The Knowledge Spillover Theory of Entrepreneurship

The International Library of Entrepreneurship

Series Editor: David B. Audretsch
Distinguished Professor and Ameritech Chair of Economic Development and Director, Institute for Development Studies, Indiana University, USA

1. Corporate Entrepreneurship and Growth
   Shaker A. Zahra
2. Women and Entrepreneurship: Contemporary Classics
   Candida G. Brush, Nancy M. Carter, Elizabeth J. Gatewood, Patricia G. Greene and Myra M. Hart
3. The Economics of Entrepreneurship
   Simon C. Parker
4. Entrepreneurship and Technology Policy
   Albert N. Link
5. Technological Entrepreneurship
   Donald S. Siegel
6. Entrepreneurship and Economic Growth
   Martin Carree and A. Roy Thurik
7. New Firm Startups
   Per Davidsson
8. International Entrepreneurship
   Benjamin M. Oviatt and Patricia Phillips McDougall
9. Entrepreneurship Education
   Patricia G. Greene and Mark P. Rice
10. Entrepreneurship and Global Capitalism (Volumes I and II)
    Geoffrey G. Jones and R. Daniel Wadhwani
11. The Political Economy of Entrepreneurship
    Magnus Henrekson and Robin Douhan
12. Financing Entrepreneurship
    Philip Auerswald and Ant Bozkaya
13. Entrepreneurial Teams and New Business Creation
    Mike Wright and Iris Vanaelst
14. Innovation and Entrepreneurship
    David Audretsch, Stephan Heblich and Oliver Falck
15. Entrepreneurship in Developing Countries
    Thorsten Beck
16. Entrepreneurship and Regional Development
    Zoltan J. Acs
17. Strategies for New Venture Development
    Ari Ginsberg
18. The Knowledge Spillover Theory of Entrepreneurship
    Zoltan J. Acs

Future titles will include:

Entrepreneurship and Public Policy
David Smallbone

Entrepreneurship in Recession
Simon C. Parker

Corporate Governance in Small and Medium Sized Firms
David B. Audretsch and Eric Lehmann

Nascent Entrepreneurship
Per Davidsson and Scott R. Gordon

Wherever possible, the articles in these volumes have been reproduced as originally published using facsimile reproduction, inclusive of footnotes and pagination to facilitate ease of reference.

The Knowledge Spillover Theory of Entrepreneurship

Edited by

Zoltan J. Acs

University Professor
and Director, Center for Entrepreneurship and Public Policy
School of Public Policy
George Mason University, USA

THE INTERNATIONAL LIBRARY OF ENTREPRENEURSHIP

An Elgar Research Collection
Cheltenham, UK • Northampton, MA, USA

Published by
Edward Elgar Publishing Limited
The Lypiatts
15 Lansdown Road
Cheltenham
Glos GL50 2JA
UK

Edward Elgar Publishing, Inc.
William Pratt House
9 Dewey Court
Northampton
Massachusetts 01060
USA

A catalogue record for this book is available from the British Library

Library of Congress Control Number: 2009941147

FSC Mixed Sources
Product group from well-managed forests and other controlled sources
www.fsc.org Cert no. SA-COC-1565
© 1996 Forest Stewardship Council

ISBN 978 1 84844 235 1

Printed and bound by MPG Books Group, UK

Contents

Acknowledgements

The editor and publishers wish to thank the authors and the following publishers who have kindly given permission for the use of copyright material.

American Economic Association for articles: F.A. Hayek (1945), 'The Use of Knowledge in Society', *American Economic Review*, **XXXV** (4), September, 519–30; William J. Baumol (1968), 'Entrepreneurship in Economic Theory', *American Economic Review, Papers and Proceedings*, **58** (2), May, 64–71; Harvey Leibenstein (1968), 'Entrepreneurship and Development', *American Economic Review*, *Papers and Proceedings*, **58** (2), May, 72–83; Zoltan J. Acs and David B. Audretsch (1988), 'Innovation in Large and Small Firms: An Empirical Analysis', *American Economic Review*, **78** (4), September, 678–90; Adam B. Jaffe (1989), 'Real Effects of Academic Research', *American Economic Review*, **79** (5), December, 957–70; Paul S. Segerstrom, T.C.A. Anant and Elias Dinopoulos (1990), 'A Schumpeterian Model of the Product Life Cycle', *American Economic Review*, **80** (5), December, 1077–91; James J. Anton and Dennis A. Yao (1994), 'Expropriation and Inventions: Appropriable Rents in the Absence of Property Rights', *American Economic Review,* **84** (1), March, 190–209; Lynne G. Zucker, Michael R. Darby and Marilynn B. Brewer (1998), 'Intellectual Human Capital and the Birth of U.S. Biotechnology Enterprises', *American Economic Review*, **88** (1), March, 290–306; Peter Thompson and Melanie Fox-Kean (2005), 'Patent Citations and the Geography of Knowledge Spillovers: A Reassessment', *American Economic Review*, **95** (1), March, 450–60.

Blackwell Publishing Ltd (Wiley-Blackwell) for articles: Claudio Michelacci (2003), 'Low Returns in R&D Due to the Lack of Entrepreneurial Skills', *Economic Journal*, **113** (484), January, 207–25; Hans K. Hvide (2009), 'The Quality of Entrepreneurs', *Economic Journal*, **119** (539), July, 1010–35.

Cambridge University Press for article: Joseph A. Schumpeter (1947), 'The Creative Response in Economic History', *Journal of Economic History*, **VII** (2), November, 149–59.

Econometric Society for articles: Boyan Jovanovic (1982), 'Selection and the Evolution of Industry', *Econometrica*, **50** (3), May, 649–70; Philippe Aghion and Peter Howitt (1992), 'A Model of Growth through Creative Destruction', *Econometrica*, **60** (2), March, 323–51.

Elsevier for articles: Luc Anselin, Attila Varga and Zoltan Acs (1997), 'Local Geographic Spillovers between University Research and High Technology Innovations', *Journal of Urban Economics*, **42** (3), 422–48; David B. Audretsch and Erik E. Lehmann (2005), 'Does the Knowledge Spillover Theory of Entrepreneurship Hold for Regions?', *Research Policy*, **34**, 1191–202.

INFORMS (Institute for Operations Research and the Management Sciences) for articles: Scott Shane (2001), 'Technological Opportunities and New Firm Creation', *Management Science*, **47** (2), February, 205–20; Thomas Hellmann (2007), 'When Do Employees Become Entrepreneurs?', *Management Science*, **53** (6), June, 919–33.

MIT Press for excerpt: David B. Audretsch (1995), 'New Firms', in *Innovation and Industry Evolution*, Chapter 3, 39–64, references.

MIT Press Journals for articles: Adam B. Jaffe, Manuel Trajtenberg and Rebecca Henderson (1993), 'Geographic Localization of Knowledge Spillovers as Evidenced by Patent Citations', *Quarterly Journal of Economics*, **108** (3), August, 577–98; Zoltan J. Acs, David B. Audretsch and Maryann P. Feldman (1994), 'R&D Spillovers and Recipient Firm Size', *Review of Economics and Statistics*, **76** (2), May, 336–40.

Oxford University Press for article: Bo Carlsson, Zoltan J. Acs, David B. Audretsch and Pontus Braunerhjelm (2009), 'Knowledge Creation, Entrepreneurship, and Economic Growth: A Historical Review', *Industrial and Corporate Change*, **18** (6), December, 1193–229.

Princeton University Press for excerpt: Kenneth J. Arrow (1962), 'Economic Welfare and the Allocation of Resources for Invention', in Universities-National Bureau Committee for Economics Research, *The Rate and Direction of Inventive Activity: Economic and Social Factors. A Conference of the Universities-National Bureau Committee for Economic Research and the Committee on Economic Growth of the Social Science Research Council*, 609–26.

RAND Corporation for articles: Robert E. Lucas, Jr. (1978), 'On the Size Distribution of Business Firms', *Bell Journal of Economics*, **9** (2), Autumn, 508–23; Zvi Griliches (1979), 'Issues in Assessing the Contribution of Research and Development to Productivity Growth', *Bell Journal of Economics*, **10** (1), Spring, 92–116.

SAGE Publications, Inc. via Copyright Clearance Center for article: Zoltan J. Acs and Attila Varga (2002), 'Geography, Endogenous Growth, and Innovation', *International Regional Science Review*, **25** (1), 132–48.

Springer Science and Business Media for articles: Boyan Jovanovic (2001), 'New Technology and The Small Firm', *Small Business Economics*, **16** (1), February, 53–5; Zoltan J. Acs and Attila Varga (2005), 'Entrepreneurship, Agglomeration and Technological Change', *Small Business Economics*, **24** (3), 323–34; Zoltan J. Acs, Pontus Braunerhjelm, David B. Audretsch and Bo Carlsson (2009), 'The Knowledge Spillover Theory of Entrepreneurship', *Small Business Economics*, **32** (1), January, 15–30; Zoltan Acs, Lawrence A. Plummer and Ryan Sutter (2009), 'Penetrating the Knowledge Filter in "Rust Belt" Economies', *Annals of Regional Science*, **43** (4), 989–1012.

University of Chicago Press via Copyright Clearance Center's Rightslink service for articles: Ariél Pakes and Shmuel Nitzan (1983), 'Optimum Contracts for Research Personnel, Research Employment, and the Establishment of "Rival" Enterprises', *Journal of Labor Economics*, **1**

(4), October, 345–65; David S. Evans and Boyan Jovanovic (1989), 'An Estimated Model of Entrepreneurial Choice under Liquidity Constraints', *Journal of Political Economy*, **97** (4), 808–27; Paul M. Romer (1990), 'Endogenous Technological Change', *Journal of Political Economy*, **98** (5, Part 2), October, S71–S102; Jarle Møen (2005), 'Is Mobility of Technical Personnel a Source of R&D Spillovers?', *Journal of Labor Economics*, **23** (1), 81–114; Edward P. Lazear (2005), 'Entrepreneurship', *Journal of Labor Economics*, **23** (4), 649–80.

Every effort has been made to trace all the copyright holders but if any have been inadvertently overlooked the publishers will be pleased to make the necessary arrangement at the first opportunity.

In addition the publishers wish to thank the Library at the University of Warwick, UK, and the Library of Indiana University at Bloomington, USA, for their assistance in obtaining these articles.

Introduction: Innovation, Entrepreneurship and the Search for Knowledge Spillovers

Zoltan J. Acs

1. Introduction

David B. Audretsch and Zoltan J. Acs were attracted to the economics of technological change by innovative prowls of new technology-based firms in the 1980s. While the conventional wisdom held that large firms had an innovative advantage over small firms, in a 1988 article in the *American Economic Review* they discovered an anomaly instead of solving a problem:

> A perhaps somewhat surprising result is that not only is the coefficient of the large-firm employment share positive and significant for small-firm innovations, but it is actually greater in magnitude than for large firms. This suggests that, *ceteris paribus*, the greater extent to which an industry is composed of large firms, the greater will be the innovative activity, but that increased innovative activity will tend to emanate more from the small firms than from the large firms (Acs and Audretsch, 1988, 686).

The anomaly of where new technology-based startups acquire knowledge was unresolved.

Building on the work of Griliches (1979, Chapter 7), Adam Jaffe (1989, Chapter 15) was the first to identify the extent to which university research spills over into the generation of commercial activity. Building on Jaffe's work, Maryann Feldman at Carnegie Mellon University expanded the knowledge production function to innovative activity and incorporated aspects of the regional knowledge infrastructure. Attila Varga at West Virginia University extends the Jaffe–Feldman approach by focusing on a more precise measure of local geographic spillovers. Varga approaches the issue of knowledge spillovers from an explicit spatial econometric perspective. The Jaffe–Feldman–Varga spillovers (from here on JFV) go a long way toward explaining the role of knowledge spillovers in technological change. Building on this foundation, the model was recently extended to identify entrepreneurship as a conduit through which knowledge spillovers take place (Acs, Braunerhjelm, Audretsch and Carlsson, 2009, Chapter 28). Finally, the role of agglomerations in knowledge spillovers represents the frontier in this scientific revolution (Clark, Feldman and Gertler, 2000). The purpose of this Introduction is to catalogue the contribution of JFV – two of them my students – that simultaneously and independently sparked a search for the mechanism of knowledge spillovers.

Section 2 outlines the main contributions of Jaffe, Feldman and Varga. Section 3 examines extensions of the model by Jaffe, Trajtenberg and Henderson as well as recent criticisms of the model by Thomson and Fox-Kean. Section 4 examines spatialized explanations of economic growth by Acs and Varga; Fujita, Krugman and Venables; and Romer. Section 5 presents work on the Knowledge Spillover Theory of Entrepreneurship. Section 6 discusses agglomerations with policy discussed in Section 7. Conclusions are in the final section.

2. Jaffe–Feldman–Varga

In his 1989 paper in the *American Economic Review* (Chapter 15), Adam Jaffe extended his path breaking 1986 study measuring the total R&D pool available for spillovers to identify the contribution of spillovers from university research to commercial innovation. Jaffe's findings were the first identifying the extent to which university research spills over into the generation of inventions and innovations by private firms. In order to relate the response of this measure to R&D spillovers from universities, Jaffe modifies the 'knowledge production function' introduced by Zvi Griliches (1979, Chapter 7) for two inputs: private corporate expenditures on R&D and research expenditures undertaken at universities.

Essentially, this is a two-factor Cobb–Douglas production function that relates an output measure for 'knowledge' to two input measures: research and development performed by industry; and research performed by universities. Formally, this is expressed as:

$$\log(K) = \beta_{K1} \log(R) + \beta_{K2} \log(U) + \varepsilon_K \tag{1}$$

where K is a proxy for knowledge measured by patent counts, R is industry R&D and U is university research, with ε_K as a stochastic error term. The analysis is carried out for US states for several points in time and disaggregated by sector. The potential interaction between university and industry research is captured by extending the model with two additional equations that allow for simultaneity between these two variables:

$$\log(R) = \beta_{R1} \log(U) + \beta_{R2} Z_2 + \varepsilon_R \tag{2}$$

and

$$\log (U) = \beta_{U1} \log(R) + \beta_{U2} Z_1 + \varepsilon_U \tag{3}$$

where U and R are as before, Z_1 and Z_2 are sets of exogenous local characteristics, and ε_R and ε_U are stochastic error terms.

Jaffe's statistical results provide evidence that corporate patent activity responds positively to commercial spillovers from university research. The lack of evidence that geographic proximity within the state matters clouds results concerning the role of geographic proximity in spillovers from university research. According to Jaffe (1989, 968), 'there is only weak evidence that spillovers are facilitated by geographic coincidence of universities and research labs within the state'. In other words, we know very little where knowledge spillovers go.

Maryann Feldman expands on the work of Jaffe in two ways (Feldman 1994, Feldman and Florida, 1994; Acs, Audretsch and Feldman, 1992, 1994, Chapter 17). First, she uses a new data source – a literature-based innovation output indicator developed by the US Small Business Administration that directly measures innovative activity (Acs and Audretsch, 1988, Chapter 10) and extends the knowledge production function (Jaffe, 1989) to account for tacit knowledge and commercialization linkages.

Griliches (1979, Chapter 7) introduced a model of technological innovation which views innovative output as the product of knowledge generating inputs. Jaffe (1989) modified this production function approach to consider spatial and technical area dimensions. However, Jaffe's model only considers what were previously defined as the elements of the formal knowledge base. Such a formulation does not consider other types of knowledge inputs, which contribute to the realization of innovative output. This is important since innovation requires both technical and business knowledge if profitability is to be the guide for making investments in research and development. Following the innovation knowledge base conceptual model, a more complete specification of innovative inputs would include

$$\log(K) = \beta_{K1}\log(R) + \beta_{K2}\log(U) + \beta_{K3}\log(BSERV) + \beta_{K4}\log(VA) + \varepsilon_K \quad (4)$$

where K is measured by counts of innovations, and R and U are as before. VA is the tacit knowledge embodied by the industry's presence in an area. $BSERV$ stands for the presence of business services that represents a link to commercialization.

The last input in the knowledge base model is the most evasive. There are a variety of producer services that provide knowledge to the market and the commercialization process. For example, the services of patent attorneys are a critical input to the innovation process. Similarly, marketing information plays an important role in the commercialization process.

Substitution of the direct measure of innovative activity for the patent measure in the knowledge production function generally strengthens Jaffe's (1989) arguments and reinforces his findings. Most importantly, use of the innovation data provides even greater support than was found by Jaffe: as he predicted, spillovers are facilitated by the geographic coincidence of universities and research labs within the state. In addition, there is at least some evidence that, because the patent and innovation measures capture different aspects of the process of technological change, results for specific sectors may be, at least to some extent, influenced by the technological regime. Thus, it is found that the importance of university spillovers relative to private-company R&D spending is considerably greater in the electronics sector when the direct measure of innovative activity is substituted for the patent measure.

However, the relative importance of industry R&D and university research as inputs in generating innovative output clearly varies between large and small firms (Acs, Audretsch and Feldman, 1994, Chapter 17). That is, for large firms, not only is the elasticity of innovative activity with respect to industry R&D expenditures more than two times greater than the elasticity with respect to expenditures on research by universities, but it is nearly twice as large as the elasticity of small-firm innovative activity with respect to industry R&D. By contrast, for small firms the elasticity of innovative output with respect to expenditures on research by universities is about one-fifth greater than the elasticity with respect to industry R&D. Moreover, the elasticity of innovative activity with respect to university research is about 50 percent greater for small enterprises than for large corporations.

These results support the hypothesis that private-corporate R&D plays a relatively more important role in generating innovative activity in large corporations than in small firms. By contrast, spillovers from the research activities of universities play a more decisive role in the innovative activity of small firms. Geographic proximity between university and corporate laboratories within a state clearly serves as a catalyst to innovative activity for firms of all sizes. However, the impact is apparently greater on small firms than on large firms.

There were two limitations of the Jaffe–Feldman research. First, the unit of analysis at the state level was too aggregate requiring a geographical coincidence index to control for co-location. Second, the research did not take into consideration the potential influence of spatial dependence that may invalidate the interpretation of econometric analyses based on contiguous cross-sectional data.

Attila Varga mitigates these limitations examining both the state and the metropolitan statistical area (MSA) levels and using spatial econometric techniques[1] (Varga, 1998, 2000; Anselin, Varga and Acs, 1997, Chapter 18, 2000a, 2000b; Acs, Anselin and Varga, 2002). These extensions yielded a more precise insight into the range of spatial externalities between innovation and R&D in the MSA and university research both within the MSA and in surrounding counties. He was able to shed some initial light on this issue for high technology innovations measured as an aggregate across five two-digit Standard Industrial Classification (SIC) industries and also at a more detailed industrial sector level. He found a positive and highly significant relationship between MSA innovations and university research indicating the presence of localized university research spillovers in innovation. In comparison to the effect of industrial knowledge spillovers (i.e., knowledge flows among industrial research laboratories) the size of the university effect is considerably smaller as it is one-third of the size of the industrial research coefficient. University knowledge spillovers follow a definite distance decay pattern as shown by the statistically significant albeit smaller size university research coefficient for adjoining counties within a 50-mile distance range from the MSA center.

There are notable differences among sectors with respect to the localized university effect as studied at the MSA level. Specifically for the four high technology sectors such as machinery, chemicals, electronics and instruments, significant localized university spillover impact was found only for electronics and instruments while for the other two industries the university research coefficient remains consistently insignificant.

Acs, Anselin and Varga (2002) test whether the patent data developed by the United States Patent and Trademark Office (PTO) is, in fact, a reliable proxy measure of innovative activity at the regional level as compared to the literature-based innovation output indicator developed by the US Small Business Administration (SBA). This is important, since the patent data are readily available over time and can be used to study the dynamics of localized knowledge flows within regional innovation systems. Before this study, there was some evidence that patents provide a reliable measure of innovative activity at the industry level (Acs and Audretsch, 1989) and some evidence that patents and innovations behave similarly at the state level (Acs, Audretsch and Feldman, 1992). However, this has not been tested at the sub-state level.

The correlation between the PTO patent and SBA innovation counts at the MSA level is reasonably high (0.79) and this could be taken as a first indication that patents might be a reliable measure of innovation at the regional level. However, this correlation coefficient value is not high enough to guarantee that the role of different regional actors in knowledge creation would turn out similar with both measures if applied in the same empirical model. We proceeded by replacing innovation counts with the patent measure in the same model as in Anselin, Varga and Acs (1997) in order to directly compare the results of the two measures of new technological knowledge and assess the extent to which patents may be used as a reliable proxy.

Sizes of all the parameters in the estimated knowledge production function are smaller for innovation than for patents, suggesting that firms in the product development stage rely on localized interactions (with universities as well as with other actors) less intensively than in earlier stages of the innovation process. The other important finding of this comparative study is that the importance of university knowledge spillovers (measured by the size of the university research parameter) compared to that of R&D spillovers among private firms is substantially less pronounced for patents than for innovations. Since patenting reflects more the earlier stages of innovation whereas the direct innovation measure accounts for the concluding stage of the innovation process, the relatively higher weight of local universities in innovation than in patenting appears to reflect the different spatial patterns of basic and applied research collaboration. To collaborate with universities in applied research, firms tend to choose local academic institutions whereas basic research collaboration can be carried out over larger distances.

3. Extensions of the JFV Model

Jaffe, Trajtenberg and Henderson (1993, Chapter 16; 2005) expand on the above work to answer the question if knowledge externalities are localized. This is important since growth theory assumed that knowledge spills over to agents within the country, but not to other countries. This implicit assumption begs the question to what extent knowledge externalities are localized. Jaffe, Trajtenberg and Henderson extend the search for knowledge spillovers by using a matching method that found that knowledge spillovers are strongly localized. Their method matches each citing patent to a non-citing patent intended to control for the pre-existing geographic concentration of production. Using patent data, they came to two conclusions, one that spillovers are particularly significant at the local level, and the other that localization fades slowly over time. These results and the large research issue are reproduced in Jaffe and Trajtenberg (2002).

Audretsch and Feldman (1996) explore the question of the geography of innovation and production. They provide evidence concerning the spatial dimension of knowledge spillovers. Their findings suggest that knowledge spillovers are geographically bounded and localized within spatial proximity to the knowledge source. Feldman and Audretsch (1999) further examine the question of knowledge spillovers by looking into the question of specialization versus industrial diversity in cities. Their research supports the ideas that diversity leads to more innovation.

Recently, Thompson and Fox-Kean (2005a, Chapter 19; 2005b) challenged the findings of Jaffe, Trajtenberg and Henderson. They suggest that the Jaffe, Trajtenberg and Henderson matched case control methodology included a serious spurious component. Controlling for the unobservable using matching methods is invariably a dangerous exercise because one can rarely be confident that the controls are doing their job. In some cases, imperfect matching may simply introduce noise and a corresponding loss of efficiency. They suggest at least two reasons why the matching method may not adequately control for existing patent activity. First, the level of aggregation might not be fine enough. Second, patents typically contain many distinct claims each of which is assigned a technological classification. These two features of the control selection process mean that there is no guarantee that the control patent has any industrial similarity with either the citing or originating patent. Of course, their conclusion that spillovers stop at the country level also needs explaining.

Empirical research done within the JFV framework and the extensions introduced so far were established and originally carried out in the United States with the use of state, MSA and county level data sets. However, the issue of the geographic extent of knowledge spillovers has definite international validity. The JFV model has been replicated and continually refined in the search for the geographical boundaries of knowledge flows in Europe, South America and Asia. Varga (2006) provides an assessment of the international literature.

4. The 'Spatialized' Explanation of Economic Growth

Building on the JFV model of knowledge spillovers, Acs and Varga (2002, Chapter 25) suggests that a 'spatialized' theoretical framework of technology-led economic growth that needs to reflect three fundamental issues. First, it should explain why knowledge-related economic activities concentrate in certain regions leaving others relatively underdeveloped. Second, it needs to answer the questions of how technological advances occur and what are the key processes and institutions involved with a particular focus on the geographic dimension. Third, it must present an analytical framework where the role of technological change in regional and national economic growth is clearly explained. In order to answer these three questions Acs and Varga examine three separate and distinct literatures: the new economic geography, the new growth theory, and the new economics of innovation.

Although the three approaches focus on different aspects, the three are at the same time complements. The 'new' theories of growth endogenize technological change and as such interlink technological change with macroeconomic growth. However, the way technological change is described is strongly simplistic and the economy investigated is formulated in an a-spatial model. On the other hand, systems of innovation frameworks are very detailed with respect to the innovation process but say nothing about macroeconomic growth. However, the spatial dimension has been introduced into the framework in the recently developed 'regional innovation systems' studies (Braczyk, Cooke and Heidenreich, 1998).

The idea behind the innovation systems approach is quite simple but extremely appealing. According to this, in most cases, innovation is a result of a collective process and this process is shaped in a systematic manner. The elements of the system are innovating firms and firms in related and connected industries (suppliers, buyers), private and public research laboratories, universities, supporting business services (like legal or technical services), financial institutions (especially venture capital), and the government. These elements are interconnected by innovation-related linkages where these linkages represent knowledge flows among them. Linkages can be informal in nature (occasional meetings in conferences, social events etc.) or they can also be definitely formal (contracted research, collaborative product development etc.). The effectiveness (i.e., productivity in terms of number of innovations) of the system is determined by both the knowledge already accumulated by the actors and the level of their interconnectedness (i.e., the intensity of knowledge flows). Ability and motivations for interactions are shaped largely by traditions, social norms, values and the countries' legal systems.

New economic geography models investigate general equilibrium in a spatial setting (Krugman, 1991). This means that they provide explanations not just for the determination of equilibrium prices, incomes and quantities in each market but also the development of the particular geographical structure of the economy. In other words, new economic geography derives economic and spatial equilibrium simultaneously (Fujita, Krugman and Venables, 1999; Fujita and Thisse, 2002). Spatial equilibrium arises as an outcome of the balance between centripetal forces working towards agglomeration (such as increasing returns to scale, industrial demand, localized knowledge spillovers) and centrifugal forces promoting dispersion (such as transportation costs). Until the latest developments in recent years, new economic geography models did not consider the spatial aspects of economic growth. However, even the recent models of technological change follow the same pattern as endogenous growth models and fail to reach the complexity inherent in innovation systems studies.

As emphasized by Acs and Varga (2002, Chapter 25), although each of the above three approaches has its strengths and weaknesses, each could serve to create the building blocks of an explanatory framework of technology-led economic growth. The three suggest that a specific combination of the Krugmanian theory of initial conditions for spatial concentration of economic activities with the Romerian theory of endogenous economic growth, complemented with a systematic representation of interactions among the actors of Nelson's innovation system, could be a way of developing an appropriate model of technology-led regional economic development.

Following Acs and Varga (2002, Chapter 5), Varga (2006) develops an empirical modeling framework of geographical growth explanation. This framework is the spatial extension of the endogenous growth model in Romer (1990, Chapter 12) and it integrates elements of the innovation systems and the new economic geography literatures. For a more formal treatment, Varga (2006) applies the generalized version of the Romer (1990) equation of macroeconomic level knowledge production developed in Jones (1995)[2]:

$$dA = \delta\, H_A^{\lambda} A^{\varphi} \tag{5}$$

where H_A stands for human capital in the research sector working on knowledge production (operationalized by the number of researchers), A is the total stock of technological knowledge available at a certain point in time whereas dA is the change in technological knowledge resulted from private efforts to invest in research and development. δ, λ and φ are parameters.

Technological change is generated by research and its extent depends on the number of researchers involved in knowledge creation (H_A). However, their efficiency is directly related to the total stock of already available knowledge (A). Knowledge spillovers are central to the growth process: the higher A is the larger the change in technology produced by the same number of researchers. Thus, macroeconomic growth is strongly related to knowledge spillovers.

Parameters in the Romer knowledge production function play a decisive role in the effectiveness of macro level knowledge production. The same number of researchers with a similar value of A can raise the level of already existing technological knowledge with significant differences depending on the size of the parameters. First, consider δ ($0 < \delta < 1$) which is the research productivity parameter. The larger δ is, the more efficient H_A is in producing economically useful new knowledge.

The size of φ reflects the extent to which the total stock of previously established knowledge impacts knowledge production. Given that A stands for the level of codified knowledge (available in books, scientific papers or patent documentations), φ is the parameter of codified knowledge spillovers. The size of φ reflects the portion of *A* that spills over and, as such, its value largely influences the effectiveness of research in generating new technologies.

The research spillover parameter is λ. A larger λ indicates a stronger impact on technological change with the same number of researchers. In contrast to φ and δ that are determined primarily in the research sector and as such their values are exogenous to the economy, λ is endogenous. Its value reflects the diffusion of (codified and tacit) knowledge accumulated by researchers. Technological diffusion depends on three interactions: first, on the intensity of interactions among researchers (H_A); second, on the quality of public research, and the extent to which the private research sector is connected to it (especially to universities) by formal and informal linkages; and third the development level of supporting/connected industries and business services and the integration of innovating firms into the system. The extensive innovation systems literature evidences that the same number of researchers contribute to different efficiencies depending on the development of the system. In the Romer equation, this is reflected in the size of λ.

Within the JFV framework, a series of papers demonstrate that a significant fraction of knowledge spillovers is bounded spatially. These findings imply that the geographic structure of R&D is a determinant of technological change and ultimately economic growth. *Ceteris paribus*, in an economy where R&D institutions are highly concentrated, intensive knowledge spillovers will result in a higher level of innovation than in a system where research is more evenly distributed over space. Thus λ is also sensitive to the spatial structure of H_A. Even with the same number of researchers, λ can have different values depending on the extent that research and development is spatially concentrated.

Finally, λ depends on the interaction of researchers and entrepreneurs. The distribution of entrepreneurs is also not evenly distributed over space. The more entrepreneurs are concentrated regionally where knowledge is produced the greater the impact of knowledge spillovers on economic growth.

5. A Knowledge Spillover Theory of Entrepreneurship

In this section, the JFV model of knowledge spillovers is extended by Acs and Audretsch who develop the Knowledge Spillover Theory of Entrepreneurship in order to answer the question, 'What is the conduit by which knowledge spillovers occur?' As a first step in this direction, the theory incorporates two of the above literatures: new growth theory (Romer) and the new economics of innovation (Nelson, 1991) to explain how entrepreneurship facilitates the spillover of knowledge.

A modern synthesis of the entrepreneur is someone who specializes in making judgmental decisions about the coordination of scarce resources (Lazear, 2005, Chapter 32). In this definition, the term 'someone' emphasizes that the entrepreneur is an individual. Judgmental decisions are decisions for which no obvious correct procedure exists – a judgmental decision cannot be made simply by plugging available numbers into a scientific formula and acting based on the resulting number. In this framework, entrepreneurial activity depends upon the interaction between the characteristics of opportunity and the characteristics of the people who exploit them. Since discovery is a cognitive process, it can take place only at the individual level. Individuals, whether working in an existing organization or unemployed, are the entities discovering opportunities. The organizations employing people are inanimate and cannot engage in *discovery*. Therefore, any explanation for the mode of opportunity discovery must be based on choices made by individuals about how they would like to exploit the opportunity that they have discovered (Hayek, 1937).

So where do opportunities come from? Today we know that the technology opportunity set is endogenously created by investments in new knowledge. The new growth theory, formalized by Romer (1986), assumes that firms exist exogenously and then engage in the pursuit of new economic knowledge as input into the process of generating endogenous growth. Technological change plays a central role in the explanation of economic growth, since on the steady state growth path the rate of per capita GDP growth equals the rate of technological change.

However, not only does new knowledge contribute to technological change, it also creates opportunities for use by third party firms, often entrepreneurial startups (Shane, 2001, Chapter 23). The creation of new knowledge gives rise to new opportunities through knowledge spillovers, therefore, entrepreneurial activity does not involve simply the arbitrage of opportunities (Kirzner, 1973) but also the exploitation of new opportunities created but not appropriated by incumbent organizations (Hellmann, 2007, Chapter 34). Thus, while the entrepreneurship literature considers opportunity to exist exogenously, in the new economic growth literature, opportunities are endogenously created through the purposeful investment in new knowledge. The theory as suggested by Audretsch (1995, 48, Chapter 22) '...proposes shifting the unit of observation away from exogenously assumed firms to individuals – agents confronted with new knowledge and the decision whether and how to act upon that new knowledge.'

The theory relaxes two central (and unrealistic) assumptions of the endogenous growth model to develop a theory that improves the microeconomic foundations of endogenous growth theory (Acs, Braunerhjelm, Audretsch and Carlsson, 2009, Chapter 28). The first is that knowledge is automatically equated with economic knowledge. In fact, as Arrow (1962, Chapter 3) emphasized, knowledge is inherently different from the traditional factors of production, resulting in a gap between knowledge (K) and what he called economic knowledge (K^c). The second involves the assumed spillover of knowledge. The existence of the factor of knowledge is equated with its automatic spillover, yielding endogenous growth. In the Knowledge Spillover Theory of Entrepreneurship, *institutions* impose a gap between new knowledge and economic knowledge ($0 < K^c/K < 1$) that results in a lower level of knowledge spillovers.

The model is one where new product innovations can come either from incumbent organizations or from entrepreneurial startups (Schumpeter, 1911 [1934]). According to Baumol (2004, 9),

> the bulk of private R&D spending is shown to come from a tiny number of very large firms. Yet, the revolutionary breakthroughs continue to come predominantly from small entrepreneurial enterprises, with large industry providing streams of incremental improvements that also add up to major contributions.

We can think of incumbent firms that rely on the *flow* of knowledge to innovate as focusing on incremental innovation, i.e. product improvements (Acs and Audretsch, 1988, Chapter 10).

Entrepreneurial startups that have access to knowledge spillovers from the *stock* of knowledge and entrepreneurial talent are more likely to be engaged in radical innovation that leads to new industries or completely replace existing products (Acs, Audretsch and Feldman, 1994, Chapter 17). Startups play a major role in radical innovations such as software, semiconductors, biotechnology (Zucker, Darby and Brewer, 1998, Chapter 21) and the information and communications technologies (Jorgenson, 2001). The presence of these activities is especially important at the early stages of the life cycle when technology is still fluid.

Equation (6) suggests that entrepreneurial startups (*E*) will be a function of the difference between expected profits (π^*) minus wages (w). Expected profits are conditioned by the knowledge stock (*K*) that positively affects startups and is negatively conditioned by knowledge commercialized by incumbent firms. Yet a rich literature suggests that there is a compelling array of financial, institutional and individual barriers to entrepreneurship, which results in a modification of the entrepreneurial choice equation:

$$E = \gamma\,(\pi^*\,(K^{\xi}) - w)/\beta \tag{6}$$

where β represents those institutional and individual barriers to entrepreneurship, spanning factors such as risk aversion, financial constraints, and legal and regulatory restrictions (Acemoglu, Johnson and Robinson, 2004). The existence of such barriers explains why economic agents might choose not to enter into entrepreneurship, even when confronted with knowledge that would otherwise generate a potentially profitable opportunity. Thus, this model shows how local differences in knowledge stocks, the presence of large firms as deterrents to knowledge exploitation, and an entrepreneurial culture might explain regional variations in the rates of entrepreneurial activity. The primary theoretical predictions of the model are:

- An increase in the stock of knowledge positively affects the level of entrepreneurship.
- The more efficient incumbents are at exploiting knowledge flows, the smaller the effect of new knowledge on entrepreneurship.
- Entrepreneurial activities decrease in the face of higher regulations, administrative barriers and governmental market intervention.

Thus, entrepreneurship becomes central to generating economic growth by serving as a conduit, albeit not the sole conduit, by which knowledge created by incumbent organizations spills over to agents who endogenously create new firms. The theory is actually a theory of *endogenous entrepreneurship*, where entrepreneurship is a response to opportunities created by investments in new knowledge that was not commercialized by incumbent firms. The theory suggests that, *ceteris paribus*, entrepreneurial activity will tend to be greater in contexts where investments in new knowledge are relatively high, since the startups will benefit from knowledge that spills over from the source actually producing that new knowledge. In a low knowledge context, the lack of new ideas will not generate entrepreneurial opportunities based on potential knowledge spillovers. A series of studies link entrepreneurship and economic growth at the regional level (Acs and Armington, 2006); Audretsch, Keilbach and Lehmann, 2006) and at the national level (Acs, Audretsch, Braunerhjelm and Carlsson, 2009) finding that entrepreneurship does in fact offer an explanation for how knowledge spillovers occur.

Acs and Varga (2005, Chapter 31) empirically test the theory within the JFV framework. They build their modeling approach on the interpretation of the Romerian equation (equation 5) provided in section 4. They start with the assumption that the value of λ bears the influence of the level of entrepreneurship because the value of new economic knowledge is uncertain. While most R&D is carried out in large firms and universities, it does not mean that the individuals that discover the opportunity will carry out the subsequent exploitation. An implication of the theory of firm selection is that new firms may enter an industry in large numbers to exploit knowledge spillovers. The higher the rate of startups the greater should be the value of λ because of knowledge spillovers.

The empirical model in which the parameter λ in equation (5) is endogenized has the following form:

$$\log(NK) = \delta + \lambda\log(H) + \varphi\log(A) + \varepsilon \quad (7)$$

$$\lambda = (\beta_1 + \beta_2\log(ENTR) + \beta_3\log(AGGL) \quad (8)$$

where *NK* stands for new knowledge (i.e., the change in *A*), *ENTR* is entrepreneurship, *AGGL* is agglomeration, *A* is the set of publicly available scientific-technological knowledge and ε is stochastic error term. Implementation of (7) into (8) results in the following estimated equation:

$$\log(NK) = \delta + \beta_1\log(H) + \beta_2\log(ENTR)\log(H) + \beta_3\log(AGGL)\log(H) + \varphi\log(A) + \varepsilon \quad (9)$$

In equation (9), the estimated value of the parameter β_2 measures the extent to which research interacted with entrepreneurship contributes to knowledge spillovers. Applied to European data, Acs and Varga (2005, Chapter 31) find a statistically significant value of β_2 that is taken as evidence supporting the Knowledge Spillover Theory of Entrepreneurship.

6. Agglomeration: The Final Frontier

The JFV model is extendable for empirically testing agglomeration effects in knowledge spillovers. Agglomeration forces are crucial in technological change and as such in economic growth explanation. Varga (2006) points out that in equation (5) the size of λ is also influenced by agglomeration. Insights from the new economic geography can help understand the dynamic effects of the spatial structure of R&D on macroeconomic growth (Baldwin and Forslid, 2000; Fujita and Thisse, 2002; Baldwin et al., 2003). If spatial proximity to other research labs, universities, firms and business services matters in innovation, firms are motivated to locate R&D laboratories in those regions where actors of the system of innovation are already agglomerated in order to decrease innovation costs.

Thus, spatial concentration of the system of innovation is a source of positive externalities and, as such, these externalities (as centrifugal forces in R&D location) determine the strength of the cumulative process that leads to a particular spatial economic structure. However, agglomeration effects can be negative as well. Increasing housing costs and travel time make innovation more expensive and might motivate labs to move out of the region. The actual balance between centrifugal and centripetal forces determines the geographical structure of the system of innovation. Through determining the size of λ in equation (5), this also influences the rate of technological progress (dA/A) and, eventually, the macroeconomic growth rate (dy/y).

Within the JFV framework Varga (2000 and 2001) estimates the magnitude of agglomeration effects. Based on a data set of 125 US metropolitan areas, he finds that spatial concentration of high technology production and business services have a definite positive relationship with the intensity of local academic knowledge transfers. Increasing returns resulting from the spatial concentration of economic activities is clearly demonstrated in the study. It is shown that the same amount of local expenditures on university research yields dramatically different levels of innovation output depending on the concentration of economic activities in the metropolitan area. A critical mass of agglomeration is found necessary for regions to experience substantial local economic effects of academic research spending. This critical mass is characterized by a city population of around three million, employment in high technology production facilities and business service firms about 160 000 and 4000, respectively. In Varga (2001), agglomeration effects in university knowledge spillovers for two 'high technology' sectors (electronics and instruments) are also demonstrated.

How can the JFV framework contribute to study empirically the dynamism of agglomeration (i.e., the dynamism of λ) described in detail by the new economic geography? To model empirically the effects of centripetal and centrifugal forces on spatial structure, researchers develop spatial computable general equilibrium (SCGE) models (e.g., Thissen, 2003). These models are empirical counterparts of the new economic geography and are extremely powerful tools for explaining spatial distribution of economic activities under different starting assumptions.

Now it is technically possible to integrate the JFV approach into SCGE modeling in order to study the dynamic effects of knowledge spillovers on geography, technological change and growth. With this step the JFV model becomes a crucial bridge between academic research on the geography of innovation and policy analysis for studying different scenarios of economic development. Varga (2008) demonstrates that incorporating the lessons of the JFV framework into development policy analysis opens up the possibility of building 'new generation models' with such simulations where both regional, inter-regional and macro effects of different policy scenarios can be studied and compared to each other. The Geographic Macro and Regional Modelling (GMR)-Hungary model (Varga, 2007) is the first one in this field.

7. Public Policy

Policy makers are interested in promoting economic growth at the national and the regional level. Politicians look to academia to help them understand the process of economic development and inform their decisions. Academics long ago identified technical change through innovation as a key process for generating long-term stable economic growth. However, that begs the question, 'What causes innovation in a region or economy?' (Acs and Sanders, 2007).

In accordance with the evidence, the entrepreneur is the agent with whom the buck stops. The creation of the knowledge he commercializes is not (necessarily) motivated by the rents that the entrepreneur receives for commercialization. Rents reward the act of commercialization, and as such should not be destroyed, to enhance static efficiency. However, the claim to these rents should also not be transferred to the generators of knowledge that may have had no intent of commercializing and/or require no incentive to create such knowledge in the first place.[3] It is not the generation of new knowledge that is is valuable to society at large but rather its utilization and subsequent economic growth. Knowledge creation, of course, is a necessary but insufficient condition for innovation and growth, and creation without implementation is clearly a waste of resources. We argue, therefore, that policy makers should stop and think about the bottlenecks in the innovative process before committing large amounts of public money and/or entitlements to profits and rent to the (formal) knowledge generation process.

These results also carry over to the regional level if considering the impact of limited geographical labor mobility, transport costs and communication costs. As the knowledge spillovers that drive economic growth are likely to be regionalized, regional policies should aim to facilitate spillovers. A first requirement is that sufficient resources are available for both knowledge creation and knowledge commercialization. And as entrepreneurial talent is a key resource in the innovation chain, regional policies should try to develop it. Moreover, the impediments to knowledge spillovers from creators to commercializers deserve attention. Legal impediments such as non-competition clauses in labor contracts should be abandoned. Investing in physical and communication infrastructures and by stimulating or enabling the exchange of knowledge, local and regional governments can support the entire innovation chain. Direct support to new entrants or R&D should be given only as long as that does not reduce the incentives to create or commercialize new knowledge.

The model outlined in this paper has important policy implications at the aggregate and regional level but also raises important questions. The presence of knowledge spillovers is well documented in the literature. But the exact channels through which such spillovers arise are a challenging area for further research. The three propositions predicting regional clustering are empirically indistinguishable in most studies due to data availability issues. Florida (2002), presents evidence in support of the third channel we have discussed, but to our knowledge, studies that try to distinguish between them have not yet been done. In addition, at the aggregate level, the theory and its underlying assumptions require further empirical scrutiny. The available evidence supports the claim that knowledge spillovers are important for (regional) economic growth but much more can be done to test the model predictions. This empirical research agenda will hopefully inspire other researchers.

8. Conclusions

In this review, the Jaffe–Feldman–Varga model is introduced and an assessment as to its relevance for economics research is made. It is highlighted that this approach has become a widely applied tool for testing the spatial extent of knowledge spillovers in different countries, different sectors and at different spatial scales. In addition, this model became a workhorse of empirical studies of entrepreneurship, agglomeration and growth. The JFV approach has also proven to become a crucial element in 'new generation development policy modeling'. Thus the JFV model of knowledge spillovers and its extensions offer an avenue to re-explain the mechanism by which knowledge spillovers operate and open the door for a new understanding of regional and macroeconomic development. If this indeed happens, we would have experienced a paradigm shift in economic science.

Notes

1. When models are estimated for cross-sectional data on neighboring spatial units, the lack of independence across these units (or, the presence of spatial autocorrelation) can cause serious problems of model misspecification when ignored (Anselin, 1988). The methodology of spatial econometrics consists of testing for the potential presence of these misspecifications and of using the proper estimators for models that incorporate the spatial dependence explicitly (for a recent review, see Anselin, 2001).
2. The functional form corresponds to the Jones (1995) version; however, the interpretation of λ and φ is different in Varga (2006).
3. This may well be the effect of stronger patent and Intellectual Property Rights Protection.

References

Acemoglu, D., S. Johnson and J. Robinson. (2004), 'Institutions as the Fundamental Cause of Long-Run Growth', in Philippe Aghion and Steven Durlauf (eds.), *Handbook of Economic Growth*, New York, NY: Elsevier North Holland, Vol. 1, Chapter 6, 405–72.

Acs, Z.J. and C. Armington. (2006), *Entrepreneurship, Geography and American Economic Growth*, Cambridge, UK: Cambridge University Press.

Acs, Z.J. and D.B. Audretsch. (1989), 'Patents as a Measure of Innovative Activity', *Kyklos*, **42**, 171–80.

Acs, Z.J and M. Sanders. (2007), 'Intellectual Property and the Knowledge Spillover Theory of Entrepreneurship', Max Planck Institute of Economics, Jena.

Acs, Z.J., L. Anselin and A. Varga. (2002), 'Patents and Innovation Counts as Measures of Regional Production of New Knowledge', *Research Policy*, **31**, 1069–85.

Acs, Z.J., D. Audretsch and M.P. Feldman. (1992), 'Real Effects of Academic Research: Comment', *American Economic Review*, **82**, 363–7.

Anselin, L. (1988), *Spatial Econometrics: Methods and Models*, Boston, MA: Kluwer Academic Publishers.

Anselin, L. (2001), 'Spatial Econometrics', in B. Baltagi (ed.), *A Companion to Theoretical Econometrics*, Oxford: Basil Blackwell, 310–30.

Anselin, L., A. Varga and Z.J. Acs. (2000a), 'Geographic and Sectoral Characteristics of Academic Knowledge Spillovers', *Papers in Regional Science*, **79**, 435–45.

Anselin, L., A. Varga and Z.J. Acs. (2000b), 'Geographic Spillovers and University Research: a Spatial Econometric Approach', *Growth and Change*, **31**, 501–15.

Audretsch, D. (1995), *Innovation and Industry Evolution*, Cambridge, MA: MIT Press.

Audretsch, D. and M.P. Feldman. (1996), 'R&D Spillovers and the Geography of Innovation and Production', *American Economic Review*, **86**, 630–40.

Audretsch, D., M.C. Keilbach and E.E. Lehmann. (2006), *Entrepreneurship and Economic Growth*, Oxford, UK: Oxford University Press.

Baldwin, R.E. and R. Forslid. (2000), 'The Core-Periphery Model and Endogenous Growth: Stabilising and De-stabilizing Integration', *Economica*, **67**, 307–24.

Baldwin, R., R. Forslid, P. Martin, G. Ottaviano and F. Robert-Nicoud. (2003), *Economic Geography and Public Policy*, Princeton, NJ: Princeton University Press.

Baumol, William. (2004), 'Entrepreneurial Enterprises, Large Established Firms and Other Components of the Free-market Growth Machine', *Small Business Economics*, **23**(1), 9–21.

Braczyk, H., P. Cooke and M. Heidenreich. (1998), *Regional Innovation Systems: The Role of Governances in a Globalized World*, London: UCL Press.

Clark, G.L., M.P. Feldman and M.S. Gertler. (2000), *The Oxford Handbook of Economic Geography*, Oxford, UK: Oxford University Press.

Feldman, M.P. (1994), *The Geography of Innovation*, New York, NY: Kluwer Academic Publishers.

Feldman, M.P. and R. Florida. (1994), 'The Geographic Sources of Innovation: Technological Infrastructure and Product Innovation in the United States', *Annals of the Association of American Geographers*, **84**, 210–29.

Feldman, M.P. and D.B. Audretsch. (1999), 'Innovation in Cities: Science-Based Diversity, Specialization and Localized Competition', *European Economic Review*, **43**, 409–29.

Florida, R. (2002), 'The Economic Geography of Talent', *Annals of American Geographers*, Routledge Taylor & Francis Group, **92**(4), 743–55.

Fujita, M., P. Krugman and A. Venables. (1999), *The Spatial Economy*, Cambridge, MA: MIT Press.

Fujita, M and J. Thisse. (2002), *Economics of Agglomeration. Cities, Industrial Location, and Regional Growth*, Cambridge, UK: Cambridge University Press.

Hayek, Frederick A. von. (1937), 'Economics and Knowledge', *Economica* (New Series), **4**, 33–54.

Jaffe, A. and M. Trajtenberg. (2002), *Patents, Citations and Innovations: A Window on the Knowledge Economy*, Cambridge, MA: MIT Press.

Jaffe, A., M. Trajtenberg and R. Henderson. (2005), 'Patent Citations and the Geography of Knowledge Spillovers: A Reassessment: Comment', *American Economic Review*, **95**(1) 461–5.

Jones, C. (1995), 'R&D Based Models of Economic Growth', *Journal of Political Economy*, **103**, 759–84.

Jorgenson, D.W. (2001), 'Information Technology and the U. S. Economy', *American Economic Review*, **91**, 1–32.

Kirzner, Isreal. M. (1973), *Competition and Entrepreneurship*, Chicago, IL: University of Chicago Press.

Krugman, Paul. (1991), 'Increasing Returns and Economic Geography', *Journal of Political Economy*, **99**, 483–99.

Nelson, J.R. (1991), *National Innovation Systems*, Cambridge, MA: Harvard University Press.

Romer, P. (1986), 'Increasing Returns and Economic Growth', *Journal of Political Economy*, **94**, 1002–37.

Schumpeter, Joseph. A. (1911) [1934], *The Theory of Economic Development*, Cambridge, MA: Harvard University Press.

Thissen, M. (2003), RAEM 2.0 A Regional Applied General Equilibrium Model for the Netherlands. Manuscript, pp. 19.

Thompson, P. and M. Fox-Kean. (2005b), 'Patent Citations and the Geography of Knowledge Spillovers: A Reassessment: Reply', *American Economic Review*, **95**(1), 465–7.

Varga, A. (1998), *University Research and Regional Innovation: A Spatial Econometric Analysis of Academic Technology Transfers*, Boston, MA: Kluwer Academic Publishers.

Varga, A. (2000), 'Local Academic Knowledge Spillovers and the Concentration of Economic Activity', *Journal of Regional Science*, **40**, 289–309.

Varga, A. (2001), 'Universities and Regional Economic Development: Does Agglomeration Matter?' in B. Johansson, C. Karlsson and R. Stough (eds.), *Theories of Endogenous Regional Growth – Lessons for Regional Policies*, 345–67. Berlin: Springer.

Varga, A. (2006), 'The Spatial Dimension of Innovation and Growth: Empirical Research Methodology and Policy Analysis', *European Planning Studies*, **9**, 1171–86.

Varga, A. (2007), GMR-HUNGARY: A Complex Macro-regional Model for the Analysis of Development Policy Impacts on the Hungarian Economy. Final Report, Project No. NFH 370/2005.

Varga, A. (2008), 'From the Geography of Innovation to Development Policy Analysis: The GMR-approach', *Annals of Economics and Statistics*, 87–88, 83–102.

Part I
Knowledge

[1]

The American Economic Review

VOLUME XXXV SEPTEMBER, 1945 NUMBER FOUR

THE USE OF KNOWLEDGE IN SOCIETY

By F. A. Hayek*

I

What is the problem we wish to solve when we try to construct a rational economic order?

On certain familiar assumptions the answer is simple enough. *If* we possess all the relevant information, *if* we can start out from a given system of preferences and *if* we command complete knowledge of available means, the problem which remains is purely one of logic. That is, the answer to the question of what is the best use of the available means is implicit in our assumptions. The conditions which the solution of this optimum problem must satisfy have been fully worked out and can be stated best in mathematical form: put at their briefest, they are that the marginal rates of substitution between any two commodities or factors must be the same in all their different uses.

This, however, is emphatically *not* the economic problem which society faces. And the economic calculus which we have developed to solve this logical problem, though an important step toward the solution of the economic problem of society, does not yet provide an answer to it. The reason for this is that the "data" from which the economic calculus starts are never for the whole society "given" to a single mind which could work out the implications, and can never be so given.

The peculiar character of the problem of a rational economic order is determined precisely by the fact that the knowledge of the circumstances of which we must make use never exists in concentrated or integrated form, but solely as the dispersed bits of incomplete and frequently contradictory knowledge which all the separate individuals possess. The economic problem of society is thus not merely a problem

* The author is Tooke professor of political economy and statistics at the University of London (London School of Economics and Political Science).

of how to allocate "given" resources—if "given" is taken to mean given to a single mind which deliberately solves the problem set by these "data." It is rather a problem of how to secure the best use of resources known to any of the members of society, for ends whose relative importance only these individuals know. Or, to put it briefly, it is a problem of the utilization of knowledge not given to anyone in its totality.

This character of the fundamental problem has, I am afraid, been rather obscured than illuminated by many of the recent refinements of economic theory, particularly by many of the uses made of mathematics. Though the problem with which I want primarily to deal in this paper is the problem of a rational economic organization, I shall in its course be led again and again to point to its close connections with certain methodological questions. Many of the points I wish to make are indeed conclusions toward which diverse paths of reasoning have unexpectedly converged. But as I now see these problems, this is no accident. It seems to me that many of the current disputes with regard to both economic theory and economic policy have their common origin in a misconception about the nature of the economic problem of society. This misconception in turn is due to an erroneous transfer to social phenomena of the habits of thought we have developed in dealing with the phenomena of nature.

II

In ordinary language we describe by the word "planning" the complex of interrelated decisions about the allocation of our available resources. All economic activity is in this sense planning; and in any society in which many people collaborate, this planning, whoever does it, will in some measure have to be based on knowledge which, in the first instance, is not given to the planner but to somebody else, which somehow will have to be conveyed to the planner. The various ways in which the knowledge on which people base their plans is communicated to them is the crucial problem for any theory explaining the economic process. And the problem of what is the best way of utilizing knowledge initially dispersed among all the people is at least one of the main problems of economic policy—or of designing an efficient economic system.

The answer to this question is closely connected with that other question which arises here, that of *who* is to do the planning. It is about this question that all the dispute about "economic planning" centers. This is not a dispute about whether planning is to be done or not. It is a dispute as to whether planning is to be done centrally, by one authority for the whole economic system, or is to be divided among many individuals. Planning in the specific sense in which the term is used in contemporary controversy necessarily means central planning—direction of the whole economic system according to one unified plan. Competition, on the other hand, means decentralized planning by many separate persons. The half-way house between the two, about which many people talk but which few like when they see it, is the delegation of planning to organized industries, or, in other words, monopoly.

Which of these systems is likely to be more efficient depends mainly on the question under which of them we can expect that fuller use will be made of the existing knowledge. And this, in turn, depends on whether we are more likely to succeed in putting at the disposal of a single central authority all the knowledge which ought to be used but which is initially dispersed among many different individuals, or in conveying to the individuals such additional knowledge as they need in order to enable them to fit their plans in with those of others.

III

It will at once be evident that on this point the position will be different with respect to different kinds of knowledge; and the answer to our question will therefore largely turn on the relative importance of the different kinds of knowledge; those more likely to be at the disposal of particular individuals and those which we should with greater confidence expect to find in the possession of an authority made up of suitably chosen experts. If it is today so widely assumed that the latter will be in a better position, this is because one kind of knowledge, namely, scientific knowledge, occupies now so prominent a place in public imagination that we tend to forget that it is not the only kind that is relevant. It may be admitted that, so far as scientific knowledge is concerned, a body of suitably chosen experts may be in the best position to command all the best knowledge available—though this is of course merely shifting the difficulty to the problem of selecting the experts. What I wish to point out is that, even assuming that this problem can be readily solved, it is only a small part of the wider problem.

Today it is almost heresy to suggest that scientific knowledge is not the sum of all knowledge. But a little reflection will show that there is beyond question a body of very important but unorganized knowledge which cannot possibly be called scientific in the sense of knowledge of general rules: the knowledge of the particular circumstances of time and place. It is with respect to this that practically every individual has some advantage over all others in that he possesses unique information of which beneficial use might be made, but of which use can be made only if the decisions depending on it are left to him or are made with his active coöperation. We need to remember only how much we have to learn in any occupation after we have completed our theoretical training, how big a part of our working life we spend learning particular jobs, and how valuable an asset in all walks of life is knowledge of people, of local conditions, and special circumstances. To know of and put to use a machine not fully employed, or somebody's skill which could be better utilized, or to be aware of a surplus stock which can be drawn upon during an interruption of supplies, is socially quite as useful as the knowledge of better alternative techniques. And the shipper who earns his living from using otherwise empty or half-filled journeys of tramp-steamers, or the estate agent whose whole knowledge is almost exclusively one of temporary opportunities, or the *arbitrageur* who gains from local differences of commodity prices, are all performing eminently useful functions based on special knowledge of circumstances of the fleeting moment not known to others.

It is a curious fact that this sort of knowledge should today be generally regarded with a kind of contempt, and that anyone who by such knowledge gains an advantage over somebody better equipped with theoretical or technical knowledge is thought to have acted almost disreputably. To gain an advantage from better knowledge of facilities of communication or transport is sometimes regarded as almost dishonest, although it is quite as important that society make use of the best opportunities in this respect as in using the latest scientific discoveries. This prejudice has in a considerable measure affected the attitude toward commerce in general compared with that toward production. Even economists who regard themselves as definitely above the crude materialist fallacies of the past constantly commit the same mistake where activities directed toward the acquisition of such practical knowledge are concerned—apparently because in their scheme of things all such knowledge is supposed to be "given." The common idea now seems to be that all such knowledge should as a matter of course be readily at the command of everybody, and the reproach of irrationality leveled against the existing economic order is frequently based on the fact that it is not so available. This view disregards the fact that the method by which such knowledge can be made as widely available as possible is precisely the problem to which we have to find an answer.

IV

If it is fashionable today to minimize the importance of the knowledge of the particular circumstances of time and place, this is closely connected with the smaller importance which is now attached to change as such. Indeed, there are few points on which the assumptions made (usually only implicitly) by the "planners" differ from those of their opponents as much as with regard to the significance and frequency of changes which will make substantial alterations of production plans necessary. Of course, if detailed economic plans could be laid down for fairly long periods in advance and then closely adhered to, so that no further economic decisions of importance would be required, the task of drawing up a comprehensive plan governing all economic activity would appear much less formidable.

It is, perhaps, worth stressing that economic problems arise always and only in consequence of change. So long as things continue as before, or at least as they were expected to, there arise no new problems requiring a decision, no need to form a new plan. The belief that changes, or at least day-to-day adjustments, have become less important in modern times implies the contention that economic problems also have become less important. This belief in the decreasing importance of change is, for that reason, usually held by the same people who argue that the importance of economic considerations has been driven into the background by the growing importance of technological knowledge.

Is it true that, with the elaborate apparatus of modern production, economic decisions are required only at long intervals, as when a new factory is to be erected or a new process to be introduced? Is it true that, once a plant has been built, the rest is all more or less mechanical, determined by the character of the plant, and leaving little to be changed in adapting to the ever-changing circumstances of the moment?

The fairly widespread belief in the affirmative is not, so far as I can ascertain, borne out by the practical experience of the business man. In a competitive industry at any rate—and such an industry alone can serve as a test—the task of keeping cost from rising requires constant struggle, absorbing a great part of the energy of the manager. How easy it is for an inefficient manager to dissipate the differentials on which profitability rests, and that it is possible, with the same technical facilities, to produce with a great variety of costs, are among the commonplaces of business experience which do not seem to be equally familiar in the study of the economist. The very strength of the desire, constantly voiced by producers and engineers, to be able to proceed untrammeled by considerations of money costs, is eloquent testimony to the extent to which these factors enter into their daily work.

One reason why economists are increasingly apt to forget about the constant small changes which make up the whole economic picture is probably their growing preoccupation with statistical aggregates, which show a very much greater stability than the movements of the detail. The comparative stability of the aggregates cannot, however, be accounted for—as the statisticians seem occasionally to be inclined to do—by the "law of large numbers" or the mutual compensation of random changes. The number of elements with which we have to deal is not large enough for such accidental forces to produce stability. The continuous flow of goods and services is maintained by constant deliberate adjustments, by new dispositions made every day in the light of circumstances not known the day before, by *B* stepping in at once when *A* fails to deliver. Even the large and highly mechanized plant keeps going largely because of an environment upon which it can draw for all sorts of unexpected needs; tiles for its roof, stationery for its forms, and all the thousand and one kinds of equipment in which it cannot be self-contained and which the plans for the operation of the plant require to be readily available in the market.

This is, perhaps, also the point where I should briefly mention the fact that the sort of knowledge with which I have been concerned is knowledge of the kind which by its nature cannot enter into statistics and therefore cannot be conveyed to any central authority in statistical form. The statistics which such a central authority would have to use would have to be arrived at precisely by abstracting from minor differences between the things, by lumping together, as resources of one kind, items which differ as regards location, quality, and other particulars, in a way which may be very significant for the specific decision. It follows from this that central planning based on statistical information by its nature cannot take direct account of these circumstances of time and place, and that the central planner will have to find some way or other in which the decisions depending on them can be left to the "man on the spot."

V

If we can agree that the economic problem of society is mainly one of rapid adaptation to changes in the particular circumstances of time and place, it would seem to follow that the ultimate decisions must be left to the people who are familiar with these circumstances, who know directly of the relevant changes and of the resources immediately available to meet them. We cannot expect that this problem will be solved by first communicating all this knowledge to a central board which, after integrating *all* knowledge, issues its orders. We must solve it by some form of decentralization. But this answers only part of our problem. We need decentralization because only thus can we ensure that the knowledge of the particular circumstances of time and place will be promptly used. But the "man on the spot" cannot decide solely on the basis of his limited but intimate knowledge of the facts of his immediate surroundings. There still remains the problem of communicating to him such further information as he needs to fit his decisions into the whole pattern of changes of the larger economic system.

How much knowledge does he need to do so successfully? Which of the events which happen beyond the horizon of his immediate knowledge are of relevance to his immediate decision, and how much of them need he know?

There is hardly anything that happens anywhere in the world that *might* not have an effect on the decision he ought to make. But he need not know of these events as such, nor of *all* their effects. It does not matter for him *why* at the particular moment more screws of one size than of another are wanted, *why* paper bags are more readily available than canvas bags, or *why* skilled labor, or particular machine tools, have for the moment become more difficult to acquire. All that is significant for him is *how much more or less* difficult to procure they have become compared with other things with which he is also concerned, or how much more or less urgently wanted are the alternative things he produces or uses. It is always a question of the relative importance of the particular things with which he is concerned, and the causes which alter their relative importance are of no interest to him beyond the effect on those concrete things of his own environment.

It is in this connection that what I have called the economic calculus proper helps us, at least by analogy, to see how this problem can be solved, and in fact is being solved, by the price system. Even the single controlling mind, in possession of all the data for some small, self-contained economic system, would not—every time some small adjustment in the allocation of resources had to be made—go explicitly through all the relations between ends and means which might possibly be affected. It is indeed the great contribution of the pure logic of choice that it has demonstrated conclusively that even such a single mind could solve this kind of problem only by constructing and constantly using rates of equivalence (or "values," or "marginal rates of substitution"), *i.e.,* by attaching to each kind of scarce resource a numerical index which cannot be derived from any property possessed by that particular thing, but which reflects, or in which is condensed, its significance in view of the whole means-end structure. In any small change he will have to consider only these quantitative indices (or "values") in which all the relevant information is concentrated; and by adjusting the quantities one by one, he can appropriately rearrange his dispositions without having to solve the whole puzzle *ab initio,* or without needing at any stage to survey it at once in all its ramifications.

Fundamentally, in a system where the knowledge of the relevant facts is dispersed among many people, prices can act to coördinate the separate actions of different people in the same way as subjective values help the individual to coördinate the parts of his plan. It is worth contemplating for a moment a very simple and commonplace instance of the action of the price system to see what precisely it accomplishes. Assume that somewhere in the world a new opportunity for the use of some raw material, say tin, has arisen, or that one of the sources of supply of tin has been eliminated. It does not matter for our purpose—and it is very significant that it does not matter—which of these two causes has made tin more scarce. All that the users of tin need to know is that some of the tin they used to consume is now more profitably employed elsewhere, and that in consequence they must economize tin. There is no need for the great majority of them even to know where the more urgent need has arisen, or in favor of what other needs they ought to husband the supply. If only some of them know directly of the new demand, and switch resources over to it, and if the people who are aware of the new gap thus created in turn fill it from still other sources, the effect will rapidly spread throughout the whole economic system and influence not only all the uses of tin, but also those of its substitutes and the substitutes of these substitutes, the supply of all the things made of tin, and their substitutes, and so on; and all this without the great majority of those instrumental in bringing about these substitutions knowing anything at all about the original cause of these changes. The whole acts as one market, not because any of its members survey the whole field, but because their limited individual fields of vision sufficiently overlap so that through many intermediaries the relevant information is communicated to all. The mere fact that there is one price for any commodity—or rather that local prices are connected in a manner determined by the cost of transport, etc.—brings about the solution which (it is just conceptually possible) might have been arrived at by one single mind possessing all the information which is in fact dispersed among all the people involved in the process.

VI

We must look at the price system as such a mechanism for communicating information if we want to understand its real function—a function which, of course, it fulfills less perfectly as prices grow more rigid. (Even when quoted prices have become quite rigid, however, the forces which would operate through changes in price still operate to a considerable extent through changes in the other terms of the contract.) The most significant fact about this system is the economy of knowledge with which it operates, or how little the individual participants need to know in order to be able to take the right action. In abbreviated form, by a kind of symbol, only the most essential information is passed on, and passed on only to those concerned. It is more than a metaphor to describe the price system as a kind of machinery for registering change, or a system of telecommunications which enables individual producers to watch merely the movement of a few pointers, as an engineer might watch the hands of a few dials, in order to adjust their activities to changes of which they may never know more than is reflected in the price movement.

Of course, these adjustments are probably never "perfect" in the sense in which the economist conceives of them in his equilibrium analysis. But I fear that our theoretical habits of approaching the problem with the assumption of more or less perfect knowledge on the part of almost everyone has made us somewhat blind to the true function of the price mechanism and led us to apply rather misleading standards in judging its efficiency. The marvel is that in a case like that of a scarcity of one raw material, without an order being issued, without more than perhaps a handful of people knowing the cause, tens of thousands of people whose identity could not be ascertained by months of investigation, are made to use the material or its products more sparingly; *i.e.*, they move in the right direction. This is enough of a marvel even if, in a constantly changing world, not all will hit it off so perfectly that their profit rates will always be maintained at the same constant or "normal" level.

I have deliberately used the word "marvel" to shock the reader out of the complacency with which we often take the working of this mechanism for granted. I am convinced that if it were the result of deliberate human design, and if the people guided by the price changes understood that their decisions have significance far beyond their immediate aim, this mechanism would have been acclaimed as one of the greatest triumphs of the human mind. Its misfortune is the double one that it is not the product of human design and that the people guided by it usually do not know why they are made to do what they do. But those who clamor for "conscious direction"—and who cannot believe that anything which has evolved without design (and even without our understanding it) should solve problems which we should not be able to solve consciously—should remember this: The problem is precisely how to extend the span of our utilization of resources beyond the span of the control of any one mind; and, therefore, how to dispense with the need of conscious control and how to provide inducements which will make the individuals do the desirable things without anyone having to tell them what to do.

The problem which we meet here is by no means peculiar to economics but arises in connection with nearly all truly social phenomena, with language and most of our cultural inheritance, and constitutes really the central theoretical problem of all social science. As Alfred Whitehead has said in another connection, "It is a profoundly erroneous truism, repeated by all copy-books and by eminent people when they are making speeches, that we should cultivate the habit of thinking what we are doing. The precise opposite is the case. Civilization advances by extending the number of important operations which we can perform without thinking about them." This is of profound significance in the social field. We make constant use of formulas, symbols and rules whose meaning we do not understand and through the use of which we avail ourselves of the assistance of knowledge which individually we do not possess. We have developed these practices and institutions by building upon habits and institutions which have proved successful in their own sphere and which have in turn become the foundation of the civilization we have built up.

The price system is just one of those formations which man has learned to use (though he is still very far from having learned to make the best use of it) after he had stumbled upon it without understanding it. Through it not only a division of labor but also a coördinated utilization of resources based on an equally divided knowledge has become possible. The people who like to deride any suggestion that this may be so usually distort the argument by insinuating that it asserts that by some miracle just that sort of system has spontaneously grown up which is best suited to modern civilization. It is the other way round: man has been able to develop that division of labor on which our civilization is based because he happened to stumble upon a method which made it possible. Had he not done so he might still have developed some other, altogether different, type of civilization, something like the "state" of the termite ants, or some other altogether unimaginable type. All that we can say is that nobody has yet succeeded in designing an alternative system in which certain features of the existing one can be preserved which are dear even to those who most violently assail it—such as particularly the extent to which the individual can choose his pursuits and consequently freely use his own knowledge and skill.

VII

It is in many ways fortunate that the dispute about the indispensability of the price system for any rational calculation in a complex society is now no longer conducted entirely between camps holding different political views. The thesis that without the price system we could not preserve a society based on such extensive division of labor as ours was greeted with a howl of derision when it was first advanced by von Mises twenty-five years ago. Today the difficulties which some still find in accepting it are no longer mainly political, and this makes for an atmosphere much more conducive to reasonable discussion. When we find Leon Trotsky arguing that "economic accounting is unthinkable without market relations"; when Professor Oscar Lange promises Professor von Mises a statue in the marble halls of the future Central Planning Board; and when Professor Abba P. Lerner rediscovers Adam Smith and emphasizes that the essential utility of the price system consists in inducing the individual, while seeking his own interest, to do what is in the general interest, the differences can indeed no longer be ascribed to political prejudice. The remaining dissent seems clearly to be due to purely intellectual, and more particularly methodological, differences.

A recent statement by Professor Joseph Schumpeter in his *Capitalism, Socialism and Democracy* provides a clear illustration of one of the methodological differences which I have in mind. Its author is preeminent among those economists who approach economic phenomena in the light of a certain branch of positivism. To him these phenomena accordingly appear as objectively given quantities of commodities impinging directly upon each other, almost, it would seem, without any intervention of human minds. Only against this background can I account for the following (to me startling) pronouncement. Professor Schumpeter argues that the possibility of a rational calculation in the absence of markets for the factors of production follows for the theorist "from the elementary proposition that consumers in evaluating ('demanding') consumers' goods *ipso facto* also evaluate the means of production which enter into the production of these goods."[1]

Taken literally, this statement is simply untrue. The consumers do nothing of the kind. What Professor Schumpeter's "*ipso facto*" presumably means is that the valuation of the factors of production is

[1] J. Schumpeter, *Capitalism, Socialism, and Democracy* (New York, Harper, 1942), p. 175. Professor Schumpeter is, I believe, also the original author of the myth that Pareto and Barone have "solved" the problem of socialist calculation. What they, and many others, did was merely to state the conditions which a rational allocation of resources would have to satisfy, and to point out that these were essentially the same as the conditions of equilibrium of a competitive market. This is something altogether different from showing how the allocation of resources satisfying these conditions can be found in practice. Pareto himself (from whom Barone has taken practically everything he has to say), far from claiming to have solved the practical problem, in fact explicitly denies that it can be solved without the help of the market. See his *Manuel d'économie pure* (2nd ed., 1927), pp. 233-34. The relevant passage is quoted in an English translation at the beginning of my article ➙ "Socialist Calculation: The Competitive 'Solution,'" in *Economica*, New Series, Vol. VIII, No. 26 (May, 1940), p. 125.

implied in, or follows necessarily from, the valuation of consumers' goods. But this, too, is not correct. Implication is a logical relationship which can be meaningfully asserted only of propositions simultaneously present to one and the same mind. It is evident, however, that the values of the factors of production do not depend solely on the valuation of the consumers' goods but also on the conditions of supply of the various factors of production. Only to a mind to which all these facts were simultaneously known would the answer necessarily follow from the facts given to it. The practical problem, however, arises precisely because these facts are never so given to a single mind, and because, in consequence, it is necessary that in the solution of the problem knowledge should be used that is dispersed among many people.

The problem is thus in no way solved if we can show that all the facts, *if* they were known to a single mind (as we hypothetically assume them to be given to the observing economist), would uniquely determine the solution; instead we must show how a solution is produced by the interactions of people each of whom possesses only partial knowledge. To assume all the knowledge to be given to a single mind in the same manner in which we assume it to be given to us as the explaining economists is to assume the problem away and to disregard everything that is important and significant in the real world.

That an economist of Professor Schumpeter's standing should thus have fallen into a trap which the ambiguity of the term "datum" sets to the unwary can hardly be explained as a simple error. It suggests rather than there is something fundamentally wrong with an approach which habitually disregards an essential part of the phenomena with which we have to deal: the unavoidable imperfection of man's knowledge and the consequent need for a process by which knowledge is constantly communicated and acquired. Any approach, such as that of much of mathematical economics with its simultaneous equations, which in effect starts from the assumption that people's *knowledge* corresponds with the objective *facts* of the situation, systematically leaves out what is our main task to explain. I am far from denying that in our system equilibrium analysis has a useful function to perform. But when it comes to the point where it misleads some of our leading thinkers into believing that the situation which it describes has direct relevance to the solution of practical problems, it is time that we remember that it does not deal with the social process at all and that it is no more than a useful preliminary to the study of the main problem.

[2]

THE JOURNAL OF ECONOMIC HISTORY

VOL. VII NOVEMBER 1947 NO. 2

The Creative Response in Economic History

I

ECONOMIC historians and economic theorists can make an interesting and socially valuable journey together, if they will. It would be an investigation into the sadly neglected area of economic change.

As anyone familiar with the history of economic thought will immediately recognize, practically all the economists of the nineteenth century and many of the twentieth have believed uncritically that all that is needed to explain a given historical development is to indicate conditioning or causal factors, such as an increase in population or the supply of capital. But this is sufficient only in the rarest of cases. As a rule, no factor acts in a uniquely determined way and, whenever it does not, the necessity arises of going into the details of its *modus operandi,* into the mechanisms through which it acts. Examples will illustrate this. Sometimes an increase in population actually has no other effect than that predicated by classical theory—a fall in per capita real income;[1] but, at other times, it may have an energizing effect that induces new developments with the result that per capita real income rises. Or a protective duty may have no other effect than to increase the price of the protected commodity and, in consequence, its output; but it may also induce a complete reorganization of the protected industry which eventually results in an increase in output so great as to reduce the price below its initial level.

What has not been adequately appreciated among theorists is the

[1] Even within the assumptions of classical theory this is not necessarily true; but we need not go into this.

distinction between different kinds of reaction to changes in "condition." Whenever an economy or a sector of an economy adapts itself to a change in its data in the way that traditional theory describes, whenever, that is, an economy reacts to an increase in population by simply adding the new brains and hands to the working force in the existing employments, or an industry reacts to a protective duty by expansion within its existing practice, we may speak of the development as *adaptive response*. And whenever the economy or an industry or some firms in an industry do something else, something that is outside of the range of existing practice, we may speak of *creative response*.

Creative response has at least three essential characteristics. First, from the standpoint of the observer who is in full possession of all relevant facts, it can always be understood *ex post*; but it can practically never be understood *ex ante*; that is to say, it cannot be predicted by applying the ordinary rules of inference from the pre-existing facts. This is why the "how" in what has been called above the "mechanisms" must be investigated in each case. Secondly, creative response shapes the whole course of subsequent events and their "long-run" outcome. It is not true that both types of responses dominate only what the economist loves to call "transitions," leaving the ultimate outcome to be determined by the initial data. Creative response changes social and economic situations for good, or, to put it differently, it creates situations from which there is no bridge to those situations that might have emerged in its absence. This is why creative response is an essential element in the historical process; no deterministic credo avails against this. Thirdly, creative response—the frequency of its occurrence in a group, its intensity and success or failure—has obviously something, be that much or little, to do (a) with quality of the personnel available in a society, (b) with relative quality of personnel, that is, with quality available to a particular field of activity relative to quality available, at the same time, to others, and (c) with individual decisions, actions, and patterns of behavior. Accordingly, a study of creative response in business becomes coterminous with a study of entrepreneurship. The mechanisms of economic change in capitalist society pivot on entrepreneurial activity.[2] Whether we emphasize opportunity or conditions, the responses of individuals or of groups, it is patently true that in capitalist society objective opportunities or conditions act through

[2] The function itself is not absent from other forms of society; but capitalist entrepreneurship is a sufficiently distinct phenomenon to be singled out.

entrepreneurial activity, analysis of which is at the very least a highly important avenue to the investigation of economic changes in the capitalist epoch.[3] This is compatible with widely different views about its importance as an "ultimate cause."

Seen in this light, the entrepreneur and his function are not difficult to conceptualize: the defining characteristic is simply the doing of new things or the doing of things that are already being done in a new way (innovation).[4] It is but natural, and in fact it is an advantage, that such a definition does not draw any sharp line between what is and what is not "enterprise." For actual life itself knows no such sharp division, though it shows up the type well enough. It should be observed at once that the "new thing" need not be spectacular or of historic importance. It need not be Bessemer steel or the explosion motor. It can be the Deerfoot sausage. To see the phenomenon even in the humblest levels of the business world is quite essential though it may be difficult to find the humble entrepreneurs historically.

Distinction from other functions with which enterpreneurship is frequently but not necessarily associated—just as "farmership" is frequently but not necessarily associated with the ownership of land and with the activity of a farm hand—does not present conceptual difficulties either. One necessary distinction is that between enterprise and management: evidently it is one thing to set up a concern embodying a new idea and another thing to head the administration of a going concern, however much the two may shade off into each other. Again, it is essential to note that the entrepreneurial function, though facilitated by the ownership of means, is not identical with that of the capitalist.[5] New light is urgently needed on the relation between the

[3] Arthur H. Cole has opened new vistas in this area in his presidential address before the Economic History Association, "An Approach to the Study of Entrepreneurship," THE TASKS OF ECONOMIC HISTORY (Supplemental Issue of THE JOURNAL OF ECONOMIC HISTORY), VI (1946), 1–15.

[4] An exact definition can be provided by means of the concept of production functions. On this, see Oscar Lange, "A Note on Innovations," *Review of Economic Statistics,* XXV (1943), 19–25.

[5] It is sometimes held that entrepreneurship, although it did not require antecedent ownership of capital (or very little of it) in the early days of capitalism, tends to become dependent upon it as time goes on, especially in the epoch of giant corporations. Nothing could be further from the truth. In the course of the nineteenth century, it became increasingly easier to obtain other people's money by methods other than the partnership, and in our own time promotion within the shell of existing corporations offers a much more convenient access to the entrepreneurial functions than existed in the world of owner-managed firms. Many a would-be entrepreneur of today does not found a firm, not because he could not do so, but simply because he prefers the other method.

two, especially because of the cant phrases that are current on this topic. In the third place, it is particularly important to distinguish the entrepreneur from the "inventor." Many inventors have become entrepreneurs and the relative frequency of this case is no doubt an interesting subject to investigate, but there is no necessary connection between the two functions. The inventor produces ideas, the entrepreneur "gets things done," which may but need not embody anything that is scientifically new. Moreover, an idea or scientific principle is not, by itself, of any importance for economic practice: the fact that Greek science had probably produced all that is necessary in order to construct a steam engine did not help the Greeks or Romans to build a steam engine; the fact that Leibnitz suggested the idea of the Suez Canal exerted no influence whatever on economic history for two hundred years. And as different as the functions are the two sociological and psychological types.[6] Finally, "getting new things done" is not only a distinct process but it is a process which produces consequences that are an essential part of capitalist reality. The whole economic history of capitalism would be different from what it is if new ideas had been currently and smoothly adopted, as a matter of course, by all firms to whose business they were relevant. But they were not. It is in most cases only one man or a few men who see the new possibility and are able to cope with the resistances and difficulties which action always meets with outside of the ruts of established practice. This accounts for the large gains that success often entails, as well as for the losses and vicissitudes of failure. These things are important. If, in every individual case, the difficulties may indeed be called transitional, they are transitional difficulties which are never absent in the economy as a whole and which dominate the atmosphere of capitalist life permanently. Hence it seems appropriate to keep "invention" distinct from "innovation."

The definition that equates enterprise to innovation is a very abstract one. Some classifications that are richer in content may be noticed because of their possible use in drawing up plans for specific pieces of research. There is the obvious classification—historical and systematic—of the phenomena of enterprise according to institutional forms, such as the medieval trading company, the later "chartered companies," the partnership, the modern "corporation," and the like, on all of which

[6] The relation between the two has attracted interest before. See, e.g., F. W. Taussig, *Inventors and Money-Makers* (New York: The Macmillan Company, 1915).

there exists a vast amount of historical work.[7] The interaction of institutional forms and entrepreneurial activity, the "shaping" influence of the former and the "bursting" influence of the latter, is, as has already been intimated, a major topic for further inquiry. Closely connected with this classification is the old one according to fields of activity—commerce, industry, finance [8]—which has been refined by the following distinctions: enterprise that introduces "new" commodities; enterprise that introduces technological novelties into the production of "old" commodities; enterprise that introduces new commercial combinations such as the opening up of new markets for products or new sources of supply of materials; enterprise that consists in reorganizing an industry, for instance, by making a monopoly out of it.[9]

But there are other classifications that may prove helpful. We may classify entrepreneurs according to origins and sociological types: feudal lords and aristocratic landowners, civil servants—particularly important, for instance, in Germany after the Thirty Years' War, especially in mining—farmers, workmen, artisans, members of the learned professions, all embarked upon enterprise as has often been noticed, and it is highly interesting from several points of view to clear up this matter. Or we may try to classify entrepreneurial performances according to the precise nature of the "function" filled and the aptitudes (some may even add motivation) involved. Since all this presumably changed significantly in the course of the capitalist epoch, economic historians are particularly qualified for work on this line.

Though the phrase "getting a new thing done" may be adequately comprehensive, it covers a great many different activities which, as the observer stresses one more than another or as his material displays one

---

[7] Gustav von Schmoller introduced the subject into his general treatise (*Grundriss*) of 1904. But the novelty consisted only in the systematic use he made of the result of historical research. Less systematically, the subject had entered general treatises before.

[8] Financial institutions and practices enter our circle of problems in three ways: they are "auxiliary and conditioning"; banking may be the object of entrepreneurial activity, that is to say, the introduction of new banking practices may constitute enterprise; and bankers (or other "financiers") may use the means at their command in order to embark upon commercial and industrial enterprise themselves (for example, John Law). See the recent book by Fritz Redlich, *The Molding of American Banking—Men and Ideas* (New York: Hafner Publishing Company, 1947).

[9] This case emphasizes the desirability, present also in others, of divesting our idea of entrepreneurial performance of any preconceived value judgment. Whether a given entrepreneurial success benefits or injures society or a particular group within society is a question that must be decided on the merits of each case. Enterprise that results in a monopoly position, even if undertaken for the sole purpose of securing monopoly gains, is not necessarily antisocial in its total effect although it often is.

more than another, may, locally, temporarily, or generally, lend different colors to entrepreneurship. In some cases, or to some observers, it may be the activity of "setting up" or "organizing" that stands out from the others; in other cases, or for other observers, it may be the breaking down of the resistances of the environment; in still other cases, or for still other observers, simply leadership or, again, salesmanship. Thus, it seems to me, there was a type of entrepreneur in early capitalist industry that is best described as a "fixer." Modern history furnishes many instances of entrepreneurship vested in a company promotor.[10] The typical industrial entrepreneur of the nineteenth century was perhaps the man who put into practice a novel method of production by embodying it in a new firm and who then settled down into a position of owner-manager of a company, if he was successful, or of stockholding president of a company, getting old and conservative in the process. In the large-scale corporation of today, the question that is never quite absent arises with a vengeance, namely, who should be considered as the entrepreneur. In a well-known book, R. A. Gordon has presented much interesting material bearing upon this question.[11]

II

The economic nature, amount, and distribution of the returns to entrepreneurial activity constitute another set of problems on which investigation may be expected to shed much-needed light. Conceptual difficulties confront us here even before we come up against the still more formidable difficulties of fact finding. For the "profit" of the English classics, which was analyzed by J. S. Mill into wages of management, premiums for risk, and interest on owned capital, was a return to normal business activity and something quite different from, though influenced by, the gain of successful enterprise in our sense of the term. What the latter is can best be explained by considering a special case. Suppose that a man, realizing the possibility of producing acceptable caviar from sawdust, sets up the Excelsior Caviar concern and makes it a success. If this concern is too small to influence the prices of either the product or the factors of production, he will sell the

[10] In a sense, the promotor who does nothing but "set up" new business concerns might be considered as the purest type of entrepreneur. Actually, he is mostly not more than a financial agent who has little, if any, title to entrepreneurship—no more than the lawyer who does the legal work involved. But there are important exceptions to this.

[11] Robert A. Gordon, *Business Leadership in the Large Corporation* (Washington, D.C.: The Brookings Institution, 1945).

former and buy the latter at current prices. If, however, he turns out the unit of caviar more cheaply than his competitors, owing to his use of a much cheaper raw material, he will for a time, that is, until other firms copy his method, make (essentially temporary) surplus gains. These gains are attributable to personal exertion. Hence they might be called wages. They may with equal justice be attributed to the fact that, for a time, his method is exclusively his own. Hence they might also be called monopoly gains. But whether we elect to call them wages or monopoly gains, we must add immediately that they are a special kind of wages or monopoly gains that differ in important respects from what we usually mean to denote by these terms. And so we had better call them simply entrepreneurial gains or profit. However, it should be observed that if this venture means a "fortune," this fortune does not typically arise from the actual net receipts being saved up and invested in the same or some other business. Essentially, it emerges as a *capital* gain, that is, as the discounted value of the stream of prospective excess returns.

In this simple case, which, however, does constitute a type, the investigator is not confronted with difficulties other than those involved in fact finding. Also, it is clear what happens with that surplus gain: in this case the entrepreneurial gain goes to the entrepreneur,[12] and we can also see, if we have the facts, how, to use a current phrase, the "fruits of the progress involved are handed to consumers and workmen." The speed of this process of "handing on" varies widely, but it would always work, in isolated cases like the one under discussion, through a fall in the price of the product to the new level of costs, which is bound to occur whenever competition steps up to the successful concern. But even here we meet the practice of innovators striving to keep their returns alive by means of patents and in other ways. The gains described above shade off into gains from purposive restriction of competition and create difficulties of diagnosis that are sometimes insurmountable.[13] Cumulation of carefully analyzed historical cases is

[12] It should be obvious that this does not mean that the whole social gain resulting from the enterprise goes to the entrepreneur. But the question of appraisal of social gains from entrepreneurship, absolute and relative to the entrepreneurial shares in them, and of the social costs involved in a system that relies on business interests to carry out its innovations, is so complex and perhaps even hopeless that I beg to excuse myself from entering into it.

[13] Still more difficult is, of course, responsible appraisal, that is to say, appraisal that is not content with popular slogans. Measures to keep surplus gains alive no doubt slow up the process of "handing on the fruits of progress." But the knowledge that such measures are available may be necessary in order to induce anyone to embark upon certain ventures. There

the best means of shedding light on these things, of supplying the theorist with strategic assumptions, and banishing slogans.

If innovations are neither individually small nor isolated events, complications crowd upon us. Entrepreneurial activity then affects wage and interest rates from the outset and becomes a factor—the fundamental factor in my opinion—in booms and depressions. This is one reason, but not the only one, why entrepreneurial gains are not net returns (1) to the whole set of people who attempt entrepreneurial ventures, (2) to the industrial sector in which innovation occurs, (3) to the capitalist interests that finance entrepreneurial activity and to the capitalist class as a whole.

Concerning the first point, I might have made my special case more realistic by assuming that several or many people try their hands at producing that caviar but that all but one fail to produce a salable product before the success of this one presents an example to copy. The gains of the successful entrepreneur and of the capitalists who finance him—for whenever capital finances enterprise the interest is paid out of the entrepreneurial gains, a fact that is very important for our grasp of the interest phenomenon—should be related not to his effort and their loan but to the effort and the loans of all the entrepreneurs and capitalists who made attempts and lost. The presence of gains to enterprise so great as to impress us as spectacular and, from the standpoint of society, irrational is then seen to be compatible with a negative return to entrepreneurs and financing capitalists as a group.[14]

It is similarly clear that entrepreneurial gain is not a net accretion to the returns of the industrial sector in which it occurs. The impact of the new product or method spells losses to the "old" firms. The competition of the man with a significantly lower cost curve is, in fact, the really effective competition that in the end revolutionizes the industry. Detailed investigation of this process which may take many forms might teach us much about the actual working of capitalism that we are but dimly perceiving as yet.

also may be other compensating advantages to such measures, particularly where rapid introduction into general use of new methods would involve severe dislocations of labor, and where entrepreneurial gains are important sources of venture capital.

14 Whether this actually is so in any particular case is, of course, extremely difficult to establish. The successes stand out, statistically and otherwise; the failures are apt to escape notice. This is one of the reasons why economists seem so much impressed by peak successes. Another reason for faulty appraisal is neglect of the fact that spectacular gains may stimulate more effectively than would the same sum if more equally distributed. This is a question that no speculation can decide. Only collection of facts can tell us how we are to frame our theory.

Concerning the third point, while we have a fair amount of information about how the working class fares in the process of economic change, in respect to both real wages and employment, we know much less about that elusive entity, capital, that is being incessantly destroyed and re-created. That the theorist's teaching, according to which capital "migrates" from declining to rising industries, is unrealistic is obvious: the capital "invested" in railroads does not migrate into trucking and air transportation but will perish in and with the railroads. Investigation into the histories of industries, concerns, and firms, including surveys of sectors in order to point out how long a typical firm stays in business and how and why it drops out, might dispel many a preconceived notion on this subject.

III

Finally, I should like to touch one more set of problems on which we may expect light from historical analysis, namely, the problems that come within the range of the question: does the importance of the entrepreneurial function decline as time goes on? There are serious reasons for believing that it does. The entrepreneurial performance involves, on the one hand, the ability to perceive new opportunities that cannot be proved at the moment at which action has to be taken, and, on the other hand, will power adequate to break down the resistance that the social environment offers to change. But the range of the provable expands, and action upon flashes or hunches is increasingly replaced by action that is based upon "figuring out." And modern milieus may offer less resistance to new methods and new goods than used to be the case. So far as this is so, the element of personal intuition and force would be less essential than it was: it could be expected to yield its place to the teamwork of specialists; in other words, improvement could be expected to become more and more automatic. Our impression to this effect is reinforced by parallel phenomena in other fields of activity. For instance, a modern commander no doubt means less in the outcome of a war than commanders meant of old, and for the same reasons; campaigns have become more calculable than they used to be and there is less scope for personal leadership.

But this is at present only an impression. It is for the historian to establish or to refute it. If, however, it should stand up under research, this would be a result of the utmost importance. We should be led to expect that the whole mechanism of economic development will change significantly. Among other things, the economy would progressively bureaucratize itself. There are, in fact, many symptoms of this. And consequences would extend far beyond the field of economic phenomena. Just as warrior classes have declined in importance ever since warfare—and especially the management of armies in the field—began to be increasingly "mechanized," so the business class may decline in importance, as its most vital figure, the entrepreneur, progressively loses his most essential function. This would mean a different social structure.

Therefore, the sociology of enterprise reaches much further than is implied in questions concerning the conditions that produce and shape, favor or inhibit entrepreneurial activity. It extends to the structure and the very foundations of, at least, capitalist society or the capitalist sector of any given society. The quickest way of showing this starts from recognition of the facts that, just as the rise of the bourgeois class as a whole is associated with success in commercial, industrial, and financial enterprise, so the rise of an individual family to "capitalist" status within that class is typically [15] associated with entrepreneurial success; and that the elimination of a family from the "capitalist" class is typically associated with the loss of those attitudes and aptitudes of industrial leadership or alertness that enter our picture of the entrepreneurial type of businessman.

Now these facts, if they are facts, might teach us a lot about such fundamental problems as the nature of the class structure of capitalist society; the sort of class civilization which it develops and which differs so characteristically from the class civilization of feudal society; its schema of values; its politics, especially its attitudes to state and church and war; its performance and failures; its degree of durability. But a great deal of work needs to be done in order to arrive at scientifically defensible opinions about all these and cognate things. First of all, these "facts" must be established. How far is it really true, for instance, that entrepreneurs, while not forming a social class themselves but originating in almost all existing strata, do "feed" or renew the capitalist stratum? To put it differently, does the latter recruit itself through entrepreneurial successes? Or, to put it still differently,

[15] That is to say, successful entrepreneurship is that method of rising in the social scale that is characteristic of the capitalist blueprint. It is, of course, not the only method. First, there are other possibilities within the economic sphere, such as possession of an appreciating natural agent (for example, urban land) or mere speculation or even, occasionally, success in mere administration that need not partake of the specifically entrepreneurial element. Secondly, there are possibilities outside the business sphere, for business success is no more the only method of rising in capitalist society than knightly service was in feudal society.

does the "typical" history of industrial families lead back to entrepreneurial performances that "created" a concern which then, for a time, yielded capitalistic surpluses by being merely "administrated" with more or less efficiency? How much statistical truth is there in the slogan: "Three generations from overalls to overalls"? Secondly, what is, as measured by observable results, the economic and cultural, also political, importance of the further fact that, though the entrepreneurial function cannot be transmitted by inheritance, except, possibly, by biological inheritance, the financial or industrial position that has been created can? How much truth is there in the contention that the industrial family interest is, in capitalist society, the guardian of the nation's economic future?

These questions, which could be readily multiplied, have often attracted attention. Every textbook of economic history contains some material about the origins of entrepreneurs of historical standing, and a number of studies have been inspired by full awareness of the importance of the answers for our understanding of capitalist society and of the ways in which it works.[16] But these studies are few and that attention has been desultory. We do not know enough in order to form valid generalizations or even enough to be sure whether there are any generalizations to form. As it is, most of us as economists have some opinions on these matters. But these opinions have more to do with our preconceived ideas or ideals than with solid fact, and our habit of illustrating them by stray instances that have come under our notice is obviously but a poor substitute for serious research. Veblen's—or, for that matter, Bucharin's—*Theory of the Leisure Class* exemplifies well what I mean. It is brilliant and suggestive. But it is an impressionistic essay that does not come to grips with the real problems involved. Yet there is plenty of material. A great and profitable task awaits those who undertake it.

Harvard University JOSEPH A. SCHUMPETER

[16] An example is the study by F. J. Marquis and S. J. Chapman on the managerial stratum of the Lancashire cotton industry in the *Journal of the Royal Statistical Society*, LXXV, Pt. III (1912), 293–306.

[3]

Economic Welfare and the Allocation of Resources for Invention

KENNETH J. ARROW
THE RAND CORPORATION

INVENTION is here interpreted broadly as the production of knowledge. From the viewpoint of welfare economics, the determination of optimal resource allocation for invention will depend on the technological characteristics of the invention process and the nature of the market for knowledge.

The classic question of welfare economics will be asked here: to what extent does perfect competition lead to an optimal allocation of resources? We know from years of patient refinement that competition insures the achievement of a Pareto optimum under certain hypotheses. The model usually assumes among other things, that (1) the utility functions of consumers and the transformation functions of producers are well-defined functions of the commodities in the economic system, and (2) the transformation functions do not display indivisibilities (more strictly, the transformation sets are convex). The second condition needs no comment. The first seems to be innocuous but in fact conceals two basic assumptions of the usual models. It prohibits uncertainty in the production relations and in the utility functions, and it requires that all the commodities relevant either to production or to the welfare of individuals be traded on the market. This will not be the case when a commodity for one reason or another cannot be made into private property.

We have then three of the classical reasons for the possible failure of perfect competition to achieve optimality in resource allocation: indivisibilities, inappropriability, and uncertainty. The first problem has been much studied in the literature under the heading of marginal-cost pricing and the second under that of divergence between social and private benefit (or cost), but the theory of optimal allocation of resources under uncertainty has had much less attention. I will summarize what formal theory exists and then point to the critical notion of information, which arises only in the context of uncertainty. The

NOTE: I have benefited greatly from the comments of my colleague, William Capron. I am also indebted to Richard R. Nelson, Edward Phelps, and Sidney Winter of The RAND Corporation for their helpful discussion.

economic characteristics of information as a commodity and, in particular, of invention as a process for the production of information are next examined. It is shown that all three of the reasons given above for a failure of the competitive system to achieve an optimal resource allocation hold in the case of invention. On theoretical grounds a number of considerations are adduced as to the likely biases in the misallocation and the implications for economic organization.[1]

Resource Allocation under Uncertainty

The role of the competitive system in allocating uncertainty seems to have received little systematic attention.[2] I will first sketch an ideal economy in which the allocation problem can be solved by competition and then indicate some of the devices in the real world which approximate this solution.

Suppose for simplicity that uncertainty occurs only in production relations. Producers have to make a decision on inputs at the present moment, but the outputs are not completely predictable from the inputs. We may formally describe the outputs as determined by the inputs and a "state of nature" which is unknown to the producers. Let us define a "commodity-option" as a commodity in the ordinary sense labeled with a state of nature. This definition is analogous to the differentiation of a given physical commodity according to date in capital theory or according to place in location theory. The production of a given commodity under uncertainty can then be described as the production of a vector of commodity-options.

This description can be most easily exemplified by reference to agricultural production. The state of nature may be identified with the weather. Then, to any given set of inputs there corresponds a number of bushels of wheat if the rainfall is good and a different number if rainfall is bad. We can introduce intermediate conditions of rainfall

[1] For other analyses with similar points of view, see R. R. Nelson, "The Simple Economics of Basic Scientific Research," *Journal of Political Economy*, 1959, pp. 297–306; and C. J. Hitch, "The Character of Research and Development in a Competitive Economy," The RAND Corporation, p. 1297, May 1958.

[2] The first studies I am aware of are the papers of M. Allais and myself, both presented in 1952 to the Colloque International sur le Risque in Paris; see M. Allais, "Généralisation des théories de l'équilibre économique général et du rendement social au cas du risque," and K. J. Arrow, "Rôle des valeurs bousières pour la répartition la meilleure des risques," both in *Économètrie*, Colloques Internationaux du Centre National de la Recherche Scientifique, Vol. XL, Paris, Centre National de la Recherche Scientifique, 1953. Allais' paper has also appeared in *Econometrica*, 1953, pp. 269–290. The theory has received a very elegant generalization by G. Debreu in *Theory of Values*, New York, Wiley, 1959, Chap. VII.

in any number as alternative states of nature; we can increase the number of relevant variables which enter into the description of the state of nature, for example by adding temperature. By extension of this procedure, we can give a formal description of any kind of uncertainty in production.

Suppose—and this is the critical idealization of the economy—we have a market for all commodity-options. What is traded on each market are contracts in which the buyers pay an agreed sum and the sellers agree to deliver prescribed quantities of a given commodity *if* a certain state of nature prevails and nothing if that state of nature does not occur. For any given set of inputs, the firm knows its output under each state of nature and sells a corresponding quantity of commodity-options; its revenue is then completely determined. It may choose its inputs so as to maximize profits.

The income of consumers is derived from their sale of supplies, including labor, to firms and their receipt of profits, which are assumed completely distributed. They purchase commodity-options so as to maximize their expected utility given the budget restraint imposed by their incomes. An equilibrium is reached on all commodity-option markets, and this equilibrium has precisely the same Pareto-optimality properties as competitive equilibrium under certainty.

In particular, the markets for commodity-options in this ideal model serve the function of achieving an optimal allocation of risk bearing among the members of the economy. This allocation takes account of differences in both resources and tastes for risk bearing. Among other implications, risk bearing and production are separated economic functions. The use of inputs, including human talents, in their most productive mode is not inhibited by unwillingness or inability to bear risks by either firms or productive agents.

But the real economic system does not possess markets for commodity-options. To see what substitutes exist, let us first consider a model economy at the other extreme, in that no provisions for reallocating risk bearing exist. Each firm makes its input decisions; then outputs are produced as determined by the inputs and the state of nature. Prices are then set to clear the market. The prices that finally prevail will be a function of the state of nature.

The firm and its owners cannot relieve themselves of risk bearing in this model. Hence any unwillingness or inability to bear risks will give rise to a nonoptimal allocation of resources, in that there will be discrimination against risky enterprises as compared with the optimum. A preference for risk might give rise to misallocation in the opposite direction, but the limitations of financial resources are likely to make underinvestment in risky enterprises more likely than the opposite. The inability of individuals to buy protection against uncertainty similarly gives rise to a loss of welfare.

In fact, a number of institutional arrangements have arisen to mitigate the problem of assumption of risk. Suppose that each firm and individual in the economy could forecast perfectly what prices would be under each state of nature. Suppose further there were a lottery on the states of nature, so that before the state of nature is known any individual or firm may place bets. Then it can be seen that the effect from the viewpoint of any given individual or firm is the same as if there were markets for commodity-options of all types, since any commodity-option can be achieved by a combination of a bet on the appropriate state of nature and an intention to purchase or sell the commodity in question if the state of nature occurs.

References to lotteries and bets may smack of frivolity, but we need only think of insurance to appreciate that the shifting of risks through what are in effect bets on the state of nature is a highly significant phenomenon. If insurance were available against any conceivable event, it follows from the preceding discussion that optimal allocation would be achieved. Of course, insurance as customarily defined covers only a small range of events relevant to the economic world; much more important in shifting risks are securities, particularly common stocks and money. By shifting freely their proprietary interests among different firms, individuals can to a large extent bet on the different states of nature which favor firms differentially. This freedom to insure against many contingencies is enhanced by the alternatives of holding cash and going short.

Unfortunately, it is only too clear that the shifting of risks in the real world is incomplete. The great predominance of internal over external equity financing in industry is one illustration of the fact that securities do not completely fulfill their allocative role with respect to risks. There are a number of reasons why this should be so, but I will confine myself to one, of special significance with regard to invention. In insurance practice, reference is made to the moral factor as a limit to the possibilities of insurance. For example, a fire insurance policy cannot exceed in amount the value of the goods insured. From the purely actuarial standpoint, there is no reason for this limitation;

the reason for the limit is that the insurance policy changes the incentives of the insured, in this case, creating an incentive for arson or at the very least for carelessness. The general principle is the difficulty of distinguishing between a state of nature and a decision by the insured. As a result, any insurance policy and in general any device for shifting risks can have the effect of dulling incentives. A fire insurance policy, even when limited in amount to the value of the goods covered, weakens the motivation for fire prevention. Thus, steps which improve the efficiency of the economy with respect to risk bearing may decrease its technical efficiency.

One device for mitigating the adverse incentive effects of insurance is coinsurance; the insurance extends only to part of the amount at risk for the insured. This device is used, for example, in coverage of medical risks. It clearly represents a compromise between incentive effects and allocation of risk bearing, sacrificing something in both directions.

Two exemplifications of the moral factor are of special relevance in regard to highly risky business activities, including invention. Success in such activities depends on an inextricable tangle of objective uncertainties and decisions of the entrepreneurs and is certainly uninsurable. On the other hand, such activities should be undertaken if the expected return exceeds the market rate of return, no matter what the variance is.[3] The existence of common stocks would seem to solve the allocation problem; any individual stockholder can reduce his risk by buying only a small part of the stock and diversifying his portfolio to achieve his own preferred risk level. But then again the actual managers no longer receive the full reward of their decisions; the shifting of risks is again accompanied by a weakening of incentives to efficiency. Substitute motivations whether pecuniary, such as executive compensation and profit sharing, or nonpecuniary, such as prestige, may be found, but the dilemma of the moral factor can never be completely resolved.

A second example is the cost-plus contract in one of its various forms. When production costs on military items are highly uncertain, the military establishment will pay, not a fixed unit price, but the cost of production plus an amount which today is usually a fixed fee. Such a contract could be regarded as a combination of a fixed-price contract with an insurance against costs. The insurance premium could be

[3] The validity of this statement depends on some unstated assumptions, but the point to be made is unaffected by minor qualifications.

regarded as the difference between the fixed price the government would be willing to pay and the fixed fee.

Cost-plus contracts are necessitated by the inability or unwillingness of firms to bear the risks. The government has superior risk bearing ability and so the burden is shifted to it. It is then enabled to buy from firms on the basis of their productive efficiency rather than their risk bearing ability, which may be only imperfectly correlated. But cost-plus contracts notoriously have their adverse allocative effects.[4]

This somewhat lengthy digression on the theory of risk bearing seemed necessitated by the paucity of literature on the subject. The main conclusions to be drawn are the following: (1) the economic system has devices for shifting risks, but they are limited and imperfect; hence, one would expect an underinvestment in risky activities; (2) it is undoubtedly worthwhile to enlarge the variety of such devices, but the moral factor creates a limit to their potential.

Information as a Commodity

Uncertainty usually creates a still more subtle problem in resource allocation; information becomes a commodity. Suppose that in one part of the economic system an observation has been made whose outcome, if known, would affect anyone's estimates of the probabilities of the different states of nature. Such observations arise out of research but they also arise in the daily course of economic life as a by-product of other economic activities. An entrepreneur will automatically acquire a knowledge of demand and production conditions in his field which is available to others only with special effort. Information will frequently have an economic value, in the sense that anyone possessing the information can make greater profits than would otherwise be the case.

It might be expected that information will be traded in, and of course to a considerable extent this is the case, as is illustrated by the numerous economic institutions for transmission of information, such as newspapers. But in many instances, the problem of an optimal allocation is sharply raised. The cost of transmitting a given body of information is frequently very low. If it were zero, then optimal allocation would obviously call for unlimited distribution of the informa-

[4] These remarks are not intended as a complete evaluation of cost-plus contracts. In particular, there are, to a certain extent, other incentives which mitigate the adverse effects on efficiency.

tion without cost. In fact, a given piece of information is by definition an indivisible commodity, and the classical problems of allocation in the presence of indivisibilities appear here. The owner of the information should not extract the economic value which is there, if optimal allocation is to be achieved; but he is a monopolist, to some small extent and will seek to take advantage of this fact.

In the absence of special legal protection, the owner cannot, however, simply sell information on the open market. Any one purchaser can destroy the monopoly, since he can reproduce the information at little or no cost. Thus the only effective monopoly would be the use of the information by the original possessor. This, however, will not only be socially inefficient, but also may not be of much use to the owner of the information either, since he may not be able to exploit it as effectively as others.

With suitable legal measures, information may become an appropriable commodity. Then the monopoly power can indeed be exerted. However, no amount of legal protection can make a thoroughly appropriable commodity of something so intangible as information. The very use of the information in any productive way is bound to reveal it, at least in part. Mobility of personnel among firms provides a way of spreading information. Legally imposed property rights can provide only a partial barrier, since there are obviously enormous difficulties in defining in any sharp way an item of information and differentiating it from other similar sounding items.

The demand for information also has uncomfortable properties. In the first place, the use of information is certainly subject to indivisibilities; the use of information about production possibilities, for example, need not depend on the rate of production. In the second place, there is a fundamental paradox in the determination of demand for information; its value for the purchaser is not known until he has the information, but then he has in effect acquired it without cost. Of course, if the seller can retain property rights in the use of the information, this would be no problem, but given incomplete appropriability, the potential buyer will base his decision to purchase information on less than optimal criteria. He may act, for example, on the average value of information in that class as revealed by past experience. If any particular item of information has differing values for different economic agents, this procedure will lead both to a nonoptimal purchase of information at any given price and also to a nonoptimal allocation of the information purchased.

WELFARE ECONOMICS AND INVENTIVE ACTIVITY

It should be made clear that from the standpoint of efficiently distributing an existing stock of information, the difficulties of appropriation are an advantage, provided there are no costs of transmitting information, since then optimal allocation calls for free distribution. The chief point made here is the difficulty of creating a market for information if one should be desired for any reason.

It follows from the preceding discussion that costs of transmitting information create allocative difficulties which would be absent otherwise. Information should be transmitted at marginal cost, but then the demand difficulties raised above will exist. From the viewpoint of optimal allocation, the purchasing industry will be faced with the problems created by indivisibilities; and we still leave unsolved the problem of the purchaser's inability to judge in advance the value of the information he buys. There is a strong case for centralized decision making under these circumstances.

Invention as the Production of Information

The central economic fact about the processes of invention and research is that they are devoted to the production of information. By the very definition of information, invention must be a risky process, in that the output (information obtained) can never be predicted perfectly from the inputs. We can now apply the discussion of the preceding two sections.

Since it is a risky process, there is bound to be some discrimination against investment in inventive and research activities. In this field, especially, the moral factor will weigh heavily against any kind of insurance or equivalent form of risk bearing. Insurance against failure to develop a desired new product or process would surely very greatly weaken the incentives to succeed. The only way, within the private enterprise system, to minimize this problem is the conduct of research by large corporations with many projects going on, each small in scale compared with the net revenue of the corporation. Then the corporation acts as its own insurance company. But clearly this is only an imperfect solution.

The deeper problems of misallocation arise from the nature of the product. As we have seen, information is a commodity with peculiar attributes, particularly embarrassing for the achievement of optimal allocation. In the first place, any information obtained, say a new method of production, should, from the welfare point of view, be available free of charge (apart from the cost of transmitting information). This insures optimal utilization of the information but of course provides no incentive for investment in research. In an ideal socialist economy, the reward for invention would be completely separated from any charge to the users of the information.[5] In a free enterprise economy, inventive activity is supported by using the invention to create property rights; precisely to the extent that it is successful, there is an underutilization of the information. The property rights may be in the information itself, through patents and similar legal devices, or in the intangible assets of the firm if the information is retained by the firm and used only to increase its profits.

The first problem, then, is that in a free enterprise economy the profitability of invention requires a nonoptimal allocation of resources. But it may still be asked whether or not the allocation of resources to inventive activity is optimal. The discussion of the preceding section makes it clear that we would not expect this to be so; that, in fact, a downward bias in the amount of resources devoted to inventive activity is very likely. Whatever the price, the demand for information is less than optimal for two reasons: (1) since the price is positive and not at its optimal value of zero, the demand is bound to be below the optimal; (2) as seen before, at any given price, the very nature of information will lead to a lower demand than would be optimal.

As already remarked, the inventor will in any case have considerable difficulty in appropriating the information produced. Patent laws would have to be unimaginably complex and subtle to permit such appropriation on a large scale. Suppose, as the result of elaborate tests, some metal is discovered to have a desirable property, say resistance to high heat. Then of course every use of the metal for which this property is relevant would also use this information, and the user would be made to pay for it. But, even more, if another inventor is stimulated to examine chemically related metals for heat resistance, he is using the information already discovered and should pay for it in some measure; and any beneficiary of his discoveries should also pay. One would have to have elaborate distinctions of partial property rights of all degrees to make the system at all tolerable. In the interests of the possibility of enforcement, actual patent laws sharply restrict the range of appropriable information and thereby reduce the incentives to engage in inventive and research activities.

[5] This separation exists in the Soviet Union, according to N. M. Kaplan and R. H. Moorsteen of The RAND Corporation (verbal communication).

WELFARE ECONOMICS AND INVENTIVE ACTIVITY

These last considerations bring into focus the interdependence of inventive activities, which reinforces the difficulties in achieving an optimal allocation of the results. Information is not only the product of inventive activity, it is also an input—in some sense, the major input apart from the talent of the inventor. The school of thought that emphasizes the determination of invention by the social climate as demonstrated by the simultaneity of inventions in effect emphasizes strongly the productive role of previous information in the creation of new information. While these interrelations do not create any new difficulties in principle, they intensify the previously established ones. To appropriate information for use as a basis for further research is much more difficult than to appropriate it for use in producing commodities; and the value of information for use in developing further information is much more conjectural than the value of its use in production and therefore much more likely to be underestimated. Consequently, if a price is charged for the information, the demand is even more likely to be suboptimal.

Thus basic research, the output of which is only used as an informational input into other inventive activities, is especially unlikely to be rewarded. In fact, it is likely to be of commercial value to the firm undertaking it only if other firms are prevented from using the information obtained. But such restriction on the transmittal of information will reduce the efficiency of inventive activity in general and will therefore reduce its quantity also. We may put the matter in terms of sequential decision making. The a priori probability distribution of the true state of nature is relatively flat to begin with. On the other hand, the successive a posteriori distributions after more and more studies have been conducted are more and more sharply peaked or concentrated in a more limited range, and we therefore have better and better information for deciding what the next step in research shall be. This implies that, at the beginning, the preferences among alternative possible lines of investigation are much less sharply defined than they are apt to be later on and suggests, at least, the importance of having a wide variety of studies to begin with, the less promising being gradually eliminated as information is accumulated.[6] At each stage the decisions about the next step should be based on all available information. This would require an unrestricted flow of informa-

[6] The importance of parallel research developments in the case of uncertainty has been especially stressed by Burton H. Klein; see his, "A Radical Proposal for R. and D.," *Fortune*, May 1958, p. 112 ff.; and Klein and W. H. Meckling, "Application of Operations Research to Development Decisions," *Operations Research*, 1958, pp. 352–363.

tion among different projects which is incompatible with the complete decentralization of an ideal free enterprise system. When the production of information is important, the classic economic case in which the price system replaces the detailed spread of information is no longer completely applicable.

To sum up, we expect a free enterprise economy to underinvest in invention and research (as compared with an ideal) because it is risky, because the product can be appropriated only to a limited extent, and because of increasing returns in use. This underinvestment will be greater for more basic research. Further, to the extent that a firm succeeds in engrossing the economic value of its inventive activity, there will be an underutilization of that information as compared with an ideal allocation.

Competition, Monopoly, and the Incentive to Innovate

It may be useful to remark that an incentive to invent can exist even under perfect competition in the product markets though not, of course, in the "market" for the information contained in the invention. This is especially clear in the case of a cost reducing invention. Provided only that suitable royalty payments can be demanded, an inventor can profit without disturbing the competitive nature of the industry. The situation for a new product invention is not very different; by charging a suitable royalty to a competitive industry, the inventor can receive a return equal to the monopoly profits.

I will examine here the incentives to invent for monopolistic and competitive markets, that is, I will compare the potential profits from an invention with the costs. The difficulty of appropriating the information will be ignored; the remaining problem is that of indivisibility in use, an inherent property of information. A competitive situation here will mean one in which the industry produces under competitive conditions, while the inventor can set an arbitrary royalty for the use of his invention. In the monopolistic situation, it will be assumed that only the monopoly itself can invent. Thus a monopoly is understood here to mean barriers to entry; a situation of temporary monopoly, due perhaps to a previous innovation, which does not prevent the entrance of new firms with innovations of their own, is to be regarded as more nearly competitive than monopolistic for the purpose of this analysis. It will be argued that the incentive to invent is less under monopolistic than under competitive conditions but even in the latter case it will be less than is socially desirable.

WELFARE ECONOMICS AND INVENTIVE ACTIVITY

We will assume constant costs both before and after the invention, the unit costs being c before the invention and $c' < c$ afterward. The competitive price before invention will therefore be c. Let the corresponding demand be x_c. If r is the level of unit royalties, the competitive price after the invention will be $c' + r$, but this cannot of course be higher than c, since firms are always free to produce with the old methods.

It is assumed that both the demand and the marginal revenue curves are decreasing. Let $R(x)$ be the marginal revenue curve. Then the monopoly output before invention, x_m, would be defined by the equation,

$$R(x_m) = c\ .$$

Similarly, the monopoly output after invention is defined by,

$$R(x'_m) = c'\ .$$

Let the monopoly prices corresponding to outputs x_m and x'_m, respectively, be p_m and p'_m. Finally, let P and P' be the monopolist's profits before and after invention, respectively.

What is the optimal royalty level for the inventor in the competitive case? Let us suppose that he calculates p'_m, the optimal monopoly price which would obtain in the postinvention situation. If the cost reduction is sufficiently drastic that $p'_m < c$, then his most profitable policy is to set r so that the competitive price is p'_m, i.e. let,

$$r = p'_m - c'\ .$$

In this case, the inventor's royalties are equal to the profits a monopolist would make under the same conditions, i.e. his incentive to invent will be P'.

Suppose, however, it turns out that $p'_m > c$. Since the sales price cannot exceed c, the inventor will set his royalties at,

$$r = c - c'\ .$$

The competitive price will then be c, and the sales will remain at x_c. The inventor's incentive will then be, $x_c(c - c')$.

The monopolist's incentive, on the other hand, is clearly $P' - P$. In the first of the two cases cited, the monopolist's incentive is obviously less than the inventor's incentive under competition, which is P', not $P' - P$. The preinvention monopoly power acts as a strong disincentive to further innovation.

ALLOCATION OF RESOURCES FOR INVENTION

The analysis is slightly more complicated in the second case. The monopolist's incentive, $P' - P$, is the change in revenue less the change in total cost of production, i.e.,

$$P' - P = \int_{x_m}^{x'_m} R(x)\,dx - c'\,x'_m + c\,x_m\,.$$

Since the marginal revenue $R(x)$ is diminishing, it must always be less than $R(x_m) = c$ as x increases from x_m to x'_m, so that,

$$\int_{x_m}^{x'_m} R(x)\,dx < c\,(x'_m - x_m)\,,$$

and,

$$P' - P < c\,(x'_m - x_m) - c'\,x'_m + c\,x_m = (c - c')\,x'_m\,.$$

In the case being considered, the postinvention monopoly price, p'_m, is greater than c. Hence, with a declining demand curve, $x'_m < x_c$. The above inequality shows that the monopolist's incentive is always less than the cost reduction on the postinvention monopoly output, which in this case is, in turn, less than the competitive output (both before and after invention). Since the inventor's incentive under competition is the cost reduction on the competitive output, it will again always exceed the monopolist's incentive.

It can be shown that, if we consider differing values of c', the difference between the two incentives increases as c' decreases, reaching its maximum of P (preinvention monopoly profits) for c' sufficiently large for the first case to hold. The ratio of the incentive under competition to that under monopoly, on the other hand, though always greater than 1, decreases steadily with c'. For c' very close to c (i.e., very minor inventions), the ratio of the two incentives is approximately x_c/x_m, i.e., the ratio of monopoly to competitive output.[7]

[7] To sketch the proof of these statements quickly, note that, as c' varies, P is a constant. Hence, from the formula for $P' - P$, we see that,

$$d(P' - P)/dc' = dP'/dc' = R(x'_m)(dx'_m/dc') - c'(dx'_m/dc') - x'_m = -x'_m\,,$$

since $R(x'_m) = c'$. Let $F(c')$ be the difference between the incentives to invent under competitive and under monopolistic conditions. In the case where $p'_m < c$, this difference is the constant P. Otherwise,

$$F(c') = x_c\,(c - c') - (P' - P),$$

so that

$$dF/dc' = x'_m - x_c\,.$$

For the case considered, we must have $x'_m < x_c$, as seen in the text. Hence, $dF/dc' \leq 0$, so

WELFARE ECONOMICS AND INVENTIVE ACTIVITY

The only ground for arguing that monopoly may create superior incentives to invent is that appropriability may be greater under monopoly than under competition. Whatever differences may exist in this direction must, of course, still be offset against the monopolist's disincentive created by his preinvention monopoly profits.

The incentive to invent in competitive circumstances may also be compared with the social benefit. It is necessary to distinguish between the realized social benefit and the potential social benefit, the latter being the benefit which would accrue under ideal conditions, which, in this case, means the sale of the product at postinvention cost, c'. Clearly, the potential social benefit always exceeds the realized social benefit. I will show that the realized social benefit, in turn, always equals or exceeds the competitive incentive to invent and, a fortiori, the monopolist's incentive.

Consider again the two cases discussed above. If the invention is sufficiently cost reducing so that $p'_m < c$, then there is a consumers' benefit, due to the lowering of price, which has not been appropriated by the inventor. If not, then the price is unchanged, so that the consumers' position is unchanged, and all benefits do go to the inventor. Since by assumption all the producers are making zero profits both before and after the invention, we see that the inventor obtains the entire realized social benefit of moderately cost reducing inventions but not of more radical inventions. Tentatively, this suggests a bias against major inventions, in the sense that an invention, part of whose costs could be paid for by lump-sum payments by consumers without making them worse off than before, may not be profitable at the maximum royalty payments that can be extracted by the inventor.

---

that $F(c')$ increases as c' decreases.

Let $G(c')$ be the ratio of the incentive under competition to that under monopoly. If $p'_m < c$, then,

$$G(c') = P'/(P' - P),$$

which clearly decreases as c' decreases. For $p'_m > c$, we have,

$$G(c') = x_c (c - c')/(P' - P).$$

Then,

$$dG/dc' = [\, -(P' - P)\, x_c + x_c (c - c')\, x'_m] \,/(P' - P)^2.$$

Because of the upper bound for $P' - P$ established in the text, the numerator must be positive; the ratio decreases as c' decreases.

Finally, if we consider c' very close to c, $G(c')$ will be approximately equal to the ratio of the derivatives of the numerator and denominator (L'Hopital's rule), which is, x_c/x'_m, and which approaches x_c/x_m as c' approaches c.

ALLOCATION OF RESOURCES FOR INVENTION

Alternative Forms of Economic Organization in Invention

The previous discussion leads to the conclusion that for optimal allocation to invention it would be necessary for the government or some other agency not governed by profit-and-loss criteria to finance research and invention. In fact, of course, this has always happened to a certain extent. The bulk of basic research has been carried on outside the industrial system, in universities, in the government, and by private individuals. One must recognize here the importance of nonpecuniary incentives, both on the part of the investigators and on the part of the private individuals and governments that have supported research organizations and universities. In the latter, the complementarity between teaching and research is, from the point of view of the economy, something of a lucky accident. Research in some more applied fields, such as agriculture, medicine, and aeronautics, has consistently been regarded as an appropriate subject for government participation, and its role has been of great importance.

If the government and other nonprofit institutions are to compensate for the underallocation of resources to invention by private enterprise, two problems arise: how shall the amount of resources devoted to invention be determined, and how shall efficiency in their use be encouraged? These problems arise whenever the government finds it necessary to engage in economic activities because indivisibilities prevent the private economy from performing adequately (highways, bridges, reclamation projects, for example), but the determination of the relative magnitudes is even more difficult here. Formally, of course, resources should be devoted to invention until the expected marginal social benefit there equals the marginal social benefit in alternative uses, but in view of the presence of uncertainty, such calculations are even more difficult and tenuous than those for public works. Probably all that could be hoped for is the estimation of future rates of return from those in the past, with investment in invention being increased or decreased accordingly as some average rate of return over the past exceeded or fell short of the general rate of return. The difficulties of even ex post calculation of rates of return are formidable though possibly not insuperable.[8]

The problem of efficiency in the use of funds devoted to research

[8] For an encouraging study of this type, see Z. Griliches, "Research Costs and Social Returns: Hybrid Corn and Related Innovations," *Journal of Political Economy*, 1958, pp. 419–431.

is one that has been faced internally by firms in dealing with their own research departments. The rapid growth of military research and development has led to a large-scale development of contractual relations between producers and a buyer of invention and research. The problems encountered in assuring efficiency here are the same as those that would be met if the government were to enter upon the financing of invention and research in civilian fields. The form of economic relation is very different from that in the usual markets. Payment is independent of product; it is governed by costs, though the net reward (the fixed fee) is independent of both. This arrangement seems to fly in the face of the principles for encouraging efficiency, and doubtless it does lead to abuses, but closer examination shows both mitigating factors and some explanation of its inevitability. In the first place, the awarding of new contracts will depend in part on past performance, so that incentives for efficiency are not completely lacking. In the second place, the relation between the two parties to the contract is something closer than a purely market relation. It is more like the sale of professional services, where the seller contracts to supply not so much a specific result as his best judgment. (The demand for such services also arises from uncertainty and the value of information.) In the third place, payment by results would involve great risks for the inventor, risks against which, as we have seen, he could hedge only in part.

There is clear need for further study of alternative methods of compensation. For example, some part of the contractual payment might depend on the degree of success in invention. But a more serious problem is the decision as to which contracts to let. One would need to examine the motivation underlying government decision making in this area. Hitch has argued that there are biases in governmental allocation, particularly against risky invention processes, and an excessive centralization, though the latter could be remedied by better policies.[9]

One can go further. There is really no need for the firm to be the fundamental unit of organization in invention; there is plenty of reason to suppose that individual talents count for a good deal more than the firm as an organization. If provision is made for the rental of necessary equipment, a much wider variety of research contracts with individuals as well as firms and with varying modes of payment, including incentives, could be arranged. Still other forms of organiza-

[9] *Op. cit.*

tion, such as research institutes financed by industries, the government, and private philanthropy, could be made to play an even livelier role than they now do.

COMMENT

C. J. HITCH, The RAND Corporation

There is, I think, one important aspect of government policy with respect to research and development that Markham failed to discuss—the government's contracting and management policies in this area. The problems here are related to Arrow's discussion of incentives, particularly his remarks concerning risk bearing and the adverse allocative effects of cost-plus contracts.

Let me illustrate one set of such problems by contrasting the management policies used by the Army Air Corps before World War II in developing new combat aircraft with the policies now used by the Air Force in developing aircraft and missiles.

Before the war the Air Corps would typically announce a competition for a new aircraft type, the desired characteristics of which would be described in general terms. Any aircraft company that wanted to enter the competition would develop and build, with its own funds, the required number of prototypes (usually one to three). The competition consisted in flying and testing, at Dayton, the prototypes entered by the companies. The prize for the winner was a production contract. The losers lost their stake.

Now, typically, the competition occurs at the design stage. The competing companies enter drawings, with their estimates of performance (and time and cost of development). The winner of the design competition is awarded a development contract which, if things go favorably, is transformed later into a production contract. Occasionally, if the program is considered very important, two competing designs may be approved for development or even production (cf. the Atlas and Titan). But usually the development of more than one model is considered too expensive.

In fact, the greatly increased cost of development is the reason given for this change in policy. When the cost of developing a new vehicle is from \$0.5 billion to \$2.0 billion, it seems evident that no aircraft company can assume the risk, and perhaps also that even the government can not finance multiple developments.

Nevertheless, the new policy appears far from ideal. The aircraft companies risk nothing (even the cost of preparing designs is usually reimbursed). Because of the great uncertainties involved in any major development, it is hard to make a wise selection at the design stage. Companies have a natural tendency to be optimistic in estimating performance, cost, and availability at this stage—sometimes much more so than at other stages as Klein and Marshall and Meckling have shown. And the Air Force has a natural tendency to favor the more optimistic proposals. As a result, the specifications in the development contract are sometimes unrealistic to the point of causing inordinate delay.

Moreover, once the winner has been selected, there is no more competition. The company may put its best team on the development for the sake of patriotism or its own long term reputation, but the powerful incentive of competition is lacking. Partly for this reason the Air Force, locked to a sole source with a cost-plus contract, has to exercise a kind and degree of control during the development process that is inconsistent with management prerogatives as they are understood and practiced in other parts of the free enterprise economy.

I believe the inefficiencies resulting from these policies and procedures are serious, and worthy of much more attention by economists. It does no good simply to inveigh against the iniquities of cost-plus contracting and government risk bearing when we are unable to propose a practical alternative. Perhaps part of the answer lies in some form of risk sharing. What is badly needed here is an economics invention or, more probably, several of them. The government is going to be in the business of supporting research and development on a large scale for a long time, and it is important that it use policies that take advantage of the incentives present in the economy.

[4]

ENTREPRENEURSHIP IN ECONOMIC THEORY*

By WILLIAM J. BAUMOL
Princeton University

The entrepreneur is at the same time one of the most intriguing and one of the most elusive characters in the cast that constitutes the subject of economic analysis. He has long been recognized as the apex of the hierarchy that determines the behavior of the firm and thereby bears a heavy responsibility for the vitality of the free enterprise society. In the writings of the classical economist his appearance was frequent, though he remained a shadowy entity without clearly defined form and function. Only Schumpeter and, to some degree, Professor Knight succeeded in infusing him with life and in assigning to him a specific area of activity to any extent commensurate with his acknowledged importance.

In more recent years, while the facts have apparently underscored the significance of his role, he has at the same time virtually disappeared from the theoretical literature. And, as we will see, while some recent theoretical writings seem at first glance to offer a convenient place for an analysis of his activities, closer inspection indicates that on this score matters have not really improved substantially.

This paper will undertake to examine three major matters. First, I will review briefly the grounds on which entrepreneurship should concern us. Second. I will seek to explain why economic theory has failed to develop an illuminating formal analysis of entrepreneurship and I shall conclude that it is unlikely to do so for the foreseeable future. Finally, I shall argue that theory can say a great deal that is highly relevant to the subject of entrepreneurship even if it fails to provide a rigorous analysis of the behavior of the entrepreneur or of the supply of entrepreneurship.

Before proceeding with the discussion I would like to make a distinction that is somewhat artificial but nevertheless important. It is necessary for us to differentiate between the entrepreneurial and the managerial functions. We may define the manager to be the individual who oversees the ongoing efficiency of continuing processes. It is his task to see that available processes and techniques are combined in proportions appropriate for current output levels and for the future outputs that are already in prospect. He sees to it that inputs are not wasted, that

* The author would like very much to thank his colleague W. A. Lewis, whose comments were used liberally in the revision of this paper. He must also thank the National Science Foundation whose grant greatly facilitated the completion of this paper.

schedules and contracts are met, he makes routine pricing and advertising outlay decisions, etc., etc. In sum, he takes charge of the activities and decisions encompassed in our traditional models.

The preceding description is not intended to denigrate the importance of managerial activity or to imply that it is without significant difficulties. Carl Kaysen has remarked that in practice most firms no doubt find themselves in a position well inside their production possibility loci and one of their most challenging tasks is to find ways of approaching those loci more closely; i.e., of increasing their efficiency even within the limits of known technology. This is presumably part of the job of the manager who is constantly on the lookout for means to save a little here and to squeeze a bit more there. But for many purposes the standard models would appear to provide an adequate description of the functions of the manager. Given an arrangement which calculation, experience, or judgment indicate to constitute a reasonable approximation to the current optimum, it is the manager's task to see that this arrangement is in fact instituted to a reasonable degree of approximation.

The entrepreneur (whether or not he in fact also doubles as a manager) has a different function. It is his job to locate new ideas and to put them into effect. He must lead, perhaps even inspire; he cannot allow things to get into a rut and for him today's practice is never good enough for tomorrow. In short, he is the Schumpeterian innovator and some more. He is the individual who exercises what in the business literature is called "leadership." And it is he who is virtually absent from the received theory of the firm.

I. *On the Significance of the Entrepreneur*

If we are interested in explaining what Haavelmo has described as the "really big dissimilarities in economic life," we must be prepared to concern ourselves with entrepreneurship. For the really big differences are most usually those that correspond to historical developments over long periods of time or to the comparative states of various economies, notably those of the developed and the underdeveloped areas.

It has long been recognized that the entrepreneurial function is a vital component in the process of economic growth. Recent empirical evidence and the lessons of experience both seem to confirm this view. For example, some empirical studies on the nature of the production function have concluded that capital accumulation and expansion of the labor force leave unexplained a very substantial proportion of the historical growth of the nation's output. Thus, in a well-known paper, Solow [6, p. 320] has suggested on the basis of American data for the period 1909–49 that "gross output per man-hour doubled over the interval, with 87½ percent of the increase attributable to technical change and the remaining 12½ percent to increase in the use of capital."[1] But any such innovation, whether it is purely technological or it consists in a modification in the way in which an industry is organized, will require entrepreneurial initiative in its introduction. Thus we are led to suspect that by ignoring the entrepreneur we are prevented from accounting fully for a very substantial proportion of our historic growth.

Those who have concerned themselves with development policy have apparently been driven to similar conclusions. If we seek to explain the success of those economies which have managed to grow significantly with those that have remained relatively stagnant, we find it difficult to do so without taking into consideration differences in the availability of entrepreneurial talent and in the motivational mechanism which drives them on. A substantial proportion of the energies of those who design plans to stimulate development has been devoted to the provision of means whereby entrepreneurs can be trained and encouraged.

The entrepreneur is present in institutional and applied discussions of a number of other economic areas. For example, his absence is sometimes cited as a significant source of the difficulties of a declining industry, and a balance-of-payments crisis is sometimes discussed in similar terms. Thus both macro problems and micro problems offer a substantial place for him in their analysis. Whether or not he is assigned the starring role he would appear in practice to be no minor character.

II. *The Entrepreneur in Formal Models*

Contrast all this with the entrepreneur's place in the formal theory. Look for him in the index of some of the most noted of recent writings on value theory, in neoclassical[2] or activity analysis models of the firm. The references are scanty and more often they are totally absent. The theoretical firm is entrepreneurless—the Prince of Denmark has been expunged from the discussion of *Hamlet*.

It is not difficult to explain his absence. Consider the nature of the model of the firm. In its simplest form (and in this respect we shall see that the more complex and more sophisticated models are no better) the theoretical firm must choose among alternative values for a small number of rather well-defined variables: price, output, perhaps adver-

[1] Solow's result and other similar conclusions have recently been challenged in an article by Jorgenson and Griliches [3]. However, their contention does not necessarily imply any denigration of the role of the entrepreneur. They argue merely that entrepreneurship and innovation have achieved growth in outputs only with the aid of corresponding increases in input quantities.

[2] There is one residual and rather curious role left to the entrepreneur in the neoclassical model. He is the indivisible and non-replicable input that accounts for the U-shaped cost curve of a firm whose production function is linear and homogeneous. How the mighty have fallen!

tising outlay. In making this choice management is taken to consider the costs and revenues associated with each candidate set of values, as described by the relevant functional relationships, equations, and inequalities. Explicitly or implicitly the firm is then taken to perform a mathematical calculation which yields optimal (i.e., profit maximizing) values for all of its decision variables and it is these values which the theory assumes to be chosen—which are taken to constitute the business decision. There matters rest, forever or until exogenous forces lead to an autonomous change in the environment. Until there is such a shift in one of the relationships that define the problem, the firm is taken to replicate precisely its previous decisions, day after day, year after year.

Obviously, the entrepreneur has been read out of the model. There is no room for enterprise or initiative. The management group becomes a passive calculator that reacts mechanically to changes imposed on it by fortuitous external developments over which it does not exert, and does not even attempt to exert, any influence. One hears of no clever ruses, ingenious schemes, brilliant innovations, of no charisma or of any of the other stuff of which outstanding entrepreneurship is made; one does not hear of them because there is no way in which they can fit into the model.[3]

It must be understood clearly that what I have been saying constitutes no criticism, not even an attempt to reprove mildly the neoclassical model of the firm. I think that model does what it was designed to do and does it well. Like any respectable analysis, one hopes that it will be modified, amended, and improved with time. But not because it cannot handle an issue for which it is irrelevant. The model is essentially an instrument of optimality analysis of well-defined problems, and it is precisely such (very real and important) problems which need no entrepreneur for their solution.

Some readers may suspect that I am subtly putting forward as more appropriate candidates for the job some alternative models of the firm with which I have to some degree been associated. But this is certainly not my intention, because it seems clear to me that these models are

[3] The problem was recognized long ago by Thorstein Veblen. One may recall the characteristic passage in which he described the economic man as "a lightning calculator of pleasures and pains, who oscillates like a homogeneous globule of desire of happiness under the impulse of stimuli that shift him about the area, but leave him intact. He has neither antecedent nor consequent. He is an isolated, definitive human datum, in stable equilibrium except for the buffets of impinging forces that displace him in one direction or another. Self-imposed in elemental space, he spins symmetrically about his own spiritual axis until the parallelogram of forces bears down upon him, whereupon he follows the line of the resultant. When the force of the impact is spent, he comes to rest, a self-contained globule of desire as before. . . . [he] is not a prime mover. He is not the seat of a process of living, except in the sense that he is subject to a series of permutations enforced upon him by circumstances external and alien to him" [7, pp. 73–74].

no better for the purpose than the most hidebound of conventional constructs. For example, consider what Oliver Williamson has described as the "managerial discretion models," in which the businessman is taken to maximize the number of persons he employs, or sales, or still another objective distinct from profits. True, this businessman has (somewhere outside the confines of the model) made a choice which was no mere matter of calculation. He has decided, in at least some sense, to assign priority to some goal other than profit. But having made this choice he becomes, no less than the profit maximizer, a calculating robot, a programmed mechanical component in the automatic system that constitutes the firm. He makes and enforces the maximizing decision and in this the choice of maximand makes no difference.

Nor can the "practical pertinence" of the decision variables make the difference in carving out a place for the entrepreneur. Maximization models have recently been developed in which, instead of prices and outputs, the decision variables are the firm's real investment program, or its financial mix (the proportion of equity and debt in its funding), or the attributes of a new product to be launched by the company. These decisions seem to smell more of the ingredients of entrepreneurship. But though the models may be powerful and serve their objective well, they take us not a whit further in the analysis of entrepreneurship, for their calculations are again mechanistic and automatic and call for no display of entrepreneurial initiative.

Finally, it must be understood that the timeless nature of these models has nothing to do with the problem. Professor Evans [2] long ago developed a model in which the firm considered the consequence of its decisions for the time path of prices and where the calculus of variations served as his instrument of analysis. In one of my own models the firm was taken to choose not a stationary, once-and-for-all output level, but selected instead an optimal growth rate. None of these alternatives helps matters. In all these models, automaton maximizers the businessmen are and automaton maximizers they remain.

And this shows us why our body of theory, as it has developed, offers us no promise of being able to deal effectively with a description and analysis of the entrepreneurial function. For maximization and minimization have constituted the foundation of our theory, and as a result of this very fact the theory is deprived of the ability to provide an analysis of entrepreneurship. The terminology of game theory has been extremely suggestive; the willingness of the behaviorists to break away from traditional formulations has been encouraging; but I see no real breakthroughs in this area even on the distant horizon. At most I hope for more brilliant observations and descriptive insights such as those provided by Schumpeter and more recently by Leibenstein, but I foresee for the immediate future no more formal, manipulatable engine of calculation and analysis.[4]

III. *On the Supply of Entrepreneurship*

There is yet another reason why a marriage between theory and policy is not easily arranged in this area. In its discussions of inputs our formal analysis deals, by and large, with the way in which these inputs are used, and tells us relatively little about where they come from. In our growth models, for example, the behavior of the labor supply exerts a critical influence on the economy's expansion path. But the determination of the growth of the labor force itself is generally taken to be an exogenous matter. Similarly, in a neoclassical or a programming analysis of production one investigates how inputs should be used in the production process, but one assumes that their supply is somehow determined outside the system. Thus even if we were to develop a model which were successful in advancing the theory of entrepreneurship to the level of sophistication of our treatment of other inputs, we would have defined more effectively the entrepreneurial role, but we would have added relatively little to our understanding of the determinants of the level of output of entrepreneurship.

From the point of view of policy, however, the priorities would seem to be reversed. The first order of business in an economy which exhibits very little business drive is presumably to induce the appearance of increased supplies of entrepreneurial skills which would then be let loose upon the area's industry. The policy-maker thus is interested primarily in what determines the supply of entrepreneurship and in the means that can be used to expand it.

But there is reason to suppose that these issues are to a very considerable extent matters of social psychology, of social arrangements, of cultural developments and the like. And perhaps this is why many of the recent discussions of the theory of entrepreneurship have been contributed by the sociologists and the psychologists.[5] This may then be no

[4] My colleague, Professor Lewis, has adduced yet another reason why the current theory does not help us to understand the entrepreneur. He remarks in a note to me that "the entrepreneur is doing something new and is therefore to some extent a monopolist. . . . We have no good theory of entrepreneurship because we have no good theory of monopoly. Our theory that monopolists [act] to maximize profit is obviously absurd, given the low elasticity of demand of most monopolized products." I agree that this observation points to a most fundamental gap in the theory of the firm.

[5] For a remarkable study of entrepreneurship by a social psychologist, see McClelland [4]. While the book is not free of somewhat distracting jargon, and is naïve in spots, particularly in its literal interpretation of the role of the profit motive in economic analysis, it does offer a number of extremely interesting hypotheses and provides in their support quantities of psychological test results relating to a great variety of cultures. In what is perhaps the most interesting part of his discussion from our point of view, the author claims to show that entrepreneurs are motivated by *n*-achievement (the need for achievement) and not by desire for money (pp. 233–37). In his tests, people with high levels of *n*-achievement do no better when offered larger

fortuitous development. The very nature of the more pressing issues relating to entrepreneurship may invite more directly the attention of the practitioners of disciplines other than theoretical economics.

IV. *A Place for Theory and Entrepreneurship*

Given these difficulties besetting any attempt to construct a relevant economic theory in the area, I can offer only one suggestion for a theoretical approach to entrepreneurship, but one which I think is not without promise. We may not be able to analyze in detail the supply of entrepreneurship, the entrepreneur's strategy choices, his attitudes to risk, or the sources of his ideas. But one can hope to examine fruitfully what can be done to encourage his activity. Here an analogy is illuminating. The Keynesian analysis really bypasses the issue of expectations which is surely at the heart of the investment decision and yet the model succeeds in coming to grips with some means that can stimulate investment. In the same way one can undertake to grapple, assisted by theoretical instruments, with the policies that encourage entrepreneurship.

This can be done by considering not the means which the entrepreneur employs or the process whereby he arrives at his decisions but by examining instead the determinants of the payoff to his activity.[6] In his operations he must bear risks, never mind just how he does this, but let the theory consider how the marginal costs of his risk bearing can be reduced. He employs the results of work in research and development; very well, let us investigate what means make it easier, economically, to undertake R and D. Theoretical analysis of the effects of alternative tax arrangements, for example, should shed some light on these matters. The role of the structure of interest rates is no doubt also pertinent and we do have a powerful body of literature which treats of

amounts of money for success, whereas people with low *n*-achievement scores do much better when offered money. However, it should be noted that while a rise in absolute income levels does not seem to stimulate *n*-achievers, a rise in marginal returns does seem likely to spur them on, according to the author because it provides a clearer measure of accomplishment. (The economist would no doubt propose a different explanation.) He also claims to show that *n*-achievers choose smaller risks than the average man: they are not gamblers, but are calculators and planners. The entrepreneur is not essentially a man who chooses to bear risks—that is the speculator, a man with quite a different personality (pp. 210–25). Another interesting McClelland claim is that the *n*-achiever is not an individualist and does not depend for his success on private enterprise (pp. 292–300). He gets just as much satisfaction from the manipulation of a committee, or from working for a government, since his interest is in results rather than in these other considerations. This is perhaps the reason huge committee-run corporations can be successful.

[6] I believe the key element of Schumpeter's contribution to the theory of entrepreneurship is precisely of this variety. In its discussion of the functions of the entrepreneur, *The Theory of Economic Development* [5] offers us little more than a taxonomy. But enormous illumination is provided by Schumpeter's analysis of the process whereby the rewards of innovation are only gradually eroded by the competitive process and the corollary observation that some imperfection in the market mechanism is essential to permit some financial reward for innovation.

these matters. On all of these fronts analysis is well advanced and it is no heroic exercise to imagine rather complex and probing theoretical formulations capable of shedding light upon them.

It should be recognized, moreover, that such a theoretical analysis can be of enormous significance for policy. In a growth-conscious world I remain convinced that encouragement of the entrepreneur is the key to the stimulation of growth. The view that this must await the slow and undependable process of change in social and psychological climate is a counsel of despair for which there is little justification. Such a conclusion is analogous to an argument that all we can do to reduce spending in an inflationary period is to hope for a revival of the Prostestant ethic and the attendant acceptance by the general public of the virtues of thrift! Surely we have learned to do better than that, in effect by producing a movement along the relevant functional paths rather than undertaking the more heroic task involved in shifting the relationships. This is precisely why I have just advocated more careful study of the rewards of entrepreneurship. Without awaiting a change in the entrepreneurial drive exhibited in our society, we can try to learn how one can stimulate the volume and intensity of entrepreneurial activity, thus making the most of what is permitted by current mores and attitudes. If the theory succeeds in no more than showing us something about how that can be done,[7] it will have accomplished very much indeed.

[7] For a crude attempt at such an analysis, see the last chapter of [1].

REFERENCES

1. W. J. Baumol, *Business Behavior, Value and Growth* (revised ed., Harcourt, Brace and World, 1967).
2. G. C. Evans, "The Dynamics of Monopoly," *Amer. Math. Monthly*, Feb., 1924, pp. 77–83.
3. D. W. Jorgenson and Z. Griliches, "The Explanation of Productivity Change," *Rev. of Econ. Studies*, July, 1967, pp. 249–83.
4. D. C. McClelland, *The Achieving Society* (Princeton, 1961).
5. J. A. Schumpeter, *The Theory of Economic Development* (Cambridge, Mass., 1936).
6. R. M. Solow, "Technical Change and the Aggregate Production Function," *Rev. of Econ. and Statis.*, Aug., 1957, pp. 312–20.
7. T. B. Veblen, "Economics and Evolution," *The Place of Science in Modern Civilization* (New York, 1919).

[5]

ENTREPRENEURSHIP AND DEVELOPMENT*

By HARVEY LEIBENSTEIN
Harvard University

I

The received theory of competition gives the impression that there is no need for entrepreneurship. If all inputs are marketed and their prices are known, and if all outputs are marketed and their prices are known, and if there is a definite production function that relates inputs to outputs in a determinate way, then we can always predict the profit for any activity that transforms inputs into outputs. If net profits are positive, then this should serve as a signal for entry into this market. The problem of marshaling resources and turning them into outputs appears to be a trivial activity. From this point of view it is hard to see why there should ever be a deficiency of entrepreneurship. But there is frequently a lack of entrepreneurship. The answer is that the standard competitive model hides the vital function of the entrepreneur.[1]

My aim in what follows is twofold: to suggest a theory of the economy and of entrepreneurship in which entrepreneurship has a unique and critical role and to use this theory to indicate why entrepreneurship is a significant variable in the development process.

In a paper published in 1966 [9] I argued that there does not exist a one-to-one correspondence between sets of inputs and outputs.[2] There are three main reasons for this: contracts for labor are incomplete, the production function is not completely specified or known, and not all factors of production are marketed. I will argue that these are the basic postulates for an economy in which entrepreneurship has a distinct and critical role.

We may distinguish two broad types of entrepreneurial activity: at one pole there is routine entrepreneurship, which is really a type of management, and for the rest of the spectrum we have Schumpeterian or "new type" entrepreneurship. (We shall refer to the latter as N-

* The author would like to thank his colleagues Sam Bowles, Albert O. Hirschman, Gustav Papanek, Nathan Rosenberg, and Ray Vernon for helpful comments that led to some revisions of an earlier version. They are not responsible for the deficiencies that remain.

[1] This point is elaborated in detail in Professor Baumol's paper [3]. His quotation from Veblen is especially apt. Professor Hirschman makes similar points in [8a, pp. 2–5].

[2] See [9] for evidence of specific cases. Econometric evidence on production functions is hard to interpret. Production functions fitted for specific industries frequently have very low values for R^2. While this is consistent with the notion that there is no one-to-one correspondence between inputs and putputs, there are also many other reasons why the fits may be poor. See Marc Nerlove, "Recent Empirical Studies on the CES and Related Production Functions," in *The Theory and Empirical Analysis of Production* (N.B.E.R., 1967), p. 78.

entrepreneurship.) By routine entrepreneurship we mean the activities involved in coordinating and carrying on a well-established, going concern in which the parts of the production function in use (and likely alternatives to current use) are well known and which operates in well-established and clearly defined markets. By N-entrepreneurship we mean the activities necessary to create or carry on an enterprise where not all the markets are well established or clearly defined and/or in which the relevant parts of the production function are not completely known. In both cases the entrepreneur coordinates activities that involve different markets; he is an intermarket operator. But in the case of N-entrepreneurship not all of the markets exist or operate perfectly and the entrepreneur, if he is to be successful, must fill in for the market deficiencies. To my mind one of the main obstacles to our understanding of the entrepreneurial role lies in the conventional theory of the production function. This theory seems so reasonable at first blush that we are likely not to notice the subtle assumptions it makes. The basic culprits are the following assumptions: that the complete set of inputs are specified and known to all actual or potential firms in the industry, and that there is a fixed relation between inputs and outputs. The first assumption is implicit. To my knowledge, it is never stated explicitly, but I have not made an exhaustive search of the literature to check this. The second assumption is explicit, but it is rarely challenged.

In its usual conception the production function is considered to be clearly defined, fully specified, and completely known. Where and to whom in the firm this knowledge is supposed to be available is never stated. In fact, there are great gaps of knowledge about the production function. Points on the production function refer to well-defined inputs. To the extent that they are not completely defined in actuality, the entrepreneur must in some way make up the deficiency. Suppose that to produce a certain commodity, a certain type of machine has to be employed. If no one in the country produces such a machine and if imports are barred, only entrepreneurs who have access to information on how to construct the machine can enter the industry. The potential entrepreneur has to make up for a market deficiency. But that is not his only major function.

Important inputs not well marketed are types of management and market knowledge. Even managers of the more routine type may not be available in well-organized markets in many developing countries. Where available, their capacities may be very difficult to assess. One of the important capacities of management is the ability to obtain and use factors of production that are not well marketed. In some countries the capacity to obtain finance may depend on family connections rather than on the willingness to pay a certain interest rate. A successful entrepreneur may, at times, have to have the capacity to operate well in the political arena connected with his economic activities.

The usual characteristics attributed to entrepreneurs involve gap-filling as one of their essential underlying qualities. For example, it may be thought desirable that entrepreneurs possess at least some of the capacities to: search and discover economic opportunities, evaluate economic opportunities, marshal the financial resources necessary for the enterprise, make time-binding arrangements, take ultimate responsibility for management, be the ultimate uncertainty and/or risk bearer,[3] provide and be responsible for the motivational system within the firm, search and discover new economic information, translate new information into new markets, techniques, and goods, and provide leadership for the work group. In a world of perfect markets, if such a world were possible, each of these characteristics would be marketed as a specific service. Thus, some firms might specialize in the discovery of economic opportunities and sell this information to others. A similar remark could be made of each of the capacities mentioned above. The reason that this is not the case is because some inputs are inherently unmarketable, and some are difficult to market and are frequently unmarketed. For example, we cannot have a perfect market in risk-taking since, among other reasons, there is a "moral risk" problem in profit insurance. (The entrepreneur can intentionally do poorly and cash in on the policy.) Similarly, if the motivational system is the sum of all the human elements and their relations to each other within the firm rather than something specifically provided from outside the firm, then this element cannot be marketed. One of our basic points is that the conditions for perfect markets and the nature of some commodities are inconsistent with each other.

It is important to stress that entrepreneurial activities do not arise only because of market structure imperfections. This view gives too shallow an interpretation of the entrepreneurial role.[4] First, some gaps in markets are inherent in all cases. Second, and what is perhaps less apparent, is that the entrepreneur has to employ some inputs that are somewhat vague in their nature (but nevertheless necessary for production), and whose output is indeterminate. The provision of leadership, motivation, and the availability of the entrepreneur to solve po-

[3] Schumpeter [12, p. 137] is very firm on the point that the entrepreneur is not a risk bearer or uncertainty bearer: "The one who gives credit comes to grief if the undertaking fails." Furthermore, in countries with highly developed stock markets some entrepreneurs can shift the risk by selling shares.

[4] A narrow "imperfect market" interpretation of the entrepreneurial role gives the impression that markets are perfectable, say by the elimination of monopolistic influences, and that by doing so, the significant aspects of the entrepreneurial role can be eliminated thereby. This is not the view taken in this paper. The ideas of this paper are not brought out fully by thinking that the entrepreneurs' role depends only on market imperfections.

tential crisis situations, the capacity to carry ultimate responsibility for the organizational structure and the major time-binding (implicit or explicit) contractual arrangements are of this sort. Third, and most important, the entrepreneur has to possess what might be called, for want of a better term, an "input completing" capacity. If six inputs are needed to bring to fruition a firm that produces a marketable product, it does no good to be able to marshal easily five of them. The gap-filling and the "input-completing" capacities are the unique characteristics of the entrepreneur.

As we have defined the entrepreneur he is an individual or group of individuals with four major characteristics: he connects different markets, he is capable of making up for market deficiencies (gap-filling), he is an "input-completer," and he creates or expands time-binding, input-transforming entities (i.e., firms).

Entrepreneurship is frequently a scarce resource because entrepreneurs are gap-fillers and input-completers and these are scarce talents. Other things equal, the amount of gap-filling and input-completing required determines the degree of scarcity. Gap-filling is necessary because information about some inputs are unmarketable; and because private information about markets cannot always be proven and made public information. Of course, gap-filling will also be necessary where universalistic markets have not been developed, or where the inputs are, in principle, marketable but for some reason such markets have not arisen. For any given economic activity there is a minimum quantum of various inputs that must be marshaled. If less than this minimum variety is universalistically available, the entrepreneur has the job of stepping into the breech to fill the lack of marketable inputs; i.e., he must be an input-completer.

In my "X-efficiency" paper [9] I argued that neither individuals nor groups (say, firms) work as hard or as effectively or search for new information and techniques as diligently as they could, nor is effort maintained at a constant level. The nature and degree of directed human effort of a given individual is not invariable in the sense in which the characteristics of some physical inputs and their capacities may be said to be invariable. The degree of directed effort depends on a variety of factors that determine the internal motivational state of the firm and the external motivational state of the appropriate segment of the economy. Thus, under some circumstances the level of directed effort of the human inputs may be low and, as a consequence, some firms operate under a considerable degree of slack [5] [9]. Persistant slack implies the existence of entrepreneurial opportunities.

The motivational state is likely to be composed of the following elements: (1) The system of financial rewards for effort, some of which may be directly related to the quantity of output but some of the rewards may not be clearly related to output. (2) There may also be a system of rewards and "punishments" related to aspects of behavior other than the productive ones. For example, promotion within a firm may be related to personality traits or kinship or personal ties unconnected to the direct pursuit of the aims of the firm. (3) Finally, there is an interpersonal mechanism of group approval and disapproval, as well as approval-disapproval relations between individuals in different relative hierarchical statuses that normally influence productive behavior. The sum of these relationships is essentially the motivational state of the system. It seems clear that the degree and nature of directed effort will depend on the motivational state. This is especially likely to be true for nonroutine aspects of directed effort such as those involved in the introduction of technological change.

There is a significant relation between the entrepreneur's perceptive capacity and the fact that firms operate under some degree of slack [9]. The existence of slack and the fact that not all inputs are marketed means that the market signals for profit opportunities are blurred. Since there is no one-to-one correspondence between inputs and outputs, a knowledge of output price and input prices can no longer yield the necessary signals. On the other hand, an error in perception can be partially counterbalanced by increased effort in marshaling resources and in operating the plant.

It is noteworthy that the traditional theory does not explain the existence of firms as time-binding entities. The theory presented here suggests that since the production function is incomplete, firms become valuable storehouses of detailed experience and knowledge. In part, this means that successful firms are entities that house successful motivational systems that can be retained only through a scheme of renewable contractual arrangements of different time durations. It is in this way that the firm captures some of the long-term benefits of previous gap-filling and input-completing conquests.

A way of looking at the essential elements is to visualize the economy as a net made up of nodes and pathways. The nodes represent industries or households that receive inputs (or consumer goods) along the pathway and send outputs (final goods and inputs for the other commodities) to other nodes. The perfect competition model would be represented by a net that is complete, that has pathways that are well marked and well defined, that has well-marked and well-defined nodes, and one in which each element (i.e., firm or household) of each node deals with every other node along the pathways on equal terms for the same commodity. In the realistic model we have in mind there are holes and tears in the net, obstructions (knots) along the pathways, and some nodes and pathways, where they exist, are poorly defined and poorly marked or entirely unmarked from the viewpoint of elements of other nodes. We may refer to this net as impeded, incomplete, and "dark" in contrast to the unimpeded and "well-lit" net that represents the competitive model. Of course, a portion of the real economy net may very loosely approximate the "unimpeded" net of the perfect competition model. Entrepreneurs working in the well-defined, non-hole, non-obstruction part of the net carry out routine entrepreneurial-managerial activities, while those that operate on the impeded, incomplete, and dark parts carry out N-entrepreneurial activities. Entrepreneurial activities will make some portions of the net less impeded through extending markets (i.e., creating new pathways) but may make others more so through the creation of monpolies, or the creation of other obstacles (e.g., high entry costs) where they previously did not exist. Inventions and the creation of new knowledge will to some extent extend the net to vague and incomplete areas, but other inventions may substitute relatively well-defined pathways and nodes for those which were ill-defined and obstruction-laden previously.

II

Although there is no universally accepted theory of development we can point to two important elements in the process: (1) Per capita income growth requires shifts from less productive to more productive techniques per worker, the creation or adopton of new commodities, new materials, new markets, new organizational forms, the creation of new skills, and the accumulation of new knowledge. (2) Part of the process is the interaction between the creation of economic capacity and the related creation of demand so that some rough balance between capacity growth and demand growth takes place. The entrepreneur as a gap-filler and input-completer is probably the prime mover of the capacity creation part of these elements of the growth process.[5]

We now know that development is not simply a process of physical and human capital accumulation in the usual sense. If that were all that were involved, then development would simply be a function of the willingness to save. Experience has shown that this is not the case. The work of Solow and others [1] [2] [13] have shown that growth cannot be explained by the contributions of the increase in standard inputs. The work of Chenery and Strout [4] emphasizes that the degree of capital absorption can be a significant constraint to growth in developing

[5] The basic idea is that firms do not operate on their production possibilities frontier. In part, the internal motivational state of the firm determines the degree to which actual output is less than the production possibilities frontier output. Thus, costs per unit of output are not minimized. The size of the difference between actual costs and true minimum costs offers opportunities for those entrepreneurs who think they can produce at lower costs.

countries. The existence of and need for gap-filling and input-completing capacities could explain why standard inputs do not account for all outputs and why capital absorption should be a problem. Economic planning experience in many countries reveals that there is frequently a considerable divergence between plan targets and results. This divergence may be partly explained by the fact that enterpreneurship is not a normal input whose contribution can be readily determined, predicted, planned for, or controlled.

We now sketch briefly some of the basic strands of a theory from which the concept of the entrepreneur as a gap-filler and input-completer derives.

The demand side is determined by the following: (1) The maximal production possibilities set in the sense of maximum knowledge. By maximum knowledge we mean that the techniques are known somewhere in the world—knowledge that is conceivably obtainable although it may be at an exceptionally high cost. (2) We deduct from the large maximal possibilities set the subset of techniques in use and those techniques that contain the following basic characteristics: they are actually known in detail without anything more than routine search activities and the inputs required for production are marketed on a routine basis. (3) What is left is that portion of the maximal production possibilities set which forms the potential opportunities for gap-fillers. Now, gap-filling and input-completing activities are usually costly. Taking these costs into account and calculating the expected prices of marketed inputs and potential outputs, each element in the gap-filling opportunity set can be associated with a set of potential profits or losses (depending on who does the gap-filling). We reduce the gap-filling opportunity set to those possibilities that are associated with expected yields of positive net profits. This set is likely to be very much larger than what will actually be pursued by entrepreneurs. The gap-filling opportunity set is likely to be non-unique since the costs associated with gap-filling depend on the specific entrepreneur that attempts to take advantage of the opportunity. The sequence in which gap-fillers choose opportunities will determine the degree to which any one turns out to be profitable. In addition, the degree of effort put forth by different enterpreneurs and the same entrepreneur at different times will vary, depending on the personality, circumstances, and the motivating influences that exist at the time. Thus, the association between gap-filling opportunities and profitable opportunities is not likely to be a unique one-to-one correspondence.

The supply side is determined by the following: the set of individuals with gap-filling and input-completing capacities, the sociocultural and political constraints which influence the extent to which entrepreneurs take advantage of their capacities, and the degree to which potential entrepreneurs respond to different motivational states, especially where nontraditional activities are involved. Clearly, the personality characteristics of entrepreneurs are important. Apart from gap-filling and input-completing capacities, the potential entrepreneurs' response to opportunities will depend on their preference for certain modes of behavior as opposed to others. Thus, the entrepreneurial personality theories developed by Hagen [6] and McClelland [10] which connect nurture to the creation of entrepreneurial drives are significant elements on the supply side. Last, but not least, the alternatives open to individuals are important, since we must take into account opportunity costs of entrepreneurial acts.

In such a theory growth would depend, in part, on the degree of routine entrepreneurship, the degree to which gaps and impediments in markets exist, and the quality, motivations, and opportunity costs of the potential gap-fillers and input-completers available.

It is not possible at this stage to develop a complete and detailed model of economic development and entrepreneurship. One reason for this is that we do not have, at present, a theory of obstructed, incomplete, and "relatively dark" economic systems. However, it may be useful to sketch briefly the broad outlines of what such a model might contain if further research proved successful.

The model, if it were successfully developed, should enable us to describe the the motivational state that arises from any given state of the impeded system and the reactions to the motivational state. That is, the model should show the links between the maximal opportunity set and those opportunities that are actually perceived and pursued by entrepreneurs. We now attempt to specify the links that are likely to be involved: (1) The input gaps are in part determined exogenously. (2) Given the input gaps, and the opportunity set, the interfirm motivational state should determine the degree to which firms expand in response to the pull of profit opportunities and the push of the fear of falling behind competitive firms. The interfirm motivational state itself is determined by the number of firms in the industry, the nature of the market structure, and the energy and aims of the entrepreneurs within these firms, which in turn determines the degree of competition between firms. The interfirm motivational state is unlikely to be sufficient to determine how any specific firm behaves. Among the intervening elements is the perceptive mechanism of the firm which determines the way in which firms receive, filter, and process market information and the degree to which firms become aware of changes of relative competitive status. (3) Thus, the intrafirm motivational state, whose constituents we have described above, determines how firms react to the activities of competitors, and to changes in the opportunities the firm faces. The intrafirm motivational state depends in part on the organizational structure of the firm and in part on the rate of change of manpower (especially managerial personnel) within the firm. The basic notion here is that as new individuals enter the firm, the existing equilibrium between decision-makers and their reactions to each other and to external opportunities may change so that the intrafirm motivational state changes accordingly. Of course, this last depends also on the degree to which new management personnel are similar or different in their capabilities and attitudes from those that they replace. (4) Finally, the input-completing and gap-filling capacities of the potential entrepreneurial pool determines the response of members of this pool to changes in opportunities and motivational states. An important aspect of the abilities involved is both the perception of economic opportunities and the capacity to assess such opportunities. These are presumably determined in part by factors exogenous to the system such as those involved in nurture, informal training, experience, as well as formal education of individuals. In sum, the model should in some way enable us to specify the relations of the links mentioned to the nature of economic states so that we can determine entrepreneurial reactions to changes in the economic state.

It might be helpful to classify N-entrepreneurs into different categories and determine each category's responsiveness to a given motivational state. Probably a significant part of such a model would be the interaction of different types of entrepreneurs to each other's activities (i.e., imitation, linkages, followers on "cleared" pathways, knowledge spread, etc.). Each period the response of the N-entrepreneurs to the motivational state creates a new state of the system and changes the motivational states in subsequent periods. At the same time it changes the supply of N-entrepreneurs in subsequent periods since some of those that enter foreclose their availability on subsequent occasions. Thus, the impulses created by entrepreneurial acts lead to sequences of entrepreneurial activities and changing opportunities which influence the pattern and rate of growth.[6] In addition, basic secular factors would have to be taken into account, since each year some potential entrepreneurs retire and others enter, while, at the same time, inventions lead to changes in the technical frontier and add new elements to the impeded and incomplete part of the market net.

III

To be of interest a theory needs some conjectures to tell us how some basic elements in the theory behave. Hence, to add some interest to this

[6] It would be interesting to see under what assumptions we could derive from such a model the growth promoting backward linkages suggested by Professor Hirschman [7] [8].

paper, I will hazard the following, all of which are on an "other things equal" basis: (1) The greater the rate of growth desired, the greater the quantum of gap-filling and input-completing capacities required. (2) The supply of active gap-fillers depends on opportunity costs. (3) The greater the assets of the group related to the gap-filler by kinship or friendship ties, the greater the gap-filling capacity of the entrepreneur involved. (4) Differential gap-filling and input-completing capacities are a critical element in explaining the differential rewards of entrepreneurs. (5) The routinization of gap-filling activities reduces the rewards of entrepreneurs.

There are a set of theories about entrepreneurship which revolve around the notions that in underdeveloped countries entrepreneurs prefer traditional industries, that their behavior is tradition-bound, and that they face overriding institutional obstacles. Yet, developing countries have periods of low growth and other periods of rapid growth. My conjecture in this connection is that in fact traditionalism is not the critical element but that the motivations present—e.g., the profit rates—are such that those with gap-filling capacities are willing and able to exert themselves under some motivational circumstances and reduce the degree of exertion under others. Thus, the ebb and flow of low and high growth rates can be explained without positing institutional rigidities that would appear to be almost impossible to overcome.

Two related elements that come to mind are the facts that entrepreneurs frequently come from groups which have fairly large extended families who are often engaged in trade and that they are disproportionately recruited from elements of the population that in some sense or other are looked upon as "outsiders." The extended family aspect can be explained by the fact that gap-filling capacities depend in part on kinship relations in which there is a much higher degree of trust and through which one can draw on more diverse capacities than exist on a universalistic basis. While there are many aspects to the outsider part of the phenomena, part of it, perhaps, can be explained by the fact that to the extent that outsiders are restricted from some economic opportunities, their opportunity costs as entrepreneurs are likely to be lower than other portions of the population, and hence they more readily engage in entrepreneurial activities compared to "insiders" whose opportunity costs are higher. However, not all outsiders become entrepreneurs since low opportunity costs can only be a facilitating and not a sufficient condition for entrepreneurship.

I realize that I run the risk of being charged, to use Professor Baumol's phrase, with offering nothing more than a taxonomy. I want to suggest that this is not the case—that the characteristics of the world described in this paper and the specified nature of the entrepreneurial role is such that it does lead to potentially interesting conclusions for development problems.

Our basic assumptions are as follows: (1) Motivation internal to the firm is a basic input that is not marketed. (2) There always exists some degree of slack (or excess capacity) due to low X-efficiency. (3) To bring any enterprise into fruition requires the marshaling of a minimum quantum of inputs. (4) Some inputs are "nonexhaustible" in the usual sense; that is, they do not necessarily decrease with use. Indeed, in some cases the opposite may be the case. Knowledge and motivation are two inputs of this type.

Some possible conclusions derivable from our assumptions are as follows:

1. While entrepreneurship may be scarce because of a lack of input-completing capacities, some entrepreneurial characteristics may in fact be in surplus supply; that is, they are unused simply because of the lack of the input completing capacity. In addition, some may be unused because the motivational state does not bring forth an adequate entrepreneurial response. As a consequence, it is possible that in some cases, small changes in the motivational state or in the reduction of market impediments may turn entrepreneurial scarcity into an abundant supply.

2. Our analysis of entrepreneurship requires us to reconsider the literature on investment criteria. Since investment can alter the market impediments and hence alter the supply of entrepreneurship, we must consider such possible side effects in our investment criteria. Thus, a lower profit investment that releases entrepreneurial energies and capacities may be more fruitful in the long run than a higher profit investment, if profit is calculated apart from the side effects we have just mentioned.

3. Some types of input creation which would normally appear to be functional may in fact be dysfunctional when the side effects are taken into account. For example, some types of higher education provided to potential entrepreneurs may be dysfunctional in that it increases the opportunity costs of potential entrepreneurs and may as a consequence decrease the supply of entrepreneurship.[7]

4. The theory suggests that training can do something to increase the supply of entrepreneurship. Obviously, not all characteristics of entrepreneurs are trainable. However, since entrepreneurship requires a combination of capacities, some of which may be vital gaps in carrying out the input-completing aspect of the entrepreneurial role, training can eliminate some of these gaps. For example, it may be difficult to train people to spot economic opportunities, but it is possible to train them to

[7] Somerset Maugham's story of the illiterate verger is an illustration of this possibility.

assess such opportunities once perceived. Similarly, certain managerial skills are trainable, but without them new firms might not survive because of their inability to overcome initial managerial difficulties.

For policy purposes, the theory suggests that development economists focus their attention when concerned with specific countries on studying the gaps, obstructions, and impediments in the market network of the economy in question and on the gap-filling and input-completing capacities and responsiveness to different motivational states of the potential entrepreneurs in the population.

References

1. M. Abramovitz, "Resources and Output Trends in the United States Since 1870," *A.E.R.*, May, 1956.
2. O. Aukrust, "Investment and Economic Growth," *Prod. Meas. Rev.*, Feb., 1959.
3. W. Baumol, "The Entrepreneurship in Economic Theory," *A.E.R.*, May, 1968.
4. H. B. Chenery and A. M. Strout, "Foreign Assistance and Economic Development," *A.E.R.*, Sept., 1966, p. 686.
5. R. M. Cyert and J. G. March, *A Behaviorial Theory of the Firm* (Prentice-Hall, 1963).
6. E. E. Hagen, *On the Theory of Social Change* (Irwin, 1962).
7. Albert O. Hirschman, *The Strategy of Economic Development* (Harvard Univ. Press, 1951).
8. ———, *Development Projects Observed* (1967).

8a. ———, *Journeys Through Progress* (1963).

9. H. Leibenstein, "Allocative Efficiency vs. X-Efficiency," *A.E.R.*, June, 1966.
10. D. McClelland, *The Achieving Society* (Princeton, 1961).
11. G. F. Papanek, *Pakistan's Development* (Cambridge, 1967).
12. J. A. Schumpeter, *The Theory of Economic Development* (Harvard Univ. Press, 1951).
13. R. Solow, "Technical Change and the Aggregate Production Function," *Rev. of Econ. and Statis.*, Aug., 1957.

Part II
Knowledge and the Firm

[6]

On the size distribution of business firms

Robert E. Lucas, Jr.
Department of Economics
University of Chicago

This paper proposes a new theory of the size distribution of business firms. It postulates an underlying distribution of persons by managerial "talent" and then studies the division of persons into managers and employees and the allocation of productive factors across managers. The implications of the theory for secular changes in average firm size are developed and tested on U.S. time series.

1. Introduction

■ The designation of Herbert A. Simon as a Distinguished Fellow of the American Economic Association provides a happy occasion for his friends, colleagues, and students to gather to talk economics. For me, it also offers the opportunity of claiming a distinction for myself, one which I feel I have earned many times over yet have no degree to show for it: that of being one of Herb's students in economics, and in social science more generally, at Carnegie Tech.

In this paper, I would like to return to an area to which I contributed, rather obliquely, ten years ago, and to which Simon has contributed substantially both before and since: the size distribution of business firms. This topic is one on which important issues of economic policy are held to hinge: in wealthy economies, "bigness" is widely viewed as a menace against which government activity should, perhaps, be directed; in poor economies, "littleness" is often viewed as a sign of backwardness to be dealt with by government policy. The lack of discipline with which such issues are typically discussed reflects, I think, the lack of an adequate understanding of the forces determining firm size.

2. Existing theory and its implications

■ For many years, the theory of firm size advanced in Jacob Viner's classic paper (1932) dominated economists' thinking on the size distribution of firms. This theory predicts a unique size distribution within an *industry* under the

I would like to thank Sherwin Rosen and José Scheinkman for very helpful discussions, and Dan Gressel for his assistance. At the Conference Honoring Herbert A. Simon's Economic Contributions at Carnegie-Mellon University in October, 1977, I received helpful suggestions and corrections from many participants—Dennis Epple, Tsu Yao, and Herbert Simon in particular—as acknowledged more specifically in the text. Finally, I would like to thank two anonymous referees and Alvin Klevorick.

assumption that individual firms have *U*-shaped long-run average cost functions. In equilibrium, each firm produces at the minimum point of this curve, with firm entry or exit adapting so as to adjust total industry production to quantity demanded at the zero-profit price. The size distribution which emerges, then, is a solution to an extremum problem: allocate production over firms so as to minimize total cost.

Viner's theory proceeds under the *assumption* of product market competition. It cannot, therefore, provide a complete guide to policy directed at reducing monopoly power. It does, however, direct attention to a major potential *cost* of policies directed against large firms. By keeping firms small, relative to their most efficient (in a productive sense) scale, such a policy results in a waste of resources which must somehow be balanced against the gains from reducing monopoly inefficiency. Not surprisingly, scale economies have been a standard and often successful defense in antitrust actions.

The evidence against this nonvacuous version of the Viner theory is by now so overwhelming that few economists accept it, except perhaps as a model of *plant* or *store* size. Perhaps most fundamentally, the fact that the typical firm sells in a large number of product markets leads to enormous difficulties in applying the Marshallian industry abstraction to the firm size problem. Second, in contradiction to the theory, most changes in product demand are met by changes in firm size, *not* by entry or exit of firms. Third, the percentage rate of firm growth appears to be independent of its size (measured by sales, employment, or assets). Finally, Simon, in collaboration with Charles Bonini (1958) and Yuji Ijiri (1964), observed that by examining the distribution of firms by size at a single point in time, one can make inferences about the stochastic process which governs firm growth. This insight has been exploited in several directions, and has led to further confirmation of the law that firm growth is independent of size.

The models of Bonini, Ijiri, and Simon make no use of the hypothesis of maximizing behavior on the part of firms. Yet it is not at all difficult to reconcile their findings with a suitably modified, *constant-returns-to-scale* version of Viner's theory (cf. Lucas, 1967). On this version of the theory, *all* size distributions are competitive equilibria, and all are solutions to the extremum problem of producing at minimum resource cost.

This conclusion that the size distribution is, under competition, a matter of indifference from a welfare point of view is more far-reaching than is commonly recognized. It means that the difficult balancing between productive efficiency on the one hand and enforcement of competition on the other, as indicated by the Viner theory, is entirely unnecessary. The possible *gains* from policies directed against large firms are, because of the possibility of inefficiencies due to monopoly, positive; the *costs* in productive efficiency losses are, under constant returns, nil. Thus even a clumsy strategy of opposition to mergers and forced dissolutions will lead to beneficial or at worst, harmless, results.

These are the policy implications of the existing theory of the size distribution of business firms, under the version of that theory which is consistent with available evidence. Yet with the exception of George Stigler's discussion (1964), I am not aware of any place where these implications have been carefully developed and seriously advocated as the basis for actual practice in antitrust policy.

I suspect that one reason so few economists have followed Stigler's advocated policies against bigness lies in a suspicion that, if the existing size distribution does not uniquely solve the problem "allocate production over firms so as to minimize cost," then it uniquely solves some *other* extremum problem. Yet simply voicing this suspicion, without spelling out what this other problem is, is merely an application of the justly discredited principle that "what is, is best."

3. An alternative theory

■ In this paper I want to develop the implications of a suggestion of Henry G. Manne (1965) that the observed size distribution is a solution to the problem: allocate productive factors over managers of different ability so as to maximize output. Manne's paper was devoted mainly to explaining how takeovers via proxy fights, share purchases, and mergers can be viewed as *mechanisms* by which talented managers acquire more productive factors. My strategy will be to assume at the outset that this allocation of assets and employees over managers is *perfectly* carried out (by some mechanism which I will not specify) and then to work out some of the implications of this hypothesis. This line will, I hope, be taken as complementary to that pursued by Manne.

As a preliminary matter, it is necessary to decide whether to follow Viner and develop a model at the *industry* level, or to theorize at an economy-wide level. The former would be appropriate if managerial ability were industry-specific, as a result of "nature" or of the accumulation of industry-specific expertise. There is no doubt that these considerations matter, at some level, but the multiproduct nature of at least the largest modern firms and the mobility of top managers across industries suggest that, if one is choosing between these extremes, it is best to attack the problem at an economy-wide level.

To keep matters simple, at least initially, I shall develop a model of a closed economy with a given quantity of homogeneous capital and a given workforce which is homogeneous with respect to productivity as an employee. Each member of the workforce is also endowed with a "talent for managing" which varies across workers. A *firm* in this economy is one manager, together with the capital and labor under his or her control.[1] Resource allocation involves, first, a division of the workforce into managers and employees and, second, the allocation of factors of production across managers. After specifying the technology of production and of management, I shall study the problem of allocating resources in an output-maximizing way: the dual of the problem studied by Viner. In common with Viner's theory, the solution to this extremum problem will be a competitive equilibrium.

As with Viner's theory, this model "predicts" the size-distribution of firms, but only *given* the distribution of persons by managerial talent. Without some means of learning about the latter, then, the theory does not place restrictions on the kind of data utilized in the Simon-Bonini-Ijiri studies. On the other hand, the theory places definite restrictions on the way in which average

[1] Richard Kihlstrom and Jean-Jacques Laffont (1977) developed an equilibrium theory in which agents differ in their attitudes toward risk with the relatively least risk averse becoming entrepreneurs. I am here adopting exactly their formulation, with "attitudes toward risk" replaced by "talent for managing."

firm size should vary across economies at different levels of development. These implications form the basis for some time-series regressions.

After developing an explicit model (Sections 4 and 5) and reporting some empirical results (Section 6), I shall consider two ways in which the model might fruitfully be elaborated (Section 7). Section 8 is a conclusion.

4. An explicit model

■ In this section I shall consider a closed economy with a workforce of size N and with K units of homogeneous capital, both inelastically supplied to the market. These factors of production may be combined, in a manner to be specified in a moment, so as to produce Y homogeneous units of output. After specifying the technology, I shall study the problem of combining productive factors so as to maximize Y; the solution to this problem will also be a competitive equilibrium.

It will be helpful (though slightly artificial) to consider separately the *production technology* and the *managerial* (or entrepreneurial) *technology*. For the former, let $f(n,k)$ be the output produced with n units of labor and k of capital, under "normal" or "representative" management. Let this technology exhibit constant returns, so that we may write $f(n,k) = n\phi(r)$, where $r = k/n$, and ϕ: $R^+ \rightarrow R^+$ is a twice differentiable function, increasing and strictly concave. If everyone in the economy were capable of "normal" management at any scale of operations, then optimal and equilibrium output would be $Y = N\phi(K/N) = N\phi(R)$, letting R denote capital per capita, and equilibrium wages and capital rentals would be $w = \phi(R) - R\phi'(R)$ and $u = \phi'(R)$, respectively. Rents to managers or entrepreneurs would be zero.

The managerial technology involves two elements: variable skill or talent, and an element of diminishing returns to scale, or to "span of control." For the first, each agent is endowed with a managerial talent level x, drawn from a fixed distribution Γ: $R^+ \rightarrow [0,1]$. If agent x manages resources n and k, his "firm" produces $xg[f(n,k)]$ units of output, where g: $R^+ \rightarrow R^+$ is twice-differentiable, increasing, and strictly concave, satisfying $g(0) = 0$. That is to say, each "firm" consists of a single manager (or entrepreneur), n homogeneous employees, and k homogeneous units of capital.[2]

This description of management is a shallow one, in at least two respects. First, it does not say anything about the nature of the tasks performed by managers, other than that whatever managers do, some do it better than others. Given this assumption, however, one is led immediately to the question: why does the best manager not run everything? Therefore, I assume concavity of the function g. Second, this technology precludes pyramidal managerial structures: managers managing other managers. One could postulate a technology for such organizations without any difficulty in a mathematical sense, but without a clear idea of where one is going, this is likely to lead to an uninformative taxonomy. Let me proceed for a while at this simple level, with the understanding that what we may hope for is not serious organization theory, but perhaps some insights into why organization theory matters economically.

[2] Notice that *no* individual x is capable of producing under what I have called above "normal management" $f(n,k)$ at all scales of operations. Thus the normal or "representative" manager is an expositional fiction, as these terms are used here.

To complete the statement of the problem, it is convenient to think of a continuum of agents, so that the entire distribution Γ of talent is always fully represented. Then an allocation of resources is described by two functions $n(x)$ and $k(x)$, giving the labor and capital managed by agent x. Since production requires both chiefs and Indians, an allocation which yields a positive output will have $n(x) = k(x) = 0$ for some x-values (those corresponding to employees) and positive for others (those corresponding to managers). For efficient or equilibrium allocations, it will be only the most talented who manage, so that there will exist a cutoff level $z > 0$ such that if $x < z$, one is an employee, and, if $x \geq z$, one is a manager. By an *allocation*, then, I shall mean a *number* z and a pair of *functions* $n(x)$, $k(x):R^+ \rightarrow R^+$ such that $n(x) = k(x) = 0$ for $x < z$ and $n(x)$, $k(x) > 0$ for $x \geq z$.

An allocation is *feasible* if it does not utilize more than the available population N and capital K. That is,

$$1 - \Gamma(z) + \int_z^{\infty} n(x)d\Gamma(x) \leq 1 \tag{1}$$

(the fraction engaged in managing plus the fraction engaged as employees sum to no more than one) and

$$\int_z^{\infty} k(x)d\Gamma(x) \leq \frac{K}{N} = R. \tag{2}$$

An *efficient* allocation is one which maximizes output,

$$\frac{Y}{N} = \int_z^{\infty} xg[f(n(x),k(x))]d\Gamma(x) \tag{3}$$

subject to (1) and (2).

In the Lagrangian associated with this variational problem, let w and u be the multipliers (both constant with respect to x) associated respectively with the constraints (1) and (2). Then the efficient allocation will also be the *competitive equilibrium*, with w and u being the equilibrium wage rate and rental price of capital. The income or rent to manager $x > z$ will be the residual

$$xg[f(n(x),k(x))] - wn(x) - uk(x). \tag{4}$$

The first-order conditions for this maximum problem include:

$$xg'(f)f_n(n(x),k(x)) = w, \quad x \geq z \tag{5}$$

and

$$xg'(f)f_k(n(x),k(x)) = u, \quad x \geq z \tag{6}$$

so that the marginal products of both factors are equated across firms. Recalling that $f(n,k) = n\phi(r)$, where $r = k/n$, (5) and (6) imply

$$\frac{\phi(r) - r\phi'(r)}{\phi'(r)} = \frac{w}{u}. \tag{7}$$

Thus, given the ratio of factor prices, all firms have a common capital-labor ratio $r(w/u)$ given implicitly by (7). The function $r(\cdot)$ is strictly increasing.

Given r from (7), the equilibrium scale $n(x)$ of firm x can be obtained from either (5) or (6). Using (6),

$$xg'[n(x)\phi(r)]\phi'(r) = u, \tag{8}$$

which gives employment as an implicit function $n(x,w,u){:}x \geq z$. The function $n(\cdot)$ is increasing in x and u, and decreasing in w.

The first-order condition for the cutoff value z is

$$zg[f(n(z),k(z))] = w[1 + n(z)] + uk(z). \tag{9}$$

For the marginal manager z, as for all, $k(z) = rn(z)$ with r given by (7), so that (9) may be written:

$$zg[n(z)\phi(r)] = w + (w + ur)n(z). \tag{10}$$

The right-hand side of (10) (or (9)) is total cost, consisting of a "fixed cost" w, the opportunity cost of the manager's own time, and a constant-per-employee variable cost, $(w + ur)n(z)$. The left side is output (or, with output serving as numeraire, the *value* of output). Thus (10) is a breakeven, or average-cost-equals-price condition for the marginal manager.

The marginal manager's employment level must also satisfy the *marginal-cost-equals-price* condition (8), evaluated at $x = z$. Then, given r, (8) and (10) are two equations in z and $n(z)$. Figure 1 displays these relationships in the (n,z) plane. In this plane, equation (10) gives the talent level needed to break even with a firm of employment n, as a function of n; equation (8) gives the talent level at which employment of n is profit-maximizing, also as a function of n. The relationship between these two curves is familiar from its analogue in the Viner theory. They determine a unique employment level $n(z,w,u)$ for the marginal manager: the level which *minimizes* average cost. The managerial talent level $z(w,u)$ which just breaks even at this employment level is the equilibrium cut-off.

One way to attempt to construct an equilibrium solution is as follows. First, think of the problem of maximizing output (3), subject to (1) and (2) for a *fixed* capital-population ratio R and a *fixed* managerial cutoff level z. This is a concave program, the first-order conditions for which are given by (1), (2), (7), and (8). It has a unique solution $n(x,w,u)$, $r(w/u)$ with factor prices $u(z,R)$, $w(z,R)$. To obtain it, one substitutes $n(x,w,u)$ and $r(w/u)$, obtained from (7) and (8), into the constraints (1) and (2) and solves for factor prices as continuous functions of z and R. Call the per capita output Y/N which results from this procedure $H(z,R)$. This much is a standard exercise.

FIGURE 1

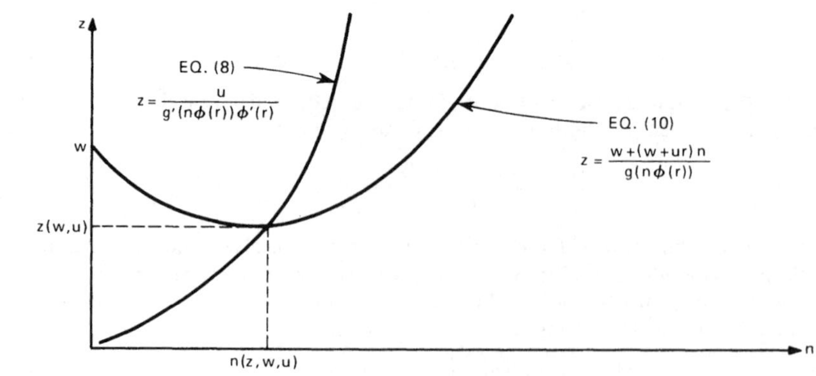

Next, we find the cutoff value z which maximizes $H(z,R)$, for each fixed R. The first-order condition for *this* problem is (10), with w and u read as the solutions $w(z,R)$, $u(z,R)$ found as above, and with $n(z)$ and r equal to $n(z,w(z,R),u(z,R))$ and $r[w(z,R)/u(z,R)]$, respectively. The derivatives H_{zz} and H_{zR} needed to check on the uniqueness of a solution for z, and on the way it varies with R, are easy enough to calculate, but not very illuminating. This is not a concave problem. More structure is needed, it appears. In the next section, I shall get it by invoking Gibrat's law.

5. Implications of Gibrat's law

■ The theory of the preceding section is static, determining the size distribution of firms given the economy-wide capital-population ratio. Taken literally, the theory has *no* implications for the pattern of growth of individual firms; a *firm* is defined empirically as a collection of assets, and the matching of managers to asset collections can, according to the theory, change arbitrarily from period to period. That is, the change in assets managed by manager x may or may not be interpreted as growth of an empirically defined firm. In this section, however, I want to think of the model of Section 4 as a limiting case of a model in which there *are* costs of rearranging assets among managers, so that one can think of "manager x" and "firm x," interchangeably. Then the percentage rate of growth of employment in firm x is

$$\frac{d}{dt} \ln [n(x,w(t),u(t))] \tag{11}$$

and of assets

$$\frac{d}{dt} \ln [r(w(t),u(t))n(x,w(t),u(t))]. \tag{12}$$

It should be clear that the theory in Section 4 is consistent with a wide variety of behavior of expressions (11) and (12). A well-known feature of observed patterns in firm growth, however, is the independence of firm growth and size: Gibrat's law, or the law of proportionate effect. Though this law is certainly not implied by the model just developed, there are thus empirical reasons for giving special attention to the special case which satisfies the law. This is the objective of this section.

In the present model, Gibrat's law is the hypothesis that the derivatives with respect to x (size) of expressions (11) and (12) are zero. Carrying out the differentiation (of either one) gives

$$0 = \frac{\partial}{\partial x} \frac{n_w(x,w,u)}{n(x,w,u)} \frac{dw}{dt} + \frac{\partial}{\partial x} \frac{n_u(x,w,u)}{n(x,w,u)} \frac{du}{dt}. \tag{13}$$

If (13) holds for *all* patterns of factor price changes, (13) implies

$$\frac{\partial}{\partial x} \frac{n_w(x,w,u)}{n(x,w,u)} = \frac{\partial}{\partial x} \frac{n_u(x,w,u)}{n(x,w,u)} = 0. \tag{14}$$

It is easily checked that these two statements are in fact the same, so I will develop the wage condition only. Differentiating (8) through with respect to w and solving for n_w yield

$$n_w(x,w,u) = -r_w\left(\frac{w}{u}\right)\frac{g'(f)\phi''(r) + g''(f)[\phi'(r)]^2 n}{g''(f)\phi(r)\phi'(r)}. \tag{15}$$

Dividing by $n(x,w,u)$ and differentiating with respect to x yield

$$\frac{\partial}{\partial x}\frac{n_w(x,w,u)}{n(x,w,u)} = -r_w\left(\frac{w}{u}\right)\frac{\phi''(r)}{\phi(r)\phi'(r)}\frac{\partial}{\partial x}\left[\frac{g'(f)}{ng''(f)}\right], \tag{16}$$

recalling that the capital-labor ratio r does not vary with x. Then (14) and (16) give:

$$0 = n_x(x,w,u)\{n\phi(r)(g'')^2 - g'g'' - n\phi(r)g'g^{(3)}\}. \tag{17}$$

Since $n_x > 0$ for $x \geq z$, (17) requires that the expression in brackets be zero. (Similar development of the capital rental condition in (14) leads to the same conclusion.)

Now $n\phi(r)$ is just the argument of the function g and its derivatives, so that (17) gives a second-order differential equation in the function $g'(v)$:

$$v[g''(v)]^2 - g'(v)g''(v) - vg'(v)g^{(3)}(v) = 0. \tag{18}$$

A convenient change of variable is

$$m(v) = \ln[g'(v)], \tag{19}$$

since (18) and (19) imply

$$\frac{vm''(v)}{m'(v)} = -1. \tag{20}$$

Integrating (20) gives

$$\ln(m'(v)) = \ln(A_1) - \ln(v)$$

or, using (19)

$$\ln\left[\frac{vg''(v)}{g'(v)}\right] = \ln(A_1).$$

Integrating again yields

$$\ln[g'(v)] = A_2 + A_1\ln(v)$$

and again

$$g(v) = \alpha v^\beta + A_3, \tag{21}$$

where $\alpha = (A_1 + 1)^{-1}e^{A_2}$ and $\beta = A_1 + 1$. We have found, in summary, that a necessary and sufficient condition for the model of Section 4 to obey Gibrat's law—in the sense of (14)—is that g takes the form given by (21). Given assumptions already imposed on g, it must also be the case that $0 < \beta < 1$ in (21), that $\alpha > 0$, and that $A_3 = 0$.

With these additional restrictions on the function g, I shall resume the construction sketched at the close of Section 4.[3] First, (8) is solved for

$$n(x,w,u) = \frac{1}{\phi(r)}\left[\frac{u}{\alpha\beta x\phi'(r)}\right]^{1/(\beta-1)}, \tag{22}$$

where r is given by (7). Next, inserting $n(x,w,u)$ from (22) and $k(x,w,u) = rn(x,w,u)$ into the constraints (1) and (2) gives

$$\left[\frac{u}{\alpha\beta\phi'(r)}\right]^{1/(\beta-1)}\frac{1}{\phi(r)}L(z) = \Gamma(z) \tag{23}$$

[3] Several errors in equations (22)–(27) and in their interpretation appeared in an earlier draft. These were pointed out to me by Dennis Epple, Tsu Yao, and a *Bell Journal* referee. In revising this section, I have made use of some notes of Tsu Yao's which, besides correcting mistakes, suggested considerable simplification.

and

$$r\Gamma(z) = R, \tag{24}$$

where

$$L(z) = \int_z^\infty x^{1/(1-\beta)} d\Gamma(x). \tag{25}$$

Now per capita output Y/N may be expressed as a function $H(z,R)$ of z and R by inserting the employment solution in (22) into equation (3):

$$\begin{aligned}\frac{Y}{N} &= \int_z^\infty xg[f(n(x),k(x))]d\Gamma(x)\\ &= \int_z^\infty x\alpha[n(x,w,u)\phi(r)]^\beta d\Gamma(x)\\ &= \int_z^\infty x\alpha\left[\frac{\alpha\beta x\phi'(r)}{u}\right]^{\beta/(1-\beta)} d\Gamma(x)\\ &= \alpha\left[\frac{\alpha\beta\phi'(r)}{u}\right]^{\beta/(1-\beta)} L(z).\end{aligned}$$

From (23),

$$\left[\frac{\alpha\beta\phi'(r)}{u}\right]^{\beta/(1-\beta)} = \left[\frac{\phi(r)\Gamma(z)}{L(z)}\right]^\beta,$$

so that

$$\frac{Y}{N} = H(z,R) = \alpha(\phi(r))^\beta(\Gamma(z))^\beta(L(z))^{1-\beta}, \tag{26}$$

where $r = R/\Gamma(z)$ from (24).

The equilibrium z will maximize $H(z,R)$ or, equivalently, $\ln H(z,R)$. The first-order condition for this problem (equivalent to (10)) is

$$\begin{aligned}0 &= \frac{\partial}{\partial z}\left[\beta \ln \phi\left(\frac{R}{\Gamma(z)}\right) + \beta \ln \Gamma(z) + (1-\beta)\ln L(z)\right]\\ &= \frac{\beta}{\phi}\phi'(r)\frac{-R}{\Gamma^2}\Gamma'(z) + \frac{\beta}{\Gamma}\Gamma'(z) + \frac{1-\beta}{L}L'(z) \qquad (27)\\ &= \frac{\beta}{\Gamma}\Gamma'(z)\left[1 - \frac{r\phi'(r)}{\phi}\right] - \frac{1-\beta}{L}z^{1/(1-\beta)}\Gamma'(z).\end{aligned}$$

The z-value satisfying (27) equates $(1-\beta)z^{1/(1-\beta)}$ and $\beta[1-(r\phi'(r)/\phi(r))] \times L(z)/\Gamma(z)$. Both functions are drawn in Figure 2. The first passes through the origin, and is increasing and convex. The ratio $L(z)/\Gamma(z)$ tends to infinity as $z \to 0$, and declines monotonically to zero as $z \to \infty$. With (24) the equilibrium r tends to infinity as $z \to 0$ and to R as $z \to \infty$. I shall assume that $1 - r\phi'(r)/\phi(r)$ remains bounded away from both 0 and 1 as these limits are taken, so that the asymptotic behavior of $\beta[1-(r\phi'(r)/\phi(r))]L(z)/\Gamma(z)$ is that of $L(z)/\Gamma(z)$. The derivative of $1 - r\phi'(r)/\phi(r)$ with respect to r is

$$\frac{d}{dr}\left(-\frac{r\phi'}{\phi}\right) = -\frac{\phi'(\phi - r\phi')}{\phi^2}\left[1 - \frac{1}{\sigma}\right],$$

where σ is the elasticity of substitution in production. Hence for the Cobb-

FIGURE 2

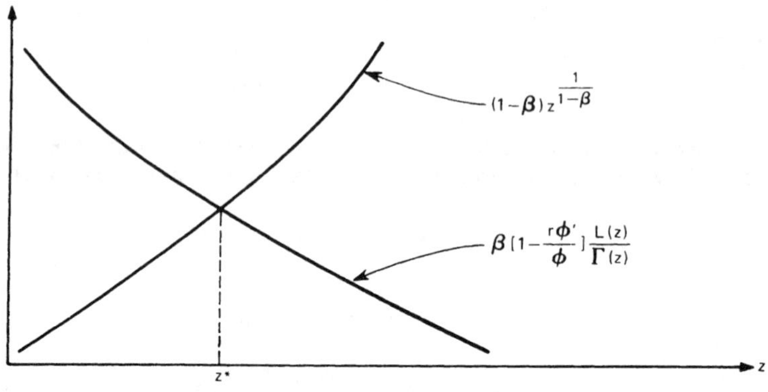

Douglas case $\sigma \equiv 1$, $1 - r\phi'/\phi$ is constant with respect to *both* z and R, and a unique solution z^* is determined as shown in Figure 2. Evidently, $\ln H(z,R)$ increases to the left of z^* and decreases to the right, so that z^* is a global maximum (though H need not be concave). In this Cobb-Douglas case, the equilibrium cutoff z^* does not vary with R, so that equilibrium average employment per firm, $[1 - \Gamma(z)]^{-1}\Gamma(z)$, would not vary systematically with the per capita wealth of an economy.

For the case $\sigma < 1$ uniformly, the curve $\beta[1 - (r\phi'/\phi)]L(z)/\Gamma(z)$ continues to slope down, as in Figure 2, and shifts to the *right* with increases in per capita capital R. In this case, which considerable evidence[4] suggests as the interesting one empirically, average firm size $[1 - \Gamma(z)]^{-1}\Gamma(z)$ is predicted to be an *increasing* function of per capita wealth. For $\sigma > 1$ this relationship is reversed (provided an equilibrium exists at all).

What is the logic underlying this connection between average firm size and the degree of factor substitutability? Think of an economy, in equilibrium, receiving an addition to its stock of capital. Holding the managerial cut-off z fixed, this new capital would be allocated over existing firms, increasing r as given by (24). This change will affect *both* the wage rate and the return to the marginal manager z. The wage rate as a function of r and z can be found from (23) and (7):

$$w = \alpha\beta\left(\frac{L(z)}{\Gamma(z)}\right)^{1-\beta}\phi^{\beta-1}(\phi - r\phi').$$

Managerial rent as a function of r, z, and employment $n(x)$ is, from (4), (21) and (23),

$$\pi = \alpha(1 - \beta)\left(\frac{L(z)}{\Gamma(z)}\right)^{1-\beta}\phi^{\beta}n(x).$$

[4] Some of this evidence is reviewed and augmented in Lucas (1969). An excellent recent study by Ernst R. Berndt (1976) obtains time series estimates of σ (for U.S. manufacturing) ranging as high as 1.2, depending on the way capital costs are measured. Though Berndt prefers the high estimates, for reasons explained in his paper, I have more confidence in those in his column D, Table 2, p. 65, for reasons given in his note 24, p. 66. These are around 0.5 to 0.8.

Now the ratio of wages to marginal rents is

$$\frac{w}{\pi} = \frac{\beta}{1-\beta}\frac{1}{n(z)}\left[1 - \frac{r\phi'}{\phi}\right].$$

With an elasticity of substitution less than one, $(d/dr)[-(r\phi'/\phi)]$ is positive, as shown above. That is, an increase in capital per capita *raises* wages relative to marginal managerial rents, or, in other words, raises the opportunity cost of managing relative to the return. This induces marginal managers to become employees, raising the equilibrium z and the average size of firms.

6. Some implications of the theory

■ The model developed in the preceding section is a variant on the production side of a Solow-type growth model. It implies an "aggregate production function" in the sense of a stable relationship between output per capita, Y/N, and capital per capita, R. To obtain this relationship, one inserts the optimal value of z as a function of R (as shown on Figure 2) into (26):

$$\frac{Y}{N} = \int_{z(R)}^{\infty} xg[n(x,R)\phi(r(R))]d\Gamma(x) = F(R). \tag{28}$$

Hence the implications of the theory for the dynamic behavior of aggregates are indistinguishable from the implications of standard, neoclassical growth theory. Whether this should be viewed as a virtue or a defect is, of course, debatable.

The model also "predicts" the full size distribution of firms, but only *given* a distribution of the (probably unobservable) managerial talent. It is more accurate to say that it predicts a particular relationship between the talent distribution and the size distribution of firms. To develop this relationship, take (7) as solved for r as a function of w/u and solve (8) for x:

$$x = \frac{u}{g'[n\phi(r)]\phi'(r)}. \tag{29}$$

Under the Gibrat's law restriction (21) and with (23), equation (29) is specialized to

$$x = \left[\frac{L(z)}{\Gamma(z)}\right]^{1-\beta} n^{1-\beta}. \tag{30}$$

The right side of (29) or (30) gives the talent level which will manage a firm with n employees at a given talent cutoff z and capital per capita R (or at given factor prices).

Now let $S(n)$ be the probability that a randomly selected firm will have fewer than n employees. Then under (30) $S(n)$ will be the probability that x is less than $(L/\Gamma)^{1-\beta}n^{1-\beta}$ *conditional on* $x \geq z$, or

$$S(n) = \frac{\Gamma\left[\left(\frac{L(z)}{\Gamma(z)}\right)^{1-\beta} n^{1-\beta}\right] - \Gamma(z)}{1 - \Gamma(z)} \tag{31}$$

for $n \geq (\Gamma/L)^{1/(1-\beta)}$ and 0 otherwise.

Any stochastic account of the development of managerial talent, leading to a limit argument motivating a particular form for Γ will, using (31), have definite implications for S. Thus, statistical and economic accounts of the size distribution need not be taken as *alternatives*, but can as well be viewed as complementary. Dennis Epple suggested the following illustration of this idea. Let Γ be a Pareto cumulative distribution function:

$$\Gamma(x) = 1 - B^{\rho}x^{-\rho}, \quad x \geq B,$$

where $B < z$. Then (31) implies that $S(n)$ is also a Pareto distribution function:

$$S(n) = 1 - z^{\rho}\left(\frac{L(z)}{\Gamma(z)}\right)^{-\rho(1-\beta)} n^{-\rho(1-\beta)} \tag{32}$$

for $n \geq (\Gamma/L)z^{1/(1-\beta)}$. An advantage of this combined statistical and economic motivation of a Pareto distribution of firm sizes is that there is a clear motivation for the value of the minimal size firm. The arbitrariness in selecting the parameter B in the talent distribution does not carry over to the predicted distribution S of firms.

A tighter implication of the Gibrat's law version of the model concerns the relationship of managerial compensation (or "rent") to firm size, as measured by employees. (This, too, I owe to Dennis Epple.) The general expression for managerial rent is given in (4). As noted in the preceding section, this implies a rent of

$$\alpha(1 - \beta)\left(\frac{L(z)}{(z)}\right)^{1-\beta} \phi^{\beta}n(x),$$

which is proportional to $n(x)$ for all x.[5]

A third implication for empirical testing is the predicted relationship between average firm size and the wealth of the economy. Average firm size is the ratio of employees to managers, or $[1 - \Gamma(z)]^{-1}\Gamma(z)$, which is an increasing function of z. According to the Gibrat's law version of the model, and provided the elasticity of substitution is less than unity, the equilibrium value of z will increase with the capital-labor ratio, R.[6]

Below, I report estimates of the parameters of

$$\ln(M_t) = \theta_0 + \theta_1 \ln(y_t) + \theta_2 t + \epsilon_t, \tag{33}$$

where M_t is employees per firm in the United States in year t; y_t is per capita GNP in constant dollars for the United States; and θ_0, θ_1, and θ_2 are fixed parameters. To obtain efficient estimators of θ_0, θ_1, and θ_2 in the presence of serially correlated residuals, the disturbances ϵ_t were assumed to follow

$$\epsilon_t = \rho_1\epsilon_{t-1} + \rho_2\epsilon_{t-2} + \eta_t, \tag{34}$$

where $\{\eta_t\}$ is a sequence of independent, normal variates with 0 mean and variance σ^2. The estimate of θ_1 will then be interpreted as an estimate of the

[5] This proportionality prediction was also noted by a referee. Available evidence on this point, obtained by D. R. Roberts (1956) and cited (and pointed out to me) by Simon (1957) is not favorable. Roberts finds managerial compensation varying with the *log* of employees. This suggests to me the necessity of incorporating hierarchical considerations, as discussed (but not carried out) in Section 7, below.

[6] Notice that this implication involves only the accuracy of the model in accounting for the *marginal*, or smallest, firm. Thus it is likely that modifications (such as the introduction of hierarchies) which affect mainly larger firms will not alter this implication.

TABLE 1

REGRESSION RESULTS*

ESTIMATES	DEPENDENT VARIABLE, TIME PERIOD		
	CONCERNS, 1900–1970	FIRMS, 1929–1963	MFG. FIRMS, 1929–1963
θ_0 (SE)	13.2 (3.80)	40.2 (7.83)	53.07 (10.59)
θ_1 (SE)	0.838 (0.071)	0.894 (0.119)	0.987 (0.172)
θ_2 (SE)	−0.004 (0.002)	−0.018 (0.004)	−0.024 (0.005)
ρ_1 (SE)	1.16 (0.12)	0.92 (0.19)	1.11 (0.17)
ρ_2 (SE)	−0.30 (0.13)	−0.25 (0.17)	−0.47 (0.14)
R_1^2	0.977	0.954	0.901
R_2^2	0.977	0.947	0.895
DW	1.84	1.78	1.85

*IN EACH COLUMN, THE INDEPENDENT VARIABLE IS $\ln$[GNP/LABOR FORCE]. THE DEPENDENT VARIABLE IS $\ln$[CONCERN WORKERS/CONCERNS] FOR COLUMN CONCERNS, AND SO FORTH.

sample average value of

$$\left[\frac{RF'(R)}{F(R)}\right]^{-1} \frac{\Gamma'(z)}{\Gamma(z)(1 - \Gamma(z))} z'(R),$$

which should (and does) turn out to be positive.[7]

The high level of abstraction of the theory of Sections 4 and 5 makes it difficult to match theoretically defined firms with firms as variously defined empirically. My strategy was to use as many measures of the number of firms in existence as I could find, and to hope that the choice did not make a great deal of difference. (The ones used are described in Table 2 below.) For each definition of firm, as conformable as possible an employment level was used.

Rather than attempt to measure the capital-labor ratio, R, directly, it seemed safer to use per capita GNP. The two variables are linked theoretically by (28), and using GNP per capita permits thinking of "capital" in the broadest possible terms, measurable or not. Some of the many possible economic motivations for including a trend are discussed in the next section.

Estimates of the parameters of (33) and (34) for annual U.S. data, with three dependent variables M_t, are reported in Table 1. Table 2 describes the sources and the construction of variables used. Estimates were obtained using Durbin's (1960) two-step method. Estimates of ρ_1 and ρ_2 are from the first step; R_1^2 is the R^2 from this step. Estimates of θ_0, θ_1, and θ_2 are from the second step (which is just regression (33) with $\ln(M_t)$ and $\ln(y_t)$ filtered by $1 - \hat{\rho}_1 L - \hat{\rho}_2 L^2$, where L is the lag operator). R_2^2 is the fraction of the variance of *original*

[7] This did not come as a surprise to me, nor will it to any economist with any familiarity with U.S. business population statistics. One might prefer to view the "test" results reported below as bearing on the size of the elasticity of substitution.

TABLE 2

SOURCES AND VARIABLE CONSTRUCTION

RAW SERIES	
CONCERNS	– FROM U.S. DEPARTMENT OF COMMERCE (1975), SERIES V–20. THOUSANDS OF FIRMS, AS DEFINED BY DUN & BRADSTREET.
	– EXCLUDES FINANCE, REAL ESTATE, INSURANCE, RAILROADS, AMUSEMENTS, MANY ONE–MAN SERVICES, PROFESSIONS, FARMERS.
FIRMS	– FROM U.S. DEPARTMENT OF COMMERCE (1975), SERIES V–13. THOUSANDS OF FIRMS, FROM SOCIAL SECURITY AND IRS SOURCES.
	– EXCLUDES SELF–EMPLOYED WITH NO EMPLOYEES, PROFESSIONS, FARMERS.
MFG. FIRMS	– FROM U.S. DEPARTMENT OF COMMERCE (1975), SERIES V–15. THOUSANDS OF MANUFACTURING FIRMS: SAME SOURCES AS "FIRMS."
GNP	– GNP IN $BILLIONS 1958.
CONCERN WORKERS	– FROM U.S. DEPARTMENT OF COMMERCE (1975), SERIES D–127 LESS D–137 LESS D–139 LESS RR WORKERS (SEE BELOW). THOUSANDS OF WORKERS, CONFORMING AS CLOSELY AS POSSIBLE TO EMPLOYMENT IN CONCERNS (DEFINED ABOVE).
RR WORKERS	– THOUSANDS OF RAILWAY WORKERS, CONSTRUCTED FROM SERIES IN U.S. BUREAU OF LABOR STATISTICS (1976) AND U.S. DEPARTMENT OF COMMERCE (1975).
FIRM WORKERS	– FROM U.S. DEPARTMENT OF COMMERCE (1975), SERIES D–127 LESS D–139. THOUSANDS OF WORKERS.
MFG. WORKERS	– FROM U.S. DEPARTMENT OF COMMERCE (1975), SERIES D–130. THOUSANDS OF WORKERS.
LABOR FORCE	– FROM U.S. DEPARTMENT OF COMMERCE (1975), SERIES D–1 FOR 1900–1947 AND SERIES D–12 X D–1 (1947)/D–12 (1947) FOR 1948–1970.

(that is, unfiltered) dependent variable which is explained by (33)–(34) with estimates obtained as just described. *DW* is the Durbin-Watson statistic from the second step.

The results with all three choices of dependent variable show a clear and accurately measured effect of GNP (or wealth) on average firm size. The estimated elasticity is in the range 0.8–1.0. The independent effect of trend on firm size is negative, with firm size declining (GNP held fixed) at an annual rate of about 0 to 2.5 percent. Together, these two variables do an excellent job of accounting for the secular behavior of firm size (however measured) and for large-scale cyclical movements (mainly the Great Depression).

7. Elaborations of the model

■ The extreme simplicity of the theory developed in Sections 4 and 5 matches very well the crude and aggregative data used in Section 6. If one were to attempt tests on finer data (say, quantitative histories of individual firms or entrepreneurs) it would be evident that refinement in several directions would be necessary. In this section, I shall briefly discuss two refinements that seem important and promising: elaborations to consider human capital and hierarchical managements.

The employment variable theoretically used in Section 4 was for labor of constant quality. The actual variable used in the regressions of Section 6 is for labor of improving quality. The simplest way to correct for this would be to imagine that employee and managerial quality are improving at possibly different, fixed exponential rates. This introduces a trend term (of indeterminate effect) into (33), but has no other effect on the model.

More interesting human capital considerations are suggested by commonly known features of managerial careers: people tend to move from employee to managerial status later in their careers (as opposed to immediately upon entry to the workforce, as predicted by the theory above); those that make this transition tend to be among the most skilled employees. These facts suggest the existence of a kind of human capital which is productive both in managing and in working for others, and which is accumulated most rapidly as an employee. Without trying to formulate a model embodying these features exactly, one can guess that in place of an equilibrium cutoff talent level z for managers, one would find an equilibrium *function* $\tau(x)$, the value of which is the age at which a person of talent x begins to manage. If so, and if $F(s)$ is the fraction of the workforce younger than age s, then the fraction of the workforce managing in equilibrium is

$$\int_{x_0}^{\infty}\int_{\tau(x)}^{\infty} d\Gamma(x)dF(s),$$

where x_0 is the lowest talent level ever to manage. The case studied in Sections 4 and 5 was $x_0 = z$, $\tau(x) \equiv 0$, in which case this expression was $1 - \Gamma(z)$: a decreasing function of capital per capita, R. It is likely that the equilibrium values of x_0 and $\tau(x)$ both would increase with R, so that the aggregative implications of this more elaborate theory would not differ from those of the simple version tested. On more interesting samples, however, a more refined theory along this line might be worth developing in full.

The managerial "technology" postulated in Section 4 was more complex than that usually postulated in economic models (where labor and capital are somehow costlessly combined) but considerably simpler than the complex hierarchies observed in actual business firms. To capture some of this complexity, and to broaden the term manager to include others than top executives, one would need to postulate a technology for managing managers. (See Tuck (1954) and Beckman (1977).) Again, without attempting to carry out such a construction, one can guess as to its likely implications for the aggregative data used in Section 6, and for other samples.

With diminishing returns on span-of-control built in at each managerial level, an equilibrium with multilevel management should consist of an allocation of capital and product workers over "level one" managers, an allocation of these units over level two managers, and so forth, until all resources are distributed over the top level managers. In such a scheme, it seems likely that the theory of firm size would be similar to that developed in Sections 4 and 5, with "managers" reinterpreted as "top level managers" and, of course, with considerably more complex substitution possibilities in production. Fruitful analysis along these lines would, I think, require more specific thinking about what it is, exactly, that managers do.

8. Conclusions

■ On a recent vacation in Quebec, my family and I stopped for lunch at a small, inexpensive restaurant on the St. Lawrence River. The decor, the menu, and the service in this family-run place were unique to it, and reflected a large number of managerial decisions, all solved in a way reflecting both the tastes of the owners and local prices of food and other materials. Even politics was involved: the flags of Canada and the United States flew out front; the flag of Quebec was absent. The Quebec economy is developing rapidly, however, and should we return in ten years we shall find, I imagine, a Poulet Frit Kentucky outlet in its place, with decor, menu, services, and politics identical to its twins in Montreal or Quebec City. This will occur, according to the theory developed above, because rising real wages will make working for some one else more lucrative than the return to making managerial decisions for a single, small restaurant.

If one could not think of dozens of examples of specific economic activities of this sort which, over time, are being carried out in larger and larger firms (though not necessarily larger plants or stores) one would, I suspect, have little confidence in theorizing and testing at the aggregative level of this paper. But anyone *can* think of many such examples, and to many observers, this trend to bigness seems ominous. In this paper, I have tried to show one way in which this trend (like so many other changes which seem ominous to people) can arise as a natural consequence of increasing wealth in a competitive economy. If correct, the theory implies that policies to modernize backward economies by consolidation and policies to resist conglomerates in advanced economies may both be misguided—for exactly the same reason.

References

Beckman, M. J. "Management Production Functions and the Theory of the Firm." *Journal of Economic Theory*, Vol. 14 (February 1977), pp. 1–18.

Berndt, E. R. "Reconciling Alternative Estimates of the Elasticity of Substitution." *Review of Economics and Statistics*, Vol. 58 (February 1976), pp. 59–68.

Durbin, J. "Estimation of Parameters in Time Series Regression Models." *Journal of the Royal Statistical Society* (Series B) (1960).

Ijiri, Y. and Simon, H. A. "Business Firm Growth and Size." *The American Economic Review*, Vol. 54, No. 2 (March 1964), pp. 77–89.

Kihlstrom, R. E. and Laffont, J. J. "A General Equilibrium Entrepreneurial Theory of Firm Formation Based on Risk Aversion." Unpublished working paper, July 1977.

Lucas, R. E., Jr. "Adjustment Costs and the Theory of Supply." *Journal of Political Economy*, Vol. 75, No. 4, Part 1 (August 1967), pp. 321–334.

———. "Labor-Capital Substitution in U.S. Manufacturing" in A. C. Harberger and M. J. Bailey, eds., *The Taxation of Income from Capital*, Washington: The Brookings Institution, 1969, pp. 223–274.

Manne, H. G. "Mergers and the Market for Corporate Control." *Journal of Political Economy*, Vol. 73, No. 2 (April 1965), pp. 110–120.

Roberts, D. R. "A General Theory of Executive Compensation Based on Statistically Tested Propositions." *Quarterly Journal of Economics*, Vol. 20 (1956), pp. 270–294.

Simon, H. A. "The Compensation of Executives." *Sociometry*, Vol. 20 (March 1957), pp. 32–35.

——— and Bonini, C. P. "The Size Distribution of Business Firms." *The American Economic Review*, Vol. 48, No. 4 (September 1958), pp. 607–617.

Stigler, G. J. "The Case against Big Business" in E. Mansfield, ed., *Monopoly Power and Economic Performance*, New York: W. W. Norton, 1964.

Tuck, R. H. *An Essay on the Economic Theory of Rank*. Oxford: Basil Blackwell, 1954.

U.S. Bureau of Labor Statistics. *Employment and Earnings Statistics for the U.S.* Washington, D.C.: U.S. Government Printing Office, 1976.

U.S. Department of Commerce. *Historical Statistics of the United States: Colonial Times to 1970*. Washington, D.C.: U.S. Government Printing Office, 1975.

Viner, J. "Cost Curves and Supply Curves." *Zeitschrift für Nationalökonomie*, Vol. 3 (1932), pp. 23–46.

[7]

Issues in assessing the contribution of research and development to productivity growth

Zvi Griliches
Professor of Economics
Harvard University

This article outlines the production function approach to the estimation of the returns to R&D and then proceeds to discuss in turn two very difficult problems: the measurement of output in R&D intensive industries and the definition and measurement of the stock of R&D "capital." The latter concept leads to a discussion of modeling of the spillover effects of R&D and to suggestions for possible measurement of such effects via the concept of technological distance between firms and industries. Somewhat more familiar econometric problems (multicollinearity and simultaneity) are taken up in the next section and another section is devoted to estimation and inference problems arising more specifically in the R&D context. Several recent studies of returns to R&D are then surveyed, and the paper concludes with a plea for a lowering of expectations as to what the available data can tell us and with suggestions for ways of expanding the current data base in this field.

1. Introduction

■ Economists have used two rather different styles of research in their attempts to assess the contribution of research and development (R&D) expenditures to economic growth: historical case studies and econometric estimates of production functions containing an R&D variable. There have been a number of detailed case studies of particular innovations tracing out their subsequent consequences (see Griliches (1958) and Mansfield *et al.* (1977) for examples and Griliches (1973) for a survey). Much can be and has been learned through such studies. They are, however, very data- and time-expensive and are always subject to attack as not being representative, since they tend to concentrate on prominent and successful innovations and fields. Thus, it is never quite clear what general conclusions one can draw on the basis of such studies.

The econometric production function approach tries to meet these objections by abandoning the interesting detail of specific events and concentrating instead on total output or total factor productivity as a function of past R&D

This paper was written during my stay at the Institute for Advanced Studies, The Hebrew University, Mount Scopus, Jerusalem. The work on this paper was supported by NASA and NSF. I am indebted to Henry Herzfeld, Carol Kitti, Ariel Pakes, and Mark Schankerman for helpful comments.

investments (and other variables). Here *all* productivity growth (to the extent that it is measured correctly) is related to *all* expenditures on R&D and an attempt is made to estimate statistically the part of productivity growth that can be attributed to R&D (and sometimes, also, to its components). While the production function approach is more general than the case study approach, it is also coarser and suffers from all the problems that beset attempts to infer causality from behavioral data on the basis of correlational techniques. Nevertheless, currently it is the only available general way of trying to answer questions about the contribution of R&D to growth. It is the purpose of this paper then to explore both the promise and the problems raised by this approach. Some of these problems are conceptual and semantic: What do we want to measure? Others are substantive: What do we measure? And others are statistical-methodological: How can we tell from the data what happened?

Productivity and its growth are best discussed in the context of a "production function," $Y = F(X, \ldots)$, which describes the relationship between various inputs X and final output Y. Productivity ($A = Y/X$) is then defined as the ratio of output (Y) to some index of the total input X and its determinants are then discussable in terms of the list of variables included in X, the mathematical form assumed for the production function $F(\ )$, the particular empirical observations chosen to represent Y and X and the statistical methods used to infer the properties of $F(\ )$ from the data.

A major conceptual issue in any such study is the definition and scope of Y. Are we talking about GNP, the change in some measure of national wealth, or an even broader concept of economic welfare? In the health sector, for example, a major product of research and development is the reduced morbidity of the population. To the extent that R&D affects the workforce and increases hours of work, it will affect X and Y in a parallel fashion and have no effect on A as it is conventionally measured. Moreover, to the extent that it affects the morbidity of children and nonworking wives, it may not show up in our measures at all or show up perversely (e.g., an improvement in the health of children would be associated with less hospitalization, a smaller demand for drugs, and a resultant decline in GNP as it is conventionally measured).[1]

A semantic issue is what do we mean by the "contribution of research and development to growth"? Is it the partial derivative of Y with respect to some measure of research and development flow (R) or stock (K) or is it the total derivative, including the indirect effects of a change in research and development on Y through the induced changes in X? That is, are we asking about the marginal effect of an additional dollar spent on R&D holding all other investments constant, or are we asking for the total effect of a particular R&D investment, including the contribution of all the other investments (in equipment and training) induced by it. To some extent this is a distinction between economic-accounting approaches and causal-historical ones. Most of the economic literature has interpreted the question in the first, partial sense, though occasionally one can find some dissenting voices (Gordon, 1968 and Rymes, 1971). It is a semantic issue, since a complete model of this process can answer both questions and derive the answer to one of them from the other. But it does reflect differences in emphasis and should be kept in

[1] Of course, some of the freed resources will be spent on something else. The point of the illustration is, however, the fact that the "productivity" in the health sector as conventionally measured may in fact decline as the result of R&D.

mind in interpreting and framing statements intended to affect research and development policy.

Measurement issues arise both in the case of output and in the case of inputs. Difficult problems exist in the measurement of output in the government and service sectors. Problems arise also in the measurement of output in the private goods sector for complex and changing goods. They are essentially the dual of the familiar "quality change" problem in the construction of price indices. Unfortunately, the more research and development intensive is an industry, the more likely is its output to be subject to such measurement problems. Problems arise also in defining the "stock" of research and development and in developing appropriate deflators for it.

Serious difficulties in econometric inference result from the fact that most of the variables of interest tend to move together over time and space, making it hard to untangle their separate effects. Moreover, it is not easy to establish causality. Research and development investments are themselves affected by the level of output and by past profits and productivity, forcing one to formulate simultaneous equations models and to turn towards much more complex estimation techniques.

Clearly the issues alluded to above exceed by far the scope of one modest survey paper. In what follows I shall, by and large, accept the existing national income accounting framework and interpret the question of the contribution of research and development to productivity as referring to aggregate or industrial productivity measures as they emerge from these accounts.[2] Also, I shall join the majority of the writers on this subject by trying to estimate the *partial* contribution of R&D to growth, holding the contribution of other inputs (labor and capital) constant. Since I have discussed the conceptual and measurement issues at some lengths earlier (Griliches, 1973, 1977), here I shall only briefly recapitulate some of the major points.

The paper proper starts with a brief outline of the production function model used in analyzing returns to R&D and then proceeds to discuss in turn two very difficult problems: the measurement of output in R&D intensive industries and the definition and measurement of the stock of R&D "capital." The latter concept leads us to a discussion and modeling of the spillover effects of R&D (Section 2) and to suggestions for possible measurement of such effects via the concept of technological distance between firms and industries (Sections 3 and 4). Somewhat more familiar econometric problems (multicollinearity and simultaneity) are taken up in Section 5, and Section 6 is devoted to problems arising more specifically in the R&D context. Conclusions and recommendations for additional data collection and new research are to be found in the last section of the paper.

2. A model

■ Let $Y = F(X,K,u)$ be the "production" function connecting some measure of output, Y, at the micro or macro level, to the "inputs" X, K, and u, where X stands for an index of conventional inputs such as labor and capital, K is

[2] There is a whole literature criticizing the aggregate production function approach. Within its own frame of reference there are serious problems of aggregation (Fisher, 1969) and one can also raise questions about the profit maximizing or cost minimizing framework that underlies much of this work (Nelson and Winter, 1974 and elsewhere).

a measure of the current state of technical knowledge, determined in part by current and past research and development expenditures, and u stands for all other unmeasured determinants of output and productivity. Define also $A = Y/X$ as the level of (total factor) productivity, and $\tau = y - x = (dY/dt)/Y - (dX/dt)/X$ as its rate of growth. Let us also assume that there exists a relationship between K, the current level of technological knowledge, and $W(B)R$, an index of current and past levels of research and development expenditures, where $W(B)$ is a lag polynomial, describing the relative contribution of past and current research and development levels to K, and B is the lag (backward shift) operator. Thus,

$$K = G[W(B)R, v],$$

where v is another set of unmeasured influences on the accumulated level of knowledge and

$$W(B)R_t = (w_0 + w_1B + w_2B^2 + \ldots)R_t = w_0R_t + w_1R_{t-1} + w_2R_{t-2} + \ldots$$

Now the various issues mentioned above can be restated in terms of what we would like conceptually Y, X, K, and R to measure, what the problems with actual measures available to us are, what we need to assume about the world for the relationships $F(\ )$, $G(\ )$, and $W(B)$ to exist and be of interest to us, what we want to know about them, and how we are going to find it out in the face of imperfect data and with the presence of unmeasured forces u and v.

For example, just to write the function $F(X,K,u)$ implies the assumption of separability of the conventional inputs X (labor and capital) from the series of past and current research and development investments R. Similarly, writing $W(B)R_t$ as a linear function of all past R&D investments implies that there are no diminishing returns or rising costs at the annual R&D level.[3] One can look at such assumptions either as a statement about the properties of the real world or as a statement about the conditions for the approximations implied in the construction of aggregate indices of total input (X) or the stock of knowledge capital (K), not to mislead us too much. Given the limited quantity and quality of the data available to us, such assumptions are not really testable (for attempts at such tests *cf.* Berndt and Christenson (1973)). In what follows I shall ignore these "garden variety" index number problems and concentrate instead on the problems peculiar to research and development.[4]

Nor will I worry much here about the functional form of $F(\ )$. For expositional simplicity I shall assume it to be Cobb-Douglas and assume that the unmeasured factors u can be considered as random after the introduction of a time trend into the equation to represent the systematic component of the unmeasured factors. Then we can rewrite F as:

$$Y = DC^{\alpha}L^{\beta}K^{\gamma}e^{\lambda t+u},$$

where D is a constant, t is a time index, e is the base of natural logarithms, and α, β, γ, and λ are some of the parameters we are interested in estimating. If we had more and better data, we could try for a more complex description of the production process, using more general functional forms such as the CES

[3] An alternative approach would complicate this model further by adding an annual knowledge production function of the form $K = H(R,K)$ and defining $\dot{K}$ accordingly.

[4] See Diewert (1977) for a recent discussion of the necessary assumptions for the existence of such composite commodities as "labor," "capital," and "research capital."

or the translog, and introducing more parameters to be estimated.[5] But for the purposes of this paper, this simple characterization will suffice.

Let us define a conventional total input index X as:

$$X = C^s L^{1-s},$$

where s is the observed factor share of physical capital. Let us assume, for a while, that s is observed correctly and that it is proportional to the true coefficient of capital, i.e., $s = \alpha/(\alpha + \beta)$, and there is no error in computing the true *relative* shares of labor and capital. Then measured total factor productivity,

$$A = Y/X = DX^{\alpha+\beta-1}K^{\gamma}e^{\lambda t+u},$$

depends not only on the contribution of research capital K, the contribution of the trend t in the other unmeasured factors, and the random factor u, but also on the level of other inputs X, as long as there are nonconstant returns to scale ($\alpha + \beta \neq 1$). Moreover, any error of measurement in one of the inputs will transmit itself directly to the productivity measure. For example, let the "true" relevant measure of L be given by $L = Q_L N$, where Q_L is the average "quality" per worker and N is the total number of workers. Now if N rather than L is used in the construction of X, then measured productivity will be given by:

$$\tilde{A} = Y/\tilde{X} = D\tilde{X}^{\alpha+\beta-1}Q_L^{\beta}K^{\gamma}e^{\lambda t+u},$$

where $\tilde{X} = C^s N^{1-s}$, and similarly for errors in the measurement of C. Since the issue of errors of measurement in C and L has been discussed at length elsewhere (Griliches, 1963; Jorgenson and Griliches, 1965), I only want to remind the reader about their potential importance in any practical context.

The issues we want to concentrate on here are the definition of Y, the measurement of K, and the estimation of γ.

3. The measurement of output in research and development intensive industries

■ One of the major difficulties in measuring the contribution of research and development to economic growth is the fact that much of it is performed in industries whose product is itself badly measured. In three major areas of public research and development investments, defense, space, and health, the output measures are based on inputs and hence cannot and do not reflect the improvements in productivity which have been achieved with the help of research and development investments. In space exploration output is measured by man-days and expenditures on equipment and does not rise or fall with the success of the venture. In defense, companies sell equipment to and build battleships for the government. Again, there are no price indices of military aircraft that take into account their improved performance in terms of speed or maneuverability nor are there price indices that reflect the improved resolution and range of

[5] To estimate a more general functional form we would need to observe firms utilizing very different combinations of factors of production. But since most firms face rather similar factor prices, why should they have very dissimilar factor ratios? In our context, the issue of the functional form $F(\ )$ is not very interesting or crucial unless we are interested in the specific interaction of K with a particular input (e.g., if we suspect a particular complementarity between physical and research capital). As far as the role of K itself is concerned, the functional form issue can be investigated by adding the square or other nonlinear functions of K to the equation.

radar equipment. Defense purchases are almost always deflated by cost indices and, except for the fact that such indices are often not exactly consistent with the industry data, the resulting productivity measures should show no growth, even though large research and development expenditures have been made successfully to improve the performance of such items.

A similar problem arises in the measurement of output in the service sector, especially in health. The output of physicians is measured basically by the number of patient visits and the output of hospitals is measured by patient days. Any improvement in the performance of these activities in the sense of a higher frequency of cure of specific ailments not only will not show up as an increase in the product of this sector, but might actually lead it to register a decline.

The problem is more complex with new or improved products which are sold directly to consumers, such as pocket calculators or drugs. Here the measurement of output will depend on the market structure of the industry and the procedures used by the statistical agencies (e.g., the Bureau of Labor Statistics in the United States) in constructing the relevant price indices. If the producer of the new item were a perfectly discriminating monopolist, he would capture all the social returns to the innovation. Since most monopolists cannot discriminate perfectly and since their market position is far from secure, the actual revenue received by them will fall short of what could have been realized by a discriminating and secure monopolist. What happens to price indices will depend on whether they allow for the "quality" improvements embedded in the new item or not. By and large they do not make such quality adjustments.[6] Instead, the new product is "linked in" at its introductory (or subsequent) price with the price indices left unchanged. Hence measured output goes up by the revenue received, costs go up by the increased utilization of inputs in production, and productivity goes up by the amount of the total return to the innovation that the innovator succeeds in appropriating for himself. Subsequently, the entry of other competitors with similar products may force him to lower the price, but that will not increase measured output, since his revenues will decrease proportionately except as the reduction in price results in growth of the overall market for this product. The social return from this erosion of the original innovator's monopoly position will be recorded in the conventional measures. More generally, the amount of social returns reflected in the productivity measures will depend, in part, on how early the price of a new item is included in the official price index. The tendency of statistical agencies is to include it rather late, after the new product has "matured" a bit, and after its price has declined to "normal" levels, thereby missing much of the social contribution of such an invention. Actual scenarios are, of course, even more complicated than outlined above, but the long and the short of it is that official productivity measures in consumer goods industries such as drugs, where no "quality" adjustments are made, reflect only the original private returns appropriated by the inventors and the consumer surplus arising from the subsequent research of competitors which erodes their original market positions.

[6] Methods for incorporating more extensive "quality" adjustments into the official indices are available but require a much more detailed data base. See Griliches (1971) for a review of the "hedonic" indices literature.

A simplified example may be of some help here. Imagine an invention of a more sensitive (faster) photographic color film. Let the demand curve for this film be as follows: at \$100 per roll only 100 rolls would be sold, at \$1 per roll 10,000 rolls will be sold, with all other prices and quantities given by the straight line passing through these points ($q = 10{,}100 - 100p$). Let the average and marginal cost of producing a roll of this film (after the invention) equal \$1, including normal markup for overhead. This is also the price of the old film on the market of which 8,000 units were being sold originally. We are assuming then that the new film substitutes for the old film, that at \$2 everyone will switch over to the new film, and that at the \$1 price the market will expand by an additional 2,000 units.

Now there are several possibilities:

(1) Imagine that this new film was developed by NASA and is licensed freely to everybody. Then the price will be \$1, and 10,000 units of the new film (and none of the old) will be sold. Total cost in the film industry will go up from 8,000 to 10,000, as will total revenue. If, as is likely, the statistical agencies do not interpret this as a fall in the real price of color film, both input and output will have been measured as having increased proportionately and *no increase* in productivity will be recorded.

(2) If this film instead has been produced by a private inventor with a fool-proof patent on it, then it would pay him as a monopolist to produce 5,000 units at a price of \$5.01 per unit and leave 3,000 units of the old film to be produced at \$1 per unit. Now industry revenue will increase to 25,050 + 3,000 = 28,050, while the total cost remains at 8,000, implying an increase in measured productivity of 250 percent (as against zero in the previous competitive, freely licensed case).

(3) Now imagine the possibility that this monopolist could discriminate in the sales of this film individually and get each individual to pay him the maximum price he would be willing to pay for the privilege of using this film. Such a monopolist could collect \$510,000 from his sales of 10,000 units at an average price of \$51 per unit. In this case revenue will have increased by 511/8 = 64-fold, while costs will have gone up by only 25 percent, which implies a 52-fold increase in productivity. This is exactly the "true" increase in productivity which occurred in case 1 but could not be measured there, since the whole benefit was passed on to the consumer without its being measured anywhere.

(4) It is possible that in situation 1 or 3 the statistical agency notices that all of the previous users of the old film (at the \$1 price) are willing to pay at least \$2 per roll for the new film and declares that the "quality" of the new film is at least double that of the old one or, in other words, that the real price of film has declined by 50 percent. Then real output will have been measured to increase from 8,000 to 20,000 (in old film units) or by 150 percent while costs (inputs) will have gone up only by 25 percent, which implies a (approximately) 100 percent growth in measured productivity. This is clearly not the right number (case 3 has the correct computation), but it is better than nothing. Unfortunately, such computations are made only in a few industries. Among the major CPI components only automobiles and, recently, housing prices are subject to quality adjustments. Lack of such adjustments can lead to very serious biases in some industries. In computers, for example, the national income accounting convention has been to show no price change whatsoever. There is independent evidence,

however, that the "real" price of computers has been falling by about 20 to 30 percent per year in the last decade, leading to a very serious downward bias in the estimate of output and productivity growth in this industry. (See Chow (1967) and Sharpe (1967), among others.)

(5) If the innovation is not a consumer product but is used instead as an input in another industry, then the missing productivity is only misplaced (provided the output of the using industry is correctly measured). If the film discussed above is used only commercially to replace elaborate lighting arrangements, then its contribution would show up in an increase in the productivity of the photographic services industry (rather than the film industry). Similarly, computer-using industries have really bought more input than has been measured and will therefore show a rise in their productivity, even though the true productivity increase occurred in the computer-producing industry. Thus, to the extent that the output of the invention-using industries is well measured, the returns to such research and development can be found in the aggregate data (though not correctly attributed). But as we have already noted, in many important invention-using sectors such as government, health, and education, output is not well measured, and hence the productivity of the purchased and mismeasured inputs will also not be reflected there.

Up to this point we have been discussing primarily product innovations. Cost reducing innovations will, of course, be reflected in the productivity measures as long as they occur in industries where output is measured independently of input. Conventional productivity measures reflect, therefore, the cost-reducing inventions made in the industry itself, the privately appropriated part of product innovations within the industry, and the social product of inventions in the input-producing industries which have not already been reflected in the price of purchased inputs.

A word also should be said here about simple cross sectional comparisons. The revenues of a firm reflect its current and past research and development activities. These private returns, however, can exceed the social ones if they occur at the expense of another firm. If, for example, the research and development of a particular firm doubled its revenue by taking it away from a preexisting rival, then, this is not a social return. It is just a transfer. It is more likely, however, that social returns exceed private ones, since only a fraction of them is appropriated by the original inventors.

To summarize, much of reported research and development is expended in areas where its direct contribution cannot be measured. In an earlier paper (Griliches, 1973), I estimated these areas to account for about half of all reported research and development. An additional large component of research and development is aimed at final consumer product rather than process innovations and is reflected in productivity measures only to the extent that producers succeed in appropriating its fruits. Since much of the product of research and development is entirely unmeasured and much of the rest is mismeasured, it is not surprising that it has been rather difficult to find its traces in the data.

4. The measurement of research and development capital

■ In Section 2 we talked about knowledge capital (K). Empirically it is too broad a concept; it aspires to and contains too much. We can, however, focus on the contribution of identified investments in advancing the state of knowledge in a particular (or related) area(s). The contribution of "science" in general to a particular industry is probably not measurable, since there is no way of knowing how much "science" is actually used in one industry versus another.

Focusing on the contribution of industrial R&D, we can rewrite the definition of K as:

$$K_t = a_0[W(B)R_t]^{\eta} e^{\mu t + v},$$

where $W(B)R$ is some lag function of past R&D investments, μt is the trend component of all other influences on the state of knowledge, and v is the random transitory component of it. Substituting this into our production function formula, one can absorb η into the γ parameter of the production function, the trend component μt into the general efficiency trend term λt, and the v term into the overall disturbance term u. This leaves us with $W(B)R$, a measure of R&D "capital," as the topic to be discussed in this section.

There are three major issues in the measurement of such "capital":

(1) The fact that the research and development process takes time and that current research and development may not have an effect on measured productivity until several years have elapsed forces one to make assumptions about the relevant lag structure $W(B)$.

(2) Past research and development investments depreciate and become obsolete. Thus the growth in the net "stock" of research and development capital is not equal to the gross level of current or recent resources invested in expanding it.

(3) The level of knowledge in any one sector or industry not only is derived from "own" research and development investments but also is affected by the knowledge borrowed or stolen from other sectors or industries. Thus, the productivity of industry i will depend also on the research and development investments of industries j and h, among others.

Before we proceed to discuss these issues in turn, two other remarks are in order. First, we are obviously aware that there are very different levels and types of knowledge and that research and development results are embodied in people, blueprints, patents, books, and oral tradition. To try to aggregate such diverse items into one notion (index) of research and development "capital" is quite presumptuous. The "sin" of aggregation, however, is not so different here from that committed when constructing measures of national output where bushels of wheat, haircuts, and striptease hours are aggregated into one GNP figure, or "physical" capital which aggregates buildings, planes, computers, and shovels. The difference is, and this is the import of the second remark, that in most of the conventional cases the components to be aggregated have an observable market counterpart to which a value can be attached. A piece of equipment is sold and can be resold at a market price. The results of research and development investments are by and large not directly observable. Research and development capital is fundamentally an "input" rather than an "output" measure. In this, however, it is not so different from, say, the plant construction component of physical capital measures, which are also largely based on cost cumulations rather than on a market valuation of the final results.[7] Never-

[7] Note that nothing tangible corresponds to this notion of R&D "capital." It is just an alternative to expensing R&D as a current input. Ideally it would equal the value of the firm's "know how" and should be related to the excess of market value over appropriately depreciated and deflated book value. See Ben-Zion (1977) for an attempt at measurement in this vein.

theless, the lack of direct measures of research and development output introduces an inescapable layer of inexactitude and randomness into our formulation.[8]

There are several lags involved in the research and development-productivity nexus. First, a particular research and development project may take more than a year to complete. Second, when complete and if successful, it may still take some time before a decision is made to use it or produce it. Once an innovation decision is made, it may show up in the firm's revenue stream only with another lag. If a process innovation, it may be introduced gradually, affecting only parts of the firm's cost structure in the beginning. If a product innovation, it may take time for consumers to find out about it and to accept it. At the aggregate firm level there are many such projects that have started at different dates and are in different stages of fruition. More generally, the convolution and aggregation of many such lag structures should lead to a rather flat but somewhat bell-shaped lag structure connecting total firm research and development to its subsequent productivity (Griliches, 1967).

There is also the issue of depreciation or obsolescence of this capital. If one distinguishes between the firm-specific knowledge capital and the general state of knowledge in the industry as a whole, then at least as far as the first is concerned, it is quite clear that its earning capacity erodes over time, both because better products and processes become available and because its own knowledge begins to lose its specificity (it leaks to other firms in the industry). Thus, from the private point of view there is depreciation of this capital, probably at a rather high rate (see Pakes and Schankerman (1977) for some scattered evidence on this point). This fact, together with the rather short-term nature of much commercial research and development (see the discussion by Mansfield and others in Williams (1973, pp. 87–90)), would imply a research and development lag structure that peaks somewhere between three to five years earlier and then declines rather rapidly, with little of the original research and development product remaining "private" past ten years or so.

The question of depreciation is much more complicated for social research and development capital measures at the industry or national level. The fact that private knowledge loses its privacy and hence its value is a private loss, not a social one. Nevertheless, there is likely also to be some depreciation in social knowledge. Some of the new products may make the knowledge about older products and processes redundant. Alternatively, much of what is measured as investment in research and development is not "net" investment from a social point of view, since it replaces already existing knowledge. If 50 percent more investment is needed to produce an alternative product that is 10 percent superior to the older version, it may be pursued from the private point of view, since with about half the investment one can appropriate most of the rent collected by the inventors of the earlier version. But from the social point of view, if research and development capital were measured by "output," it would only go up by 10 percent rather than 50. We can say then, equivalently, that either only a fraction of current research and development flow is to be thought of as a net addition to the social stock of knowledge capital or that some fraction of the preexisting stock of this capital is replaced (depreciated) annually. The real problem here is our lack of information about the possible rates of such

[8] I shall ignore here the possibility of using patents or publication counts as indices of research and development output. They are only available for a limited range of sectors and industries.

depreciation. The only thing one might be willing to say is that one would expect such social rates of depreciation to be lower than the private ones.

The problem is even more difficult as far as the measurement of public R&D "capital" (such as is generated by NASA, the Department of Agriculture, universities, and other similar organizations) is concerned. Components of this type of R&D capital that contribute to productivity growth in the industrial sector are likely to be subject to both lengthier lags and lower depreciation rates. It is hard to see, however, where one could get relevant evidence on this topic.[9]

The last major issue is that of "spillovers," the effect of "outside" knowledge capital—outside the firm or industry in question—on the within-industry productivity. The level of productivity achieved by one firm or industry depends not only on its own research efforts but also on the level of the pool of general knowledge accessible to it. Looking at a cross section of firms within a particular industry, one will not be able to distinguish such effects. If the pools of knowledge differ for different industries or areas, some of it could be deduced from interindustry comparisons over time and space. Moreover, the productivity of own research may be affected by the size of the pool or pools it can draw upon. This would lead to the formulation of models allowing for an interaction between the size of individual and aggregate research and development effort.[10]

A simple model of such within-industry spillover effects is given by

$$Y_i = BX_i^{1-\gamma}K_i^{\gamma}K_a^{\mu},$$

where Y_i is the output of the ith firm which depends on its index of conventional inputs X_i, its specific knowledge capital K_i, and on the state of aggregate knowledge in this industry K_a. Note that I have assumed constant returns in the firm's own inputs, X_i and K_i. This simplifies the example greatly. Now let us assume that: (1) the aggregate level of knowledge capital $K_a = \sum_i K_i$ is simply the sum of all specific firm research and development capital levels; and (2) own resources are allocated optimally and all firms in the industry face the same relative factor prices. Then we know that the individual K_i to X_i ratios will be given by

$$\frac{K_i}{X_i} = \frac{\gamma}{1-\gamma}\frac{P_x}{P_k} = r,$$

where P_x and P_k are the prices of X and K, respectively, and r, the K/X ratio does not depend on i. We can aggregate our individual production functions:

$$\sum_i Y_i = \sum_i BX_i(K_i/X_i)^{\gamma}K_a^{\mu} = \sum_i BX_ir^{\gamma}K_a^{\mu} = Br^{\gamma}K_a^{\mu}\sum_i X.$$

Now since the K_i/X_i ratios are equal to r, so also is $\sum K_i/\sum X_i$, which we can substitute back into this equation, yielding:

$$\sum_i Y_i = B(\sum_i K_i/\sum_i X_i)^{\gamma}K_a^{\mu}\sum_i X_i = B(\sum_i X)^{1-\gamma}K_a^{\mu+\gamma},$$

where, by assumption, $\sum_i K_i = K_a$. Thus, we get an aggregate production function with the coefficient of aggregate knowledge capital being higher

[9] Some work has been done on the contribution of publicly supported research in agriculture to productivity. See Griliches (1973) for a review of this literature.

[10] See Evenson and Kislev (1975, Chapter 4) for an example of such a "borrowing" function involving scientific publications.

$(\gamma + \mu)$ than at the micro level (γ only), since at the aggregate level it reflects not only the private but also the social returns to research and development.[11] The above formula provides a framework for reconciling micro and macro results in this area.[12]

Of course, this formula is rather simplistic and is based on a whole string of untenable assumptions, the major ones being: the assumption of constant returns to scale with respect to X_i and K_i and the assumption of common factor prices for all firms within an industry. These assumptions could be relaxed. This would add a number of "mix" terms to the equation, indicating how aggregate productivity would shift if the share of, say, the larger firms, were to increase (as in the case of economies of scale). If the mix of firms stays stable, such refinements do not add much. In any case, the above formula was presented to suggest the nature of a class of results that one could get from such assumptions, rather than to set out a final model of such phenomena.

The problem is much more complicated when we realize that we do not deal with one closed industry, but with a whole array of firms and industries which "borrow" different amounts of knowledge from different sources according to their economic and technological distance from them. The concept of such a "distance" is very hard to define empirically. If we return to our previous example and now interpret the index i as referring to industries rather than firms, it makes little sense to define K_a as $\sum_i K_i$. Rather

$$K_{a_i} = \sum_j w_{ij} K_j$$

is the amount of aggregate knowledge borrowed by the ith industry from all available sources. K_j measures the levels available in these sources, while w_{ij}, the "weighting" function, can be interpreted as the effective fraction of knowledge in j borrowed by industry i. Presumably w_{ij} becomes smaller as the "distance," in some sense, between i and j increases. Thus we need an additional distributed (lag) over space function to construct a measure of the stock of borrowed knowledge.

On what should such a weighting function be based? There have been earlier suggestions based on "vertical borrowing" concepts: Brown and Conrad (1967) used the input-output table to measure the "closeness" of industries proportional to their purchases from each other, while Terleckyj (1974) used the capital and intermediate inputs purchases matrix weights, assuming that "borrowed" research and development is embodied in purchased inputs. Raines (1968) used the "horizontal" product field classification of NSF to include inputs to an industry's research and development and also the research and development expenditures of other industries which were reported as belonging to its product field.

Actually, there are two distinct notions of research and development "spillovers" here which are often confused in the literature. The first one, research and development intensive inputs purchased from other industries at

[11] The effect of aggregated private "knowledge" is K^{γ}. The total spillover effect of this knowledge is K^{μ}. Since we have assumed that all private knowledge spills over to some extent and have measured it as $K_a = \sum K_i$, the total effect of all private knowledge at the aggregate level is given by $\gamma + \mu$, rather than just by γ.

[12] The first time I saw this type of formulation was in an unpublished note by Grunfeld and Levhari (1962).

less than their full "quality" price, is related to issues in the measurement of capital equipment and materials and their prices and is not really a case of pure knowledge spillover. If capital equipment purchase price indices reflected fully the improvements in their quality, i.e., were based on hedonic calculations (Griliches, 1971), there would be no need to deal with this problem. As currently measured, however, total factor productivity in industry *i* is affected not only by its own research and development but also by productivity improvements in industry *j* to the extent of its purchases from that industry and to the extent that the improvements in *j* have not been appropriated by its producers and/or have not been incorporated in the official price indices of that (*i*) industry by the relevant statistical agencies. The use of purchase flow weighted research and development measures assumes that social returns in industry *j* are proportional to its research and development investment levels and that the amount of such returns transferred to industry *i* is proportional to its purchases (or stocks) from industry *j*.

A good example of such productivity transfers would be the computer industry. It has had a tremendous real productivity growth, most of it unmeasured in its official indices, and most of it unappropriated within the industry itself (because of rather intensive competitive pressures). Different industries have benefited differentially from it, depending on their rate of computer purchases. One way of accounting for it would be to adjust upward the relevant capital equipment figures by their computer content. The alternative is to "import" the computer industry's research and development in proportion to the purchases from it.

But these are not real knowledge spillovers. They are just consequences of conventional measurement problems. True spillovers are the ideas borrowed by the research teams of industry *i* from the research results of industry *j*. It is not clear that this kind of borrowing is particularly related to input purchase flows. The photographic equipment industry and the scientific instruments industry may not buy much from each other but may be, in a sense, working on similar things and hence benefiting much from each other's research. One could argue that this is what the SIC classification is for. Presumably, the usefulness of somebody else's research to you is highest if he is in the same four-digit SIC classification as you are; it is still high if he is in the same three-digit industry group as you are; and, while lower than before, the results of the research by a firm in your own two-digit classification (but not three-digit) is likely to be more valuable to you than the average results of research outside of it.[13] The problem arises when we want to extend this scale across the other two-digit industries. Here there is no natural order of closeness (e.g., is "leather" closer to "food" or to "textiles"?).

The following alternatives appear reasonable and worth trying: (1) grouping three-digit SIC categories into clusters based on *a priori* notions about the extent of commonality in their technological and scientific base; (2) using the NSF's applied research and development product field by industry table to induce a distance metric, on the assumption that if an industry is doing research and development on some other industry's products, it is in some sense closer to it

[13] The situation is further complicated by the fact that the major R&D performers are conglomerates, spanning several SIC four-, three-, and even two-digit classifications. The NSF's applied R&D product field data help here a little but not enough. Complete within-firm product line breakdowns along the lines suggested by the FTC would be very useful in this context.

technologically than if it does not; (3) using company industrial diversification data from the Census of Enterprises to compute an alternative technological closeness metric; and (4) using information on rates of cross referencing of patents across product fields to infer the technological distance between them.[14] In each of these cases one will have to assume some simple weighting functions (e.g., the influence declining exponentially with the particular concept of distance) or group the data into a few categories: immediate neighborhood, related fields, and the rest. The available data will not support very refined approaches. There just are not enough degrees of freedom or independent variations in such productivity and research and development series to allow us to estimate very fancy distributed lag schemes over both time and all other industries.

The alternative to the search for such a concept of technological closeness or distance is to use the research investments of different industries as separate variables. But that is not really feasible. At best we would have about 20 years of data for each of about 20 industries. We cannot include 20 separate R&D variables in each of the industry equations; there simply will not be enough degrees of freedom there. We therefore have to aggregate somehow and that is what the idea of technological distance is for: to tell us how to weight the different research series and collapse them into one or a few variables so that we can estimate and assess the empirical importance of R&D spillovers. With such estimates we would compute not only the return to particular R&D in its "own" industry but also the total returns to R&D including the spillovers beyond its borders.

In trying to construct a research and development capital stock measure at the firm or industry level we face, thus, two major tasks: deciding on the appropriate lag structure and finding the right weights for "outside" research and development to represent borrowed knowledge and spillovers. Unfortunately, we have not enough theoretical or factual knowledge to guide us in these tasks. They will have to be solved empirically. The available data base, however, does not inspire much confidence in our ability to do so.

Before we conclude this section, we have to mention one other garden variety measurement problem: the lack of a research and development deflator. There is no official research and development deflator index available currently. NSF is using the implicit GNP deflator which is not so high-skill labor intensive as would be appropriate for a research and development input price index. Battelle publishes a research and development "cost" index, based on Milton (1972), but that appears to be an index of total cost (including other inputs) per scientist. It does not hold either the composition of the research and development labor force or the quantity of other inputs purchased constant. In short, it is not a price index. S. Jaffe reviewed the problem in NSF (1972) and suggested the use of a weighted index of labor compensation and the implicit price index in the nonfinancial corporate sector. That seems the best that one can do at the moment from secondary sources, but there is no reason why a better research and development input price index could not be constructed on the basis of primary sources. The NSF surveys ask for quite a bit of detail on the composition of research and development expenditures while the Battelle (previously Los Alamos) research and development scientists' compensation

[14] Work along lines (2) and (4) is being pursued by Schankerman (1979) and myself.

surveys have a great deal of data on the composition and compensation of the research and development labor force. This detail could be used to construct a whole set of separate research and development deflators for each of the major research and development performing industries.

5. Econometric problems

■ Besides the measurement problems discussed above, there are two other serious econometric problems facing the analyst in this area: multicollinearity and simultaneity. Although both are common, "garden variety" econometric problems, each has serious consequences. The problem of multicollinearity arises from the fact that many of the series we are interested in moved very much together over the period of observation. That being the case, it is then difficult (often impossible) to infer their separate contributions with any precision. There are no cheap solutions to this problem. It requires either less collinear data, more prior information, or a reduction in the aspiration level of the questions to be asked of the data.

Micro-time-series data at the individual firm or establishment level are probably the best way to reduce the multicollinearity that plagues such series at the industry- or economy-wide level. There is much more variability in the R&D histories of particular firms than in the R&D histories of the corresponding industries. This variability can be used to answer questions about the R&D lag structure, the relative effects of government- versus privately-financed R&D, or of basic versus applied research, and so on. Such "gains," however, come only at a serious cost. As noted above, at the micro level one can ask only questions about private returns to R&D. Very little of their social returns is detectable at this level.[15]

The analysis of industry-level data can be facilitated somewhat by the imposition of reasonable *a priori* restrictions. Thus, the use of total factor productivity measures as dependent variables already imposes on the data a set of implicit assumptions about the functional form of the production function and about specific values of its parameters; for example, it sets the output elasticities equal to their observed factor shares.

A similar problem will arise in trying to determine the exact shape of the R&D lag structure. To do that one has to have many years of R&D data and treat each of these years as a separate variable. But, in fact, R&D expenditures are highly correlated from year to year. Thus, it is unlikely that one can estimate their separate contribution with any precision (see Griliches (1967) for a more detailed discussion of such problems). Here it is probably best *to assume* a functional form for the lag distribution on the basis of prior knowledge and general considerations and not to expect the data to answer such fine questions. That is, a "solution" to the multicollinearity problem is a moderation of our demands on the data—our desires have to be kept within the bounds of our means.

The simultaneity problem refers to the possible confusion in causality:

---

[15] This does not mean that one cannot study the social returns to R&D at the micro level, but that would require a much more detailed data base with information on the magnitude of the actual technological breakthroughs and estimates of the relevant demand elasticities. See Mansfield *et al.* (1977) for details of such an approach. It goes far beyond the production function framework considered here. It is based, however, largely on confidential private information and is thus difficult to reproduce, extend, and evaluate.

future output and its profitability depend on past R&D, while R&D, in turn, depends on both past output and the expectation about its future. With long time series and detailed lag assumptions one may be able to analyze a recursive equations system with current output depending on past R&D, and past R&D depending on past rather than current output. In cross sectional data with only a few observations per firm, it is much harder to make such distinctions, particularly since current expectations about the future are based on current and past data.

It may be useful to outline the problem in somewhat more detail. In the context of simple static profit maximization a complete production and input choice model would consist of the production function,

$$y = \alpha c + \beta l + \gamma k + u,$$

assuming a Cobb-Douglas functional form and denoting the logarithms of corresponding variables by lower case letters, and marginal revenue product equals marginal cost conditions of the form,

$$y - c = a_1 + \pi_1 + v_1,$$

$$y - l = a_2 + \pi_2 + v_2,$$

$$y - k = a_3 + \pi_3 + v_3,$$

where the π's are the respective real factor prices of the various inputs, the v's are random errors in factor demand conditions, and a's are constants which depend on the coefficients of the production function and the product demand and factor supply elasticities (in the case of imperfect competition). (See Nerlove (1965) for more detail on this type of model.) We can solve this system of equations for k in terms of the π's, v's, and u, and write the resulting "reduced form" equation as

$$k = D\{-\alpha(\pi_1 + v_1) - \beta(\pi_2 + v_2) - (1 - \alpha - \beta)(\pi_3 + v_3) + u\},$$

where

$$D = 1/(1 - \alpha - \beta - \gamma).$$

Under these assumptions, it is clear that k is a function of u and that a simple OLS estimate of the production function would result in a biased estimate of γ, since the "independent" variable of interest (k) is in fact correlated with the disturbance in the same equation.

There are several "solutions" to the simultaneity problem. First, if one has good series on the π's, the real factor costs of the various inputs, one could use them as instrumental variables for the estimation of the production function. Unfortunately, in the R&D context one is unlikely to have good factor price series.[16] Even if one had the prices, they are likely to be highly collinear over time. Second, if one is willing to assume that disturbances in the input choice equations, the v's, are independent of the production function disturbance u, one could use input-output ratios, say $k - y = \pi_3 + v_3$, as instruments, since by hypothesis they are independent of u. This is known in the literature as the

[16] First, there are no published R&D deflators at the two-digit SIC level; second, if they were available, they would still be very highly correlated with the cost of labor and cost of capital indices, which are likely to be major ingredients of such indices. What we will not have are changes in "real" R&D costs in a field caused by various technological breakthroughs.

"indirect least squares" method. The implicit assumption of certainty about the future underlying such a static model makes little sense, however, in the R&D context. What is maximized here is the present value of all future profits, and the relevant output concept is an expected one and not the current one, especially if current output (and demand) is subject to special and transitory circumstances. In such a context not all of the "u" is transmitted to the factor demand equations (see Mundlak and Hoch (1965) and Mundlak (1963) for a more detailed exposition of such "partial transmission" models) and the indirect least squares method alluded to above is not consistent anymore. One needs a way of estimating the ratio of the permanent (transmitted) variance in u before one can use such methods again. Third, if both time series and cross sectional data are available and one is willing to assume a simple permanent-transitory transmission model: $u = \mu + t$, where μ is the permanent component which affects input demand choice while t, the transitory component, does not, then consistent estimates can be had from the within-firm covariance estimates of the parameter. This is equivalent to allowing a separate constant term (dummy variable) for each firm, which would absorb the μ term in it. Unfortunately, such data sets are rare. Moreover, the covariance approach may exacerbate other problems, such as errors in the variables, which may also afflict these kinds of data. Fourth, one may be able to find other "indicator" variables which may be related to the permanent components of the variables of interest and hence may help to solve the identification problem in such models.

For example, data may be available on the stock market value of the firm S. If this measure is proportional to the permanent output of the firm, then taking logarithms and ignoring constants, we can write

$$s = y - t + \epsilon,$$

where ϵ is a random term uncorrelated with either μ or t. Now one can use input to value of the firm ratios ($k - s$, etc.) as components of the indirect least squares procedure alluded to above.[17] More generally, a more explicit formulation of the expectation formation mechanism, together with the use of additional indicator variables such as the value of the firm, its physical capital investment policy, and the number of patents it has received, may allow us to formulate and estimate a more general "unobservables" type model (Griliches, 1974).

To recapitulate, the possible mutual dependency of R&D investments on past and future expected output requires careful attention to model formulation and specification and better and more detailed data to support the application of more sophisticated estimation techniques. Without a careful consideration of such issues, one may wind up reporting something as an estimate of the effect of R&D on output which may be mostly a reflection of the effect of output on R&D rather than *vice versa*.

6. Special problems

■ There are several special questions that one would like to have answered in this context:

[17] The actual model will have to be more complicated, since the value of the firm depends not on expected output but on expected profits, hence also indirectly on planned input expenditures. See Ben-Zion (1977) for an attempt to use the market value of the firm to infer the depreciation rate of R&D expenditures.

What are the relative returns to basic versus applied research?

Are the returns to government-financed R&D similar to those of company-financed R&D?

What are the "spillover" effects of government-financed R&D in sectors where the direct effects are almost unmeasurable (e.g., NASA R&D spending)?

How can one distinguish between economies of scale and R&D-induced productivity growth?

It would be interesting to distinguish between returns to basic and applied research, especially at the aggregate level. At the micro level, if firms allocate their resources rationally, the *private ex post* rates of return on different types of research should be about the same (except, possibly for some differences due to their differential riskiness) and we should not be able to distinguish among them. At the industry level one could conceivably find significant differences if, say, the appropriability of the results of basic research is more difficult than the appropriability of applied research. Also, the definition of "basic" implies a potentially wider range of applicability and an *a priori* higher externality component in its results. Thus, one might expect that the discrepancy between social and private returns would be higher, other things equal, in industries with a higher basic to total R&D ratio. Unfortunately, it may prove very difficult to isolate such effects. Basic research accounts for only about 3 percent of the total R&D expenditures in industry. Also, the lag structure of its effects is likely to be flatter and more variable, and hence more difficult to estimate. Therefore, one should not be too surprised if attempts to estimate the separate contribution of basic research fail. Nor should this be interpreted as implying the unimportance of basic research. It may only reflect the inability of our data to reveal such effects.[18]

There is much more basic research done in universities and government institutes than in industry, but it is almost impossible to assess its independent effect on productivity. At the industry level all industries are to some extent beneficiaries of the same research effort. There are no data, however, (except for agriculture) which could connect different expenditures on basic research to specific industries. At best there is a breakdown by fields of science but there is little quantitative information on the differential importance of "scientific fields" for different industries. One could try to construct differentially weighted "relevant" basic research series for different industries, but the results are likely to be quite collinear, and this approach does not appear to be too promising. At the aggregate level one has only one time series to explain and the basic research time series is likely to be no different from many other trending aggregate series. Without much more detail on the structure of basic research and on how it is used (if at all) in industry, it is unrealistic to expect to uncover its effects by econometric methods.[19]

Similar issues arise in the attempt to distinguish between the effects of publicly- and privately-financed R&D. Within any firm one would not expect to find much of a difference. A dollar is a dollar irrespective of source (unless

---

[18] In a recent still unpublished paper Mansfield (1978) does find a significant separate impact of basic R&D expenditures on industry productivity growth rates. This indicates that the paragraph above may be too pessimistic.

[19] There is always the route of detailed historical case studies but they are subject to problems of selection bias and incomplete coverage.

there are explicit expenditure and accounting rules connected with the use of federal R&D money which lead to inefficiencies). But a concentration of federally supported R&D expenditures in one area may lead to an overall decline in the rate of return to all R&D there. This may explain the difference between the results of Griliches (1975a), who found no differential effect of federal versus private company R&D dollars on the levels and rates of growth of total factor productivity at the *firm* level, and Terleckyj (1975), who found a significant discount of federal R&D dollars as far as their effect on productivity at the two-digit aggregated industry level is concerned.[20] Within an industry, at the firm level there should not be much of a difference between the effectiveness of different dollars, while at the aggregate level different rates of federal R&D investment must imply differences in *ex post* private and social returns, unless these expenditures are directed primarily at areas where there is private under-investment in R&D (a large *ex ante* gap between social and private returns) and concentrated on the less appropriable portions of it.

It might be interesting to look at this more from the "R&D investment function" side. Does federal R&D substitute for or complement private R&D investments? This would depend, in part, on the specific conditions and rules which accompany such federally financed expenditures. Holding federal funds constant, a firm that invests more of its own money than another presumably faces better investment opportunities or is more certain of appropriating a larger fraction of the total return.

Both in this case and in the previously discussed basic versus applied research context, questions are raised as to whether the different types of research are "substitutes" or "complements." Technically, this is a question about the functional form of the production function. In the Cobb-Douglas case all inputs are "complements." This is true of inputs that are considered separately (e.g., labor and capital, and basic and applied research if they have been entered separately with different exponents). When inputs are just added together into one aggregate total R&D measure, the implicit assumption is made that they are infinitely substitutable at the dollar-per-dollar rate. It is rather difficult to investigate questions of functional form with the usual kinds of data (see Griliches and Ringstad (1971) for a more detailed discussion of such difficulties). If one knew more about the differential price (real capital costs) of federal versus company R&D investments, one could investigate this problem through an analysis of input demand equations. Functional form differences are difficult to detect at the production function level, since the different curvature parameters are of second-order importance there. But at the input demand structure level these same parameters (such as the elasticity of substitution) have a first-order effect and thus may be easier to estimate. This, unfortunately, requires us to have the "right" prices for the different types of input—a feasible but very difficult task.

The third question, the detection of the "spillovers" from governmental R&D was discussed earlier in the context of defining a notion of "borrowed" R&D. A model was outlined which "weighed" the contribution of the jth in-

[20] Both studies are based on cross sections of growth rates. Terleckyj covers the years 1948–1966 while Griliches uses only 1957–1965. Terleckyj uses 1958 R&D intensity data while Griliches uses 1957–1965 R&D levels and growth rates. Griliches also finds that the estimated rate of return is lower in industries with high levels of federal R&D support (e.g., electrical equipment, aircraft, and missiles).

dustry's R&D to the stock of borrowed knowledge in industry i in proportion to some notion of the relevant technological distance between them. Besides the "borrowability" of a particular piece of knowledge of industry i from industry j, there is also the question of differential social productivity of R&D in different industries. There may be effectively more to borrow from some industries than others, per dollar of their R&D expenditures. We would like to be able to test hypotheses of the form "NASA R&D was more productive in the sense of having higher spillovers into the civilian economy than, say, DOD-financed R&D." Statistically this would consist of separating our measures of the total stock of borrowed R&D into components (NASA R&D, DOD R&D, other federal R&D, industry R&D, etc.) and asking whether they have consistently different effects on the productivity growth of other industries. It is feasible to try such an approach. Whether the available data can sustain such an effort is unclear, however.

The last question, the role of economies of scale in all of this, refers really to two somewhat distinct phenomena. The usual measures of total factor productivity are very much affected by short-term fluctuations in capacity utilization. To get a correct measure of the shift in technological opportunities of an economy (or industry), some adjustments or allowances have to be made for it. Also, there are industries where the longer-run trends in productivity are affected by changes in the scale of consumption. For example, part of the productivity growth in the telephone communications industry is attributable to the rise in calls per subscriber. Similarly, in the electric utility industry, spreading the peak load may lead to significant gains in measured productivity. Some of these changes may be related to earlier R&D expenditures but most are not. The data should be adjusted, if possible, so as not to confuse the issues.

A second type of interaction of R&D with scale occurs in industries where the technological results have been biased towards larger-scale enterprises. Here the payoff to R&D may be delayed if the market does not grow so quickly as anticipated. Moreover, the growth in the market and the growth in the number of "large" customers may, in turn, affect the direction of the R&D effort itself.

There is no satisfactory theoretical framework yet for discussing both the R&D decision process and the returns to it in a world of increasing returns to scale. These few cryptic remarks can at best serve as a placemark for future research. To pursue this topic further here would take us, however, too far afield.

7. Previous work

■ In this section I shall review briefly several recent studies of R&D returns which fall within the framework of this paper in the sense that they use both the production function framework and econometric methodology to derive their results.

The study which deals with the most aggregate level is the Chase Econometrics Associates' (1975) report, "The Economic Impact of NASA R&D Spending." It is a time-series study (14 to 18 years) of aggregate total factor productivity growth as it relates to distributed lag terms of past NASA R&D expenditures and "other" R&D expenditures. It finds extremely high returns to NASA R&D (on the order of 40 percent per year in perpetuity and more than double the returns to all other R&D). There are several difficulties with this study (see Griliches (1975) for detailed comments) and it is difficult to accept its results at face value. First, it looks for returns to NASA R&D at the total GNP level, rather than in the sectors where one could have hoped to measure the effect of the spillover of such R&D nontautologically. Second, there is a relatively short time series available to the authors for this purpose. Finally, the results are brittle with respect to the various assumptions and data adjustments made. Minor changes in procedure and variable definitions lead to substantial changes in the results. The study is interesting in that it disaggregates total R&D into two components (NASA *vs.* other), uses a distributed lag formulation to capture the delays in the effects of R&D over time and makes various adjustments for changes in capacity utilization over time. Unfortunately, the shortness of the time series and the use of a problematic dependent variable make their results both imprecise and difficult to interpret.[21] In addition, they impose the same lag structure on both NASA R&D and all other R&D. But as I have argued above, only the spillovers of NASA R&D should show up in such aggregate productivity figures, and surely they must have a different lag structure from most of the other R&D which is short term and has relatively quick direct returns.

At the aggregate two-digit industry level, there is a series of studies by Terleckyj (1974, 1975, and 1977) which relate a cross section of estimated rates of total factor productivity growth (1948–1966) for 20 to 33 two-digit industries to a number of R&D intensity ratios and other variables. Two findings stand out in these studies: (1) variables that "work" (at this level) are functions of private R&D investments and not of federally financed ones (as contrasted with the CEA study of the effects of NASA R&D); and (2) an average of R&D in "other" industries, weighted in proportion to the industry's purchases from them, has a significant influence on productivity growth. The significance of the "own" R&D intensity ratio in such studies is not surprising, and was observed in earlier studies by others (e.g., see Griliches (1972)). The major new finding is the large and statistically significant coefficient of his "borrowed" R&D measure. Terleckyj does not distinguish, however, between whether this variable works because of the mismeasurement of the inputs bought by the respective industries and whether the variable works because of the use of the input-output matrix weights as measures of technological "distance" between industries. If the first reason is true, the results are less interesting than they appear at first sight. For example, it is quite likely that his productivity measures (taken from Kendrick (1973)) underestimate the contribution of equipment capital and hence overestimate productivity growth in industries which had a relatively high investment in equipment. This occurs because the Kendrick measures weight capital by stock rather than flow weights and overdeflate them (*cf.* Jorgenson and Griliches (1972) for additional discussion of these and related issues). Since equipment industries are relatively R&D intensive, the resultant variable may be nothing more than a correction to the originally erroneously constructed productivity series. If the measure is supposed to represent technological distance and capture the pure spillover effects of R&D elsewhere, its logic would require the use not of the direct input-output matrix weights but of the corresponding "total" weights, which would take into account

[21] They use total GNP instead of the more relevant private civilian economy concept. Also their use of the Council of Economic Advisers' "potential" output series is debatable. The results do not survive the switch to Denison's potential output or Jorgenson's real output series.

the flows of information both directly from *i* to *j* and indirectly from *i* through *h* and *g* to *j*, and so forth. Here, too, the sample is small and the evidence is brittle.[22]

At the micro (firm) level Griliches (1975) related total factor productivity growth measures for 883 large U.S. companies during the 1957–1965 period to various measures of the growth in R&D capital and found a sizeable and significant effect of R&D on productivity growth. This study is also subject to serious limitations. First and most importantly, because of the nature of the data base (individual company reports), only questions about *private* returns to R&D could be asked. Second, because of the shortness of the available time series, no effort was made to investigate the lag structure of R&D effects. And third, only rudimentary attention was paid to the simultaneity problem.

8. Some conclusions and recommendations

■ One of the major points of this paper is the plea for realism as to what the production function approach can and cannot accomplish. Given good data, it can tell us something about average returns to R&D investments in the past and whether they appear to be changing over time. It may be able to indicate industries where returns have been especially high or low, but it will not be able to tell us whether a particular proposed R&D project is a good bet or not.

Given the kind of data we have and are likely to have in the near future, there are questions that one is unlikely to be able to answer from such data: e.g., questions about the exact time structure of the effects of R&D on productivity or the role of "science" and "basic" research in all of this. The level of "science" in the aggregate changes only very slowly, and we have no good way of assessing its differential impact on different industries. Unless a cause (stimulus) has varied much in its intensity over time or in its effects over industries, one is unlikely to be able to isolate it reliably by standard econometric techniques. In studying the contribution of NASA's R&D to the growth of the U.S. economy, we are faced with the fact that at the aggregate level we have only one cycle to work with: a rapid and continuous rise in NASA's R&D expenditures to the mid-1960s and then a more or less continuous decline to the mid-1970s. Given reasonably long and variable lags in the effects of such investments, we have, at best, only two observations, one up and one down. Given all else that is happening at the aggregate level at the same time, we are unlikely to be able to estimate reliably the contribution of a cyclical factor whose one cycle roughly corresponds to the time range of the available data.

This does not mean that the topic cannot be studied at all, only that the production function framework is not the suitable avenue for it. One could presumably identify a list of specific *products* of NASA research and trace out their subsequent impact on productivity in other industries. This, however, would require a much more micro-oriented approach.

At the moment, the lack of relevant data and the conceptual poverty of our models are the major impediments to progress in this area. As far as data are concerned, we have needs at both the macro (industry) and micro levels. At the

[22] A recent study by Mathtech (1978) using similar data but annual observations (instead of growth rates) is unable to detect a significant effect of such input-output table weighted borrowed R&D measures.

macro (industry) level a consistent set of total factor productivity measures corresponding to the 24 two- and three-digit industry detail given in the NSF's R&D publications (or the 32 industries corresponding to the applied R&D detail) is needed. These productivity measures should use consistent output and capital deflation methods, be clear about the treatment of sales to government, if possible exclude the R&D expenses and labor force from its measures of current inputs, and be up to date (i.e., go through 1976 or 1977 rather than end in the late 1960s as most of the studies have done). This will require a review and construction of industry price indices which should be consistent, as much as possible, in their treatment of quality change.

A price deflator is needed for the conversion of the R&D expenditures to "real" magnitudes. The earlier work of Jaffe (NSF, 1972) can and should be updated and extended.

The data on R&D collected by the NSF could also be improved. Industry codes use 1967 definitions which lack detail in some high technology areas and may be also out of date by now. Conversion to 1972 codes will have to be made. Coverage of subcontractors of R&D is almost nonexistent. Also, the basic unit surveyed is a firm, not an establishment. Many of the large R&D performing companies are conglomerates. An effort should be initiated to collect R&D data either at the establishment level (which has a more clearly defined industrial identity) or at the "product line" level, as has been recently suggested by the FTC. More effort should also be invested in improving the "Applied R&D by Product Field" estimates.[23]

At the micro (company) level the Census-Griliches-NSF Large-Company panel (1957–1965) should be reconstructed and updated to 1975. Efforts in this direction are now in progress.

Because of confidentiality problems and difficulties in access to census-based data, another "open" firm panel should be created from the published (mainly SEC) record. See, for example, the recent compilations in *Business Week* on this topic. This data set could be matched to other data for the same firms available in such sources as the Compustat tape, and the Patent Office records on patents applied for and granted by product field.

At the conceptual level we need more research on two very difficult topics: (1) how to measure the public product (and hence also the returns to R&D) in such important sectors as space, defense, and health; and (2) how to conceptualize and estimate technological distance between firms and industries and the associated notions of externalities and spillovers in research.

The first really corresponds to efforts to improve and expand the national income accounts towards more welfare-oriented measures. There do appear to be data and there is no insuperable conceptual problem in constructing a more relevant health sector accounting scheme. Whether something comparable could be done in space and defense is not clear at the moment. Suggestions for possible attacks on the second question, measuring the technological distance between firm and industries, have been advanced in the body of this paper and will not be repeated here. They too appear to be feasible.

At the empirical-econometric level we need studies that compare and carefully contrast results at different levels of aggregation in an attempt to detect

[23] NSF-Census should also write up and publish the results of their "Response Analysis" efforts of the Industrial R&D Survey rather than just allude to them (see p. 20 of NSF 77-324).

and measure the importance of social returns to R&D. We also need studies that pay much more attention to the estimation of the various lag structures between R&D expenditures and productivity growth, and at the same time recognize the problem of simultaneity and tackle it in a nontrivial manner. In addition, with patent data now available for 55 SIC product fields, a serious effort should be mounted to determine whether they actually measure something of interest and exactly what it may be. Especially, one should be able to tell whether there is any connection between the rate of patent activity and subsequent productivity growth in the various industries.

This is a large order, but we are talking about a major source of economic growth and about one of the few variables (R&D) which public policy can affect in the future (and has affected in the past). Hence it is of the utmost importance that we improve our knowledge in this area. It will not be easy, but it can be done, and it should not take a very large increase in research resources to accomplish at least some of it.

References

AGNEW, C.E. AND WISE, D.E. "The Impact of R&D on Productivity: A Preliminary Report." Paper presented at the Southern Economic Association Meetings. Princeton: Mathtech, Inc., 1978.

BEN-ZION, U. "The Investment Aspect of Nonproduction Expenditures: An Empirical Test." Unpublished manuscript, 1977.

BERNDT, E. AND CHRISTENSEN, L. "Testing for a Consistent Aggregate Index of Labor Input." *The American Economic Review*, Vol. 44 (1974), pp. 391–404.

BROWN, M. AND CONRAD, A. "The Influence of Research on CES Production Relations" in M. Brown, ed., *The Theory and Empirical Analysis of Production*, Studies in Income and Wealth, Vol. 3, New York: Columbia University Press for NBER, 1967, pp. 275–340.

CHASE ECONOMETRICS ASSOCIATES INC. "The Economic Impact of NASA R&D Spending." Bala Cynwyd, Pa.: 1975.

CHOW, G. "Technological Change and the Demand for Computers." *The American Economic Review*, Vol. 57 (1967), pp. 1117–1130.

DIEWERT, E. "Aggregation Problems in the Measurement of Capital." University of British Columbia Discussion Paper No. 77-09, 1977.

EVENSON, R.E. AND KISLEV, Y. *Agricultural Research and Productivity*. New Haven: Yale University Press, 1975.

FISHER, F.M. "The Existence of Aggregate Production Functions." *Econometrica*, Vol. 37, No. 4 (1969), pp. 553–577.

GORDON, R.J. "The Disappearance of Productivity Change." Harvard ED Report No. 105. Mimeographed, 1968.

GRILICHES, Z. "The Sources of Measured Productivity Growth, U.S. Agriculture, 1940–1960." *Journal of Political Economy* (August 1963).

———. "Distributed Lags: A Survey." *Econometrica* (January 1967).

———. "Research Expenditures and Growth Accounting" in B.R. Williams, ed., *Science and Technology in Economic Growth*, London: MacMillan, 1973, pp. 59–95.

———. "Errors in Variables and Other Unobservables." *Econometrica*, Vol. 42, No. 6 (1974), pp. 971–998.

———. "Returns to Research and Development Expenditures in the Private Sector" (1975a) in J.W. Kendrick and B. Vaccara, eds., *New Developments in Productivity Measurement*, Studies in Income and Wealth, Vol. 41, forthcoming.

———. "Comments on CEA: The Economic Impact of NASA R&D Spending." Mimeographed, 1975b.

———. "Economic Problems of Measuring Returns to Research" in Y. Elkana *et al.*, eds., *Toward a Metric of Science: The Advent of Science Indicators*, New York: John Wiley & Sons, Inc., 1977.

———, ED. *Price Indices and Quality Change*. Cambridge: Harvard University Press, 1971.

GRILICHES, Z. AND RINGSTAD, V. *Economies of Scale and the Form of the Production Function*. Amsterdam: North-Holland Publishing Company, 1971.

GRUNFELD, Y. AND LEVHARI, D. "A Note on External Economies." Mimeographed, 1962.

JORGENSON, D.W. AND GRILICHES, Z. "The Explanation of Productivity Change." *Review of Economic Studies*, Vol. 34, No. 3 (1967), pp. 249–283.

——— AND ———. "Issues in Growth Accounting: A Reply to E.F. Denison." *Survey of Current Business* (May, Part II, 1972).

KENDRICK, J.W. *Postwar Productivity Trends in the U.S., 1948–1969*. New York: Columbia University Press for NBER, 1973.

MANSFIELD, E., RAPAPORT, J., ROMERO, A., WAGNER, S., AND BEARDSLEY, G. "Social and Private Rates of Returns from Industrial Innovations." *Quarterly Journal of Economics*, Vol. 91, No. 2 (1977), pp. 221–240.

MILTON, H.S. "Cost-of-Research Index, 1920–1970." *Operations Research* (1972), pp. 1–17.

MUNDLAK, Y. "Estimation of Production and Behavioral Functions from a Combination of Cross Section and Time Series Data" in C. Christ *et al.*, eds., *Measurement in Economics*, Studies in Mathematical Economics and Econometrics in Memory of Yehuda Grunfeld, Stanford: Stanford University Press, 1963.

——— AND HOCH, I. "Consequences of Alternative Specifications of Cobb-Douglas Production Functions." *Econometrica*, Vol. 33 (1965), pp. 814–828.

NATIONAL SCIENCE FOUNDATION. "A Price Index for Deflation of Academic R&D Expenditures." NSF 72-310. Washington, D.C., 1972.

———. *Research and Development in Industry, 1975*. NSF 77-324. Washington, D.C.: 1977.

NELSON, R. AND WINTER, S. "Neoclassical vs. Evolutionary Theory of Economic Growth: Critique and Prospectus." *Economic Journal*, Vol. 84, No. 336 (1974), pp. 886–905.

NERLOVE, M. *Estimation and Identification of Cobb-Douglas Production Functions*. Chicago: Rand McNally & Company, 1965.

PAKES, A. AND SCHANKERMAN, M.A. "The Rate of Obsolescence of Knowledge, Research Gestation Lags, and the Private Rate of Return to Research Resources." Harvard Institute of Economic Research, Discussion Paper 659, October 1978.

RAINES, F. "The Impact of Applied Research and Development on Productivity." Washington University Working Paper No. 6814, 1968.

RYMES, T.K. *On the Concept of Capital and Technical Change*. Cambridge: Cambridge University Press, 1971.

SCHANKERMAN, M. "The Determinants, Rate of Return, and Productivity Impact of Research and Development." Unpublished Ph.D. dissertation, Harvard University, 1979.

SHARPE, W.F. *The Economics of Computers*. New York: Columbia University Press, 1967.

TERLECKYJ, N.E. *Effects of R&D on the Productivity Growth of Industries: An Exploratory Study*. Washington, D.C.: National Planning Association, 1974.

———. "Direct and Indirect Effects of Industrial Research and Development on the Productivity Growth of Industries" (1975) in J.W. Kendrick and B. Vaccara, eds., *New Developments in Productivity Measurement*, Studies in Income and Wealth, Vol. 41, forthcoming.

———. "Output of Industrial Research and Development Measured as Increments to Production of Economic Sectors." Paper given at the 15th Conference of the International Association for Research in Income and Wealth, York, England, 1977.

WILLIAMS, B.R. *Science and Technology in Economic Growth*. London: MacMillan, 1973.

[8]

Econometrica, Vol. 50, No. 3 (May, 1982)

SELECTION AND THE EVOLUTION OF INDUSTRY

BY BOYAN JOVANOVIC[1]

Recent evidence shows that within an industry, smaller firms grow faster and are more likely to fail than large firms. This paper provides a theory of selection with incomplete information that is consistent with these and other findings. Firms learn about their efficiency as they operate in the industry. The efficient grow and survive; the inefficient decline and fail. A perfect foresight equilibrium is proved by means of showing that it is a unique maximum to discounted net surplus. The maximization problem is not standard, and some mathematical results might be of independent interest.

1. THEORY AND EVIDENCE ON THE GROWTH AND SURVIVAL OF FIRMS

DO SMALL FIRMS grow faster than large firms? Are they less likely to survive? Early studies found no relation between the size of firms and their growth rates [**8**, **14**, **16**]. The growth of firms seemed to be proportional to their size. In later work, adjustment costs with constant returns to scale were shown to imply that firms *should* grow in proportion to their size [**10**, **11**].

Recent evidence from larger samples tells a different story. Mansfield [**13**] finds that smaller firms have *higher* and *more variable* growth rates. Du Rietz [**6**], in a sample of Swedish firms, again finds that smaller firms grow faster, and that they *are* less likely to survive [**6**, **8**, **13**]. These findings conflict with the adjustment costs theory in which all firms grow at the same rate, and in which failure does not happen.

To explain these deviations from the proportional growth law, I propose a theory of "noisy" selection. Efficient firms grow and survive; inefficient firms decline and fail. Firms differ in size not because of the fixity of capital, but because some discover that they are more efficient than others. The model gives rise to entry, growth, and exit behavior that agrees, broadly, with the evidence.

The model also agrees with some more tentative findings. First, firm size and concentration seem to be positively related to rates of return.[2] Second, the correlation *over time* of rates of return is higher for larger firms and in the concentrated industries [**15**, **17**]. Third, the variability of rates of return *at a point in time* is higher in the concentrated industries [**17**]. Finally, higher concentration is associated with higher profits for the larger firms, but *not* for the smaller firms [**4**].

Enduring differences in size and in growth are no doubt caused in part by the fixity of capital. This paper shows, I think, that selection matters too.

[1] The original draft is dated December, 1979. I would like to thank Carl Futia, Roy Radner, and Ed Green for helpful discussion.

[2] Weiss [**18**] summarizes a number of studies that report this finding. Two exceptions are Stigler [**17**] who found no relation and Samuels and Smyth [**15**] who found a negative relation.

2. A BRIEF DESCRIPTION OF THE MODEL

The model deals with a small industry to which factors are supplied at a constant price. The product is homogeneous and the time-path of the demand for the product is deterministic and known.

Costs are random, and different among firms. For each firm, the mean of its costs may be thought of as the firm's "true cost." The *distribution* of true costs among the potential firms is known to all, but no firm knows what *its* true cost is. All firms have the same prior beliefs, and each firm regards itself as a random draw from the population distribution of true costs. This "prior" distribution is then updated as evidence comes in.

If the firm has low true costs, it is likely that the evidence will be favorable, and the firm will survive. If its costs are high and the evidence adverse, the firm may not wait too long before withdrawing from the industry.

The number of firms in the industry is always infinite—each firm is of measure zero so that it is too small to affect price. With uncertainty at the individual level but with no *aggregate* uncertainty, the path of output prices is deterministic and is assumed to be self-fulfilling in equilibrium.

Firms and potential entrants know the entire equilibrium price sequence, and based on it, they make entry, production, and exit decisions. A one-time entry cost is borne at the time of entry. Thereafter, only production costs are incurred. In equilibrium, the net present value of entry cannot be positive, for if it were, more firms would enter.

In the next section, the model is presented, and the firm's optimization problem is defined. Some of the properties of the model then become clear. Figure 1 portrays them concisely: efficient firms grow and survive; the inefficient decline and fail. Toward the end of the section, results are described which are obtained in the later, mathematical sections of the paper. The implications are

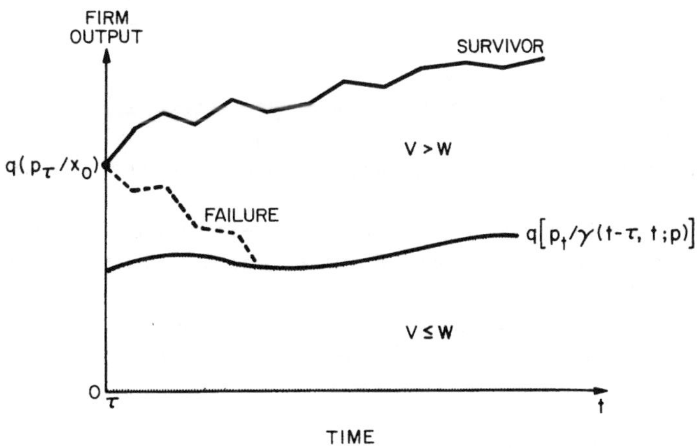

FIGURE 1

compared to the empirical evidence. The perfect foresight equilibrium is defined in Section 4. Section 5 is devoted to proving that equilibrium exists, is unique, and is, in a sense, socially optimal. The proof involves showing that equilibrium coincides with the unique maximum of the discounted sum of consumer plus producer surplus. Two theorems that describe the equilibrium appear in Section 6, and some of the longer proofs are contained in the Appendix.

A curious feature of the paper is that proofs of "obvious" results are complicated. The reader may be eager—on the first reading at least—to take such results on trust. So, the remainder of the paper is arranged into two distinct parts. The next section contains a discussion of all the results. If the reader is not interested in the mathematics, he can stop there, for the discussion is self-contained. Sections 4 and beyond are devoted to the formal development, which is interesting in its own right. Proving that equilibrium is a maximum is not new, of course. But the problem of maximizing discounted surplus here does not fit a regular mold. Similar problems will no doubt arise again and results proved here could then be useful. A description of the nature of the problem being solved and of how it differs from other work on stochastic optimization is contained at the beginning of Section 5.

3. THE MODEL

In an industry with a homogeneous output, firms differ in efficiency. Some are more efficient than others at *all* levels of output. Let q be the output of a firm, and $c(q)$ a cost function which satisfies

$$c(0) = 0, \quad c'(0) = 0, \quad c'(q) > 0, \quad c''(q) > 0 \quad \text{and} \quad \lim_{q\to\infty} c'(q) = \infty.$$

Total costs are $c(q_t)x_t$ where x_t is a random variable independent across firms. For the firm of type θ, let $x_t = \xi(\eta_t)$ where $\xi(\cdot)$ is a positive, strictly increasing, and continuous function with $\lim_{\eta_t\to-\infty}\xi(\eta_t) = \alpha_1 > 0$ and $\lim_{\eta_t\to\infty}\xi(\eta_t) = \alpha_2 \leq \infty$, and where

$$\eta_t = \theta + \epsilon_t, \qquad \epsilon_t \sim N(0, \sigma^2) \text{ iid}.$$

Firms with large values of θ will generate larger x_t's, and be less efficient at all levels of output. The ϵ_t are firm-specific shocks, independent over time and across firms.

Among potential firms, θ is normally distributed with mean $\bar{\theta}$ and variance σ_θ^2. An entrant does not know his own θ, but he knows that he is a random draw from $N(\bar{\theta}, \sigma_\theta^2)$. He also knows the variance of ϵ_t, as well as the exact form of $\xi(\cdot)$ so that observing his own costs at t allows him to infer η_t.

The firm is too small to affect price. It chooses q_t so as to maximize expected profits:

$$\max_{q_t} \left[p_t q_t - c(q_t) x_t^* \right]$$

where x_t^* is the expectation of x_t conditional upon information received prior to t. The output decision is made *before* x_t is observed, and is denoted by $q(p_t/x_t^*)$. As one would expect, it is decreasing in x:

$$\frac{\partial q}{\partial x_t^*} = \frac{-c'}{x_t^* c''} < 0 \quad \text{and} \quad \frac{\partial^2 q}{\partial x_t^{*2}} = \frac{1}{x_t^*}\left[\frac{c'c'''}{(c'')^2} - 2\right]\frac{\partial q}{\partial x_t^*}. \tag{3.1}$$

THE EXIT DECISION: Let $W > 0$ be the expected present value of the firms's fixed factor (its "managerial ability" or "advantageous location") if it is employed in a different activity. The value of W is the same for all firms in the industry *regardless of how successful they are in that industry*. In other words, if the firm learns that it is efficient in this industry, this does not increase its estimated efficiency anywhere else. This assumption may seem restrictive, but it could be relaxed to allow for correlation in firms' efficiencies in different industries, without changing the nature of the results. What really matters is that if favorable information about a firm's costs in an industry raises its expected earnings *in that industry* by one dollar, its expected earnings elsewhere increase by *less* than a dollar. Here it is assumed that new information about θ leaves expected alternative earnings unchanged.

A cost of entry, k, is borne by the firm when it enters—the cost of establishing a particular location for example. And θ might be the degree of suitability of the location. The firm learns about θ with the passage of time.

The firm has an infinite horizon and a constant discount rate, r. At time t, if the firm is in the industry, it will have a pair of sufficient statistics $(\bar{\eta}_n, n)$ which characterize its beliefs about its parameter θ. Here n is the number of periods that the firm has been in the market (the age of the firm) and $\bar{\eta}_n = \sum^n \eta_i/n$. These two statistics are sufficient for the posterior distribution on θ, as this distribution is normal [**20**, p. 15].

In spite of the infinite horizon and the constant discount rate, the present value of earnings will depend on t too, because the price path, $\{p_t\}_0^\infty$, treated as given by the firm, is in general not constant over time. Therefore, once $\{p_t\}_0^\infty$ is given, t determines where one is along the price sequence.

Since $x_t^* = \int \xi(\eta) P^0(d\eta \mid \bar{\eta}_n, n)$, (where $P^0(\cdot \mid \bar{\eta}_n, n)$ is the normal posterior distribution of η_t with variance which depends only on n), and since $\xi(\eta)$ is strictly increasing, x_t^* is strictly increasing in $\bar{\eta}_n$ for each n. Therefore the pair (x_t^*, n) is also a sufficient statistic.

Let $\pi(p_t, x) \equiv p_t q(p_t/x) - c[q(p_t/x)]x$ be the expected value of profits maximized with respect to q when $x_t^* = x$. For a bounded price sequence $p \equiv \{p_t\}_0^\infty$, let $V(x, n, t;\, p)$ be the value, at t, of staying in the industry for one period and then behaving optimally, when the information is (x, n) and when the price sequence is p. Then V satisfies[3]

$$V(x,n,t;\,p) = \pi(p_t,x) + \beta\int \max\left[W, V(z,n+1,t+1;\,p)\right]P(dz \mid x,n). \tag{3.2}$$

[3] In equation (3.2), $P(z \mid x, n)$ is the probability that $x_{t+1}^* \leq z$ given that $x_t^* = x$, and given that the firm has been in the industry for n periods.

At entry, when the firm has only its prior information, $x = x_0 \equiv$ prior mean of x_t, and $V(x_0, 0, t;\, p) - k$ is the net value of entry at t. The following simple result is proved in the Appendix.

THEOREM 1: (i) *A unique, bounded and continuous solution for* V *in equation* (3.2) *exists and* (ii) V *is strictly decreasing in* x.

Thus firms with higher expected costs have a lower value of staying in the industry. Let $\gamma(n, t;\, p)$ be the level of x_t^* at which the firm is *indifferent* between staying in the industry and leaving it. Then $\gamma(\cdot)$ is the solution for x to the equation[4]

$$(3.3) \qquad V(x, n, t;\, p) = W.$$

As V is strictly decreasing in x, $\gamma(\cdot)$ is uniquely defined. Consequently, the level of output below which the firm will not produce (but will exit instead) is $q[p_t/\gamma(n, t;\, p)]$. It is drawn in Figure 1. Here τ is the time at which the firm enters the industry, so $n = t - \tau$. For any price sequence, the boundary defines an "exit" region in which $V \leq W$ (shaded area) and a "continuation" region, in which $V > W$.

The firm's output sequence, $q(p_t/x_t^*)$, is a random process which starts from $q(p_\tau/x_0)$. The x_t^* sequence is a Martingale: $E_t x_{t+k}^* = x_t^*$ for any $k \geq 0$ [**7**, p. 212]. If output remains above the shaded area, the firm stays in the industry. Therefore the firms that survive are larger than the firms that fail—at the point of failure, the firm is smaller than all surviving members of its cohort.

The x_t^* sequences are independent across firms. They tend to *diverge*, as do the output sequences. Each firm is of measure zero, so the number of firms is always infinite, and the concentration *ratio* always zero. But a popular measure of concentration is the Gini coefficient. The greater the dispersion of x_t^* across firms, the greater the dispersion of firm size. At $t = 0$, all firms are of the same size, and the Gini coefficient is zero. As the x_t^* diverge, so do the outputs, and the Gini coefficient increases over time. But the increase need not be monotonic. And this is *exactly* the type of increase the Gini coefficient has exhibited at the economy-wide level in the U.K. [**8**].

Average profits also rise as the industry matures—at least they do if the equilibrium price sequence is constant (the latter possibility is the subject of Theorem 3 and is discussed below as well). The reason is simple. Since the unprofitable firms leave while profitable firms stay, the profits of the survivors as a group will increase so long as the price of the product does not fall. If it does not, profits increase with the age of the industry, as does concentration. In this sense then, the model predicts a positive relation between profits and concentration. But if the equilibrium price sequence falls over time, it may offset the upward "selection effect" on profits, and nothing can then be said about the time path of profits.

[4]If $V < W$ for all $x \in [\alpha_1, \alpha_2]$, we set $\gamma = \alpha_1$, while if $V > W$ for all $x \in [\alpha_1, \alpha_2]$, we set $\gamma = \alpha_2$.

Why does concentration increase the profits of the large firms but not of the small firms? A high Gini coefficient results from high inequality in firm size in the industry. And the latter is caused by more inequality in efficiency. This means that some firms—large firms—earn higher rents. But the marginal firms—small firms—do not earn rents; there is no reason to expect a positive relation between concentration and the profits of these firms.

As for the positive relation between concentration and the *variability* of profits, take the case where $\sigma_t^2 = 0$—all firms are equally efficient, and all are equal in size. The variability of profits is zero, as is the Gini coefficient. Since both profits and output decrease with x, there is a *one-to-one* relation between the relative dispersion in profits and in size.

Another implication is that unusually high profits today lead to unusually high growth between today and tomorrow. The reason is that the firm's *revision* of x_t^* depends on *realized* profits. Since $\pi_t = p_t q_t - c(q_t)x_t$,

$$\pi_t - E_t\pi_t = -c(q_t)(x_t - x_t^*).$$

So if profits at t are large compared to average profits *for that size of firm*, it means that x_t is unusually low. And this leads to a downward revision of expected marginal costs: $x_{t+1}^* < x_t^*$. But then next period's output—and growth—will be higher than usual. So, high profits are transformed into high growth. Of course, the standard "explanation" has relied on imperfect capital markets: constraints on borrowing lead to higher growth for firms that can finance it internally.

Fluctuations in output occur in any industry. What fraction is due to changes in the output of existing firms as opposed to changes—through entry or exit—in the number of firms? If demand changes are erratic and unforeseen, one might expect the existing firms to meet a large proportion of such changes. But in this paper, all the demand changes are foreseen. Holding constant the behavior of demand, a lot hinges on whether $q(\cdot)$ is concave in x or not (see (3.1) for the concavity condition). Why should this matter? Suppose for the moment that price is constant, say at $\bar{p}$. Since x_t^* is a Martingale, concavity and Jensen's inequality imply that $E_t q(\bar{p}/x_{t+1}^*) < q(\bar{p}/x_t^*)$. So when the output price does not increase, the existing firms would produce *less* in each successive period. This is only re-inforced by the reduction of output due to exit of some of the existing firms. So, if $q(\cdot)$ is concave in x, increases in demand should be met by new entrants. Theorem 3 proves that when demand is non-decreasing and $q(\cdot)$ concave in x, the unique equilibrium is one in which price is constant over time and in which entry and exit occur in every period.

No technological progress takes place in this model. Yet it seems possible (but I cannot prove it) that even if demand is constantly shifting to the right, equilibrium price will constantly decline. Since the efficient survive while the inefficient fail, the average efficiency of the survivors improves from period to period. Convexity of $q(\cdot)$ in x is a necessary condition for such an equilibrium to occur. On the other hand, prices cannot monotonically increase, as the net value to entry would become positive.

What can be proved is that equilibrium *always* coincides with the unique maximum to the discounted consumer surplus (Theorem 2). This is not surprising —firms take prices as given and there are no externalities. Still, as background to the current work on oligopolistic entry-deterrence, it is useful to know that at least when numbers are large, entry—and exit—occurs neither too early nor too late.

Some specialized results can also be proved. The variability of growth rates will be largest among the young (and therefore smaller) firms. But for mature firms that have survived for a long time, x_t^* converges to a constant. Therefore, if growth rates are to be equal among *mature* firms (and there is some empirical basis for this: samples of *large* firms were used in [**14** and **16**]), one must have, for each x,

$$\frac{1}{q}\frac{dq}{dp_t} = k(p_t)$$

where $k(p_t)$ is some function which does not depend on firm size and is therefore independent of x.

Solving this differential equation under the constraint that q is a function only of the ratio p_t/x,

$$q(p_t/x) = \delta_1\left[\frac{p_t}{x}\right]^{\delta}, \tag{3.4}$$

where δ_1 and δ are positive constants. This can only happen if $c(q)$ assumes the Cobb-Douglas form

$$c(q) = \beta_1 q^{\beta_2} \qquad \text{with} \qquad \beta_2 = 1/\delta + 1$$

and with

$$\beta_1 = \delta_1^{-1/\delta}\left[\frac{\delta}{1+\delta}\right].$$

Since the restriction in (3.4) applies to *all* firms, the growth rate of any given firm is

$$\left[\frac{p_{t+1}}{p_t}\right]^{\delta}\left[\frac{x_t^*}{x_{t+1}^*}\right]^{\delta} - 1. \tag{3.5}$$

In the sections that follow, we shall *not* restrict the cost function to be Cobb-Douglas. But it is instructive to pursue further the implications of the existence of approximately equal expected growth rates of firms of a given vintage. Let $z_t \equiv x_t^*/x_{t+1}^*$. If (3.5) holds, a weak form of the proportional growth law requires that $E_t z_t^{\delta}$ be the same for all x_t^*. A strong form requires that the entire distribution of z_t (conditional on information at t) be the same for all firms. The strong form of the law has been empirically rejected, and we now show that it cannot hold within the framework of this model: since the Martingale x_t^* has decreasing incremental variance as the precision on θ grows, two firms with the same x_t^* but different precisions could not have the same distribution for z_t. However, z_t can have the same distribution for firms in the same age cohort. Since one may write $x_{t+1}^* = x_t^*(1 + u_t)$ where u_t has mean zero due to the Martingale property, $z_t = 1/(1 + u_t)$ will be identically distributed for each firm in the cohort so long as the distribution of u_t does not depend on x_t^*. Then $\xi(\eta)$ may be chosen so that this property holds. Let $x_t - \alpha_1$ be log-normally distributed so that $\xi(\eta) = \alpha_1 + e^{\eta}$. Let θ_n and ν_n denote the posterior mean and variance on θ when n observations $\{\eta_i\}_{i=0}^{n-1}$ are available. Then,

$$x_t^* = \alpha_1 + \exp\left\{\theta_{t-\tau} + \tfrac{1}{2}\left(\nu_{t-\tau} + \sigma^2\right)\right\}.$$

As $\alpha_1 \to 0$,

$$z_t \to \frac{1}{1+u_t} = \frac{x_t^*}{x_{t+1}^*} = \exp\left\{(\theta_{t-\tau} - \theta_{t+1-\tau}) + \frac{1}{2}(\nu_{t-\tau} - \nu_{t+1-\tau})\right\}$$

and since the distribution of $\theta_{t-\tau} - \theta_{t+1-\tau}$ (conditional on information available at t) is normal with mean zero and variance that does *not* depend on $\bar{\eta}_n$, only on $t - \tau$, the variable $1/(1 + u_t)$ does not depend on x_t^*.

Since the variance of u_t declines as the firm becomes more mature, younger firms have more variability in their growth rates. They will also grow faster than the older firms. This follows from the convexity of x_t^*/x_{t+1}^* and the application of Jensen's inequality: $E_t(x_t^*/x_{t+1}^*) > x_t^*/E_t x_{t+1}^* = 1$. So even the weak form of the law cannot hold except within a single age cohort.

The implication that smaller firms should have higher and more variable growth rates is in accord with evidence. But there is a selection bias in the data. Smaller firms are more likely to fail and the model implies (see Figure 1) that the firms which fail are exactly those which otherwise would have grown more slowly. If, as is done in practice, all failures are omitted from the sample, one overestimates the growth rate of small firms relative to that of large firms. The model implies that even if one were to eliminate this bias by an appropriate choice of statistical technique, the results should show a higher growth-rate for smaller firms.

4. EQUILIBRIUM

Before defining equilibrium, the industry supply and demand functions are defined. Let

$$\begin{aligned}(4.1)\qquad \Psi(x \mid t, \tau; p) = \text{Prob}\big[&x_s^* < \gamma(s-\tau, s; p),\ s = \tau+1, \ldots, t-1, \\ &\text{and } x_t^* < \min\left[x, \gamma(t-\tau, t; p)\right] \quad \text{given } x_\tau^* = x_0 \\ &\text{and given that entry occurred at } \tau\ (\tau < t)\big].\end{aligned}$$

Ψ is the probability that the firm which enters at τ, and follows its optimal stopping policy, is still in the industry at t, at which time its $x_t^* \le x$. (Note that since s runs from $\tau + 1$, *immediate* exist is not feasible—it is not feasible for the firm to exit without having spent at least one period in the industry.) Then,

$$\phi(t, \tau; p) \equiv \int q(p_t/x)\Psi(dx \mid t, \tau; p) \tag{4.2}$$

is the expected output at t of a firm of vintage τ. If y_τ is the measure of entrants at time τ, the output of these firms at t is $y_\tau \phi(t, \tau; p)$. This output is deterministic, as each firm is of measure zero. Let Q_t be the aggregate industry output at t. Then, the industry *supply function* is

$$Q_t = \sum_{\tau=0}^{t} y_\tau \phi(t, \tau; p) \equiv Q_t(p, y) \tag{4.3}$$

where $y \equiv \{y_\tau\}_0^\infty$ is a sequence of entry. So $Q_t(p, y)$ is the industry output at t which results if the firms are faced with an *arbitrary* pair of price and entry sequences (p, y), and if they make optimal output and exit decisions in response to the price sequence p.

A deterministic *demand function* $D[Q_t, t]$ is given for each t. For each t, $D(\cdot)$ is strictly decreasing in Q_t.

A perfect foresight equilibrium has the property that if firms and prospective entrants behave on the assumption that a particular price sequence will occur, then their behavior does in fact give rise to this price sequence. In other words, the equilibrium price sequence is self-fulfilling.

Definition of Equilibrium: Equilibrium is of a pair of functions $q(\cdot)$ and $\Psi(\cdot)$ that characterize optimal output and exit behavior of firms, and a pair of nonnegative sequences (p, y) such that for all $t = 0, 1, \ldots,$

$$p_t = D\{Q_t(p, y), t\}, \tag{D.1}$$

$$\begin{aligned} V(x_0, 0, t; p) - k &= W \quad \text{if } y_t > 0, \\ V(x_0, 0, t; p) - k &\leq W \quad \text{if } y_t = 0, \end{aligned} \tag{D.2}$$

where $Q_t(p, y)$ is defined in (4.3).

Condition D.1 expresses the self-fulfilling property of the equilibrium price sequence. Condition D.2 states that at each t, the net present value to entry cannot be positive, for if it were, more firms would enter, so that this could not be an equilibrium. The value to entry may be negative in some of the periods, in which case no firm would enter.

5. EXISTENCE, UNIQUENESS, AND OPTIMALITY OF EQUILIBRIUM

We prove the existence and uniqueness of the equilibrium by showing that it is a unique maximum to a particular functional—the discounted consumer plus producer surplus. The equilibrium is an optimum in this sense.

Before plunging into the algebra, let us review the elements of the argument. A benevolent planner chooses an industry output sequence $\{Q_t\} \equiv Q$ so as to maximize the discounted sum of surplus. He has exactly the same amount of information as is collectively available to firms. He assigns the entry and exit in each time period, as well as the output of each firm. The benevolent planner is concerned only with one industry and shadow prices in other industries are held constant both through time and with respect to the industry output level.

One possible approach could have been to use dynamic programming. But the state space is too large here: a firm is characterized by its expected cost and its age (this is the basis for (3.1)) while the industry is described by a *measure* over (x_t^*, n)—the measure of firms, at t, with those characteristics. Because of the dimensionality of the state space, we take the route of direct optimization subject to constraints. There are problems with this approach too, because the horizon is infinite. The constraint space is ℓ_∞, and linear functionals in the dual of ℓ_∞ do not assume a simple form (see (5.8)). The key result is in Lemma 5 (which is probably of independent interest) where it is shown that—essentially because of discounting—the class of linear functionals in the dual of ℓ_∞ can be reduced to a "manageable" subclass.

The cheapest way (for the planner) of producing an aggregate output sequence Q, is defined as $K(Q)$. The hard part of the proof is showing that $K(Q)$ is well-defined (Lemma 2), convex, and differentiable (Lemma 7).

Having established the relevant properties of $K(Q)$, we proceed to compare the necessary conditions for $S(Q)$ to be at a maximum, with the necessary conditions for the actions of firms to be optimal in the perfect foresight equilibrium. This comparison turns out to be easier if an alternative representation for the value of entry is used. This representation is derived (at the outset) in equation (5.2). We now proceed with the analysis.

In view of equation (4.1), another representation for the value of entry at τ, $V(x_0, 0, \tau; p)$ is

$$V(x_0,0,\tau;p) = \sum_{t=\tau}^{\infty} \beta^{t-\tau} \int_{\alpha_1}^{\alpha_2} \pi(p_t, x)\Psi(dx \mid t,\tau;p) + \sum_{t=\tau}^{\infty} \beta^{t-\tau} W\{\Psi(\alpha_2 \mid t-1,\tau;p) - \Psi(\alpha_2 \mid t,\tau;p)\} \tag{5.1}$$

because the expression in curly brackets is just the probability of exit exactly at t. (Here we define $\Psi(\alpha_2 \mid \tau - 1, \tau; p) = 1$.) Writing $\psi_t \equiv \Psi(\alpha_2 \mid t, \tau; p)$ (since α_2, τ, and p are fixed), the second part of equation (5.1) may be written as

$$W\sum_{t=\tau}^{\infty} \beta^{t-\tau}(\psi_t - \psi_{t-1}) = W\left[\psi_{\tau-1} + \beta \sum_{t=\tau}^{\infty} \beta^{t-\tau}\psi_t - \sum_{t=\tau}^{\infty} \beta^{t-\tau}\psi_t\right]$$

$$= -(1-\beta)W\sum_{t=\tau}^{\infty} \beta^{t-\tau}\psi_t + W$$

(because $\psi_{\tau-1} = 1$). Therefore,

$$(5.2)\qquad V(x_0, 0, \tau; p) = W + \sum_{t=\tau}^{\infty} \beta^{t-\tau} \int_{\alpha_1}^{\alpha_2} [\pi(p_t, x) - (1-\beta)W] \Psi(dx \mid t, \tau; p).$$

(The intuition here is that $(1-\beta)W$ is the per-unit time expected foregone income for the firm while it is in the industry.)

Now let $\gamma \equiv \{\gamma_t\}_{t=1}^{\infty}$ be a particular sequence bounded above by α_2 and below by α_1. Any such sequence γ represents a *feasible* exit policy. Let Γ be the set of all such sequences. That is,

$$\Gamma = \{\gamma : \gamma_t \in [\alpha_1, \alpha_2], t = 1, \ldots\}.$$

Then, analogously to equation (4.1), for any $\gamma \in \Gamma$, let

$$\hat{\Psi}_\gamma(x \mid t, \tau) = \text{Prob}\left[x_s^* < \gamma_{s-\tau}, s = \tau + 1, \ldots, t, \text{ and } x_t^* < \min(x, \gamma_{t-\tau}) \right.$$
$$\left. \text{given } x_\tau = x_0 \text{ and entry occurred at } \tau \right].$$

In words, $\hat{\Psi}_\gamma(\cdot)$ is the distribution of x at t (for vintage $-\ \tau$ firms) if they follow the feasible policy characterized by γ.

Lemma 1: *The density* $\hat{\psi}_\gamma(x \mid t, \tau) = \partial \hat{\Psi}_\gamma(x \mid t, \tau)/\partial x$ *exists for all* γ, *and* $t > \tau$, *and any* $x \in (\alpha_1, \alpha_2)$. *Furthermore,* $\hat{\psi}_\gamma$ *is differentiable in each element,* γ_j, *of the sequence* γ, *for all* $j \geq 1$.

Proof: Contained in the Appendix.

Now let $\gamma(p, \tau) \equiv \{\gamma(t-\tau, t; p)\}_{t=\tau}^{t=\infty}$ [the latter is defined in (3.3)]. Then $\gamma(p, \tau) \in \Gamma$. Therefore $\Psi(x \mid t, \tau; p)$ [defined in (4.1)] has density $\psi(x \mid t, \tau; p)$. Then $\psi(x \mid t, \tau; p)$ is the solution to the problem

$$(5.3)\qquad \max_{\gamma \in \Gamma} \left\{ W + \sum_{t=\tau}^{\infty} \beta^{t-\tau} \int_{\alpha_1}^{\alpha_2} [\pi(p_t, x) - (1-\beta)W] \hat{\psi}_\gamma(x \mid t, \tau)\, dx \right\}$$

because the optimal "cutoff" sequence $\gamma(p, \tau)$ is contained in Γ. Of course, $\psi(x \mid t, \tau; p) = \hat{\psi}_{\gamma(p,\tau)}(x \mid t, \tau)$. Writing $\gamma_j(p, \tau)$ for the jth member of the optimal sequence, since $\gamma_j(p, \tau) \in [\alpha_1, \alpha_2]$, the necessary conditions for optimality may be written for $j = 1, \ldots,$

$$(5.4)\qquad \left\{ \sum_{t=\tau}^{\infty} \beta^{t-\tau} \int_{\alpha_1}^{\alpha_2} [\pi(p_t, x) - (1-\beta)W] \frac{\partial \psi}{\partial \gamma_j}(x \mid t, \tau; p)\, dx \right\}$$
$$\times [\gamma_j - \gamma_j(p, \tau)] \leq 0$$

for all $\gamma_j \in [\alpha_1, \alpha_2]$.

Let $Q \equiv \{Q_t\}_0^\infty$ be a bounded output sequence. Let ℓ_∞ denote the set of bounded infinite sequences. We now define the social benefit functional $S: \ell_\infty \to R$ as

$$S(Q) = \sum_{t=0}^{\infty} \beta^t \int_0^{Q_t} D(z,t)\,dz - K(Q) \tag{5.5}$$

where $K: \ell_\infty \to R$ is the present value of the minimized cost of producing the output sequence Q.

The problem of attaining $K(Q)$ will be referred to as the "planner's cost minimization problem." We shall assume that the planner possesses all of the information available to each firm. He chooses an entry sequence[5] $y \equiv \{y_t\}_0^\infty \in A_1$, where $A_1 = \{y : 0 \leq y_t \leq \bar{y}_t$, where $\sum_{t=0}^{\infty} \bar{y}_t < \infty\}$, and for firms of vintage τ he chooses an exit policy $\gamma(\tau) \equiv \{\gamma_j(\tau)\}_{j=1}^\infty \in \Gamma$. Finally he assigns a nonnegative output level $\hat{q}(x,t)$ to a firm which at t has $x_t^* = x$. Let $A_2 = \{\hat{q}(x,t)$: each t and $x \in [\alpha_1, \alpha_2]$, $0 \leq \hat{q}(x,t) \leq b(x)$, with $b(x)$ Lebesgue integrable$\}$. By Tychonoff's theorem on product spaces, A_1, Γ, and A_2 are compact sets in their product topology. Let $\Gamma^\infty = \Gamma \times \Gamma \times \cdots$, and let $\gamma \equiv \{\gamma(\tau)\}_{\tau=0}^\infty$, so that $\gamma \in \Gamma^\infty$. Similarly, let $A_2^\infty = A \times A \times \cdots$, and let $\hat{q} \equiv \{\hat{q}(\cdot, t)\}_{t=0}^\infty$, so that $\hat{q} \in A_2^\infty$. Finally let $s \equiv (y, \gamma, \hat{q}) \in A_1 \times \Gamma^\infty \times A_2^\infty \equiv \Omega$. Again, Ω is a product of compact spaces and is therefore compact in its product topology. Then one may write

$$K(Q) = \inf_{s \in \Omega} f(s) \qquad \text{subject to} \qquad G_t(s, Q_t) \leq 0$$

for $t = 0, 1, \ldots$, where

$$f(s) = \sum_{t=0}^{\infty} \beta^t \left\{ \sum_{\tau=0}^{t} y_\tau \int_{\alpha_1}^{\alpha_2} (c[\hat{q}(x,t)]x + (1-\beta)W)\hat{\psi}_{\gamma(\tau)}(x \mid t, \tau)\,dx + k y_t \right\}$$

[5]The restriction that $\sum y_t$ be bounded is stated for analytical convenience. The optimal entry sequence is bounded because at the optimum one must have $G_t = 0$ for all t. For if $G_t < 0$, the planner could reduce $\hat{q}(x,t)$ for some firms, thereby reducing f without violating any of the constraints. Rewriting G_t as $-\sum_{\tau=0}^{t} y_\tau \hat{\phi}(t,\tau) + Q_t \leq 0$, one then has that $\sum_{\tau=0}^{t} y_\tau \hat{\phi}(t,\tau)$ is bounded because Q_t is bounded. Then, if $\sum^t y_\tau$ did not converge, one would have output per entrant

$$\lim_{\substack{t \to \infty \\ (t > T)}} \frac{\sum_{\tau=T}^{t} \hat{\phi}(t,\tau) y_\tau}{\sum_{\tau=T}^{t} y_\tau} = 0$$

for every $T < \infty$, which cannot be optimal. It is of relevance, however, that one may have entry at each point in time ($y_t > 0$ each t) while $\sum^\infty y_t < \infty$. In fact, this may well happen in equilibrium (Theorem 3).

and where

$$G_t(s, Q_t) = -\sum_{\tau=0}^{t} y_\tau \int_{\alpha_1}^{\alpha_2} \hat{q}(x,t)\hat{\psi}_{\gamma(\tau)}(x \mid t,\tau)\,dx + Q_t$$

so that $f:\Omega \to R^1$ and $G_t:\Omega \times \ell_\infty \to R^1$. The meaning of the constraint $G_t(s, Q_t) \leq 0$ is that the planner is required to produce *no less* than the target output Q_t at t. This constraint holds at each t.

Let $G(s, Q) \equiv \{G_t(s, Q_t)\}_{t=0}^{\infty}$. Since G_t is bounded (uniformly in t), $G:\Omega \times \ell_\infty \to \ell_\infty$. The constraint may then be written as $G(s, Q) \leq 0$.

LEMMA 2: *There exists a constrained minimum for f on* Ω.

PROOF: Contained in the Appendix.

Simplifying the notation slightly, we write $G(s) \leq 0$ for the constraint, dropping Q from the notation. We now define the Gateaux differentials $\delta f(s_0, s_1) = df[s_0 + \alpha(s_1 - s_0)]/d\alpha$ and $\delta G(s_0, s_1) = dG[s_0 + \alpha(s_1 - s_0)]/d\alpha$ for $s_0, s_1, \in \Omega$. They are differentials at the point s_0, with increment $(s_1 - s_0)$. We have $\delta f:\Omega \to R$ and $\delta G:\Omega \to \ell_\infty$. Since Ω is convex, the point $s_0 + \alpha(s_0 - s_0)(= (1-\alpha)s_0 + \alpha s_1) \in \Omega$ for all $\alpha \in [0, 1]$.

DEFINITION: A point $s_0 \in \Omega$ is said to be a regular point of the inequality $G(s) \leq 0$ if $G(s_0) \leq 0$ and there exists a point $s_1 \in \Omega$ such that $G(s_0) + \delta G(s_0, s_1) < 0$ [**12**, p. 248]. The inequality is strict at each coordinate, i.e., $G_t(s_0) + \delta G_t \cdot (s_0, s_1) < 0$ for each t.

LEMMA 3: *Let* $f(s_0) = \min f(s)$ *subject to* $G(s) \leq 0$. *Then, if* $Q_t > 0$ *(each t),* s_0 *is a regular point.*

PROOF: Contained in the Appendix. Of course, the condition that the maximum occurs at a regular point is the analogue of the Kuhn-Tucker constraint qualification in the finite dimensional case.

Let P be the positive cone of ℓ_∞. Then P contains an interior point (any bounded, strictly positive sequence which is bounded away from zero is an interior point of P). Let ℓ_∞^* be the dual space of ℓ_∞, that is, the space of bounded linear functionals on ℓ_∞. Let $\lambda \in \ell_\infty^*$ be a particular functional. Its value at the point $z \in \ell_\infty$ will be denoted by $\lambda(z)$. We then have the following lemma.

LEMMA 4: *If* s_0 *minimizes f on* Ω *subject to* $G(s) \leq 0$, *then there exists a functional* $\lambda^* \in \ell_\infty^*$ *such that if* $Q > 0$,

$$\delta f(s_0; s) + \lambda^*[\delta G(s_0; s)] \geq 0 \qquad \textit{for all } s \in \Omega \tag{5.6}$$

and such that $\lambda^* > 0$ *and*

$$\lambda^*[G(s_0)] = 0. \tag{5.7}$$

PROOF: [**12**, pp. 249–250]: (His proof relies on s_0 being regular (and this we have established in Lemma 3) and on the interior point property of P.)

We now proceed to characterize λ^*. Since $\lambda^* \in \ell_\infty^*$, it is representable in the following way: for any sequence $\eta \in \ell_\infty$:

$$\lambda^*(\eta) = \sum_{t=0}^{\infty} \lambda_t^* \eta_t + \int \eta \, d\hat{\lambda}^* \tag{5.8}$$

where $\hat{\lambda}^*$ is a bounded and *purely finitely additive* measure on the set of all possible subsets of the positive integers [**9**, p. 870; **19**, p. 52] and where $\sum|\lambda_t^*| < \infty$. The purely finitely additive property implies $\int \eta \, d\hat{\lambda}^* = 0$ for all $\eta \in \ell_\infty$ such that only finitely many elements of the sequence η are nonzero. We now prove that $\lambda_t^* > 0$ (each t) and that $\int \eta \, d\hat{\lambda}^* = 0$, for all $\eta \in \ell_\infty$.

At the optimal solution, s_0, $G_t(s_0) = 0$. For if $G_t(s_0) < 0$, one could reduce f by reducing $\hat{q}^0(x,t)$ (for t only) on a set of positive measure without violating the constraint at t, and leaving all the other contraints unaffected. If s_1 differs from s_0 only in as much as $\hat{q}^1(x,t) \neq \hat{q}^0(x,t)$ for some x and for *fixed t*, only the constraint at t is affected. More precisely $\delta G_i(s_0;s_1)$ is zero except for $i = t$. Hence, for such variations, $\lambda^*[\delta G(s_0;s_1)] = \lambda_t^* \delta G_t(s_0;s_1)$ (in view of equation (5.8)). Equation (5.6) then implies

$$\sum_{\tau=0}^{t} y_\tau^0 \int \left\{ \beta^t c'\left[\hat{q}^0(x,t)\right]x - \lambda_t^* \right\} \left[\hat{q}^1(x,t) - \hat{q}^0(x,t)\right] \hat{\psi}_{\gamma^0(\tau)}(x \mid t,\tau)\, dx \geq 0 \tag{5.9}$$

for all $\hat{q}^1$ which are integrable. (We shall refer to "sets on which

$$\sum_{\tau=0}^{t} y_\tau^0 \hat{\psi}_{\gamma^0(\tau)}(x \mid t,\tau)$$

is of positive Lebesgue measure" as sets of positive measure. Clearly, variations in $\hat{q}$ on sets of measure zero in this sense are immaterial since they affect neither f nor G.) Since $Q_t > 0$, $\hat{q}^0(x,t) > 0$ on a set of positive measure, so that $c'[\hat{q}^0(\cdot)] > 0$ on these sets. But then $\lambda_t^* > 0$, or else there would exist an admissible variation which would violate equation (5.9). Also, if $\hat{q}^0(x,t) = 0$ on some set of positive measure, it is readily seen that since $c'(0) = 0$, equation (5.9) would again be violated by an admissible variation. Therefore $\hat{q}^0 > 0$, which implies that except on sets of measure zero, for each t,

$$\lambda_t^* = \beta^t c'\left[\hat{q}^0(x,t)\right]x > 0. \tag{5.10}$$

We are now able to prove the following lemma.

LEMMA 5: $\int \eta \, d\hat{\lambda}^* = 0$ *for all* $\eta \in \ell_\infty$.

PROOF: Contained in the Appendix.

Lemma 5 establishes that the second expression on the right-hand side of equation (5.8) is equal to zero. Now let $s_1 = s_0$ except that $y_t^1 \neq y_t^0$ for some t. Then equation (5.6) reads (in view of Lemma 5)

$$\left\{ \frac{\partial f}{\partial y_t} + \sum_{j=t}^{\infty} \lambda_j^* \frac{\partial G_j}{\partial y_t} \right\} (y_t^1 - y_t^0) \geq 0$$

for all y_t^1. Therefore, if $y_t^0 > 0$, $\partial f/\partial y_t + \sum_{j=t}^{\infty} \lambda_j^* \partial G_j/\partial y_t = 0$ while if $y_t^0 = 0$, then $\partial f/\partial y_t + \sum_{j=t}^{\infty} \lambda_j^* \partial G_j/\partial y_t \geq 0$. This implies that

$$(5.11) \quad \sum_{j=t}^{\infty} \int \{ \beta^j [c(\hat{q}^0)x + W(1-\beta)] - \lambda_j^* \hat{q}^0 \} \hat{\psi}_{\gamma^0(t)}(x \mid j,t)\, dx + \beta^t k \geqq 0 \quad \text{as} \quad y_t^0 \underset{>}{=} 0.$$

Finally we consider the variation for which $s_1 = s_0$, except that $\gamma_i^1(\tau) \neq \gamma_i^0(\tau)$ for some $i \geq 1$ and some $\tau \geq 0$. Applying Lemma 5, equation (5.6) reads

$$(5.12) \quad \left\{ \frac{\partial f}{\partial \gamma_i(\tau)} + \sum_{j=\tau+i}^{\infty} \lambda_j^* \frac{\partial G_j}{\partial \gamma_i(\tau)} \right\} [\gamma_i^1(\tau) - \gamma_i^0(t)] \geq 0$$

(note that a change in $\gamma_i(\tau)$ does not affect G_t for $t < \tau + i$). Then equation (5.12) implies that for all $\gamma_i^1(\tau) \in [\alpha_1, \alpha_2]$,

$$(5.13) \quad \left\{ \sum_{j=\tau+i}^{\infty} \int \{ \beta^j [c(\hat{q}^0)x + W(1-\beta)] - \lambda_j^* q^0 \} \frac{\partial \hat{\psi}_{\gamma^0(\tau)}}{\partial \gamma_i(\tau)} (x \mid t,\tau)\, dx \right\} \times y_\tau^0 [\gamma_i^1(\tau) - \gamma_i^0(\tau)] \geq 0.$$

If $y_\tau^0 = 0$, this inequality is automatically satisfied—there are no firms of vintage τ.

LEMMA 6: *The Lagrangean* $L(s,\lambda^*) = f(s) + \lambda^*[G(s)]$ *is minimized over* s *at* (s_0, λ^*) *satisfying equations* (5.10), (5.11), *and* (5.13). *The point* $f(s_0)$ *is a global constrained minimum.*

PROOF: Contained in the Appendix. Lemma 6 ensures that the minimum attained at s_0 is indeed the global minimum.

LEMMA 7: $K(Q)$ *is convex in* Q, *and is differentiable in the elements of* Q, *with* $\partial K/\partial Q_t = \lambda_t^*$.

PROOF: Contained in the Appendix.

The social marginal cost is λ_t^*, in present value terms. From equation (5.10) this is also the discounted expected marginal cost of each firm.

We recall that the derived properties of $K(Q)$ hold only for sequences Q which are bounded away from zero by some positive ϵ. However, for each such ϵ, λ_t^* is bounded uniformly in $Q \in \ell_\infty$. Let $\bar{\lambda}$ be the upper bound on λ. The following assumption on demand ensures that marginal benefits exceed marginal costs at some sufficiently small level of output at each t: we assume that for some $\epsilon > 0$ sufficiently small,

$$(5.14) \qquad D(\epsilon, t) > \bar{\lambda} \qquad \text{all } t.$$

Secondly, in order to ensure that the socially optimal output sequence is bounded, we assume that for each t

$$(5.15) \qquad \int_0^\infty D(z,t)\,dz < A$$

for some A sufficiently large and independent of t.

PROPOSITION 1: *There is exactly one bounded sequence* $\{Q_t^*\}$ *which maximizes* $S(Q)$ *in equation* (5.5), *and it satisfies*

$$(5.16) \qquad \beta^t D[Q_t^*, t] = \lambda_t^*, \qquad t = 0, 1, \ldots .$$

PROOF: Equation (5.15) implies that Q_t^* is bounded. Equation (5.14) ensures that Q_t^* is bounded away from zero. Each decision Q_t then belongs to a compact set, and $S(Q)$ is strictly concave in Q, because $D(Q_t, t)$ is downward sloping. Therefore the maximum exists and is unique. Since $S(Q)$ is also differentiable, one has $\partial S / \partial Q_t = 0$ at the optimum, which implies that equation (5.16) holds, and the proof is complete.

Let $\hat{q}^*(x,t)$ and $\hat{\psi}_{Y^*(\tau)}(x \mid t, \tau)$ denote the optimal $\hat{q}$ and $\hat{\psi}$ for minimizing f when $Q = Q^*$. We may now state:

THEOREM 2 (Existence, Uniqueness, Optimality): *There is exactly one equilibrium price sequence* $\tilde{p}$ *and entry sequence* $\tilde{y}$ *satisfying* D.1. *and* D.2. *in Section* 4. *The price sequence is given by* $\tilde{p}_t = \beta^{-t}\lambda_t^*$. *The social benefit functional* $S(Q)$ *is at a maximum with* $Q_t^* = Q_t(\tilde{p}, \tilde{y})$, *where* $\hat{q}^*(x,t) = q(\tilde{p}_t / x)$ *and where* $\hat{\psi}_{Y^*(\tau)}(x \mid t, \tau) = \hat{\psi}_{Y(\tilde{p},\tau)}(x \mid t, \tau)$.

PROOF: From equation (5.10) it is seen that $\hat{q}^*(x,t) = q(\tilde{p}_t / x)$. Substituting this for $\hat{q}^*$ into equation (5.13), this condition becomes identical to equation (5.4) when the latter is evaluated at $\psi(x \mid t, \tau; \tilde{p})$. Therefore $\hat{\psi}_{Y^*(t)} = \psi(x \mid t, \tau; \tilde{p})$ as asserted. Substituting for $\hat{q}^*$ and $\hat{\psi}_{Y^*(t)}$ in equation (5.11), this condition is identical to Condition D.2. Finally, from Lemma 8, $D[Q_t(\tilde{p}, \tilde{y}), t] = p_t^*$ so that Condition D.1. is satisfied. Therefore, the maximum is necessarily an equilibrium. This establishes existence. For uniqueness and optimality, note that since any candidate equilibrium price and entry sequence, $(\tilde{p}, \tilde{y})$ satisfies $\tilde{p}_t = \beta^{-t}(\partial K/\partial Q_t)[Q_t(\tilde{p}, \tilde{y})]$, Condition (D.1) requires that $\tilde{p}_t = D[Q_t(\tilde{p}, \tilde{y}), t]$. This requires that $S(Q)$ be maximized by $Q_t(\tilde{p}, \tilde{y})$. By lemma 7 this maximum is unique. Therefore $Q(\tilde{p}, \tilde{y}) = Q^*$, and $\tilde{p}_t = \beta^{-t}\lambda_t^*$, so that the equilibrium $(\tilde{p}, \tilde{y})$ is necessarily a maximum. This completes the proof of the theorem.

6. TWO CHARACTERIZATION THEOREMS

This section gives some characterization of the behavior of entry and prices in equilibrium. Much depends on whether $q(p_t/x)$ is convex in x. This is made precise in the following theorem.

THEOREM 3: *Assume that $D[Q_t, x]$ is nondecreasing in t (demand grows monotonically). If $q(p/x)$ is a strictly concave function of x [see equation (3.1)], then the equilibrium price sequence is constant, with $p_t^* = \bar{p}$ for each t, and entry occurs at each t ($y_t^* > 0$) while $\sum_0^\infty y_t^* < \infty$, and $V(x_0, 0, t; \bar{p}) - k = W$ for each t.*

PROOF: Since (D.2) must hold, profits to entry must be zero at each t. Suppose the optimal exit policies of firms are such that no firm exits under any circumstance. Since x_t^* is a martingale, strict concavity of $q(\cdot)$ in x implies (by Jensen's inequality) $E_t q(\bar{p}/x_{t+1}^*) < q(\bar{p}/x_t^*)$. This implies that $\phi(t, \tau; p^*) > \phi(t+1, \tau; p^*)$ for each $t \geq \tau$. This inequality is only reinforced by firms making an exit under some contingencies. In other words, if the price is constant at $\bar{p}$, the output of the survivors (of any given vintage) would decline with time *even if none were to exit*. It follows that it will decline even more if some do exit. Define $A_t(\bar{p})$ by $\bar{p} = D[A_t(\bar{p}), t]$. In other words, $A_t(\bar{p})$ is the amount of output which must be forthcoming at t if price is to be kept constant at $\bar{p}$. Then since demand is nondecreasing in t, the sequence $\{A_t(\bar{p})\}$ is also nondecreasing in t. Since for each t, ϕ declines in t, the output of the survivors is declining. Therefore $y_t^* > 0$ for each t because if $y_t^* = 0$ for some t, price at t would be strictly greater than in the previous period. Finally, since $A_t(\bar{p})$ is bounded, one has $\sum_0^\infty y_t^* < \infty$ if $\inf_{t,\tau}\phi(t, \tau; p^*) > 0$. For this is sufficient to show that for each τ, $\lim_{t\to\infty}\Psi(\alpha_2 \mid t, \tau; p^*) > 0$, that is, the probability of permanent survival is positive for firms of each vintage. To prove that this is so, consider the policy: "stay in for one more period, then exit no matter what." The expected reward from this policy is $\pi(\bar{p}, x) + \beta W$. Therefore the firm will not exit if this reward is greater than W, or, equivalently, if $\pi(\bar{p}, x) > (1-\beta)W$. Let $\hat{x}$ be such that $\pi(\bar{p}, \hat{x}) = (1-\beta)W$, so that the firm stays in if $x_t^* < \hat{x}$. Of course, $\hat{x} > \alpha_1$, otherwise the net value of entry at all t would be negative. For each firm of age n (see Section 3), $x_t^* = E(x \mid \bar{\eta}_n, n)$. Let $\bar{\eta}_n$ be such that $\hat{x} = E(x \mid \bar{\eta}_n, n)$. Since $\hat{x} > \alpha_1, \hat{\eta}_n$ is bounded from below. Therefore exit does not occur so long as $\bar{\eta}_n < \hat{\eta}_n$. But the probability that a normal posterior mean does not ever reach a boundary that is *bounded* is strictly positive (Chernoff [3]). Therefore $\lim_{t\to\infty}\Psi(\alpha_2 \mid t, \tau; p^*) > 0$ for all τ, and $\sum_0^\infty y_t^* < \infty$. This completes the proof of the theorem.

The concavity of $q(\cdot)$ in x is (under conditions of increasing demand) sufficient to imply a constant-price equilibrium. But it is not necessary. The concavity of q in x was, in the proof of Theorem 3, used only to prove that $\phi(t+1,\tau;\,p^*) < \phi(t,\tau;\,p^*)$. If q is convex in x, this means that the collective output of the *surviving* firms is larger at $t+1$ than it was at t. But the subtraction from the total output due to firms' exit might still be larger than this increase, resulting, on net, in a decrease in ϕ.

It seems possible that after withdrawals have occurred, the surviving firms will increase their output by *more* than the increase in the quantity demanded (at an unchanged price), thereby making it necessary for price to fall over time. In view of Theorem 3, convexity of q in x is a *necessary* condition for such an equilibrium to occur.

THEOREM 4: *$K(Q)$ is homogeneous to degree one: for any $\alpha > 0$, $\alpha K(Q) = K(\alpha Q)$. An equiproportional shift to the right of demand (for all t) brings about an increase in Q_t in the same proportion for all t. Equilibrium prices remain unchanged.*

PROOF: If $(y^0, \hat{q}^0, \gamma^0)$ minimizes f when the constraint is Q_0, then $(\alpha y^0, \hat{q}^0, \gamma^0)$ minimizes it when the constraint is αQ_0: Since f is linear homogeneous in y, this proves $\alpha K(Q) = K(\alpha Q)$ for all Q. Now introduce the demand-shift-parameter μ, such that $D[Q_t, t, u] = D[\mu Q_t, t]$. Equilibrium requires that $\beta^t D[\mu Q_t(\mu), t] = \partial K[Q(\mu)]/\partial Q_t$, where $\{Q_t(\mu)\}$ is the equilibrium quantity sequence indexed by μ. We need to show that $Q(\mu) = Q(1)/\mu$. By homogeneity of K, $\alpha \partial K(Q)/\partial Q_t = \alpha \partial K(\alpha Q)/\partial Q_t$ for all $\alpha > 0$ and Q, so that $\partial K(Q)/\partial Q_t = \partial K(\alpha Q)/\partial Q_t$. Letting $Q = Q(1)$ and $\alpha = 1/\mu$, the result follows.

Bell Laboratories, Murray Hill, New Jersey

Manuscript received January 1981; revision received May, 1981.

APPENDIX

PROOF OF THEOREM 1: Part (i): The proof of this part consists of an application of Theorem 5 of [2]. Let T denote the operator which defines V as the fixed point of the equation (3.2) $V = TV$. First it needs to be shown that T transforms continuous, bounded functions into other continuous, bounded functions. Boundedness follows if $\pi(p_t, x_t^*)$ is bounded—the latter is true because (a) p_t is bounded, (b) $x_t^* \geq \alpha_1 > 0$, and (c) $\lim_{q\to\infty} c'(q) = \infty$, so that the necessary condition $p_t - c'(q_t)x_t^*$ is always satisfied at a bounded output level. To prove that T preserves continuity, note firstly that $\pi(\cdot)$ is continuous in x_t^*, as is P. Therefore T preserves boundedness and continuity.

Secondly, T is a monotone (increasing) operator. That is, for two functions f_1 and f_2, if $f_1 \geq f_2$ everywhere, then $Tf_1 \geq Tf_2$ everywhere.

Finally, for any function f and any constant $c > 0$, $T(f+c) \leq Tf + \beta c$.

Therefore T is a contraction operator with modulus $\beta < 1$, and the Banach fixed-point theorem may be applied to yield assertion (i) of the theorem.

Part (ii): The proof is in two steps. We first prove that V is nonincreasing in x and then use this to prove that it is strictly decreasing.

Let T denote the operator in equation (3.2) so that V is the unique solution to the equation $V = TV$. Since T is a contraction operator, we have $V = \lim_{n\to\infty} T^n g$ for *any* bounded, continuous function g [where $T^n g \equiv T(T^{n-1}g)$]. Any monotonicity properties of *each* member of the sequence $T^n g$ are preserved (weakly) by the limit function V. We shall now show that if g is decreasing in x, so is Tg. If $g(z)$ is decreasing in z, then so is the function $\max[W, g(z)]$. Then, since $P(z \mid x, n)$ is increasing in x, $\int \max[W, g(z)]P(dz \mid x, n)$ is decreasing in x. By induction, these properties carry over to the limit function V which is therefore nonincreasing in x. But then $\int \max[W, V(z, n+1, t+1; p)]P(dz \mid x, n)$ is nonincreasing in x. But π is strictly decreasing in x and therefore so is V.

Proof of Lemma 1: One has

$$\hat{\Psi}_\gamma(x \mid t, \tau) = \int_{\alpha_1}^{\alpha_2} H[\min(x, \gamma_{t-\tau}) \mid z, t-1-\tau]\hat{\Psi}_\gamma(dz \mid t-1, \tau)$$

where $H(z' \mid z, n)$ is the (one-step) transition probability that $x_t^* \le z'$ given the pair of sufficient statistics $x_{t-1}^* = z$ and $n = t - 1 - \tau$. But H is a continuous transform of the normal CDF, and is differentiable in z' for $z' \in (\alpha_1, \alpha_2)$. Therefore if $\hat{\Psi}$ is continuous at t, then it is also differentiable at $t+1$. Thus if $\hat{\Psi}$ is differentiable when $t = \tau + 1$, it is differentiable for all $t > \tau$. But $\hat{\Psi}(x \mid \tau+1, \tau) = H[\min(x, \gamma_1 \mid x_0, 0]$ if $x_0 < \gamma_0$, while if $x_0 \ge \gamma_0$, $\hat{\Psi}(x \mid \tau+1, \tau) = 0$, so that $\hat{\Psi}$ is differentiable when $t = \tau + 1$. Furthermore since H is continuously differentiable, so is $\hat{\Psi}$, except at the point $x = \gamma_{t-\tau}$, (where the derivative is in general discontinuous as it becomes zero for $x > \gamma_{t-\tau}$) so that the density $\hat{\psi}$ exists and is piecewise continuous. Turning to differentiability of $\hat{\psi}$ in γ_j, note that $\hat{\psi}_\gamma(x \mid t, \tau)$ does not depend on γ_j if $j > t - \tau$. For any $j < t - \tau$, $H[\cdot]$ does not depend on γ_j, and therefore if $\hat{\psi}$ is differentiable at t, it is differentiable at $t+1$. Since this is true for any t, it is sufficient to show that $\hat{\psi}$ is differentiable at $t = \tau + j$, where

$$\hat{\Psi}_\gamma(x \mid \tau+j, \tau) = \int_{\alpha_1}^{\alpha_2} H[\min(x, \gamma_j) \mid z, j+\tau-1]\hat{\psi}_\gamma(z \mid \tau+j-1, \tau)\,dz$$

and where $\hat{\psi}_\gamma(z \mid \tau+j-1, \tau)$ does not depend on γ_j. Then the cross-derivative $\partial^2\hat{\Psi}/\partial x\,\partial\gamma_j$ exists, and this completes the proof of the lemma.

Proof of Lemma 2: Let $\Omega(Q) = \{s : G(s, Q) \le 0\} \subseteq \Omega$. We need to show that f is continuous on $\Omega(Q)$ and that $\Omega(Q)$ is compact [**1**, p. 69]. In the product topology, sequences $\{s_t\}_0^\infty$ $(s_t \in \Omega)$ converge to s_0 *weakly* [**5**, p. 32]. That is, $s_1 \to s_0$ if and only if $y_t^1 \to y_t^0$ each t, $\hat{q}^1(x, t) \to \hat{q}^0(x, t)$ for each t except on a set of Lebesgue measure zero, and $\gamma_j^1(\tau) \to \gamma_j^0(\tau)$ each j. Suppose then that $|f(s_1) - f(s_0)| \to \delta > 0$ as $s_1 \to s_0$, so that f is not continuous at s_0. Write $f(s)$ as $\sum_{t=0}^\infty \beta^t f_t(s)$. Since $f_t(s)$ is bounded on Ω, one may choose $T(\delta) < \infty$ such that $\sum_{t=T(\delta)}^\infty \beta^t |f_t(s_1) - f_t(s_0)| < (\delta/2)\ \forall s_1 \in \Omega$. Then

$$|f(s_1) - f(s_0)| \le \sum_{t=0}^{T(\delta)-1} \beta^t |f_t(s_1) - f_t(s_0)| + \frac{\delta}{2}$$

$$\le [T(\delta) - 1] \max_{0 \le t \le T(\delta)-1} |f_t(s_1) - f_t(s_0)| + \frac{\delta}{2}.$$

But for any $t < \infty$, $\hat{\psi}_{\gamma(\tau)}(x \mid t, \tau)$ depends only upon $\gamma_j(\tau)$ for $j = 1, \cdots t - \tau$ and is continuous in γ_j, so that for each $t > \tau$, and each x, $\hat{\psi}_{\gamma^1(\tau)} \to \hat{\psi}_{\gamma^0(\tau)}$ as $s_1 \to s_0$ so that $|f_t(s_1) - f_t(s_0)| \to 0$ as $s_1 \to s_0$. But then $|f(s_1) - f(s_0)| \to (\delta/2) < \delta$, a contradiction. Therefore f is continuous on Ω.

Compactness of $\Omega(Q)$ is assured if $\Omega(Q)$ is closed, because a closed subset of a compact space is compact [**1**, p. 68]. $\Omega(Q)$ is closed if its complement is open. Consider then a point $s_0 \in \Omega$ such that $G_t(s_0, Q_t) > 0$ for each t. One must show that there exists an open neighborhood of s_0, $N(s_0)$, such that $G(s, Q) > 0$ for all $s \in N(s_0)$. Let $G_t(s, Q_t) = -\sum_0^t y_i z_i(\hat{q}, \gamma) + Q_t$. Then let $G(s_0, Q) > 0$. For any $s \in \Omega$, since y_t and z_t are bounded,

$$G_t(s, Q_t) - G_t(s_0, Q_t) = -\sum_0^t y_i(z_i - z_i^0) - \sum_0^t (y_i - y_i^0)z_i^0.$$

Since $\sum_0^t y_i < \infty$, and $y_i \geq 0$, one has $\sum_T^\infty y_t \to 0$ as $T \to \infty$. Then one may choose $T = T(\delta)$ sufficiently large such that $\sum_{T(\delta)}^t y_i|z_i - z_i^0| + \sum_{T(\delta)}^t |y_t - y_i^0| z_i^0 < (\delta/2)$ for all $t \geq T(\delta)$. Then for any $t \geq T(\delta)$,

$$|G_t(s, Q_t) - G_t(s_0, Q_t)| \leq \sum_{i=0}^{T(\delta)-1} (y_i|z_t - z_i^0| + |y_i - y_i^0| z_i^0) + \frac{\delta}{2}$$

$$\leq [T(\delta) -]\Big(c_1 \max_{0 \leq i \leq T(\delta)-1} |z_i - z_i^0|$$

$$+ c_2 \max_{0 \leq i \leq T(\delta)-1} |y_t - y_i^0|\Big) + \frac{\delta}{2}$$

where the second inequality follows as both y_t and z_t are bounded. But for any $t < \infty$, $|z_t - z_i^0| \to 0$ and $|y_t - y_t^0| \to 0$ as $s \to s_0$. Since this is true for any $\delta > 0$, one has, for any t, that $G_t(s, Q_t) \to G_t(s_0, Q_t)$ as $s \to s_0$. This is true for each t; therefore $G(s, Q)$ converges to $G(s_0, Q)$, so that G is a continuous mapping. Therefore the set $\{s : G(s, Q) > 0\}$ is open. But s_0 is a member of this set (since by assumption $G(s_0, Q) > 0$). Therefore $\Omega(Q)$ is closed and the proof of Lemma 2 is complete.

PROOF OF LEMMA 3: The $s_1 = (y^1, \gamma^0, \hat{q}^0)$ and $s_0 = (y^0, \gamma^0, \hat{q}^0)$. That is s_1 differs from s_0 only in the entry sequence y. Then,

$$\delta G_t(s_0; s_1) = -\sum_{\tau=0}^{t} (y_\tau^1 - y_\tau^0) \int \hat{q}^0(x, t) \hat{\psi}_{\gamma^0(\tau)}(x \mid t, \tau)\, dx.$$

Since $G_t(s_0) \leq 0$, one need only show that $\delta G_t < 0$ (each t) for some $s_1 \in \Omega$. Since $Q_t > 0$, one has $\sum_0^t y_\tau^0 \int q^0 \hat{\psi}_{\gamma^0}\, dx > 0$ for each t. But then one may set $y_\tau^1 - y_\tau^0 = \alpha y_\tau^0$ $(\alpha > 0)$ in which case $\delta G_t(s_0; s_1) < 0$ for each t. Since $\sum^\infty y_\tau^0$ is finite, so is $(1 + \alpha)\sum^\infty y_\tau^0$ so that $s_1 \in \Omega$ and Lemma 3 has been proved.

PROOF OF LEMMA 5: Suppose $\exists \eta \in \ell_\infty$ such that $|\int \eta\, d\hat{\lambda}^*| = \epsilon_1 > 0$. Then for any ϵ_2 no matter how small, $|\int \epsilon_2 \eta\, d\hat{\lambda}^*| = \epsilon_2 \epsilon_1 > 0$. Let $T > 0$ be given, and let

$$\eta_t^T = \begin{cases} \eta_t & \text{if } t \geq T, \\ 0 & \text{if } 0 \leq t < T, \end{cases} \qquad \text{and} \qquad \eta_t^1 = \begin{cases} 0 & \text{if } t \geq T, \\ \eta_t & \text{if } 0 \leq t < T. \end{cases}$$

By linearity, $\int \epsilon_2 \eta\, d\hat{\lambda}^* = \epsilon_2 \int (\eta^1 + \eta)\, d\hat{\lambda}^* = \epsilon_2 \int \eta^T\, d\hat{\lambda}^*$, where the second equality is due to the purely finitely additive property of $\hat{\lambda}^*$. Next we show that $\exists s_1, s_2 \in \Omega$ such that $\delta G(s_0; s_1) = \epsilon_2 \eta^T$ and $\delta G(s_0; s_2) = -\epsilon_2 \eta^T$. For the first equality one may choose s_1 the same as s_0 except that $\hat{q}^1 = \hat{q}^0(1 + \epsilon_2 \eta_t^T / Q_t)$ for each x, t. This is a feasible variation, since $\exists \epsilon > 0$ such that $Q_t \geq \epsilon \forall t$ so that one may set ϵ_2 sufficiently small such that $\epsilon_2 \bar{\eta}/\epsilon \leq 1$ (where $\bar{\eta}$ is such that $\eta_t \leq \bar{\eta}$ $\forall \eta \in \ell_\infty$). Similarly, to obtain the second equality one sets $\hat{q}^1 = \hat{q}^0(1 - \epsilon_2 \eta_t^T / Q_t)$. For each η, $\|\eta^T\| \not\to 0$ unless $\lim_{t\to\infty} \eta_t = 0$. But if $\lim_{t\to\infty} \eta_t = 0$, then the lemma is true [**9**, p. 872]. Suppose then that $\|\eta^T\| \to \epsilon_3 > 0$. (The limit must exist because $\|\eta^T\|$ is nonincreasing in T and positive.) Clearly, $\lim_{T\to\infty} \delta f(s_0; s_0 + \epsilon_2 \eta^T) = 0$, and $\lim_{T\to\infty} \sum_{t=0}^\infty \lambda_t^* \delta G_t(s_0; s_0 + \epsilon_2 \eta^t) = 0$, while for T no matter how large, $|\int \epsilon_2 \eta^T\, d\hat{\lambda}^*| = \epsilon_2 \epsilon_1 = |\int -\epsilon_2 \eta^T\, d\hat{\lambda}^*| > 0$. By linearity, therefore, there is a feasible variation for which $\delta f + \lambda^*(\delta G) < 0$, a contradiction to equation (5.6). This completes the proof of the lemma.

PROOF OF LEMMA 6: If the Lagrangean is minimized at s_0, then the global minimum property follows from [**12**, p. 220, Theorem 1], and from the fact that $G(s_0) = 0$. It then remains to be shown that $L(s, \lambda^*) \geq L(s_0, \lambda^*)$ for $s \in \Omega$. Since $c(\cdot)$ is convex, one has $c(\hat{q}) \geq c(\hat{q}^0) + c'(\hat{q}^0)(\hat{q} - \hat{q}^0)$

$= c(\hat{q}^0) + x^{-1}\beta^{-t}\lambda_t^*(\hat{q} - \hat{q}^0)$. Therefore, for $s \in \Omega$,

$$f(s) + \lambda^*[G(s)]$$

$$\geq \sum_{t=0}^{\infty}\left[\int \{\beta^t[c(\hat{q}^0)x + \beta^{-t}\lambda_t^*(\hat{q} - \hat{q}^0) + (1-\beta)W] - \lambda_t^*\hat{q}\}\right.$$

$$\left.\times \sum_{\tau=0}^{t} \psi_{\gamma(\tau)}y_\tau\, dx + \beta^t k y_t + \lambda_t^* Q_t\right]$$

$$= \sum_{t=0}^{\infty}\int \beta^t[c(\hat{q}^0)x - \beta^{-t}\lambda_t^*\hat{q}^0 + (1-\beta)W]\sum_{\tau=0}^{t}\hat{\psi}_{\gamma(\tau)}y_\tau\, dx + \beta^t k y_t + \lambda_t^* Q_t$$

$$= \sum_{\tau=0}^{\infty} y_\tau\left\{\sum_{t=\tau}^{\infty}\beta^t\int[c(\hat{q}^0)x - \beta^{-t}\lambda_t^*\hat{q}^0 + (1-\beta)W]\hat{\psi}_{\gamma(\tau)}\, dx\right\} + \beta^\tau k y_\tau + \lambda_\tau^* Q_\tau.$$

However, $c(\hat{q}^0)x - \beta^{-t}\lambda_t^*\hat{q}^0 = \pi(\beta^{-t}\lambda_t^*, x)$. Therefore, in view of equation (5.3), the expression in curly brackets is minimized by $\gamma(\{\beta^{-t}\lambda_t^*\}_0^\infty, \tau)$ (see the first line following equation (5.3)). But the necessary conditions for this to occur, in equation (5.4) are identical to those of equation (5.13) (when one sets $p_t = \beta^{-t}\lambda_t^*$ in equation (5.4)). Therefore, the expression in curly brackets is minimized for each τ, by $\hat{\psi}_{\gamma^0(\tau)}$. Therefore,

$$f(s) + \lambda^*[G(s)] \geq \sum_0^\infty y_\tau r_\tau + \sum_0^\infty \lambda_\tau^* Q_\tau,$$

where

$$r_\tau = \sum_{t=\tau}^{\infty}\beta^t\int[c(\hat{q}^0)x - \beta^{-t}\lambda_t^*\hat{q}^0 + (1-\beta)W]\hat{\psi}_{\gamma^0(\tau)}\, dx + \beta^t k.$$

Therefore,

$$f(s) + \lambda^*[G(s)] \geq \sum y_t^0 r_t + \sum (y_t - y_t^0) r_t + \sum \lambda_t^* Q_t$$

$$= f(s_0) + \lambda^*[G(s_0)] + \sum (y_t - y_t^0) r_t.$$

But equality implies that $r_t(y_t - y_t^0) \geq 0\ \forall t$, implying that $f(s) + \lambda^*[G(s)] \geq f(s_0) + \lambda^*[G(s_0)]$, for $s \in \Omega$. The lemma has been proved.

Proof of Lemma 7: Let s_0 be optimal when $Q = Q_0$, and let s_1 be optimal when $Q = Q_1$. $f(s_0) + \lambda^*[G(s_0, Q_0)] \leq f(s_1) + \lambda^*[G(s_1, Q_0)]$ (using Lemma 6). Therefore $f(s_1) - f(s_0) \geq \lambda^*[G(s_0, Q_0) - G(s_1, Q_0)]$. But $G(s_0, Q_0) = 0$. Also, $G(s_1, Q_1) = 0$. Therefore $G(s_1, Q_0) = -Q_1 + Q_0$. Therefore $K(Q_1) - K(Q_0) = f(x_1) - f(x_0) \geq \lambda^*[Q_1 - Q_0]$. Next let Q_1, Q_2 be given and let $Q_\rho \equiv \rho Q_1 + (1-\rho)Q_2$. Then

$$K(Q_1) - K(Q_\rho) \geq \lambda^*[Q_1 - Q_\rho], \quad \text{and} \quad K(Q_2) - K(Q_\rho) \geq \lambda^*[Q_2 - Q_\rho].$$

Multiplying these expressions by ρ and by $1 - \rho$ respectively, and adding them together, one obtains

$$\rho K(Q_1) + (1-\rho)K(Q_2) - K(Q_\rho) \geq \lambda^*\left[\rho(Q_1 - Q_\rho) + (1-\rho)(Q_2 - Q_\rho)\right]$$

$$= \lambda^*[\rho Q_1 + (1-\rho)Q_2 - Q_\rho] = 0$$

so that K is convex in Q. Turning to differentiability, let $Q_1 = Q_0$ except that $Q_t^1 = (1+\epsilon)Q_t$ where $\epsilon > 0$. Let s_1 be optimal when $Q = Q_1$. Then $f(s_1) \leq f(\tilde{s}_1)$, where $\tilde{s}_1 = s_0$ except that $\hat{q}^1(x, t) = (1+\epsilon)$

$\hat{q}^0(x, t)$. Then $\tilde{s}_1$ meets all the constraints, and

$$f(s_1) - f(s_0) \leq f(\tilde{s}_1) - f(s_0) \tag{A.1}$$

$$= \beta^t \sum_{\tau=0}^{t} y_\tau \int \{c'[\hat{q}^0(x,t)]x\epsilon\hat{q}^0(x,t)\} \hat{\psi}_{\gamma^0(\tau)}(x \mid t, \tau)\, dx + o(\epsilon)$$

$$= Q_t^0 \lambda_t^* \epsilon + o(\epsilon)$$

where the last equality follows from equation (5.10) and the definition of $G_t(s_0, Q)$. But $K(Q_1) - K(Q_0) \geq \lambda^*[Q_1 - Q_0] = \lambda_t^* \epsilon Q_t^0$. Together with equation (A.1) this implies $Q_t^0 \lambda_t^* \epsilon + o(\epsilon) \geq K(Q_1) - K(Q_0) = \lambda_t^* \epsilon Q_t^0$. Dividing by ϵ and taking the limit as $\epsilon \to 0$ establishes that the right-derivative $\partial K / \partial Q_t$ exists and is equal to λ_t^*. A parallel argument taking $\epsilon < 0$ establishes the same property for the left-derivative. The proof is complete.

REFERENCES

[1] BERGE, C.: *Topological Spaces*. New York: MacMillan, 1963.

[2] BLACKWELL, D.: "Discounted Dynamic Programming," *Annals of Mathematical Statistics*, 36(1965), 226–235.

[3] CHERNOFF, H.: "Stochastic Control," *Sankhyä*, Ser. A, 43(1968), 111–142.

[4] DEMSETZ, H.: "Industry Structure, Market Rivalry and Public Policy," *Journal of Law and Economics*, 16(1973), 1–9.

[5] DUNFORD, N., AND J. T. SCHWARTZ: *Linear Operators*, Part I. New York: Interscience Publishers, 1958.

[6] DU REITZ, G.: "New Firm Entry in Swedish Manufacturing Industries during the Post-War Period," Doctoral Dissertation, Stockholm, 1975.

[7] FELLER, W.: *Introduction to Probability*, II (2nd ed.). New York: John Wiley and Sons, 1971.

[8] HART, P. E., AND S. J. PRAIS: "The Analysis of Business Concentration: A Statistical Approach," *Journal of the Royal Statistical Society*, 119, pt. 2 (1956), 150–191.

[9] HILDEBRANDT, T. H.: "On Bounded Functional Operations," *Transactions of the American Mathematical Society*, 64(1948), 868–875.

[10] ———: "On the Size Distribution of Business Firms," *Bell Journal of Economics*, 9(1978), 508–523.

[11] LUCAS, R. E., AND E. C. PRESCOTT: "Investment Under Uncertainty," *Econometrica*, 39(1971), 659–681.

[12] LUENBERGER, D. G.: *Optimization by Vector Space Methods*. New York: John Wiley and Sons, 1969.

[13] MANSFIELD, E.: "Entry, Gibrat's Law, Innovation, and the Growth of Firms," *American Economic Review*, 52(1962), 1023–1051.

[14] PASHIGIAN, P., AND S. HYMER: "Firm Size and Rate of Growth," *Journal of Political Economy*, 52(1962), 556–569.

[15] SAMUELS, J. M., AND D. J. SMYTH: "Profits, Variability of Profits and Firm Size," *Economica*, 35(1968), 127–139.

[16] SIMON, H. E., AND C. P. BONINI: "The Size Distribution of Business Firms," *American Economic Review*, 48(1958), 607–617.

[17] STIGLER, G. J.: *Capital and Rates of Return in Manufacturing Industries*. Princeton: Princeton University Press (for NBER), 1963.

[18] WEISS, L.: "Quantitative Studies of Industrial Organization," in *Frontiers of Quantitative Economics*, ed. by M. D. Intrilligator. Amsterdam: North-Holland, 1971.

[19] YOSIDA, K., AND E. HEWITT: "Finitely Additive Measures," *Transactions of the American Mathematical Society*, 72(1952), 46–66.

[20] ZELLNER, A.: *An Introduction to Bayesian Inference in Econometrics*. New York: John Wiley and Sons, 1971.

[9]

Optimum Contracts for Research Personnel, Research Employment, and the Establishment of "Rival" Enterprises

Ariél Pakes, *Hebrew University of Jerusalem and the Falk Institute*

Shmuel Nitzan, *Hebrew University of Jerusalem*

This paper considers the problem of hiring scientists for research and development projects when one takes explicit account of the fact that the scientist may be able to use the information acquired during the project in a rival enterprise. Management's problem is to determine an optimum labor policy for its project. The policy consists of an employment decision and a labor contract. Given optimum behavior, it is straightforward to analyze the effect of the potential for mobility of scientific personnel on project profitability and on research employment. We also formalize conditions under which one would expect to observe a scientist leaving his employer to set up or join a rival.

> However, no amount of legal protection can make a thoroughly appropriable commodity of something so intangible as information. The very use of information in any productive way is bound to reveal it, at least in

We are indebted to Morris Teubal and two referees for their insights into the problems discussed here, and to the participants in seminars at the Hebrew University, the National Bureau of Economic Research, the University of Toronto, and the University of Chicago for their comments. Ariél Pakes acknowledges the financial assistance of BSF grant 1881/79 and NSF grant PRA79-13740 and SOC78-04279.

[*Journal of Labor Economics*, 1983, vol. 1, no. 4]

0734-306X/83/0104-0005$01.50

> part. Mobility of personnel among firms provides a way of spreading information. [Arrow (1962), p. 615]

At least since the work of Schumpeter (1942), economists have stressed the fact that the private rate of return to research resources (and hence research employment) is determined, in part, by the degree to which a firm can maintain proprietary rights (monopoly power) over the information produced in its research laboratories. Given the rapid increase in industrial research expenditure since World War II and the increasingly convincing empirical evidence on their impact, it is a bit disconcerting that so little work has been done on how firms facing this appropriability problem ought to behave. That is, how should a firm act in order to protect its innovations? How does the action it takes affect research demand and the structure of science-based industry?

One frequently cited mechanism for the spread of the information produced in industrial research laboratories is the mobility of scientific personnel. It is not a difficult mechanism to explain. Research projects often require a good idea and substantial investments before they produce an output which can be embodied in a marketable good or service. The output of the project is a new piece of information which has no (or a very low) cost of reproduction; and any scientist working on the project is likely to be at least as aware of this information as the management. The scientist can therefore wait until the information has been developed and then decide to leave and use it in a rival enterprise. Indeed (as will be discussed below), there are many well-documented cases of scientists leaving their employers to set up or join a rival.

The purpose of this paper is to investigate the effects of the mobility of scientific personnel on project profitability, research employment, and the structure of science-based industry. A simple framework capable of analyzing these issues is provided in Section I. We begin with an entrepreneur who has an idea which could lead to a successful innovation. There is no (or a very imperfect) market for the idea as such, and in order to profit from it, the entrepreneur finds it in his interest to hire a scientist to help develop the idea into a marketable output. The entire project takes two periods. At the end of the first period the random outcome of the research stage of the project is known with certainty. The second period is used to market the project's output.

The entrepreneur's problem is to determine an optimal labor policy for the project that takes explicit account of the fact that at the end of the first period the scientist will have the option of leaving to join a rival. The labor policy consists of an employment decision (the quality of the scientist to hire) and a labor contract. This problem is a particular example of the principal's problem in the theory of agency (for a general description of the principal's problem, see Ross [1973]). There is a growing literature in labor economics on the characteristics of optimal labor contracts in an assortment of environments (related references are provided in the text). To focus on those aspects of our problem that are a direct consequence of the mobility of scientific personnel, we assume risk neutrality, scientists of known qualities, an absence of other sources of uncertainty (other than that involved in the research project), and that the effort the scientist devotes to the project is independent of the terms of the contract. The contract must, however, control, to the extent profitable, the mobility of the scientist in the second period (a problem related to Becker and Stigler's [1974] discussion of appropriate compensation schemes for law enforcers when malfeasance is possible; see also Harris and Raviv [1978] and Lazear[1979]). We develop the conditions that an optimal contract must satisfy (and consider the type of contractual agreements that are likely to satisfy them); the implications of the potential for mobility of scientific personnel both on project profitability and on research employment; and the conditions under which optimal behavior by all agents can result in the scientist's actually leaving in the second period to set up (or join) a rival.

Sections II and III differ in the assumptions made on the economics of the research and marketing process. Due to this difference the model of Section III can generate situations where optimal behavior by all agents does lead to the establishment of a rival, while the model of Section II cannot. The Appendix provides formal proofs for arguments made in the text, and the paper concludes with a short summary.

I. The Basic Framework

Consider an entrepreneur with an idea which could lead to a successful innovation. There is no (or a very imperfect) market for the idea as such, and to profit from it fully the entrepreneur must develop the idea into a marketable output. This in turn requires the commitment of resources to a research project (and, if the project is successful, to production and marketing). In particular, because there are costs of acquiring skills and a premium on time (lest someone else develop an innovation which renders the entrepreneur's idea obsolete), it is in the entrepreneur's interest to hire a scientist. Initially assume that the only costly resource employed in the research project is this scientist. The outcome of the research project will be known after a single period. The period after the research is completed (the second period) will be used to market its output. At the end of the research project the scientist will possess information which enables him, if he so desires, to market the new information himself. As a result, the entrepreneur realizes that no matter what the project's technical outcome, at least two different market situations could arise. If the scientist does not leave the firm to set up (or join) a rival enterprise, the entrepreneur's firm may monopolize the information that resulted from the project. Alternatively, if the scientist sets up a rival, there will be at least two firms with the information. In this section we assume that these are the only two market situations that can arise. That is, either a single firm or two rivals will market the output from the project.

The entrepreneur's objective is to maximize the expected profits from the project. He can choose the skill level, say s, of the scientist he will employ. Given s, a series of exogenous events (associated with the realization of a particular state of nature) will determine the project's outcome. Let $\epsilon \in R$ (R is the set of real numbers) be a random variable indicating the possible states of nature and denote by $G(\epsilon)$ the distribution function which embodies the entrepreneur's prior beliefs about the likelihood of the alternative states of nature. For any $\epsilon \in R$ let $f(\epsilon, s)$ be the net revenues to be generated by the project, when the scientist chosen has skill level s and if he does not set up a rival after the project is completed. We assume that $f(\epsilon, s)$ is finite and nonnegative for all ϵ and s.[1] A scientist who does not set up a rival has two alternatives. He can either stay with the firm or return to the labor market to engage in some other activity. To simplify the analysis, we assume that if he should decide to stay with the firm during the second period the entrepreneur can always employ him in some other activity in which his marginal productivity equals his alternative wage.[2] If, on the other hand, the scientist does set up (or join) a rival, let the entrepreneur's net revenues be $f(\epsilon, s) - l(\epsilon, s)$, where $l(\epsilon, s)$, the loss due to the establishment of the rival, is assumed to be nonnegative and finite for all ϵ and s.

In order to hire a scientist, management specifies a labor contract, which is assumed to consist of a flat wage, $w_0(s)$, to be paid before the realization of ϵ and a bonus payment, $w_1(\epsilon, s)$, which is paid in the second period and therefore need only be paid if the scientist decides to stay with the firm. Thus, for any s, a labor contract is a couple consisting of a number, $w_0(s)$, and a random function, $w_1(\epsilon, s)$. The choice of the triple $\{s, w_0(s), w_1(\epsilon, s)\}$ is referred to as a labor policy.

Consider now the behavior of scientists of different skill levels. The wages they can receive in alternative places of employment are set by the labor market and are an increasing function of their skill. We normalize s so that it equals these alternative wages and assume $s \in [0, \bar{s}] = U$. It

[1] That is, if the net revenues to be generated from marketing the output were expected to be negative, the firm would simply not market that output. Note that "net revenues" is used to denote the sales minus the costs incurred during the marketing period. "Profits" refers to these net revenues minus the costs of the research project.

[2] A similar assumption (constant marginal productivity over time) is used in most of the literature on labor turnover and quits; see the discussion in Nickell (1976) or in Salop and Salop (1976). It simplifies the analysis considerably by allowing one to abstract from the effect of possible changes in the future demand for employees on current employment decisions.

is assumed that scientists choose their place of employment by maximizing a utility function which is linear in the money income they receive, that is, scientists are expected-payment maximizers (risk aversion is discussed briefly below). A scientist who becomes interested in the project is on inquiry informed of the terms of the contract. These terms include the probability of distribution of the bonus—that is, of $w_1(\epsilon, s)$ or, as would seem more likely, that of $f(\epsilon, s)$ and its relationship to $w_1(\epsilon, s)$ (see the next section);[3] they also include the stipulation that the bonus will be paid only if the scientist decides to stay with the firm during the second period. The scientist is aware that by joining the project he will acquire information of potential value, and he realizes that it may not be in his interest to stay for both periods. If he does in fact join, he will face three alternatives at the end of the first period. He can stay and obtain the bonus, $w_1(\epsilon, s)$. Alternatively, he can try to capitalize on the information he has acquired by setting up or joining a rival. To evaluate this possibility, the scientist uses his knowledge of science-based industry and the information he has been given on the project to compute the possible payoffs from using the information he has acquired on the project in a rival enterprise, say $p(\epsilon, s)$. Assume that if the scientist does join (or set up) a rival he receives $p(\epsilon, s)$ plus the market value of his labor during the second period s. (All that is required is for the amount the scientist earns in a rival to be less than or equal to $p(\epsilon, s) + s$.) Finally, deciding that neither of these alternatives is particularly attractive, the scientist may return to the labor market to engage in another activity and earn the going rate on his skill level, s. Being an expected-payment maximizer, he will accept the bonus and stay with the firm if $\epsilon \in A = \{\epsilon | w_1(\epsilon, s) \geq \max [p(\epsilon, s) + s, s]\}$. He will quit to set up a rival if $\epsilon \in Q = \{\epsilon | p(\epsilon, s) + s > \max [w_1(\epsilon, s), s]\}$;

[3] Thus, if the distribution of $w_1(\epsilon, s)$ is approximately normal, the scientist is told so and provided with its mean and variance. In a broader framework one would also ask whether the entrepreneur could, by providing the scientist with the wrong distribution function, increase expected profits. It can be shown that, provided the scientist does not have sufficient information on the distribution of $f(\epsilon, s)$ from other sources (say, from the stock market or from a community of people involved in science-based industry who have considered similar projects in the past), and provided nothing else in the problem changes as a result of the entrepreneur's actions, the entrepreneur could increase his expected profits by persuading the scientist of an overly optimistic distribution function (see Sec. III). In this case, however, the other aspects of the problem are likely to change. In particular, the entrepreneur is likely to want to hire additional scientists later, and they are not likely to believe in an entrepreneur who has cheated before. Moreover, the scientist may realize that he has been cheated early on in the project and leave before its completion. Similar issues arise in much of the literature on labor contracts (see, e.g., Lazear 1979) and in this section we assume, as is customary, that the considerations outlined above induce the entrepreneur to reveal the correct information. A deeper analysis of this issue may, however, be warranted.

and if $\epsilon \in R - A - Q$ he will return to the labor market.[4] Clearly, then, a scientist of skill level s will accept the offer to join the project only if $w_0(s) + {}_A\int w_1(\epsilon, s)dG + {}_Q\int[p(\epsilon, s) + s]dG + {}_{R-A-Q}\int sdG \geq 2s$, that is, if the expected payments from joining the project are at least equal to the payments he would receive over the two periods in alternative employment (for simplicity we shall ignore discount factors).

The economic structure of the problem set out in this section suggests one more constraint (Sec. III differs in this respect). If the scientist's evaluation of $p(\epsilon, s)$ and the entrepreneur's beliefs on $l(\epsilon, s)$ are at all close to the gains the scientist would actually make from setting up a rival and the entrepreneur's consequent losses, respectively, we would expect $p(\epsilon, s) - l(\epsilon, s) \leq 0$, for all $\epsilon \in R$. The difference between $p(\epsilon, s)$ and $l(\epsilon, s)$ is likely to be a result of two factors: the additional costs incurred if a rival is set up, and the fact that postproject competitive behavior is likely to divert to consumers some of the gains that would otherwise accrue to the monopolist. Thus, the sum of the gains that would accrue to the two agents in a situation involving rivalry could never exceed those that would accrue in the monopolistic situation. This would be true even if neither of them incurs any additional costs as a result of the rival appearing. The rival, however, is likely to incur some startup costs; and in situations involving rivalry, both agents may incur additional costs due to product differentiation and advertising. Although, as we show in the next section, the condition that $p(\epsilon, s) - l(\epsilon, s) \leq 0$ is sufficient for our purposes, we begin by restricting the problem further and assuming that $l(\epsilon, s) = \alpha f(\epsilon, s)$ for $0 \leq \alpha \leq 1$, and $p(\epsilon, s) = \alpha f(\epsilon, s) - c(s)$. Here, if the scientist sets up a rival, the entrepreneur loses a fraction α of the net revenues that he would otherwise have obtained. The scientist, on setting up or joining a rival, gains all these net revenues but must incur setup costs, $c(s) \geq 0$.

We can now formalize the entrepreneur's expected-profit-maximization problem.

Problem A

$$\max_{\{s \in U, w_0(s), w_1(\epsilon, s)\}} \Pi = {}_R\int f(\epsilon, s)\, dG - w_0(s) - {}_A\int[w_1(\epsilon, s) - s]dG - {}_Q\int \alpha f(\epsilon, s)dG\,,$$

where $A = \{\epsilon | w_1(\epsilon, s) \geq \max[\alpha f(\epsilon, s) - c(s) + s, s]\}$ and $Q = \{\epsilon | \alpha f(\epsilon, s) -$

[4] Let B and C be two sets. To simplify the notation we denote the union of B and C when B and C are disjoint by $B + C$ and the set of all elements in B which are not contained in C by $B - C$.

$c(s) + s > \max [w_1(\epsilon, s), s]\}$, subject to

$$2s \leq w_0(s) + {}_A\int w_1(\epsilon, s)dG + {}_Q\int[\alpha f(\epsilon, s) - c(s) + s]dG + {}_{R-A-Q}\int s dG \,. \quad (1)$$

We shall impose the usual regularity conditions on $f(\epsilon, s)$. That is, for each ϵ, $f(\epsilon, s)$ will be assumed to be increasing and concave in s $[f'_s(\epsilon, s) > 0, f''_{ss}(\epsilon, s) < 0]$, while $f''_{ss}(\epsilon, s)$ is assumed to be bounded. (The regularity conditions are required for the existence of a unique optimum s, but not for our derivation of the optimum contract.)

II. Optimum Labor Policies

> Semiconductor firms attract and reward valuable employees not so much by offering high salaries as by giving stock options. [Braun and MacDonald (1978), p. 132]

Proposition 1 and the explanation which follows it provide the solution to problem A. (The App. gathers the formal proofs for the propositions and corollaries presented in the text.)

PROPOSITION 1: Any labor policy consisting of a contract, $\{w_0(s), w_1(\epsilon, s)\}$, satisfying $w_1(\epsilon, s) \geq \alpha f(\epsilon, s) - c(s) + s$, for all $\epsilon \in A^* = \{\epsilon | \alpha f(\epsilon, s) - c(s) \geq 0\}$, and $w_0(s) = s - {}_A\int[w_1(\epsilon, s) - s]dG$; and the unique skill level s^* which satisfies ${}_R\int f'_s(\epsilon, s^*)dG = 1$ (if no such $s^* \in U$ exists, then the optimum s is at one of the boundaries, 0 or $\bar{s}$); is a solution for problem A.

The knowledge that initiates the project (and that the scientist acquires by virtue of joining) generates two random variables: the value of the knowledge to the entrepreneur, and its value to the scientist. The realizations of these random variables are determined both by the outcome of the research (i.e., by ϵ) and by the scientist's action in the second period. For any ϵ in which the sum of the two values is at least as large if the project's output is utilized in a single firm—that is, for which $p(\epsilon, s) < l(\epsilon, s)$—the entrepreneur can always pay the scientist as much as the scientist would earn in a rival and still earn more than the entrepreneur would earn if the rival were set up. (These are outcomes in which the knowledge becomes capital with characteristics similar to those of firm-specific human capital as defined by Becker [1975]).[5] Since problem A assumes that such outcomes occur with probability 1, contracts which are optimal for it must always deter the scientist from setting up a rival.

[5] The difference is that we are dealing with a type of capital that is not embodied in the scientist (the scientist need not actually use it in the second period for either agent to earn returns).

If we note that A^* is the set of all ϵ for which the scientist would actually set up a rival were he to leave (instead of engage in some other activity), it follows that an optimal contract must have $w_1(\epsilon, s) \geqslant \alpha f(\epsilon, s) - c(s) + s$ for $\epsilon \in A^*$ (this condition ensures that $Q^* = \emptyset$, the null set). Given this condition and any s, the entrepreneur is interested in minimizing the expected payments to the scientist subject to the constraint that these payments must be sufficient to draw a scientist of quality s to the job (condition [1]). Substituting $Q^* = \emptyset$ into (1), this argument sets the initial wage at $w_0(s) = s - {}_A\int[w_1(\epsilon, s) - s]dG$. The scientist's initial wage is set lower than his alternative wage by an amount just equal to the premium over the alternative wage the scientist can expect to earn in the second period. Optimal contracts for problem A are therefore similar to the bond-pension schemes discussed in several places in the literature on labor contracts (see the references in n. 8); with the difference that a "pension" $[w_1(\epsilon, s)]$ which satisfies proposition 1 is a random variable whose realization depends on the fortunes of the firm. Finally, substituting $Q^* = \emptyset$ and the optimal $w_0(s)$ into the profit function, it is clear that the only problem left to an optimizing entrepreneur is to choose an $s \in U$ to maximize ${}_R\int f(\epsilon, s)dG - s$. Clearly s^*, as defined in the proposition, solves this problem. Corollary 1.1 is an immediate consequence of proposition 1.

COROLLARY 1.1: The labor policy consisting of s^*, and the contract $w_1^*(\epsilon, s) = \max\,\{\alpha f(\epsilon, s) - c(s) + s, s\}$, for all $\epsilon \in R$, and $w_0^*(s) = s - {}_R\int[w_1^*(\epsilon, s) - s]dG = s - {}_{A^*}\int[\alpha f(\epsilon, s) - c(s)]dG$, is optimal for problem A.

Of the contracts satisfying proposition 1, we emphasize $\{w_0^*(s), w_1^*(\epsilon, s)\}$ for two reasons; first, it is exceedingly easy to implement. The hiring agreement need only specify that during the second period the scientist will be offered, in addition to his alternative wage, a stock option. The stock option specifies that if the scientist decides to stay with the firm in the second period he will be allowed to purchase, at a cost of $c(s)$, a fraction α of the firm's equity.[6] Of course the offer of a profit-sharing scheme of the form $w_1(\epsilon, s) = \beta f(\epsilon, s) + s$, with $\beta \geqslant \max_{\epsilon\in R}\,\{[\alpha f(\epsilon, s) - c(s)]/f(\epsilon, s)\}$ and with $w_0(s)$ as defined in the corollary, will do equally well. In fact, a stock option (or profit-sharing) scheme of this form will be optimal for arbitrary gain, $p(\epsilon, s)$, and loss, $l(\epsilon, s)$, functions provided that $\beta \geqslant \max_{\epsilon\in R}\,\{p(\epsilon, s)/f(\epsilon, s)\}$.[7] The second, and perhaps more important,

[6] This, of course, assumes that the project is financed by the issue of equity and that the stock market provides a reasonable approximation of the value of the project. Note that if $\alpha f(\epsilon, s) - c(s) < 0$, it does not matter whether the scientist stays with the firm or engages in some other activity.

[7] Note that this assumes either that the scientist cannot sell the information without leaving the firm or that if this did happen the entrepreneur, whose profits would fall, could stop the payment of the bonus. If neither of these conditions is satisfied, then the entrepreneur ought to choose a contract (from those satisfying proposition 1) which ensures that it is not in the scientist's interest to both stay

reason for emphasizing stock option agreements is that they tie the second-period payment to the scientist directly to the fortunes of the firm—that is, the realization of ϵ—and to those aspects of the structure of the industry that determine the gains and losses that would result from setting up a rival—here, α and $c(s)$. As will be shown below, when we complicate problem A the set of optimal contracts narrows (often to a single one), but an optimal contract must always have these characteristics. These are also the characteristics that distinguish contracts designed to help solve the appropriability problem in a firm's research laboratory from contracts designed to help solve other agency problems that arise in the context of employer-employee relationships (which explains the quotation at the beginning of this section).[8]

Proposition 1 implies that the entrepreneur can maximize expected profits by following a simple two-stage decision procedure. First, choose s to maximize expected profits, ignoring the fact that the scientist may be able to use the information he acquires on the project to set up a rival; then choose a labor contract that ensures both that a rival never appears and that the expected costs of employing the scientist just equal his alternative wage. There will always exist a particularly simple labor contract consisting of a flat rate and a profit-sharing scheme which satisfies these conditions.

These conclusions have two implications of particular interest. First, provided all agents act optimally, input decisions (s) and the expected profitability of research projects are independent of the institutional, market, and technological conditions which determine the gains and the losses agents would incur if a rival were set up. Second, again provided all agents act optimally, we should not observe a scientist breaking away from an established firm to join or set up a rival. These results would not be altered if one allowed for the fact that many different factors of production are typically used in research projects (in this case, all input quantities should be chosen to maximize expected profits, ignoring the possibility that a rival will appear).[9]

---

with the firm and sell the information in the second period (the contract would then have to provide the scientist with all the net revenues in the second period apart from a sum equal to the cost of selling the information).

[8] The latter include monitoring on-the-job effort, the allocation of uncertainty, and the unraveling of unobserved characteristics (see, among others, Stiglitz [1975], Ackerloff [1976], Nickell [1976], Salop and Salop [1976], Mortensen [1978], Lazear [1979], and Lazear and Rosen [1981]). Though allowing for these other issues complicates the problem significantly (and this probably accounts for the at least implicit complexity of the contractual agreements that we actually observe), none of them (to our knowledge) leads to contracts with the characteristics noted above.

[9] In particular, the entrepreneur will never be induced to substitute any other factor for skill beyond the point where the ratio of the expected marginal products of all factors equals the ratio of their costs, regardless, of course, of whether an

It should be noted, however, that if there are bounds on contracts or if agents have nonlinear utility functions, then the first of these implications is modified but the second need not be. It is easiest to illustrate this point by adding a nonnegativity constraint to problem A (see also n. 10 below). Note that if the project has some probability of generating large net revenues, or if α is large while $c(s)$ is small, then to implement an optimal contract for problem A, $w_0(s)$ may have to be very small or negative, that is, $w_0(s) \leqslant s - {}_{A^*}\int\{\alpha f(\epsilon, s) - c(s)\}dG$ (from proposition 1). If one adds the constraint that, due to convention or legal restrictions (minimum wage laws), $w_0(s)$ has to be above some minimum value (which we take for simplicity to be zero), then the potential mobility of scientific personnel can affect both project profitability and input decisions; but, provided all agents act optimally, the scientist will still never actually set up or join a rival. To see this, note that the addition of the nonnegativity constraint, $w_0(s) \geqslant 0$, does not alter the fact that for any ϵ the entrepreneur can always do better by deterring the scientist from setting up a rival; and this leads to the conclusion that for a contract to be optimal it must never induce the scientist to do so.[10] The nonnegativity constraint will,

---

increase in skill decreases the cost of setting up a rival. The problem does get more complicated in situations where it may be profitable to hire more than one scientist who will (at the end of the first period) possess information that enables them to set up a rival. The reason is that the gains any one of the scientists would make from setting up a rival may, in this case, depend on the behavior of the others. Provided the total gains from setting up a rival are always less than the entrepreneur's losses and that constraints analogous to (1) continue to hold, the solution does not, however, differ substantively. The entrepreneur ought still to ensure that a rival never appears by paying each scientist a high enough bonus (say, higher than the maximum gain he could make as a result of a rival appearing) to lower the initial wage as much as possible and thus to ensure that he (the entrepreneur) garners all the expected net revenues from the project other than the scientists' alternative cost. Of course, for many (usually all but one or two) of the individuals working on a research project, these considerations are irrelevant. Most personnel do not work on tasks enabling them to gather information that would be of much value to a rival.

[10] The argument that, given ϵ, the entrepreneur can always do better by deterring the scientist from setting up a rival does not depend on the linearity of either agent's utility function. Allowing for nonlinear utility functions, however, complicates the analysis significantly. The source of the complication is that with nonlinear utility functions there will be cases in which an optimal contract from the set of contracts considered in this paper would induce the establishment of a rival even though, because of the argument given above, there are contracts not in this set which are preferred by both agents and which do not induce the appearance of a rival. Broadening the set of feasible contracts complicates the technical formulation of the problem significantly without changing the basic thrust of the arguments presented here (which is the reason we have chosen to illustrate the points in this paragraph with the aid of the nonnegativity constraint). It should also be noted that the discussion in this paragraph assumes that the scientist cannot both receive the bonus in the second period and sell the information to a third party (see n. 7).

however, tend to make contracts with higher initial wages and lower bonus payments preferable. The lowest bonus function that still satisfies the condition that it will never induce the appearance of a rival is given by $w_1^*(\epsilon, s)$, as defined by corollary 1.1. Given $w_1^*(\epsilon, s)$, the lowest initial wage which satisfies both the nonnegativity and feasibility conditions is $w_0^+(s) = \max\,\{0, w_0^*(s)\}$; and, as a result, the contract $\{w_0^+(s), w_1^*(\epsilon, s)\}$ will always constitute an optimal solution. Clearly in this (second-best) situation the appropriability problem in a firm's research laboratory can only decrease project profitability; but, and this is the interesting point, it may actually *increase* research employment. Consider, for example, the case where the nonnegativity constraint is binding at $\{s^*, w_0^+(s^*), w_1^*(\epsilon, s^*)\}$, but the feasibility constraint (1) is not—that is, $w_0^+(s^*) = 0 > w_0^*(s^*)$; if this were not the case the policy of corollary 1.1 could still be implemented, and would be optimal—and assume $c(s) = c$ (a constant). Since the feasibility constraint is not binding a marginal increase in skill from s^* results in an increase in expected costs just equal to the increase in expected bonus required to ensure that a rival will not be established, and this is just a fraction, α, of the increase in expected benefits. The entrepreneur will, therefore, be induced to increase skill beyond s^*; in fact, the same argument implies that the entrepreneur will increase skill until the expected benefits to the scientist over the two periods just equal his alternative wage. After that point, say s^+, the increase in expected costs from further increases in skill is unity, while the increase in expected benefit is ${}_R\int f_s'(\epsilon, s^+)dG$, which, since $s^+ > s^*$ and $f(\cdot, \cdot)$ is concave in s, is less than unity.[11] Finally, we stress that the conditions discussed in this paragraph ought not, in themselves, lead to scientists actually breaking away from their employer to set up rivals. Thus to find situations that may induce scientists to do so, we examine the economics of the research and marketing processes more closely.

III. Quits and the Establishment of "Rivals"

> Mobility within the semiconductor industry has been aided by the very tolerant attitude of most firms towards movement of personnel. [Braun and MacDonald (1978), p. 135]

Though it may well be true that in most science-based industries one

[11] If no such $s \in U$ exists, the entrepreneur will choose the maximum value of s, i.e., $\bar{s}$. Formal proofs of the statements in this paragraph are provided in proposition 1A and corollary 1A of the Appendix. Barzel (1968) and Dasgupta and Stiglitz (1980) have also argued that the public-good characteristics of knowledge can lead to overinvestment in research, but in their models this result occurs because there can be many different entrepreneurs who search, simultaneously, for the innovations emanating from a given idea. The next section considers the implications of the appropriability problem in a firm's research laboratories when this can happen.

seldom observes a scientist breaking away from his employer to set up a rival, there is at least one industry, the semiconductor industry, where this has happened quite often.[12] Indeed, many studies of the structure of the semiconductor industry begin by constructing a "family tree" of firms that were offshoots from each other (see, for example, Freeman [1974], pp. 147–49). As the discussion of the last section indicates (and corollary 2.2 will formalize), to accommodate this phenomenon in our framework there must be possible outcomes in which $p(\epsilon, s) > l(\epsilon, s)$; that is, outcomes in which the scientist believes that he will gain more by setting up a rival than what the entrepreneur believes he (the entrepreneur) will lose as a result of the rival's appearance. There are, therefore, at least two questions of interest. First, when would we expect such outcomes to occur? Second, how do they affect project profitability and research employment? In this section we discuss two sets of conditions which have the potential of actually inducing the scientist to set up a rival. In both, the actions of all agents are a result of optimum behavior based on correct information.[13] Moreover, these conditions are not unlike those which have developed in the semiconductor industry. Our discussion will focus on their implications for project profitability and on optimum labor policies, leaving several interesting questions concerning the environments which generate these conditions for subsequent work. In particular, it will be shown that, provided the entrepreneur uses an optimum labor contract, he need not be troubled by the possibility that scientific personnel may move. In fact, all else equal, the entrepreneur should prefer situations in which it is relatively easy for the scientist to do so.

Third-Party Entrants

Sections I and II have assumed that the entrepreneur and the scientist whom he employs jointly retain perfect monopoly power over the information produced during the research project. There are, however, several ways in which third parties might gain access to this information. The output of the project may spread as a result of inspection of either the product in which it is embodied or the patents resulting from the project (indeed, patents would not exist if there were no possibility of

[12] One would need a very detailed history of an industry to determine that it does not exhibit this phenomenon. On the other hand, many science-based industries seem to operate with a fairly stable number of firms; see Temin (1979) on the drug industry since World War II, Phillips (1974) on the aircraft industry, and Freeman's (1974) description of the synthetic materials industry after World War I.

[13] Of course, the scientist may wrongly believe that $p(\epsilon, s) > l(\epsilon, s)$; or the entrepreneur may err in his evaluation of $l(\epsilon, s)$. Provided one precisely specifies the entrepreneur's information on $p(\epsilon, s)$ in the first period, this case can be analyzed in a model similar to those developed below, and it has the same qualitative implications on project profitability and research employment.

third-party entrants), or a third party may complete a similar research project after the leading firm. If there is a possibility of third-party entrants the entrepreneur's optimum strategy will take into account his beliefs on how third parties are likely to act (for a formal model of entrepreneurial behavior in a related context, see Gaskins [1971]). This optimum strategy will include a labor policy. Clearly, however, if there are potential third-party entrants, the logic that led us to the optimum labor policies of the preceding section is no longer appropriate. There the argument was that since the benefits that would accrue to the scientist from setting up a rival were smaller than the consequent losses of the firm, the entrepreneur would always devise a contract deterring the scientist from setting up the rival. If other agents have access to the information, however, then the scientist, by setting up a rival, may also be able to break into profits that would otherwise accrue to third parties. In this case, both the scientist and the entrepreneur may be able to gain, if the scientist can set up a rival.

To analyze this situation we must add some notation. Again $f(\epsilon, s)$ represents the net revenues that would accrue to the entrepreneur if the scientist did not set up a rival (given the expected behavior of third parties). If the scientist does set up a rival, the payoffs to entrepreneur, scientist, and third parties all change. For simplicity, we maintain the assumption that if the scientist does set up a rival, the entrepreneur loses a fraction α of the net revenues—to allow for the possibility that his revenues actually increase we permit $\alpha < 0$ in this section, though we continue to refer to $\alpha f(\epsilon, s)$ as the entrepreneur's losses; and that the costs of setting up a rival are $c(s)$. Depending on the realization of ϵ, on technological and market conditions, and on the oligopolistic strategies pursued by the agents, part of the losses incurred by the firm may now go to agents other than the entrepreneur and the scientist (to the third parties or to consumers); alternatively, by setting up a rival, the scientist may be able to break into the net revenues that would otherwise accrue to the third parties. Accordingly, in this section we assume that the gains that would accrue to the scientist if he set up a rival are $\alpha f(\epsilon, s) - c(s) + s + r(\epsilon, s)$, where the only restriction placed on $r(\epsilon, s)$ is that it be finite for all $\epsilon \in R$ and $s \in U$.

At the end of the first period the scientist will be induced to set up a rival if this is the best of the three alternatives available to him, that is, if $\alpha f(\epsilon, s) - c(s) + s + r(\epsilon, s) > w_1(\epsilon, s)$ and $\alpha f(\epsilon, s) - c(s) + s + r(\epsilon, s) > s$. He will stay with the firm if the first inequality is reversed and $w_1(\epsilon, s) \geqslant s$, while if neither of these two conditions is met he will return to the labor market and engage in some other activity.

Given the scientist's decision rules (and assuming contracts are unbounded), the entrepreneur's expected profit-maximization problem is given by problem B.

Problem B

$$\max_{\{s \in U, w_0(s), w_1(\epsilon,s)\}} \Pi = {}_R\int f(\epsilon, s)dG - w_0(s)$$

$$- {}_{A_r}\int [w_1(\epsilon, s) - s]dG - {}_{Q_r}\int \alpha f(\epsilon, s)dG \ ,$$

where $A_r = \{\epsilon | w_1(\epsilon, s) \geqslant \max [\alpha f(\epsilon, s) - c(s) + s + r(\epsilon, s), s]\}$ and $Q_r = \{\epsilon | \alpha f(\epsilon, s) - c(s) + s + r(\epsilon, s) > \max [w_1(\epsilon, s), s]\}$, subject to

$$2s \leqslant w_0(s) + {}_{A_r}\int w_1(\epsilon, s)dG + {}_{Q_r}\int [\alpha f(\epsilon,s) - c(s) + s + r(\epsilon, s)]dG + {}_{R-A_r-Q_r}\int s dG \ . \tag{2}$$

Proposition 2 provides the optimal contract for problem B.

PROPOSITION 2: Regardless of the choice of s, an optimum labor contract for problem B is $w_1^r(\epsilon, s) = \max \{\alpha f(\epsilon, s) + s, s\}$, $w_0^r(s) = s - {}_R\int \{\max [\alpha f(\epsilon, s), 0]\}dG - {}_{Q_r^*}\int [r(\epsilon, s) - c(s)]dG$; where $Q_r^* = \{\epsilon | r(\epsilon, s) - c(s) > 0$ and $\alpha f(\epsilon, s) - c(s) + r(\epsilon, s) > 0\}$.

The bonus scheme, $w_1^r(\epsilon, s)$, is implemented by the entrepreneur offering the scientist a profit-sharing agreement plus his alternative wage. The share is the fraction of the firm's net revenues the scientist would take away if the scientist were to set up a rival. The scientist only accepts the profit-sharing agreement if $\alpha f(\epsilon, s) > 0$. To see why this bonus scheme is optimal, note that, since the scientist can always return to the labor market to engage in some other activity, the scientist will not set up a rival unless $\alpha f(\epsilon, s) - c(s) + r(\epsilon, s) > 0$. Now consider those realizations of ϵ where this condition is satisfied. If $\alpha f(\epsilon, s) - c(s) + r(\epsilon, s) > \alpha f(\epsilon, s)$, the entrepreneur would have to pay the scientist more to induce him to stay than the entrepreneur would lose if the scientist set up a rival. For those realizations of ϵ, then, it will pay the entrepreneur to allow the employee to leave. The set Q_r^* of proposition 2 is defined as the set of all ϵ in which the last two inequalities are met, and the bonus scheme $w_1^r(\epsilon, s)$ will induce the scientist to set up a rival if and only if $\epsilon \in Q_r^*$. The initial wage, $w_0^r(\epsilon, s)$, is the minimum flat wage acceptable (satisfying [2]), given that the scientist will receive $\alpha f(\epsilon, s) - c(s) + s + r(\epsilon, s)$ if $\epsilon \in Q_r^*$ and $w_1^r(\epsilon, s)$ otherwise.

To derive the optimum skill level for problem B, we require additional assumptions on $T(s) = {}_{Q_r^*}\int [r(\epsilon, s) - c(s)]dG$, which is likely to increase in s, since one would expect $r_s'(\epsilon, s) \geqslant 0$ and $c_s'(s) \leqslant 0$. If in addition we assume that $T(s)$ is concave for all s, or at least not sufficiently convex to offset the concavity of ${}_R\int f(\epsilon, s)dG$, and that $T_s''(s)$ is bounded, we have the following corollary.

COROLLARY 2.1: If there exists an $s^r \in U$ which satisfies ${}_R\int f_s'(\epsilon, s^r)dG + {}_{Q_r^*(s^r)}\int [r_s'(\epsilon, s^r) - c_s'(s^r)]dG = 1$, then s^r is the optimum skill level for problem B. Otherwise the optimum s is at one of the boundaries, 0 or $\bar{s}$.

The sum of the entrepreneur's and scientist's gains from setting up a rival is $r(\epsilon, s) - c(s)$, and $Q_r^*(s)$ is the set of all ϵ for which this sum is positive and for which the scientist, if he quits, would actually set up the rival (rather than engage in some other activity). By using a labor contract which induces the scientist to set up a rival if $\epsilon \in Q_r^*(s)$, and then setting the flat wage at the minimum acceptable given optimum behavior by the scientist, the entrepreneur ensures that his expected gain includes the maximum expected gain that could possibly arise as a result of the mobility of scientific personnel, $T(s)$. The entrepreneur will choose s to maximize these gains plus ${}_R\int f(\epsilon, s)dG - s$. All else equal, then, the entrepreneur ought clearly to prefer environments in which it is easier for the scientist to set up a rival and where the scientist can earn larger profits from the rival. Corollary 2.2 is an immediate consequence of the definition of $Q_r^*(s^r)$.

COROLLARY 2.2: Given optimum behavior of all agents, the necessary and sufficient condition for quits and the establishment of rivals to be possible is that $r(\epsilon, s^r) - c(s^r) > 0$ and $\alpha f(\epsilon, s^r) + r(\epsilon, s^r) - c(s^r) > 0$ for some ϵ with positive probability.

Thus, one should observe scientists setting up rivals when setup costs are low and when other agents (third parties) are likely to enter similar activities. Indeed, one would expect these two conditions to be mutually supportive, that is, low setup costs may induce third-party entrants, while third-party entrants may attract venture capital to the particular field of activity (see the description of the evolution of Santa Clara, or "Silicon" Valley, in Bylinsky [1976], chap. 4).

To summarize, it is clear that the inability of the firm to maintain perfect monopoly power over its discoveries may induce third-party entrants, thereby creating conditions which result in the scientist's breaking away from the entrepreneur to set up a new enterprise. Nonetheless, the mobility of scientific personnel is not, in itself, a source of concern to entrepreneurs. Indeed, an optimum labor contract ensures that the scientist will only set up a new enterprise when it would help the entrepreneur to gain some of the revenues that would otherwise accrue to third parties.

Spin-Offs

One way of stating the general result underlying the discussion of this section is that an optimum labor contract could induce the scientist to set up a rival if there are possible states of nature in which the establishment of the rival increases the sum of the returns accruing to the two agents involved in the project. The classic economic reason for total profits to increase when two firms (rather than one) engage in an activity is increasing costs (here interpreted to include the costs of gathering and processing information; see the review by Spence [1975] and, more recently, Rosen [1981] and Keren and Levhari [1983]). There is one characteristic of research projects which deserves emphasis in this context. Research projects, especially those in relatively young, science-based industries, often produce multiple outputs, that is, discoveries leading to several different innovations, some of which are only marginally related to the project's original goals.[14] In cases with multiple discoveries, in general, there will be costs and benefits associated with the coordination of subsequent financial, development, and marketing decisions. The particular aspects of the process worth coordinating may be best served by different organizational structures (the possibilities include, in addition to the formation of a separate enterprise, the establishment of a holding company, and the setting up of separate research and production departments within a given firm); which one is economic will depend on technological, market, and (particularly when antitrust laws are a factor) institutional considerations. The analysis and implications of situations in which there exists the possibility of multiple discoveries and increasing costs is similar to that of the last subsection and therefore will not be developed here. It should be noted, however, that provided setup costs are not too high, this argument leads us to expect to observe scientists' breaking away from their employers to set up rivals when the economic possibilities embodied in the applied science being used in research projects are relatively unexplored and particularly plentiful (see the historical discussion by Rosenberg [1963, 1969]).

IV. Summary

The basic conclusion of this paper is that the potential mobility of scientific personnel need not, in itself, have an adverse effect on project profitability. This conclusion follows from the fact that an optimizing entrepreneur who is free to choose among alternative contracts will always choose one which only induces the scientist to leave and join a rival if the sum of the benefits to the two agents increases as a result of the scientist's leaving. Contracts which specify labor payments in the form of a combination of a flat wage rate and stock option (or other profit-sharing agreement) ought to be able to induce a close approximation to this behavior. Given the expected value of the profit-sharing agreement, the entrepreneur should set the initial wage, w_0, to a level which is low

[14] See, e.g., the discussion in Nelson (1959). Mueller's (1962) study of the origin of DuPont's major innovations lists nine of them as originating in DuPont's own laboratory and of these, two (the discoveries of teflon and f ducose lacquers) resulted from research accidents.

enough to ensure that the total of the expected payments to the scientist is just equal to the scientist's alternative wage. The expected profits of an entrepreneur who follows this policy will include all the monetary gains expected to accrue to either of the two agents other than the scientist's alternative wage. That is, the entrepreneur extracts all possible rents accruing to his idea.

Clearly, then, if the sum of the benefits to the two agents cannot be increased by the scientist's leaving, the entrepreneur should specify a contract which will never induce him to do so. In this case the best the entrepreneur can do is to choose all factors to maximize the expected profitability of the research project, ignoring the possibility of quits, and then to choose a labor contract which ensures that quits will never occur. It follows that project profitability and research employment will be independent of the technological, market, and institutional factors which determine the gains the scientist would make from setting up a rival. On the other hand, if there are possible outcomes from the research project in which the sum of the monetary benefits to the two agents can be increased by the scientist leaving, the entrepreneur ought to choose a contract which will induce the scientist to leave if such an outcome materializes. In this case, the entrepreneur's expected profits will, in fact, be an increasing function of the gains the scientist could make in a rival.

We have focused on two characteristics of research projects which could lead to situations in which optimum behavior by all agents would result in the scientist breaking away from his employer to establish a rival. First, it may be possible for third parties to gain access to the information generated from the project, in which case the scientist's setting up a rival may enable the entrepreneur to recoup some of the profits that would otherwise accrue to third parties. Second, the research project may lead to several discoveries, in which case costs of coordination may result in its being economic to develop and market them in separate enterprises. In either of these cases an optimizing entrepreneur will be led to make research employment decisions to maximize all the profits to be generated by the project, whether the associated net revenues accrue to the rival or to the original firm.

In short, though we are quite sure that there are mechanisms which, because of the fact that they can be used to spread the information produced in a firm's research laboratories, reduce the profitability of research projects and (perhaps) of employment in them, the potential mobility of scientific personnel need not be one of them. The reason is straightforward. Provided the firm is free to choose among alternative labor contracts, it can provide an incentive structure which controls the mobility of the scientist—only inducing him to leave and set up a rival when it is in the firm's interest for him to do so.

Appendix

This Appendix provides formal proofs for statements made in the text.

PROPOSITION 1: We prove that no matter the choice of s, a contract $\{\underline{w}_0(s), \underline{w}_1(\epsilon, s)\}$ will be optimal for problem A provided that $\underline{w}_1(\epsilon, s) \geq \alpha f(\epsilon, s) - c(s) + s$ for $\epsilon \in A^*$ and $\underline{w}_0(s) = s - {}_A\!\int[\underline{w}_1(\epsilon, s) - s]dG$. Given this contract, the proof that s^* is the optimal choice of s follows directly from the argument given in the text.

To prove this we assume, to the contrary, that there exists an s, say $\tilde{s}$, and a feasible contract $\{\tilde{w}_0(\tilde{s}), \tilde{w}_1(\epsilon, \tilde{s})\}$ such that $\Pi\{\tilde{s}, \tilde{w}_0(\tilde{s}), \tilde{w}_1(\epsilon, \tilde{s})\} > \Pi\{\tilde{s}, \underline{w}_0(\tilde{s}), \underline{w}_1(\epsilon, \tilde{s})\}$ and show that this involves a contradiction. Since $\{\tilde{w}_0(\tilde{s}), \tilde{w}_1(\epsilon, \tilde{s})\}$ must satisfy (1); $\tilde{w}_0(\tilde{s}) \geq 2s - {}_A\!\int\tilde{w}_1(\epsilon, s)dG - {}_{\tilde{Q}}\!\int[\alpha f(\epsilon, {}_{\tilde{Q}}\tilde{s}) - c(\tilde{s}) + \tilde{s}]dG - {}_{R-\tilde{A}-\tilde{Q}}\!\int \tilde{s}dG$, with $\tilde{A}$ and $\tilde{Q}$ defined as in problem A. Substituting this expression into the profit function and comparing the profits earned from this contract to those earned from $\{\tilde{s}, \underline{w}_0(\tilde{s}), \underline{w}_1(\epsilon, \tilde{s})\}$, we find $\Pi\{\tilde{s}, \tilde{w}_0(\tilde{s}), \tilde{w}_1(\epsilon, \tilde{s})\} - \Pi\{\tilde{s}, \underline{w}_0(\tilde{s}), \underline{w}_1(\epsilon, \tilde{s})\} = -{}_{\tilde{Q}}\!\int c(\tilde{s})dG \leq 0$, which is a contradiction. Q.E.D.

Proposition 1A and corollary 1.1A underlie the discussion in the last paragraph of Section II.

PROPOSITION 1A: If the constraint

$$w_0(s) \geq 0\ , \tag{A1}$$

is added to problem A, then no matter the choice of s, the contract $\{w^+_0(s), w^*_1(\epsilon, s)\}$ is optimal, where $w^+_0(s) = \max\{0, w^*_0(s)\}$ and $\{w^*_0(s), w^*_1(\epsilon, s)\}$ is defined by corollary 1.1.

PROOF: We first show that for any contract $\{\tilde{w}_0(s), \tilde{w}_1(\epsilon, s)\}$ satisfying (1) and (A1) to be optimal, $\tilde{Q}$ must equal the null set. We then show that of all contracts satisfying $\tilde{Q} = \emptyset$ and (1)–(A1), $\{w^+_0(s), w^*_1(\epsilon, s)\}$ is an optimum. Thus consider comparing the expected profits from the contract $\{\tilde{w}_0(s), \tilde{w}_1(\epsilon, s)\}$ to those from $\{\hat{w}_0(s), w^*_1(\epsilon, s)\}$, where $\hat{w}_0(s) = \tilde{w}_0(s) + {}_{\tilde{A}}\!\int[\tilde{w}_1(\epsilon, s) - w^*_1(\epsilon, s)]dG$. Note that if $\{\tilde{w}_0(s), \tilde{w}_1(\epsilon, s)\}$ satisfies (1)–(A1), so does $\{\hat{w}_0(s), w^*_1(\epsilon, s)\}$. On direct substitution into the profit function we have $\Pi\{s, \tilde{w}_0(s), \tilde{w}_1(\epsilon, s)\} - \Pi\{s, \hat{w}_0(s), w^*_1(\epsilon, s)\} = -{}_{\tilde{A}}\!\int[w^*_1(\epsilon, s) - s]dG - {}_{\tilde{Q}}\!\int\alpha f(\epsilon, s)dG + {}_{A^*}\!\int[w^*_1(\epsilon, s) - s]dG = -{}_{\tilde{Q}}\!\int c(s)dG \leq 0$, where the last equality follows from the fact that $A^* = \tilde{Q} + \tilde{A} - D[D = \{\epsilon | \epsilon \in \tilde{A} \text{ and } \epsilon \notin A^*\}]$, and $w^*_1(\epsilon, s) = s$ for $\epsilon \in D$. Thus we restrict our discussion to contracts which result in $Q = \emptyset$, which implies that $\tilde{w}_1(\epsilon, s) - s \geq \alpha f(\epsilon, s) - c(s)$ for all $\epsilon \in A^*$. Among these contracts the feasibility condition implies that $\tilde{w}_0(s) \geq s - {}_{A^*}\!\int[\tilde{w}_1(\epsilon, s) - s]dG$, from which it follows that for given s expected profits must be less than equal to the number $\Pi_1(s) = {}_R\!\int f(\epsilon, s)dG - s$. Note that $\Pi_1(s)$ would be the optimum profit if there were no nonnegativity constraint; see proposition 1. On the other hand, since $\tilde{w}_1(\epsilon, s) - s \geq \alpha f(\epsilon, s) - c(s)$ for $\epsilon \in A^*$, the nonnegativity constraint implies that expected profits must be less than or equal to $\Pi_2(s) = {}_R\!\int f(\epsilon, s)dG - {}_{A^*}\!\int[\alpha f(\epsilon, s) - c(s)]dG$. The optimum profit level would be $\Pi_2(s)$ if there were no feasibility constraint, that is, if expected payments to the employee could be made as small as we like provided that $w_0(s) \geqslant 0$. But $\Pi[s, w_0^+(s), w_1^*(\epsilon, s)] = \min[\Pi_1(s), \Pi_2(s)]$. Q.E.D.

COROLLARY 1.1A: Let s^+ be the optimal value of s for problem A when the nonnegativity constraint (A1) is added, and assume $c(s) = c \geqslant 0$. Note that for $s \geqslant s^*$, $w_0^*(s)$ is a monotonic increasing function of s. Given this fact, s^+ is defined as follows. If $w_0^*(s^*) \geqslant 0$, which implies that the nonnegativity constraint is not binding at s^*, then $s^+ = s^*$; if $w_0^*(\bar{s}) < 0$, which implies that there does not exist an $s \in (s^*, \bar{s}]$ that satisfies $w_0^*(s) \geqslant 0$, then $s^+ = \bar{s}$; and if $w_0^*(s^*) < 0$ while $w_0^*(\bar{s}) \geqslant 0$, then s^+ is the unique value of $s \in (s^*, \bar{s}]$ that satisfies $w_0^*(s^+) = 0$.

PROOF: From proposition A, s^+ is that value of s which satisfies $\max_{s \in U} \{\min[\Pi_1(s), \Pi_2(s)]\}$. Note that $w_0^*(s) = \Pi_2(s) - \Pi_1(s)$ and that $\Pi_2'(s) \geqslant 0$ for all $s \in U$, while $\Pi_1(s)$ is concave and has a maximum at s^*. Thus, if $w_0^*(s^*) \geqslant 0$, $\Pi_1(s^*)$ is attainable, $\leqslant \Pi_2(s^*)$, and since $\Pi_1(s^*) \geqslant \Pi_1(s)$ for all $s \in U$, $s^+ = s^*$. If $w_0^*(s^*) < 0$ while $w_0^*(\bar{s}) > 0$, then, since $w_0^{*\prime}(s) > 0$ for all $s \in (s, \bar{s}]$, there exists a unique s, say $\hat{s} \in (s^*, \bar{s}]$, with $w_0^*(\hat{s}) = 0$. For $s \geqslant \hat{s}$, $\Pi_1(s) \leqslant \Pi_1(\hat{s})$, while for $s \leqslant \hat{s}$, $\Pi_2(s) \leqslant \Pi_2(\hat{s})$, which proves that in this case $s^+ = \hat{s}$. Finally, if $w_0^*(\bar{s}) < 0$, then $\Pi_2(\bar{s})$ is attainable, $\leqslant \Pi_1(\bar{s})$, and since $\Pi_2(\bar{s}) \geqslant \Pi_2(s)$ for all $s \in U$, $s^+ = \bar{s}$. Q.E.D.

Recall that proposition 2 and corollary 2.1 provide the optimal contract and the optimal skill level for problem B.

PROPOSITION 2: To prove this proposition we assume, to the contrary, that there exists an s, say $\tilde{s}$, and a feasible contract $\{\tilde{w}_0(\tilde{s}), \tilde{w}_1(\epsilon, \tilde{s})\}$ such that $\Pi\{\tilde{s}, \tilde{w}_0(\tilde{s}), \tilde{w}_1(\epsilon, \tilde{s})\} > \Pi\{\tilde{s}, w_0^r(\tilde{s}), w_1^r(\epsilon, \tilde{s})\}$, and then show that this involves a contradiction. Since $\{\tilde{w}_0(\tilde{s}), \tilde{w}_1(\epsilon, \tilde{s})\}$ must satisfy (2), $\tilde{w}_0(\tilde{s}) \geqslant 2\tilde{s} - {}_{A_r}\!\int \tilde{w}_1(\epsilon, \tilde{s})dG - {}_{\tilde{Q}_r}\!\int[\alpha f(\epsilon, \tilde{s}) - c(\tilde{s}) + \tilde{s} + r(\epsilon, \tilde{s})]dG - {}_{R-\tilde{A}_r-\tilde{Q}_r}\!\int \tilde{s}dG$ where $\tilde{Q}_r = \{\epsilon | \alpha f(\epsilon, \tilde{s}) - c(\tilde{s}) + \tilde{s} + r(\epsilon, \tilde{s}) > \max[\tilde{w}_1(\epsilon, \tilde{s}), \tilde{s}]\}$. Substituting this condition and the optimum labor contract into the definition of Π we have $\Pi\{\tilde{s}, \tilde{w}_0(\tilde{s}), \tilde{w}_1(\epsilon, \tilde{s})\} - \Pi\{\tilde{s}, w_0^r(\tilde{s}), w_1^r(\epsilon, \tilde{s})\} \leqslant {}_{\tilde{Q}_r}\!\int[r(\epsilon, \tilde{s}) - c(\tilde{s})]dG - {}_{Q_r^*}\!\int[r(\epsilon, \tilde{s}) - c(\tilde{s})]dG \leqslant 0$, a contradiction. To prove the last inequality, let Ω be the set of all subsets of the set $M = \{\epsilon | \alpha f(\epsilon, \tilde{s}) - c(\tilde{s}) + r(\epsilon, \tilde{s}) \geqslant 0\}$. Clearly, Q_r^*, $\tilde{Q}_r$, $\subset \Omega$, and Q_r^* maximizes ${}_Q\!\int[r(\epsilon, \tilde{s}) - c(\tilde{s})]dG$ (over $Q \subset \Omega$). Q.E.D.

COROLLARY 2.1: To prove this corollary, substitute the optimal contract, $\{w_0^r(s), w_1^r(\epsilon, s)\}$, into the profit function for problem B to obtain $\Pi(s) = {}_R\!\int f(\epsilon, s)dG - s + {}_{Q_r^*(s)}\!\int[r(\epsilon, s) - c(s)]dG$. Taking the derivative of this expression with respect to s and setting it equal to zero, one obtains s^r as defined in the corollary.[15]

References

Ackerloff, George. "The Economics of Caste and of the Rat Race and Other Woeful Tales." *Quarterly Journal of Economics* 90 (November 1976): 599–617.

[15] To see this partition $Q_r^*(s)$ into J subsets, that is $Q_r^*(s) = \Sigma_{j=1}^J q_j(s)$, where $q_j(s) = \{\epsilon | \kappa_{0j}(s) < \epsilon < \kappa_j(s)\}$. Then $\Pi(s) = {}_R\!\int f(\epsilon, s)dG - s + \Sigma_{j=1}^J \int_{\kappa_{0j}(s)}^{\kappa_j(s)}[r(\epsilon, s) - c(s)]dG$. Noting that the value of the integrand at both its upper and lower limits for each of the J groups is zero, we obtain the derivative in the corollary (see Kolmogorov and Fomin 1970, section 31.3).

Arrow, Kenneth J. "Economic Welfare and the Allocation of Resources for Invention." In *The Rate and Direction of Inventive Activity: Economic and Social Factors,* edited by R. R. Nelson. Universities–National Bureau Conference Series no. 13. Princeton, N.J.: Princeton University Press (for NBER), 1962.

Barzel, Yoram. "Optimal Timing of Innovations." *Review of Economics and Statistics* 5 (August 1968): 348–55.

Becker, G. S. *Human Capital.* 2d ed. Human Behavior and Social Institutions no. 5. New York and London: Columbia University Press (for NBER), 1975.

Becker, G. S., and Stigler, George J. "Law Enforcement, Malfeasance, and Compensation of Enforcers." *Journal of Legal Studies* 3 (January 1974): 1–18.

Braun, Ernest, and MacDonald, Stuart. *Revolution in Miniature: The History and Impact of Semiconductor Electronics.* Cambridge: Cambridge University Press, 1978.

Bylinsky, Gene. *The Innovation Millionaires: How They Succeed.* New York: Scribner's, 1976.

Dasgupta, Partha, and Stiglitz, Joseph. "Uncertainty, Industrial Structure, and the Speed of R & D." *Bell Journal of Economics* 11 (Spring 1980): 1–28.

Freeman, Christopher. *The Economics of Industrial Innovation.* Harmondsworth: Penguin Books, 1974.

Gaskins, Darius W., Jr. "Dynamic Limit Pricing: Optimal Pricing under Threat of Entry." *Journal of Economic Theory* 3 (September 1971): 306–22.

Harris, Milton, and Raviv, Artur. "Some Results on Incentive Contracts with Applications to Education and Employment, Health Insurance, and Law Enforcement." *American Economic Review* 68 (March 1978): 20–30.

Keren, Michael, and Levhari, David. "The Internal Organization of the Firm and the Shape of Average Costs." *Bell Journal of Economics* (Autumn 1983): in press.

Kolomogorov, A. N., and Fomin, S. V. *Introductory Real Analysis.* Rev. English ed., translated and edited by Richard A. Silverman. New York: Dover, 1970.

Lazear, Edward P. "Why Is There Mandatory Retirement?" *Journal of Political Economy* 87 (December 1979): 1261–84.

Lazear, Edward P., and Rosen, Sherwin. "Rank-Order Tournaments as Optimum Labor Contracts." *Journal of Political Economy* 89 (October 1981): 841–64.

Mortensen, Dale T. "Specific Capital and Labor Turnover." *Bell Journal of Economics* 9 (Autumn 1978): 572–86.

Mueller, Willard F. "The Origins of the Basic Inventions Underlying Du Pont's Major Product and Process Innovations, 1920 to 1950." In *The Rate and Direction of Inventive Activity: Economic and Social Factors,* edited by R. R. Nelson. Universities–National Bureau Conference Series no. 13. Princeton, N.J.: Princeton University Press (for NBER), 1962.

Nelson, Richard R. "The Simple Economics of Basic Scientific Research." *Journal of Political Economy* 67 (June 1959): 297–306.

Nickell, Steve J. "Wage Structures and Quit Rates." *International Economic Review* 17 (February 1976): 191–203.

Phillips, A. *Technology and Market Structure: A Study of the Aircraft Industry.* Lexington, Mass.: Lexington, 1974.

Rosen, Sherwin. "Output, Income and Rank in Hierarchical Firms." Economics Research Center/NORC Discussion Paper No. 81-10, University of Chicago, 1981.

Rosenberg, Nathan. "Technological Change in the Machine Tool Industry, 1840–1910." *Journal of Economic History* 23 (December 1963): 414–43.

———. "The Direction of Technological Change: Inducement Mechanisms and Focusing Devices." *Economic Development and Cultural Change* 18 (October 1969, pt.1): 1–24.

Ross, Stephen, A. "The Economic Theory of Agency: The Principal's Problem." *American Economic Review, Papers and Proceedings* 63 (May 1973): 134–39.

Salop, Joanne, and Salop, Steven. "Self-Selection and Turnover in the Labor Market." *Quarterly Journal of Economics* 90 (November 1976): 619–27.

Schumpeter, Joseph A. *Capitalism, Socialism, and Democracy.* 3d ed. New York: Harper, 1950. (Originally published, 1942).

Spence, A. Michael. "The Economics of Internal Organization: An Introduction." *Bell Journal of Economics* 6 (Spring 1975): 163–72.

Stiglitz, Joseph E. "Incentives and Risk Sharing in Sharecropping." *Review of Economic Studies* 51 (April 1974): 219–56.

———. "Incentives, Risk and Information: Notes Towards a Theory of Hierarchy." *Bell Journal of Economics* 6 (Autumn 1975): 552–79.

Temin, Peter. "Technology, Regulation, and Market Structure in the Modern Pharmaceutical Industry." *Bell Journal of Economics* 10 (Autumn 1979): 429–46.

[10]

Innovation in Large and Small Firms: An Empirical Analysis

By Zoltan J. Acs and David B. Audretsch*

We present a model suggesting that innovative output is influenced by R&D and market structure characteristics. Based on a new and direct measure of innovation, we find that (1) the total number of innovations is negatively related to concentration and unionization, and positively related to R&D, skilled labor, and the degree to which large firms comprise the industry; and (2) these determinants have disparate effects on large and small firms.

As Simon Kuznets (1962) observed, perhaps the greatest obstacle to understanding the role of innovation in economic processes has been the lack of meaningful measures of innovative inputs and outputs. More recently, there has been the development of new data sources measuring different aspects of technical change. These new sources of data have included measures of patented inventions from the computerization by the U.S. Patent Office (Bronwyn Hall et al., 1986; Adam B. Jaffe, 1986; Ariel Pakes and Zvi Griliches, 1980), better measures of research and development (John Bound et al., 1984, and F. M. Scherer, 1982), and stock market values of inventive output (Pakes, 1985). While several of these new and improved data sources have been used to examine the relationship between innovative activity and firm size, there have been virtually no studies able to apply a more direct measure of the innovative output. For example, the limitations of using patent data were significant enough to supplement them with renewal data (Pakes and Mark Schankerman, 1984). Further, while most of the empirical research has examined only the innovative activity contributed by relatively large firms, the innovative output of the smallest firms has received only scant attention and quantification.[1] Thus, most of the inferences which have been made about the causes of innovative activity have been based on observing only the behavior of larger firms.[2] Such inferences may be misleading since, as we show, almost half of the number of innovations are contributed by firms which employ fewer than 500 workers.

The purpose of this paper is to add to the literature on new measures examining technical change by introducing a more direct measure of innovative activity, to determine some of its basic properties, and to illustrate its use with a reduced form empirical model. We present a model which investigates the degree to which innovative output is affected by different industry characteristics, and the extent to which small and large firms respond differently to various stimuli. The econometric analysis enables the testing of

*Research Fellows, Wissenschaftszentrum Berlin, Reichpietschufer 50, D-1000 Berlin 30, Federal Republic of Germany. We wish to thank George Bittlingmayer, Bo Carlsson, Paul Geroski, Albert N. Link, Richard R. Nelson, William K. Scheirer, Joachim Schwalbach, J.-Matthias Graf von der Schulenburg, Hideki Yamawaki, Klaus Zimmermann, two anonymous referees, and seminar participants at the U.S. Small Business Administration, Case Western Reserve University, and the University of Bradford for helpful comments. We are especially grateful for the suggestions by F. M. Scherer and the computational assistance of Michael Karge and Jianping Yang. All errors and omissions remain our responsibility.

[1]For a thorough review of the literature relating technical change to innovation activity, see Morton I. Kamien and Nancy L. Schwartz (1975), F. M. Scherer (1980), and Richard C. Levin et al. (1985).

[2]For example, Scherer (1965) related market structure to the number of patents for fewer than 500 of the largest U.S. corporations.

two hypotheses: (1) the degree to which *R&D* expenditures produce innovative output is conditioned by the market structure characteristics; and (2) as Sidney G. Winter (1984) suggests, small- and large-firm innovative activity respond to distinct technological and economic regimes.

In Section I of this paper, we introduce the new innovation data and compare them with the more traditional measures of technical change, including patented inventions and *R&D* expenditures. After presenting the model in Section II, cross-section regressions estimating the logs of the 1982 total number of innovations, large-firm innovations, small-firm innovations, and the small-firm innovation share for 247 four-digit SIC (Standard Industrial Classification) industries are presented in Section III. Finally, in Section IV, a summary and conclusion of the model and empirical results are provided. We find that innovative output increases with industry *R&D* expenditures at a less than proportional rate. While some of the appropriability measures, such as market concentration and unionization, are negatively related to innovation activity, the extent to which an industry is comprised of large firms is positively related to the total number of innovations. Similarly, we find considerable evidence supporting the hypothesis by Winter (1984) that innovation activity for small firms responds to a different technological and economic environment than does innovation activity for large firms.

I. The Innovation Data

The measure of innovative activity used in this paper is the number of innovations in each four-digit SIC industry recorded in 1982. The data, which were only recently released by the U.S. Small Business Administration,[3] consist of 8,074 innovations introduced into the United States in 1982. Of these innovations, 4,476 were identified as occurring in manufacturing industries. The Small Business Administration constructed this data base by examining over 100 technology, engineering, and trade journals, covering each manufacturing industry. From the sections in each trade journal listing innovations and new products, a data base consisting of the innovations by four-digit SIC industry was formed. The entire list of trade journals used to compile these data is available from the authors. The Small Business Administration defines an innovation as "a process that begins with an invention, proceeds with the development of the invention, and results in the introduction of a new product, process or service to the marketplace." (Keith L. Edwards and Theodore J. Gordon, 1984, p. 1) Because the innovations recorded in 1982 were the result of inventions made, on average, 4.2 years earlier, in some sense the innovation data base represents the inventions made around 1978 that were subsequently introduced to the market in 1982. The data were also checked for duplication. In fact, 8,800 innovations were actually recorded, but it was subsequently found that 726 of them appeared either in separate issues of the same journal or else in different journals. Thus, double counting was avoided. While the Small Business Administration does not claim that every single innovative activity is included in the data base, it does consider the data to be "comprehensive and reliable."

The innovation data were classified according to the industry of origin based on the SIC code of the innovating enterprise. The data were then classified into innovations by large firms, defined as firms with at least 500 employees, and innovations by small firms, defined as firms with fewer than 500 employees. For example, an innovation made by a subsidiary of a diversified firm would be classified by industry according to the SIC industry of the innovating subsidiary (enterprise) and not by the SIC industry of the parent firm. However, the innovation would be classified by size according to the size of the entire firm and not just by the size of the subsidiary.[4]

[3]For a more complete explanation of the data, see the Appendix.

[4]Because 67 innovations could not be classified according to firm size, the number of total innovations does not always equal the sum of large- and small-firm innovations.

TABLE 1—NUMBER OF INNOVATIONS AND THE RELATIVE INNOVATIVE ACTIVITY OF LARGE AND SMALL FIRMS IN THE MOST INNOVATIVE INDUSTRIES, 1982

Industry	Total Innovations	Large-Firm Innovations/ Small-Firm Innovations	Large Firm Employment/ Small-Firm Employment
Electronic Computing Equipment	395	0.696	16.544
Process Control Instruments	165	0.731	1.632
Radio and TV Communication Equipment	157	1.153	4.814
Pharmaceutical Preparations	133	9.231	19.408
Electronic Components	128	0.740	0.894
Engineering and Scientific Instruments	126	0.518	1.096
Semiconductors	122	3.138	5.757
Plastics Products	107	0.268	0.332
Photographic Equipment	88	8.778	18.231
Office Machinery	77	6.700	11.658
Instruments to Measure Electricity	77	0.596	2.077
Surgical Appliances and Supplies	67	4.154	3.566
General Industrial Machinery	67	4.154	0.976
Surgical and Medical Instruments	66	0.833	3.566
Special Industry Machinery	64	2.048	2.717
Industrial Controls	61	0.326	1.732
Toilet Preparations	59	2.278	5.289
Valves and Pipe Fittings	54	0.606	2.436
Electric Housewares and Fans	53	7.833	17.182
Measuring and Controlling Devices	52	0.067	0.866
Food Products Machinery	50	3.083	1.976
Motors and Generators	49	3.900	14.625
Plastic Materials and Resins	45	2.000	6.937
Industrial Inorganic Chemicals	40	4.000	4.917
Radio and TV Receiving Sets	40	8.750	13.925
Hand and Edge Tools	39	2.455	2.125
Fabricated Platework	38	3.222	1.597
Fabricated Metal Products	35	0.706	0.495
Pumps and Pumping Equipment	34	1.125	3.484
Optical Instruments and Lenses	34	0.571	1.179
Polishes and Sanitation Goods	33	0.684	2.937
Industrial Trucks and Tractors	33	0.650	6.404
Medicinals and Botanicals	32	5.400	4.376
Aircraft	32	31.000	111.111
Environmental Controls	32	2.200	3.566
Total	2,617	1.272	8.440

Source: U.S. Small Business Administration.

There are several other important qualifications which should be made concerning the innovation data. The trade journals report relatively few process, service, and management innovations and tend to capture mainly product innovations. The most likely effect of this bias is to underestimate the number of innovations emanating from large firms, since larger enterprises tend to produce more process innovations than do their smaller counterparts. However, because it was found that the large-firm innovations are more likely to be reported in trade journals than are small-firm innovations, the Small Business Administration considers the biases to be at least somewhat offsetting.

Table 1 lists the total number of innovations in those industries which had the greatest number of innovations in 1982, along with the corresponding ratios of large-firm innovation to small-firm innovation and large-firm employment to small-firm employment (also for 1982). Large firms are defined as firms with at least 500 employees and small firms are defined as firms with fewer than 500. In some industries, the large firms exhibit considerably more innovative activity than do their smaller counterparts, while in other industries the small firms are apparently more innovative. One potential concern might be that the significance and "quality" of the innovations vary considerably between large and small firms. Based on 4,938 of the innovations, the Small Business Administration classified each innovation according to one of the following levels of significance: (1) the innovation established an entirely new category of product; (2) the innovation is the first of its type on the market in a product category already in existence; (3) the innovation represents a significant improvement in existing technology; and (4) the innovation is a modest improvement designed to update an existing product. While none of the innovations in the sample were in the highest level of significance, 80 were in the second level, 576 in the third level, and 4,282 were classified in the fourth level. Within each level of significance, the distribution between large- and small-firm innovations proved to be remarkably constant. In both the second and third significance categories, the large firms accounted for 62.5 percent of the innovations and the small firms for the remaining 37.5 percent. In the fourth significance category, the large firms accounted for just a slightly smaller share of the innovations, 56.6 percent, while the small firms contributed the remaining 43.4 percent. A chi-square test for the hypothesis that there is no difference in the frequency of innovation with respect to innovation significance and firm size cannot be rejected at the 99 percent level of confidence (Edwards and Gordon, 1984).[5] Thus, based on the U.S. Small Business Administration's classification of the significance level of innovations, there does not appear to be a great difference in the "quality" and significance of the innovations between large and small firms. However, the extent of innovative activity does not necessarily correspond to the market values of the innovations. It is conceivable that larger firms may tend to focus on innovations with a higher market value.

The second column of Table 1 shows that while the number of large-firm innovations exceeded that of small firms in twenty-one of the thirty-five most innovative industries, the small firms were more innovative in the remaining fourteen. By contrast, the third column indicates that large firms accounted for a greater share of 1982 industry employment in all but five of these highly innovative industries. For the entire sample of manufacturing industries the simple correlation between these two measures is only 0.316. While the ratio of the innovations-per-employee is 6.64 times greater for small firms than for large firms for the most innovative industries, small firms were only 43 percent more innovative than their larger counterparts for the entire sample of manufacturing industries.[6]

Table 2 shows the distribution of manufacturing industries according to innovation frequency. About one-quarter of the innovating industries contributed at least sixteen innovations, while slightly more than one-half of these industries had fewer than six innovations.

In order to compare the new innovation measures with the more traditional measures of technical change, Table 3 provides a correlation matrix of input and output measures of innovative activity. There is a striking difference in the simple correlation of 0.481

[5] Unfortunately, no further information regarding the distribution of innovations by significance category has been released by the U.S. Small Business Administration. Therefore, no comparisons across industries can be made.

[6] It should be noted that while there were a total of 8,074 innovations identified in the data base, some innovations occurred outside of manufacturing and others were contributed from firms which could not be found in publishing company directories. While some of these were subsidiaries of large firms, many more were small companies. A random sample of 600 firms (with 375 responses) was used to allocate the entire set of innovations into 55 percent from small firms and 45 percent from large firms, resulting in an innovation-per-employee ratio 2.38 times greater in small firms than in large firms.

TABLE 2—THE DISTRIBUTION OF MANUFACTURING INDUSTRIES ACCORDING TO INNOVATION FREQUENCY

	Number of Innovations[a]							
	1–5	6–10	11–15	16–20	21–25	26–30	31 >	Total
Total Innovations	149	36	21	13	15	6	36	276
(percent)	(53.99)	(13.04)	(7.61)	(4.71)	(5.43)	(2.17)	(13.04)	(99.99)
Large-Firm Innovations	127	28	24	6	7	8	18	218
(percent)	(58.26)	(12.84)	(11.01)	(2.75)	(3.21)	(3.67)	(8.26)	(100.00)
Small-Firm Innovations	129	33	16	9	4	1	13	205
(percent)	(62.93)	(16.10)	(7.80)	(4.39)	(1.95)	(0.49)	(6.34)	(100.00)

Source: U.S. Small Business Administration.
[a]Only industries in which there is at least some innovation activity are included in this table.

TABLE 3—CORRELATION MATRIX OF INPUT AND OUTPUT MEASURES OF INNOVATIVE ACTIVITY

	Total Innovations	Large-Firm Innovations	Small-Firm Innovations	Total *R&D* Expenditures	Company *R&D* Expenditures
Large-Firm Innovations	0.920	–			
Small-Firm Innovations	0.922	0.698	–		
Total *R&D* Expenditures	0.481	0.532	0.379	–	
Company *R&D* Expenditures	0.746	0.737	0.672	0.764	–
Patents	0.467	0.482	0.382	0.327	0.440

between total *R&D* expenditures (also from the 1977 FTC Line of Business) and total innovations, and of 0.746 between company *R&D* expenditures and total innovations. That the correlation between company *R&D* and innovative activity is stronger than the correlation between total *R&D* and innovative activity is consistent with the findings by Griliches (1986) that privately financed *R&D* has a larger effect on private productivity and profitability than does government-financed *R&D*. The somewhat lower correlation between the *R&D* measures and the total number of patented inventions[7] between 1970 and 1972 is not surprising, as Pakes and Griliches (1980, p. 378) observe that "patents are a flawed measure (of innovative output); particularly since not all new innovations are patented and since patents differ greatly in their economic impact." Still, the correlation between the output measures, total innovations, and patents is greater than that between the *R&D* measures and patents. Also of interest is that, just as the correlation between innovations and *R&D* is greater for large firms than for small ones, the correlation between patents and innovations is greater for large firms than for small ones.

The relationships between the input and output measures of innovative activity are somewhat sensitive to the total amount of innovative activity. That is, in industries in which there is little innovative activity, defined as those with fewer than seven innovations, the correlation between all of the measures of technical change become considerably weaker. In particular, the correla-

[7]The correlations involving the patents are at the three-digit SIC level. (The corresponding variables were also aggregated to the three-digit SIC level.) All other correlations are for the four-digit SIC level.

TABLE 4—COMPARISON OF INNOVATION DATA WITH *R&D* AND PATENT MEASURES[a]

Industry Group	Total Innovations	Patents	Patents/ Innovation	Company *R&D* (millions)	*R&D* (millions)/ Innovation
Food and Tobacco	206	311	1.51	272	1.32
Textiles and Apparel	29	147	5.07	65	2.24
Lumber and Furniture	83	50	0.60	37	0.45
Paper	61	292	4.79	150	2.46
Chemicals (excluding drugs)	332	3,492	10.52	1,260	3.80
Drugs	170	868	5.11	449	2.64
Petroleum	24	1,046	43.58	360	15.00
Rubber and Plastics	129	637	4.94	287	2.22
Stone, Clay, and Glass	59	477	8.09	149	2.53
Primary Metals	74	424	5.73	239	3.23
Fabricated Metal Products	340	450	1.32	246	0.72
Machinery (excluding office)	612	1,657	2.71	852	1.39
Computers and Office Equipment	566	1,045	1.85	1,054	1.86
Industrial Electrical Equipment[b]	444	836	1.88	210	0.47
Household Appliances	64	232	3.63	78	1.22
Communications Equipment	262	2,384	9.10	1,136	4.34
Motor Vehicles and Other Transportation Equipment[c]	152	809	5.32	1,791	11.78
Aircraft and Engines	48	501	10.44	653	13.60
Guided Missiles and Ordnance	16	173	10.81	103	6.44
Instruments	736	1,351	1.84	652	0.89
Total[d]	4,407	17,182	3.90	10,043	2.28

[a]Company *R&D* (1974) and patent (June 1976–March 1977) data are from Scherer (1983).
[b]Includes SIC 361, 362, 364, and 367.
[c]Includes SIC 371, 373, 374, 375, and 379.
[d]Includes only industries in this table.

tion between large-firm innovations and patents falls to 0.107, while that between small-firm innovations and patents is nearly twice as great. Similarly, the correlation between each of the *R&D* measures and small-firm innovations falls to less than 0.070 while that between each of the *R&D* measures and the large-firm innovations is not so greatly affected. In addition, the correlation between patents and each of the *R&D* measures falls to below 0.100 for the low-innovative industries. In the high-innovative industries, defined as those with at least fifteen innovations, the correlation between small-firm innovations and patents is also considerably weaker, as is that between the *R&D* measures and small-firm innovations.

Table 4 provides a more detailed comparison of an input measure—company *R&D* expenditures (from the FTC Line of Business Survey)—and two output measures, the total number of innovations and the total number of patented inventions between June 1976 and March 1977 (from Scherer, 1983). As the simple correlation of 0.74 indicates, there is a fairly strong relationship between the ratio of patents-per-innovation and *R&D* expenditures (millions of dollars)-per-innovation.[8] Of particular interest is that the patent-per-innovation ratio in drugs is about one-half of that in the rest of the broader chemical sector. This is remarkably similar to the pattern of patents-per-scientists and engineers identified by Robert E. Evenson (1984). The relatively high ratios of patents-per-innovation in the chemical and petroleum sectors, and the relatively low ratios in the computers, electrical machinery, lumber, and instruments sectors is explained by Edwin Mansfield (1984, p. 462): "The value

[8]Patents represent 59 percent of all invention patents issued to U.S. corporations during the sample period and 61 percent of patents issued to industrial corporations (Scherer, 1983).

TABLE 5—WEIGHTED MEAN NUMBER OF TOTAL INNOVATIONS, SMALL-FIRM INNOVATION, AND EMPLOYMENT SHARES BY TWO-DIGIT SECTOR[a]

Sector	Total Innovations	Small-Firm Innovation Share	Small-Firm Employment Share
Food	3.739	0.228	0.220
	(1.977)	(0.052)	(0.175)
Textiles	0.333	0.133	0.325
	(0.062)	(0.161)	(0.156)
Apparel	0.576	0.099	0.565
	(0.605)	(0.159)	(0.220)
Lumber	0.647	0.088	0.564
	(0.338)	(0.228)	(0.264)
Furniture	5.539	0.365	0.596
	(0.422)	(0.074)	(0.264)
Paper	3.588	0.161	0.234
	(0.103)	(0.022)	(0.199)
Printing	1.471	0.191	0.524
	(0.368)	(0.048)	(0.245)
Chemicals	17.929	0.313	0.135
	(0.014)	(0.041)	(0.163)
Petroleum	4.800	0.400	0.113
	(2.400)	(0.200)	(0.163)
Leather	0.546	0.273	0.422
	(0.776)	(0.230)	(0.182)
Stone,	2.185	0.219	0.347
Clay, and Glass	(0.036)	(0.013)	(0.264)
Primary Metals	2.846	0.276	0.227
	(1.631)	(0.043)	(0.167)
Fabricated Metal	9.556	0.388	0.524
Products	(4.301)	(0.038)	(0.192)
Machinery	25.886	0.485	0.289
(nonelectrical)	(1.355)	(0.014)	(0.199)
Electrical	20.946	0.411	0.164
Equipment	(1.824)	(0.008)	(0.155)
Transportation	9.000	0.149	0.075
Equipment	(1.000)	(0.029)	(0.325)
Instruments	56.615	0.506	0.221
	(16.344)	(0.017)	(0.121)

[a]The sector means are weighted averages (by 1977 value-of-shipments) of the four-digit SIC industry values. The small-firm innovation share is defined as the percentage of total innovations contributed by small firms. The small-firm employment share is defined as the percentage of industry employment accounted for by firms with fewer than 500 employees. The data come from the U.S. Small Business Administration, Small Business Data Base, 1982. Standard deviations are listed in parentheses.

and cost of individual patents vary enormously within and across industries... Many inventions are not patented. And in some industries, like electronics, there is considerable speculation that the patent system is being bypassed to a greater extent than in the past. Some types of technologies are more likely to be patented than others."

A broader comparison of the distribution of the innovations classified according to firm size, both between and within two-digit SIC sectors is presented in Table 5. For example, the first column shows that the mean number of total innovations in the 47 industries (weighted by the 1977 value-of-shipments) comprising the food sector was 3.74. The second column shows the small-firm innovation share, measured as the percentage of total innovations contributed by small firms, and indicates that in certain sectors the mean number of innovations contributed by the large firms far exceeded the mean number contributed by small firms, while in other sectors the pattern is reversed. As the third column indicates, in about half of the sectors the small-firm employment share exceeds the small-firm innovation share, and in the other half the innovation share of small firms exceeds the employment share.

II. The Empirical Model

We follow others (Jaffe, 1986; Scherer, 1982; and Mansfield, 1981) and assume innovative output is related to innovation-inducing inputs in the previous period according to a log relationship. The independent variables we expect to be important are: *R&D* expenditures (either total or company), 1977; capital intensity, measured as gross assets divided by value-of-shipments, 1977; the four-firm concentration ratio, 1977; the mean percentage of employees belonging to a union between 1973 and 1975; advertising expenditures divided by value-of-shipments; the share of 1977 industry employment accounted for by large firms (enterprises with more than 500 employees); a measure of skilled labor, defined as the percentage of employment consisting of professional and kindred workers, plus managers and administrators, plus craftsmen and kindred workers, 1970. Finally, to control for industry size, the 1977 value-of-shipments is included in the empirical model. All variable sources are listed in the Appendix.

Concentration, advertising, and capital intensity have all been hypothesized to facilitate appropriability and therefore encourage innovation (William S. Comanor, 1967; and Kamien and Schwartz, 1975). More recently, several models have been developed arguing that unions capture rents from intangible capital investments, and, in particular, those accruing from innovation-producing *R&D* (Robert A. Connolly, Barry Hirsch, and Mark Hirschey, 1986; and Hirsch and Albert N. Link, 1986). To the extent to which unions are successful in such rent-seeking activities, the ease of appropriability by the innovative firm is clearly reduced. Therefore, a negative relationship between unionization and total innovative activity is expected.

By substituting the number of large-firm innovations and small-firm innovations as dependent variables into the regression model, the hypothesis by Winter (1984) that two technological regimes exist can be examined. According to Winter (1984, p. 297), "An entrepreneurial regime is one that is favorable to innovative entry and unfavorable to innovative activity by established firms; a routinized regime is one in which the conditions are the other way round." To the extent to which innovations from small and large firms emanate from different technological regimes, a difference in the parameters in the small- and large-firm innovation equations would offer support for Winter's hypothesis that large- and small-firm innovations are promoted under differing economic and technological conditions.[9] The relative small-firm innovative advantage "is likely to be roughly proportional to the number of people exposed to the knowledge base from which innovative ideas might derive." (Winter, 1984, p. 297)

While the exact dynamic relationship between the endogenous and exogenous variables is not known, this should not pose a significant problem in the context of cross-sectional analysis where each of the industry-specific attributes remains relatively invariant over time. While this model assumes the traditional view that technical change is endogenous to market structure, Partha Dasgupta and Joseph Stiglitz (1980) argue that, in fact, certain aspects of market structure are endogenous to technical change.

III. Regression Results

Using the logged value of the number of total innovations in 1982 as the dependent

[9]While we have seen a major shift in the twentieth century, with the large firm and the industrial research laboratory playing a much more important role in innovation, individual entrepreneurs, who are not associated with established firms (Joseph Schumpeter, 1934, p. 62), continue to survive and thrive in a number of industries (Kenney Martin, 1986).

TABLE 6—REGRESSIONS OF LOG OF TOTAL NUMBER OF INNOVATIONS
(t-statistics in parentheses)

	1	2
Log Total R&D	0.364	–
	(6.434)[a]	
Log Company R&D	–	0.409
		(7.424)[a]
Log Capital/Output	0.007	−0.040
	(0.045)	(−0.278)
Log Concentration	−0.613	−0.631
	(−3.853)[a]	(−4.061)[a]
Log Unionization	−0.460	−0.425
	(−3.078)[a]	(−2.907)[a]
Log Advertising	−0.061	−0.056
	(−1.169)	(−1.096)
Log Large-Firm Employment Share	0.438	0.400
	(2.170)[a]	(2.023)[a]
Log Skilled Labor	0.505	0.559
	(2.257)[a]	(2.620)[a]
Log Industry Size	0.032	0.621
	(0.372)	(4.301)[a]
Constant	1.883	1.821
	(1.251)	(1.492)
Sample Size	247	247
R^2	0.457	0.482
F	25.040[a]	27.730[a]

[a]Statistically significant at the 95 percent level of confidence, two-tailed test.

variable, the cross-section regressions for 247 four-digit SIC manufacturing industries are shown in Table 6. While equation (1) uses the measure of total expenditures on *R&D*, equation (2) uses company expenditures on *R&D*. There are two major results from this table. First, the negative coefficient of concentration suggests that lower, and not higher, levels of concentration tend to be associated with increased innovation activity. While the emergence of a negative and significant relationship is consistent with the findings of Connolly and Hirschey (1984), it is a much stronger result than found by Scherer (1965). Second, the elasticity of both company and total *R&D* with respect to the total number of innovations is considerably less than one. Even when the measure of company *R&D*, which apparently explains innovative activity better than total *R&D*, is used in equation (2), the elasticity of *R&D* remains below 0.5. This is somewhat less than the elasticity of about one found between firm *R&D* and patents by Bound et al. (1984).

The positive coefficient of the large-firm employment share indicates that the greater the proportion of an industry consisting of firms with more than 500 employees, the greater is the innovation activity. Of the other variables representing appropriability, only unionization apparently is negatively associated with the total number of innovations, providing support for the Hirsch and Link (1986) hypothesis.[10]

Table 7 shows the results for analogous regressions where the log of the number of large-firm innovations, log of the number of small-firm innovations, and the small-firm innovation share are the dependent variables. Since the error terms in the large- and small-firm innovation regressions are presumably interrelated, the generalized least squares method for seemingly unrelated regressions was estimated using the Aitken procedure (Jan Kmenta, 1971). Skilled labor and advertising apparently have disparate relationships with small- and large-firm innovations. While skilled labor is positively and advertising negatively associated with small-firm innovations, neither is statistically associated with large-firm innovations. Although concentration is negatively associated with both large- and small-firm innovation, the elasticity of concentration with respect to small-firm innovations is more than double that for large firms. A perhaps somewhat surprising result is that not only is

[10]Regressions analogous to those in Table 6 were estimated substituting the patent measure used in Table 3 as the dependent variable. The results are quite similar to those using the innovation measure in Table 6. Total *R&D*, capital/output, the large-firm employment share, and skilled labor are found to be positively related to the number of patents, while concentration is negatively related to patents. Neither advertising, which has a negative coefficient, nor unionization, which has a positive coefficient, can be considered statistically significant. Thus, substituting the patent measure for the innovation measure yields very similar results. Only unionization and capital/output have different qualitative relationships between the patent and innovation measures.

TABLE 7—REGRESSIONS OF THE LOG OF THE NUMBER OF LARGE- AND SMALL-FIRM INNOVATIONS, AND THE SMALL-FIRM INNOVATION SHARE (*t*-statistics listed in parentheses)[a]

Equation	Total *R&D*	Company *R&D*	Concentration	Capital/Output	Unionization	Advertising	Large-Firm Employment Share	Skilled Labor	Industry Size	Constant	Sample Size	R^2	F
1 Large-Firm Innovations	0.251 (4.711)[c]	-	−0.276 (−1.836)[b]	0.107 (0.763)	−0.415 (−2.937)[c]	−0.042 (−0.854)	0.371 (1.943)[b]	0.294 (1.388)	0.150 (1.825)[b]	0.020 (0.017)	247	0.372	-
2 Small-Firm Innovations	0.278 (5.850)[c]	-	−0.705 (2.036)[c]	−0.060 (−0.481)	−0.442 (−3.517)[c]	−0.082 (−1.866)[b]	0.437 (2.572)[c]	0.383 (2.036)[c]	−0.158 (−2.155)[c]	3.802 (1.056)	247	0.366	-
3 Large-Firm Innovations	-	0.294 (5.612)[c]	−0.294 (−1.987)[c]	0.078 (0.568)	−0.388 (−2.785)[c]	−0.038 (−0.787)	0.339 (1.802)[b]	0.318 (1.567)	0.110 (1.361)	−0.022 (−0.019)	247	0.393	-
4 Small-Firm Innovations	-	0.324 (7.015)[c]	−0.723 (−5.556)[c]	−0.093 (−0.769)	−0.412 (−3.356)[c]	−0.078 (−1.813)[b]	0.401 (2.423)[c]	0.411 (2.300)[c]	−0.202 (−2.842)[c]	3.756 (3.669)[c]	247	0.398	-
5 Small-Firm Innovation Share	−0.389 (−1.765)[b]	-	−0.004 (−2.811)[c]	−0.505 (−3.077)[c]	−0.001 (0.569)	−7.306 (−0.456)	0.355 (2.535)[c]	0.785 (2.881)[c]	−0.154 (−0.367)	0.168 (1.459)	247	0.117	3.94[b]
6 Small-Firm Innovation Share	-	−0.138 (−0.463)	−0.004 (−2.837)[c]	−0.484 (−2.934)[b]	−0.001 (0.637)	−7.194 (−0.446)	0.330 (2.327)[c]	0.666 (2.482)[c]	−0.400 (−0.903)	0.209 (1.834)[b]	247	0.116	3.92[b]

[a] The method of generalized least squares (Aitken procedure) was used to estimate large-firm innovations and small-firm innovations. The logs for all explanatory variables are used to estimate log large-firm innovations and log small-firm innovations but not for the small-firm innovation share.
[b] Statistically significant at the 90 percent level of confidence, two-tailed test.
[c] Statistically significant at the 95 percent level of confidence, two-tailed test.

the coefficient of the large-firm employment share positive and significant for small-firm innovations, but it is actually greater in magnitude than for large firms. This suggests that, *ceteris paribus*, the greater extent to which an industry is composed of large firms, the greater will be the innovative activity, but that increased innovative activity will tend to emanate more from the small firms than from the large firms. Perhaps this indicates that, in industries composed predominately of large firms, the existing small firms must resort to a strategy of innovation in order to remain viable. This is consistent with the findings of Richard E. Caves and Thomas A. Pugel (1980) that smaller firms in an industry tend to perform better if they use different strategies from those followed by the larger firms.[11]

Equations (5) and (6) address a slightly different question. Rather than considering what determines the extent of innovative activity in small firms, small-firm innovative activity is examined relative to that in the entire industry.[12] The small-firm innovation share, or the number of small-firm innovations divided by the number of total innovations, is used as the dependent variable. The negative coefficients of total *R&D*, concentration, and capital/output, suggest that the number of small-firm innovations tends to be small relative to that in the entire industry in industries which are high in *R&D*, capital-intensive, and which are concentrated. The positive coefficients of the large-firm employment share and skilled labor imply that the innovation activity of small firms is high relative to the industry level in in-

[11] The disparate effect of the large-firm employment share on small- and large-firm innovation rates is perhaps implied by Scherer (1980, p. 422), "...the most favorable industrial environment for rapid technological progress would appear to be a firm size distribution that includes a predominance of companies with sales below $500 million, pressed on one side by a horde of small, technologically oriented enterprises bubbling over with bright new ideas and on the other by a few larger corporations with the capacity to undertake exceptionally ambitious developments."

[12] For further analysis of the difference between large- and small-firm innovation rates, see our 1987 paper.

TABLE 8—HYPOTHESES TESTS FOR THE DIFFERENCES IN THE DETERMINANTS OF INNOVATION BETWEEN LARGE AND SMALL FIRMS (BASED ON REGRESSIONS 3 AND 4 IN TABLE 7)

Variable Hypothesized[a]	SSR[b]	F
Company *R&D*	118.21	3.665[d]
Capital/Output	120.80	0.279
Concentration	118.25	3.625[d]
Unionization	120.71	0.398
Advertising	120.73	0.359
Large-Firm Employment Share	120.73	0.359
Skilled Labor	118.23	3.630[d]
All Variables[c]	104.276	2.731[d]

[a] The variable hypothesized and therefore constrained to have an identical effect on large-firm innovations and small-firm innovations. Each hypothesis is tested by comparing the sum of the squares explained by the regression from the regression with the restricted coefficient to that from the unrestricted regression. For further explanation of the hypothesis test, see Kmenta (1971).

[b] SSR refers to the sum of squares explained by the estimated regressions for small-firm innovations with the restricted coefficient.

[c] The coefficients of all the explanatory variables are constrained to be identical for the large-firm innovation and small-firm innovation equations.

[d] Statistically significant at the 95 percent level of confidence.

dustries in which there is only a small share of small firms, and in which skilled labor plays an important role.

The results from Table 7 do not support an unequivocal conclusion regarding the exact differences in innovation behavior between large and small firms. While the first four equations imply that skilled labor, advertising, and perhaps concentration and the large-firm employment share, account for the different patterns in innovative activity between large and small firms, the last two equations suggest that total *R&D*, concentration, capital/output, large-firm employment share, and skilled labor account for the differences in relative innovative activity. Thus, in Table 8 the hypotheses that each variable has an identical effect on small- and large-firm innovation activity are tested. The resulting *F*-test is statistically significant for skilled labor, company *R&D*, and concentration, as well as for the hypothesis that all of the variables jointly have an identical effect on small- and large-firm innovation activity. Thus, while there still remains ambiguity regarding the exact patterns between *R&D*, appropriability, and innovation activity for large and small firms, there is considerable support for Winter's (1984) hypothesis that different economic and technological regimes may account for at least some of the differences between the innovation activity of large and small firms.

IV. Conclusion

We have introduced a new and more direct measure of innovative activity in this paper and found that, while the total number of innovations is closely related to *R&D* expenditures (particularly after excluding federal *R&D*), and patented inventions, especially at the more aggregated level, the exact relationship between *R&D* and innovation is somewhat different from that between *R&D* and patents. While Bound et al. (1984) found that, based on firm data, the elasticity of *R&D* expenditures with respect to patented inventions is about one, we find that the number of innovations increases with increased industry *R&D* expenditures but at a decreasing rate. Similarly, while the literature has found a somewhat ambiguous relationship between concentration and various measures of technical change, our results are unequivocal—industry innovation tends to decrease as the level of concentration rises. In addition, there is evidence supporting the hypothesis that unionization is negatively related to innovation activity. Of course, these results are at best an introduction of the innovation data and need to be qualified, particularly in light of the likely simultaneity between industry technology and market structure. Future research may be able to incorporate the endogeneity of market structure in a manner suggested by Levin and Peter Reiss (1984).

This paper has also provided evidence supporting Winter's (1984) prediction that the innovation activity of small and large firms responds to considerably different technological and economic environments.

These findings are not without ambiguity. Further research should apply the patent measure in a similar framework to determine whether its performance differs from that of the innovation measure used here.

APPENDIX: DATA SOURCES AND FURTHER EXPLANATIONS

The data on innovations and employment for small and large firms come from the U.S. Small Business Administration (Edwards and Gordon, 1984). These data can be obtained upon request from the authors. The U.S. Department of Commerce, Bureau of the Census, *Annual Survey of Manufactures*, 1977, *Industry Profiles*, Washington: USGPO, issued 1981, is the source for capital/output (gross assets divided by value-of-shipments), concentration, and value-of-shipments. The advertising measure was derived by using the value-of-shipments data described above and advertising expenditures, from the 1972 U.S. Input-Output Table.

The unionization data are from Richard B. Freeman and James L. Medoff (1979). The percentage of total employment that is unionized for three-digit SIC industries between 1973 and 1975 is reported. We repeat these three-digit SIC values at the four-digit level. The measure of skilled labor is from the U.S. Department of Commerce, Bureau of the Census, Census Population 1970, Subject Report PC (2)-7C, *Occupation by Industry*, Washington: USGPO, issued 1972.

The total *R&D* and company *R&D* data are from the 1977 Federal Trade Commision Line of Business Report. The patent measure used for the disaggregated three-digit SIC correlations in Table 3 is from the Office of Technology Assessment and Forecast of the U.S. Patent Office. Both the patent and skill measures are taken from the *U.S. International Trade Commission's Industrial Characteristics and Trade Performance Data Bank*, June 1975, Office of Economic Research, Washington, D.C. The USITC data bank is described by Edward J. Ray (1981) and these measures are available from the authors on request.

REFERENCES

Acs, Zoltan J. and Audretsch, David B., "Innovation, Market Structure, and Firm Size," *Review of Economics and Statistics*, November 1987, *69*, 567–75.

Bound, John, Cummins, Clint, Griliches, Zvi, Hall, Bronwyn H. and Jaffe, Adam, "Who Does *R&D* and Who Patents?," in Zvi Griliches, ed., *R&D, Patents, and Productivity*, Chicago: University of Chicago, 1984, 21–54.

Caves, Richard E. and Pugel, Thomas A., *Intraindustry Differences in Conduct and Peformance: Viable Strategies in U.S. Manufacturing Industries*, New York University Press, 1980.

Comanor, William S., "Market Structure, Product Differentiation, and Industrial Research," *Quarterly Journal of Economics*, November 1967, *81*, 639–57.

Connolly, Robert A., Hirsch, Barry T. and Hirschey, Mark, "Union Rent Seeking, Intangible Capital, and the Market Value of the Firm," *Review of Economics and Statistics*, November 1986, *68*, 567–77.

______ **and Hirschey, Mark,** "*R&D*, Market Structure and Profits: A Value-Based Approach," *Review of Economics and Statistics*, November 1984, *66*, 682–86.

Dasgupta, Partha and Stiglitz, Joseph, "Industrial Structure and the Nature of Innovative Activity," *Economic Journal*, June 1980, *90*, 266–93.

Edwards, Keith L. and Gordon, Theodore, J., "Characterization of Innovations Introduced on the U.S. Market in 1982," The Futures Group, U.S. Small Business Administration, Contract No. SBA-6050-0A-82, March 1984.

Evenson, Robert E., "International Invention: Implications for Technology Market Analysis," in Zvi Griliches, ed., *R&D, Patents, and Productivity*, Chicago: University of Chicago, 1984, 55–70.

Freeman, Richard B. and Medoff, James L., "New Estimates of Private Sector Unionism in the United States," *Industrial and Labor Relations Review*, January 1979, *32*, 143–74.

Griliches, Zvi, "Productivity, *R&D*, and Basic Research at the Firm Level in the 1970's," *American Economic Review*, March 1986, *76*, 141–54.

Hall, Bronwyn H., Griliches, Zvi and Hausman, Jerry A., "Patents and R&D: Is There a Lag?," *International Economic Review*, June 1986, *27*, 265–302.

Hirsch, Barry T. and Link, Albert N., "Labor Union Effects on Innovative Activity," unpublished manuscript, University of North Carolina-Greensboro, 1986.

Jaffe, Adam B., "Technological Opportunity and Spillovers of *R&D*: Evidence from Firms' Patents, Profits, and Market Value," *American Economic Review*, December 1986, *76*, 984–1001.

Kamien, Morton I. and Schwartz, Nancy L., "Market Structure and Innovation: A Survey," *Journal of Economic Literature*, March 1975, *13*, 1–37.

Kmenta, Jan, *Elements of Econometrics*, New York: Macmillan, 1971.

Kuznets, Simon, "Inventive Activity: Problems of Definition and Measurement," in R. R. Nelson, ed., *The Rate of Direction of Inventive Activity*, Princeton: Princeton University Press, 1962, 19–43.

Levin, Richard C., Cohen, Welsley M. and Mowery, David C., "*R&D* Appropriability, Opportunity, and Market Structure: New Evidence on Some Schumpeterian Hypotheses," *American Economic Review*, May 1985, *75*, 20–24.

______, and Reiss, Peter C., "Tests of a Schumpeterian Model of *R&D* and Market Structure," in Zvi Griliches, ed., *R&D, Patents, and Productivity*, Chicago: University of Chicago Press, 1984, 462–64.

Mansfield, Edwin, "Comment on Using Linked Patent and *R&D* Data to Measure Interindustry Technology Flows," in Zvi Griliches, ed., *R&D, Patents, and Productivity*, Chicago: University of Chicago, 1984, 175–204.

______, "Composition of R and D Expenditures: Relationship to Size of Firm, Concentration, and Innovative Output," *Review of Economics and Statistics*, November 1981, *63*, 610–15.

Martin, Kenney, "Schumpeterian Innovation and Entrepreneurs in Capitalism," *Research Policy*, February 1986, *1*, 21–33.

Pakes, Ariel, "On Patents, *R&D*, and the Stock Market Rate of Return," *Journal of Political Economy*, April 1985, *93*, 390–409.

______, and Griliches, Zvi, "Patents and *R&D* at the Firm Level: A First Report," *Economics Letters*, 1980, *4*, 377–81.

______, and Griliches, Zvi, "Patents and *R&D* at the Firm Level: A First Look," in Zvi Griliches, ed., *R&D, Patents, and Productivity*, Chicago: University of Chicago, 1984, 55–70.

______, and Schankerman, Mark, "The Rate of Obsolescence of Patents, Research Gestation Lags, and the Private Rate of Return to Research Resources," in Zvi Griliches, ed., *R&D, Patents, and Productivity*, Chicago: University of Chicago, 1984, 73–88.

Ray, Edward J., "The Determinants of Tariffs and Nontariff Trade Restrictions in the United States," *Journal of Political Economy*, February 1981, *89*, 105–21.

Scherer, F. M., "The Propensity to Patent," *International Journal of Industrial Organization*, 1983, *1*, 107–28.

______, "Inter-Industry Technology Flows and Productivity Growth," *Review of Economics and Statistics*, December 1982, *64*, 627–34.

______, *Industrial Market Structure and Economic Performance*, Chicago: Rand McNally College Publishing, 1980.

______, "Market Structure and the Employment of Scientists and Engineers," *American Economic Review*, June 1967, *57*, 524–30.

______, "Firm Size, Market Structure, Opportunity, and the Output of Patented Inventions," *American Economic Review*, December 1965, *55*, 1097–125.

Schumpeter, Joseph, *The Theory of Economic Development*, Cambridge: Harvard University Press, 1934.

Winter, Sidney G., "Schumpeterian Competition in Alternative Technological Regimes," *Journal of Economic Behavior and Organization*, September-December 1984, *5*, 287–320.

[11]

An Estimated Model of Entrepreneurial Choice under Liquidity Constraints

David S. Evans

National Economic Research Associates

Boyan Jovanovic

New York University

Is the capitalist function distinct from the entrepreneurial function in modern economies? Or does a person have to be wealthy before he or she can start a business? Knight and Schumpeter held different views on the answer to this question. Our empirical findings side with Knight: Liquidity constraints bind, and a would-be entrepreneur must bear most of the risk inherent in his venture. The reasoning is roughly this: The data show that wealthier people are more inclined to become entrepreneurs. In principle, this could be so because the wealthy tend to make better entrepreneurs, but the data reject this explanation. Instead, the data point to liquidity constraints: capital is essential for starting a business, and liquidity constraints tend to exclude those with insufficient funds at their disposal.

I. Introduction

Do liquidity constraints hinder people from starting businesses? The answer to this question is important for several reasons. First, liquidity

We thank seminar participants at Columbia University, at the University of Florida at Gainesville, and at the National Bureau of Economic Research at Cambridge for helpful comments; the C. V. Starr Center for Applied Economics at New York University for technical assistance; and Francesco Goletti and Zheng Gu for excellent research assistance. We are greatly indebted to Linda Leighton for developing the data set used in this paper and to the referee for many valuable comments and suggestions. This work was done while Evans was a visitor at the C. V. Starr Center.

[*Journal of Political Economy*, 1989, vol. 97, no. 4]

constraints may explain the finding by Evans and Leighton (1989) that the hazard into self-employment is constant in age.[1] This finding is not consistent with the occupational choice stories told by Johnson (1978), Jovanovic (1979), and Miller (1984), which imply that individuals will try riskier occupations such as entrepreneurship when they are younger. But entrepreneurship may in fact not be an option for younger workers because they will have had less time to build up the capital needed to start a business and, with liquidity constraints, will have difficulty borrowing enough start-up funds.

Second, the belief that capital markets do not provide adequate funds for new businesses is one of the rationales for government assistance programs to small business.[2] The U.S. Small Business Administration provides subsidized loans and loan guarantees to small businesses for start-up and expansion. Great Britain, France, Belgium, and the Netherlands have recently adopted financial assistance programs for unemployed workers who start businesses (see Bendick and Egan 1987). The U.S. Department of Labor plans to conduct an experiment in which a sample of unemployment insurance recipients will be given the option of receiving business start-up funds instead of unemployment benefits (see U.S. Department of Labor 1988).

Third, the liquidity constraint is central to the dispute between Frank Knight and Joseph Schumpeter over the nature of entrepreneurship. Knight (1921) argues that bearing risk is one of the essential characteristics of entrepreneurship.[3] He apparently recognizes that capital markets provide too little capital to entrepreneurs because of moral hazard and adverse selection problems (see LeRoy and Singell 1987). Consequently, entrepreneurs must finance themselves and bear the risk of failure. Schumpeter (1934, 1950), on the other hand, argues that the functions of the entrepreneur and the capitalist are quite separate:[4] The role of the entrepreneur is to identify arbitrage opportunities in the economy, while modern capital markets generally enable him to find a capitalist to bear the risks for him.[5]

[1] Data from the National Longitudinal Survey of Young Men indicate that the probability of entering self-employment is independent of both age and labor market experience for men who are under 40. Data from a much larger sample of individuals from the *Current Population Survey* indicate that the probability of entering self-employment is independent of age for those under 50.

[2] This is not to say that these programs are cost effective. As far as we know, there is no evidence on whether the Small Business Administration loan programs are efficient. Bendick and Egan (1987) suggest that the unemployment programs may not be efficient.

[3] See Kanbur (1979) and Kihlstrom and Laffont (1979) for models based on Knight.

[4] This view continues among the Austrians (see Kirzner 1973).

[5] For example, Schumpeter (1934, p. 77) says that "most economists up to the time of the younger Mill failed to keep capitalist and entrepreneur distinct because the manufacturer of a hundred years ago was both; and certainly the course of events since then

This paper examines the importance of liquidity constraints by estimating a model of entrepreneurial choice in which the tightness of the liquidity constraint is a parameter. People have endowments of entrepreneurial ability and assets that may be correlated. The financial capital that they can devote to a business is a multiple of these assets. This multiple is a measure of the degree of liquidity constraints. In Section II we present the model and discuss evidence for our assumption that individuals face an L-shaped liquidity constraint.

The model is estimated with data on roughly 1,500 white males who were wage workers in 1976 and either wage workers or self-employed workers in 1978. The data were drawn from the National Longitudinal Survey of Young Men. After discussing the data, in Section III we report reduced-form results and other evidence consistent with the model.

We present maximum likelihood estimates of the structural parameters of the model in Section IV. The key parameters are the degree of the liquidity constraint, the returns to capital in entrepreneurship, the mean and variance of the distribution of entrepreneurial ability in the population, and the correlation between entrepreneurial ability and assets. It is important to note that we find that a person cannot use more than 1.5 times his or her initial assets for starting a new venture. Thus we reject Schumpeter in favor of Knight: Most individuals who enter self-employment face a binding liquidity constraint and as a result use a suboptimal amount of capital to start up their businesses. Further work is needed to find out if these structural estimates are robust to alternative specifications and samples. Moreover, our findings have no normative implications without added assumptions on why liquidity constraints are there in the first place.

In Section V we summarize the results and present suggestions for further research. While we focus on liquidity constraints, we make several contributions of more general interest. First, we estimate the parameters of the entrepreneurial ability distribution, which we find to be nondegenerate. Thus we confirm Knight's conjecture (1921)—recently elaborated on by Lucas (1978) and Jovanovic (1982)—that one of the key determinants of entrepreneurship is the distribution of "business acumen" in the population. To our knowledge, this paper reports the first structural estimates of the distribution of this key parameter.

Second, the selection problem analyzed in this paper is not stan-

---

has facilitated the making of this distinction." He notes that the entrepreneur and the capitalist are sometimes one and the same. But in his extensive discussions of the relationship between the entrepreneur and the capitalist, he shows no awareness that market forces might require the entrepreneur to be his own capitalist.

dard and may have applications in other areas. The interaction of selection on ability and liquidity constraints gives rise to a complicated set of selection conditions for which we derive the likelihood function. A similar problem would arise in, for example, the demand for durables with heterogeneous consumers subject to liquidity constraints.

II. The Model

The model is static. At the start of the period, the individual must decide whether to work for himself (i.e. become an entrepreneur) or continue to work for someone else (i.e. remain a wage worker). At the end of the period, self-employment opportunity will yield him a gross payoff equal to y, while wage work will yield him a wage of w. We assume away unemployment or withdrawal from the labor force.

A person takes up self-employment if his expected net income is higher there. Otherwise he chooses wage work. The wage equation is

$$w = \mu x_1^{\gamma_1} x_2^{\gamma_2} \xi. \tag{1}$$

Here x_1 is the person's previous experience as a wage worker, x_2 is his education, μ is a constant, and ξ is a disturbance whose logarithm has variance σ_ξ^2, independent across workers. Entrepreneurial earnings are[6]

$$y = \theta k^\alpha \epsilon. \tag{2}$$

Here θ is "entrepreneurial ability," k is the amount of capital invested in the business, ϵ is a lognormal disturbance whose logarithm has variance σ_ϵ^2, independent across entrepreneurs, and $\alpha \in (0, 1)$.[7] An abler entrepreneur has a higher total product and a higher marginal product of capital at all levels of capital. This restriction on the relationship between average and marginal products over entrepreneurs can be found in Lucas (1978) and Jovanovic (1982). It is consistent with the empirical finding that *total profits* increase with firm size when the latter is measured by assets.[8]

[6] Two exogenous observables were included in the wage equation but not in the entrepreneurial earnings equation; see the last paragraph of app. C (available from the authors).

[7] We assume that $E(\epsilon) = 1$ so that $E[\log(\epsilon)] = -\sigma_\epsilon^2/2$. The disturbance term ϵ reflects an independent and identically distributed productivity shock.

[8] Most studies of the relationship between firm size and the rate of return find that the rate of return either increases with firm size or is constant in firm size. See Scherer (1980) for a review of studies. Therefore, total profits must increase with firm size. A potentially important assumption here is that θ and μ are uncorrelated. We tried to model the case in which they are correlated but found that the resulting model was analytically intractable. A correlated θ and μ should not affect the theoretical implications of the model, although it would affect the characteristics of the constrained entrepreneurs and the likelihood estimates. If θ and μ were positively (negatively) correlated, we would expect that liquidity constraints would be less (more) empirically

Now a word about ϵ. This is the unforeseen component of his income from choosing to become an entrepreneur. Its time-series properties are, strictly speaking, of no import for the analysis. It could, for instance, be a permanent component that affects the entrepreneur's productivity for as long as he is self-employed, or it could represent an independent and identically distributed disturbance to his productivity over time. If ϵ did have a permanent component, revealed only ex post, then we would have a theory of exit as in Jovanovic (1982). The present paper will not explore this intuition any farther because of the great complexity that it would impose on the empirical analysis.

An entrepreneur's net income is

$$y + r(z - k). \tag{3}$$

Here r is *one plus* the rate of interest, and z is the entrepreneur's beginning-of-period wealth. If $z < k$, the entrepreneur is a net borrower, and $r(z - k)$ is the amount he repays at the end of the period. He cannot default, no matter how his business turns out. The latter assumption is not too implausible if people can borrow only a limited amount or have to put up collateral (as indeed appears to be the case as discussed below).[9] We shall assume that each person can borrow up to an amount that is proportional to his wealth; the factor of proportionality is denoted by $\lambda - 1$. Since the amount borrowed cannot exceed $(\lambda - 1)z$, the most that a person can invest in the business is $z + (\lambda - 1)z = \lambda z$.[10] The entrepreneur therefore faces the constraint

$$0 \leq k \leq \lambda z, \tag{4}$$

important. Entrepreneurs will tend to be high- (low-) wage people who are more (less) likely to have accumulated sufficient start-up funds. The results below provide some evidence that entrepreneurs may be relatively poor wage workers. The argument is that individuals who have accumulated a lot of assets will tend to be relatively good wage workers. Therefore, assets can be viewed as a proxy for wage ability. We find that entrepreneurial ability and assets are negatively correlated, which suggests that entrepreneurial and wage ability may be negatively correlated.

[9] The default rate among business exits appears to be fairly small. In 1976 the fraction of all concerns that defaulted was one-third of 1 percent. While comparable estimates on exits are not available, Evans (1987*b*) found that about 21 percent of manufacturing firms exited between 1976 and 1981, for an annual exit rate of about 4 percent per year. His sample was weighted toward larger manufacturing firms. The exit rate of all firms is likely to be higher.

[10] In a more realistic model, λ would depend not only on z but also on observed characteristics, x, since the latter affect y and, hence, the probability of repayment. Ando (1985) finds, e.g., that, for established businesses, the probability of having a loan application accepted increases with the amount of business experience of the applicant, the size of the firm, and the past credit record of the firm. These factors are probably less important for the new businesses considered here since most people starting a business have no previous business experience or credit history. Nevertheless, in future work it would be useful to explore whether the liquidity constraint and the interest rate depend on demographic characteristics.

where the parameter λ satisfies

$$\lambda \geq 1, \tag{5}$$

and it is equal for everyone. The constraint tells us the maximum amount of capital that the entrepreneur can control. The interest factor, r, is also equal for everyone. Moreover, for simplicity we assume that the lending rate equals the borrowing rate.[11]

This form for the liquidity constraint is analytically convenient. Moreover, evidence indicates that the liquidity constraint facing businesses is similar to the L-shaped constraint (4). Ando (1985, p. C.5) finds that most new businesses are likely to face even more severe constraints:[12] "Several conclusions emerge from these studies. One is the critical role of personal savings and loans from friends and relatives, particularly in business formation. It is by far the largest source of capital in new firms and in firms beginning to grow. Once the firm is established, the role of personal savings diminishes as institutional investors perceive less risk and become more willing to provide capital." A study by Scott and Dunkelberg (cited by Ando) for the National Federation of Independent Businessmen found that firms that had been in existence for less than 4 years reported that, on average, only 50 percent of their initial loan request was met. These findings suggest that new businesses are liquidity constrained and that the amount of capital available to them is limited by their personal assets.[13]

[11] The assumption of a constant interest rate is at least a crude approximation to reality. A significant portion of loan applications are rejected, and many new businesses claim difficulty in obtaining financing (see below). This suggests that the supply curve of capital is not upward sloping over a wide range. Banks do not appear to fine-tune risk premia to individual borrowers. For a theoretical discussion of credit rationing, see Stiglitz and Weiss (1981). For a recent study of consumer credit rationing, see Mariger (1987), who finds that almost 20 percent of his sample of families were liquidity constrained. Loans from friends and relatives may be one means to evade the liquidity constraint. There is some evidence, e.g., that the entrepreneurial success of some immigrant groups is due to their access to capital from family and communal networks. Koreans, e.g., participate in rotating credit associations for business start-ups. See Light (1972) for a discussion. Among the established male business owners surveyed by Ando (1985), loans from friends and relatives constituted about 10 percent of the start-up capital, while bank loans constituted about 41 percent.

[12] She reports that, of loans taken in the previous three years, 40 percent required personal collateral and 39.5 percent required business collateral. Although there is some ambiguity, if loans required either personal or business collateral, then 80 percent of the loans would have required some collateral. Of loans that had been rejected, 70.8 percent required personal collateral and 64.5 percent business collateral.

[13] Ham and Melnik (1987) find evidence of liquidity constraints even for firms much larger than those considered here. They report that most credit agreements place an upper limit on borrowing and that about 20 percent of the firms in their sample reached the maximum amount of their commitment size. See their discussion for an alternative interpretation, however.

The Entrepreneur's Investment Decision

At the time that the investment decision is made, the entrepreneur cannot foretell the realization of ϵ, although he does know θ. Under the assumption that he is risk neutral, his investment decision, k, solves

$$\max_{k \in [0, \lambda z]} [\theta k^{\alpha} + r(z - k)]. \tag{6}$$

At an interior maximum, the first-order condition is

$$\theta \alpha k^{\alpha - 1} - r = 0, \tag{7}$$

which leads to the solution

$$k = \left(\frac{\theta \alpha}{r}\right)^{1/(1-\alpha)}, \tag{8}$$

which is valid as long as the right-hand side is no greater than λz. When the latter is true, we say that the entrepreneur is not constrained with respect to how much he can borrow, or is simply "unconstrained." For the entrepreneur to be unconstrained, his θ must satisfy

$$\theta \leq (\lambda z)^{1-\alpha} \frac{r}{\alpha}. \tag{9}$$

Otherwise he is constrained.

Since our data do not contain precise enough information on how much is invested, we shall substitute out from the entrepreneurial earnings equation (2) the optimal capital invested (as given by [8] for the unconstrained and by λz for the constrained). This leads to the following expression for entrepreneurial income:

$$y = \begin{cases} \theta^{1/(1-\alpha)} \left(\dfrac{\alpha}{r}\right)^{\alpha/(1-\alpha)} \epsilon & \text{if } \theta \text{ satisfies (9)} \\ \theta (\lambda z)^{\alpha} \epsilon & \text{otherwise.} \end{cases} \tag{10}$$

Clearly, if θ satisfies (9), y does not depend on z. If, on the other hand, θ does not satisfy (9), then $\partial y / \partial z = \alpha y / z$ and $\partial y / \partial \theta = y / \theta$, so that the slope of the indifference curves in figure 1 is $d\theta / dz|_{y \text{ const}} = -\alpha \theta / z$.

Selection into Entrepreneurship

The would-be entrepreneur, we assume, knows θ *before* he commits himself on whether to start a business. In this we deviate from Jovanovic (1982) and instead follow Lucas (1978). Since the individual knows his θ, he will choose to start a business if and only if his ex-

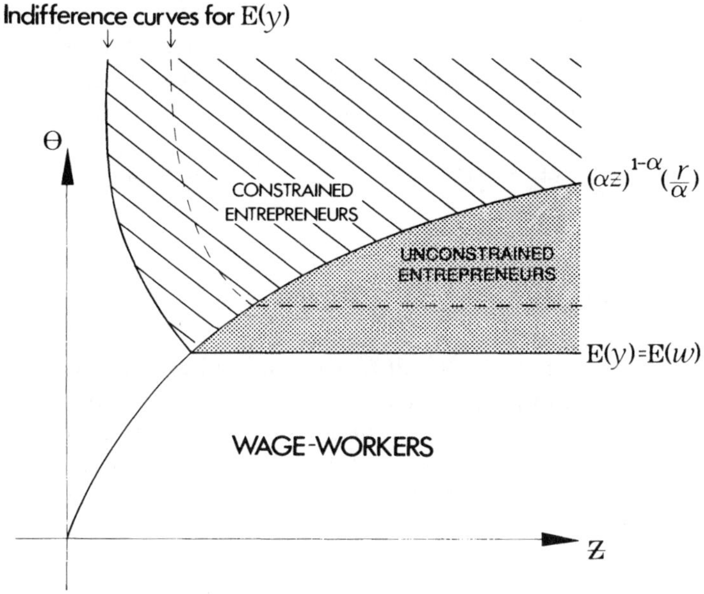

FIG. 1.—Nature of the selection into entrepreneurship

pected net income from doing so exceeds that from wage work:

$$\max[\theta k^{\alpha} + r(z - k)] \geq \mu x_1^{\gamma_1} x_2^{\gamma_2} + rz. \tag{11}$$

Since the left-hand side of (11) increases with θ while the right-hand side does not, we have the following proposition.

PROPOSITION 1: A person with characteristics (θ, z, x) chooses entrepreneurship if $\theta \in (f(z, x), \infty)$, where $f(z, x)$ is the value of θ that solves (11) as an exact equality.

That higher-θ people select into entrepreneurship and that the preferred region in θ-space is a half-open interval is a property that does not depend on the nature of the production function. All that matters is that output is increasing in θ at all levels of the capital stock. On the other hand, the shape of the three regions in figure 1 depends on our assumption that an increase in θ raises the marginal product of capital everywhere. That is, k and θ are gross complements.

Canceling the term rz from both sides of (11) and substituting the optimal k (from [8] for the unconstrained and from λz for the constrained) leaves us with the following *selection conditions*: (1) Equation (9) holds and

$$\theta^{1/(1-\alpha)}\left(\frac{\alpha}{r}\right)^{\alpha/(1-\alpha)} - r\left(\frac{\alpha}{r}\right)^{1/(1-\alpha)}\theta^{1/(1-\alpha)} \geq \mu x_1^{\gamma_1} x_2^{\gamma_2},$$

or (2) equation (9) does not hold and $\theta(\lambda z)^{\alpha} - r\lambda z > \mu^{\gamma}$. Appendix B (available from the authors on request, together with apps. A and C) shows that selection conditions 1 and 2 are equivalent to the following selection conditions:

$$\mu^{1-\alpha}\left(\frac{r}{\alpha}\right)^{\alpha}(1-\alpha)^{\alpha-1}x_1^{\gamma_1(1-\alpha)}x_2^{\gamma_2(1-\alpha)} \leq \theta \leq (\lambda z)^{1-\alpha}\left(\frac{r}{\alpha}\right) \qquad (1')$$

or

$$\theta > \max\left[(\lambda z)^{1-\alpha}\left(\frac{r}{\alpha}\right), \mu x_1^{\gamma_1}x_2^{\gamma_2}(\lambda z)^{-\alpha} + r(\lambda z)^{1-\alpha}\right]. \qquad (2')$$

If θ satisfies either constraint, the individual chooses self-employment. The complicated nature of these constraints stems from the wealth constraint on investment. The special case of *no* wealth constraints obtains when $\lambda = \infty$. Then the right-hand side of condition 2′ is ∞ as is the right-hand side of condition 1′, so that the selection rule takes on a simple one-sided form. Figure 1 shows the nature of the selection.

The Joint Distribution of θ and z

Since θ and z may be correlated in the population, we take the following steps to control for this. The parameter θ is assumed to be independent of everything except possibly z. We write

$$\ln \theta = \delta_0 + \delta_1 \ln z + \eta, \qquad (12)$$

where $\eta \sim N(0, \sigma_{\eta}^2)$, so that $\ln \theta \sim N(\delta_0 + \sigma_1 \ln z, \sigma_{\eta}^2)$.

We shall not dwell too long on δ_1—the parameter determining the population correlation between θ and z. It may reflect greater past savings by those high-θ people who, knowing their θ, expected to become entrepreneurs one day. Or, if we stretch the interpretation a bit, it may reflect lower absolute risk aversion of wealthy people, making them more inclined to become entrepreneurs (as in Kihlstrom and Laffont [1979]).[14] Whatever the source of the correlation between θ and z, the parameter δ_1 will not, in general, be invariant to changes in other parameters of the model. We include it so as to avoid interpreting the apparent correlation between z on the one hand and entrepreneurial choice and success on the other as being caused entirely by the presence of liquidity constraints: A positive δ_1 would imply that, even in the absence of liquidity constraints (i.e., $\lambda = \infty$), wealthier people would be more likely to become entrepreneurs sim-

[14] We thank Larry Samuelson for this interpretation.

ply because of a "spurious" correlation between z and the unobservable θ. As it turns out, we find no evidence that δ_1 is positive.[15]

III. The Data and Reduced-Form Results

We use data from the National Longitudinal Survey of Young Men (NLS) to estimate the structural parameters of the model. This survey consists of a sample of 5,225 men who were between the ages of 14 and 24 in 1966 and who were queried periodically between 1966 and 1981. Individuals reported net family assets, the key variable for our purposes, in 1971, 1976, and 1981. We look at a subsample of 1,949 white males who were wage workers in 1976, who either were wage workers or were self-employed in 1978, and who were not unemployed, out of the labor force, in the military, or in school full-time in either 1976 or 1978.[16] Of these individuals, 89 (4.5 percent) entered self-employment. Unfortunately, we lost a number of other individuals who failed to report information we needed. Asset and income variables were often missing. For the maximum likelihood analysis we also had to delete some individuals who reported negative net worth or negative self-employment income (since we take the logs of these variables), leaving us with 1,443 white men for the actual estimation. Given that this subsample consists of less than half of the 3,918 white males who started in the survey in 1966, our results may not be representative of the population at large. For the reduced-form results reported below, we include all individuals for whom the relevant information is available. Table 1 reports variable definitions, summary statistics, and sample selection. Further details are discussed in Evans and Leighton (1987).

The men in our sample were between the ages of 24 and 34 in 1976. This age range covers a portion of the life cycle during which a large proportion of the men who will eventually become self-employed make the transition. According to *Current Population Survey* data, approximately 2.5 percent of 21–25-year-old males were self-

[15] In fact, δ_1 is negative and statistically significant. This result suggests that high-asset people tend to be relatively poor entrepreneurs. But it is also possible that this is a sign that the model is misspecified. Some of the results for the model with δ_1 negative are implausible, as we discuss below.

[16] Note that we have eliminated workers who were unemployed in 1976. Ongoing work by Evans and Leighton finds that unemployed workers are about twice as likely to switch into self-employment than other workers. A few of the individuals in our sample, however, did experience spells of unemployment between the 1976 and 1978 survey weeks and were not eliminated. Of the entrants, six (10 percent of the 59) experienced some unemployment, and of the nonentrants, 97 (7 percent of the 1,384 nonentrants) experienced some unemployment. In the reduced-form results reported below, we included a dummy variable for workers who experienced some unemployment between survey weeks but found that it had no substantive effect on the results.

TABLE 1

DEFINITION OF VARIABLES

Name	Mean (Standard Deviation)	Definition
Enter	.0409 (.2022)	Equals 0 if wage worker in 1976 and 1978 survey weeks and 1 if wage worker in 1976 survey week and self-employed in 1978 survey week; defined only for individuals who were employed in both survey weeks and who were not full-time students
Assets	20.0092 (50.0533)	Net family assets in 1976 survey week (in thousands of dollars)
Income	17.6553 (19.1731)	Individual wage earnings in 1978 for wage workers and individual self-employment earnings in 1978 for self-employed workers (in thousands of dollars)*
Work experience	11.9959 (12.5343)	Years of wage experience as of 1978 survey week†
Education	13.9390 (14.1546)	Years of education as of 1978 survey week

NOTE.—Data drawn from the National Longitudinal Survey of Young Men, 1966–81. The initial sample size consisted of 3,918 white young men aged 14–24 in 1966. Data were available for 2,536 in both 1976 and 1978. Of these men, 1,986 were employed in both survey weeks and were not enrolled in school full-time. The resulting sample consisted of 126 entrants and 1,860 nonentrants on the basis of the self-employment indicator variable available from the NLS. Of the 126 entrants, 37 were deleted because in 1978 they reported having been self-employed more than 2 years. Most of the deletions were incorporated self-employed who apparently reported themselves as wage workers in 1976 even though they were running their own businesses. Approximately 110 individuals had missing asset data, and 60 individuals had missing income data. The maximum likelihood estimation also deleted individuals who reported nonpositive assets or self-employment earnings.

* Two adjustments were made to self-employment income: (1) Some individuals who run incorporated businesses report self-employment income as wage and salary income. If an individual was incorporated, reported no self-employment income, and reported wage and salary income, we used his wage and salary income in place of self-employment income. (2) For individuals who had operated their businesses for less than a year (as measured by tenure as of the 1978 survey week), we prorated self-employment income by the length of time in business to obtain an estimate of annual self-employment earnings.

† Wage experience accumulated from 1966. An estimate of pre-1966 job experience was used for workers who had jobs prior to the start of the survey. See app. A to Evans and Leighton (1987) for details.

employed, 5.7 percent of 26–30-year-old males, 9.0 percent of 31–35-year-old males, and 11.2 percent of 36–40-year-old males. The population fraction of self-employment remains fairly constant between age 40 and retirement age.[17]

In 1978 the average self-employed man in our sample earned $15,746 compared with $16,760 for wage earners. The average increase in earnings was smaller for those who switched into self-employment than for those who remained wage workers.[18] Of the 89

[17] These results are based on averages for 1976–86. See Evans and Leighton (1987) for details on the data and Evans and Leighton (1989) for a discussion of the dynamics of changes in the stock of self-employed workers over the life cycle.

[18] A possible reason for the discrepancy is that self-employment income gets more favorable tax treatment.

TABLE 2

ESTIMATED PROBABILITY OF ENTERING SELF-EMPLOYMENT: REDUCED-FORM RESULTS, 1976–78

	PROBIT ESTIMATES*	
VARIABLE	(1)	(2)
Assets	.0053	.0075
	(.0282)	(.0032)
Assets²/100	−.0010	−.0015
	(.0008)	(.0011)
Wage experience	.0248	.0320
	(.0144)	(.0148)
Education	−.0128	−.0038
	(.0163)	(.0174)
Married	−.4799	−.4491
	(.1162)	(.1183)
Urban	−.1511	−.0879
	(.1124)	(.1166)
Handicapped	−.1376	−1730
	(.2131)	(.2152)
Wage income in 1976/1,000	...	−.0222
		(.0104)
Constant	−1.3915	−1.3991
	(.3057)	(.3143)
Log likelihood	−330.0562	−327.4762
Observations	1,839	1,835

NOTE.—Standard errors are in parentheses.
* The F-statistic for assets and assets^2 was 2.24 (p-value of .0107) for the regression in the first column and 3.64 (p-value of .0264) in the second column.

individuals who switched into self-employment, 20 percent formed incorporated businesses.

Before we estimate the structural parameters of the model, it is useful to consider some simple tests of the model. If one assumes a zero correlation between assets and entrepreneurial ability, our model has two implications. First, there is a positive correlation between the probability of starting a business and assets *if and only if* there are liquidity constraints: Such constraints mean that a wealthier person can start a business with a more efficient capital level and thereby realize a greater return than a poorer one. Table 2 reports probit estimates of the relationship between the probability of entering self-employment and initial assets. The estimates in the first column control for education, experience, and several demographic characteristics. The effect of assets on the probability of starting a business is positive (at the sample mean) and is statistically significant at the 10 percent level. The estimates in the second column also condition on wage earnings in the previous period. The effect of

TABLE 3

ESTIMATED EFFECT OF LOG ASSETS ON LOG SELF-EMPLOYMENT EARNINGS: REDUCED-FORM RESULTS

	REGRESSION ESTIMATES	
VARIABLE	1978	1981
Log assets	.1424	−.0683
	(.0538)	(.1239)
Log experience	.2290	1.0010
	(.3534)	(1.0273)
Log education	−.3344	.4627
	(.5589)	(1.1865)
Married	−.1422	.1532
	(.2694)	(.5536)
Urban	−.0054	.2044
	(.2059)	(.4828)
Handicapped	.9884	−.9083
	(.5423)	(1.0332)
Constant	8.8066	1.2021
	(3.1708)	(8.6413)
R^2	.2267	.0806
F-statistic	2.54	.45
Observations	59	37

NOTE.—Dependent variable is log self-employment earnings. Semilog regressions yielded qualitatively similar results. Standard errors are in parentheses.

assets on the probability of starting a business is positive (at the sample mean) and is statistically significant at the 2 percent level.[19]

The second implication is that the correlation between entrepreneurial earnings and initial assets is positive since wealthier people will have started businesses with more efficient capital levels. Table 3 reports regression estimates that show a statistically significant positive correlation between log self-employment earnings and log assets after controlling for education, experience, and several demographic characteristics.[20]

A third implication is that people with smaller assets will be forced to devote a larger proportion of their assets to their businesses. Ando (1985) estimates that a 10 percent increase in assets leads to a 4 percent increase in the amount of personal savings devoted to a business. Thus poorer individuals do devote a larger proportion of their wealth to their businesses, just as our model implies.

Although the model is static, it suggests several dynamic implications. Constrained firms will start with a suboptimal amount of capital and therefore will be smaller than unconstrained firms. As a result,

[19] Evans and Leighton (1989) report similar results for 1980–81.

[20] We obtain similar results using a semilog regression and using only levels.

constrained businesses will have a greater tendency to reinvest earnings back into the business than unconstrained firms since the return on capital invested in the business is higher for the constrained firms. Therefore, smaller firms will grow faster than larger firms that entered at the same time. Evans (1987*a*, 1987*b*) and Dunne, Roberts, and Samuelson (1987) confirm that smaller firms do indeed grow faster than larger firms when business age and other characteristics are held constant.[21]

As firms grow over time, the importance of the initial liquidity constraint will diminish. We have checked this implication by comparing the effect of initial (1976) assets on entrepreneurial earnings in 1978 and 1981. The results reported in table 3 are consistent with the diminishing importance of assets over time.[22] The elasticity of earnings with respect to initial assets is positive and statistically significant for 1978 and negative and highly insignificant for 1981. This suggests that assets are not acting as a proxy for entrepreneurial ability, whose effect should persist over time.[23]

Recent attempts to subsidize small business start-ups by unemployed workers also provide some weak evidence for our model. Relaxing binding liquidity constraints by providing unemployment benefits in a lump sum or otherwise providing start-up assistance should increase business formations. Bendick and Egan (1987) report that the British Enterprise Allowance Program roughly doubled the formation of businesses among unemployed workers.[24]

IV. Maximum Likelihood Estimates

The likelihood function is derived in the appendix (available from the authors). The regressions reported above indicate that work experi-

[21] Another related implication of the model is that firms that face a tighter liquidity constraint will grow more rapidly. We tested this implication by regressing the growth of income between 1978 and 1981 for entrants who were self-employed in both years against the log of assets (35 observations were available). The relationship was negative, as predicted, but statistically insignificant.

[22] We might also expect that failures will experience smaller growth in assets than successes. The average change in nominal assets between 1976 and 1981 was 250 percent for successes and 111 percent for failures. The price level increased 159.7 percent between these years. Therefore, failures had negative asset growth and successes had positive asset growth.

[23] We also estimated the 1978 earnings equation using the 1981 sample to determine whether the change in the sign of the asset variable is due to self-selection. The estimated coefficient on assets was smaller for the restricted sample (about .6) but remained statistically significant at the 5 percent level.

[24] The Enterprise Allowance Program gave workers an operating subsidy for up to 1 year instead of unemployment benefits. Workers had to have $1,500 of their own to invest in the business before they could be eligible. Bendick and Egan note that workers received more from participating in the program than if they had remained unemployed.

ence and education were not substantively or statistically significant determinants of entrepreneurial earnings. Therefore, entrepreneurial earnings are assumed to depend only on assets. Wage earnings are assumed to depend on education and experience.[25]

Table 4 reports the maximum likelihood estimates. The estimated correlation between entrepreneurial ability and assets is negative and statistically significant. Thus we can reject the hypothesis that assets are a positive proxy for entrepreneurial ability. The key finding is that there are binding capital constraints. In the second column, which constrains δ_1 to be zero, the point estimate of λ is 1.44 with a 99 percent (plus or minus three standard deviations) confidence interval of (1.31, 1.59). Our results support Knight ($\lambda = 1$) over Schumpeter ($\lambda = \infty$).

The estimate of α in the second column means that a 10 percent increase in the capital devoted to a business leads to a 2.2 percent increase in earnings. This coefficient is highly significant.[26] The estimate of σ_θ is almost the same as the estimate of σ_ϵ. This means that entrepreneurs face considerable risk, not just as perceived by outsiders (who can see neither ϵ nor θ), but also as perceived by themselves since they cannot see ϵ until after they have committed their investment. Surprisingly, the distribution of unanticipated earnings shocks seems to be more variable for wage workers than for entrepreneurs.

[25] As mentioned in the appendix, α is constrained to lie in the unit interval (i.e., marginal product of capital is positive and diminishing) and λ is constrained to be positive. Our rough expectation is that α would be about a third—the share of capital in income (see n. 26 for more on this point). The assumption that $0 < \alpha < 1$ implies that there are diminishing returns to capital and that capital and ability are complements. The assumption that $\lambda > 0$ precludes the implausible implication that liquidity decreases with assets. Both assumptions are necessary because $\log(\alpha)$, $\log(1 - \alpha)$, and $\log(\lambda)$ appear in the likelihood function. It was not possible to keep all the parameter estimates within the permissible range during the likelihood estimation without these constraints. In order to ensure that the estimates reported are for a global maximum of the likelihood function, we perturbed the values of the key parameters and checked to see that the likelihood function was indeed lower and that the parameters moved toward the estimates at the maximum. We tried several high and low combinations of α and λ in this process. We also tried several alternative functional forms for the capital constraint (e.g., including higher-order terms in assets and allowing the capital constraint to depend on some of the exogenous variables) but had difficulty obtaining convergence.

[26] The estimates of α of .39 and .22 are not too far from those obtained elsewhere using the Cobb-Douglas production function. In models without liquidity constraints, α should equal capital's share in income, which is about one-third. In our model, however, α is the share of capital in the income of *unconstrained* entrepreneurs only. As for a constrained entrepreneur, he spends $r\lambda z$ on capital and expects $\theta(\lambda z)^\alpha$ in income. This yields a capital share of $r(\lambda z)^{1-\alpha}/\theta$. But since a constrained entrepreneur's θ violates (9), this means that $1/\theta < [(\lambda z)^{1-\alpha} r/\alpha]^{-1}$, which means that the empirical share of capital *exceeds* α for the constrained. Therefore, our estimate of α should be *less* than one-third, and on these grounds, again, the estimates in the second column of table 4 are more plausible than those in the first column.

TABLE 4

MAXIMUM LIKELIHOOD ESTIMATES OF ENTREPRENEURIAL SELECTION UNDER LIQUIDITY CONSTRAINTS FOR 1976–78

COEFFICIENT	NAME	ESTIMATE (1)	ESTIMATE (2)
Capital returns	α	.3862	.2186
		(.0822)	(.0200)
Capital constraint	λ	1.7263	1.4401
		(.0782)	(.0429)
Log entrepreneurial ability—constant	δ_0	2.3388	2.2732
		(.0498)	(.0404)
Log entrepreneurial ability—assets	δ_1	−.1160	. . .
		(.0343)	
Wage constant	μ	.2098	.1947
		(.2168)	(.2115)
Experience returns	γ_1	.3225	.3281
		(.0390)	(.0386)
Education returns	γ_2	.7200	.7207
Standard deviations for:			
Entrepreneurial ability	σ_θ	.2682	.2849
		(.0139)	(.0118)
Entrepreneurial earnings	σ_ϵ	.2666	.2797
		(.0222)	(.0163)
Wage earnings	σ_ξ	.4114	.4114
		(.0043)	(.0043)
Log likelihood		−954.0183	−959.8647
Observations		1,443	1,443

NOTE.—Estimation performed with the MAXLIK procedure in Gauss. Asymptotic standard errors are in parentheses.

Given the stringent functional form assumptions and the extensive attrition from our sample, the reader should regard the estimated structural parameters merely as suggestive of what more refined research will reveal. Nonetheless, let us take the estimates seriously and use them to evaluate the impact of liquidity constraints on business start-ups. We shall use the estimates from column 2 of table 4 for these calculations.[27]

The optimal amount of start-up capital for a white person with average education and experience is estimated to be $16,739 for someone with log entrepreneurial ability three standard deviations above the mean, $8,603 for two standard deviations above the mean, and $4,421 for one standard deviation above the mean. The geometric mean assets in our sample were $7,433. Someone with these assets could control business capital of $10,704. Thus only the

[27] Most of the results were similar if col. 1 is used instead, with the exception of the effect of removing the constraint. The correlated case predicts that removing the constraints leads to roughly a sixfold increase in the rate of entry, an implausible inference. Complete results are available from the authors on request.

ablest would not be able to obtain sufficient capital if they decided to start a business.[28] Most people are not constrained in the sense that, if they did decide to start a business, they would have sufficient capital according to our estimates.

Only high-ability/low-asset people are affected by the wealth constraint. But it is precisely these people who are most likely to want to switch to self-employment. Those with high ability can earn more in self-employment than in wage work, especially if they have poor wage earnings. But those with poor wage earnings are also likely to have accumulated relatively few assets. Indeed, compared with nonentrants, entrants had lower average wage earnings ($12,339 for entrants vs. $13,592 for nonentrants in 1976) and lower geometric mean assets ($5,927 for entrants and $7,504 for nonentrants in 1976).

Using the estimated structural parameters, we can calculate the fraction of the population that has values of θ (and of the other characteristics) satisfying constraints 1′ and 2′.[29] The average probability of being a constrained entrepreneur (i.e., satisfying 1′) is 3.75 percent, and the average probability of being an unconstrained entrepreneur (i.e., satisfying 2′) is 0.06 percent. Thus the liquidity constraint is binding for virtually all the individuals who are likely to start a business.

The liquidity constraint reduces the amount of capital flowing to entrepreneurship in two ways.[30] First, it will prevent some people from trying entrepreneurship. If the liquidity constraint were removed, the estimates indicate that the average probability of becoming an entrepreneur over the period would increase to 5.11 percent from 3.81 percent. Thus we estimate that the liquidity constraint deters 1.3 percent of the population from trying entrepreneurship.[31] Second, individuals who do try entrepreneurship use less capital because of the constraint. For example, an individual with $5,927 of assets (the geometric mean for entrants) and with log entrepreneurial ability two standard deviations above the mean uses $68 too little capital ($8,603 − 1.44 × $5,927). The difference between optimal and actual capital is lower for people with lower ability—an individual

[28] In fact, for individuals with average characteristics and assets, only those with log entrepreneurial ability 2.33 standard deviations above the mean would be constrained.

[29] These figures are obtained by taking the logarithmic values of the elements of the constraints and using the assumption that ln θ is normally distributed to calculate the probability that an individual with observed characteristics will satisfy the constraint. The sample mean of these individual probabilities is then taken.

[30] The figures reported below are in 1976 dollars. Prices have roughly doubled since 1976.

[31] This, of course, is a partial equilibrium result. Increasing the supply of entrepreneurs would decrease entrepreneurial rewards and thereby reduce the number of individuals who would like to switch.

with \$5,927 of assets and log entrepreneurial ability three standard deviations above the mean would use the optimal amount of capital—and higher for people with less assets—an individual with \$3,717 of assets and log entrepreneurial ability two standard deviations above the mean uses \$3,250 too little capital.

The structural parameters also give a rough idea of the aggregate self-employment investment that might not be made because of the liquidity constraint. There were 50.3 million white male members of the labor force in 1976. The average annual entry rate into self-employment for the NLS sample between 1976 and 1978 was 2.25 percent. Therefore, 1.26 million whites are estimated to have entered self-employment annually. If we assume that each of these people used \$68 too little capital (as would be the case for an individual whose log entrepreneurial ability is two standard deviations above the mean and who has assets of \$5,917—the geometric mean for entrants), the total lost investment because of suboptimal investment by initial entrants is \$86 million a year. Additionally, 1.30 percent of our sample were deterred from entering self-employment for a 2-year period (0.65 percent annually) or 0.3 million people annually. If we assume that these individuals would have invested the optimal start-up capital of \$8,603 (the amount for a person with log entrepreneurial ability two standard deviations above the mean), the total lost investment because of deterred entry is \$2,580 million. We thus obtain a total "lost" investment of \$2.7 billion in 1976 dollars. By way of comparison, gross private domestic investment in 1976 was \$174 billion. Again, it is important to emphasize that (1) these calculations are meant to be illustrative[32] and (2) without a better understanding of the nature of the liquidity constraints, the welfare implications of these calculations are unclear.

V. Conclusions

This paper develops and estimates a behavioral model of entrepreneurial choice under liquidity constraints. Several interesting results emerge. One can reject Schumpeter's view that capital markets allow a separation of the entrepreneurial and capitalist functions: Liquidity constraints bind. It is estimated that entrepreneurs are limited to a capital stock that is no more than about one and one-half times their wealth. As a result, almost all the entrepreneurs in our sample are estimated to devote less capital to their businesses than they would like to. Our findings are consistent with those of Fazzari, Hubbard,

[32] More accurate calculations would require integrating over the joint distribution of ability, assets, education, and experience.

and Petersen (1987), who find evidence of significant capital market imperfections even for publicly traded manufacturing corporations.

We conclude with several words of caution and some suggestions for further research. Our results are based on a simple static model of entrepreneurship. Further work on entrepreneurial selection and asset accumulation over the life cycle would be useful. The estimates rely on a sample that is subject to several potential selection biases whose correction could alter our conclusions. Because of sample attrition and missing values, only about half of the initial sample was used in the estimation. An interesting extension would have only a fraction q of 1976 wage workers observing a self-employment opportunity, and then q could be estimated; currently q is assumed to be one. The model also relies on a simple formulation of the liquidity constraints. Estimating more general liquidity constraints in which income and demographic characteristics play a role would enrich our understanding of the role that liquidity constraints play in entrepreneurial choice.

References

Ando, Faith. *Access to Capital by Subcategories of Small Business.* Report prepared for the U.S. Small Business Administration. Fort Washington, Pa.: JACA Corp., December 1985.

Bendick, Marc, Jr., and Egan, Mary Lou. "Transfer Payment Diversion for Small Business Development: British and French Experience." *Indus. and Labor Relations Rev.* 40 (July 1987): 528–42.

Dunne, Timothy; Roberts, Mark; and Samuelson, Laurence. "Plant Failure and Employment Growth in the U.S. Manufacturing Sector." Manuscript. University Park: Pennsylvania State Univ., Dept. Econ., January 1987.

Evans, David S. "The Relationship between Firm Growth, Size, and Age: Estimates for 100 Manufacturing Industries." *J. Indus. Econ.* 35 (June 1987): 567–81. (*a*)

———. "Tests of Alternative Theories of Firm Growth." *J.P.E.* 95 (August 1987): 657–74. (*b*)

Evans, David S., and Leighton, Linda S. *Self-Employment Selection and Earnings over the Life Cycle.* Washington: Small Bus. Admin., December 1987.

———. "Some Empirical Aspects of Entrepreneurship." *A.E.R.* 79 (June 1989).

Fazzari, Steven; Hubbard, R. Glenn; and Petersen, Bruce C. "Financing Constraints and Corporate Investment." Working Paper no. 2387. Cambridge, Mass.: NBER, September 1987.

Ham, John C., and Melnik, Arie. "Loan Demand: An Empirical Analysis Using Micro Data." *Rev. Econ. and Statis.* 69 (November 1987): 704–9.

Johnson, William R. "A Theory of Job Shopping." *Q.J.E.* 92 (May 1978): 261–78.

Jovanovic, Boyan. "Job Matching and the Theory of Turnover." *J.P.E.* 87, no. 5, pt. 1 (October 1979): 972–90.

———. "Selection and the Evolution of Industry." *Econometrica* 50 (May 1982): 649–70.

Kanbur, Steven M. "Of Risk Taking and the Personal Distribution of Income." *J.P.E.* 87 (August 1979): 769–97.

Kihlstrom, Richard E., and Laffont, Jean-Jacques. "A General Equilibrium Entrepreneurial Theory of Firm Formation Based on Risk Aversion." *J.P.E.* 87 (August 1979): 719–48.

Kirzner, Israel M. *Competition and Entrepreneurship.* Chicago: Univ. Chicago Press, 1973.

Knight, Frank H. *Risk, Uncertainty and Profit.* New York: Houghton Mifflin, 1921.

LeRoy, Stephen F., and Singell, Larry D., Jr. "Knight on Risk and Uncertainty." *J.P.E.* 95 (April 1987): 394–406.

Light, Ivan H. *Ethnic Enterprise in America: Business and Welfare among Chinese, Japanese, and Blacks.* Berkeley: Univ. California Press, 1972.

Lucas, Robert E., Jr. "On the Size Distribution of Business Firms." *Bell J. Econ.* 9 (Autumn 1978): 508–23.

Mariger, Randall P. "A Life-Cycle Consumption Model with Liquidity Constraints: Theory and Empirical Results." *Econometrica* 55 (May 1987): 533–57.

Miller, Robert A. "Job Matching and Occupational Choice." *J.P.E.* 92 (December 1984): 1086–1120.

Scherer, Frederic M. *Industrial Market Structure and Economic Performance.* 2d ed. Boston: Houghton Mifflin, 1980.

Schumpeter, Joseph A. *The Theory of Economic Development.* Cambridge, Mass.: Harvard Univ. Press, 1934.

———. *Capitalism, Socialism, and Democracy.* 3d ed. New York: Harper and Row, 1950.

Stiglitz, Joseph E., and Weiss, Andrew. "Credit Rationing in Markets with Imperfect Information." *A.E.R.* 71 (June 1981): 393–410.

U.S. Department of Labor. *Self Employment Demonstration Project Request for Proposal RFP-DAA-88-23.* Washington: Employment and Training Admin., U.S. Dept. Labor, January 15, 1988.

Part III
Endogenous Technical Change

[12]

Endogenous Technological Change

Paul M. Romer

University of Chicago

Growth in this model is driven by technological change that arises from intentional investment decisions made by profit-maximizing agents. The distinguishing feature of the technology as an input is that it is neither a conventional good nor a public good; it is a nonrival, partially excludable good. Because of the nonconvexity introduced by a nonrival good, price-taking competition cannot be supported. Instead, the equilibrium is one with monopolistic competition. The main conclusions are that the stock of human capital determines the rate of growth, that too little human capital is devoted to research in equilibrium, that integration into world markets will increase growth rates, and that having a large population is not sufficient to generate growth.

I. Introduction

Output per hour worked in the United States today is 10 times as valuable as output per hour worked 100 years ago (Maddison 1982). In the 1950s, economists attributed almost all the change in output per hour worked to technological change (Abramovitz 1956; Kendrick 1956; Solow 1957). Subsequent analysis raised our estimates of

Prepared for the conference "The Problem of Economic Development: Exploring Economic Development through Free Enterprise," held at the State University of New York at Buffalo, May 1988. I have benefited from the comments of many seminar and conference participants and two discussants (Rob Vishny, Buffalo, May 1988, and Dale Jorgenson, National Bureau of Economic Research Economic Fluctuations meeting, July 1988). Discussions with Gary Becker, Karl Shell, Robert Lucas, Gene Grossman, and Elhanan Helpman were especially helpful. Research assistance was provided by Danyang Xie. The original work was supported by National Science Foundation grant SES-8618325. It was revised while I was a visitor at the Center for Advanced Study in the Behavioral Sciences and supported by NSF grant BNS87-00864.

[*Journal of Political Economy*, 1990, vol. 98, no. 5, pt. 2]

the importance of increases in the effective labor force and the effective stock of capital in generating growth in output per worker (Jorgenson, Gollop, and Fraumeni 1987), but technological change has surely been important as well. The raw materials that we use have not changed, but as a result of trial and error, experimentation, refinement, and scientific investigation, the instructions that we follow for combining raw materials have become vastly more sophisticated. One hundred years ago, all we could do to get visual stimulation from iron oxide was to use it as a pigment. Now we put it on plastic tape and use it to make videocassette recordings.

The argument presented in this paper is based on three premises. The first is that technological change—improvement in the instructions for mixing together raw materials—lies at the heart of economic growth. As a result, the model presented here resembles the Solow (1956) model with technological change. Technological change provides the incentive for continued capital accumulation, and together, capital accumulation and technological change account for much of the increase in output per hour worked.

The second premise is that technological change arises in large part because of intentional actions taken by people who respond to market incentives. Thus the model is one of endogenous rather than exogenous technological change. This does not mean that everyone who contributes to technological change is motivated by market incentives. An academic scientist who is supported by government grants may be totally insulated from them. The premise here is that market incentives nonetheless play an essential role in the process whereby new knowledge is translated into goods with practical value. Our initial understanding of electromagnetism arose from research conducted in academic institutions, but magnetic tape and home videocassette recorders resulted from attempts by private firms to earn a profit.

The third and most fundamental premise is that instructions for working with raw materials are inherently different from other economic goods. Once the cost of creating a new set of instructions has been incurred, the instructions can be used over and over again at no additional cost. Developing new and better instructions is equivalent to incurring a fixed cost. This property is taken to be the defining characteristic of technology.

Most models of aggregate growth, even those with spillovers or external effects, rely on price-taking behavior. But once these three premises are granted, it follows directly that an equilibrium with price taking cannot be supported. Section II of the paper starts by showing why this is so. It also indicates which of the premises is dropped in growth models that do depend on price-taking behavior. The argument in this section is fundamental to the motivation for the particular model of monopolistic competition that follows, but it is more general than the model itself.

In the specific model outlined in Section III, a firm incurs fixed design or research and development costs when it creates a new good. It recovers those costs by selling the new good for a price that is higher than its constant cost of production. Since there is free entry into this activity, firms earn zero profit in a present value sense.

The conclusions of the model follow directly from this specification. On the basis of results from the static theory of trade with differentiated goods (see, e.g., Helpman and Krugman 1985), one should expect that fixed costs lead to gains from increases in the size of the market and therefore to gains from trade between different countries. Perhaps the most interesting feature of the equilibrium calculated for the model constructed here is that increases in the size of the market have effects not only on the level of income and welfare but also on the rate of growth. Larger markets induce more research and faster growth.

The analysis also suggests why population is not the right measure of market size and why the presence of a large domestic market in countries such as China or India is not a substitute for trade with the rest of the world. The growth rate is increasing in the stock of human capital, but it does not depend on the total size of the labor force or the population. In a limiting case that may be relevant for historical analysis and for the poorest countries today, if the stock of human capital is too low, growth may not take place at all.

These implications of the model are taken up briefly in the final sections of the paper. Section III describes the functional forms that are used to describe the preferences and the technology for the model. It defines an equilibrium that allows for both monopolistic competition and external effects arising from knowledge spillovers. Section IV offers a brief intuitive description of a balanced growth equilibrium for the model. Section V formally characterizes the equilibrium. Section VI describes the welfare properties of the equilibrium. Section VII discusses the connection implied by the model between trade, research, and growth. Algebraic details of the derivations are placed in the Appendix.

II. Rivalry, Excludability, and Nonconvexities

Economists studying public finance have identified two fundamental attributes of any economic good: the degree to which it is rivalrous and the degree to which it is excludable (Cornes and Sandler 1986). Rivalry is a purely technological attribute. A purely rival good has the property that its use by one firm or person precludes its use by another; a purely nonrival good has the property that its use by one firm or person in no way limits its use by another. Excludability is a function of both the technology and the legal system. A good is excludable if the owner can prevent others from using it. A good such as the code for a computer program can be made excludable by means of a legal system that prohibits copying or by means of encryption and copy protection schemes.

Conventional economic goods are both rivalrous and excludable. They are privately provided and can be traded in competitive markets. By definition, public goods are both nonrival and nonexcludable. Because they are nonexcludable, they cannot be privately provided or traded in markets. Public goods can be introduced into a model of price-taking behavior by assuming the existence of a government that can levy taxes. Basic scientific research is an example of a public good that could be provided in this way and that is relevant for modeling growth.

Rivalry and excludability are closely linked because most rival goods are excludable. (A parking space in a shopping center parking lot is an example of a good that is effectively nonexcludable because the cost of enforcing excludability is too high relative to the value of the good.) The interesting case for growth theory is the set of goods that are nonrival yet excludable. The third premise cited in the Introduction implies that technology is a nonrival input. The second premise implies that technological change takes place because of the actions of self-interested individuals, so improvements in the technology must confer benefits that are at least partially excludable. The first premise therefore implies that growth is driven fundamentally by the accumulation of a partially excludable, nonrival input.

To evaluate these claims, it helps to have a specific case in mind. The example of a nonrival input used in what follows is a design for a new good. The vast majority of designs result from the research and development activities of private, profit-maximizing firms. A design is, nonetheless, nonrival. Once the design is created, it can be used as often as desired, in as many productive activities as desired.

In this sense, a design differs in a crucial way from a piece of human capital such as the ability to add. The design is nonrival but the ability to add is not. The difference arises because the ability to add is inherently tied to a physical object (a human body) whereas the design is not.[1] The ability to add is rivalrous because the person who

[1] The original version of this paper used the terms "embodied" and "disembodied" to refer to the difference between an intangible such as the ability to add, which is tied to a specific person, and an intangible such as a design, which is not. This choice of terminology is not used in this revision because embodiment has another meaning in growth theory and because the notion of rivalry already exists in the public finance literature.

possesses this ability cannot be in more than one place at the same time; nor can this person solve many problems at once. As noted above, rivalry leads to a presumption that human capital is also excludable. Thus human capital can be privately provided and traded in competitive markets. In contrast, the design is nonrival because it is independent of any physical object. It can be copied and used in as many different activities as desired.

Like any scientific concept, nonrivalry is an idealization. It is sometimes observed that a design cannot be a nonrival good because it is itself tied to the physical piece of paper or the physical computer disk on which it is stored. What is unambiguously true about a design is that the cost of replicating it with a drafter, a photocopier, or a disk drive is trivial compared to the cost of creating the design in the first place. This is not true of the ability to add. Training the second person to add is as costly as training the first. For simplicity, the arguments here will treat designs as idealized goods that are not tied to any physical good and can be costlessly replicated, but nothing hinges on whether this is literally true or merely close to being true.

Nonrivalry has two important implications for the theory of growth. First, nonrival goods can be accumulated without bound on a per capita basis, whereas a piece of human capital such as the ability to add cannot. Each person has only a finite number of years that can be spent acquiring skills. When this person dies, the skills are lost, but any nonrival good that this person produces—a scientific law; a principle of mechanical, electrical, or chemical engineering; a mathematical result; software; a patent; a mechanical drawing; or a blueprint—lives on after the person is gone. Second, treating knowledge as a nonrival good makes it possible to talk sensibly about knowledge spillovers, that is, incomplete excludability. These two features of knowledge—unbounded growth and incomplete appropriability—are features that are generally recognized as being relevant for the theory of growth. What thinking about nonrivalry shows is that these features are inextricably linked to nonconvexities.

If a nonrival input has productive value, then output cannot be a constant-returns-to-scale function of all its inputs taken together. The standard replication argument used to justify homogeneity of degree one does not apply because it is not necessary to replicate nonrival inputs. Suppose that a firm can invest 10,000 hours of engineering time to produce a design for a 20-megabyte hard disk drive for computers. Suppose that it can produce a total of 2 trillion megabytes of storage per year (i.e., 100,000 units of the drive) if it builds a $10 million factory and hires 100 workers. If it merely replicates the rival inputs—the factory and the workers—it can double its output to 4 trillion megabytes of storage per year.

Suppose that the firm could have invested 20,000 hours of engineering time in the design work instead of 10,000 hours and, by doing so, could have designed a 30-megabyte hard disk drive that could be manufactured with the same factory and workers. When the firm doubles all its inputs, it uses a 20,000-hour design, two factories, and 200 workers and produces 6 trillion megabytes of storage per year, three times the original output.

More formally, if $F(A, X)$ represents a production process that depends on rival inputs X and nonrival inputs A, then by a replication argument, it follows that $F(A, \lambda X) = \lambda F(A, X)$. This replication argument assumes that X is an exhaustive list of the rival inputs. Because the focus here is on national economies, the argument neglects integer problems that may be relevant for a small market that gets stuck between n and $n + 1$ plants. The fact that it may not be possible to actually replicate all the inputs in the list X has no bearing on this argument about the properties of $F(\cdot)$.

If A is productive as well, it follows that F cannot be a concave production function because $F(\lambda A, \lambda X) > \lambda F(A, X)$. Because of the properties of homogeneous functions, it also follows that a firm with these kinds of production possibilities could not survive as a price taker. If disk drives sold for marginal cost, annual revenue for the firm would just equal interest payments on the capital and wage payments to workers. More generally, since $F(A, X) = X \cdot (\partial F/\partial X)(A, X)$, it follows that

$$F(A, X) < A \cdot \frac{\partial F}{\partial A}(A, X) + X \cdot \frac{\partial F}{\partial X}(A, X).$$

If all inputs were paid their value marginal product, the firm would suffer losses.

This point has been made many times before (Schumpeter 1942; Arrow 1962*b*; Shell 1966, 1967, 1973; Nordhaus 1969; Wilson 1975). Previous growth models have avoided this difficulty in various ways. Solow (1956) treats A as an exogenously provided public input (i.e., an input that is both nonexcludable and nonrival). Shell (1966, 1967) treats it as a public input that is provided by the government. In each case, the factor A receives no compensation, and every individual firm is assumed to be free to exploit the entire stock of A. These models are consistent with the first premise, that technological change drives growth, and the third, that the technology is a nonrival good, but they are inconsistent with the second premise. They both deny the role that private, maximizing behavior plays in generating technological change.

In an attempt to make the evolution of A responsive to market incentives, Arrow (1962*a*) assumed that an increase in K necessarily leads to an equiproportionate increase in knowledge through "learning by doing," but he still treats knowledge as a public good. Lucas (1988) assumed in effect that it is production of human capital rather than physical capital that generates this nonrival, nonexcludable good. Both of these papers make the production of a nonrival, nonexcludable good an unintentional side effect of the production of a conventional good.

The learning-by-doing formulation has the advantage that it makes the rate of accumulation of nonrival knowledge endogenous, but it is unsatisfactory because it takes the strict proportionality between knowledge and physical capital or knowledge and education as an unexplained and exogenously given feature of the technology. It preserves the public-good character of knowledge assumed by Solow and Shell but makes it a public good that is privately provided as a side effect. Like the other public-good formulations, it rules out the possibility that firms make intentional investments in research and development.

This formulation has the additional difficulty that it is not robust. The nonrival input produced through learning by doing must be completely nonexcludable. If it were even partially excludable, Dasgupta and Stiglitz (1988) show that decentralized equilibrium with many firms would not be sustainable.

In a partial equilibrium model of an industry in which firms face upward-sloping cost curves, Shell (1973) proposed a model with price taking in which expenditure on research was compensated out of quasi rents. Griliches (1979), again in an industry setting, made this formulation more explicit. He assumed that the production function takes the form $F(A_N, A_E, X)$, where A_E represents an excludable part of the benefits of research and development and A_N represents the nonexcludable part. Since A_E is excludable, it is accumulated intentionally. The nonexcludable part A_N is created as a side effect of producing A_E. He also assumed that the function $F(\cdot)$ is homogeneous of degree one in X and A_E taken together.

In an aggregate model of growth, I made the same kind of assumption (Romer 1986). To make the dynamic analysis in this paper simple, I reduced the dynamic model to one with a single-state-variable model by assuming that the excludable good A_E that the firm produces intentionally is used in fixed proportions with physical capital. As a result, the model ends up having dynamics similar to those of Arrow's learning-by-doing model, and the mathematical equations can be interpreted equally well in terms of learning by doing that is incidental to capital production.

The advantage of the interpretation that knowledge is compensated out of quasi rents is that it allows for intentional private investments in research and development. The difficulty is that it violates the logic of the replication argument. If the input A_E is truly the result of research and development, it is a nonrival good. In this case, the function $F(\cdot)$ must be homogeneous of degree one in X alone and the Dasgupta and Stiglitz (1988) argument applies. If a nonrival input is even partially excludable, nonconvexities are present and a decentralized equilibrium cannot be sustained. Stated in terms of cost curves, the replication argument implies that long-run cost curves are horizontal when all rival inputs are treated as variable. What appear to be quasi rents are merely competitive returns to rival factors that are in fixed supply. These quasi rents cannot be used to compensate both the innovation activity and the rival fixed factors. Even if these factors are in fixed supply at the aggregate level, they presumably have alternative uses and will not be supplied to an activity if they are not paid their marginal product.

An alternative approach to growth theory with price-taking competition is to dispute the first and third premises, that technological change drives growth and that knowledge about the technology is a nonrival input. Human capital models such as those presented by King and Rebelo (1987), Jones and Manuelli (1988), Rebelo (1988), and Becker, Murphy, and Tamura (this issue) treat all forms of intangible knowledge as being analogous to human capital skills that are rivalrous and excludable. There is no nonrival input like the technology and, hence, no nonconvexities or spillovers.

The only way to accept all three premises described in the Introduction is to return to the suggestion of Schumpeter (1942) and explicitly introduce market power. Shell (1973) described a model with a single monopolist who invests in technological change, but as he recognizes, it is difficult to give an aggregate interpretation to a model dominated by a single firm. In a recent paper (Romer 1987), I presented a model with market power but also with free entry and many firms. It builds on the model of monopolistic competition in consumption goods formulated by Dixit and Stiglitz (1977), applied in a dynamic setting by Judd (1985), used in a dynamic model of trade by Grossman and Helpman (1989*c*), and extended to differentiated inputs in production by Ethier (1982). The specification used here differs from that in my earlier paper primarily because it emphasizes the importance of human capital in the research process. The earlier model showed that scale is an important determinant of the rate of growth. The analysis here shows that the correct measure of scale is not population but human capital.

III. Description of the Model

The four basic inputs in this model are capital, labor, human capital, and an index of the level of the technology. Capital is measured in units of consumption goods. Labor services L are skills such as eye-hand coordination that are available from a healthy physical body. They are measured by counts of people. As used here, human capital H is a distinct measure of the cumulative effect of activities such as formal education and on-the-job training.

The concept of human capital as years of education or training that are person specific is close to the one used in labor market contexts (e.g., Heckman 1976; Rosen 1976). It corresponds to the practice in growth accounting applications that take account of changes in the quality of the labor force due to changes in observables such as the level of education and experience (see, e.g., Gollop and Jorgenson 1980). This concept of human capital is more limited than the notion used in theoretical models of growth based on unlimited human capital accumulation such as those presented by King and Rebelo (1987), Lucas (1988), and Becker et al. (this issue). These models implicitly combine a notion of knowledge that can outlive any individual with a labor market notion of human capital that does not.

The model used here separates the rival component of knowledge, H, from the nonrival, technological component, A. Because it has an existence that is separate from that of any individual, A can grow without bound. In the specific formulation used below, each new unit of knowledge corresponds to a design for a new good, so there is no conceptual problem measuring A. It is a count of the number of designs.

The formal model of the economy has three sectors. The research sector uses human capital and the existing stock of knowledge to produce new knowledge. Specifically, it produces designs for new producer durables. An intermediate-goods sector uses the designs from the research sector together with forgone output to produce the large number of producer durables that are available for use in final-goods production at any time. In practice, one might expect research on a new design and the production of the new good to take place within the same firm, and nothing in the analysis here rules this out; design work can take place either internally or in a separate firm that sells its patent to the firm that will produce the actual good. A final-goods sector uses labor, human capital, and the set of producer durables that are available to produce final output. Output can be either consumed or saved as new capital.

To keep the dynamic analysis simple and highlight the effects of interest, several simplifying assumptions are used. The first is that the population and the supply of labor are both constant. This rules out an analysis of fertility, labor force participation, or variation in hours worked per worker. The second is that the total stock of human capital in the population is fixed and that the fraction supplied to the market is also fixed. Thus the supply of the aggregate factors L and H

is fixed. The assumption on H is made largely for technical reasons. The dynamic analysis is greatly simplified by restricting attention to equilibria with constant growth rates. In a stationary population in which people have finite lives, the only feasible constant growth rate for total years of education or experience is zero. A more complicated dynamic analysis could consider the effects of the kinds of increases in H and L that have been observed historically, but even in this kind of analysis, H must ultimately approach an upper bound.

The other simplifying assumptions are extreme assumptions on factor intensities. One has already been made implicitly. Assuming that capital can be accumulated as forgone output is equivalent to assuming that capital goods are produced in a separate sector that has the same technology as the final-output sector. Forgoing consumption is then equivalent to shifting resources from the consumption sector into the capital sector. Also, the plausible assertion that research is relatively human capital– and knowledge-intensive is translated into an extreme specification in which only knowledge and human capital are used to produce new designs or knowledge. Labor and capital do not enter at all. These kinds of restrictions will reduce the analysis of the dynamics of this system to a system of equations that can be explicitly solved by doing algebra. Presumably, a relaxation of these assumptions that preserves the factor intensity orderings used here would not change the basic dynamics of the model.

Final output Y in this model is expressed as a function of physical labor L, human capital devoted to final output H_Y, and physical capital. The unusual feature of the production technology assumed here is that it disaggregates capital into an infinite number of distinct types of producer durables. For now, let these durables be indexed by an integer i. (Soon, the index i will be assumed to be a continuous variable instead of a discrete one to avoid integer constraints.) Only a finite number of these potential inputs, the ones that have already been invented and designed, are available for use at any time. Thus if $x = \{x_i\}_{i=1}^{\infty}$ is the list of inputs used by a firm that produces final output, there is some value A such that $x_i = 0$ for all $i \geq A$. Because A changes as new producer durables are invented, it is important to be able to describe final output as a stationary function of all conceivable input lists.

In this kind of environment, a simple functional form for output is the following extension of the Cobb-Douglas production function:

$$Y(H_Y, L, x) = H_Y^{\alpha} L^{\beta} \sum_{i=1}^{\infty} x_i^{1-\alpha-\beta}. \tag{1}$$

This production function differs from the usual production function only in its assumption about the degree to which different types of capital goods are substitutes for each other. In the conventional specification, total capital K is implicitly defined as being proportional to the sum of all the different types of capital. This definition implies that all capital goods are perfect substitutes. One additional dollar of capital in the form of a truck has the same effect on the marginal productivity of mainframe computers as an additional dollar's worth of computers. Equation (1) expresses output as an additively separable function of all the different types of capital goods so that one additional dollar of trucks has no effect on the marginal productivity of computers.

Among the various types of capital goods, one can imagine pairs of inputs that are close substitutes (trucks and trains), pairs that are complements (computers and communications networks), and many pairs that fall somewhere in between. The conventional formulation explores the case in which all durables are perfect substitutes. The model here considers the case in which all durables have additively separable effects on output. An investigation of complementarity as well as of mixtures of types of substitutability is left for future work.

Because the production function in equation (1) is homogeneous of degree one, output in the final-goods sector can be described in terms of the actions of a single, aggregate, price-taking firm. The sector that produces producer durables, however, cannot be described by a representative firm. There is a distinct firm i for each durable good i. A firm must purchase or produce a design for good i before commencing production. Once it owns the design, the firm can convert η units of final output into one durable unit of good i. As in the standard one-sector model, the formal specification here describes the sector that produces capital goods as a black box that takes final output in on one side and gives capital goods out of the other side. The correct interpretation of this formal description is that the forgone consumption is never manufactured. The resources that would have been used to produce the forgone output are used instead to manufacture capital goods. It is possible to exchange a constant number of consumption goods for each unit of capital goods if the production function used to manufacture capital goods has exactly the same functional form as the production function used to manufacture consumption goods.

Once a firm has produced a design for durable i, it can obtain an infinitely lived patent on that design. If the firm manufactures $x(i)$ units of the durable, it rents those durables to final-output firms for a rental rate $p(i)$. Since firm i will be the only seller of capital good i, it will face a downward-sloping demand curve for its good. Since the durables are assumed not to depreciate, the value of one unit of durable i is the present discounted value of the infinite stream of rental income that it generates.

There are many equivalent institutional arrangements that can support any given equilibrium. In what follows, the firm that owns the patent on the design for durable i is assumed to be the only one that manufactures it, but this is only a convenience. Whether the owner of the patent manufactures the good itself or licenses others to do so, it can extract the same monopoly profit. Design of new durables and manufacturing could take place within the same firm, but it is easier to describe the equilibrium if the research and development department is treated as a separate firm and designs are transferred for an explicit price. It is also easier to assume that the firm that buys a design and manufactures a differentiated producer durable rents its durables instead of selling them outright. In particular, this shows that there are market mechanisms that avoid the usual durable-goods-monopoly problem. Provided that the manufacturer of the durable could commit to levels of output, nothing would change if the durables were sold instead of rented. The analysis is further simplified by assuming that the durables do not depreciate. Adding depreciation would merely add a familiar term to the user cost of capital.

In parallel with the usual one-sector model and in conformity with national income accounting conventions, it is useful to define an accounting measure of total capital K as cumulative forgone output. Thus $K(t)$ evolves according to the rule

$$\dot{K}(t) = Y(t) - C(t), \tag{2}$$

where $C(t)$ denotes aggregate consumption at time t. Because it takes η units of forgone consumption to create one unit of any type of durable, this accounting measure K is related to the durable goods that are actually used in production by the rule $K = \eta \sum_{i=1}^{\infty} x_i = \eta \sum_{i=1}^{A} x_i$.

Thus H and L are fixed, and K grows by the amount of forgone consumption. It remains to specify the process for the accumulation of new designs, that is, for the growth of $A(t)$. As noted above, research output depends on the amount of human capital devoted to research. It also depends on the stock of knowledge available to a person doing research. If designs were treated as discrete indivisible objects that are not produced by a deterministic production process, the production technology for designs would have to take explicit account of both integer constraints and uncertainty. There is no doubt that both indivisibility and uncertainty are important at the micro level and over short periods of time. The simplifying assumption made here is that neither is crucial to a first-pass analysis of technological change at the aggregate level. Henceforth, the index i

for the different types of goods is treated as a continuous variable, and the sum in equation (1) is replaced by an integral:

$$Y(H_Y, L, x) = H_Y^{\alpha} L^{\beta} \int_0^{\infty} x(i)^{1-\alpha-\beta}\, di. \qquad (1')$$

(Here x must be interpreted as a function from an appropriately defined function space.) With this formal structure, the output of new designs produced by researcher j can be written as a continuous, deterministic function of the inputs applied. If the researcher possesses an amount of human capital H^j and has access to a portion A^j of the total stock of knowledge implicit in previous designs, the rate of production of new designs by researcher j will be $\delta H^j A^j$, where δ is a productivity parameter.

This formulation with a continuum of goods is close to that used by Judd (1985) in his discussion of patents. He also studies the process whereby goods are introduced as new patents are produced. The model here corresponds to Judd's discussion of the case in which patents are infinitely lived. The main differences are that Judd treats the differentiated goods as consumption goods rather than producer durables and uses a form of exogenous technological change to generate growth of productivity in the research sector. Here, growth in A by itself increases the productivity of human capital in the research sector.

Although other assumptions about secrecy and property rights could be considered, the equilibrium here is based on the assumption that anyone engaged in research has free access to the entire stock of knowledge. This is feasible because knowledge is a nonrival input. All researchers can take advantage of A at the same time. The output of researcher j is therefore $\delta H^j A$. If we sum across all people engaged in research, the aggregate stock of designs evolves according to

$$\dot{A} = \delta H_A A, \qquad (3)$$

where H_A has the obvious interpretation of total human capital employed in research.

Equation (3) contains two substantive assumptions and two functional form assumptions. The first substantive assumption is that devoting more human capital to research leads to a higher rate of production of new designs. The second is that the larger the total stock of designs and knowledge is, the higher the productivity of an engineer working in the research sector will be. According to this specification, a college-educated engineer working today and one working 100 years ago have the same human capital, which is measured in terms of years of forgone participation in the labor market. The engineer working today is more productive because he or she can take advantage of all the additional knowledge accumulated as design problems were solved during the last 100 years.

The two functional form assumptions are that the output of designs is linear in each of H_A and A when the other is held constant. These assumptions, like the exclusion of the inputs L and $x(\cdot)$, are made largely for analytical convenience. Linearity in H_A is not important for the dynamic properties of the model, but weakening this assumption would require a more detailed specification of how income in the research sector is allocated to the participants.

Linearity in A is what makes unbounded growth possible, and in this sense, unbounded growth is more like an assumption than a result of the model. In what follows, it will become clear that the marginal product of human capital H_Y employed in the manufacturing sector grows in proportion to A. If A were replaced in equation (3) by some concave function of A—that is, if the marginal productivity of human capital in the research sector does not continue to grow in proportion to A—human capital employed in research would shift out of research and into manufacturing as A becomes larger. This will cause the rate of growth to slow down. Whether opportunities in research are actually petering out, or will eventually do so, is an empirical question that this kind of theory cannot resolve. The specification here, in which unbounded growth at a constant rate is feasible, was chosen because there is no evidence from recent history to support the belief that opportunities for research are diminishing. Moreover, linearity in A is convenient analytically, and assumptions about what will happen in the far future for values of A that are very large relative to the current level have very little effect on the question of interest: How do other variables in the model affect the rate of growth of A?

The crucial feature of the specification used here is that knowledge enters into production in two distinct ways. A new design enables the production of a new good that can be used to produce output. A new design also increases the total stock of knowledge and thereby increases the productivity of human capital in the research sector. The owner of a design has property rights over its use in the production of a new producer durable but not over its use in research. If an inventor has a patented design for widgets, no one can make or sell widgets without the agreement of the inventor. On the other hand, other inventors are free to spend time studying the patent application for the widget and learn knowledge that helps in the design of a wodget. The inventor of the widget has no ability to stop the inventor of a wodget from learning from the design of a widget. This means that the benefits from the first productive role for a design are completely excludable, whereas the benefits from the second are completely non-excludable. In an overall sense, this means that the nonrival design inputs are partially excludable.

In theory and in practice, there is always some ambiguity about what constitutes a design for a new and different good and what constitutes a copy of an existing design. In the model, this ambiguity is artificially resolved by the form of the production function Y. This functional form implies that new goods are never close substitutes for existing goods since all the producer durables enter into production in an additively separable fashion. (See Pascoa [1986] for a discussion of this general property of "no neighboring goods.") Although it greatly simplifies the analysis, this is not a realistic feature of the model. In particular, it rules out the possibility of obsolescence. Nonetheless, the general results here should be robust to more careful modeling of the nature of the interaction between different specialized producer durables. What matters for the results is that the knowledge is a nonrival good that is partially excludable and privately provided.

At the aggregate level, H_A and H_Y are related by the constraint $H_Y + H_A = H$. According to these equations, any person can devote human capital to either the final-output sector or the research sector. Implicitly this formulation neglects the fact that L and H are supplied jointly. To take the equations used here literally, one must imagine that there are some skilled persons who specialize in human capital accumulation and supply no labor.

To fix notation for prices, let spot prices at any point in time be measured in units of current output and let r denote the interest rate on loans denominated in goods. Let P_A denote the price of new designs, and let w_H denote the rental rate per unit of human capital. Because goods can be converted into capital one for one, the spot price for capital is one and its rate of return is r. Because of the assumption that anyone engaged in research can freely take advantage of the entire existing stock of designs in doing research to produce new designs, it follows from equation (3) that P_A and w_H are related by $w_H = P_A \delta A$.

Once a design has been produced, a large number of potential suppliers of the new good bid for the right to do so. Each of these firms takes the price P_A for designs, the price of one for capital goods, and the interest rate as given, but if it begins production it sets prices to maximize profits. Formally, it helps to let the rental price $p(i)$ for the ith durable lie in the range $\mathbb{R}_+ \cup \{\infty\}$ for its durable good. If no firm produces good i, its price can be understood to be $p(i) = \infty$.

Faced with a price list $\{p(i): i \in \mathbb{R}_+\}$ for all the producer durables, including infinite prices for the durables that have not been invented yet, the representative final-output firm chooses a profit-maximizing quantity $x(i)$ for each durable. Because it is a constant-returns-to-scale firm, its input demands are defined only after the scale of operation is pinned down. Let L and H_Y be the total amounts of labor and human capital that are used in the production of final-output goods. (The split of total H between H_Y and H_A remains to be determined.) Given values for H_Y and L, it is possible to derive the aggregate demand for the durables from a maximization problem that is conditional on them:

$$\max_x \int_0^\infty [H_Y^\alpha L^\beta x(i)^{1-\alpha-\beta} - p(i)x(i)]di.$$

Differentiating under the integral sign leads to an inverse demand function

$$p(i) = (1 - \alpha - \beta)H_Y^\alpha L^\beta x(i)^{-\alpha-\beta}. \tag{4}$$

(There is an important technical issue about what it means to the final-goods producer if prices change on a set of values of i that has measure zero. Because of the symmetry in this model, eq. [4] can readily be derived by a limiting argument. For a general discussion of this issue, see Pascoa [1986, 1990].)

The demand curve in equation (4) is what the producer of each specialized durable takes as given in choosing the profit-maximizing price to set. Faced with given values of H_Y, L, and r, a firm that has already incurred the fixed-cost investment in a design will choose a level of output x to maximize its revenue minus variable cost at every date:

$$\begin{aligned} \pi &= \max_x p(x)x - r\eta x \\ &= \max_x (1 - \alpha - \beta)H_Y^\alpha L^\beta x^{1-\alpha-\beta} - r\eta x. \end{aligned} \tag{5}$$

The flow of rental income is $p(x)$ times x. The cost is the interest cost on the ηx units of output needed to produce x durables. To keep the analysis simple, it is convenient to assume that the capital is putty-putty, so that the firm can solve this problem at every point in time, converting units of durables back into general capital, and avoid the interest cost if it decides to supply fewer units. At any date, the only sunk cost is the initial expenditure on the design. This assumption is harmless because the demand for durables is stationary in equilibrium, so no disinvestment ever takes place.

The monopoly pricing problem specified in equation (5) is that of a firm with constant marginal cost that faces a constant elasticity demand curve. The resulting monopoly price is a simple markup over marginal cost, where the markup is determined by the elasticity of demand, $\bar{p} = r\eta/(1 - \alpha - \beta)$. The flow of monopoly profit is $\pi = (\alpha + \beta)\bar{p}\bar{x}$, where $\bar{x}$ is the quantity on the demand curve (4) implied by the price $\bar{p}$.

Each producer of specialized durables must rent its output to a large number of final-goods producers that can operate at any scale. By assumption, it is not possible for the producer to monitor the use of its durables. As a result, price discrimination is not feasible. The best the firm can do is charge the simple monopoly price.

The decision to produce a new specialized input depends on a comparison of the discounted stream of net revenue and the cost P_A of the initial investment in a design. Because the market for designs is competitive, the price for designs will be bid up until it is equal to the present value of the net revenue that a monopolist can extract. At every date t, it must therefore be true that

$$\int_t^\infty e^{-\int_t^\tau r(s)ds}\pi(\tau)d\tau = P_A(t). \tag{6}$$

If P_A is constant (as it will be in the equilibrium described below), this condition can be put in a more intuitive form. Differentiating with respect to time t yields

$$\pi(t) - r(t)\int_t^\infty e^{-\int_t^\tau r(s)ds}\pi(\tau)d\tau = 0.$$

Substituting in the expression for P_A from equation (6) yields

$$\pi(t) = r(t)P_A. \tag{6'}$$

This equation says that at every point in time, the instantaneous excess of revenue over marginal cost must be just sufficient to cover the interest cost on the initial investment in a design. (This formulation of the intertemporal zero profit constraint is taken from Grossman and Helpman [1989*c*].)

The solution of the model for a balanced growth equilibrium given in Section V shows that the technology described above implies a negatively sloped linear relation between the rate of growth of output and the rate of return on investment. To close the model, it remains to specify preferences that imply a parallel relation between the rate of growth of consumption and the marginal rate of intertemporal substitution. This relation is easily derived for Ramsey consumers with discounted, constant elasticity preferences:

$$\int_0^\infty U(C)e^{-\rho t}dt, \quad \text{with } U(C) = \frac{C^{1-\sigma}-1}{1-\sigma} \quad \text{for } \sigma \in [0, \infty).$$

The implied intertemporal optimization condition for a consumer faced with a fixed interest rate r is that $\dot{C}/C = (r - \rho)/\sigma$. Preferences enter the solution of the model only through this relation between the consumption growth rate and the interest rate.

The consumers are endowed with fixed quantities of labor L and human capital H that are supplied inelastically. At time 0, consumers own the existing durable-goods-producing firms, and the net revenues of these firms are paid to consumers as dividends. Final-goods firms earn zero profits and own no assets, so they can be ignored in the specification of endowments.

An equilibrium for this model will be paths for prices and quantities such that (i) consumers make savings and consumption decisions taking interest rates as given; (ii) holders of human capital decide whether to work in the research sector or the manufacturing sector taking as given the stock of total knowledge A, the price of designs P_A, and the wage rate in the manufacturing sector w_A; (iii) final-goods producers choose labor, human capital, and a list of differentiated durables taking prices as given; (iv) each firm that owns a design and manufactures a producer durable maximizes profit taking as given the interest rate and the downward-sloping demand curve it faces, and setting prices to maximize profits; (v) firms contemplating entry into the business of producing a durable take prices for designs as given; and (vi) the supply of each good is equal to the demand.

IV. Discussion of the Model

A reasonable intuition for the behavior of this model can be inferred by considering the Solow (1956) model, in which the evolution of A is given exogenously, and the Uzawa (1965) model, in which the evolution of A is determined by the allocation of resources between a research sector and a final-goods sector. For a fixed amount of A, and therefore a fixed set of producer durables, the model is almost identical to the Solow model. Because of the symmetry in the model, all the durable goods that are available are supplied at the same level, henceforth denoted as $\bar{x}$. If they were not, it would be possible to increase profits in the producer durable sector by reducing the output of high-output firms and diverting the capital released in this way to low-output goods. Since A determines the range of durables that can be produced and since η units of capital are required per unit of durable goods, it is possible to solve for $\bar{x}$ from the equation $K = \eta A\bar{x}$. Then output Y can be written as

$$\begin{aligned} Y(H_A, L, x) &= H_Y^{\alpha} L^{\beta} \int_0^{\infty} x(i)^{1-\alpha-\beta}\, di \\ &= H_Y^{\alpha} L^{\beta} A \bar{x}^{1-\alpha-\beta} \\ &= H_Y^{\alpha} L^{\beta} A \left(\frac{K}{\eta A}\right)^{1-\alpha-\beta} \\ &= (H_Y A)^{\alpha} (LA)^{\beta} (K)^{1-\alpha-\beta} \eta^{\alpha+\beta-1}. \end{aligned} \tag{7}$$

The last line of this equation shows that the model behaves just like the neoclassical model with labor and human capital augmenting technological change. In particular, it exhibits the usual diminishing returns to capital accumulation. Given the assumed form of preferences, a fixed level of A will lead to an equilibrium with a steady state in which the level of K is determined by the requirement that the marginal product of capital is equal to the discount rate. If A grew at an exogenously specified exponential rate, the economy would converge to a path on which K grows at the same exponential rate as A, just as it does in the Solow model. Along the transition path, the ratio of K to A would change, which implies that r and $\bar{x}$ would change as well. Along the balanced growth path, r, $\bar{x}$, and the ratio of K to A are all constant.

The nonconvexity evident in the expression for final output as a function of the primary inputs of the model (H, L, K, and A) is supported in a decentralized equilibrium that relies on monopolistic competition. In contrast, the nonconvexity present in equation (3) describing output of designs is supported through competition with external effects that arise from knowledge spillovers. In each case, the nonconvexity arises because the nonrival good A is an input in production. In the final-output sector, A matters indirectly because of its effects on the availability of the new $x(\cdot)$ goods. In the research sector, A enters directly.

Both spillovers and price setting seem essential to capturing the features of knowledge in a model of growth. There is little doubt that much of the value to society of any given innovation or discovery is not captured by the inventor, and any model that missed these spillovers would miss important elements of the growth process. Yet it is still the case that private, profit-maximizing agents make investments in the creation of new knowledge and that they earn a return on these investments by charging a price for the resulting goods that is greater than the marginal cost of producing the goods.

V. Solution of the Model for a Balanced Growth Equilibrium

The strategy for characterizing the model that is followed here is to solve for an equilibrium in which the variables A, K, and Y grow at constant exponential rates. This is generally referred to as a balanced growth equilibrium. The intuition from the Solow model suggests that such an equilibrium will exist if A grows at a constant exponential rate. The intuition from the Uzawa model suggests that it is possible for A to grow at an exponential rate because equation (3) for $\dot{A}$ is linear in A. It will grow at a constant rate if the amount of human capital H_A that is devoted to research stays constant. Verifying that a balanced growth equilibrium exists therefore reduces to the problem of showing that prices and wages are such that H_Y and H_A remain constant as Y, K, C, and A grow.

By focusing only on balanced growth paths, the analysis neglects the transient dynamics that arise when the economy starts from a ratio of K to A that differs from the ratio that is maintained along the balanced growth path. One should be able to study convergence to the balanced growth ratio of K to A using the tools used for studying the Solow and Uzawa models, but this analysis is not attempted here.

The first step in the solution of the model is to derive the relation between the growth rate of output and the rate of return on investment. Coupled with the relation between interest rates and growth rates implied by the preference side of the model, this will determine the rate of growth and interest rate. Given the intuition from the Solow model and the results derived in my earlier model of differentiated inputs (Romer 1987), it follows that along the balanced growth path, the ratio of K to A should be constant, which implies that $\bar{x}$ is constant as well. Because of the accumulation of both K and A, the wage paid for human capital in the final-output sector will grow in proportion to A, but by equation (3), the productivity of human capital in research also grows in proportion to A. Since the productivity of human capital grows at the same rate in both sectors, H_Y and H_A will remain constant if the price P_A for new designs is constant.

Figure 1 illustrates the behavior of the inputs in the model. All the producer durables that have been designed up to time t are used at the level $\bar{x}$ that indicates the height of the rectangle. The width is the measure of the number of designs or durables in use, $A(t)$. The area $A(t)\bar{x}$ is equal to total capital divided by η. Over time, $\bar{x}$ remains constant and A grows at a constant exponential rate.

It remains to check that this description of a balanced growth path is consistent with all the equilibrium conditions. As noted in the dis-

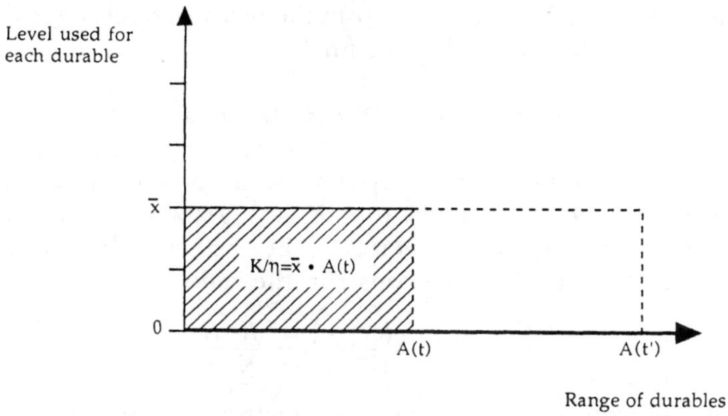

FIG. 1.—Producer durables used in production at time t and $t' > t$

cussion following the monopoly profit problem (5), the flow of profit that can be extracted by the seller of any particular durable input is $\pi = (\alpha + \beta)\bar{p}\bar{x}$. Since the present discounted value of this stream of profit must equal the price P_A of the design, it follows that

$$P_A = \frac{1}{r}\pi = \frac{\alpha + \beta}{r}\bar{p}\bar{x} = \frac{\alpha + \beta}{r}(1 - \alpha - \beta)H_Y^{\alpha}L^{\beta}\bar{x}^{1-\alpha-\beta}. \quad (8)$$

The last equality in this expression follows by using equation (4) to evaluate $\bar{p}$ in terms of $\bar{x}$.

The condition determining the allocation of human capital between the final-output and research sectors says that the wages paid to human capital in each sector must be the same. In the final-output sector, the wage for human capital is its marginal product. Since human capital receives all the income from the research sector, the wage there is $P_A\delta A$. To equalize returns to human capital in both sectors, $H_Y = H - H_A$ must be chosen so that

$$\begin{aligned} w_H = P_A\delta A &= \alpha H_Y^{\alpha-1}L^{\beta}\int_0^{\infty}\bar{x}^{1-\alpha-\beta}\,di \\ &= \alpha H_Y^{\alpha-1}L^{\beta}A\bar{x}^{1-\alpha-\beta}. \end{aligned} \quad (9)$$

Substituting P_A from equation (8) into equation (9) and simplifying yield

$$H_Y = \frac{1}{\delta}\frac{\alpha}{(1 - \alpha - \beta)(\alpha + \beta)}r. \quad (10)$$

For a fixed value of $H_A = H - H_Y$, the implied exponential growth rate for A is δH_A. From the monopoly pricing problem, we know that

$\bar{x}$ is constant if r is. The intuition from the Solow model suggests, and examination of the expression for final output

$$Y = H_Y^{\alpha} L^{\beta} \int_0^{\infty} \bar{x}^{1-\alpha-\beta}\, di = H_Y^{\alpha} L^{\beta} A \bar{x}^{1-\alpha-\beta}$$

from equation (7) shows, that output grows at the same rate as A if L, H_Y, and $\bar{x}$ are fixed. If $\bar{x}$ is fixed, then K must grow at the same rate as A, because total usage of capital is $A\bar{x}\eta$. Let g denote the growth rate of A, Y, and K. Since K/Y is a constant, the ratio

$$\frac{C}{Y} = 1 - \frac{\dot{K}}{Y} = 1 - \frac{\dot{K}}{K}\frac{K}{Y}$$

must also be constant. The common growth rate g for all these variables is therefore

$$g = \frac{\dot{C}}{C} = \frac{\dot{Y}}{Y} = \frac{\dot{K}}{K} = \frac{\dot{A}}{A} = \delta H_A.$$

Together with equation (10), the constraint $H_Y = H - H_A$ implies a relation between the growth rate g and the interest rate r:

$$g = \delta H_A = \delta H - \frac{\alpha}{(1 - \alpha - \beta)(\alpha + \beta)} r, \tag{11}$$

which can be simplified as

$$g = \delta H_A = \delta H - \Lambda r, \tag{11'}$$

where Λ is a constant that depends on the technology parameters α and β,

$$\Lambda = \frac{\alpha}{(1 - \alpha - \beta)(\alpha + \beta)}. \tag{12}$$

Implicitly, the allocation of H between the two sectors is constrained by the requirement that H_A must be nonnegative and that H_Y can be no larger than H. This implies that g is nonnegative. If this constraint is binding, equation (10) will hold as an inequality.

To close the model, it remains to impose the relation between the growth rate g and the interest rate r implied by the preference side of the model, $g = \dot{C}/C = (r - \rho)/\sigma$. Combined with equation (11), this gives an expression for g in terms of the fundamentals of the model,

$$g = \frac{\delta H - \Lambda\rho}{\sigma\Lambda + 1}, \tag{13}$$

where Λ is as defined in equation (12).

The expression for the growth rate suggests a minor technical restriction. For the integral in the consumer's preferences to be finite, the rate of growth of current utility $(1 - \sigma)g$ must be less than the discount rate ρ. Thus, for $\sigma \in [0, 1)$, $(1 - \sigma)\delta H/(\Lambda + 1)$ must be less than the discount rate ρ. If this does not hold, the integral can be infinite and some kind of overtaking criterion must be used to describe the behavior of the consumer.

VI. Interpretation and Welfare Properties of the Solution

Almost all the content of the model is contained in equation (11), which summarizes the effects of the technological side of the model, including the effects of imperfect competition in the market for producer durables. Part of the intuition behind equation (11) is easy to grasp. The opportunity cost of human capital is the wage income that can be earned instantaneously in the manufacturing sector. The return to investing human capital in research is a stream of net revenue that a design generates in the future. If the interest rate is larger, the present discounted value of the stream of net revenue will be lower. Less human capital will be allocated to research, and the rate of growth will be lower.

Other aspects of the intuition behind equation (11) are more subtle. The surprising feature of equation (11) is that neither L nor the parameter η is present. Equation (4) shows that an increase in L increases the demand faced by each monopolistic firm selling a durable good. From the monopoly problem (5), it is clear that a reduction in η reduces the costs of the monopolist and increases output $\bar{x}$. In either case, the stream of net revenue generated by a new design increases. Nonetheless, the amount of human capital devoted to research increases neither when L increases nor when η falls.

In a partial equilibrium analysis, one would always expect any change that increases the return to an activity to increase the allocation of resources to that activity. What a general equilibrium analysis emphasizes is that any intervention that increases the return to one activity can very well increase the return to some other activity that competes with the first activity for resources. An increase in L or a reduction in η (which increases $\bar{x}$) raises the return to human capital employed in manufacturing at the same time that it raises the return to human capital in research. For the functional forms used here, these two effects exactly cancel. The allocation of human capital between research and manufacturing does not change when L and η change.

This kind of exact cancellation is not a robust feature of the model.

If slightly different functional forms were used, the net effect of an increase in L or a decrease in η could be either to increase or to reduce the amount of human capital used in the research sector. For example, in an extension of this model (Romer 1990), I show how an increase in L could reduce research effort and the rate of growth. The correct inference from this model is that the effect of an increase in L or a decrease in η on the rate of growth is ambiguous, something that a priori theorizing cannot resolve.

This ambiguity contrasts with the strong and robust implication that reductions in the interest rate will speed up growth. Equivalently, from equation (12), it follows that any change in the preference parameters that acts to reduce the interest rate (an increase in patience captured by a decrease in the discount rate ρ, or an increase in the intertemporal rate of substitution, as captured by a decrease in σ) will increase research and growth. This implication follows directly from the assumption that the benefits of research come largely in the future and that the costs are incurred immediately.

From a policy point of view, the difference between a reduction in η and a reduction in the equilibrium interest rate is very important. A direct subsidy to investment in physical capital financed by a lump-sum tax is mathematically equivalent to a reduction in η. (An illustration of this easy result is given in Romer [1989].) On the basis of an intuition developed for the one-sector model, economists often identify the marginal product of physical capital with the market interest rate. If they were the same, anything that increases the capital stock and reduces the marginal product of physical capital would have the same growth-enhancing effect as a reduction in the interest rate. The intuition that subsidies to physical capital accumulation will also speed up growth in A was reinforced by the first generation of endogenous growth models. In Arrow's (1962*a*) model of learning by doing and in my first model (Romer 1986), the rate of growth of A was forced by assumption to be the same as the rate of growth of K. As a result, an intervention such as an investment tax credit that increased the accumulation of K necessarily increased the accumulation of A as well.

The model presented here shows that when the decision to invest in physical capital is uncoupled from the decision to invest in research, the effects of a subsidy to physical capital are quite different from the effects of a reduction in the market interest rate. If the fundamental policy problem is that we have too many lawyers and MBAs and not enough engineers, a subsidy to physical capital accumulation is a weak, and possibly counterproductive, policy response.

In the previous paper with differentiated producer durables (Romer 1987), I found that an increase in scale measured by L would increase the rate of growth. Here, an increase in L has no effect.

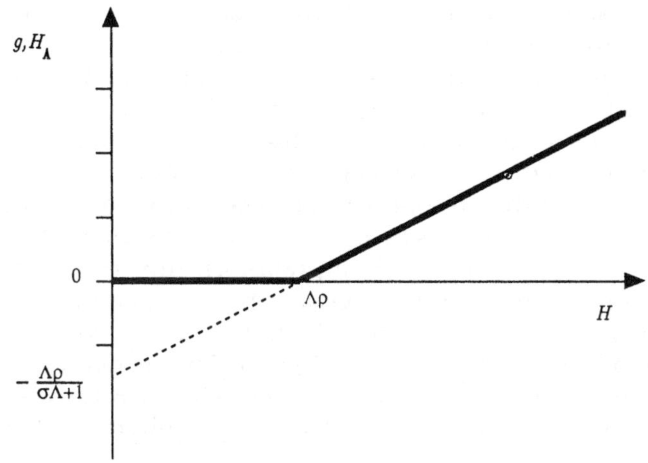

FIG. 2.—Growth rate and amount of human capital in research as a function of total human capital (for $\delta = 1$).

Instead, an increase in scale as measured by total human capital H has the effect of speeding up the rate of growth. This effect is illustrated in figure 2, which plots the rate of growth and the amount of human capital used in research as a function of total human capital. Both of these models have an underlying form of increasing returns in research. As a result, an increase in a scale variable induces an increase in the rate of growth. Human capital is the relevant scale variable in this model because it is the input that is used most intensively in research.

That the research sector in this model exhibits increasing returns is clear from equation (4). If the research technology exhibited constant returns to scale, a doubling of both the human capital and the stock of knowledge would leave the marginal product of human capital in research unchanged. Under the specification used here, a doubling of both leads to an increase in the marginal product of human capital in research. As a result, a permanent increase in the total stock of human capital in the population leads to an increase in the ratio of A to K and a more than proportional increase in the amount of human capital that is devoted to the research sector, as illustrated in figure 2. This implication is of interest from both a historical and a cross-sectional point of view. It is surely the case that the total level of human capital and the fraction of human capital devoted to research are higher now than they were at any time in the past. Moreover, the fraction of human capital devoted to research is apparently highest in the most developed countries of the world.

An extreme possibility illustrated in figure 2 is that if the total level of human capital is too small, stagnation may arise. If H is too low, the nonnegativity constraint on H_A is binding and growth does not take place. In this case, all the feasible growth rates for A are too small relative to the discount rate to justify the sacrifice in current output necessary for growth to take place. This result offers one possible way to explain the wide variation in growth rates observed among countries and the fact that in some countries growth in income per capita has been close to zero. This explanation is reminiscent of the explanation for the absence of growth in prehistoric time that is offered by some historians and anthropologists: civilization, and hence growth, could not begin until human capital could be spared from the production of goods for immediate consumption. This model cannot offer a complete explanation for these observations because it treats the stock of H (and of L) as given, but it does suggest directions for further work.

The effects of a subsidy to capital can be contrasted with a policy designed to encourage research. A subsidy to employment in the research sector that is financed through lump-sum taxes has the same effects on growth as an increase in the productivity parameter δ in equation (3). In the long run, the subsidy will cause an increase in the growth rate, a fall in P_A, and a reduction in $\bar{x}$ and in the ratio of K to A.

There are two reasons to expect that too little human capital is devoted to research. The most obvious reason is that research has positive external effects. An additional design raises the productivity of all future individuals who do research, but because this benefit is nonexcludable, it is not reflected at all in the market price for designs. The second and an equally important reason why too little human capital is devoted to research is that research produces an input that is purchased by a sector that engages in monopoly pricing. The markup of price over marginal cost forces a wedge between the marginal social product of an input used in this sector and its market compensation. Equation (7) shows that a new design (i.e., an additional unit of A) increases output at every date by an amount $H_Y^{\alpha}L^{\beta}\bar{x}^{1-\alpha-\beta}$; from the calculation in (5), the producer of a design captures only a fraction $1 - \alpha - \beta$ of this net benefit to society.

Both of these effects cause human capital to be undercompensated. The marginal product of capital in the manufacturing sector is equal to the wage w_H, but the marginal product in the research sector is higher than the wage because the price of the patent captures only part of the social value of the patent. From equation (9), (10), or (11), it follows that an increase in total human capital would lead to an increase in the amount of human capital employed in the research sector. As a result, in equilibrium, the marginal value of an additional unit of human capital is higher than the market wage. For simplicity, the model here has treated the stock of total human capital as being exogenously fixed, so undercompensating human capital has no welfare effects. In a more realistic model in which human capital is accumulated endogenously, the supply will be too low. As a result, a second-best policy for a government that cannot affect the allocation of human capital between different sectors would be to subsidize the production of human capital.

Within the confines of the model, the social optimum can be achieved by subsidizing the accumulation of A. Demonstrating this result rigorously starting from arbitrary initial conditions forces the analysis to depart from consideration of balanced growth paths. Any intervention designed to move an economy from one balanced growth path to another must consider the transition dynamics along the way, and an explicit analysis of these dynamics is beyond the scope of this paper. It is easy, however, to compare the growth rate along the balanced growth path that emerges from the equilibrium with the one that would emerge from the solution to a social planning problem. Because of the symmetry between the different producer durables in this economy, the optimal level of $x(i)$ is the same for all i between zero and A. This level is related to K and A by the constraint that $K = \eta x A$. Using this to express x in terms of K and A, we can therefore write the social planning problem for this economy as

$$\max \int_0^\infty \frac{C^{1-\sigma} - 1}{1 - \sigma} e^{-\rho t} dt,$$

subject to

$$\dot{K} = \eta^{\alpha+\beta-1} A^{\alpha+\beta} H_Y^{\alpha} L^{\beta} K^{1-\alpha-\beta} - C,$$

$$\dot{A} = \delta H_A A,$$

$$H_Y + H_A \leq H.$$

As is shown in the Appendix, the balanced growth solution to the first-order necessary conditions for this problem has a growth rate g^* that is given by

$$g^* = \frac{\delta H - \Theta\rho}{\Theta\sigma + (1 - \Theta)}, \tag{14}$$

where $\Theta = \alpha/(\alpha + \beta)$. The coefficient Λ from equation (12) is equal to this coefficient Θ times the markup from the monopoly sector, $1/(1 - \alpha - \beta)$. This accounts for part of the difference between the equilibrium and the socially optimal rate of growth. However, there is an additional effect because the term in the denominator in equation (14) also replaces the constant 1 from equation (13) with the expression $1 - \Theta$. This change reflects the effect of correcting for the external effects associated with the production of new ideas. Both of these changes in the expression for the rate of growth—the substitution of Θ for Λ and the substitution of $1 - \Theta$ for 1—cause the socially optimal allocation of human capital to research to be higher, and this causes the socially optimal rate of growth to be higher.

VII. Growth, Trade, and Research

The final observations about this model pertain to its implications for growth, trade, and research. These can be seen most simply by comparing the balanced growth equilibrium for two identical closed economies that operate in isolation with the balanced growth equilibrium that would obtain if the economies had always been fully integrated. In isolation, the common growth rate is given by g from equation (13), with H set equal to the amount of human capital in each country. In the second case, the growth rate is found by replacing H by total worldwide human capital, $2H$. Both the fraction H_A/H of total worldwide human capital devoted to research and the rate of growth increase.

This thought experiment suggests why a decision to engage in trade may be important even for a country that has a large population, such as China or India. If access to a large number of workers or consumers were all that mattered, having a large population would be a good substitute for trade with other nations. The model here suggests that what is important for growth is integration not into an economy with a large number of people but rather into one with a large amount of human capital. Many of the details of trade between different economies of this kind remain to be worked out,[2] but since growth seems to be correlated with the degree of integration into worldwide markets but not closely related to population size or density, the results from this model seem promising.

The most direct test of this implication of the model would come from a controlled experiment in which the level of research activity was monitored both before and after a country was opened to trade with the rest of the world. Sokoloff (1988) reports historical data on a natural experiment that comes close to this test. He finds cross-sectional variation that supports the model: counties in the United States in the early nineteenth century that had access to navigable waterways had higher rates of patenting than counties that did not.

[2] Since the first draft of this paper was written, Grossman and Helpman (1989*a*, 1989*b*) have explored detailed models of trade and growth with this kind of underlying technology.

More convincingly, he shows that over time, the introduction of water transportation (because of either the construction of a new canal or the dredging of a river) was followed by a sharp increase in the rate of patenting in counties adjacent to the waterway.

In subsequent work, Sokoloff and Khan (1989) examine the time-series variability in patenting for specific individuals. They find that there was a substantial group of people with broad general knowledge who moved in and out of research in response to aggregate disturbances (primarily business cycles and a trade embargo by the British) just as one would expect if human capital that could be used to do research had alternative uses in manufacturing and commercial trade. They conclude that at least in the nineteenth century, there was a fairly elastic short-run supply of human capital for use in research. Combined with the prior evidence that patenting activity does respond to changes in the size of the market, this offers some assurance that the basic mechanisms described in the model are relevant for historical experience.

VIII. Conclusions

The model presented here is essentially the one-sector neoclassical model with technological change, augmented to give an endogenous explanation of the source of the technological change. The most robust welfare conclusion from the model is that because research projects exchange current costs for a stream of benefits in the future, the rate of technological change is sensitive to the rate of interest. Although all the research is embodied in capital goods, a subsidy to physical capital accumulation may be a very poor substitute for direct subsidies that increase the incentive to undertake research. In the absence of feasible policies that can remove the divergence between the social and private returns to research, a second-best policy would be to subsidize the accumulation of total human capital.

The most interesting positive implication of the model is that an economy with a larger total stock of human capital will experience faster growth. This finding suggests that free international trade can act to speed up growth. It also suggests a way to understand what it is about developed economies in the twentieth century that permitted rates of growth of income per capita that are unprecedented in human history. The model also suggests that low levels of human capital may help explain why growth is not observed in underdeveloped economies that are closed and why a less developed economy with a very large population can still benefit from economic integration with the rest of the world.

Appendix

Calculation of the Balanced Growth Social Optimum

To derive the necessary conditions for the social optimization problem given in the text, construct the current-valued Hamiltonian:

$$\mathcal{H} = \frac{C^{1-\sigma}}{1-\sigma} + \lambda[\eta^{\alpha+\beta-1}A^{\alpha+\beta}(H - H_A)^{\alpha}L^{\beta}K^{1-\alpha-\beta} - C] + \mu\delta H_A A.$$

The necessary conditions follow by maximizing $\mathcal{H}$ with respect to the control variables C and H_A, and from the equations for the evolution of the multipliers λ and μ:

$$\dot{\lambda} = \rho\lambda - \frac{\partial\mathcal{H}}{\partial K}, \quad \dot{\mu} = \rho\mu - \frac{\partial\mathcal{H}}{\partial A}.$$

The first-order condition for maximizing $\mathcal{H}$ with respect to C gives the usual expression relating marginal utility and the multiplier λ:

$$C^{-\sigma} = \lambda. \quad \text{(A1)}$$

If the symbol Δ is used to represent the term $\eta^{\alpha+\beta-1}A^{\alpha+\beta}(H - H_A)^{\alpha}L^{\beta}K^{1-\alpha-\beta}$ from the Hamiltonian, the first-order condition for maximizing $\mathcal{H}$ with respect to H_A can be written as

$$\Delta = (H - H_A)\frac{\delta\mu}{\alpha\lambda}A. \quad \text{(A2)}$$

Then with equation (A2), the evolution equation for μ can be simplified to yield

$$\frac{\dot{\mu}}{\mu} = \rho - \delta\left(\frac{\alpha+\beta}{\alpha}H - \frac{\beta}{\alpha}H_A\right). \quad \text{(A3)}$$

For a balanced growth equilibrium, it must be the case that $\dot{\mu}/\mu = \dot{\lambda}/\lambda$ and that $\dot{C}/C = \dot{A}/A$. With equation (A1), these can be combined to yield $-\sigma(\dot{A}/A) = \dot{\mu}/\mu$. Then combining the evolution equation for A, $\dot{A}/A = \delta H_A$, with equation (A3) gives an equation in H, H_A, and the basic parameters of the model:

$$-\sigma\delta H_A = \rho - \delta\left(\frac{\alpha+\beta}{\alpha}H - \frac{\beta}{\alpha}H_A\right). \quad \text{(A4)}$$

Equation (14) in the text can be derived from this equation by solving for H_A and using the fact that the growth rate g is given by $g = \delta H_A$.

References

Abramovitz, Moses. "Resource and Output Trends in the United States since 1870." *A.E.R. Papers and Proc.* 46 (May 1956): 5–23.

Arrow, Kenneth J. "The Economic Implications of Learning by Doing." *Rev. Econ. Studies* 29 (June 1962): 155–73. (*a*)

———. "Economic Welfare and the Allocation of Resources for Invention." In *The Rate and Direction of Inventive Activity: Economic and Social Factors,* edited by Richard Nelson. Princeton, N.J.: Princeton Univ. Press (for NBER), 1962. (*b*)

Becker, Gary S.; Murphy, Kevin M.; and Tamura, Robert. "Human Capital, Fertility, and Economic Growth." *J.P.E.*, this issue.

Cornes, Richard, and Sandler, Todd. *The Theory of Externalities, Public Goods, and Club Goods.* Cambridge: Cambridge Univ. Press, 1986.

Dasgupta, Partha, and Stiglitz, Joseph E. "Learning-by-Doing, Market Structure and Industrial and Trade Policies." *Oxford Econ. Papers* 40 (June 1988): 246–68.

Dixit, Avinash K., and Stiglitz, Joseph E. "Monopolistic Competition and Optimum Product Diversity." *A.E.R.* 67 (June 1977): 297–308.

Ethier, Wilfred J. "National and International Returns to Scale in the Modern Theory of International Trade." *A.E.R.* 72 (June 1982): 389–405.

Gollop, Frank M., and Jorgenson, Dale W. "U.S. Productivity Growth by Industry." In *New Developments in Productivity Measurement and Analysis,* edited by John W. Kendrick and Beatrice N. Vaccara. Chicago: Univ. Chicago Press (for NBER), 1980.

Griliches, Zvi. "Issues in Assessing the Contribution of Research and Development to Productivity Growth." *Bell J. Econ.* 10 (Spring 1979): 92–116.

Grossman, Gene M., and Helpman, Elhanan. "Comparative Advantage and Long Run Growth." Working Paper no. 2809. Cambridge, Mass.: NBER, January 1989. (*a*).

———. "Growth and Welfare in a Small Open Economy." Working Paper no. 2970. Cambridge, Mass.: NBER, May 1989. (*b*)

———. "Product Development and International Trade." *J.P.E.* 97 (December 1989): 1261–83. (*c*)

Heckman, James J. "A Life-Cycle Model of Earnings, Learning, and Consumption." *J.P.E.* 84, no. 4, pt. 2 (August 1976): S11–S44.

Helpman, Elhanan, and Krugman, Paul R. *Market Structure and Foreign Trade: Increasing Returns, Imperfect Competition, and the International Economy.* Cambridge, Mass.: MIT Press, 1985.

Jones, Larry, and Manuelli, Rodolfo. "A Model of Optimal Economic Growth." Working paper. Stanford, Calif.: Stanford Univ., 1988.

Jorgenson, Dale W.; Gollop, Frank M.; and Fraumeni, Barbara M. *Productivity and U.S. Economic Growth.* Cambridge, Mass.: Harvard Univ. Press, 1987.

Judd, Kenneth L. "On the Performance of Patents." *Econometrica* 53 (May 1985): 567–85.

Kendrick, John W. "Productivity Trends: Capital and Labor." *Rev. Econ. and Statis.* 38 (August 1956): 248–57.

King, Robert G., and Rebelo, Sergio. "Business Cycles with Endogenous Growth." Working paper. Rochester, N.Y.: Univ. Rochester, 1987.

Lucas, Robert E., Jr. "On the Mechanics of Economic Development." *J. Monetary Econ.* 22 (July 1988): 3–42.

Maddison, Angus. *Phases of Capitalist Development.* Oxford: Oxford Univ. Press, 1982.

Nordhaus, William D. *Invention, Growth, and Welfare: A Theoretical Treatment of Technological Change.* Cambridge, Mass.: MIT Press, 1969.

Pascoa, Mario. "Monopolistic Competition and No-Neighboring-Goods." Working Paper no. 86-14. Philadelphia: Univ. Pennsylvania, Center Analytic Res. Econ. and Soc. Sci., 1986.

———. "Noncooperative Equilibrium and Chamberlinian Monopolistic Competition." *J. Econ. Theory* (1990), in press.

Rebelo, Sergio. "Long Run Policy Analysis and Long Run Growth." Working paper. Rochester, N.Y.: Univ. Rochester, 1988.

Romer, Paul M. "Increasing Returns and Long-Run Growth." *J.P.E.* 94 (October 1986): 1002–37.

———. "Growth Based on Increasing Returns Due to Specialization." *A.E.R. Papers and Proc.* 77 (May 1987): 56–62.

———. "Capital Accumulation in the Theory of Long Run Growth." In *Modern Business Cycle Theory,* edited by Robert J. Barro. Cambridge, Mass.: Harvard Univ. Press, 1989.

———. "Capital, Labor, and Productivity." *Brookings Papers Econ. Activity* (1990), in press.

Rosen, Sherwin. "A Theory of Life Earnings." *J.P.E.* 84, no. 4, pt. 2 (August 1976): S45–S67.

Schumpeter, Joseph A. *Capitalism, Socialism, and Democracy.* New York: Harper, 1942.

Shell, Karl. "Toward a Theory of Inventive Activity and Capital Accumulation." *A.E.R. Papers and Proc.* 56 (May 1966): 62–68.

———. "A Model of Inventive Activity and Capital Accumulation." In *Essays on the Theory of Optimal Economic Growth,* edited by Karl Shell. Cambridge, Mass.: MIT Press, 1967.

———. "Inventive Activity, Industrial Organisation and Economic Growth." In *Models of Economic Growth,* edited by James A. Mirrlees and Nicholas H. Stern. New York: Wiley, 1973.

Sokoloff, Kenneth L. "Inventive Activity in Early Industrial America: Evidence from Patent Records, 1790–1846." *J. Econ. Hist.* 48 (December 1988): 813–50.

Sokoloff, Kenneth L., and Khan, Zorina. "The Democratization of Invention during Early Industrialization: Evidence from the United States, 1790–1846." Working paper. Los Angeles: Univ. California, 1989.

Solow, Robert M. "A Contribution to the Theory of Economic Growth." *Q.J.E.* 70 (February 1956): 65–94.

———. "Technical Change and the Aggregate Production Function." *Rev. Econ. and Statis.* 39 (August 1957): 312–20.

Uzawa, Hirofumi. "Optimum Technical Change in an Aggregative Model of Economic Growth." *Internat. Econ. Rev.* 6 (January 1965): 18–31.

Wilson, Robert. "Informational Economies of Scale." *Bell J. Econ.* 6 (Spring 1975): 184–95.

[13]

A Schumpeterian Model of the Product Life Cycle

By PAUL S. SEGERSTROM, T. C. A. ANANT, AND ELIAS DINOPOULOS*

This paper presents a dynamic general equilibrium model of North–South trade in which research and development races between firms determine the rate of product innovation in the North. Tariffs designed to protect dying industries in the North from Southern competition reduce the steady-state number of dominant firms in the North, reduce the rate of product innovation, and increase the relative wage of Northern workers. (*JEL* 411, 111)

In his celebrated "product life cycle" paper Raymond Vernon (1966) argued that many products experience cycles. These products are initially discovered and produced in developed countries (the North), and exported to less developed countries (the South). As the techniques of production become more standardized, production shifts to less developed countries due to lower labor costs. These older products are then exported back to developed countries.

The product-life-cycle hypothesis has attracted considerable attention among international-trade theorists in recent years. In this literature, the rate at which an individual firm discovers and successfully markets new products is either treated as exogenously given (Paul Krugman, 1979; David Dollar, 1986, 1987) or as a "deterministic" function of the firm's expenditures on new product development (Robert Feenstra and Kenneth Judd, 1982; Thomas Pugel, 1982; Barbara Spencer and James Brander, 1983; Leonard Cheng, 1984; Richard Jensen and Marie Thursby, 1986, 1987). Thus, from the individual firm's perspective, successful product innovation is either effortless or guaranteed by large expenditures on new product development. In contrast, Joseph Schumpeter (1942) stressed that firms compete with each other to successfully introduce new products. The recent industrial-organization literature has followed Schumpeter's lead (see, e.g., Glen Loury, 1979; Tom Lee and Louis Wilde, 1980; Jennifer Reinganum, 1982). In these research and development (R&D) models, there are losers as well as winners because a firm can spend substantial resources on new product development only to find that another firm has discovered and patented the new product first.

In this paper, we construct a dynamic, general equilibrium model of North–South trade that combines the product-life-cycle hypothesis with Schumpeter's (1942) description of product innovation. We model each R&D race as an "invention lottery" in which the probability of winning the race is proportional to resources devoted to R&D by each firm. The duration of each R&D race is a deterministic decreasing function of the amount of aggregate resources devoted to R&D. Every time a new product is discovered, a new R&D race between firms in the North begins. The winner of each R&D race earns dominant firm profits for an exogenously given patent period, after which perfect competition prevails. Firms in the North choose how much labor to hire for R&D by maximizing expected discounted profits, and consumers maximize their discounted lifetime utility.

*Department of Economics, Michigan State University, East Lansing, MI 48824; Department of Economics, Delhi School of Economics, Delhi University, New Delhi, India; and Department of Economics, University of Florida, Gainesville, FL 32611, respectively. We thank Richard Brecher, Robert Feenstra, Gene Grossman, Paul Krugman, Dave Richardson, Barbara Spencer, Marie Thursby, and the other participants at the NBER summer institute on international studies, July 1987, for helpful comments. The paper has also benefited from the comments of Earl Grinols, Haideh Salehi-Esfahani, and three anonymous referees. Any errors that remain are our own responsibility.

We show that a unique steady-state equilibrium exists in which the number of new products, consumer expenditures, and assets are all constant over time. In the steady state, Northern workers earn higher wages than their Southern counterparts if the South has a sufficiently large fraction of the world labor force. Moreover, the pattern of trade continuously changes with each product initially being exported and then later imported by the North.

Endogenizing the rate of technological change generates some surprising comparative steady-state results in our model. When wages in the North and in the South are equal, an increase in the patent length (or a decrease in the rate of technology transfer to the South) increases the rate of product innovation in the North. This result is consistent with the partial-equilibrium industrial-organization literature on R&D competition, because an increase in the patent length increases the reward for winning an R&D race. However, when Northern workers earn higher wages than Southern workers, an increase in the patent length *decreases* the rate of product innovation in the North. The increase in the patent length, by itself, increases the reward for innovative activity, but Northern wages rise more than enough to offset this effect.

Unlike the previously mentioned studies, we also examine the effects of tariffs designed to protect dying industries in the North from Southern competition. When Northern workers earn higher wages than Southern workers, we find that an increase in the number of industries being protected in the North leads to higher relative wages for Northern workers and a slower rate of innovation. Thus, we are able to theoretically link protectionist trade policies with slower economic growth.

The rest of this paper is organized as follows: In Section I, the dynamic general equilibrium model of North–South trade is presented. In Section II, we characterize the steady-state equilibrium of the model. The relationship between relative labor endowments and steady-state relative wages is explained in Section III. Section IV analyzes the effects of changes in patent length and tariff protection on the steady-state equilibrium. Finally, our conclusions are presented in Section V.

I. The Model

Consider a world consisting of two countries: the North and the South. Let $\overline{L}^{\mathrm{N}}$ and $\overline{L}^{\mathrm{S}}$ be the aggregate endowments of labor in the North and in the South, respectively. These labor endowments do not change over time.

In this world, there is a countably infinite set of product $N \equiv \{1,2,3,\dots\}$. At any point in time $t \in [0,\infty)$, these products can be partitioned into three sets: the set of products that any firm in the world knows how to produce, the set of products that only one firm in the world knows how to produce, and the set of products that no firm in the world knows how to produce. Firms that produce products in the second set are called dominant firms.

At time t, every firm that knows how to produce product $j \in N$ has the same production technology. Constant returns to scale prevail in production, with one unit of labor producing one unit of product j. Labor is the only factor of production, and all workers in the world are equally productive. Factors of production are not mobile internationally.

At time $t=0$, there are n products for which the production technology is common knowledge, and the remainder of the products have unknown production technology. Time $t=0$ represents the beginning of a sequence of R&D races between firms in the North. Only workers in the North are capable of doing R&D-type work, and therefore only firms in the North compete in these R&D races. Every time an R&D race in the North ends, a new R&D race immediately begins. At the beginning of the jth R&D race, each firm i in the North must decide how much labor L_{ij}^{R} to devote to R&D. This choice by firm i represents a commitment to employing L_{ij}^{R} units of R&D labor for the duration of the jth R&D race. The firm that wins the jth R&D race becomes the sole producer in the world of product $n+j$ for a time period of length

$T > 0$. After this "patent" expires, the production technology for product $n+j$ becomes common knowledge.[1]

We model each R&D race as an "invention lottery." Each of the L_j^R worker-researchers participating in the jth R&D race is equally likely to discover product $n+j$ at time $\hat{t}_j = h(L_j^R)$ after the beginning of the jth R&D race, where $L_j^R \equiv \Sigma_i L_{ij}^R$ is the aggregate labor devoted to R&D. Thus, at the end of the jth race, nature draws one of the L_j^R "lottery tickets" and one of the L_j^R worker-researchers discovers the new product. The firm that employs the winner earns dominant firm profits until its patent expires. Firm i wins the jth race with probability L_{ij}^R / L_j^R.

By modeling the R&D process as an "invention lottery," we capture two features of R&D races that we feel are important. First, individual firms investing in R&D face an uncertain return; there are winners and losers. In contrast, with other models of technological change and international trade (see, e.g., Jensen and Thursby, 1986; Gene Grossman and Elhanan Helpman, 1989; Feenstra and Judd, 1982), a firm is not guaranteed success in developing a new product by spending some fixed sum of money on product development. Secondly, new products tend to be discovered faster and at a greater discounted cost, as more resources are devoted to R&D [this is implied by properties of the $h(\cdot)$ function to be specified shortly]. There is an intertemporal trade-off associated with R&D activity. From the individual firm's perspective, as it spends more money on R&D (a flow expenditure), its possibility of winning the race increases, other firm's probabilities of success decline, and the race ends sooner.[2]

In this model, at each point in time t, wages for workers in each country are determined by competitive market forces. We set the equilibrium wage rate for workers in the South equal to one and let w denote the equilibrium relative wage of workers in the North. Both production workers and R&D workers in the North are paid the same wage w.

In addition to the absence of international labor mobility, we assume that production of goods protected by patents takes place only in the North. In other words, Southern firms can produce a good only after its patent protection has expired. Possible institutional justification for this assumption would be that enforcement of patent laws in the South is considerably weaker than in the North and the labor market within the South constitutes an effective channel of technology diffusion. Thus, in the absence of effective patent protection in the South, a Northern dominant firm producing in the South faces the risk that some of its workers might establish another firm manufacturing the same product.[3]

Infinitely lived consumers maximize total lifetime utility. Each consumer has an identical time-separable utility function

$$U \equiv \int_0^\infty e^{-\rho t} \log u(\cdot)\, dt \tag{1}$$

where $\rho > 0$ is the constant subjective discount rate and $u(\cdot)$ is an instantaneous utility function.[4] We adopt a particular

[1] T is inversely related to the rate of technology transfer in Krugman (1979) and Jensen and Thursby (1986). "Patents" need not be given a literal interpretation. The patent length T serves as a proxy for all relevant factors that impede technology transfer between the North and South.

[2] The deterministic length $\hat{t} = h(L^R)$ of each R&D race is admittedly artificial, but as will become clear, the main results in the paper are driven by factor market constraints, which would be present whether the length of each R&D race were deterministic or stochastic. To analyze tractably the effects of commercial policy in a general equilibrium setting, we also abstract from certain interesting features of R&D races (established-firm advantages, variable R&D expenditures over time, imitation in spite of patent protection, etc.) that have been extensively studied in the partial-equilibrium industrial-organization literature.

[3] Allowing dominant firms to produce in the South could generate multinational firms along the lines proposed by Grossman and Helpman (1989). Multinational activity in our model results in the wage being equalized between the North and South unless all Northern labor is engaged in R&D.

[4] The same form of total lifetime utility is used by Grossman and Helpman (1989).

form of $u(\cdot)$,

$$(2)\quad u(x_1, x_2, x_3, \dots) \equiv \prod_{j=1}^{n} \left(\sum_{i=0}^{\infty} \alpha^i x_{j+ni} \right).$$

This is a generalized symmetric Cobb-Douglas utility function where n can be interpreted as the number of product groups. Since products within each group are perfect substitutes, we call this the CDP (Cobb-Douglas with perfect substitutes) utility substitutes) utility function. Product group j $(j = 1,2,3,\dots,n)$ consists of products j, $n+j$, ...; and $\alpha > 1$ represents the extent to which each new product improves upon existing products in the same product group.

To illustrate the effect of product innovation on consumer utility, suppose that initially there are n products available for consumption. Given time separability, consumers are, in effect, maximizing the utility function $\tilde{U} \equiv x_1 x_2 x_3 \dots x_n$ at that instant in time. The discovery of product $n+1$ means that consumers are now, in effect, maximizing the utility function $\tilde{U} \equiv (x_1 + \alpha x_{n+1}) x_2 x_3 \dots x_n$. If the equilibrium prices of products 1 and $n+1$ both equal one, which would be the case if both products 1 and $n+1$ were produced competitively, then no consumer would purchase product 1 (given $\alpha > 1$), and it would become obsolete. Thus, new products substitute perfectly for old products, and product innovation in our model takes the form of superior products replacing inferior products.[5]

We assume that there is a capital market in the North which supplies the savings of Northern consumers to firms engaged in R&D. The equilibrium interest rate $r(t)$ clears the capital market at each point in time t. Firms borrow funds from this market to pay workers as the R&D is done. Each firm issues a risky security that yields a positive return if it wins and a negative return if it loses an R&D race. Assuming risk neutrality and perfect competition in R&D, Northern firms enter each R&D race until expected discounted profits are driven to zero. Southern consumers are not allowed to participate in the capital market, and therefore at each instant of time, their income equals their expenditure. If this assumption is relaxed, then Southern savings would end up financing part of the North's R&D expenditure, a result that is contrary to empirical evidence.[6]

Free trade is assumed to exist between the North and the South throughout time, and products are assumed to be nonstorable. Furthermore, at any time t, perfect competition prevails in the market for each product whose patent has expired. Thus, the market price for all such products equals the marginal cost of production in the South (one). Given the consumer preferences, one unit of product j gives each consumer as much utility as α units of product $j-n$. When both products are competitively produced and sell at the same market prices (equal to one), the competitive market for product j renders product $j-n$ obsolete.

The endowment of labor in the North $\bar{L}^{N}$ is assumed to be sufficiently small so that, even if all the workers in the North did R&D-type work, the number of dominant firms would be less than the number of product groups n. That is, $h(\bar{L}^{N})n > T$. This condition guarantees that there are never two dominant firms producing products in the same product group.[7] At time t, the dominant firm producing product j must

[5]Nancy Stokey (1988) models product replacement in a different context. The rest of the product-life-cycle literature has treated product innovation as being the introduction of greater variety.

[6]All the comparative steady-state results concerning the effects of changes in labor endowments, patents, and tariffs on the number of dominant firms, the rate of innovation, relative wages, and world assets would be unaffected if the capital market were international. Nor would the magnitudes of these variables be affected. However, the distribution of world assets between the North and the South and the pattern of trade in the steady state would change.

[7]Even if innovations did not occur in the previously described sequence, all the results in the paper would be unaffected if product j is only discovered when product $j-n$ is competitively produced. For example, it is possible that certain groups do not experience any innovation at all.

compete only against a competitive fringe of firms producing product $j-n$.

The dominant firm and firms in the competitive fringe simultaneously set prices, and we solve for a Bertrand-type Nash equilibrium. Let E^{W} denote instantaneous world expenditure. Given the instantaneous CDP utility function, world expenditure at time t on products in product group j is E^{W}/n. With the equilibrium price of one being charged by firms in the competitive fringe, the dominant firm has zero sales if it charges a price p^{d} greater than α. On the other hand, the competitive fringe has zero sales if the dominant firm charges a price p^{d} less than α. If $p^{\mathrm{d}}=\alpha$, then consumers are indifferent between spending E^{W}/n on product j and spending E^{W}/n on product $j-n$. We assume that all the indifferent consumers buy from the dominant firm (all rules for rationing the demand of indifferent consumers among firms are somewhat arbitrary). Then dominant-firm profits are

$$\text{(3)} \quad \pi^{\mathrm{d}}(p^{\mathrm{d}}) = \begin{cases} 0 & \text{if } p^{\mathrm{d}}>\alpha \\ (p^{\mathrm{d}}-w)E^{\mathrm{W}}/p^{\mathrm{d}}n & \text{if } p^{\mathrm{d}}\leq\alpha. \end{cases}$$

These profits are clearly maximized where $p^{\mathrm{d}}=\alpha$. Thus, in the Nash noncooperative equilibrium in prices that we examine in the rest of this paper, each dominant firm produces $q^{\mathrm{d}}\equiv E^{\mathrm{W}}/\alpha n$ and earns profits

$$\text{(4)} \qquad \pi^{\mathrm{d}} \equiv \frac{E^{\mathrm{W}}}{n}\left(\frac{\alpha-w}{\alpha}\right).$$

The competitive fringe constrains each dominant firm from charging prices higher than α.

Several restrictions are placed on the $h(\cdot)$ function that defines the R&D technology. First, $h(\cdot)$ is assumed to be continuously differentiable with $h'(\cdot)<0$. This guarantees that product innovation occurs at a faster rate when firms in the North devote more resources to R&D. Second $\bar{h}\equiv h(0)< +\infty$; that is, some product innovation occurs even if no resources are devoted to R&D. Third, $h(L^{\mathrm{R}})>0$ and $h''(L^{\mathrm{R}})\geq 0$ for all $L^{\mathrm{R}}>0$; that is, no matter how much labor is devoted to R&D, innovation never occurs instantaneously. Fourth,

$$\text{(5)} \qquad \frac{d}{dL^{\mathrm{R}}}L^{\mathrm{R}}(e^{\rho h(L^{\mathrm{R}})}-1)>0.$$

Notice that $\int_{-h(L^{\mathrm{R}})}^{0} wL^{\mathrm{R}}e^{-\rho t}dt = wL^{\mathrm{R}}(e^{\rho h(L^{\mathrm{R}})}-1)/\rho$ is the discounted labor cost of developing a new product (discounted to the end of the R&D race) when the market interest rate $r(t)$ equals each consumer's subjective discount rate ρ. We show in the next section that $r(t)=\rho$ in the steady-state equilibrium. Thus, this condition states that the appropriately discounted labor costs of developing a new product rise as firms try to speed up the process by devoting more resources to R&D. Equation (5) will hold if the $h(\cdot)$ function is downward sloping but sufficiently flat. Fifth, we make a technical restriction

$$\text{(6)} \qquad -h'(0) < \frac{(1-e^{-\rho T})h(\bar{L}^{\mathrm{N}})}{\bar{L}^{\mathrm{N}}(e^{\rho h(\bar{L}^{\mathrm{N}})}-1)T}$$

which will also hold if the $h(\cdot)$ function is sufficiently flat. As shown in Appendix A, condition (6) is sufficient but hardly necessary for the steady-state equilibrium to be unique.

Finally, we assume that the labor force in the North ($\bar{L}^{\mathrm{N}}$) is sufficiently large relative to the labor force in the South ($\bar{L}^{\mathrm{S}}$) so that

$$\text{(7)} \qquad \frac{n\alpha\bar{L}^{\mathrm{N}}}{\bar{L}^{\mathrm{S}}+\alpha\bar{L}^{\mathrm{N}}} > \frac{T}{\bar{h}}.$$

As will become clear in Section II, inequality (7) guarantees that, in any steady-state equilibrium, aggregate R&D expenditures are strictly positive.

II. The Steady-State Equilibrium

In this section, we show that a unique steady-state equilibrium exists for the dynamic, general equilibrium model of North–South trade. In this steady state, the number of dominant firms m, the relative wage of Northern workers w, the profit flow of each dominant firm π^{d}, the aggregate labor devoted to research and development L^{R}, world expenditure E^{W}, world wage income I^{W}, Northern assets A^{N}, and the market interest rate r are all positive constants over time and are interrelated in several specific ways.[8]

World expenditure E^{W}, Northern assets A^{N}, and the equilibrium interest rate r must be consistent with the consumer's savings–consumption decisions over time. Let t_0 represent the beginning of an R&D race where J innovations have already occurred. The representative consumer's discounted future utility from expenditure path $E(t)$, $t \in [t_0, \infty)$ is

$$(8)\quad U = \sum_{i=0}^{\infty} \int_{t_0 + i\hat{t}}^{t_0+(i+1)\hat{t}} e^{-\rho t} \log\left[\frac{\alpha^{i} E(t)^{n}}{n^{n}\alpha^{m}}\right] dt + \Gamma(J, t_0)$$

where $\hat{t} = h(L^{R})$ is the length of each steady-state R&D race and m is the number of products produced by dominant firms. Because the m products produced by dominant firms are sold at price $p^{d} = \alpha$ and the $n - m$ competitively produced goods are sold at price $p^{c} = 1$, every time an innovation occurs, the consumer's instantaneous utility increases by factor α. From the point of view of future decision making $\Gamma(J, t_0) = \int_0^{t_0} e^{-\rho t} \log u(\cdot)\, dt$ is a constant. Appendix B shows that optimal consumer behavior is characterized by a constant expenditure path over time when the steady-state interest rate equals ρ, the consumer's subjective discount rate. Furthermore, it is shown in Appendix B that the relationship among steady-state expenditure E^{W}, assets A^{N} and wage income I^{W} is

$$(9)\quad E^{W} = \rho A^{N} + I^{W}.$$

Each consumer spends his wage income and interest earnings from his assets at each instant in time. These assets have been accumulated before the economy reaches the steady-state equilibrium. In other words, equation (9) is just a breakdown of steady-state expenditure by income source.

It is perhaps surprising that, although the instantaneous utility function is characterized by periodic jumps (caused by innovations), there nevertheless exists a steady state with constant expenditures and a constant interest rate over time. However, the optimal expenditure path is derived from the *marginal* utility of expenditure function (see Appendix B), which is invariant with respect to jumps caused by innovations.

Wage income in the world consists of income from production work and income from R&D work:

$$(10)\quad I^{W} = w\bar{L}^{N} + \bar{L}^{S}.$$

Notice that equation (10) implies that R&D workers are paid concurrently. Because goods are nonstorable, world GNP must equal world expenditure E^{W}:

$$(11)\quad E^{W} = m\pi^{d} + w(\bar{L}^{N} - L^{R}) + \bar{L}^{S}.$$

The first two terms represent the value of Northern production, and the last term equals the value of Southern production.

Expected discounted profits of firm i in a typical R&D race are

$$(12)\quad -wL_i^{R}\left[1 - e^{-\rho h(L^{R})}\right] + \frac{L_i^{R}}{L^{R}} \pi^{d} e^{-\rho h(L^{R})}\left[1 - e^{-\rho T}\right].$$

Firm i must pay each of L_i^{R} workers the wage w for the duration $h(L^{R})$ of the R&D race. With probability L_i^{R}/L^{R} the ith firm wins the race and earns profits π^{d} until its patent expires. Perfect competition and free entry in each R&D race drives expected discounted profits of each firm to zero. Summing over all firms in the R&D race, we obtain

$$(13)\quad -wL^{R}\left[1 - e^{-\rho h(L^{R})}\right] + \pi^{d} e^{-\rho h(L^{R})}\left[1 - e^{-\rho T}\right] = 0.$$

[8] World assets equal Northern assets and Southern expenditure equals Southern income, because Southern consumers do not participate in the capital market.

In other words, aggregate profits discounted to the beginning of an R&D race are equal to zero.

Using equations (10), (11), and (13), it can be shown that the value of aggregate assets is

$$(14) \qquad A^{\mathrm{N}} = \frac{m\pi^{\mathrm{d}} - wL^{\mathrm{R}}}{\rho} = \frac{\pi^{\mathrm{d}}}{\rho} \sum_{j=1}^{m} (1 - e^{-\rho j\hat{t}}).$$

Equation (14) implies that $A^{\mathrm{N}} > 0$. Assets must be positive in the steady state, because Northern consumers saved in the past to finance the innovation process that led to m dominant firms.

We use geometric techniques to derive the solution and perform comparative steady-state analysis. Combining equations (4), (11), and (13) yields

$$(15) \qquad m = Z(L^{\mathrm{R}}, w) \equiv \frac{n\alpha}{\alpha - w} - \frac{\left[w(\overline{L}^{\mathrm{N}} - L^{\mathrm{R}}) + \overline{L}^{\mathrm{S}}\right](1 - e^{-\rho T})}{wL^{\mathrm{R}}(e^{\rho h(L^{\mathrm{R}})} - 1)}.$$

With w fixed, Z can be interpreted as the steady-state zero-profit condition expressed in (m, L^{R}) space. Given the assumptions about $h(\cdot)$ in Section I, the partial derivatives are unambiguously signed: $\partial Z/\partial L^{\mathrm{R}} > 0$, $\partial Z/\partial w > 0$, and $\lim_{L \to 0} Z(L, w) = -\infty$ for any $w \geq 1$.

The function $m = Z(L^{\mathrm{R}}, w)$ increases in L^{R} for the following reason: from condition (5), when L^{R} increases, the discounted cost of developing an innovation $L^{\mathrm{R}}(e^{\rho h(L^{\mathrm{R}})} - 1)$ increases. Equation (13) implies that to maintain zero discounted profits, dominant firm profits π^{d}, which represent the reward for winning an R&D race, must be higher. For a given wage w, π^{d} is higher only if world expenditure E^{W} is higher [eq. (4)], and from equation (11), there must be more dominant firms for world expenditure to be higher. In other words, firms can only afford to devote more resources to R&D if there are more dominant firms earning positive economic profits and, thus, higher world income and expenditure.

For a given L^{R}, an increase in w leads to a proportionate increase in π^{d} by equation (13). From equation (4), world expenditure must increase more than proportionately. Since world wage income $w(\overline{L}^{\mathrm{N}} - L^{\mathrm{R}}) + \overline{L}^{\mathrm{S}}$ increases less than proportionately with an increase in w, equation (11) implies that m must increase. Thus, for a given L^{R}, $m = Z(L^{\mathrm{R}}, w)$ increases in w. In other words, for firms to justify paying their R&D workers higher wages, world expenditure on their products must increase, and this can only occur if there are more dominant firms earning profits.

To maintain m dominant firms in the steady state, each time a patent expires, a new product must be discovered. Thus, during the period of time T, m new products must be discovered. This generates the steady-state R&D supply function in (m, L^{R}) space:

$$(16) \qquad m = R(L^{\mathrm{R}}) \equiv T/h(L^{\mathrm{R}}).$$

Given the properties of $h(\cdot)$, the R&D supply increases in L^{R}; $\partial R/\partial L^{\mathrm{R}} > 0$ and $R(0) = T/\overline{h} > 0$, because some product innovation occurs even if no resources are devoted to R&D. Finally, $R(\overline{L}^{\mathrm{N}}) = T/h(\overline{L}^{\mathrm{N}}) < n$, which guarantees that the number of dominant firms is less than the number of product groups.

The relative wage w is determined by competitive market forces. Depending on the distribution of labor endowments between the North and South, the steady-state relative wage w can exceed or be equal to one. The relative wage w cannot be less than one, since Northern workers are as productive as Southern workers and only Northern workers can do R&D-type work.

If $w > 1$ in the steady-state equilibrium, then m products are produced exclusively in the North by dominant firms and $n - m$ products are produced exclusively in the South by competitive firms, with one product from each product group being produced at any point in time. By symmetry, for each product that the South produces, the aggregate output is $\overline{L}^{\mathrm{S}}/(n - m)$, and the equilibrium price is one. Since this production must satisfy world demand, $\overline{L}^{S}/(n-m)=E^{W}/n$. Thus,

$$L^{R}+\frac{m\overline{L}^{S}}{\alpha(n-m)}=\overline{L}^{N} \tag{17}$$

whenever $w>1$. Equation (17) states that all workers in the North are either engaged in R&D or manufacturing for dominant firms. It implicitly defines $m=F(L^{R})$. F can be interpreted as the steady-state labor market constraint in (m,L^{R}) space. Clearly $\partial F/\partial L^{R}<0$. Given equation (17), any increase in L^{R} must be matched by an equal decrease in the aggregate output of dominant firms. This can only happen if both world expenditure and the number of dominant firms decrease. When $w>1$ in the steady-state equilibrium, the three graphs $Z(\cdot)$, $R(\cdot)$, and $F(\cdot)$ must simultaneously intersect. This case is illustrated by point A in Figure 1.

On the other hand, if a competitive product is produced in the North, it must be that Northern and Southern production workers get paid the same wage $w=1$; that is, $L^{R}+mE^{W}/n\alpha\le\overline{L}^{N}$. Then, the aggregate output for the typical Southern-produced product is $\overline{L}^{S}(n-m)$. However, this production does not have to satisfy demand; that is, $\overline{L}^{S}/(n-m)\le E^{W}/n$. Thus, when $w=1$, the graphs $Z(\cdot)$ and $R(\cdot)$ must intersect at a point where the labor market constraint $m\le F(L^{R})$ is satisfied. This case is illustrated by point B in Figure 2. It is proved in Appendix A that a unique steady-state equilibrium exists in both cases.

In this steady-state equilibrium, each product $j\in N$ experiences a Vernon-type product life cycle. Once discovered, each product is produced in the North by a dominant firm for a period of length T. Then production shifts to a competitive industry in the South (if $w>1$). Eventually each product becomes obsolete, and world production ceases.

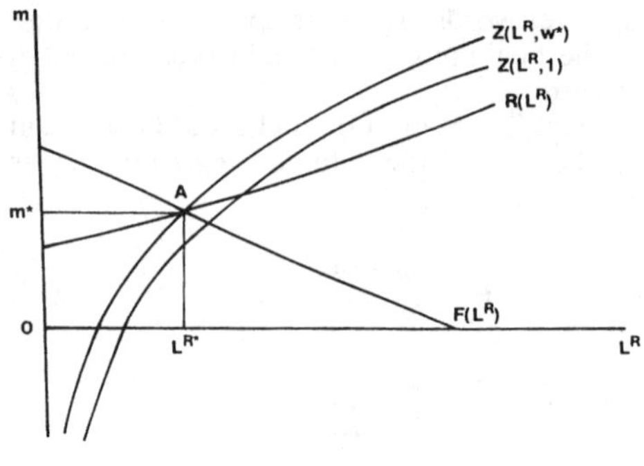

FIGURE 1. THE STEADY-STATE EQUILIBRIUM WITH $w>1$

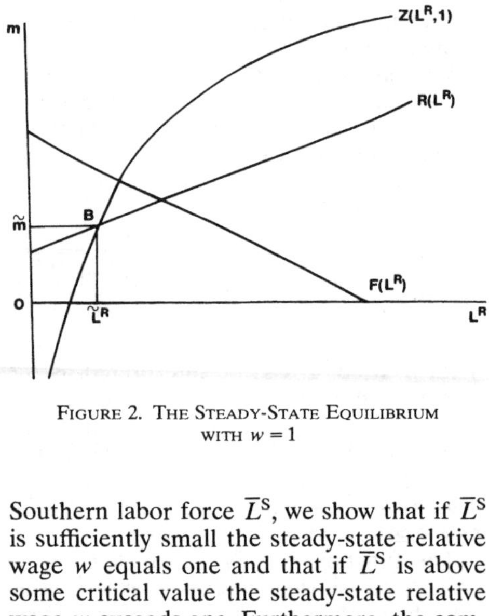

FIGURE 2. THE STEADY-STATE EQUILIBRIUM WITH $w=1$

III. Labor Endowments and Wages in the Steady-State Equilibrium

In this section, we examine the relationship between relative labor endowments and steady-state relative wages. By varying the Southern labor force $\overline{L}^{S}$, we show that if $\overline{L}^{S}$ is sufficiently small the steady-state relative wage w equals one and that if $\overline{L}^{S}$ is above some critical value the steady-state relative wage w exceeds one. Furthermore, the comparative steady-state effects of an increase in the Southern labor endowment depend critically on which case we are in. To see this, consider the steady-state effect of a once-and-for-all increase in the Southern labor force $\overline{L}^{S}$. Increasing $\overline{L}^{S}$ causes both the zero profit condition $Z(L^{R},1)$ and the labor market constraint $F(L^{R})$ to shift down. If the steady-state relative wage w^{*} equals one [and eq. (17) holds with strict inequality], then an increase in $\overline{L}^{S}$ increases the steady-state labor force engaged in R&D (L^{R}) and increases the steady-state number

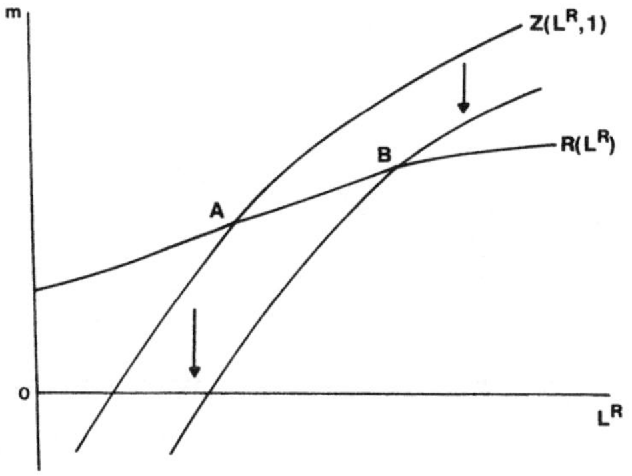

FIGURE 3. EFFECTS OF INCREASE IN SOUTHERN LABOR ENDOWMENT WHEN $w = 1$

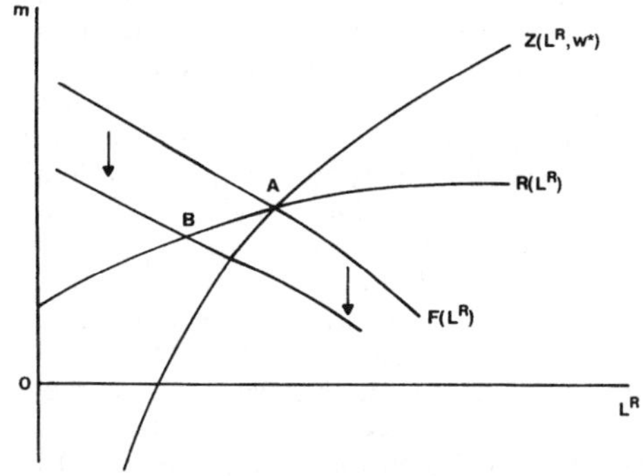

FIGURE 4. EFFECTS OF INCREASE IN SOUTHERN LABOR ENDOWMENT WHEN $w > 1$

of dominant firms in the North (m). This is illustrated by the movement from point A to point B in Figure 3. However, for sufficiently large $\overline{L}^S$, the labor market constraint becomes binding, and the steady-state relative wage begins to rise. With $w^* > 1$, an increase in $\overline{L}^S$ decreases the steady-state labor force engaged in R&D, decreases the steady-state number of dominant firms in the North, and increases the relative wage of Northern workers.[9] This case is illustrated by the movement from point A to point B in Figure 4.

The intuition behind this set of results is easy to explain. When the steady-state relative wage w^* equals one and equation (17) holds with strict inequality, at each point in time t, some Northern workers produce competitive products that are also produced in the South. An increase in the Southern labor force $\overline{L}^S$ increases Southern income I^S and Southern expenditure and thus increases world expenditure E^W. As a result of the increase in world expenditure, dominant firms want to produce more ($q^d = E^W/n\alpha$) and to hire more production workers. Since dominant-firm profits [$\pi^d = (\alpha - w)E^W/\alpha n$] increase, perfect competition in each R&D race induces firms to devote more labor L^R to R&D. With a fixed endowment of labor $\overline{L}^N$ in the North, the increased employment of production workers by dominant firms and the increased employment of R&D labor are exactly balanced by a decreased employment of production workers by competitive firms in the North. However, when the steady-state relative wage w^* exceeds one, this reallocation of labor within the North in response to an increase in $\overline{L}^S$ is not possible, because there are no workers in the North producing competitive goods. Without any change in w^*, an increase in $\overline{L}^S$ leads to excess demand for labor by firms in the North. It is still true that an increase in the Southern labor force $\overline{L}^S$ increases world expenditure E^W, and as a result, dominant firms want to hire more production workers ($q^d = E^W/n\alpha$). Thus, the relative wage w^* must rise enough [and dominant-firm profits $\pi^d = (\alpha - w)E^W/n\alpha$ fall enough] so that firms in the North hire fewer R&D workers; when firms hire fewer R&D workers, this leads to a new steady-state equilibrium with fewer dominant firms.

[9]An increase in $\overline{L}^S$, given w, shifts down $Z(\cdot)$ (not shown in Fig. 4). Since the final equilibrium is at point B, w and $Z(\cdot)$ must appropriately shift up.

IV. The Effects of Patents and Tariffs

First, consider the steady-state effect of a once-and-for-all increase in the patent length T (or a decrease in the rate of technology transfer to the South). Increasing T causes the zero profit condition $Z(L^R,1)$ to shift down and causes the R&D supply function $R(L^R)$ to shift up but leads to no change in the labor market constraint $F(L^R)$. If the steady-state relative wage w^* equals one [and eq. (17) holds with strict inequality], then an increase in T *increases* both the steady-state labor force engaged in R&D and the steady-state number of dominant firms in the North. This case is illustrated in Figure 5. Increasing T increases the reward for innovative activity. Firms respond to this incentive by increasing the resources they devote to R&D. However, if the steady-state relative wage w^* exceeds one, then an increase in T *decreases* the steady-state labor force engaged in R&D, *increases* the steady-state number of dominant firms in the North, and *increases* the relative wage of Northern workers. This is illustrated in Figure 6. This counterintuitive result is explained as follows: increasing T increases the demand for production workers by dominant firms in the North, because dominant firms have longer lives. Since $\bar{L}^N$ is fixed, the increased employment of production workers in the North must be exactly balanced by a decreased employment of R&D workers. The relative wage w^* must rise enough and dominant-firm profits π^d fall enough so that profit-maximizing firms in the North appropriately reduce their R&D expenditures.

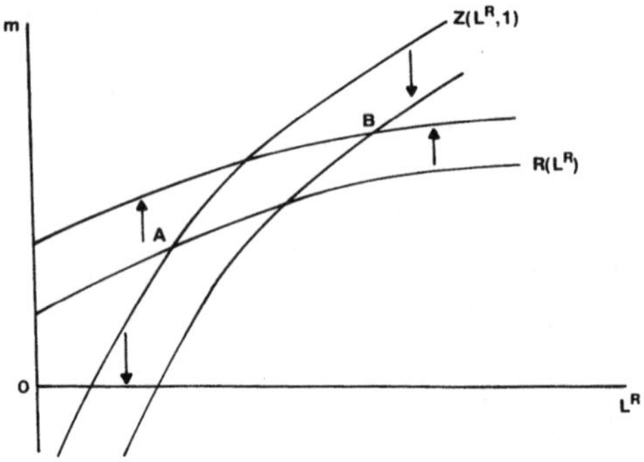

FIGURE 5. EFFECTS OF INCREASE IN PATENT LENGTH WHEN $w = 1$

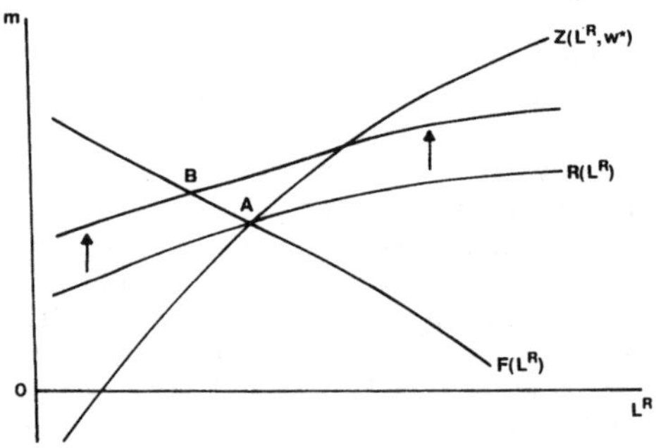

FIGURE 6. EFFECTS OF INCREASE IN PATENT LENGTH WHEN $w > 1$

Because each product experiences a Vernon-type product life cycle in the steady-state equilibrium, the international trade pattern repeatedly changes over time. Industries in the North die in the sense that production of particular products ceases. Other industries in the North are born when new products are discovered. Thus, in this steady-state equilibrium, production workers in the North repeatedly lose their jobs and must find employment in other sectors of the economy. Given this scenario, tariffs designed to save the jobs of production workers in dying industries would have considerable political support.

We now relax the previous assumption that free trade prevails between the North and the South throughout time and explore the comparative steady-state effects of tariffs designed to protect dying industries in the North from Southern competition. We will assume that the labor force $\bar{L}^N$ in the North is sufficiently small so that the steady-state relative wage of Northern production workers w^* exceeds one. This assumption is supported by empirical evidence.[10] At each time t, the government in the North imposes per-unit tariffs on the importation of the $\hat{m}$ products whose patents have most recently expired. The government chooses $\hat{m}$ as its policy instrument; that is, it chooses how many industries in the North to protect from Southern competition. For these tariffs to have any

[10]Keith Maskus (1989) and Christopher Clague (1988) present evidence showing that the wage of unskilled workers in some LDC's is about one-tenth that of unskilled Northern workers.

effects on employment in the $\hat{m}$ Northern industries, each per-unit tariff must be at least $w^* - 1$. In other words, each tariff must be prohibitive. We will assume that this holds for each protected industry. As a result, there is no international trade in any of these $\hat{m}$ protected products, and no government revenue is generated by the tariffs.

In the steady-state equilibrium with $\hat{m}$ products protected, m products are produced exclusively in the North by dominant firms, $\hat{m}$ products are produced both in the North and in the South, and $n - m - \hat{m}$ products are produced exclusively in the South by competitive firms. Since Southern income I^S still equals $\bar{L}^S$ in equilibrium, the South must produce $\bar{L}^S/n$ units of each of the $\hat{m}$ protected products for domestic consumption. Since Southern production must satisfy world demand for each of the $n - m - \hat{m}$ products, by symmetry,

$$(18) \qquad \frac{E^W}{n} = \frac{\bar{L}^S - \hat{m}\left(\frac{\bar{L}^S}{n}\right)}{n - m - \hat{m}}$$

must be satisfied. The right-hand side of equation (18) represents how much Southern labor must be used to produce each of the $n - m - \hat{m}$ products that are produced exclusively in the South.

Steady-state equations (15) and (16) remain unchanged by the introduction of tariffs. Given that $w > 1$, equation (17) becomes

$$(19) \qquad L^R + \frac{mE^W}{n\alpha} + \frac{\hat{m}E^N}{nw} = \bar{L}^N.$$

Using the identity $E^W = E^N + I^S$ and substituting equation (18) into (19) we get

$$(20) \qquad L^R + \left[\frac{m}{\alpha} + \frac{\hat{m}}{w}\right] \frac{\bar{L}^S\left(1 - \frac{\hat{m}}{n}\right)}{n - m - \hat{m}} - \frac{\hat{m}}{n}\frac{\bar{L}^S}{w} = \bar{L}^N.$$

This equation implicitly defines the new $m \equiv F(L^R, \hat{m}, w)$ function. It is easily verified that $\partial F/\partial L^R < 0$, $\partial F/\partial \hat{m} < 0$, and $\partial F/\partial w > 0$. Furthermore, when $\hat{m} = 0$, the new $F(\cdot)$ function coincides with the old $F(\cdot)$ function.

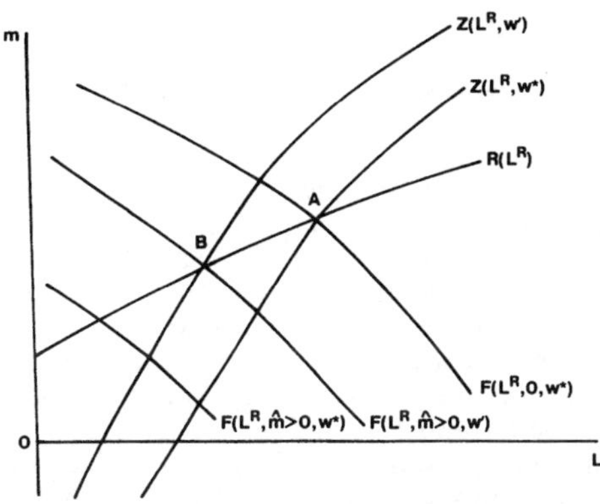

FIGURE 7. EFFECTS OF INCREASE IN THE NUMBER OF PROTECTED INDUSTRIES WHEN $w > 1$

Suppose that in the initial steady-state equilibrium $\hat{m} = 0$ and $w = w^* > 1$. This is illustrated by the intersection of the $m = Z(L^R, w)$, $m = R(L^R)$, and $m = F(L^R, \hat{m}, w)$ functions at point A in Figure 7. An increase in $\hat{m}$ shifts down the labor market constraint $F(\cdot)$. Now there is an excess demand for Northern labor at point A. The steady-state relative wage w must rise above w^*, shifting up both the zero profit condition $Z(\cdot)$ and $F(\cdot)$ until a new intersection is established (at point B, with $w = w' > w^*$).

Thus, we can conclude that when Northern production workers earn higher wages than their Southern counterparts, an increase in the number of protected industries in the North ($\hat{m}$) *decreases* the steady-state rate of product innovation in the North, *decreases* the steady-state number of dominant firms in the North, and *increases* the steady-state relative wage of Northern production workers.

The intuition behind this result is easy to explain. An increase in the number of protected industries in the North raises the demand for Northern production workers. With the Northern labor endowment fixed, the labor devoted to R&D must decline. This necessitates an increase in Northern relative wages, which raises production costs for dominant firms and lowers dominant-firm profit flows. As a result, R&D becomes less profitable, and the steady-state number of dominant firms (winners of R&D races) declines. Notice that in the case of $w=1$, increasing the number of protected industries in the North does not have any effect on the steady-state number of dominant firms and the rate of product innovation.

The introduction of a nontraded sector in the North does not affect the impact of protectionism on the rate of product innovation. The effects of patents on the rate of innovation are robust to the introduction of a nontraded good but may not be robust to the introduction of many nontraded goods. If a sufficiently large fraction of the economy involves nontraded goods, then an increase in T could lead to an increase in L^{R} when $w>1$.[11]

V. Conclusion

We have analyzed a dynamic model of innovation, technology transfer, and international trade. Although highly stylized and in some respects unrealistic, this model nevertheless captures some of the forces that are shaping the pattern of trade in the real world today—forces that are not easily captured in the traditional Heckscher-Ohlin trade model.

In our model, sustained product innovation in the North enables Northern workers to earn higher wages than comparable workers in the South. We have carefully analyzed how the rate of product innovation and the relative wage of Northern workers are affected not only by changes in the rate of technology transfer and the world labor endowment, but also by protectionist government policies. What results is an explanation for the small number of new industries in the North that have arisen to replace older, dying industries as employers of Northern labor. By artificially inflating the wages of Northern workers, protectionist government policies induce sluggish innovative performance in the North.

By focusing on the steady-state equilibrium, we have necessarily abstracted from examining the welfare implications of comparative steady-state exercises. This important task constitutes a nontrivial extension of our model, and it represents a topic for future research.

[11]An appendix containing an algebraic analysis of the effects of patents and tariffs in the presence of nontraded goods is available from the first author upon request.

APPENDIX A

Existence and Uniqueness of Steady-State Equilibrium

Proving that a steady-state equilibrium exists reduces to showing that either (i) $m=Z(L^{\mathrm{R}},w)$, $m=R(L^{\mathrm{R}})$, and $m=F(L^{\mathrm{R}})$ simultaneously intersect at some point $(\hat{L}^{\mathrm{R}},\hat{m},\hat{w})\in R^3_+$ for some $\hat{w}>1$ or (ii) $m=Z(L^{\mathrm{R}},1)$ and $m=R(L^{\mathrm{R}})$ intersect at some point $(\tilde{L}^{\mathrm{R}},\tilde{m})\in R^2_+$ where $\tilde{m}\leq F(\tilde{L}^{\mathrm{R}})$. The graph of $R(\cdot)$ is upward sloping in L^{R}, and the graph of $F(\cdot)$ is downward sloping in L^{R}. Moreover, $R(0)<F(0)$ and $R(\bar{L}^{\mathrm{N}})>F(\bar{L}^{\mathrm{N}})=0$ [given the properties of the $h(\cdot)$ function]. Consequently the functions $R(\cdot)$ and $F(\cdot)$ must have a unique intersection, which we will denote $(L^{\mathrm{R}*},m^*)$. (See Fig. 1).

Suppose that the intersection of $R(\cdot)$ and $F(\cdot)$ lies above the $Z(\cdot)$ graph evaluated at $w=1[Z(L^{\mathrm{R}*},1)<m^*]$. Then, increasing the wage shifts the $Z(\cdot)$ graph upward, and since $\lim_{w\to\alpha}Z(L^{\mathrm{R}*},w)=+\infty$, there exists a wage $w^*>1$ such that all three graphs intersect simultaneously at $(L^{\mathrm{R}*},m^*,w^*)$. This intersection is unique, corresponds to the case-(i) steady-state equilibrium, and is illustrated by point A in Figure 1.

Suppose that the intersection of $R(\cdot)$ and $F(\cdot)$ does not lie above the $Z(\cdot)$ graph evaluated at $w=1[Z(L^{\mathrm{R}*},1)\geq R(L^{\mathrm{R}*})=m^*]$. Notice that $R(0)-Z(0,1)>0$ and that $R(L^{\mathrm{R}*})-Z(L^{\mathrm{R}*},1)\leq 0$ $[\lim_{L^{\mathrm{R}}\to 0}Z(L^{\mathrm{R}},1)=-\infty<0<R(0)=T/\bar{h}]$. This guarantees that the $Z(\cdot)$ function evaluated at $w=1$ and the $R(\cdot)$ function must intersect for

some $\tilde{L}^{R}$ and $\tilde{m}$ which are less than or equal to L^{R*} and m^{*}, respectively. Consequently $w=1$, $\tilde{L}^{R}$ and $\tilde{m}$ are steady-state equilibrium values satisfying case (ii). This steady-state equilibrium is illustrated by point B in Figure 2.

The steady-state equilibrium is unique provided that the functions $Z(L^{R},1)$ and $R(L^{R})$ have at most one intersection in the interval $(0,\bar{L}^{N}]$. It suffices to show that

$$\text{(A1)}\quad \frac{\partial Z(L,1)}{\partial L} = [(\bar{L}-L)(1-e^{-\rho T})Lh'(L)\rho e^{\rho h(L)} + \bar{L}(1-e^{-\rho T})(e^{\rho h(L)}-1)] \times\left[L^{2}(e^{\rho h(L)}-1)^{2}\right]^{-1} > \frac{\partial R(L)}{\partial L} = \frac{-Th'(L)}{[h(L)]^{2}}$$

for all $L\in(0,\bar{L}^{N})$. From equation (5), it follows that

$$\text{(A2)}\quad 0 > e^{\rho h(L)}\rho Lh'(L) > 1-e^{\rho h(L)}.$$

Substituting (A2) into (A1), it suffices to show that

$$\text{(A3)}\quad \frac{1-e^{\rho T}}{L(e^{\rho h(L)}-1)} > \frac{-Th'(L)}{[h(L)]^{2}}$$

for all $L\in(0,\bar{L}^{N})$. Since $h''(L)\geq 0$ and $h'(L)<0$ for all $L\in(0,\bar{L}^{N})$,

$$\text{(A4)}\quad \frac{-Th'(L)}{[h(L)]^{2}} \leq \frac{-Th'(0)}{[h(\bar{L}^{N})]^{2}}.$$

Also, by equation (5)

$$\text{(A5)}\quad \frac{1-e^{-\rho T}}{L(e^{\rho h(L)}-1)} \geq \frac{1-e^{-\rho T}}{\bar{L}^{N}(e^{\rho h(\bar{L}^{N})}-1)}$$

for all $L\in(0,\bar{L}^{N})$. Combining equations (A3), (A4), and (A5) yields equation (6). Thus, the steady-state equilibrium is unique.

APPENDIX B

Steady-State Consumer Behavior

With time-separable utility, the representative Northern consumer's maximization problem can be solved in two stages.[12] First, for given total expenditure at time t, $E(t)$, and prices of available products, we find the allocation of expenditure that maximizes the consumer's CDP utility function. Then we solve for the time path of expenditures that maximizes U.

The first stage of the consumer optimization problem was analyzed in detail in Section II. The second stage involves choosing the optimum expenditure path $E(t)$, $t\in(0,\infty)$. The representative Northern consumer's assets $A(t)$ evolve according to the equation[13]

$$\text{(B1)}\quad \dot{A}(t) = r(t)A(t) + I(t) - E(t)$$

where $r(t)$ is the instantaneous interest rate and $I(t)$ is the consumer's income at time t. Dots denote time derivatives. Furthermore, assets and income must satisfy the feasibility condition

$$\text{(B2)}\quad \lim_{t\to\infty}\inf\left[A(t) + \int_{t}^{\infty}\exp\left(-\int_{t}^{\tau}r(x)\,dx\right)\times I(\tau)\,d\tau\right] \geq 0.$$

This feasibility condition states that the sum of assets and the discounted value of income is nonnegative in the limit as t approaches infinity.

Infinitely lived consumers maximize total lifetime utility [eq. (1)] subject to equations (B1) and (B2). After some algebraic manipulation, equation (8), which is the lifetime

[12]See Hal Varian (1984 p. 148).

[13]See Kenneth Arrow and Mordecai Kurz (1970 Ch. 7).

utility function, reduces to

$$(B3)\quad U = n\int_{t_0}^{\infty} e^{-\rho t}\log E(t)\,dt + \left(\frac{e^{-\rho t_0}}{\rho}\right)\log\frac{1}{n^n\alpha^m} + (\log\alpha)\left(\frac{e^{-\rho t_0}}{\rho}\right)\left(\frac{e^{-\rho\hat{t}}}{(1-e^{-\rho\hat{t}})}\right) + \Gamma(J,t_0).$$

Notice that the second, third, and fourth terms are all constants from the point of view of the consumer [who is choosing $E(t)$]. The third term represents the discounted value of all future innovations in the steady state. If $\alpha = 1$, then new products are identical to old products, and this term disappears.

We conjecture that, in the steady state, the market interest rate $r(t)$ is constant over time and equal to the consumer's discount parameter ρ. We will subsequently verify that, given $r = \rho$, the optimum path of consumer expenditures and assets will also be constant over time in the steady state. Furthermore all markets will clear at each instant in time.[14]

Consequently, the consumer solves the following optimal control problem in the steady state at time t_0:

$$(B4)\quad \max_{E(t)\geq 0;\, t\in[t_0,\infty)} n\int_{t_0}^{\infty} e^{-\rho t}\log E(t)\,dt + \text{constant}$$

subject to $A(t_0) = A_0$,

$$(B5)\quad \dot{A}(t) = \rho A(t) + I - E(t)$$

and

$$(B6)\quad \lim_{t\to\infty}\inf\left[A(t) + \frac{I}{\rho}\right] \geq 0.$$

The constraints (B5) and (B6) follow from (B1) and (B2), assuming that $r(t) = \rho$ for all $t \geq t_0$.

The current-value Hamiltonian for this optimal-control problem is

$$(B7)\quad H = \log E(t) + \lambda(t)\{\rho A(t) + I - E(t)\}$$

and the necessary conditions are $1/E(t) = \lambda(t)$, $\dot{\lambda}(t)/\lambda(t) = 0$, and equations (B5) and (B6). Thus,

$$(B8)\quad E(t) = E_0.$$

Solving (B5), we get

$$(B9)\quad A(t) = e^{\rho t}\left\{A_0 + \frac{I - E_0}{\rho}\right\}\frac{E_0 - I}{\rho}.$$

If $A_0 + (I - E_0)/\rho < 0$, then the dominant term in (B9) approaches $-\infty$ as t approaches $+\infty$, and the feasibility condition (B6) is not satisfied. If $A_0 + (I - E_0)/\rho > 0$, then the dominant term approaches $+\infty$, and (B6) is clearly satisfied. However, (B6) would still be satisfied if E_0 were increased slightly. Hence higher consumption at every instant in time would be feasible, and therefore the expenditure path $E(t) = E_0$ would not be optimal. Thus, the optimal expenditure path must satisfy

$$(B10)\quad A_0 + \frac{I - E_0}{\rho} = 0$$

and therefore

$$(B11)\quad A(t) = A_0 \equiv \frac{E_0 - I}{\rho}$$

which obviously satisfies (B6).

To summarize, we have shown that, if the steady-state interest rate $r(t) = \rho$, then the consumer optimizes by choosing a constant expenditure path over time E_0 where

$$(B12)\quad E_0 = \rho A_0 + I.$$

[14]Alternatively, one could use (6) and (7), allowing the instantaneous interest rate $r(t)$ to vary over time. In this case, if one assumes that expenditure does not change at the steady state, then $r(t) = \rho$; that is, the interest rate is constant in the steady state. For more details, see Grossman and Helpman (1989), in particular their equation 5.

That is, the consumer spends his wage income and interest earning on his assets, at each instant in time.

REFERENCES

Arrow, Kenneth J. and Kurz, Mordecai, *Public Investment, The Rate of Return and Optimal Fiscal Policy*, Baltimore: Johns Hopkins Press, 1970.

Cheng, Leonard, "International Competition in R&D and Technological Leadership: An Examination of the Posner-Hufbauer Hypothesis," *Journal of International Economics*, August 1984, *17*, 15–40.

Clague, Christopher K., "Comparative Costs and Economic Development," unpublished manuscript, University of Maryland, 1988.

Dollar, David, "Technological Innovation, Capital Mobility and the Product Cycle in North–South Trade," *American Economic Review*, March 1986, *76*, 177–90.

______, "Import Quotas and the Product Cycle," *Quarterly Journal of Economics*, August 1987, *102*, 615–32.

Feenstra, Robert C. and Judd, Kenneth L., "Tariffs, Technology Transfer and Welfare," *Journal of Political Economy*, December 1982, *90*, 1142–65.

Grossman, Gene M. and Helpman, Elhanan, "Product Development and International Trade," *Journal of Political Economy*, December 1989, *97*, 1261–83.

Jensen, Richard A. and Thursby, Marie C., "A Strategic Approach to the Product Life Cycle," *Journal of International Economics*, November 1986, *21*, 269–84.

______ **and** ______, "A Decision Theoretic Model of Innovation, Technology Transfer and Trade," *Review of Economic Studies*, October 1987, *54*, 631–47.

Krugman, Paul, "A Model of Innovation, Technology Transfer and the World Distribution of Income," *Journal of Political Economy*, April 1979, *87*, 253–66.

Lee, Tom and Wilde, Louis L., "Market Structure and Innovation: A Reformulation," *Quarterly Journal of Economics*, March 1980, *94*, 429–36.

Loury, Glenn C., "Market Structure and Innovation," *Quarterly Journal of Economics*, August 1979, *93*, 395–410.

Maskus, Keith E., "A Nonparametric Approach to Testing for International Technological Equivalence in Manufacturing," unpublished manuscript, University of Colorado, 1989.

Pugel, Thomas A., "Endogenous Technological Change and International Technology Transfer in a Ricardian Trade Model," *Journal of International Economics*, November 1982, *13*, 321–35.

Reinganum, Jennifer F., "A Dynamic Game of R&D: Patent Protection and Competitive Behavior," *Econometrica*, May 1982, *50*, 671–88.

Schumpeter, Joseph A., *Capitalism, Socialism and Democracy*, New York: Harper, 1942.

Spencer, Barbara J. and Brander, James A., "International R&D Rivalry and Industrial Strategy," *Review of Economic Studies*, October 1983, *50*, 707–22.

Stokey, Nancy L., "Learning by Doing and the Introduction of New Goods," *Journal of Political Economy*, August 1988, *96*, 701–17.

Varian, Hal R., *Microeconomic Analysis*, New York: Norton, 1984.

Vernon, Raymond, "International Investment and International Trade in the Product Cycle," *Quarterly Journal of Economics*, May 1966, *80*, 190–207.

[14]

Econometrica, Vol. 60, No. 2 (March, 1992), 323–351

A MODEL OF GROWTH THROUGH CREATIVE DESTRUCTION

BY PHILIPPE AGHION AND PETER HOWITT[1]

A model of endogenous growth is developed in which vertical innovations, generated by a competitive research sector, constitute the underlying source of growth. Equilibrium is determined by a forward-looking difference equation, according to which the amount of research in any period depends upon the expected amount of research next period. One source of this intertemporal relationship is creative destruction. That is, the prospect of more future research discourages current research by threatening to destroy the rents created by current research. The paper analyzes the positive and normative properties of stationary equilibria, in which research employment is constant and GNP follows a random walk with drift, although under some circumstances cyclical equilibria also exist. Both the average growth rate and the variance of the growth rate are increasing functions of the size of innovations, the size of the skilled labor force, and the productivity of research as measured by a parameter indicating the effect of research on the Poisson arrival rate of innovations; and decreasing functions of the rate of time preference of the representative individual. Under laissez faire the economy's growth rate may be more or less than optimal because, in addition to the appropriability and intertemporal spillover effects of other endogenous growth models, which tend to make growth slower than optimal, the model also has effects that work in the opposite direction. In particular, the fact that private research firms do not internalize the destruction of rents generated by their innovations introduces a business-stealing effect similar to that found in the partial-equilibrium patent race literature. When we endogenize the size of innovations we find that business stealing also makes innovations too small.

KEYWORDS: Endogenous growth, innovations, creative destruction.

1. INTRODUCTION

THE MAIN CONTRIBUTION of the literature on endogenous growth pioneered by Romer (1986) and Lucas (1988) has been to endogenize the underlying source of sustained growth in per-capita income, namely the accumulation of knowledge. There are many channels through which societies accumulate knowledge, including formal education, on-the-job training, basic scientific research, learning by doing, process innovations, and product innovations. This paper examines a channel that has received little attention in the endogenous growth literature, namely that of industrial innovations which improve the quality of products.

This channel introduces into endogenous growth theory the factor of obsolescence; better products render previous ones obsolete. Obsolescence exemplifies an important general characteristic of the growth process, namely that progress creates losses as well as gains. It also embodies Schumpeter's idea of creative

[1] The authors wish to acknowledge the helpful comments and criticisms of Roland Bénabou, Olivier Blanchard, Patrick Bolton, Louis Corriveau, Mathias Dewatripont, Dick Eckaus, Zvi Griliches, Elhanan Helpman, Rebecca Henderson, Louis Phaneuf, William Scarth, Nancy Stokey, Patrick Rey, and the Co-Editor and referees of this journal.

destruction (1942, p. 83, his emphasis):

> The fundamental impulse that sets and keeps the capitalist engine in motion comes from the new consumers' goods, the new methods of production or transportation, the new markets,.... [This process] incessantly revolutionizes the economic structure *from within*, incessantly destroying the old one, incessantly creating a new one. This process of Creative Destruction is the essential fact about capitalism.

The present paper constructs a simple model of growth through creative destruction, by modelling the innovation process as in the patent-race literature surveyed by Tirole (1988, Ch. 10) and Reinganum (1989). The expected growth rate of the economy depends upon the economy-wide amount of research. The paper shows that equilibrium in such an economy is determined by a forward-looking difference equation, according to which the amount of research in any period depends upon the expected amount of research next period, similar to the difference equation that defines equilibrium in the two-period overlapping-generations model of money (Azariadis (1981), Grandmont (1985)).

More specifically, the model assumes, following Schumpeter, that individual innovations are sufficiently important to affect the entire economy. A period is the time between two successive innovations. The length of each period is random, because of the stochastic nature of the innovation process, but the relationship between the amount of research in two successive periods can be modelled as deterministic. The amount of research this period depends negatively upon the expected amount next period, through two effects.

The first effect is that of creative destruction. The payoff from research this period is the prospect of monopoly rents next period. Those rents will last only until the next innovation occurs, at which time the knowledge underlying the rents will be rendered obsolete. Thus the expected present value of the rents depends negatively upon the Poisson arrival rate of the next innovation. The expectation of more research next period will increase that arrival rate, and hence will discourage research this period.

The second effect is a general equilibrium effect working through the wage of skilled labor, which can be used either in research or in manufacturing. In order to be consistent with the conditions for labor-market equilibrium, the expectation of more research next period must correspond to an expectation of higher demand for skilled labor in research next period, which implies the expectation of a higher real wage of skilled labor. Higher wages next period will reduce the monopoly rents that can be gained by exclusive knowledge of how to produce the best products. Thus the expectation of more research next period will discourage research this period by reducing the flow of rents expected to accrue to a successful innovator.

This functional relationship between research in two successive periods has a unique fixed point, which defines a stationary equilibrium. The stationary equilibrium exhibits balanced growth, in the sense that the allocation of skilled labor between research and manufacturing remains unchanged with each innovation; the log of GNP follows a random walk with drift. This is not always, however, the only equilibrium in the model. As in the overlapping-generations literature the functional relationship can also be satisfied by cyclical trajectories.

One noteworthy implication of the negative dependency of current research upon future research is the possible existence of what we call a "no-growth trap," a cyclical equilibrium in which the level of research oscillates deterministically between two levels each period, and in which the lower of these two levels is zero. An economy in such an equilibrium would stop growing in finite time, because with no research there would be no innovation, and hence the period with no research would never come to an end. The (rational) expectation that the next innovation would be followed by a very high level of research would discourage anyone from undertaking that innovation.

Another implication is that the average growth rate of the economy is not necessarily increased by an increase in the productivity of research. In particular, a parameter change that makes research more productive in some states of the world can discourage research in other states, by increasing the threat of obsolescence faced by the product of research in those other states, to such an extent that the average growth rate is reduced.

From a normative point of view, the average growth rate in stationary equilibrium may be more or less than socially optimal because of the presence of conflicting distortionary effects. Specifically, although the model includes the appropriability and intertemporal spillover effects which generate a less than optimal growth rate in Romer's (1990) model, it also has effects that work in the opposite direction. In particular, there is a "business-stealing" effect of the sort familiar from the patent-race literature (Tirole (1988, p. 399)). That is, researchers do not internalize the destruction of existing rents created by their innovations. When the size of innovations is taken as given, the business stealing effect can lead to too much growth. In addition, we find that when the size of innovations is endogenized, the business stealing effect tends to make innovations too small.

Other papers in the endogenous growth literature that model vertical product innovations include Segerstrom, Anant, and Dinopoulos (1990), who assume that the time between successive innovations is deterministic. They have a richer intersectoral structure than the present paper, and address a different set of questions. Stokey (1988) models vertical product innovations and obsolescence in a perfectly competitive model where innovations are the unintentional by-product of learning by doing. The cost-reducing innovations of Shleifer (1986) can also be interpreted as vertical product innovations. His model does not endogenize growth, however, except in a limited dichotomous sense; that is, the long-run average growth rate is fixed by the exogenously specified rate of invention, except in the singular case where no inventions are ever implemented and the economy stops growing.

Corriveau (1988) has a discrete-time analysis of endogenous growth based on cost-reducing innovations as in Shleifer, in which the possibility of simultaneous discoveries creates a different kind of "business-stealing" effect. In his model the payoff to current research is independent of future research because rents to innovations are assumed to accrue only in the same period as the research from which they resulted. Grossman and Helpman (1991) construct a model of vertical product innovation that explicitly integrates the analysis of Segerstrom, Anant, and Dinopoulos (1990) with that of the present paper.

Judd (1985) and Romer (1990) model growth based on horizontal product innovations, using the Dixit-Stiglitz (1977) model of product variety. These models involve no obsolescence; new products are no better than existing ones. They also involve no uncertainty. King and Rebelo (1988) introduce uncertainty into an endogenous growth model by assuming a random rate of return to the accumulation of human capital under conditions of perfect competition.

Within the patent-race literature the paper closest to the present is that of Reinganum (1985), which also emphasizes the affinity to creative destruction. The present paper adds to Reinganum's model the general equilibrium effects of future research on the rents created by current research, and of the level of manufacturing employment on the cost of research. The paper also generalizes the Reinganum model by allowing the stream of innovations to continue forever, and by explicitly analyzing the effect of the future level of research on the prospective reward to current research.[2]

Section 2 below presents the basic model of the paper. This basic model assumes for simplicity that each innovation creates an economy-wide monopoly in the production of intermediate goods. Section 3 derives the functional relationship between research in two successive periods that defines equilibrium. It then analyzes the determinants of the average growth rate and the variability of the growth rate in stationary equilibrium. One of those determinants is the degree of market power possessed by an intermediate-good monopolist, which is parameterized in the model. Section 4 characterizes the welfare properties of stationary equilibria in the basic model, under the assumption of a fixed size of innovations. Section 5 introduces the possibility of nondrastic innovations. Section 6 generalizes the model by allowing research firms to choose the size of innovations as well as their arrival rate. Section 7 deals with a strategic monopsony effect that has been ignored until this point in the argument, by which an intermediate firm can extend the expected lifetime of its monopoly by hiring more than the short-run profit-maximizing amount of skilled labor, at the cost of a higher real wage. Section 8 relaxes the assumption of a single economy-wide monopoly in the production of intermediate goods. Section 9 contains brief concluding remarks.

[2] That is, Reinganum's comparative-statics analysis follows the common practice of the patent-race literature in taking the reward to a successful innovation as given, whereas the following analysis shows that the effect of a parameter change on the time path of research involves feedback from future research to current research working through the two above-mentioned channels: creative destruction and the general equilibrium wage effect on profits, both of which flow through the reward to a successful innovation.

2. THE BASIC MODEL

2.A. *Assumptions*

There are three classes of tradeable objects: labor, a consumption good, and an intermediate good. There is a continuum of infinitely-lived individuals, with identical intertemporally additive preferences defined over lifetime consumption, and the constant rate of time preference $r > 0$. The marginal utility of consumption is assumed constant; thus r is also the rate of interest.

There is no disutility from supplying labor. There are three categories of labor: unskilled labor, which can be used only in producing the consumption good; skilled labor, which can be used either in research or in the intermediate sector; and specialized labor, which can be used only in research. Each individual is endowed with a one-unit flow of labor. Let M, N, and R denote respectively the mass of unskilled, skilled, and specialized individuals.

The consumption good is produced using the fixed quantity M of unskilled labor, and the intermediate good, subject to constant returns. Since M is fixed, the production function can be written as

$$(2.1) \qquad y = AF(x),$$

where $F' > 0$, $F'' < 0$, y is the flow output of consumption good, x the flow of intermediate input, and A a parameter indicating the productivity of the intermediate input.

The intermediate good is produced using skilled labor alone, according to the linear technology

$$(2.2) \qquad x = L,$$

where L is the flow of skilled labor used in the intermediate sector.

Research produces a random sequence of innovations. The Poisson arrival rate of innovations in the economy at any instant is $\lambda\phi(n, R)$, where n is the flow of skilled labor used in research, λ a constant parameter, and ϕ a constant-returns, concave production function. Both λ and ϕ are given by the technology of research. There is no memory in this technology, since the arrival rate depends only upon the current flow of input to research, not upon past research. Assume that skilled labor is an essential factor in research: $\phi(0, R) = 0$. Then an economy that allocates no skilled labor to research will not grow, because it will experience no innovations. (The "linear" case where $\phi(n, R) = n$, that is, where $R = 0$, will be used frequently.)

Time is continuous, and indexed by $\tau \geqslant 0$. The subscript $t = 0, 1\ldots$ denotes the interval starting with the tth innovation (or with $\tau = 0$ in the case of $t = 0$) and ending just before the $t+1$st. The length of each interval is random. All prices and quantities are assumed to remain constant within each interval. If n_t is applied to research in interval t, the length of the interval will be exponentially distributed with parameter $\lambda\phi(n_t, R)$.

Each innovation consists of the invention of a new intermediate good, whose use as input allows more efficient methods to be used in producing the consumption good. Real-world examples include such "input" innovations[3] as the steam engine, the airplane, and the computer, whose use made possible new methods of production in mining, transportation, and banking, with economy-wide effects. An innovation need not, however, be as revolutionary as these examples, but might consist instead of a new generation of intermediate good, similar to the old one.

Specifically, use of the new intermediate good increases the productivity parameter A in (2.1) by the factor $\gamma > 1$. There are no lags in the diffusion of technology.[4] The most modern intermediate good is always produced, so that:

$$A_t = A_0 \gamma^t \qquad (t = 0, 1, \ldots), \tag{2.3}$$

where A_0 is the initial value given by history. (Of course, it is always possible to produce the consumption good using an old technology, with a correspondingly old intermediate good.)

A successful innovator obtains a patent which it can use to monopolize the intermediate sector. (Section 8 relaxes this assumption by allowing for a finite number of monopolistic competitors.) The patent is assumed to last forever. However, the monopoly lasts only until the next innovation, at which time the intermediate good is replaced by the next vintage. All markets are perfectly competitive except that for intermediate goods.

2.B. *The Intermediate Monopolist's Decision Problem*

For ease of presentation the analysis starts by assuming that innovations are always drastic; that the intermediate monopolist is unconstrained by potential competition from the previous patent. This assumption will be relaxed in Section 5 below. The intermediate monopolist's objective is to maximize the expected present value of profits over the current interval. When the interval ends so do the profits. The only uncertainty concerns the length of the interval. Except in Section 7 below, the monopolist is assumed to take as given the amount of research at each time, and hence also takes as given the length of the interval.

[3] Scherer (1984) combines process- and input-oriented R and D into a measure of "used" R and D, which he distinguishes from pure product R and D. He estimates that during the period 1973–1978 in U.S. industry the social rate of return to "used" R and D lay between 71% and 104%, whereas the return to pure product R and D was insignificant.

[4] Gradual diffusion could be introduced by allowing the productivity parameter after each innovation to follow a predetermined but gradual path asymptotically approaching the limit A_t, and then to jump to A_t upon the next innovation and follow a gradual path approaching A_{t+1}. This would produce a cycle in research within each interval, as the gradual rise in productivity would induce manufacturing firms to hire more and more workers out of research until the next innovation occurs.

Let x_t be the flow of the intermediate good produced by the monopolist during interval t. By (2.2), x_t also equals employment of skilled labor in manufacturing. The inverse demand curve facing a monopolist charging the price p_t (relative to the numéraire consumption good) is the marginal product

$$(2.4) \qquad p_t = A_t F'(x_t).$$

Thus the monopolist chooses x_t to maximize $[A_t F'(x_t) - w_t]x_t$, taking as given A_t and the wage w_t of skilled labor.

Define the "productivity-adjusted wage" as $\omega_t \equiv w_t/A_t$, and the "marginal-revenue function" as $\tilde{\omega}(x) \equiv F'(x) + xF''(x)$. Assume that marginal revenue is downward-sloping and satisfies Inada-type conditions.

ASSUMPTION 1: $\tilde{\omega}'(x) < 0$ *for all* $x > 0$, $\lim_{x \to 0} \tilde{\omega}(x) = \infty$, $\lim_{x \to \infty} \tilde{\omega}(x) = 0$.

Then for any positive ω_t the monopolist's choice of output x_t is given by the first-order condition

$$(2.5) \qquad \omega_t = \tilde{\omega}(x_t),$$

or

$$(2.6) \qquad x_t = \tilde{x}(\omega_t),$$

where $\tilde{x}$ is the function $\tilde{\omega}^{-1}$. The flow of monopoly profits is

$$(2.7) \qquad \pi_t = A_t \tilde{\pi}(\omega_t),$$

where $\tilde{\pi}(\omega) \equiv -(\tilde{x}(\omega))^2 F''(\tilde{x}(\omega))$. Note that $\tilde{x}$ and $\tilde{\pi}$ are each strictly positive-valued and strictly decreasing for all positive ω_t.

An example satisfying Assumption 1 is the Cobb-Douglas example, in which the consumption-good technology is $F(x) = x^\alpha$, $0 < \alpha < 1$, which yields

$$(2.8) \qquad p_t = w_t/\alpha, \quad \pi_t = \left(\frac{1-\alpha}{\alpha}\right) w_t x_t, \quad x_t = \left(\omega_t/\alpha^2\right)^{1/(\alpha-1)}.$$

2.C. *Research*

There are no contemporaneous spillovers in research; that is, a firm employing the amounts z, s of the two factors in research will experience innovations with a Poisson arrival rate of $\lambda\phi(z,s)$, independently of the inputs of other firms. The objective of a firm in choosing z and s at each date is to maximize the flow of expected profits from research:

$$(2.9) \qquad \lambda\phi(z,s)V_{t+1} - w_t z - w_t^s s,$$

where V_{t+1} is the value of the $t+1$st innovation, and w_t^s is the wage rate of specialized labor.

Because ϕ has constant returns, and because the total flow of specialized labor must equal R in equilibrium, it follows from the Kuhn-Tucker conditions for maximizing (2.9) that

$$(2.10) \quad w_t \geqslant \varphi'(n_t)\lambda V_{t+1}, \quad n_t \geqslant 0, \quad \text{with at least one equality,}$$

where $\varphi(n_t) \equiv \phi(n_t, R)$, and n_t is the economy-wide flow of skilled labor used in research during interval t. Note that

$$(2.11) \quad \varphi(0) = 0, \quad \text{and} \quad \varphi'(n) > 0, \quad \varphi''(n) \leqslant 0 \quad \text{for all} \quad n \geqslant 0.$$

As we shall see, all research is conducted by outside research firms rather than by the incumbent monopolist. Because of constant returns to scale the number of research firms is indeterminate. The value V_{t+1} to an outside research firm is the expected present value of the flow of monopoly profits π_{t+1} generated by the $t+1$st innovation over an interval whose length is exponentially distributed with parameter $\lambda\varphi(n_{t+1})$:

$$(2.12) \quad V_{t+1} = \frac{\pi_{t+1}}{r + \lambda\varphi(n_{t+1})}.$$

The reason why the monopolist chooses to do no research is that the value to the monopolist of making the next innovation would be $V_{t+1} - V_t$, which is strictly less than the value V_{t+1} to an outside firm. This is an example of the Arrow effect, or replacement effect (see Tirole (1988, p. 392)). The efficiency effect, or rent-dissipation effect, according to which an outside firm might receive a smaller payoff from an innovation than would the present incumbent, because of having to compete with the present incumbent, is absent in the case of drastic innovations because the flow of profit π_{t+1} in (2.12) is independent of whether the firm earning those profits has access to the previous patent.[5]

There is an important intertemporal spillover in this model. An innovation raises productivity forever. It allows each subsequent innovation to raise A_t by the same multiple γ, and with the same probability $\lambda\varphi(n_t)$, but from a starting value that is higher by the multiple γ than it would otherwise have been. The producer of an innovation captures (some of) the rents from that productivity gain, but only during one interval. After that the rents are captured by other innovators, building upon the basis of the present innovation, but without compensating the present innovator.[6] This intertemporal spillover plays a role in the welfare analysis of Section 4 below.

[5] If, instead of a constant-returns research technology, each firm had an identical research technology with rising marginal cost, then the monopolist might do some research, but the Arrow effect would imply that the monopolist would do less research than each outside research firm, as shown by Reinganum (1985, Proposition 2) in a similar context.

[6] This is the spillover identified by Romer (1990, pp. S83–S85).

The model also embodies Schumpeter's idea of "creative destruction." Each innovation is an act of creation aimed at capturing monopoly rents. But it also destroys the monopoly rents that motivated the previous creation. Creative destruction accounts for the term $\lambda\varphi(n_{t+1})$ in the denominator of V_{t+1} in (2.12). More research reduces the expected tenure of the current monopolist, and hence reduces the expected present value of its flow of rents.

2.D. *Capital Markets*

The structure of capital markets can be specified in many different ways. One is to suppose that there is a frictionless Walrasian credit market in which future expected consumption can be discounted at the constant rate r. Another is to suppose that there is no credit market. According to the latter specification all nonresearch workers consume their wage income at each instant, the owners of the monopoly intermediate firm consume their flow of profits at each instant, and research workers receive no pay unless their firms innovate, at which time they are paid in shares of the next intermediate firm. According to either specification, all research firms could be assumed to be owned by their workers, and (2.9) would represent the expected flow of surplus to be divided among them. The crucial assumption that utility is linear in consumption makes these different specifications all equivalent, by removing any motive to use capital markets for risk-sharing.

3. PERFECT FORESIGHT DYNAMICS AND BALANCED GROWTH

3.A. *Equilibrium*

At any point in time there is only one decision for society to make; namely, how to allocate the fixed flow N of skilled labor between manufacturing and research. Combining (2.5), (2.7), (2.10), (2.12) and the equilibrium condition $N = n_t + x_t$ yields

$$\frac{\tilde{\omega}(N-n_t)}{\lambda\varphi'(n_t)} \geqslant \frac{\gamma\tilde{\pi}(\tilde{\omega}(N-n_{t+1}))}{r+\lambda\varphi(n_{t+1})}, \quad n_t \geqslant 0, \quad \text{with at least one equality.} \tag{3.1}$$

Condition (3.1) determines research employment at t as a function of research employment at $t+1$:

$$n_t = \psi(n_{t+1}), \tag{3.2}$$

where ψ: $[0, N) \to \mathbb{R}_+$ is a strictly decreasing function wherever it is positive-valued.

The functional relationship ψ between research employment in two successive periods is illustrated in Figure 1, where $c(n_t)$ is the "marginal cost of

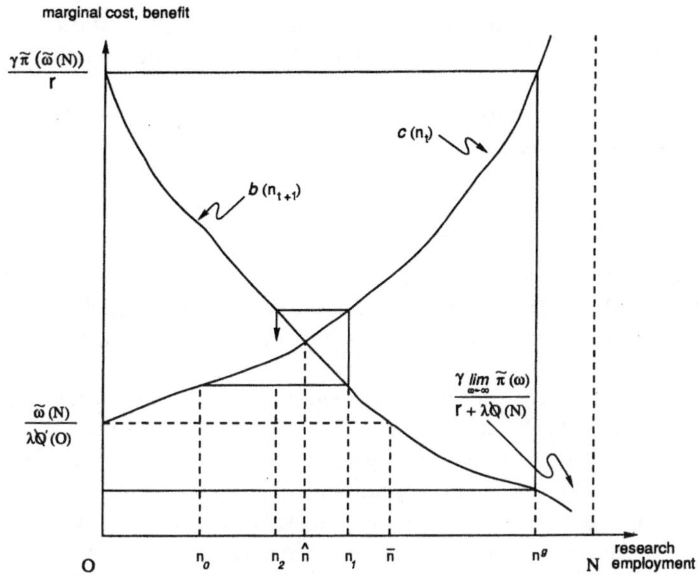

FIGURE 1.—The effect of future research on current research: $n_0 = \psi(n_1)$ and $n_1 = \psi(n_2)$. The pair $(0, n^g)$ constitutes a no-growth trap.

research" and $b(n_{t+1})$ the "marginal benefit of research," defined by

$$c(n_t) \equiv \frac{\tilde{\omega}(N - n_t)}{\lambda \varphi'(n_t)},$$

$$b(n_{t+1}) \equiv \frac{\gamma \tilde{\pi}(\tilde{\omega}(N - n_{t+1}))}{r + \lambda \varphi(n_{t+1})}.$$

By (2.11) and Assumption 1, c is strictly increasing, b is strictly decreasing, and $c(n_t) \to \infty$ as $n_t \to N$. It follows that in the case illustrated in Figure 1, where $c(0) < b(0)$, $\psi(n_{t+1})$ is well defined on $[0, N)$, and is positive and decreasing if and only if[7] $n_{t+1} < \bar{n}$. In the case where $c(0) \leqslant b(0)$, $\psi(n_{t+1})$ is identically zero.

In economic terms, there are two reasons for the negative dependency of current research on future research, corresponding to the two places in which n_{t+1} enters the expression for the marginal benefit of research, $b(n_{t+1})$. That is, a foreseen increase in research next period discourages research this period (a) by raising future wages and hence reducing the flow of profits $\tilde{\pi}(\tilde{\omega}(N - n_{t+1}))$ to be captured from the next innovation, and (b) by raising the rate of creative destruction $\lambda\varphi(n_{t+1})$ next period and hence shortening the expected lifetime of the monopoly to be enjoyed by the next innovator.

A perfect foresight equilibrium (PFE) is defined as a sequence $\{n_t\}_0^\infty$ satisfying (3.2) for all $t \geqslant 0$. In Figure 1 the sequence $\{n_0, n_1, n_2, \dots\}$ constructed from the

[7] The critical value $\bar{n}$ is defined by $c(0) = b(\bar{n})$, unless $\lim_{n \to N} b(n) > c(0)$, in which case $\bar{n} = N$.

counterclockwise spiral starting at n_0 constitutes a PFE. A stationary equilibrium corresponds to a PFE with n_t constant. It is defined as the solution to $\hat{n} = \psi(\hat{n})$.

There exists a unique stationary equilibrium. As Figure 1 shows, if $c(0) < b(0)$ then $\hat{n}$ is positive, and is defined by

$$(3.3) \qquad \frac{\tilde{\omega}(N-\hat{n})}{\lambda\varphi'(\hat{n})} = \frac{\gamma\tilde{\pi}(\tilde{\omega}(N-\hat{n}))}{r+\lambda\varphi(\hat{n})}.$$

In this case growth is positive because innovations arrive at the Poisson rate $\lambda\varphi(\hat{n}) > 0$. If $c(0) \geqslant b(0)$ then $\hat{n} = 0$ and there is no growth, because the Poisson arrival rate of innovations is $\lambda\varphi(0) = 0$. Henceforth, assume that $c(0) < b(0)$ and $\hat{n} > 0$.

Other equilibria may also exist. A two-cycle is a pair (n^0, n^1) such that $n^0 = \psi(n^1)$ and $n^1 = \psi(n^0)$. It defines a PFE of period two. If both n^0 and n^1 are positive, the PFE is a "real" two-cycle. If either n^0 or n^1 is zero, it is a "no-growth trap." In a real two-cycle, the prospect of high research in odd intervals discourages research in even intervals, and the prospect of low research in even intervals stimulates research in odd intervals.[8] A no-growth trap is the extreme case in which the prospect of high research in odd intervals shuts down research completely in even intervals. Although the no-growth trap defines an infinite sequence $\{n_t\}_0^\infty$, the oscillation will cease after one innovation. From then on no growth will occur because no innovations will occur. It is clear from Figure 1 that a no-growth trap exists if $\lim_{\omega\to\infty}\tilde{\pi}(\omega) = 0$ and r is small enough. A real two-cycle will exist if a no-growth trap exists and[9] $c'(\hat{n}) + b'(\hat{n}) > 0$.

Consider the Cobb-Douglas example: $F(x) = x^\alpha$, with a linear research technology $\varphi(n) \equiv n$. From (2.8), the equation (3.3) defining a positive $\hat{n}$ is

$$(3.4) \qquad 1 = \frac{\lambda\gamma\left(\dfrac{1-\alpha}{\alpha}\right)(N-\hat{n})}{r+\lambda\hat{n}},$$

[8] Shleifer (1986) also finds deterministic cycles in a model of multiple equilibria with innovations. The source of multiplicity in Shleifer's model is a contemporaneous strategic complementarity, whereby the incentive to innovate this period is stronger the more innovations are occurring elsewhere in the economy this period. No such strategic complementarity exists in the present model, in which more research this period raises the marginal cost of research without affecting the marginal benefit. Because Shleifer assumes that imitation destroys the return from innovations after one period, his model does not exhibit the dependency of current research upon future research which underlies the cycles, as well as the other equilibria, in the present model. Deneckere and Judd (1986) also generate cycles in a model of innovations. Their cycles arise from local instability of a unique equilibrium rather than from multiple equilibria. Like Shleifer, they also do not allow for any effect of future research upon current research.

[9] Consider the second-iterate function $\psi^2(n)$. Geometrically, ψ^2 is defined by reversing the spiral illustrated in Figure 1; thus $n_0 = \psi^2(n_2)$. Suppose a no-growth trap exists. Then $0 = \psi^2(n) < n$ for small but positive n (see Figure 1). Because $c'(\hat{n}) + b'(\hat{n}) > 0$, therefore the counterclockwise spiral in Figure 1 spirals out in the neighbourhood of $\hat{n}$, so that $\psi^2(n) > n$ for n close to but less than $\hat{n}$. By the continuity of ψ^2 there must exist an n^0 strictly between 0 and $\hat{n}$ such that $\psi^2(n^0) = n^0$. Evidently $n^0, \psi(n^0)$ constitute a real two-cycle.

and the condition for $\hat{n}$ to be positive is

$$1 < \lambda\gamma\left(\frac{1-\alpha}{\alpha}\right)N\Big/r. \tag{3.5}$$

Since $\tilde{\pi}(\infty) = 0$, a no-growth trap exists when r is small. Since[10] $b'(\hat{n})/c'(\hat{n}) \to 0$ uniformly in r as $\alpha \to 0$, a real two-cycle exists when α and r are small enough.

3.B. *Research in Stationary Equilibrium*

The rest of the paper focuses on stationary equilibria. Comparative-statics analysis of (3.3) shows the following proposition.

PROPOSITION 1: *The amount of research employment $\hat{n}$ in a stationary equilibrium increases with: (a) a decrease in the rate of interest r; (b) an increase in the size γ of each innovation; (c) an increase in the total endowment N of skilled labor; or (d) an increase in the arrival parameter λ.*

This proposition is intuitive: (a) A decrease in the rate of interest increases the marginal benefit to research, by raising the present value of monopoly profits. (b) An increase in the size of each innovation also increases the marginal benefit to research, by raising the size of next interval's monopoly profits relative to this interval's productivity. (c) An increase in the endowment of skilled labor both increases the marginal benefit and reduces the marginal cost of research, by reducing the wage of skilled labor. (d) An increase in the

[10] In this example

$$b(n) \equiv \frac{\gamma\tilde{\pi}(\tilde{\omega}(N-n))}{r+\lambda n} \quad \text{and} \quad c(n) \equiv \tilde{\omega}(N-n)/\lambda.$$

Therefore

$$b'(n) = \frac{-\gamma\tilde{\pi}'(\tilde{\omega}(N-n))\tilde{\omega}'(N-n) - \lambda b(n)}{r+\lambda n} \quad \text{and} \quad c'(n) = -\tilde{\omega}'(N-n)/\lambda.$$

Because $\lambda b(\hat{n}) = \lambda c(\hat{n}) = \tilde{\omega}(N-\hat{n})$, therefore

$$b'(\hat{n})/c'(\hat{n}) = \frac{\lambda\gamma\tilde{\pi}'(\tilde{\omega}(N-\hat{n})) + \lambda\tilde{\omega}(N-\hat{n})/\tilde{\omega}'(N-\hat{n})}{r+\lambda\hat{n}}.$$

But $\tilde{\pi}'(\tilde{\omega}(N-\hat{n})) = -(N-\hat{n})$ and $\tilde{\omega}(N-\hat{n}) = \alpha^2(N-\hat{n})^{\alpha-1}$. Therefore

$$b'(\hat{n})/c'(\hat{n}) = -\frac{\lambda(N-\hat{n})}{r+\lambda\hat{n}}\left(\gamma+\frac{1}{1-\alpha}\right).$$

From this and (3.4),

$$b'(\hat{n})/c'(\hat{n}) = -\left(\frac{\alpha}{\gamma(1-\alpha)}\right)\left(\gamma+\frac{1}{1-\alpha}\right) \to 0 \quad \text{uniformly in } r \text{ as } \alpha \to 0.$$

arrival parameter decreases both the marginal cost and the marginal benefit of research, because on the one hand it results in more "effective" units of research for any given level of employment, but on the other hand it also increases the rate of creative destruction during the next interval. The former effect dominates.

The above discussion of result (d) suggests an interesting implication of creative destruction that could arise if the arrival parameter λ were permitted to vary from one interval to the next. Suppose, for example, that with each successful innovation a new value of λ was drawn from the finite set $\{\lambda_1, \ldots, \lambda_m\}$ according to a fixed transition matrix B. Transition into a high-λ state could represent a fundamental breakthrough leading to a Schumpeterian wave of innovations, whereas transition to a low-λ state could represent the exhaustion of a line of research. Then a stationary equilibrium would involve not one level of research employment but one for each state. Now consider the effects of a *ceteris paribus* increase in λ_2. This parameter change might raise research employment in state 2, but it would tend to reduce research employment in other states, by increasing the rationally expected rate of creative destruction during the next interval. Furthermore, even though the parameter change represents an unambiguous improvement in the productivity of the research technology, it might reduce the average level of research employment in stationary equilibrium. Indeed, Appendix 1 works out a numerical example in which, in the limit, as λ_2 becomes infinite, average research employment falls to zero.

The linear Cobb-Douglas example of (3.4) and (3.5) above yields an additional comparative-statics result by parameterizing the degree of market power enjoyed by an intermediate monopolist. Specifically, $1-\alpha$ is the Lerner (1934) measure of monopoly power (price minus marginal cost divided by price), $(1-\alpha)^{-1}$ is the elasticity of demand faced by an intermediate monopolist, and $1-\alpha$ is the fraction of equilibrium revenue in the intermediate sector accruing to the monopolist, $\pi_t/(\pi_t + w_t x_t)$. Thus, by all measures, the degree of market power is measured inversely by the parameter α.

According to (3.4), an increase in the degree of market power (decrease in α) increases the stationary-equilibrium amount of research $\hat{n}$ whenever $\hat{n}$ is positive. According to (3.5), given fixed values of the parameters γ, λ, r, and N, the stationary-equilibrium amount of research will be positive if and only if there is at least some minimal degree of market power; that is, if and only if α is less than the critical value

$$\alpha^* \equiv \frac{\lambda\gamma N}{\lambda\gamma N + 1} < 1.$$

If the degree of market power falls short of this minimal value, then the flow of monopoly profits from the next innovation would not be enough to induce positive research aimed at capturing those rents even if they could be retained forever, with no creative destruction in the next interval.

3.C. *Balanced Growth*

Real output (i.e. the flow of consumption goods) in the economy during interval t is

(3.6) $$y_t = A_t F(N - \hat{n}),$$

which implies

(3.7) $$y_{t+1} = \gamma y_t.$$

Thus the time path of the log of real output $\ln y(\tau)$ will be a random step-function starting at $\ln y_0 = \ln F(N - \hat{n}) + \ln A_0$, with the size of each step equal to the constant $\ln \gamma > 0$, and with the time between each step $\{\Delta_1, \Delta_2, \ldots\}$ a sequence of iid variables exponentially distributed with parameter $\lambda\varphi(\hat{n})$. This statement and (3.3) fully specify the stochastic process driving output, in terms of the parameters of the model.

Not surprisingly, this stochastic process is nonstationary. Suppose observations were made at discrete points in time 1 unit apart. Then from (3.7),

(3.8) $$\ln y(\tau + 1) = \ln y(\tau) + \varepsilon(\tau) \qquad (\tau = 0, 1, \ldots)$$

where $\varepsilon(\tau)$ is $\ln \gamma$ times the number of innovations between τ and $\tau + 1$. From the above analysis

$$\left\{ \frac{\varepsilon(0)}{\ln \gamma}, \frac{\varepsilon(1)}{\ln \gamma}, \ldots \right\}$$

is a sequence of iid variables distributed Poisson with parameter $\lambda\varphi(\hat{n})$. Thus (3.8) can be written as

(3.9) $$\ln y(\tau + 1) = \ln y(\tau) + \lambda\varphi(\hat{n}) \ln \gamma + e(\tau) \qquad (\tau = 0, 1, \ldots),$$

where $e(\tau) \equiv \varepsilon(\tau) - \lambda\varphi(\hat{n}) \ln \gamma$. Note that $e(\tau)$ is iid., with

(3.10) $$E(e(\tau)) = 0, \quad \text{var}\, e(\tau) = \lambda\varphi(\hat{n})(\ln \gamma)^2.$$

From (3.9) and (3.10), the discrete sequence of observations on the log of output follows a random walk with constant positive drift. It also follows that the economy's average growth rate (AGR) and the variance of the economy's growth rate (VGR) are given by

(3.11) $$AGR = \lambda\varphi(\hat{n}) \ln \gamma, \quad VGR = \lambda\varphi(\hat{n})(\ln \gamma)^2.$$

Combining (3.11) with Proposition 1 allows one to sign the impact of parameter changes on the average growth rate. Increases in the arrival parameter, the size of innovations, the size of skilled labor endowment, and (in the Cobb-Douglas example) the degree of market power all raise AGR. Increases in the rate of interest lower it. The parameter changes have the same qualitative effect on VGR as on AGR. The effects are intuitive and straightforward. The effect of market power, combined with the finding that a minimal degree of market power is needed before growth is even possible, underlines the importance of imperfect competition for the growth process.

The example of Appendix 1 shows, however, that in a more general setting where the arrival parameter λ can vary from state to state, it is not always true that an unambiguous improvement in the productivity of the research technology will increase the economy's average growth rate. Instead, an increase in the arrival parameter in one state can discourage research in other states by increasing the rationally expected rate of creative destruction to such an extent that the economy's average growth rate falls.

4. WELFARE PROPERTIES OF THE STATIONARY EQUILIBRIUM

This section compares the laissez-faire average growth rate derived above with the *AGR* that would be chosen by a social planner whose objective was to maximize the expected present value of consumption $y(\tau)$. Since every innovation raises $y(\tau)$ by the same factor γ, the optimal policy consists of a fixed level of research. Expected welfare is

$$U = \int_0^{\infty} e^{-r\tau} \sum_{t=0}^{\infty} \Pi(t,\tau) A_t F(N-n)\, d\tau, \tag{4.1}$$

where $\Pi(t,\tau)$ equals the probability that there will be exactly t innovations up to time τ. Given that the innovation process is Poisson with parameter $\lambda\varphi(n)$, we have

$$\Pi(t,\tau) = (\lambda\varphi(n)\tau)^t e^{-\lambda\varphi(n)\tau}/t! \tag{4.2}$$

From (4.1) and (4.2),

$$U = \frac{A_0 F(N-n)}{r - \lambda\varphi(n)(\gamma-1)}. \tag{4.3}$$

Equation (4.3) identifies U as the initial flow of output $A_0F(x)$ discounted at the rate $r - \lambda\varphi(n)(\gamma - 1)$. This "social discount rate" is less than the rate of interest r because the stream of output will be growing over time. More specifically, the social discount rate is the rate at which each risk-neutral individual in the economy would capitalize a stream that was perpetually subject to increases by the factor $(\gamma - 1)$ with a Poisson arrival rate of $\lambda\varphi(n)$, and constant otherwise.

The socially optimal level of research n^* maximizes U. The first-order condition for an interior maximum is

$$\frac{F'(N-n^*)}{\lambda\varphi'(n^*)} = \frac{(\gamma-1)F(N-n^*)}{r - \lambda\varphi(n^*)(\gamma-1)}. \tag{4.4}$$

(If no solution exists to (4.4) then $n^* = 0$.) This level of research would produce an average growth rate of $\lambda\varphi(n^*)\ln\gamma$. Accordingly laissez-faire produces an average growth rate more (less) than optimal if $\hat{n} > (<) n^*$. Which way these inequalities go can be checked by comparing (4.4) with the equation determining the stationary equilibrium level of research $\hat{n}$:

$$\frac{\tilde{\omega}(N-\hat{n})}{\lambda\varphi'(\hat{n})} = \frac{\gamma\tilde{\pi}(\tilde{\omega}(N-\hat{n}))}{r+\lambda\varphi(\hat{n})}. \tag{3.3}$$

There are four differences between (4.4) and (3.3). The first is that the social discount rate $r-\lambda\varphi(n)(\gamma-1)$ appears in (4.4) instead of the "private discount rate" $r+\lambda\varphi(n)$. The social rate is less than the rate of interest, whereas the private rate is greater. This difference corresponds to the intertemporal spillover effect discussed in Section 2. The social planner takes into account that the benefit to the next innovation will continue forever, whereas the private research firm attaches no weight to the benefits that accrue beyond the succeeding innovation.[11]

The second difference is that total output $F(N-n^*)$ appears on the right-hand side of (4.4) instead of the flow of profits $\tilde{\pi}(\tilde{\omega}(N-\hat{n}))$ that appears in the marginal benefit in (3.3). This is the "appropriability" effect familiar from the patent-race literature.

The third difference is that the factor $(\gamma-1)$ in (4.4) replaces γ in the marginal benefit of (3.3). This corresponds to a "business-stealing" effect. The private research firm does not internalize the loss to the previous monopolist caused by an innovation. In contrast, the social planner takes into account that an innovation destroys the social return from the previous innovation.[12]

The fourth difference is that the marginal product $F'(N-n^*)$ appears in (4.4) in place of the wage $\tilde{\omega}(N-\hat{n})$ in the marginal cost in (3.3). This is a "monopoly-distortion" effect which does not appear in the partial-equilibrium patent-race literature. Specifically, the social cost of research employment exceeds the private cost because in laissez-faire the alternative user of skilled research labor is a monopolist.

[11] Two additional spillovers could easily be included. First, researchers could benefit from the flow of others' research, so that an individual firm's arrival rate would be a constant-returns function of its own and others' research. Second, there could be an exogenous Poisson arrival rate μ of imitations that costlessly circumvent the patent laws and clone the existing intermediate good. Both would have the effect of lowering AGR. Also, as shown in Aghion and Howitt (1988), the inclusion of μ would introduce another source of cycles in the economy, since each imitation would make the intermediate industry perfectly competitive, which would raise manufacturing employment, until the next innovation arrives.

[12] In the patent-race literature the business-stealing effect is usually derived in a symmetric model with no incumbent, in which all research firms enjoy some positive surplus because there is no free entry. An example is Mortensen (1982), who identifies the business-stealing effect with the comment: "Wasteful competition arises because none of the contestants takes account of the loss of the prospect that others suffer when the former's discovery ends the game" (p. 970). In the present paper the loss accrues not to the other research firms, whose value remains equal to zero, by free entry, even after an innovation by another firm, but to the incumbent monopolist who, because of the replacement effect, has chosen not to participate in the patent race. As Tirole (1988, p. 399) notes, there is another negative externality that would occur if the research technology had memory. Specifically, a firm might engage in research in order to reduce the probability that its rivals will win the race. This effect is absent from the present model, in which one firm's research has no effect on the others' probabilities of innovating.

The intertemporal-spillover and appropriability effects tend to make the laissez-faire average growth rate less than optimal, whereas the business-stealing and monopoly-distortion effects tend to make it greater than optimal. Because these effects conflict with each other, the laissez-faire average growth rate may be more or less than optimal. This can be seen in the linear Cobb-Douglas example, where n^* and $\hat{n}$ satisfy

$$(4.5) \qquad 1 = \frac{\lambda(\gamma - 1)\left(\frac{1}{\alpha}\right)(N - n^*)}{r - \lambda n^*(\gamma - 1)},$$

$$(3.4) \qquad 1 = \frac{\lambda\gamma\left(\frac{1-\alpha}{\alpha}\right)(N - \hat{n})}{r + \lambda\hat{n}}.$$

In this example, the appropriability and monopoly-distortion effects are combined in the presence of the factor $(1 - \alpha)$ in (3.4); together they tend to make the laissez-faire AGR less than optimal. This combined effect together with the intertemporal-spillover effect dominates when the size of innovations γ is large, in which case[13] $\hat{n} < n^*$. However, when there is much monopoly power (α close to zero) and innovations are not too large, the business-stealing effect dominates, in which case[14] $\hat{n} > n^*$.

5. NONDRASTIC INNOVATIONS

Until this point the analysis has assumed that innovations are drastic; that the intermediate monopolist is not constrained by potential competition from owners of previous patents. The present section shows that the analysis of stationary equilibria can be generalized to the case where innovations are nondrastic.

Innovations are nondrastic if and only if the previous incumbent could make a positive profit when the current one was charging the price $p_t = A_t F'(\tilde{x}(\omega_t))$ which yields an unconstrained maximum to the current incumbent's profit. If innovations are nondrastic, the current incumbent sets the maximum price that gives the previous incumbent nonpositive profits, and satisfies all the demand at that price, leaving none to the previous incumbent.

The previous incumbent could make a positive profit if and only if a competitive producer of consumption goods could produce at an average cost of less than unity by combining unskilled labor with the previous incumbent's good, buying the latter at a price equal to its average cost of production w_t; that

[13] From (4.5), as γ rises to the upper limit $1 + r/\lambda N$, n^* approaches N while, from (3.4), $\hat{n}$ is bounded below N.

[14] If $1/\alpha > 1 + r/\lambda N$, then $\hat{n} > 0$ for all γ, whereas if $\gamma \leqslant 1 + \alpha r/\lambda N$, then $n^* = 0$. These inequalities are compatible with the condition derived below for innovations to be drastic, namely that $\gamma \geqslant \alpha^{-\alpha}$, as can be verified with the example: $\alpha = 1/2$, $\gamma = \sqrt{2}$, $r/\lambda N = 2(\sqrt{2} - 1)$.

is, if and only if the condition

$$(5.1)\qquad C\left(w_t^M, w_t\right) \leqslant A_{t-1}$$

were to hold with strict inequality, where w_t^M is the equilibrium wage of unskilled labor and C is the unit-cost function associated with the production function F.

In equilibrium all the unskilled labor is combined with the current incumbent's intermediate good. Thus the unskilled wage must satisfy the competitive equilibrium condition:

$$(5.2)\qquad C\left(w_t^M, p_t\right) = A_t.$$

It follows that innovations are nondrastic if and only if (5.1) holds with strict inequality when w_t^M satisfies (5.2) together with $p_t = A_t F'(\tilde{x}(\omega_t))$. It also follows that if innovations are nondrastic then p_t and w_t^M satisfy (5.2) and (5.1) with equality.

In the Cobb-Douglas example, where the unconstrained optimal price for the current incumbent is w_t/α, innovations are nondrastic if and only if

$$(5.3)\qquad \gamma < \alpha^{-\alpha},$$

in which case[15]

$$(5.4)\qquad p_t = \gamma^{1/\alpha} w_t, \quad \pi_t = \left(\gamma^{1/\alpha} - 1\right) w_t x_t, \quad x_t = \left(\gamma^{1/\alpha}\omega_t/\alpha\right)^{1/(\alpha-1)}.$$

The rest of the analysis of this section will focus on stationary equilibria with positive growth in the linear Cobb-Douglas example.

The analysis assumes that the monopolist chooses to do no research, as in the case of drastic innovations. This implicitly places a lower bound on the size of innovations, because it requires the efficiency effect to be smaller than the replacement effect. Appendix 2 shows that the condition

$$(5.5)\qquad \gamma^{1/\alpha}\left(1 + \gamma^{-1}\right) \geqslant \gamma^{-1} + \min\left\{\gamma^{2/\alpha}, \alpha^{-1}\right\}$$

is sufficient for the monopolist to do no research. Note that this condition is satisfied when γ is close to the value $\alpha^{-\alpha}$ at which innovations become drastic.

If there is positive research during interval t, then, as before,

$$(5.6)\qquad \frac{w_t}{\lambda\varphi'(n_t)} = \frac{\pi_{t+1}}{r + \lambda\varphi(n_{t+1})}.$$

[15] In the Cobb-Douglas example, the unit-cost function is

$$c\left(w^M, p\right) \equiv (1-\alpha)^{-(1-\alpha)}\alpha^{-\alpha}p^{\alpha}\left(w^M\right)^{1-\alpha}.$$

It follows from this and (5.2) that if the incumbent charged the unconstrained profit-maximizing price $\alpha^{-1}w_t$, the unskilled wage would be $w_t^M = (1-\alpha)\alpha^{2\alpha/(1-\alpha)}A_t(w_t/A_t)^{-\alpha/(1-\alpha)}$. Putting this into (5.1) yields the condition $\gamma \leqslant \alpha^{-\alpha}$. Treating (5.1) as an equality and solving it and (5.2) simultaneously for (w_t^M, p_t) yields $p_t = \gamma^{1/\alpha}w_t$.

In a stationary equilibrium with positive growth, (5.4) and (5.6) imply

$$(5.7) \qquad 1 = \frac{\lambda\gamma(\gamma^{1/\alpha} - 1)(N - \hat{n})}{r + \lambda\hat{n}}.$$

Equation (5.7) defines the stationary equilibrium level of research $\hat{n}$. It is the same as the equation (3.4) that applies in the drastic case, except that the markup $\gamma^{1/\alpha}$ in (5.7) replaces the markup α^{-1} in (3.4). The comparative-statics results of Proposition 1 apply to the solution of (5.7). In addition, the stationary equilibrium level of research defined by (5.7) is increased by an increase in market power (a decrease in α) as in the drastic case. Thus in the linear Cobb-Douglas example all the comparative statics results derived for the case of drastic innovations are valid also when innovations are nondrastic.

Comparison of (5.7) and (4.5) shows that the same welfare effects analyzed in Section 4 operate in the case where innovations are nondrastic, again with the result that research and growth under laissez-faire may be more or less than optimal.[16]

As is customary in the patent-race literature this analysis has ruled out the possibility that the current and previous incumbent might contract to share the higher monopoly profits that could be earned if the previous incumbent agreed never to compete. For example, the previous incumbent might sell its patent to the current one; in the extreme case where the previous incumbent always had no bargaining power in negotiation with the current one, competition from previous vintages of the intermediate good would never constrain the monopolist, and the above analysis of drastic innovations would apply no matter how small the innovations were.

6. ENDOGENOUS SIZE OF INNOVATIONS

This section generalizes the analysis of stationary equilibria by allowing research firms to choose not only the frequency but also the size of innovations. It shows that under laissez-faire, innovations will be too small if they are drastic. In the nondrastic case, the tendency to make innovations too small is at least partly mitigated by the incentive for innovators to move away from their competitive fringe, which they can do by increasing the size of innovations.

Assume that the arrival rate of innovations to a firm employing the factor combination (z, s) and aiming for innovations of size γ is $\lambda\phi(z, s)v(\gamma)$, where $v'(\gamma) < 0$; the bigger the innovation, the harder it is to discover. Assume $v''(\gamma) < 0$; the marginal cost (in terms of lower arrival rate) of aiming for larger innovations increases with the size of innovations. Then the product $\gamma v(\gamma)$ is a concave function of γ.

[16] Suppose $\gamma = \sqrt{2}$ and $\alpha = 1/2$. To get $\hat{n} < n^*$ let r approach $(\sqrt{2} - 1)\lambda N$ from above; then n^* approaches N whereas $\hat{n}$ is bounded below N. To get $\hat{n} > n^*$ let $r = 2(\sqrt{2} - 1)\lambda N$; then $n^* = 0$ and $\hat{n} > 0$. In either case $\gamma = \alpha^{-\alpha}$, but the example is robust to a small decrease in γ that would satisfy the necessary and sufficient condition (5.3) for innovations to be nondrastic without violating the sufficient condition (5.5) for the monopolist to do no research.

The analysis focuses again on stationary equilibria with positive growth. Consider first the case of drastic innovations. By the same logic as before, the payoff to the $t+1$st innovator is

$$V_{t+1} = \frac{A_{t+1}\tilde{\pi}(\hat{\omega})}{r + \lambda\varphi(\hat{n})\upsilon(\hat{\gamma})}, \tag{6.1}$$

where $\hat{\gamma}$ is the stationary-equilibrium value of γ. If the $t+1$st innovation has size γ, not necessarily equal to $\hat{\gamma}$, then $A_{t+1} = \gamma A_t$ and $V_{t+1} = \gamma V_t$. Therefore the expected flow of profits to the research firm in interval t is

$$\lambda\phi(z,s)\upsilon(\gamma)\gamma V_t - w_t z - w_t^s s. \tag{6.2}$$

The firm takes V_t as given. Thus its profit-maximizing choice of γ also maximizes the product $\upsilon(\gamma)\gamma$. Because this product is a concave function of γ, therefore $\hat{\gamma}$ is defined by the condition[17]

$$\upsilon(\hat{\gamma}) + \hat{\gamma}\upsilon'(\hat{\gamma}) = 0. \tag{6.3}$$

The first-order condition for profit maximization with respect to skilled labor, together with (6.1) produces an expression analogous to (3.3):

$$\frac{\tilde{\omega}(N-\hat{n})}{\lambda\varphi'(\hat{n})} = \frac{\upsilon(\hat{\gamma})\hat{\gamma}\tilde{\pi}(\tilde{\omega}(N-\hat{n}))}{r + \lambda\varphi(\hat{n})\upsilon(\hat{\gamma})}. \tag{6.4}$$

The comparative statics analysis of Section 3 carries through unchanged, since $\hat{\gamma}$ is determined by (6.3) independently of all parameters that do not enter the function υ, with the obvious exception that it is no longer permissible to investigate the effects of a change in γ.

As in Section 4, the expected present value of consumption equals

$$U = \frac{A_0 F(N-n)}{r - \lambda\varphi(n)\upsilon(\gamma)(\gamma-1)}, \tag{6.5}$$

where the denominator is the social discount rate. Therefore, independently of the choice of n, the social planner will choose γ so as to maximize the expression $\upsilon(\gamma)(\gamma-1)$. The socially optimal value γ^* is then defined by

$$\upsilon(\gamma^*) + \gamma^*\upsilon'(\gamma^*) - \upsilon'(\gamma^*) = 0. \tag{6.6}$$

By concavity of $\gamma\upsilon(\gamma)$,[18] $\hat{\gamma} < \gamma^*$. Innovations are too small under laissez-faire. This result is another manifestation of the business-stealing effect. The social planner chooses γ so as to maximize the arrival rate multiplied by the net size $(\gamma-1)$ of innovations, whereas the private research firm, which does not internalize the loss of the existing vintage of intermediate good, maximizes the arrival rate times the gross size γ.

[17] Note that it is always possible to choose the function υ so that the solution to (6.3) satisfies the condition for innovations to be drastic in the Cobb-Douglas example: $\gamma \geqslant \alpha^{-\alpha}$.

[18] Since $\upsilon' < 0$, (6.6) implies that $\upsilon(\gamma)\gamma$ is locally decreasing at γ^*, so that γ^* exceeds the point $\hat{\gamma}$ at which $\upsilon(\gamma)\gamma$ is maximized.

The socially optimal level of research employment n^* satisfies the condition

$$\frac{F'(N-n^*)}{\lambda\varphi'(n^*)} = \frac{v(\gamma^*)(\gamma^*-1)F(N-n^*)}{r-\lambda\varphi(n^*)v(\gamma^*)(\gamma^*-1)}. \tag{6.7}$$

Comparison between (6.4) and (6.7) reveals the same welfare effects as in the analysis of Section 4. In addition, the fact that $\hat{\gamma} < \gamma^*$ in itself makes $\hat{n} < n^*$. This is because, as we have seen, $v(\hat{\gamma})(\hat{\gamma}-1) < v(\gamma^*)(\gamma^*-1)$. So if the other four effects were absent, and both $\hat{n}$ and n^* were determined by (6.7), the effect on research employment would be the same as if the laissez-faire economy had a smaller arrival parameter λ, which would reduce $\hat{n}$ below n^*.

The economy's average growth rate $\lambda\varphi(\hat{n})v(\hat{\gamma})\ln\hat{\gamma}$ is affected by the fact that innovations are too small under laissez-faire, although the direction of the overall effect is ambiguous. The direct effect on $\ln\hat{\gamma}$ is to decrease AGR. The direct effect on the arrival rate $\lambda\varphi(\hat{n})v(\hat{\gamma})$ is to increase AGR. The indirect effect on the arrival rate working through $\varphi(\hat{n})$ is to decrease AGR.

In the nondrastic case, the above business-stealing effect whereby innovations are too small under laissez-faire is mitigated by an additional effect, namely that private innovators tend to increase the size of innovations in order to increase their profit margins. This margin is independent of the size γ in the drastic case but it increases with γ in the nondrastic case. (In the Cobb Douglas example the profit margin is $\alpha^{-1}-1$ if the innovation is drastic and $\gamma^{1/\alpha}-1$ if nondrastic.) The following example shows, however, that this additional "profit-margin" effect does not necessarily overturn our earlier result to the effect that innovations are too small.

Example: Let $\varphi(n)=n$, $F(x)=x^{1/2}$. From (5.3) innovations are nondrastic if $\gamma<\sqrt{2}$. From (5.4), the payoff to the $(t+1)$st innovator in a stationary equilibrium is

$$V_{t+1} = \frac{(\gamma^2-1)x(\hat{\omega},\gamma)w_{t+1}}{r+\lambda v(\hat{\gamma})(N-x(\hat{\omega},\gamma))}, \tag{6.8}$$

where $x(\hat{\omega},\gamma)=(2\hat{\omega}\gamma^2)^{-2}$. In (6.8), γ is the size of innovation to be chosen by the innovator during interval t, whereas $\hat{\gamma}$ and $\hat{\omega}$ denote stationary-equilibrium values which the innovator takes as given. As in the drastic case, $w_{t+1}=\gamma w_t$. Therefore $\hat{\gamma}$ must solve the equation

$$\hat{\gamma} = \underset{\{\gamma\}}{\operatorname{arg\,max}} \frac{v(\gamma)(\gamma^2-1)x(\hat{\omega},\gamma)\gamma w_t}{r+\lambda v(\hat{\gamma})(N-x(\hat{\omega},\gamma))}.$$

Since $\partial x/\partial\gamma<0$, we have

$$\hat{\gamma} < \operatorname{arg\,max} v(\gamma)(\gamma^2-1)x(\hat{\omega},\gamma)\gamma = \bar{\gamma}.$$

From the above analysis we know that

$$\operatorname{arg\,max} v(\gamma)(\gamma-1) = \gamma^*.$$

Therefore a sufficient condition for $\hat{\gamma} < \gamma^*$ is that $(\gamma^2 - 1)x(\hat{\omega}, \gamma)\gamma = (\gamma - 1)g(\gamma)$ with $g'(\gamma) < 0$. The latter is true,[19] with $g(\gamma) \equiv (\gamma + 1)\gamma/4\hat{\omega}^2\gamma^4$.

7. STRATEGIC MONOPSONY EFFECT

In this section the intermediate firm is assumed to take into account its influence on the amount of current research and thereby on the probability of its replacement. In particular, by increasing its demand x_t for skilled labor, the monopolist can raise the wage rate that must also be paid to skilled workers in research. The effect is to reduce the equilibrium amount of research n_t and consequently to delay the arrival of the $(t+1)$st innovation. The monopolist will trade this gain off against the higher wages it must pay its own skilled labor.

The analysis focuses on stationary equilibria with positive growth. The monopolist during interval t chooses x_t to maximize the expected present value of profits:

$$V_t = \frac{(A_t F'(x) - w_t)x}{r + \lambda\varphi(N - x)},$$

subject to

$$(*) \qquad w_t = \lambda\varphi'(N - x)V_{t+1},$$

where (*) follows from (2.10). The magnitudes x_t, n_t, V_t/A_t, and w_t/A_t are constant, at the equilibrium values $\bar{x}$, $\bar{n}$, $\bar{V}$, and $\bar{\omega}$. Therefore $\bar{x}$ solves

$$(7.1) \qquad \bar{V} = \max_x \frac{\left[F'(x) - \lambda\gamma\varphi'(N - x)\bar{V}\right]x}{r + \lambda\varphi(N - x)}.$$

The first-order condition is

$$(7.2) \qquad F'(\bar{x}) + \bar{x}F''(\bar{x}) = \lambda\bar{V}\left[(\gamma - 1)\varphi'(N - \bar{x}) - \bar{x}\gamma\varphi''(N - \bar{x})\right].$$

From the constraint (*),

$$(7.3) \qquad \bar{\omega} = \lambda\gamma\varphi'(N - \bar{x})\bar{V}.$$

From (7.2), (7.3), and the definition of $\tilde{\omega}$,

$$(7.4) \qquad \bar{\omega} = \left[\frac{\gamma}{\gamma - 1 - \gamma(N - \bar{n})\varphi''(\bar{n})/\varphi'(\bar{n})}\right]\tilde{\omega}(N - \bar{n}) \equiv \bar{\omega}(\bar{n}).$$

It follows that the stationary-equilibrium level of research $\bar{n}$ is given by the analogue to (3.3):

$$(7.5) \qquad \left[\frac{\tilde{\omega}(N - \bar{n})}{\lambda\varphi'(\bar{n})}\right]\left[\frac{\gamma}{\gamma - 1 - \gamma(N - \bar{n})\varphi''(\bar{n})/\varphi'(\bar{n})}\right] = \frac{\gamma\bar{\pi}(\bar{n})}{r + \lambda\varphi(\bar{n})},$$

[19] Because it compares γ^* with $\bar{\gamma}$, this analysis would apply even if the research firm ignored the negative effect that its choice of γ has on the value of an innovation by reducing the equilibrium level of x_{t+1} and hence raising the rate of creative destruction next period.

where the function $\bar{\pi}$ is defined as

$$\bar{\pi}(\bar{n}) \equiv [F'(N-\bar{n}) - \bar{\omega}(\bar{n})](N-\bar{n}). \tag{7.6}$$

Assume that the expression $(N - n)\varphi''(n)$ is nondecreasing in n. Then the left-hand side of (7.5) is increasing in $\bar{n}$. If the right-hand side is still decreasing, then the solution to (7.5) is unique. It is straightforward to verify that all the comparative statics results in Proposition 1 apply to this solution, and that in the Cobb-Douglas example the solution is an increasing function of the degree of market power.

Welfare analysis of stationary equilibria is somewhat affected by the strategic monopsony effect, but the overall result remains, namely that the laissez-faire average growth rate may be more or less than optimal. Comparison of (7.5) with (4.4) reveals the same intertemporal-spillover, appropriability, business-stealing, and monopoly-distortion effects as before, although the monopoly-distortion effect will be quantitatively different because $\bar{\pi}(n) \neq \tilde{\pi}(\tilde{\omega}(N-n))$. There is an additional effect, however, from the presence of the term

$$\left[\frac{\gamma}{\gamma - 1 - \gamma(N-\bar{n})\varphi''(\bar{n})/\varphi'(\bar{n})}\right]$$

on the left-hand side of (7.5). This additional effect is the "monopsony-distortion" effect.

In the linear case, where φ'' is zero, the constraint (*) indicates that the intermediate firm's wage rate is independent of the amount of skilled labor it hires, so the monopsony-distortion effect induces it to hire more skilled workers in order to reduce the amount of current research, and hence the amount of creative destruction. The effect just cancels the business-stealing effect, as can be seen by multiplying both sides of (7.5) by $(\gamma - 1)/\gamma$. Thus in the linear Cobb-Douglas example, research and growth are unambiguously less than optimal.

In the general case where $\varphi'' < 0$ the monopsony-distortion effect is ambiguous, because hiring more skilled labor increases the intermediate firm's wage rate at the same time that it reduces creative destruction. Because of these conflicting tendencies it is straightforward to construct examples in which the overall monopsony-distortion effect vanishes. More specifically, given any specification of the model it is possible to perturb the research function φ in such a way[20] that $\hat{n}$ and n^* remain unchanged and the solution $\bar{n}$ to (7.5) becomes equal to $\hat{n}$. Since $\hat{n}$ can be more or less than n^* it follows that $\bar{n}$ can be more or less than n^*.

The rest of the paper ignores the strategic monopsony effect, by assuming that intermediate firms take as given the wage of skilled labor and the amount

[20] Just perturb φ in such a way that $\varphi(\hat{n})$, $\varphi(n^*)$, $\varphi'(\hat{n})$, and $\varphi'(n^*)$ remain unchanged, but $\varphi''(\hat{n})$ is made equal to $-\varphi'(\hat{n})/\gamma(N-\hat{n})$. This can be done without altering the sign of φ' and φ'' on $[0, N)$. According to (3.3) and (4.4), $\hat{n}$ and n^* will be unchanged. This perturbation makes the second factor on the left-hand side of (7.5) equal to unity when $\bar{n} = \hat{n}$, and makes $\bar{\pi}(\hat{n}) = \tilde{\pi}(\tilde{\omega}(N-\hat{n}))$. Since $\hat{n}$ solves (3.3) it will now also solve (7.5).

of research. This assumption is based on our belief that the effect is not important, because it derives from the simplifying assumption that there is only one intermediate firm in the economy. If there were many competing intermediate firms, as in the next section, each might plausibly regard itself as too small to affect the skilled wage or the amount of research.

8. MANY INTERMEDIATE GOODS

This section relaxes the simplifying assumption of a single economy-wide monopoly in intermediate goods. Suppose instead that there are m different intermediate sectors. Output of the consumption good is $y_t = \Sigma_{i=1}^{m} A_{it} F(x_{it})$, where x_{it} denotes the flow of output of the ith intermediate good during interval t, and where F has all the properties assumed above. (This requires that each sector have its own specialized brand of unskilled labor.)

Following Shleifer (1986), suppose that innovations arrive in different sectors in a deterministic order.[21] Specifically, the innovating sector is always the one with the lowest productivity parameter A_{it}. Each innovator becomes a local monopolist in that sector for a period of m successive innovations, and is replaced by the last of those m innovations. Let A_t denote the productivity parameter in the leading sector, where an innovation has arrived most recently. Assume that $A_{it} = \gamma^{(1-k)/m} A_t$ when i is the kth most advanced sector. Then γ is again the size of each innovation relative to the previous vintage of good in the same sector.

Let $\hat{x}_k$ denote the stationary-equilibrium employment of skilled labor in the kth most advanced sector. Then $\hat{x}_k$ maximizes the flow of profits:

$$\left[A_t \gamma^{(1-k)/m} F'(x_k) - w_t\right] x_k.$$

Therefore,

$$\hat{x}_k = \tilde{x}\left(\gamma^{(k-1)/m} \omega\right),$$

where $\tilde{x}$ is defined as above and ω is again the stationary-equilibrium productivity-adjusted wage, w_t/A_t. The productivity-adjusted flow of profits in the kth-leading sector is

$$\pi_t^k / A_t = \gamma^{(1-k)/m} \tilde{\pi}\left(\gamma^{(k-1)/m} \omega\right).$$

[21] The alternative of allowing innovations to be randomly distributed across sectors is analyzed in an Appendix to an earlier version of this paper, available from the authors upon request. This Appendix assumes a continuum of sectors and a continual flow of innovations. Whoever innovates at date τ is thereby allowed to enter a randomly chosen sector with the "leading technology" $A(\tau)$, where $A(\tau)$ grows continuously at the exponential rate $\sigma \lambda n$. The possibility that the same sector might receive two innovations in rapid succession, before the leading technology has advanced by much, implies that in stationary equilibrium some positive (and endogenous) fraction of innovations will be nondrastic. The model yields all the comparative-statics results of Proposition 1 above, except possibly for result (b).

The productivity-adjusted wage is the solution to the equilibrium condition

$$N = n + \sum_{k=1}^{m} \tilde{x}(\gamma^{(k-1)/m}\omega),$$

which can be written as

$$\omega = \hat{\omega}(N - n),$$

where the function $\hat{\omega}$ satisfies Assumption 1 above.

Let the technology of research be the same as before, with an arrival parameter equal to $m\lambda$. The length of each interval is distributed exponentially with parameter $m\lambda\varphi(n)$, and each local monopoly lasts for m intervals. Let τ_t denote the time of the tth innovation. Then the value of the tth innovation is

$$V_t = \sum_{k=1}^{m} E(e^{-r(\tau_{t+k-1}-\tau_t)})\left(\frac{\pi_{t+k-1}^{k}}{r + m\lambda\varphi(n)}\right)$$

$$= A_t \sum_{k=1}^{m}\left(\frac{m\lambda\varphi(n)}{r + m\lambda\varphi(n)}\right)^{k-1}\left(\frac{\tilde{\pi}(\gamma^{(k-1)/m}\omega)}{r + m\lambda\varphi(n).}\right),$$

and the condition for a positive stationary-equilibrium level of research is the analogue to (3.3):

$$(8.1)\qquad \frac{\hat{\omega}(N-\hat{n})}{m\lambda\varphi'(\hat{n})} = \sum_{k=1}^{m}\left(\frac{m\lambda\varphi(\hat{n})}{r + m\lambda\varphi(\hat{n})}\right)^{k-1}\left(\frac{\gamma^{1/m}\tilde{\pi}\left[\gamma^{(k-1)/m}\hat{\omega}(N-\hat{n})\right]}{r + m\lambda\varphi(\hat{n})}\right).$$

In the linear Cobb-Douglas example this condition is the analogue to (3.4):

$$(8.2)\qquad 1 = m\lambda\left(\frac{1-\alpha}{\alpha}\right)(N-n)\frac{\sum_{1}^{m}\gamma^{(k-1)/m(\alpha-1)}\left[\frac{\gamma^{k/m}(m\lambda\hat{n})^{k-1}}{(r+m\lambda\hat{n})^{k}}\right]}{\sum_{1}^{m}\gamma^{(k-1)/m(\alpha-1)}}.$$

Since the right-hand side of (8.1) is a decreasing function[22] of n and the left-hand side an increasing function, the solution to (8.1), if it exists, is unique. If no solution exists, then the equilibrium level of research is zero. All the comparative-statics results of Proposition 1 apply to the solution of (8.1), with the possible exception of (b), the effect of γ, the size of innovations. In the

[22] Note that

$$\frac{\partial}{\partial n}\sum_{k=1}^{m}\frac{(m\lambda\varphi(n))^{k-1}\tilde{\pi}_k}{(r+m\lambda\varphi(n))^{k}}$$

$$= m\lambda\varphi'(n)\sum_{k=1}^{m}\left\{\frac{(k-1)(m\lambda\varphi(n))^{k-2}\tilde{\pi}_k}{(r+m\lambda\varphi(n))^{k}} - \frac{k(m\lambda\varphi(n))^{k-1}\tilde{\pi}_k}{(r+m\lambda\varphi(n))^{k+1}}\right\}$$

$$= m\lambda\varphi'(n)\sum_{k=2}^{m+1}\frac{(k-1)(m\lambda\varphi(n))^{k-2}(\tilde{\pi}_k - \tilde{\pi}_{k-1})}{(r+m\lambda\varphi(n))^{k}} < 0,$$

where $\tilde{\pi}_1 > \tilde{\pi}_2 > \cdots > \tilde{\pi}_m > \tilde{\pi}_{m+1} \equiv 0$.

linear Cobb-Douglas example, however, it can be shown that (b) remains true, provided that the social discount rate (the denominator of U below) is positive.[23]

In stationary equilibrium each innovation raises the entire cross-sectional profile of productivity parameters by the factor $\gamma^{1/m}$. It therefore raises GNP by $\gamma^{1/m}$, and raises the log of GNP by the factor $(1/m)\ln\gamma$. Since the Poisson arrival rate of innovation is $m\lambda\hat{n}$, therefore the economy's average growth rate is $\lambda\hat{n}\ln\gamma$, exactly as before. The variance of the growth rate is $\lambda\hat{n}(\ln\gamma)^2/m$; aggregation across many sectors reduces variability through a law of large numbers.

By the same logic as in Section 4, social welfare is measured by

$$U = A_0 \frac{\sum_1^m \gamma^{(1-k)/m} F(x_k)}{r - m\lambda\varphi(n)(\gamma^{1/m} - 1)}.$$

A social planner would choose $(x_1, \ldots, x_m, n)$ to maximize U subject to the constraint,

$$N - n = x_1 + \cdots + x_m, \tag{8.3}$$

and $n \geqslant 0$. The first-order conditions for an interior maximum are

$$F'(x_k) = \mu\gamma^{(k-1)/m} \qquad (k = 1, \ldots, m), \tag{8.4}$$

$$m\lambda\varphi'(n)(\gamma^{1/m} - 1)U = \mu A_0, \tag{8.5}$$

where μ is a Lagrange multiplier. Let $\hat{\mu}(N-n)$ denote the value of μ such that the solutions $(x_1, \ldots, x_m)$ to (8.4) solve (8.3). Then (8.5) can be expressed as:

$$\frac{\hat{\mu}(N-n^*)}{m\lambda\varphi'(n^*)} = \sum_{k=1}^m (\gamma^{-1/m})^{k-1} \left(\frac{(\gamma^{1/m} - 1)F(x_k^*)}{r - m\lambda\varphi(n^*)(\gamma^{1/m} - 1)} \right). \tag{8.6}$$

In the linear Cobb-Douglas example,

$$1 = \frac{m\lambda\left(\dfrac{1}{\alpha}\right)(\gamma^{1/m} - 1)(N - n^*)}{r - m\lambda n^*(\gamma^{1/m} - 1)}. \tag{8.7}$$

Comparison of (8.6) with (8.1) reveals the same four effects as before. The monopoly-distortion effect is still present because $\hat{\mu}(N-n) > \hat{\omega}(N-n)$. The

[23] It suffices to show that the right-hand side of (8.2) is increasing in γ. The ratio of sums in this expression can be regarded as the expected value of the discrete random variable

$$z_k \equiv \left[\frac{\gamma^{1/m}}{1 + r/m\lambda n} \right]^k \Big/ m\lambda n,$$

under the truncated geometric distribution with parameter $\gamma^{1/m(\alpha-1)} < 1$. The effect on this ratio of a marginal increase in γ is the sum of the effect on each z_k and the effect of changing the parameter of the distribution. The former is positive. The latter would also be positive, by first-order stochastic dominance, if z_k were decreasing in k, which it is if the social discount rate is positive.

appropriability effect applies sector-by-sector, and is amplified by the fact that

$$\gamma^{-1/m} > \left(\frac{m\lambda\varphi(n)}{r + m\lambda\varphi(n)}\right)$$

if the social discount rate is positive. As before, research and growth under laissez-faire may be more or less than optimal; in the linear Cobb-Douglas example, $\hat{n} < n^*$ if γ is large,[24] but $\hat{n} > n^*$ if α is small and γ is not too large.[25]

9. CONCLUSION

The paper has presented a model of economic growth based on Schumpeter's process of creative destruction. Growth results exclusively from technological progress, which in turn results from competition among research firms that generate innovations. Each innovation consists of a new intermediate good that can be used to produce final output more efficiently than before. Research firms are motivated by the prospect of monopoly rents that can be captured when a successful innovation is patented. But those rents in turn will be destroyed by the next innovation, which will render obsolete the existing intermediate good.

It would be useful to generalize and extend the analysis in several directions, such as assuming that technology is ultimately bounded, thereby requiring the size of innovations eventually to fall. The model would gain richness and realism if capital were introduced, either physical or human capital embodying technical change, or R and D capital that affects the arrival rate of innovations. Allowing unemployment, by introducing search into the labor market, would facilitate study of the reciprocal interaction between technological change and the business cycle. All these extensions seem feasible because of the simplicity and transparency of the basic model.

European Bank for Reconstruction and Development, 122 Leadenhall St., London EC3 V4QL, U.K.

and

Department of Economics, The University of Western Ontario, London, Ontario N6A 5C2, Canada

Manuscript received June, 1989; final revision received July, 1991.

APPENDIX 1

AN EXAMPLE WITH A RANDOM ARRIVAL PARAMETER

Let λ follow a two-state Markov process on the space $\{\lambda_1, \lambda_2\}$ with all transition probabilities equal to 1/2. A stationary equilibrium is an equilibrium in which the productivity-adjusted wage rate depends only on the state of the world, not on time. Let $A_t V_j$ be the value of making the tth

[24] From (8.7), as $\gamma^{1/m}$ rises to the upper limit $1 + r/m\lambda N$, n^* approaches N while, from (8.2), $\hat{n}$ is bounded below N.

[25] If $1/\alpha > 1 + r/\lambda N$ then $\hat{n} > 0$ for all γ, whereas if $\gamma^{1/m} \leqslant 1 + \alpha r/m\lambda N$, then $n^* = 0$.

innovation and moving into state j. Assume the linear case of $\varphi(n)=n$. In any state i, the marginal expected return to research in interval t is $\lambda_i A_{t+1}(V_1+V_2)/2$. This will equal the wage if positive research occurs in state i. If research occurs in all states, the V_i's must satisfy the analogue to (2.12):

$$V_i=\frac{\tilde{\pi}(\lambda_i\gamma(V_1+V_2)/2)}{r+\lambda_i[N-\tilde{x}(\lambda_i\gamma(V_1+V_2)/2)]} \qquad (i=1,2).$$

Define $n_i \equiv N-\tilde{x}(\lambda_i\gamma(V_1+V_2)/2)$. Then the average level of research employment is

$$\bar{n}=n_1q_1+n_2q_2,$$

where q_i is the asymptotic fraction of time spent in state i. It is easily verified that

$$q_1=1-q_2=\lambda_2n_2/(\lambda_1n_1+\lambda_2n_2).$$

To complete the example, take the Cobb-Douglas case ($F(x)=x^\alpha$), and suppose $\alpha=\gamma^{-1}=1/2$ and $r=N=\lambda_1=1$. Since $\tilde{x}(\omega)=(\omega/\alpha^2)^{1/(\alpha-1)}$ and $\tilde{\pi}(\omega)=((1-\alpha)/\alpha)\omega\tilde{x}(\omega)$, the formula for each V_i can be rewritten as:

$$V_1=[16(V_1+V_2)]^{-1}-\{1-[4(V_1+V_2)]^{-2}\}V_1,$$

and

$$V_2=[16\lambda_2(V_1+V_2)]^{-1}-\lambda_2\{1-[4\lambda_2(V_1+V_2)]^{-2}\}V_2.$$

When $\lambda_2=1$, the solution to these equations is $V_1=V_2=\sqrt{3}/8\sqrt{2}$, which implies $n_1=n_2=\bar{n}=1/3$. As $\lambda_2\to\infty$, the solution approaches $V_1=1/4$, $V_2=0$, which implies $n_1=0$, $n_2=1$, and $q_1=n_2/(n_2+\lambda_1n_1/\lambda_2)=1$; hence $\bar{n}=0$.

The economy's average growth rate equals $f\ln\gamma$, where f is the asymptotic frequency of innovations:

$$f=\lambda_1n_1q_1+\lambda_2n_2q_2=\left[\frac{2\lambda_1\lambda_2}{\lambda_1+\lambda_2}\right]\bar{n}.$$

Thus when $\lambda_2=1$, $f\ln\gamma=(1/3)\ln 2>0$; and as $\lambda_2\to\infty$, $f\ln\gamma$ approaches 0.

APPENDIX 2

DERIVATION OF CONDITION (5.5)

In the stationary equilibrium described in Section 5, the monopolist has no incentive to do research if

$$\text{(A.1)} \qquad w_t \geqslant \lambda(V_{t+1}^M-V_t),$$

where

$$V_t=\frac{(\gamma^{1/\alpha}-1)w_t(N-\hat{n})}{r+\lambda\hat{n}}$$

is the value of the monopolist's current patent and

$$V_{t+1}^M=\frac{[\min(\gamma^{2/\alpha},\alpha^{-1})-1]w_{t+1}(N-\hat{n})}{r+\lambda\hat{n}}$$

would be the gross value of the next innovation to the current monopolist, for whom the innovation would be drastic if $\gamma^2>\alpha^{-\alpha}$, if next period the level of research was the stationary-equilibrium value $\hat{n}$. In fact more than this level of research would be conducted if the monopolist were to innovate, because the monopolist could then charge a markup higher than $\gamma^{1/\alpha}$, so the value would actually be less than V_{t+1}^M. Substituting these expressions for V_t and V_{t+1}^M into (A.1) and using the fact that $w_{t+1} = \gamma w_t$ produces

$$1 \geqslant \frac{\lambda[\gamma \min(\gamma^{2/\alpha}, \alpha^{-1}) - \gamma - (\gamma^{1/\alpha} - 1)](N - \hat{n})}{r + \lambda \hat{n}}. \tag{A.2}$$

Condition (5.5) follows immediately from (A.2) and (5.7).

REFERENCES

AGHION, P., AND P. HOWITT (1988): "Growth and Cycles through Creative Destruction," Unpublished, University of Western Ontario.

AZARIADIS, C. (1981): "Self-Fulfilling Prophecies," *Journal of Economic Theory*, 25, 380–396.

CORRIVEAU, L. (1988): "Entrepreneurs, Growth, and Cycles," Unpublished, University of Western Ontario.

DENECKERE, R. J., AND K. L. JUDD (1986): "Cyclical and Chaotic Behavior in a Dynamic Equilibrium Model, with Implications for Fiscal Policy," Unpublished, Northwestern University.

DIXIT, A., AND J. STIGLITZ (1977): "Monopolistic Competition and Optimum Product Diversity," *American Economic Review*, 67, 297–308.

GRANDMONT, J.-M. (1985): "On Endogenous Competitive Business Cycles," *Econometrica*, 53, 995–1045.

GROSSMAN, G. M., AND E. HELPMAN (1991): "Quality Ladders in the Theory of Growth," *Review of Economic Studies*, 58, 43–61.

JUDD, K. L. (1985): "On the Performance of Patents," *Econometrica*, 53, 567–585.

KING, R. G., AND S. T. REBELO (1988): "Business Cycles with Endogenous Growth," Unpublished, University of Rochester.

LERNER, A. P. (1934): "The Concept of Monopoly and the Measurement of Monopoly Power," *Review of Economic Studies*, 1, 157–175.

LUCAS, R. E. JR. (1988): "On the Mechanics of Economic Development," *Journal of Monetary Economics*, 22, 3–42.

MORTENSEN, D. T. (1982): "Property Rights and Efficiency in Mating, Racing and Related Games," *American Economic Review*, 72, 968–979.

REINGANUM, J. (1985): "Innovation and Industry Evolution," *Quarterly Journal of Economics*, 100, 81–99.

——— (1989): "The Timing of Innovation: Research, Development and Diffusion," in *Handbook of Industrial Organization*, Vol. I, ed. by R. Schmalensee and R. Willig. Amsterdam: North-Holland.

ROMER, P. M. (1986): "Increasing Returns and Long-Run Growth," *Journal of Political Economy*, 94, 1002–1037.

——— (1990): "Endogenous Technological Change," *Journal of Political Economy*, 98, S71-S102.

SCHERER, F. M. (1984): *Innovation and Growth: Schumpeterian Perspectives*. Cambridge, MA: MIT Press.

SCHUMPETER, J. A. (1942): *Capitalism, Socialism and Democracy*. New York: Harper and Brothers.

SEGERSTROM, P. S., T. C. A. ANANT, AND E. DINOPOULOS (1990): "A Schumpeterian Model of the Product Life Cycle," *American Economic Review*, 80, 1077–1091.

SHLEIFER, A. (1986): "Implementation Cycles," *Journal of Political Economy*, 94, 1163–1190.

STOKEY, N. L. (1988): "Learning by Doing and the Introduction of New Goods," *Journal of Political Economy*, 96, 701–717.

TIROLE, J. (1988): *The Theory of Industrial Organization*. Cambridge, MA: M.I.T. Press.

Part IV
Intra-temporal Knowledge Spillovers

[15]

Real Effects of Academic Research

By ADAM B. JAFFE*

The existence of geographically mediated "spillovers" from university research to commercial innovation is explored using state-level time-series data on corporate patents, corporate R&D, and university research. A significant effect of university research on corporate patents is found, particularly in the areas of Drugs and Medical Technology, and Electronics, Optics, and Nuclear Technology. In addition, university research appears to have an indirect effect on local innovation by inducing industrial R&D spending.

It is conventional wisdom that "Silicon Valley" near San Jose, California and Route 128 around Boston owe their status as centers of commercial innovation and entrepreneurship to their proximity to Stanford and MIT. Several case studies of these two areas confirm the important roles played there by universities. (Nancy Dorfman, 1983; Daniel Shimshoni, 1966; P. Teplitz, 1965; and Herbert Wainer, 1965). Several other areas of the country, such as San Antonio/Austin in Texas and Raleigh/Durham in North Carolina have tried explicitly to build new centers of high-technology industry around their universities. It is certainly plausible that the pool of talented graduates, the ideas generated by faculty, and the high quality libraries and other facilities of research universities facilitate the process of commercial innovation in their neighborhood, but there has been very little systematic empirical analysis of this phenomenon.[1]

Knowledge is, after all, a public good. There has been much recent interest in "spillovers" of research among firms (Jaffe, 1986). There is even more reason to believe that spillovers exist from universities to firms, since the former have less incentive to try to keep research secret. Richard Nelson (1986), using surveys of research managers, found university research to be an important source of innovation in some industries, particularly those related to the biological sciences.

For none of these spillover phenomena are the "transport" mechanisms understood. If the mechanism is primarily journal publications, then geographic location is probably unimportant in capturing the benefits of spillovers. If, however, the mechanism is informal conversations, then geographic proximity to the spillover source may be helpful or even necessary in capturing the spillover benefits.

As a first cut at this issue, I examine the production of patents assigned to corporations by state over time, and relate this to industry R&D and university research. I interpret an influence of university research on these patents at the state level (after controlling for industry R&D) as evidence of the existence of geographically mediated spillovers. I also investigate the related hy-

*Department of Economics, Harvard University, Cambridge, MA 02138. I am indebted to Mike Fogarty and Neil Bania for conversations which led to the conception of this project, and for ongoing advice; to Neil Bania and the staff of the Center for Regional Economic Issues at Case-Western Reserve University for preparing the data; to Zvi Griliches, Ariel Pakes, Mark Schankerman, and two referees for helpful advice; to Jim Hines for suggesting the paper's title; and to the Ameritech Foundation for financial support via a grant to the Center for Regional Economic Issues under the Ameritech Fellows Program. Seminar participants at Princeton, Case-Western Reserve, Yale, and the NBER provided helpful comments.

[1]A recent paper by Neil Bania, Randall Eberts, and Michael Fogarty (1987) finds a significant effect of university research spending on openings of new firms in a cross section of Standard Metropolitan Statistical Areas. I use the same university research data, and look for effects of university research on corporate patenting activity.

pothesis (implicit in the Silicon Valley story) that university research induces the location of industry R&D spending nearby.

Section I provides an overview of the model. Section II discusses data and estimation issues; Section III presents empirical results. Concluding observations follow.

I. Modeling Spillovers from University Research

I model commercial spillovers from university R&D in the "knowledge production function" framework articulated by Zvi Griliches (1979) and implemented by Ariel Pakes and Griliches (1984), Jaffe (1986) and others. I use a modified Cobb–Douglas model with two inputs:

$$\log(P_{ikt}) = \beta_{1k}\log(I_{ikt}) + \beta_{2k}\log(U_{ikt}) + \beta_{3k}[\log(U_{ikt})\log(C_{ikt})] + \epsilon_{ikt}, \tag{1}$$

where i indexes the unit of observation (states, in this case), k indexes technological areas, and t indexes time. P is corporate patents, a proxy for new economically useful knowledge;[2] I is R&D performed by industry and U is university research. ϵ_{ikt} is a stochastic error whose properties will be discussed below. The variable C is a measure of the geographic coincidence of university and industrial research activity within the state. Allowing the potency of university spillovers to depend on C is intended to mitigate the arbitrariness of states as observation units.

Equation (1) is (an approximation to) a technological relationship. In the background are behavioral relationships governing industry R&D and university research. There are two reasons to be concerned about these as well. First, they may be interesting themselves. Part of the "Silicon Valley story" may be that the presence of Stanford-induced firms to locate research labs nearby. At least from a local policy perspective, this effect of university research is as important as the direct effect on innovation holding industry R&D constant. Second, these behavioral relationships may have econometric implications for estimating equation (1), which we would like to account for even if they were not interesting themselves.

I will model these relationships as a simultaneous system. University research I allow to depend on industry R&D and certain state characteristics which for now I will call Z_1:

$$\log(U_{ikt}) = \beta_{4k}\log(I_{ikt}) + \delta_{1k}Z_1 + \xi_{ikt}. \tag{2}$$

Industry R&D I assume to depend on an overlapping but not identical set of characteristics Z_2, and on university research:

$$\log(I_{ikt}) = \beta_{5k}\log(U_{ikt}) + \delta_{2k}Z_2 + \mu_{ikt}. \tag{3}$$

Some university research is industry funded, and firms may favor local universities in this funding, or otherwise contribute to local universities' ability to attract funding. Conversely, surveys of R&D managers reveals the presence of research universities to be a factor in industrial R&D lab location decisions (Leonard Lund, 1986).

The only remaining specification issue is the structure and relations among the error terms ϵ, ξ, and μ in equations (1), (2), and (3). Since substantive issues of stochastic structure interact with measurement problems, I postpone this until after discussing the data in the next section.

II. Data and Estimation Issues

The major issues to be confronted in implementing a model consisting of equations (1), (2), and (3) are: (1) identifying an appro-

[2]Equation (1) can be derived from a production function for unobserved "knowledge" by assuming proportionality between knowledge and patents. Since, however, knowledge is not observed, little is gained by that formulation unless *other* indicators of new knowledge besides patents are employed. For a general discussion of the usefulness of patents as indicators of new knowledge see Zvi Griliches, 1987 and Griliches, Bronwyn Hall, and Ariel Pakes, 1988.

priate unit of observation; (2) measuring "geographic coincidence"; (3) specifying the nature of the "technical areas" (indexed by k); (4) considering the likely structure of the errors and the implications for estimation; and (5) considering the implications of the fact that patents are an integer variable that is sometimes zero.

States were chosen as the unit of observation on the basis of convenience. Both patents and industry R&D are easily available on this basis, the former from the patent office and the latter from NSF.[3] In particular, state-by-state breakdowns of industry R&D based on the NSF R&D Census are available for at least some states for the years 1969–77, 1979, 1981, and 1983. For universities, any geographic aggregation is possible, because the NSF tapes identify the individual institutions. The university R&D data are available for 1972–85. Unfortunately, the industry R&D numbers are missing for many states, particularly in 1983. There are 39 states for which an industry number appears at least once; only 22 states report it for the 9 years 1972–77, 1979, 1981, and 1983. Because a balanced data set is needed for some desired estimation methods, and because several of the states with partial data are missing only 1983, the estimation was carried out on 29 states for which data are complete for 1972–77, 1979, and 1981. These states are listed in Appendix B.

Though convenient, the use of states as the unit of observation is conceptually problematic for our purposes. Thinking of geographic spillovers as occurring similarly in Rhode Island and California strains credulty. The absence of detailed R&D spending data at a more micro level makes it difficult to solve this problem.[4] For this paper, I improve on the state totals only slightly. Using data from a commercial directory of industrial R&D labs, we have constructed totals of industry labs and employees by SMSA. I then use these to construct an index of geographic coincidence of industry R&D and university research for each state.[5] My hypothesis is that an average dollar of university research will yield more spillovers in Illinois, where both university research and industry labs are concentrated in Chicago, than in Indiana, where the bulk of the university research is in Bloomington and Lafayette, where there are almost no labs.

Classification by technical area is available in the patent data on the basis of the patent class to which the patent is assigned. There are currently over 300 such classes, and they represent a fairly detailed technological breakdown. On the research side, the university data are broken down by academic department. These two schemes can be matched only at a fairly broad level of aggregation. The patent classes and NSF's academic departments were each grouped into 5 broad areas: (1) Drugs and Medical Technology; (2) Chemical Technology; (3) Electronics, Optics, and Nuclear Technology; (4) Mechanical Arts; and (5) All other. Appendix A gives the classification of academic departments into these areas; the assignment of patent classes is available from the author.

Table 1 gives the 1981 data for total university research and the breakdown of this

[3]In principle, patents can be counted on almost any geographic basis desired, since the patent contains the precise address of the inventor. Further, all recent patents are available on-line from various commercial data bases.

[4]In the conclusion, I discuss the possibilities that do exist for estimation at a more micro level.

[5]The R&D lab survey reports the existence of all commercial R&D labs, with number of doctorates, number of professional employees, and number of technicians reported for some but not all labs. I estimated total professional employees by SMSA by multiplying total labs times the average number of professional employees per lab for reporting labs in the SMSA. The geographic coincidence index C_i was then calculated as

$$C_i = \frac{\sum_s U_{is} TP_{is}}{\left[\sum_s U_{is}^2\right]^{\frac{1}{2}} \left[\sum_s TP_{is}^2\right]^{\frac{1}{2}}},$$

which is the uncentered correlation of the vectors U_i and TP_i across SMSA's. For this purpose, university research or industry labs that are not in a recognized SMSA are excluded from the numerator (but not the denominator) of the calculation. For estimation purposes, the index is rescaled so that the $\log(C_i)$ has mean 0.

TABLE 1—UNIVERSITY RESEARCH BY SOURCE OF FUNDING AND TYPE OF INSTITUTION IN 1981 (MILLIONS OF DOLLARS)

	Public Universities	Private Universities	Federally Funded Research & Development Centers	Total
Federally Funded	2,609	1,941	2,457	7,006
Other Government Funded	493	470	2	542
Industry Funded	181	110	1	292
Institution Funded	780	188	2	969
All Other Sources	261	184	18	464
Total	4,324	2,470	2,480	9,274
Number of Institutions	326	230	18	574

TABLE 2—UNIVERSITY RESEARCH BY TECHNICAL AREA AND TYPE OF INSTITUTION IN 1981 (MILLIONS OF DOLLARS)

	Public Universities	Private Universities	Federally Funded Research & Development Centers	Total
Drugs and Medical	1,606	1,176	75	2,857
Chemical	225	143	215	582
Electronic, Optical, and Nuclear	372	378	1,370	2,120
Mechanical	453	289	525	1,266
All Other	1,669	484	296	2,449
Total	4,324	2,470	2,480	9,274

spending by the source of the funds and the type of institution. Table 2 shows the distribution of university research by technical area and type of institution. With this level of detail, a plethora of potential specification searches present themselves. I pursued only three. First, it could be argued that patents associated with industry- and institution-funded R&D represent the direct output of the research, rather than any spillover. For this reason, estimation was originally carried out using only federally funded research. Such results are, however, very similar to those using all research, so the latter are reported below. Second, federally funded research and development centers (such as Oak Ridge and Brookhaven) are arguably different from universities, so some experimentation was performed with allowing their funding to have different coefficients than university research, and with excluding them entirely. With the former, one could not reject the null hypothesis of equality, and results in the latter case were quite similar to results when they are included. Hence, all results herein do not distinguish FFRDC's from universities. Finally, existence of industry funding could indicate close ties between the university and local industry that would facilitate spillovers. To test this, the proportion of funding coming from industry was added to equation (1), both alone and interacted with university research, but it was completely insignificant.

The astute reader will have noticed that I have not discussed the breakdown of industrial R&D into technical areas. Data of this sort do not exist at the state level. In principle, the raw data compiled by NSF contain the necessary information. Even if NSF were inclined to publish it, however, the number of labs that would occupy a typical state/technical area cell would frequently require suppression of the total to preserve confidentiality. Thus equations (1)–(3) cannot be estimated as written because I_{ikt} is not observed. We will have to make do with expressions involving

$$I_{it} = \sum_k I_{ikt}.$$

Consider equation (1) first. As a definition, we can write:

$$I_{ikt} = \exp(\rho_{ikt}) I_{it} \tag{4a}$$

or

$$\log(I_{ikt}) = \log(I_{it}) + \rho_{ikt}. \tag{4b}$$

The (log of the) proportion of industry R&D in area k, ρ_{ikt} is not observed. Substituting equation (4b) into equation (1) yields:

$$\begin{aligned} \log(P_{ikt}) = {} & \beta_{1k}\log(I_{it}) + \beta_{2k}\log(U_{ikt}) \\ & + \beta_{3k}[\log(U_{ikt})\log(C_{ikt})] \\ & + \beta_{1k}\rho_{ikt} + \epsilon_{ikt}, \end{aligned} \tag{5}$$

where the unobserved proportion is now absorbed in the error term. Note that this is *not* the standard measurement error problem. There is little reason to expect the error ρ_{it} to be correlated with $\log(I_{it})$. There is reason, however, for a different concern: if the proportion of industry R&D in area k is correlated with the proportion of university R&D in that area, then the substitution of I_{it} for I_{ikt} introduces correlation of $\log(U_{ikt})$ with the error. There are two possible solutions to this problem. Under certain assumptions, it could be solved by instrumental variable estimation of equation (5). As will be discussed below, the likely structure of the errors in the system will suggest the necessity of IV anyway.

A second alternative is to model the relationship between ρ_{ikt} and U_{ikt}. Suppose

$$\exp(\rho_{ikt}) = (I_{ikt}/I_{it}) = (U_{ikt}/U_{it})^{\gamma} \tag{6a}$$

or

$$\rho_{ikt} = \gamma[\log(U_{ikt}) - \log(U_{it})]. \tag{6b}$$

Substituting (6b) into (5) yields:

$$\begin{aligned} \log(P_{ikt}) = {} & \beta_{1k}\log(I_{it}) \\ & + (\beta_{2k} + \beta_{1k}\gamma)\log(U_{ikt}) \\ & + \beta_{3k}[\log(U_{ikt})\log(C_{ikt})] \\ & - \beta_{1k}\gamma\log(U_{it}) + \epsilon_{ikt}. \end{aligned} \tag{7}$$

Thus, under the simple assumption of (6), adding total university R&D to the patent equation will allow us to "correct" the coefficient estimate for $\log(U_{ikt})$ to yield the parameter β_{2k}. While the assumption (6) is restrictive, it may be more plausible than the assumptions necessary to estimate (5) consistently by IV methods. Both possibilities are explored empirically below.

Turning now to the equation explaining industrial R&D itself, we can proceed to substitute from equation (6) into equation (3) and rearrange to yield:

$$\begin{aligned} \log(I_{it}) = {} & \delta_{2k}Z_2 + (\beta_{4k} - \gamma)\log(U_{ikt}) \\ & + \gamma\log(U_{it}) + \mu_{ikt}, \end{aligned} \tag{8}$$

where, again, the inclusion of both university research by area and the total allows us to back out the structural parameter β_{4k}. The only problem with equation (8) is that it is inconsistent with the identity

$$\sum_k I_{ikt} = I_{it}.$$

This is the standard problem that a log-log

relationship cannot hold simultaneously at two different levels of aggregation. I do not have a solution to it. In recognition of it, I will also estimate a version of equation (3) which simply substitutes I_{it} and U_{it} for I_{ikt} and U_{ikt}. I will also not impose the cross-equation constraints implied by the presence of γ in multiple equations.

Turning to consideration of the stochastic structure, there are two issues: the likely correlation among the errors (ϵ_{ikt}, ξ_{ikt}, and μ_{ikt}) and the choice of observables that can be assumed to be uncorrelated with them so the system can be identified. As for the first, I will estimate separate intercepts for each year in each technological area in each equation. This will take care of effects due to variations across technologies and time in the "propensity to patent," the national state of "technological opportunity" in a given technological area at any time, and the effects of inflation. I will then allow for an arbitrary pattern of correlation of the remaining error across time and across technical areas, but I will assume independence across states after allowing for the time-technical area intercepts. Thus the California, Drugs, 1972 error can be correlated with the Oregon, Drugs, 1972 error because of what was happening in Drugs in 1972, but they are assumed independent after allowing for that. On the other hand, the California, Drugs, 1972 can be correlated with the California, Drugs, 1973 error, whether because of a permanent "California effect" or because of serial correlation. Further, I will allow for an arbitrary pattern of correlation across the patent, industrial R&D, and university research equations. This implies that IV techniques are necessary despite the partially recursive nature of the system. I interpret the correlation across equations as being due to the unobserved "quality" of the universities. Good universities attract more federal money; they attract industrial labs even after controlling for their greater budgets, and they produce more spillovers per research dollar.

For the instrument set, Z_1 (for the university research equation), I will use the log of population and the numbers of private universities, public universities, and FFRDC's. For the instrument set Z_2, I will use the log of population and the log of manufacturing value added. I also include log of population in the patent equations to eliminate any effects due to pure size of state. Thus my identifying assumptions are that all of these are uncorrelated with the unobserved "quality," that industry R&D does not depend on the number of institutions of each type, and that patents and university research do not depend on manufacturing value added per se.

To summarize, I am estimating a 3-equation (partially recursive) simultaneous system on a 3-dimensional data panel. I am allowing arbitrary correlation across the equations, implying the need for instrumental variables. I am allowing for time, technical area, and time-technical area interaction effects that may be correlated with the instruments, and state and state-technical area interaction effects that are assumed independent of the instruments. With 8 years and 4 technical areas, this implies estimation of a 96-equation system by 3-stage least squares.

Finally, we have the problem that the log-log functional form cannot be strictly true for patents, which can take only integer values and may be zero. The seriousness of this problem depends qualitatively on the fraction of the data in the region near zero. (John Bound et al., 1984; A. Colin Cameron and P. K. Trivedi, 1986). In the 928 (29 states $\times$ 8 years $\times$ 4 areas)) observations used below, there are 6 zeros, 13 ones, and 7 twos. The integer problem can be solved by estimating an equation of the form of $P = \exp(X\beta)$ by nonlinear methods. This has been shown to give consistent estimates of the parameters for a wide range of assumptions about the process (though it need not be efficient. See C. Gourieroux, A. Monfort, and A. Trognon, 1984). The next section of the paper presents estimates of the log-log form, for which I set the log of patents to -1 for the six observations with zero patents. I have replicated most of the results reported here using nonlinear estimation on the exponential form; these results are qualitatively similar to those in the next section.

TABLE 3—SIMPLE STATISTICS

	Regression Sample				All States
	Mean	Standard Deviation	Minimum	Maximum	Mean
University R&D					
Drugs&Medical	28.50	35.34	1.39	203.53	20.01
Chem. Exc. Drugs	5.69	9.73	0.16	55.45	4.12
Electronics, etc.	20.95	49.16	0.08	385.39	14.02
Mechanical	12.68	25.63	0.15	174.45	9.12
Total	98.76	143.95	7.33	980.68	69.60
Total R&D Performed By Industry	582.89	823.52	3.76	4,328.87	n.a.
Geographic Coincidence Index	0.63	0.35	0.03	1.00	n.a.
Population (Thousands)	5,955.92	4,853.05	946	24,265	4,318.03
Total Number of Public Universities	7.88	6.02	1	30	4.60
Total Number of Private Universities	6.63	8.16	0	37	3.27
Total Number of FFRDC's	0.55	0.94	0	4	0.29
Corporate Patents					
Drugs&Medical	71.70	99.43	0	549	43.92
Chem. Exc. Drugs	201.24	249.00	1	948	131.55
Electronics, etc.	225.01	295.29	2	1,222	142.88
Mechanical	300.78	319.90	15	1,144	191.83
Total	879.37	975.70	36	3,447	564.70

Note: All dollar figures are millions of 1972 dollars. The "All States" column corresponds to the mean for the 50 states and D.C. for the 8 years (1972–77, 1979, and 1981) of the regression sample.

III. Results

As noted above the sample for estimation consists of 8 years (1972–77, 1979, and 1981) for 29 states. In 1981, these states had 78 percent of the U.S. population and performed about 76 percent of industry R&D and 81 percent of university research; 85 percent of patents assigned to corporations were generated within these states. Simple statistics for all the model variables are presented in Table 3. Appendix A describes the construction of all variables from the underlying sources. The complete listing by state of the means for the key variables is given in Appendix B.

Table 4 presents the simplest regression results for the patent equation. Here, the structure of the errors and the endogeneity of I and U are ignored; a pooled state-time OLS regression is simply run for each of the 4 technical areas. Panel A of Table 4 shows the results when the interaction between geographic coincidence and university research is excluded; Panel B includes it. The results in Table 4 also ignore any correction for correlation between U_{ikt} and the unobserved industrial R&D technical area distribution.

The coefficients on university research and industry R&D are quite similar in both sets of results. The elasticity of patents with respect to industry R&D ranges from 0.6 for Mechanical Arts to almost 0.9 for Drugs; it is about 0.7 in the aggregate. The university research elasticities are all statistically significant, ranging in magnitude from about 0.04 for Mechanical Arts to 0.28 for Drugs.[6] Interestingly, the university research effect is much smaller when we do not distinguish technical areas (column 1).[7] The geographic

[6] Note that the estimation is carried out after normalizing $\log(C_i)$ to have mean 0, so the coefficient on $\log(U_{ikt})$ is equal to elasticity at the mean of the data.

[7] The reported results include both research and patents in the "All Other" category in the "All Cate-

TABLE 4—OLS ON THE SIMPLEST PATENT MODELS; POOLED DATA

Variable	All Areas	Drugs	Chemicals	Electronics Etc.	Mechanical Arts
	Dependent Variable: Log of Corporate Patents by Area				
A.					
$\log(I_{it})$	0.703	0.870	0.740	0.704	0.597
	(0.035)	(0.057)	(0.061)	(0.051)	(0.041)
$\log(U_{ikt})$	0.005	0.277	0.174	0.246	0.043
	(0.050)	(0.095)	(0.070)	(0.037)	(0.0061)
$\log(Pop_{it})$	0.209	−0.128	0.478	0.078	0.297
	(0.066)	(0.131)	(0.110)	(0.083)	(0.072)
$\hat{\sigma}$	0.383	0.696	0.673	0.525	0.445
R^2	0.915	0.828	0.792	0.869	0.870
B.					
$\log(I_{it})$	0.713	0.871	0.728	0.713	0.609
	(0.035)	(0.057)	(0.061)	(0.051)	(0.042)
$\log(U_{ikt})$	0.084	0.275	0.176	0.238	0.045
	(0.047)	(0.096)	(0.070)	(0.038)	(0.021)
$\log(U_{ikt})$	0.109	0.016	0.151	0.138	0.074
$*\log(C_i)$	(0.041)	(0.080)	(0.056)	(0.064)	(0.040)
$\log(Pop_{it})$	0.179	0.135	0.399	−0.104	0.280
	(0.068)	(0.135)	(0.113)	(0.086)	(0.072)
$\hat{\sigma}$	0.380	0.697	0.663	0.524	0.443
R^2	0.915	0.828	0.801	0.879	0.870

Notes: 232 observations. All equations include year dummies. Standard errors are in parentheses.

coincidence effect is significant only for Chemicals and Electronics.

Table 5 presents results from the application of the same estimation approach to the 4 equations with total university R&D added. Though the added variable is only marginally significant statistically, the implied correction to the estimated spillover effect is important and in the predicted direction. The elasticity at the mean is reduced slightly for Drugs and Chemicals, about 40 percent for Electronics, etc., and it is essentially eliminated for Mechanical Arts.[8]

The estimates in Tables 4 and 5 are not valid in the context of the general model discussed above. The coefficient estimates are potentially biased due to the endogeneity of the research variables, and the standard errors are biased due to the lack of independence of the errors in the pooled regression.[9]

gories" total. Restricting these totals to the 4 well-defined categories does not change the results.

[8]An alternative approach to industry R&D is to try to impute the totals by technical area on the basis of an industry breakdown of value added. The best such data I could find were the BEA Gross State Product series, which give value-added by 2-digit SIC, thus not distinguishing drugs from chemicals. I tried using this data by apportioning the reported state R&D total across all manufacturing SIC's. I then used the imputed Chemical (SIC 28) portion for my Drugs and Medical Technology category, the combined Chemicals (SIC 28) and Rubber and Miscellaneous Plastics (SIC 30) for my Chemicals category, Electric and Electronic Equipment and Instruments (SIC's 36 and 38) for my Electronics, etc., and Machinery and Transportation Equipment (SIC's 35 and 37) for my Mechanical Arts category. When I substitute these imputed technical area numbers for the industry total, the coefficients on industry R&D are affected only slightly, and the coefficient on total university R&D still implies an adjustment to the university research elasticity. Therefore, I concluded that it was better simply to stick with the unimputed total.

[9]Heteroscedasticity does not seem to be a problem; robust standard errors were calculated for some equations and generally differed from those reported by only a few percent.

TABLE 5—OLS ON THE FULL PATENT MODEL; POOLED DATA

Variable (Parameters)	Dependent Variable: Log of Corporate Patents by Area				
	All Areas	Drugs	Chemicals	Electronics Etc.	Mechanical Arts
$\log(I_{it})$	0.713	0.892	0.758	0.714	0.628
(β_1)	(0.035)	(0.064)	(0.062)	(0.051)	(0.042)
$\log(U_{ikt})$	0.084	0.332	0.207	0.305	0.085
$(\beta_2+\beta_1\gamma)$	(0.047)	(0.123)	(0.108)	(0.067)	(0.061)
$\log(U_{ikt})^*$	0.109	0.153	0.167	0.043	0.080
$\log(C_i)$	(0.041)	(0.079)	(0.056)	(0.034)	(0.040)
(β_3)					
$\log(U_{it})$		−0.095	−0.103	−0.161	−0.122
$(-\beta_2\gamma)$		(0.127)	(0.149)	(0.135)	(0.089)
(β_2)	0.084	0.237	0.104	0.144	−0.042
	(0.047)	(0.125)	(0.113)	(0.071)	(0.111)
(γ)		0.107	0.136	0.225	0.194
		(0.155)	(0.172)	(0.164)	(0.151)
$\log(Pop_{it})$	0.179	0.131	0.476	−0.033	0.348
	(0.068)	(0.136)	(0.117)	(0.104)	(0.076)
$\hat{\sigma}$	0.380	0.698	0.657	0.524	0.437
R^2	0.915	0.828	0.801	0.879	0.870

Notes: 232 observations. All equations include year dummies. Standard errors are in parentheses.

The first problem is solved by instrumental variables, the second by Generalized Least Squares (GLS) estimation. Before turning to those, let me note that results analogous to Table 5 were also produced by nonlinear least squares estimation of the exponential form of the model. The parameter estimates are quite similar, though they are less precisely estimated.

Table 6 presents instrumental variable/GLS estimates (3SLS). The 3-stage least squares estimates are obtained by viewing the data as a 29-observation cross section, and estimating a separate equation for each year-technical area combination (32 altogether). Slope coefficients are constrained to equality across time, but not across technical areas. Each equation has a free intercept, which is equivalent to the estimation of the 4 technical area equations with year dummies as in Tables 4 and 5. The "first stage" uses contemporaneous values of the instruments.[10] For the third stage, the covariance matrix of residuals across the 32 patent equations is used. In principle, even greater power could be achieved by using the full 96 equation residual covariance, but this increases the computation complexity dramatically.

Comparing Tables 5 and 6, the effect of the instrumental variables estimation[11] is to increase generally both the industry and university research coefficients. The net effect on the parameter β_2 (after correction via the coefficient on total university R&D) varies

[10]Standard statistical packages use all instruments in all equations when estimating by 3-stage least squares. In this case that would correspond to using past and future values, and would result in far more instruments than observations. Thus, to do 3-stage least squares, it was necessary to run the first stage separately, then run GLS using predicted values of the endogenous variables. The standard errors in Table 6 have been corrected for the difference between the variance of the residuals when calculated with actual and predicted endogenous variables.

[11]Examination of the 2SLS results confirms that the change in the coefficients is due to the IV and not the GLS. A Hausman test comparing all slope coefficients in the 4 2SLS equations to the system results presented yields a chi-square (16) of 4.2.

TABLE 6—3SLS ON THE FULL PATENT MODEL: SEPARATE EQUATIONS FOR EACH YEAR

Variable (Parameters)	Dependent Variable: Log of Corporate Patents by Area				
	All Areas	Drugs	Chemicals	Electronics Etc.	Mechanical Arts
$\log(I_{it})$	0.940	0.989	0.895	0.892	0.844
(β_1)	(0.052)	(0.050)	(0.058)	(0.044)	(0.039)
$\log(U_{ikt})$	0.103	0.385	0.295	0.368	0.124
$(\beta_2+\beta_1\gamma)$	(0.041)	(0.090)	(0.098)	(0.058)	(0.054)
$\log(U_{ikt})^*$	0.105	0.152	0.160	0.054	0.093
$\log(C_i)$	(0.033)	(0.081)	(0.052)	(0.032)	(0.048)
(β_3)					
$\log(U_{it})$		−0.194	−0.123	−0.243	−0.151
$(-\beta_2\gamma)$		(0.134)	(0.127)	(0.138)	(0.084)
(β_2)	0.103	0.191	0.172	0.125	−0.027
	(0.041)	(0.082)	(0.104)	(0.050)	(0.104)
(γ)		0.196	0.137	0.272	0.179
		(0.142)	(0.168)	(0.150)	(0.144)
$\log(Pop_{it})$	0.181	0.125	0.494	−0.074	0.351
	(0.051)	(0.104)	(0.098)	(0.094)	(0.085)
$\hat{\sigma}$	0.420	0.699	0.659	0.529	0.438

Notes: Standard errors are in parentheses. Instruments are contemporaneous values of the log of the geographic coincidence index, log of population, log of manufacturing value added, the numbers of public universities, private universities, and FFRDC's, and interactions between the coincidence index and the other instruments. For further estimation details, see text.

across areas but is fairly small. The geographic coincidence effect is only marginally significant statistically, though the point estimates imply an effect that would be quantitatively important if it is real. The interquartile range for the university research elasticity, taking the interaction into account, is 0.10 to 0.30 for Drugs, 0.08 to 0.28 for Chemicals, and 0.09 to 0.16 for Electronics. For Mechanical Arts, the effect is slightly negative, but insignificant. The parameter γ remains in the range 0.1 to 0.2 and is not significant.[12]

Table 7 presents estimates of the industry R&D equation. The interpretation of the effect of university research on industry R&D in terms of quantity-inducement (the parameters β_{4k}) and a separate distribution effect (the parameter γ) does not seem to work very well. The estimates for γ range from -4 to 1 across the equations, though they should be the same. As a result, the implied area-specific inducement elasticities vary from -0.1 to 0.8. It may be more sensible to confine our attention to the aggregate version in the first column, which indicates an overall elasticity of industry R&D with respect to university research of 0.70. Manufacturing value added has an effect of the same magnitude, while the effect of population is smaller and insignificant.

Table 8 presents estimates of the University R&D equations. Industry R&D is fairly small and insignificant throughout. Population has a generally positive effect, though it is negative for Chemicals and not very significant. Not surprisingly, having more private universities and FFRDC's generates more research activity. Interestingly, there is a systematic negative effect of increasing the number of public universities. Since all of

[12]If the unobserved industry technical area fraction ρ_{ikt} is independent of the instruments, then an alternative strategy would be to omit the total research variable once we go to instrumental variables. This had the effect, however, of increasing the estimated coefficients on $\log(U_{ikt})$, so it would seem that the IV estimation is not an adequate solution to the omitted variable problem.

TABLE 7—3SLS RESULTS FOR THE INDUSTRY R&D EQUATION: SEPARATE EQUATIONS FOR EACH YEAR

	Dependent Variable: Log of Industry R&D				
Variable (Parameters)	All Areas	Drugs	Chemicals	Electronics Etc.	Mechanical Arts
$\log(U_{ikt})$	0.704	−0.730	3.645	0.720	1.316
$(\beta_{4k}-\gamma)$	(0.129)	(0.264)	(2.955)	(0.172)	(0.456)
$\log(U_{it})$		1.078	−4.234	−0.842	−0.994
(γ)		(0.188)	(4.013)	(0.389)	(0.607)
$\log(VA)$	0.725	0.787	−0.614	0.654	0.697
	(0.162)	(0.166)	(1.145)	(0.156)	(0.189)
$\log(Pop)$	0.131	0.421	2.586	0.782	0.311
	(0.246)	(0.271)	(2.064)	(0.282)	(0.293)
β_{4k}	0.704	0.348	−0.589	−0.122	0.822
	(0.129)	(0.195)	(3.052)	(0.258)	(0.714)
$\hat{\sigma}$	0.744	0.755	1.657	0.710	0.864

Note: Standard errors are in parentheses. Instruments are contemporaneous values of log of manufacturing value added, log of population, and the numbers of public universities, private universities, and FFRDC's. For further estimation details, see text.

TABLE 8—3SLS ESTIMATES OF UNIVERSITY RESEARCH EQUATION: SEPARATE EQUATIONS FOR EACH YEAR

	Dependent Variable: Log of University Research by Area				
Variable (Parameters)	All Areas	Drugs	Chemicals	Electronics Etc.	Mechanical Arts
$\text{Log}(I_{it})$	0.136	0.193	0.006	0.084	0.204
	(0.141)	(0.174)	(0.243)	(0.261)	(0.222)
Number of Public Universities	−0.041 (0.015)	−0.081 (0.039)	−0.088 (0.051)	−0.066 (0.027)	−0.032 (0.022)
Number of Private Universities	0.032 (0.008)	0.029 (0.010)	0.041 (0.014)	0.050 (0.015)	0.024 (0.013)
Number of FFRDC's	0.308 (0.046)	−0.035 (0.056)	0.300 (0.079)	0.827 (0.085)	0.502 (0.072)
$\text{Log}(Pop)$	0.450	0.407	−0.674	0.280	0.468
	(0.256)	(0.315)	(0.439)	(0.472)	(0.400)
$\hat{\sigma}$	0.377	0.464	0.646	0.695	0.589

Notes: Standard errors are in parentheses. Instruments are contemporaneous values of log of manufacturing value added, log of population, and the numbers of public universities, private universities, and FFRDC's. For further estimation details, see the text.

the sample states have at least one, I interpret this to indicate that states that concentrate their higher education resources in a small number of institutions (relative to population) have better institutions and hence attract more research money than those that spread their resources over a larger number of campuses.

IV. Conclusion

The analysis of state-level corporate patent activity provides some evidence of the importance of geographically mediated commercial spillovers from university research. The effect is statistically strongest in Drugs, slightly smaller and less significant in Chemicals, and smaller but quite significant in Electronics, etc. There is only weak evidence that spillovers are facilitated by geographic coincidence of universities and research labs within the state. It is interesting that the effect comes through more clearly within technical areas than it does in the total across areas. This suggests that the spillovers are limited to specific areas and not just the diffuse effect of a large research university.

From the point of view of regional economic policy, the indirect effect of university research is also important. After controlling for population and economic activity, there is an association between industry R&D and university research. Establishing causality with statistics is a tricky business, but it appears that university research causes industry R&D and not vice versa. Thus, a state that improves its university research system will increase local innovation both by attracting industrial R&D and augmenting its productivity.

Note that these results do not relate directly to the question of the social rate of return to university research. They underestimate that return, to the extent that spillovers flow beyond state boundaries.[13] On the other hand, effects within the state overestimate the social return to the extent that they reflect competitive success at the expense of other states.

It is important to emphasize that the spillover *mechanisms* have not been modeled. Despite the attempt to control for unobserved "quality" of universities, one cannot really interpret these results structurally, in the sense of predicting the resulting change in patents if research spending were exogenously increased. In future work, I intend to look in more detail at the activities of universities, moving toward an understanding of their relationships with the private sector that would really be structural. I also intend to investigate what these relationships look like at the level of the SMSA, which is arguably a more meaningful economic unit. Doing this will require using the R&D lab data (which is used herein only in constructing the geographic coincidence index) as the basis for a measure of industry R&D. This is difficult because of frequently missing data. On the other hand, the lab data do contain some information on technical areas.

With all of these caveats, it is worth noting that the effects we are discussing appear to be large. Even the overall elasticity of corporate patents with respect to university R&D (which is smaller than the area-by-area elasticities) is about 0.1 (Table 6). Given that university research is only about one-sixth of industry R&D, this is a potent effect. The indirect or inducement effect is even larger; combining the elasticity of industry R&D with respect to total university research (0.704 in Table 7) and the elasticity of corporate patents with respect to corporate research (0.814 in Table 6), the implied elasticity of induced corporate patents with respect to university research is almost 0.6.

A final interesting policy implication is the apparent disadvantage of dispersed public university systems in attracting research money. Again, the statistical relationship is only suggestive; detailed examination of the university data will be necessary to determine if this effect is real.

[13]Any systematic effect of changes in the *overall* level of federal research spending have been removed through the use of time dummies.

APPENDIX A: DATA CONSTRUCTION

1. Corporate patents: A tape was obtained from the U.S. Patent Office containing the following information for each patent granted between 1972–1986: The application year, the state (or foreign country) of the applicant, the patent class, an assignee code number, and an "assignment code." A list of assignee code numbers that correspond to organizations that are actually universities was also obtained. Corporate patents were taken to be all patents that, based on their assignment code, were assigned by the applicant to either a domestic or foreign corporation, excluding those whose assignee was a university based on the patent office list. Patents were assigned to the five technological areas used herein based on their patent class; patent classes were assigned to the areas based on their names. The totals for each state by year used herein are based on application year.

2. Industry R&D is total industry R&D by state taken directly from Research and Development in In-

dustry 1957–1983 prepared by the Bureau of the Census. (The underlying information is derived from surveys conducted by NSF).

3. University Research spending by institution by department by source of funds was taken from the *National Science Foundation Academic Science and Engineering Statistics Survey*. Departments were assigned to the five technological areas as follows: Drugs and Medical Technology includes biology and medicine; Chemical Technology includes chemistry and chemical engineering; Electronics, Optics, and Nuclear Technology includes electrical engineering, astronomy, physics, and computer science; Mechanical Arts includes mechanical engineering, civil engineering, aeronautical and astronautical engineering, other engineering, NEC, and other physical sciences, NEC; All Other includes environmental sciences, mathematics, psychology, economics, political science, sociology, other social science, and all other, NEC.

4. The geographic coincidence index was calculated as described in fn. 5 from the Directory of American Research and Technology published by Bowker (1986). These data are available in machine-readable form from Pergamon On-Line Data Services.

APPENDIX B. VARIABLE MEANS BY STATE[a]

	University Research by Area					Ind.	Geog.		Number Of:			Corporate Patents by Area;				
State	1	2	3	4	Total	R&D	Coinc.	Pop.	Pub. U.	Priv. U.	FFRDC	1	2	3	4	Total
Ala.	17.0	0.89	1.0	1.3	28.3	54	0.74	3718	7	2	0	4	14	12	31	69
Ariz.	7.9	1.79	13.0	2.9	37.4	201	0.25	2359	3	0	1	6	27	134	67	245
Ark.	5.7	0.54	0.3	0.9	12.0	9	0.27	2160	5	0	0	1	6	7	20	39
Calif.	140.7	46.74	239.9	126.1	710.4	3883	0.78	21996	25	21	4	247	545	1142	945	3230
Colo.	15.0	2.94	11.3	7.0	77.2	167	0.93	2649	5	2	1	14	59	113	103	332
Conn.	28.3	2.64	5.1	2.1	54.7	650	0.41	3087	2	9	0	88	233	281	400	1090
Fla.	20.7	2.62	4.3	6.3	70.1	375	0.13	8694	8	5	0	28	60	143	149	424
Ga.	17.0	3.33	8.8	5.8	57.8	78	0.82	5133	7	2	0	5	37	34	89	192
Ill.	45.5	22.77	74.9	49.3	254.9	894	0.96	11345	11	16	2	161	459	489	993	2451
Ind.	16.1	5.23	5.9	6.9	51.3	398	0.06	5383	6	4	0	117	137	149	319	784
Ia.	16.3	4.22	6.8	5.2	46.4	135	0.03	2891	3	5	1	10	15	48	130	243
Kans.	8.6	1.08	0.9	2.2	26.6	66	0.07	2302	5	0	0	21	21	15	65	145
Ky.	5.8	0.54	0.6	1.7	17.5	72	0.23	3502	6	0	0	4	41	28	111	209
La.	12.9	1.50	1.6	1.7	33.4	65	0.98	3958	8	3	0	9	69	10	41	163
Mass.	44.4	12.10	86.5	33.1	245.3	954	1.00	5760	6	21	2	74	230	578	364	1339
Mich.	34.1	4.00	8.0	13.8	103.2	1815	0.23	9131	10	3	0	148	392	211	859	1725
Minn.	29.7	1.89	3.1	3.3	55.7	399	1.00	3958	5	6	0	55	92	176	232	628
Miss.	30.7	1.20	3.3	5.6	61.4	420	0.84	4826	9	4	0	38	128	52	150	423
Neb.	9.9	0.57	0.5	1.0	20.4	9	0.96	11547	3	2	0	4	6	11	22	50
N.J.	7.5	2.86	34.1	6.4	70.8	1361	0.96	7352	5	5	1	418	908	658	480	2631
N.Y.	142.3	21.92	61.5	26.9	371.0	1859	0.97	17960	28	34	2	231	682	970	764	2857
N.C.	29.0	1.87	2.7	2.8	64.6	193	0.90	5587	7	2	0	16	56	63	158	327
Ohio	32.8	4.56	7.3	13.0	76.2	926	0.94	10772	13	11	0	92	586	318	858	1999
Okla.	6.4	0.93	1.4	2.8	19.9	93	0.33	2828	4	2	0	14	225	57	96	453
Pa.	51.9	7.95	13.8	20.5	139.2	1293	0.85	11884	10	21	1	208	651	561	769	2329
R.I.	2.2	0.84	1.8	1.8	14.9	32	1.00	959	1	2	0	2	10	25	44	94
Utah	9.3	2.09	2.6	6.0	32.5	72	1.00	1282	2	1	0	12	23	23	36	103
Va.	13.9	1.77	3.5	6.2	45.9	207	0.57	5109	10	4	2	22	46	89	101	297
Wisc.	25.0	3.69	3.8	4.7	65.0	224	0.48	4590	14	6	0	34	79	132	328	634

[a]All dollar figures are in millions of 1972 dollars. Population is in thousands.

REFERENCES

Bania, Neil, Eberts, Randall and Fogarty, Michael, "The Role of Technical Capital in Regional Growth," presented at the Western Economic Association Meetings, July 1987.

Bound, John, Cummins, Clint, Griliches, Zvi, Hall, Bronwyn and Jaffe, Adam, "Who Does R&D and Who Does Patents?," in Z. Griliches, ed., *R&D, Patents and Productivity*, Chicago: University of Chicago Press, 1984, ch. 2.

Cameron, A. Colin and Trivedi, Pravin, "Econometric Models Based on Count Data: Comparisons and Applications of Some Estimators and Tests," *Journal of Applied Econometrics*, January 1986, *1*, 29–63.

Dorfman, Nancy S., "Route 128: The Development of a Regional High Technology Economy," *Research Policy*, December 1983, *12*, 299–316.

Gourieroux, C., Monfort, A. and Trognon, A., "Pseudo Maximum Likelihood Methods: Applications to Poisson Models," *Econometrica*, May 1984, *52*, 701–20.

Griliches, Zvi, "Issues in Assessing the Contribution of R&D to Productivity Growth," *Bell Journal of Economics*, Spring 1979, *10*, 92–116.

______, "Patent Statistics as Economic Indicators, A Survey," mimeo., 1987.

______, "Hall, Bronwyn and Pakes, Ariel, "R&D Patents and Market Value Revisited: Is There a Second (Technological Opportunity) Factor?," National Bureau of Economic Research Working Paper No. 2624, June 1988.

Hausman, Jerry A., Hall, B. and Griliches, Z., "Econometric Models for Count Data with an Application to the Patents-R&D Relationship," *Econometrica*, July 1984, *52*, 909–38.

Jaffe, Adam B., "Technological Opportunity and Spillovers of R&D: Evidence from Firms' Patents Profits and Market Value," *American Economic Review*, December 1986, *76*, 984–1001.

Lund, Leonard, "Locating Corporate R&D Facilities," Conference Board Report No. 892, Conference Board, New York, 1986.

Nelson, Richard R., "Institutions Supporting Technical Advance in Industry," *American Economic Review*, May 1986, *76*, 186–89.

Pakes, Ariel and Griliches, Z., "Patents and R&D at the Firm Level: A First Look," in Z. Griliches, ed., *R&D, Patents, and Productivity*, Chicago: University of Chicago Press, 1984, ch. 3.

Shimshoni, Daniel, "Aspects of Scientific Entrepreneurship," unpublished doctoral dissertation, Harvard University, Cambridge, 1986.

Teplitz, P., "Spin-Off Enterprises from a Large Government-Sponsored Laboratory," unpublished doctoral dissertation, Sloan School, MIT, Cambridge, 1965.

Wainer, Herbert A., "The Spin-Off Technology from Government-Sponsored Research Laboratories: Lincoln Laboratory," unpublished doctoral dissertation, Sloan School, MIT, Cambridge, 1965.

[16]

GEOGRAPHIC LOCALIZATION OF KNOWLEDGE SPILLOVERS AS EVIDENCED BY PATENT CITATIONS*

ADAM B. JAFFE
MANUEL TRAJTENBERG
REBECCA HENDERSON

We compare the geographic location of patent citations with that of the cited patents, as evidence of the extent to which knowledge spillovers are geographically localized. We find that citations to domestic patents are more likely to be domestic, and more likely to come from the same state and SMSA as the cited patents, compared with a "control frequency" reflecting the pre-existing concentration of related research activity. These effects are particularly significant at the local (SMSA) level. Localization fades over time, but only very slowly. There is no evidence that more "basic" inventions diffuse more rapidly than others.

The last decade has seen the development of a significant body of empirical research on R&D spillovers.[1] Generally speaking, this research has shown that the productivity of firms or industries is related to their R&D spending, and also to the R&D spending of other firms or other industries. In parallel, economic growth theorists have focused new attention on the role of knowledge capital in aggregate economic growth, with a prominent modeling role for knowledge spillovers (e.g., Romer [1986, 1990] and Grossman and Helpman [1991]).

We know very little, however, about where spillovers go. Is there any advantage to nearby firms, or even firms in the same country, or do spillovers waft into the ether, available for anyone around the globe to grab? The presumption that U. S. international competitiveness is affected by what goes on at federal laboratories and U. S. universities, and the belief that universities and other research centers can stimulate regional economic growth[2] are predicated on the existence of a geographic component to the

*We gratefully acknowledge support from the Ameritech Foundation, via the Ameritech Fellows program of the Center for Regional Economic Issues at Case Western Reserve University, and from the National Science Foundation through grant SES91-10516. We thank Neil Bania, Ricardo Caballero, Michael Fogarty, Zvi Griliches, Frank Lichtenberg, Francis Narin, seminar participants at NBER and Case Western Reserve University, and two anonymous referees for helpful comments. Any errors are the responsibility of the authors.

1. E.g., Jaffe [1986] and Bernstein and Nadiri [1988, 1989]. For a recent survey and evaluation of this literature, see Griliches [1991].

2. See, e.g., Minnesota Department of Trade and Economic Development [1988]; Dorfman [1988]; Feller [1989]; and Smilor, Kozmetsky, and Gibson [1988].

The Quarterly Journal of Economics, August 1993

spillover mechanism. The existing spillover literature, however, is virtually silent on this point.[3]

In the growth literature it is typically assumed that knowledge spills over to other agents within the country, but not to other countries.[4] This implicit assumption begs the question of whether and to what extent knowledge externalities are localized. As emphasized recently by Krugman [1991], acknowledging the importance of spillovers and increasing returns requires renewed attention by economists to issues of economic geography. Krugman revives and explores the explanations given by Marshall [1920] as to why industries are concentrated in cities. Marshall identified three factors favoring geographic concentration of industries: (1) the pooling of demands for specialized labor; (2) the development of specialized intermediate goods industries; and (3) knowledge spillovers among the firms in an industry. Krugman believes that economists should focus on the first two of these, partially because he perceives that "[k]nowledge flows, by contrast, are invisible; they leave no paper trail by which they may be measured and tracked, and there is nothing to prevent the theorist from assuming anything about them that she likes" [Krugman, p. 53].

But knowledge flows do sometimes leave a paper trail, in the form of citations in patents. Because patents contain detailed geographic information about their inventors, we can examine where these trails actually lead. Subject to caveats discussed below relative to the relationship between citations and spillovers, this allows us to use citation patterns to test the extent of spillover localization. We examine citations received by patents assigned to universities, and also the citations received by a sample of domestic corporate patents. If knowledge spillovers are localized within countries, then citations of patents generated within the United States should come disproportionately from within the United States. To the extent that regional localization of spillovers is

3. Jaffe [1989] provides evidence that corporate patenting at the state level depends on university research spending, after controlling for corporate R&D. Mansfield [1991] surveyed industrial R&D employees about university research from which they benefited. He found that they most often identified major research universities, but that there was some tendency to cite local universities even if they were not the best in their field.

4. The existence of this implicit assumption was noted by Glaeser, Kallal, Scheinkman, and Shleifer [1991]: "After all, intellectual breakthroughs must cross hallways and streets more easily than oceans and continents." Grossman and Helpman [1991] consider international knowledge spillovers explicitly.

important, citations should come disproportionately from the same state or metropolitan area as the originating patent.[5]

The most difficult problem confronted by the effort to test for spillover-localization is the difficulty of separating spillovers from correlations that may be due to a pre-existing pattern of geographic concentration of technologically related activities. That is, if a large fraction of citations to Stanford patents comes from the Silicon valley, we would like to attribute this to localization of spillovers. A slightly different interpretation is that a lot of Stanford patents relate to semiconductors, and a disproportionate fraction of the people interested in semiconductors happen to be in the Silicon valley, suggesting that we would observe localization of citations even if proximity offers no advantage in receiving spillovers. Of course, the ability to receive spillovers is probably one reason for this pre-existing concentration of activity. If it were the *only* possible reason, then, under the null hypothesis of no spillover localization we should still see no localization of citations. As discussed above, however, there are other sources of agglomeration effects that could explain the geographic concentration of technologically related activities without resort to localization of knowledge spillovers. For this reason, we construct "control" samples of patents that are not citations but have the same temporal and technological distribution as the citations. We then calculate the geographic matching frequencies between the citations and originating patents, and between the controls and originating patents. Our test of localization is whether the citation matching frequency is significantly greater than the control matching frequency. Since the "control" matching frequency is, itself, likely to be partly the result of spillover-localization, we believe this to be a conservative test for the existence of localization.

The first section of the paper describes patents and patent citations, explains the construction of the control samples, and

5. Glaeser, Kallal, Scheinkman, and Shleifer [1991] characterize the "Marshall-Arrow-Romer" models as focusing on knowledge spillovers within the firms in a given industry. They examine the growth rate of industries in cities as a function of the concentration of industrial activity across cities, within-city industrial diversity, and within-city competition. They find that within-city diversity is positively associated with growth of industries in that city, while concentration of an industry within a city does not foster its growth. They interpret this contrast to mean that spillovers across industries are more important than spillovers within industries. As is discussed below, there is evidence from the R&D spillover literature to suggest that across-industry knowledge spillovers are, indeed, important. In this study, we do not consider the *industrial* identity of either generators or receivers of spillovers, though we do have some information on their *technological* similarity.

considers more carefully how citations might be used to infer spillovers. The second section presents the results of the tests of geographic localization. The following section examines whether the probability of geographic localization of any given citation can be explained by attributes of the originating or citing patents, or of relationships between them. A concluding section follows.

I. Experimental Design

A. Patents and Patent Citations[6]

A patent is a property right in the commercial use of a device.[7] For a patent to be granted, the invention must be nontrivial, meaning that it would not appear obvious to a skilled practitioner of the relevant technology, and it must be useful, meaning that it has potential commercial value. If a patent is granted,[8] a public document is created containing extensive information about the inventor, her employer, and the technological antecedents of the invention, all of which can be accessed in computerized form. Among this information are "references" or "citations." It is the patent examiner who determines what citations a patent must include. The citations serve the legal function of delimiting the scope of the property right conveyed by the patent. The granting of the patent is a legal statement that the idea embodied in the patent represents a novel and useful contribution over and above the previous state of knowledge, as represented by the citations. Thus, in principle, a citation of Patent X by Patent Y means that X represents a piece of previously existing knowledge upon which Y builds.

The examiner has several means of identifying potential citations. The applicant has a legal duty to disclose any knowledge of the prior art that she may have. In addition, the examiner is supposed to be an expert in the technological area and be able to

6. All of the data we use relate to patents granted by the U. S. patent office. About 40 percent of U. S. patents are currently granted to foreigners. Other countries, of course, also grant patents, leading to some ambiguity in the meaning of phrases like "U. S. patent" and "foreign patent." We shall use the phrase "U. S. patent" to mean a patent granted by the U. S. patent office, regardless of the residence of the inventor. We shall use the phrase "domestic patent" to refer to a patent granted (by the U. S. patent office) to an inventor residing in the United States. We shall use the phrase "foreign patent" to refer to a patent granted by the U. S. patent office to non-U. S. residents.

7. Ideas are not patentable; nor are algorithms or computer programs, though a chip with a particular program coded into it might be. The definition of a device was recently broadened to include genetically engineered organisms.

8. There is no public record of unsuccessful patent applications.

identify relevant prior art that the applicant misses or conceals. The framework for the search of the prior art is the patent classification system. Every patent is assigned to a nine-digit patent class (of which there are about 100,000) as well as an unlimited number of additional or "cross-referenced" classes. An examiner will typically begin the search of prior art using her knowledge of the relevant classes. For the purpose of identifying distinct technical areas, we utilize aggregations of subclasses to a three-digit level; at this level there are currently about 400 technical classes.[9]

For this study, we begin with two cohorts of "originating" patents, one consisting of 1975 patent applications and the other of 1980 applications. In each cohort we include all patents granted to U. S. universities and two samples of U. S. corporate patents[10] chosen to match the university patents by grant date and technological distribution. These sets of originating patents were chosen because we conjectured that the extent of geographic localization might differ depending on the nature of the originating institution. As discussed below, such differences turn out to be minor. The 1975 originating cohort contains about 950 patents that had received a total of about 4750 citations by the end of 1989. The 1980 originating cohort contains about 1450 that had received about 5200 citations by the same time.

B. Construction of "Control" Samples

The main idea of this paper is to compare the geographic location of the citations with the originating patent that they cite. But to make such a comparison meaningful, we have to consider how often we would expect them to match under some "null" hypothesis. That is, we need to compare the probability of a patent matching the originating patent by geographic area, *conditional* on its citing the originating patent, with the probability of a match *not conditioned on the existence of a citation link*. This noncitation-conditioned probability gives us a baseline or reference value against which to compare the proportions of citations that match.

9. Examples of three-digit patent classes are "Batteries, Thermoelectric and Photoelectric"; "Distillation: Apparatus"; "Robots"; seventeen distinct classes of "Organic Compounds"; and the ever-popular "Whips and Whip Apparatus."

10. The "top corporate" sample consists of patents granted to the 200 top-R&D-performing firms in the United States, as reported in S.E.C. 10-k forms and compiled by Compustat. The "other corporate" sample contains patents assigned to U. S. corporations that are not universities and not in the "top corporate" sample.

We call this baseline or reference probability the "control frequency."

Two considerations drove our choice for constructing the control frequency. First, the fraction of U. S. patents granted to foreigners has been climbing steadily during the period under study here. We do not want to conclude that citations are initially localized, but that this localization fades over time, simply because of this aggregate trend. Second, countries (and cities and states) differ in their areas of technological focus. Although such technological specialization is probably due, in part, to geographic localization of spillovers, we want to be conservative and test whether spillovers are localized *relative to what would be expected given the existing distribution of technological activity.*

To derive a control frequency that would be immune to contamination from either aggregate movements over time or localization based on the pre-existing concentration of technological activity, we went back to the patent data base and found a "control patent" to correspond to each of the citing patents. For each citing patent, we identified all patents in the *same patent class* with the *same application year* (excluding any other patents that cited the same originating patent). We then chose from that set a control patent whose grant date was as close as possible to that of the citing patent. This process yielded, for each set of citing patents, a corresponding control sample of equal size, whose distribution across time and technological areas is essentially identical to that of the citation data set. Each control patent is paired with a particular citing patent, allowing us to compare the geographic location of the control patent with that of the originating patent cited by its counterpart in the citing dataset. The frequency with which these control patents match geographically with the originating patent is an estimate of the frequency with which a randomly drawn patent that is not a citation, but has the same technological and temporal profile as the citation, matches geographically.

To put it slightly differently, when we calculate the frequency with which the citations match the geographic location of the cited patents, we are estimating the probability of geographic match for two patents, *conditional on there being a citation link and also conditional on the technological nature and timing of the citation.* When we calculate the frequency with which the "control" patents match geographically with the cited patents, we are estimating the probability of geographic match for two patents, *conditional only on the technological nature and timing of the citation.* If the citation match frequency is significantly higher, then that implies that citations are localized even after controlling for timing and technology.

C. Issues Relating to the Use of Citations to Infer Spillovers

With the construction of the control samples, we believe that we have designed a very clean test of the extent to which patent citations are geographically localized. Before going on, we must address the validity of drawing inferences about knowledge spillovers from patent citations. For discussion purposes, we can classify the links that might exist between two inventions into one of three groups: spillovers accompanied by citations, citations that occur where there was no spillover, and spillovers that occur without generating a citation. Our experiment uses the first set, but clearly the other two are non-empty. The key question is whether and to what extent we expect that either of the latter two groups would be systematically more or less localized than the group we examine. Though there are a number of considerations, all difficult to quantify, we believe that on balance it is reasonable to draw inferences about spillovers from citations.

As a general consideration, it is important to keep in mind that any analogy between patent citations and academic article citations cannot be taken too far. Academics may cite a friend (or neighbor) just to be nice, since the price of doing so is infinitesimal, or even negative if a longer list of references is perceived as making the research look more thorough. An inventor who did the same in a patent application is, in effect, leaving money lying on the table: if those citations are included in the final patent the inventor has reduced the scope of her monopoly. Further, the patent examiner should not include such citations in the patent even if the inventor did put them there. Thus, it does not seem that "gratuitous" citations are a serious concern.

A deeper problem is created by "real" citations that are not spillovers. For example, suppose that a firm gets a patent on an invention and then contracts with another firm to make some part of it, or a machine necessary to make it, or any other aspect of the downstream development. It is possible that such a contractor might later get a patent on a related technology. To the extent that the flow of rents between these parties is governed by a complete contract, there could conceivably be no externality running from the original inventor to the contractor. If we now add to this hypothetical contract the assumption that such contracted development is relatively likely to be localized, we have the potential for the observed localization of citations to be greater than the true localization of knowledge spillovers.[11]

Although such "internalized spillovers" surely exist, it is likely that most citations that are not spillovers are of a different sort: citations (added by the examiner) to previous patents of which the citing inventor was unaware. Clearly, no spillover occurs in this case. Further, it seems likely that citations of this sort should not be any more geographically localized than the control patents. If many citations are in this category, it introduces "noise" into the citations as a measure of spillovers, and biases the results *away* from finding significant localization. Our a priori belief is that this category is much larger than the previous one, suggesting that spillovers are, on balance, probably more localized than citations, but readers with different beliefs should interpret our results accordingly.

Finally, there are an enormous number of spillovers with no citations, since only a small fraction of research output is ever patented. In particular, much of the results of very basic research cannot be patented. It is plausible that basic research generates the largest spillovers,[12] and also that basic research is communicated via mechanisms that are less likely to be localized, such as international journals. For this reason, it is probably appropriate to view our results as related to applied research, and to exercise care in extrapolating to the localization of spillovers from extremely basic research.

D. Geographic Assignment of Patents

The preceding discussion has presumed that the "location" of a patent is an unambiguous construct. The patent data contain the country of residence of each inventor, and the city and state of

11. As discussed below, we focus on tests of localization that exclude citing patents that are owned by the same organization as the originating patent, precisely because such "self-citations" do not represent an externality. Citations by other organizations that have an economic relationship with the original inventor could be viewed as similar to self-citations. There is, however, a significant difference: we expect that, in general, the contract between the two parties will be quite incomplete, making it more likely than not that the citing organization could capture some rents from the original invention and hence benefit from at least a partial spillover.

12. This question is analyzed in detail in Trajtenberg, Henderson, and Jaffe [1992].

residence for U. S. inventors.[13] Use of this information is complicated by the fact that patents can have multiple inventors who can live in different places. The following procedure was followed:

1. For U. S. *inventors,* city/state combinations were placed in counties using a commercially available city directory; each U. S. inventor was then assigned to an SMSA[14] based on state and county. For this purpose an additional "phantom" SMSA was created in each state, encompassing all counties in the state outside of defined SMSAs. Approximately 98 percent of inventors were successfully assigned to SMSAs.

2. Assignments of each *patent* to a country, a state, and an SMSA were then made based on pluralities of inventors. So, for example, a patent with one inventor living in Bethesda, MD, one in Alexandria, VA, and one in rural Virginia would be assigned VA for its state and Washington, DC, for its SMSA. Ties were assigned arbitrarily, except that ties between true SMSAs and phantom SMSAs were resolved for the true one and ties between United States and foreign were resolved in favor of foreign.[15]

II. Results on Extent of Localization

As a prelude to the geographic analysis, Table I and Figure I present some descriptive data about the citations and their relationship to the originating patents. Table I shows that about 80–90 percent of the 1975 patents and 70–80 percent of the 1980 patents had received at least one citation by the end of 1989, with the higher proportion in each case applying to the university patents. Mean citations received (including zeros) were four–six for 1975 and three–four for 1980, again with the higher numbers corresponding to the university patents.[16] The average lag between the originating application year and the application year of the citing

13. Published data on the geographic distribution of U. S. patents (including those cited in the popular press) are based on the location of the organization or individual to which the patent is "assigned" by the inventor (usually her employer). For analysis of localization, the location of the assignee is not a desirable datum. There is an ambiguity in such data relating to the way employees of multinational corporations make their assignments. An employee of "Honda" in the United States could assign her patent to "Honda U.S.A., Inc." or she could assign it to the parent company in Japan. In the former case, the patent office would call it a domestic patent, and in the latter case a foreign patent; similarly for IBM Switzerland.

14. These assignments were made based on the 1981 SMSA definitions. In areas where Consolidated Metropolitan Statistical Areas were defined in 1981, these were used; elsewhere Metropolitan Statistical Areas were used. Hence we use the generic term "SMSA."

15. At the country level, 98 percent of patents were assigned unanimously. At the state level, 90 percent were assigned unanimously; an additional 4 percent had more than half of inventors in a single state. At the SMSA level, 86 percent were assigned unanimously, and an additional 6 percent had a clear majority. Overall, 4.5 percent of patents were assigned to "phantom" SMSAs.

16. Our companion paper [Trajtenberg, Henderson, and Jaffe, 1992] explores in detail the use of citation intensity and related measures for measuring the basicness and appropriability of inventions.

TABLE I
DESCRIPTIVE STATISTICS

Originating dataset	Percent receiving citations	Total no. of citations	Mean citations received	Average citation lag[a,b]	Percent self-citations[b]	Percent same patent class[b,c]
1975						
University	88.6	1933	6.12	6.53	5.6	54.3
Top corporate	84.2	1476	4.70	7.17	18.6	55.7
Other corporate	82.3	1341	4.22	7.82	9.1	57.5
1980						
University	79.9	2093	4.34	4.36	8.9	56.3
Top corporate	79.9	1701	3.54	4.41	24.6	58.3
Other corporate	74.1	1424	2.95	4.46	12.6	57.2

a. Application year of citing patent minus application year of originating patent.
b. For those patents receiving any citations.
c. Comparison is at the three-digit level (see text).

patent is 6.5 to 8 years for the 1975 cohort, and a little over 4 years for the 1980 cohort.

The inference that a citation indicates a possible knowledge spillover is much less clear in the case where the citing patent is owned by the same organization as the originating patent. For this reason, we distinguish what we call "self-citations." A self-citation is defined as a citing patent assigned by its inventors to the same party as the originating patent, which is, by construction, either a university or a domestic corporation. Not surprisingly, the self-citation rate differs for the different sources of originating patents, with universities having the lowest and top corporations the highest rates.[17] Finally, Table I shows that 55 to 60 percent of citations have a primary patent class that is the same as the primary patent class of the originating patent, indicating that the originating and citing patents are *technologically* close to one another.

Figure I provides additional detail on the distribution of lags between originating and citing patents, again defined as the difference in application years. The figure shows that citations are few in the early years,[18] and reach a plateau after about three years.

17. The apparent increase in self-citation rates between 1975 and 1980 is probably spurious; self-citations tend to come earlier than other citations. See Trajtenberg, Henderson, and Jaffe [1992] for more on this issue.

18. Patents are typically granted one to three years after application; thus, a citation lag of zero or one implies that the citing patent may well have been applied for before the originating patent was actually granted. Pending applications are not public, so in this case the citation would almost surely have been identified by the examiner.

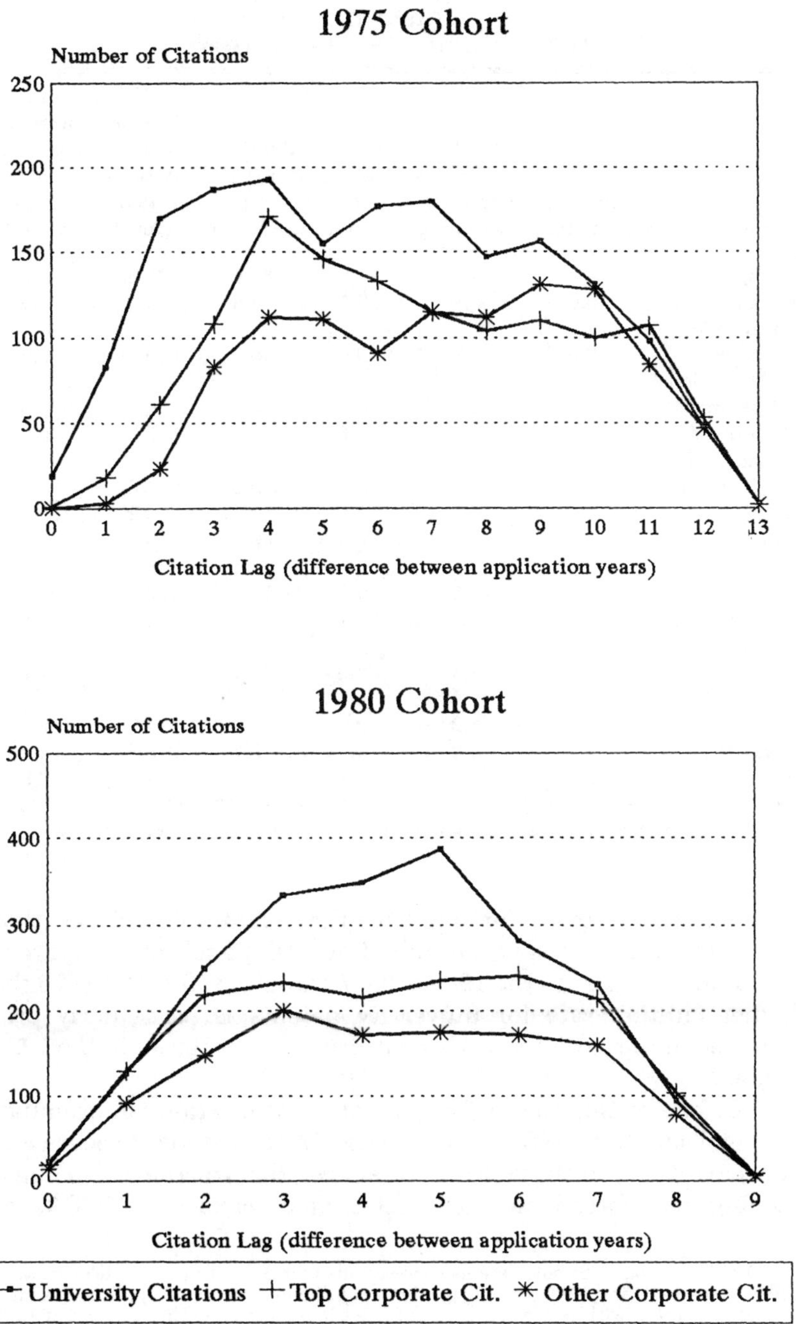
1975 Cohort
Number of Citations
250
200
150
100
50
0
0 1 2 3 4 5 6 7 8 9 10 11 12 13
Citation Lag (difference between application years)
1980 Cohort
Number of Citations
500
400
300
200
100
0
0 1 2 3 4 5 6 7 8 9
Citation Lag (difference between application years)
University Citations
Top Corporate Cit.
Other Corporate Cit.

FIGURE I

TABLE II
SMSA DISTRIBUTIONS FOR SOME DATASETS

Location	1975 University originating	1975 Top corporate originating	Citations to 1975 university	Citations to 1975 top corporate	Controls for citations to 1975 university	All citations to patents from in NY SMSA
Foreign	—	—	31.8	31.4	35.8	31.2
Boston	15.0	3.1	7.5	4.6	5.1	4.0
Los Angeles/ Anaheim	7.0	4.8	9.0	5.7	6.1	3.9
San Francisco/ Oakland	5.1	1.4	3.8	3.7	6.1	3.5
Madison, WI	4.2	—	1.6	—	0.5	0.6
Philadelphia/ Wilmington	4.2	9.3	5.4	8.2	4.5	9.1
Rural Iowa	3.8	—	1.6	0.6	0.2	—
San Jose	3.5	2.8	4.0	3.4	—	1.9
New York/ NJ/CT	3.2	13.5	9.7	11.7	13.7	28.5
Salt Lake City	3.2	—	2.1	—	0.5	0.4
Detroit/ Ann Arbor	2.6	2.4	2.6	1.7	1.7	1.2
Minneapolis/ St. Paul	1.3	5.2	2.8	2.9	1.9	2.1
Chicago	1.9	4.2	3.9	5.7	5.6	4.2
Albany	0.6	3.1	1.9	2.1	1.3	0.8

All figures are percentages. SMSA percentages for citations and controls are relative to *domestic* total.

It is not possible to tell for sure from these data when (if ever) that plateau tails off; the apparent tail-off in both panels of the figure is due at least in part to the 1989 observational cutoff.[19] For 1975 the higher citation rate for university patents is particularly pronounced in the early years; this pattern is not apparent in the 1980 cohort.

Before getting to our formal test of localization, an examination of Table II is useful to get a sense for the extent of geographic concentration in these data. It shows the fraction of patents coming from abroad and from a selection of major U. S. SMSAs for

19. The dropoff in both panels corresponds approximately to application year 1987 (1975 + 12 and 1980 + 7). Typically, a significant fraction of applications have not been granted within two years, so when we looked in 1989 this fraction of 1987 applications were not yet granted.

several of the datasets. Not surprisingly, a measurable fraction of university patents comes from Madison, WI; this is not true for corporate patents. A measurable (though smaller) fraction of the *citations* of university patents comes from Madison, and this fraction is larger than that for the controls. Indeed, the controls for the university citations look generally "more like" corporate patents than do the citations, suggesting that localization may be present. Other qualitative evidence of localization is apparent in the table, including the high percentage of NY SMSA citations that come from the NY SMSA.

The basic test of localization is presented in Table III. For each geographic area and each originating dataset, it presents the proportion of citations that geographically matched the originating patent. These proportions are shown both with and without self-citations. The matching proportions for the control samples are then shown, as well as a *t*-statistic testing the equality of the control proportions and the citation proportions (excluding self-citations).[20]

We focus first on the 1975 results on the left of the table. Starting with the country match, we find that citations *including self-citations* are domestic about 6 or 7 percent more often than the controls. Excluding self-citations eliminates this difference for the top corporate citations and cuts it roughly in half for the others. The remaining difference between the citations excluding self-cites and the controls is only marginally significant statistically.

Looking at the 1975 results for states, we find that citations of university patents come from the same state about 10 percent of the time; this rises to 15 percent for other corporate and 19 percent for top corporate. Excluding self-citations, however, makes a big difference. The university and top corporate proportions are cut to 6–7 percent, and the other corporate to just over 10. For the university and other corporate cohorts, the matching frequencies excluding self-citations are significantly greater than the matching control proportions.

20. Let p_c be the probability that a citation comes from the same geographic unit as the originating patent; let p_o be the corresponding probability for a randomly drawn patent in the same patent class (control). We test $H_o{:}p_c = p_o$ versus $H_a{:}p_c > p_o$ using the test statistic:

$$t = \frac{\hat{p}_c - \hat{p}_o}{\sqrt{[\hat{p}_c(1 - \hat{p}_c) + \hat{p}_o(1 - \hat{p}_o)]/n}},$$

where $\hat{p}_c$ and $\hat{p}_o$ are the sample proportion estimates of p_c and p_o. This statistic tests for the difference between two independently drawn binomial proportions; it is distributed as t.

TABLE III
GEOGRAPHIC MATCHING FRACTIONS

	1975 Originating cohort			1980 Originating cohort		
	University	Top corporate	Other corporate	University	Top corporate	Other corporate
Number of citations	1759	1235	1050	2046	1614	1210
		Matching by country				
Overall citation matching percentage	68.3	68.7	71.7	71.4	74.6	73.0
Citations excluding self-cites	66.5	62.9	69.5	69.3	68.9	70.4
Controls	62.8	63.1	66.3	58.5	60.0	59.6
t-statistic	2.28	−0.1	1.61	7.24	5.31	5.59
		Matching by state				
Overall citation matching percentage	10.4	18.9	15.4	16.3	27.3	18.4
Citations excluding self-cites	6.0	6.8	10.7	10.5	13.6	11.3
Controls	2.9	6.8	6.4	4.1	7.0	5.2
t-statistic	4.55	0.09	3.50	7.90	6.28	5.51
		Matching by SMSA				
Overall citation matching percentage	8.6	16.9	13.3	12.6	21.9	14.3
Citations excluding self-cites	4.3	4.5	8.7	6.9	8.8	7.0
Controls	1.0	1.3	1.2	1.1	3.6	2.3
t-statistic	6.43	4.80	8.24	9.57	6.28	5.52

Number of citations is less than in Table I because of missing geographic data for some patents. The *t*-statistic tests equality of the citation proportion excluding self-cites and the control proportion. See text for details.

At the SMSA level, 9 to 17 percent of total citations are localized. This again drops significantly when self-citations are excluded, but 4.3 percent of university citations, 4.5 percent of top corporate citations, and 8.7 percent of other corporate citations are localized excluding self-cites. This compares with control matching proportions of about 1 percent, and these differences are highly significant.

The results for citations of 1980 patents (right side of Table III) are even stronger and more significant. For every dataset, for every geographic level, the citations are quantitatively and statistically significantly more localized than the controls. The general increase in the proportion of U. S. patents taken by foreigners is reflected in a decline of 3 to 6 percent in the control percentages matching by country. The citation matching percentages actually rise, however, particularly for top corporate citations. It is impossible to tell from this comparison whether this represents a real change, or whether it is the result of the 1980 citations having shorter average citation lags. Since this gets to the issue of explaining which citations are localized, we postpone discussion until the next section.

Before moving on, the results on the extent of localization can be summarized as follows. For citations observed by 1989 of 1980 patents, there is a clear pattern of localization at the country, state, and SMSA levels. Citations are five to ten times as likely to come from the same SMSA as control patents; two to six times as likely excluding self-citations. They are three to four times as likely to come from the same state as the originating patent; roughly twice as likely excluding self-cites. Whereas about 60 percent of control patents are domestic, 70 to 75 percent of citations and 69 to 70 percent of citations excluding self-cites are domestic. Once self-cites are excluded, universities and firms have about the same domestic citation fraction; at the state and SMSA level there is weak evidence that university citations are less localized. For citations of 1975 patents, the same pattern, but weaker, emerges for citations of university and other corporate patents. For top corporate there is no evidence of localization at the state or country levels, though the SMSA fraction is significantly localized. Thus, we find significant evidence that citations are even more localized than one would expect based on the pre-existing concentration of technological activity, particularly in the early years after the originating patent.

III. Factors Affecting the Probability of Localization

The contrast between the 1975 and 1980 results suggests that localization of early citations is more likely than localization of later ones. This accords with intuition, since whatever advantages are created by geographic proximity for learning about the work of others should fade as the work is used and disseminated. Another hypothesis that is implicit in the previous discussion is that citations that represent research that is technologically similar to the originating research are more likely to be localized, because the individuals pursuing these related research lines may be localized. In addition, attributes of the originating invention or the institution that produced it may affect the probability that its spillovers are localized.

To explore these issues, we pooled the citations (excluding self-cites) to university and corporate patents for each cohort, and ran a probit estimation with geographic match/no match between the originating and citing patents as the dependent variable. As independent variables we included the log of the citation lag (set to zero for lags of zero), dummy variables for top corporate and other corporate originating patents, interactions of the lag and these dummies, and a dummy variable equal to unity if the citation has the same primary class as the originating patent. We also included a dummy variable that is unity if the *control* patent corresponding to this citation matches geographically with the originating patent, to control for the general increase over time in the fraction of U. S. patents granted to foreigners.

We also included two variables relating to the originating patent suggested by our work on basicness and appropriability of inventions [Trajtenberg, Henderson, and Jaffe, 1992]. The first, "generality" is one minus the Herfindahl index across patent classes of the citations received.[21] It attempts to capture the extent to which the technological "children" of an originating patent are diverse in terms of their own *technological* location. Thus, an originating patent with generality approaching one has citations that are very widely dispersed across patent classes; generality of zero corresponds to all citations in a single class. We argue elsewhere that generality is one aspect of the "basicness" of an invention. One might hypothesize that basic research results are less likely to be localized, because their spread is more likely to be through communication mechanisms (e.g., journals) that are not localized. The other variable characterizing the originating invention is the fraction of the originating patent's citations that were

21. Let C_{ik} be the number of citations received by patent i from subsequent patents whose primary patent class is k, and let C_i be the total number of citations received by patent i. The measure of generality is then

$$G_i = 1 - \sum_k \left(\frac{C_{ik}}{C_i}\right)^2.$$

TABLE IV
GEOGRAPHIC PROBIT RESULTS

	Country match		State match		SMSA match	
	1975	1980	1975	1980	1975	1980
Dummy for control	0.139	0.085	0.396	0.300	*	0.283
Sample match	(0.045)	(0.041)	(0.124)	(0.102)		(0.172)
Log of citation	−0.078	0.094	−0.264	0.198	−0.123	0.037
lag	(0.049)	(0.056)	(0.073)	(0.079)	(0.057)	(0.086)
Dummy for top	−0.114	−0.010	−0.383	0.013	−0.234	−0.208
corporate	(0.168)	(0.127)	(0.249)	(0.177)	(0.288)	(0.200)
Dummy for other	0.069	0.053	−0.214	−0.007	0.325	−0.042
corporate	(0.209)	(0.134)	(0.277)	(0.189)	(0.291)	(0.207)
Log-lag	0.046	−0.016	0.226	0.007	0.102	0.156
*top corp. dummy	(0.091)	(0.086)	(0.138)	(0.115)	(0.156)	(0.131)
Log-lag	0.008	−0.026	0.307	0.036	0.037	0.039
*other corp. dummy	(0.108)	(0.091)	(0.147)	(0.124)	(0.155)	(0.138)
Dummy for matching	−0.085	0.069	−0.013	0.034	−0.057	−0.016
patent class	(0.050)	(0.045)	(0.073)	(0.058)	(0.080)	(0.068)
Generality of origin	0.092	0.177	0.026	−0.140	0.013	−0.298
patent	(0.091)	(0.088)	(0.136)	(0.111)	(0.150)	(0.130)
Origin fraction	−0.813	0.162	0.815	0.883	1.174	0.828
Self-citations	(0.180)	(0.124)	(0.246)	(0.134)	(0.237)	(0.154)
# of observations	3581	4217	3573	4215	3566	3972
# of matches	2363	2925	256	490	197	298
Log likelihood	−2269	−2559	−894	−1459	−736	−1022

*The number of observations for which the control patent matched at the SMSA level was so small that this parameter could not be estimated.

Standard errors are in parentheses. All equations also included five technological field dummies.

self-cites. We take a high proportion of self-cites as evidence of relatively successful efforts by the original inventor to appropriate the invention. We expect that the nonself-citations to such a patent are more likely to be confined to suppliers, customers, or other firms that the inventing firm has a relationship with, and may therefore tend to be localized.

Finally, the extent of localization depends fundamentally on the mechanisms by which information flows, and these mechanisms may be different in different technical fields. For this reason, we also included dummy variables for broad technological fields.[22]

The results are presented in Table IV. Because of the presence

22. (1) Drugs and Medical Technology; (2) Chemicals and Chemical Processes Excluding Drugs; (3) Electronics, Optics, and Nuclear Technologies; (4) Mechanical Arts; and (5) All Other.

of the interaction terms between the lag and the corporate dummies, the coefficient on the lag itself corresponds to the fading of localization of citations of university patents. There is evidence in the 1975 results of such fading. This effect is statistically significant at the state and SMSA levels; its quantitative significance is discussed further below. For the citations of corporate patents, the interaction terms measure the difference between their fading rates and those of university citations. These terms are generally not statistically significant. In only one case (other corporate, 1975) could we reject the hypothesis of equality of fading rates at traditional confidence levels. There is, however, weak evidence that the corporate citations do not fade as rapidly as those of university patents, at least at the state and SMSA levels. The coefficients on the corporate dummies themselves capture differences in the predicted probability of localization for citations with lags of zero or one year. These are all insignificant, and there is no clear pattern.

The matching-patent-class and generality variables do not work well. The effects are generally insignificant, and show no consistent pattern. The effect of the self-citation fraction, however, is strong and puzzling. At the state and local level, there is a very significant effect in the predicted direction: citations of patents with a high self-citation fraction are more likely to be localized. This is *not* just saying that self-citations are localized, since they are excluded; it is the other citations that are more localized. At the country level, at least in 1975, this effect is reversed and is significant. Taking all results together, it suggests that for patents with a lot of self-citations, the nonself-citations are more likely to be foreign, but those that are domestic are more likely to be in the same state and SMSA as the originating patent.

The 1980 results are disappointing. The coefficient on the time lag term switches sign, though it is generally insignificant. One possibility is that these citations span too short a time period to capture the lag effect well. To test this possibility, we reran the estimation in Table IV on the 1975 citations, excluding all that were granted after 1984. This analysis tells us what we would have believed about citations of 1975 patents if we had looked for them only as long as we have looked for the citations of the 1980 patents. The results (not reported here) looked "more like" the 1980 results than the original 1975 results did. In particular, the coefficient on the lag term was insignificant, and was positive at the SMSA level.

TABLE V
PREDICTED LOCALIZATION PERCENTAGES OVER TIME (BASED ON 1975 PROBIT RESULTS FOR CITATIONS OF UNIVERSITY PATENTS)

Lag	Predicted percentage for: Same country	Same state	Same SMSA
0 or 1 year	67.1	9.7	4.8
5 years	65.5	6.5	4.0
10 years	64.6	5.3	3.7
25 years	63.5	4.0	3.3

Thus, it may be that the "perverse" results for the 1980 sample would go away if we had later citations to include.

A probit coefficient does not have an economically meaningful magnitude, because of the need to standardize the variance of the underlying error distribution. However, we can calculate what the coefficients imply about changes in the predicted probabilities. This is done in Table V, using the 1975 lag coefficient.[23] Table V was constructed by calculating the predicted localization probability using the results of Table IV, evaluating the citation lag at different values, and evaluating the other independent variables at the mean of the data. It shows that the estimates correspond to a reduction in the localization fraction after, for example, ten years, from 67.1 percent to 64.6 percent at the country level, 9.7 percent to 5.3 percent at the state level, and 4.8 percent to 3.7 percent at the SMSA level.

IV. DISCUSSION AND CONCLUSION

Despite the invisibility of knowledge spillovers, they do leave a paper trail in the form of citations. We find evidence that these trails, at least, are geographically localized. The results, particularly for the 1980 cohort, suggest that these effects are quite large and quite significant statistically. Because of our interest in true externalities, we have focused on citations excluding self-cites. For some purposes, however, this is probably overly conservative. From the point of view of the Regional Development Administra-

23. As discussed above, this is the point estimate of the lag coefficient for citations of university patents. The point estimates are different for the corporate originating patents, but we have not performed separate calculations for each dataset.

tor, it may not matter whether the subsequent development that flows from an invention is performed by the inventing firm, as long as it is performed in her state or city. Our results are also conservative because we attribute none of the localization present in the control samples to spillovers, despite the likelihood that spillovers are, indeed, one of the major reasons for the pre-existing concentration of research activity.

We also find evidence that geographic localization fades over time. The 1980 citations, which have shorter average citation lags, are systematically more localized than the 1975 citations. By using a probit analysis, we produced estimates of the rate of fading. These estimates seem to suggest a rate of fading that is both smaller than one would expect, and smaller than would be necessary to explain the difference between the 1975 and 1980 overall matching fractions. One possibility is that the difficulty of measuring the rate of fading is due to the "contamination" of citations by the patent examiner. As noted in footnote 18, it is particularly likely that citations with very short lags were added by the examiner. If we believe that such citations are less likely to represent spillovers and less likely to be localized, then this would tend to bias toward zero our measure of the effect of time on localization.[24]

We find less evidence of the effect of technological area on the localization process. Citations in the same class are no more likely to be localized. Overall, there is not really any evidence in these data that the probability of coming from a given geographic location conditional on patent class is different from the unconditional probability. This may be due to the arbitrary use of the "primary" patent class, to the exclusion of the "cross-referenced" classes. There is no legal difference in significance between the primary and cross-referenced classes, and in many cases the examiners do not place any significance on which class is designated primary. In future work we hope to explore whether using the full range of information contained in the cross-referenced classes provides a better technological characterization of the patents.

In this context it is worth noting that part of what is going on is probably that knowledge spillovers are not confined to closely related regions of technology space. Approximately 40 percent of

24. Attempts to test for this possibility by including a dummy variable for age 0 or 1, as well as other explorations of possible nonlinearity in the match-lag relationship, were inconclusive.

citations do not come from the same primary patent class; even at the level of the five broad "technological fields" listed in footnote 21, 12 to 25 of percent of citations are across fields. This is consistent with Jaffe [1986], which found that a significant fraction of the total "flow" of spillovers affecting firms' own research productivity comes from firms outside of the receiving firm's immediate technological neighborhood.

We find surprisingly little evidence of differences in localization between the citations of university and corporate patents. The largest difference is that corporate patents are more often self-cited, and self-cites are more often localized. The probit results do not allow rejection of the hypothesis that the initial localization rates for nonself-citations are indistinguishable for the different groups. They do provide some weak evidence that this initial localization is more likely to fade for the university patents, at least at the state and local levels.

In order to provide a true foundation for public policy and economic theorizing, we would ultimately like to be able to say more about the mechanisms of knowledge transfer, and about something resembling social rates of return at different levels of geographic aggregation. The limitations of patent and citation data make it difficult to go much farther with such questions within this research approach. Ex post, the vast majority of patents are seen to generate negligible private (and probably social) returns. In future work we plan to identify a small number of patents that are extremely highly cited. It is likely that such patents are both technologically and economically important [Trajtenberg, 1990]. Case studies of such patents and their citations could prove highly informative about both the mechanisms of knowledge transfer, and the extent to which citations do indeed correspond to externalities in an economic sense.

HARVARD UNIVERSITY
TEL-AVIV UNIVERSITY
MASSACHUSETTS INSTITUTE OF TECHNOLOGY

REFERENCES

Bernstein, J., and M. Nadiri, "Interindustry R&D Spillovers, Rates of Return, and Production in High-Tech Industries," *American Economic Review Papers and Proceedings*, LXXVIII (May 1988), 429–34.

Bernstein, J., and M. Nadiri, "Research and Development and Intra-industry Spillovers: An Empirical Application of Dynamic Duality," *Review of Economic Studies*, LVI (April 1989), 249–67.

Dorfman, N., "Route 128: The Development of a Regional High-Technology

Economy," D. Lampe, ed., *The Massachusetts Miracle: High Technology and Economic Revitalization* (Cambridge, MA: MIT Press, 1988).
Feller, Irwin, "R&D Theories and State Advanced Technology Programs," Paper prepared for the American Association for the Advancement of Science Annual Meeting, January 1989.
Glaeser, Edward, H. D. Kallal, J. A. Scheinkman, and A. Shleifer, "Growth in Cities," NBER Working Paper No. 3787, July 1991.
Griliches, Z., "The Search for R&D Spillovers," NBER Working Paper No. 3768, July 1991.
Grossman, G., and E. Helpman, *Innovation and Growth in the Global Economy* (Cambridge, MA: M.I.T. Press, 1991).
Jaffe, Adam, "Technological Opportunity and Spillovers of R&D: Evidence from Firms' Patents, Profits and Market Value," *American Economic Review,* LXXVI (1986), 984–1001.
——, "Real Effects of Academic Research," *American Economic Review,* LXXIX (1989), 957–70.
Krugman, P., *Geography and Trade* (Cambridge: M.I.T. Press, 1991).
Mansfield, E., "Sources and Characteristics of Academic Research Underlying Industrial Innovations," mimeo, University of Pennsylvania, 1991.
Marshall, A., *Principles of Economics* (London: Macmillan, 1920).
Minnesota Department of Trade and Economic Development, Office of Science and Technology, *State Technology Programs in the United States,* 1988.
Romer, P., "Increasing Returns and Long-Run Growth," *Journal of Political Economy,* XCIV (1986), 1002–37.
——, "Endogenous Technological Change," *Journal of Political Economy,* XCVIII (1990), S71–S102.
Smilor, R., G. Kozmetsky, and D. Gibson, *Creating the Technopolis: Linking Technology, Commercialization and Economic Development* (Cambridge, MA: Ballinger Publishing Co., 1988).
Trajtenberg, M., "A Penny for Your Quotes: Patent Citations and the Value of Innovations," *Rand Journal of Economics,* XXI (1990), 172–87.
Trajtenberg, M., R. Henderson, and A. Jaffe: "Ivory Tower Versus Corporate Lab: An Empirical Study of Basicness and Appropriability," NBER Working Paper No. 4146, 1992.

[17]

R&D SPILLOVERS AND RECIPIENT FIRM SIZE

Zoltan J. Acs, David B. Audretsch, and Maryann P. Feldman*

Abstract—The findings in this paper provide some insight into how small firms are able to innovate. Using a production function approach to relate knowledge generating inputs to innovative output, the empirical results suggest that small firms are the recipients of R&D spillovers from knowledge generated in the R&D centers of their larger counterparts and in universities. Such R&D spillovers are apparently more decisive in promoting the innovative activity of small firms than of large corporations.

I. Introduction

A recent series of studies for both the United States (Acs and Audretsch, 1987, 1988, and 1990) and Great Britain (Pavitt, Robson, and Townsend, 1987) found that while large corporations have the innovative advantage in certain industries, in other markets small firms are more innovative. This finding posed something of a paradox, because it is well known that the bulk of R&D is concentrated among the largest industrial corporations (Scherer, 1991). And, according to one of the most prevalent models of technological change (Griliches, 1979), innovative output is the product of knowledge generating inputs, most notably R&D. That small enterprises are the engine of innovative activity in certain industries despite an obvious lack of formal R&D activities,[1] raises the question, "Where do small firms get the innovation producing inputs?"

One possible answer is from other firms and institutions investing in R & D (Link and Rees, 1990; and Dorfman, 1983). An important finding of Jaffe (1986 and 1989) was that investment in R & D made by private corporations and universities "spills over" for economic exploitation by third party firms. However, while Jaffe (1986 and 1989) identified the extent to which corporate and university research spills over onto the generation of inventions by other firms, he

Received for publication August 5, 1991. Revision accepted for publication October 16, 1992.

* University of Baltimore, Wissenschaftszentrum Berlin für Sozialforschung, and Carnegie-Mellon University, respectively.

[1] Kleinknecht (1987 and 1989) has found that informal R & D activity plays a larger role in small firms than in large enterprises.

shed little light on which type of firm is the recipient of such spillovers. The purpose of this paper is to explicitly identify the degree to which university and corporate R & D spills over to small firms, and the extent to which university R & D spills over to large firms. By focusing on the recipients of the spillovers, rather than just upon the source, the manner by which large and small firms produce innovations can be more accurately identified.

We find substantial evidence that corporate R & D is a relatively more important source for generating innovations in large firms, while spillovers from university research laboratories are more important in producing innovative activity in small firms. In the following section, the "knowledge production function" model of innovation activity is applied for large and small firms. The data are introduced in the third section. In the fourth section, the empirical results from Tobit estimation are presented.

II. R & D Inputs and Innovative Output

According to the model of technological change introduced by Griliches (1979), innovative output is the product of knowledge generating inputs. Jaffe modified this production function approach for adoption to a model specified for spatial and product dimensions:

$$I_{si} = CRD_{si}^{\beta 1} * UR_{si}^{\beta 2} * (UR_{si} * GC_{si})^{\beta 3} * e_{si} \qquad (1)$$

where *I* is innovative output, *CRD* is private corporate expenditures on R & D, *UR* the research expenditures undertaken at universities, and *e* represents a stochastic error term. The unit of observation is at the spatial level, *s*, and product level, *i*.[2] To overcome the conceptual problem of using state level data to measure geographic spillovers, Jaffe constructs an index of the geographic coincidence of university and industrial research labs, *GC*.[3] Jaffe argued that the proximity of university research to corporate laboratories should raise the potency of spillovers from the universities.

Using the number of patented inventions to proxy innovative activity, Jaffe (1989) estimated equation (1) and found that university research does, in fact, spill over to promote corporate patent activity. That is, not only does patent activity increase in the presence of high private corporate expenditures on R & D, but also from research expenditures undertaken by universities within the state.

Jaffe's results made it clear that spillovers play an important role as an input in generating innovations in third party firms,[4] but they shed little light on the recipients of the spillovers. By estimating equation (1) for the innovative activity contributed by large firms separately from the innovative activity contributed by small firms, and directly comparing the magnitude of the coefficients of *CRD*, *UR*, and *UR* * *GC*, the relative importance of R & D spillovers from universities and corporate laboratories for large and small firms can be explicitly ascertained.

III. Measurement

Although Jaffe (1989) used patent counts to proxy innovative activity, we use a direct measure—the number of innovations in 1982. This measure comes from the U.S. Small Business Administration's Innovation Data Base and is explained and analyzed in considerable detail in Acs and Audretsch (1987, 1988, and 1990). Griliches (1990), Scherer (1983), and Mansfield (1984) have all observed that patent counts measure an intermediate output in the entire process of producing an innovation. According to Griliches (1990, p. 1669), "Ideally, we might hope that patent statistics would provide a measure of the (innovative) output.... The reality, however, is very far from it. The dream of getting hold of an output indicator of inventive activity is one of the strong motivating forces for economic research in this area."

In contrast to patent data, which mark the certification of an invention, innovation citations announce the market introduction of a commercially viable product.[5] Thus, the resulting data base provides a more direct measure of innovative activity than does the number of patented inventions. Not only are inventions which were not patented but ultimately introduced into the market included in the data base, but those inventions which were patented but never developed into innovations are excluded.

[2] Jaffe (1989) used "technological areas" for the product dimension.

[3] The geographic coincidence index is calculated as

$$GC_S = \frac{\sum_c UNIV_{ci} TP_{ci}}{\left[\sum_c UNIV_{ci}^2\right]^{1/2} \left[\sum_c TP_{ci}^2\right]^{1/2}};$$

where TP_{ci} is the total number of industrial R & D lab workers in a city or SMSA. The geographic coincidence index is calculated as the uncentered correlation of the vectors U_i and TP_i across SMSAs within a state. Jaffe's (1989) hypothesis is that research will yield more innovative activity if university and industrial labs are geographically concentrated. For example, more patents would be expected in Illinois, where industrial and research labs are concentrated around Chicago, than in Indiana where university labs are located in different SMSAs from industrial labs.

[4] Acs, Audretsch, and Feldman (1992) confirm that Jaffe's results for R & D spillovers onto patenting activity hold using the direct measure of innovative activity used in this paper.

[5] Another distinction is that an innovating firm did not necessarily make the invention.

TABLE 1.—INNOVATIVE OUTPUT IN LARGE AND SMALL FIRMS AND R & D INPUTS BY STATE[a]

State	Total Innovations	Large Firm Innovations	Small Firm Innovations	Industry R & D Expenditures	University Research
CALF	974	315	659	3883	710.4
NY	456	180	276	1859	371.0
NJ	426	162	264	1361	70.8
MASS	360	148	212	954	245.3
PA	245	104	141	1293	139.2
ILL	231	100	131	894	254.9
OHIO	188	76	112	926	76.2
CONN	132	77	55	650	54.7
MICH	112	61	51	1815	103.2
MINN	110	64	46	399	55.7
WISC	86	33	53	224	65.0
FLA	66	21	45	375	70.1
GA	53	20	33	78	57.8
IND	49	20	29	398	51.3
COLO	42	13	29	167	77.2
ARIZ	41	23	18	201	37.4
VA	38	19	19	207	45.9
NC	38	16	22	193	64.6
RI	24	4	20	32	14.9
OKLA	20	12	8	93	19.9
IOWA	20	12	8	135	46.4
KANS	15	3	12	66	26.6
UTAH	11	2	9	72	32.5
NEB	9	1	8	9	20.4
KY	9	6	3	72	17.5
LA	5	0	5	65	33.4
ARK	5	5	0	9	12.0
ALA	5	0	5	54	28.3
MISS	4	1	3	420	61.4

[a] Industry R & D and University Research expenditures are in millions of 1972 dollars and are taken from Jaffe (1989). Data are available for 29 states.

The innovation data are classified according to the size of the innovating firm. Large firms are defined as having at least 500 employees, and small firms are defined as having fewer than 500 employees. Innovations are attributed to the state in which the establishment responsible for the commercial development of the innovation is located. The data base distinguishes between the location of the innovating establishment and the location of the parent firm. In the case of a multi-establishment firm, the innovation would be classified by size on the basis of the employment of the entire parent firm, but would be located at the innovating establishment and not at the headquarters of the parent firm.

Table 1 ranks the states according to innovative activity and shows the distribution of innovations across large and small firms, along with the corresponding industry R & D expenditures and expenditures on research by universities. The two innovative inputs included in table 1 are taken from Jaffe (1989). The industry and university expenditures are based upon the average value measured (in constant 1972 million dollars) over the eight year period, 1972–1977, 1979, and 1981. We follow Jaffe (1989) by including the population of each state to control for size differentials across the geographic unit of observation. We also include the geographic coincidence index to control for within state variation. The geographic coincidence index is explained and documented by Jaffe (1989, p. 959) and our footnote 3. While Jaffe rescaled the index "for estimation purposes" so that its log had a mean of zero, we see no advantage in doing so and leave the geographic coincidence index unscaled.

IV. Empirical Results

The dependent variable is the number of innovations (alternatively by all firms, by large firms, and by small firms) in a specific technological area and within a particular state. Since data are available for five technological areas over 29 states, the entire sample size is 145 observations.[6] There are no innovations registered for certain technological areas within particular states. Thus, equation (1) is estimated using the Tobit model. The results for the innovative output for all firms and

[6] For a detailed description of these technological areas as well as the manner by which they were selected, see Jaffe (1989, appendix A).

TABLE 2.—TOBIT REGRESSIONS OF INNOVATIVE ACTIVITY BY STATE AND TECHNOLOGICAL AREA
(t-statistics listed in parentheses)

	(1) All firms	(2) Large firms	(3) Small Firms
Log Industry R & D	0.615	0.950	0.550
	(5.457)[b]	(7.133)[b]	(4.184)[b]
Log University Research	0.550	0.446	0.661
	(5.635)[b]	(4.569)[b]	(5.848)[b]
Log Geographic Coincidence	0.089	0.033	0.111
× University Research	(1.802)[a]	(0.687)	(1.965)[b]
Log Population	−0.246	−0.554	−0.314
	(−3.653)[b]	(−6.558)[b]	(−3.914)[b]
Log-Likelihood	−233.21	−202.19	−225.86
Sample Size	145	145	145

[a] Statistically significant at the 90% level of confidence for a two-tailed test.
[b] Statistically significant at the 95% level of confidence for a two-tailed test.

separately for large and small firms are shown in table 2.[7]

The first equation estimates the innovative activity of all firms, the second equation estimates the innovative activity of large firms, and the third equation estimates the innovative activity of small firms. Regardless of firm size, the knowledge production function for innovative output holds. Additional inputs in knowledge generating R & D, both by private corporations and by universities, lead to increases in innovative output.

However, the relative importance of industry R & D and university research as inputs in generating innovative output clearly varies between large and small firms. That is, for large firms, not only is the elasticity of innovative activity with respect to industry R & D expenditures more than two times greater than the elasticity with respect to expenditures on research by universities, but it is nearly twice as large as the elasticity of small-firm innovative activity with respect to industry R & D. By contrast, for small firms the elasticity of innovative output with respect to expenditures on research by universities is about one-fifth greater than the elasticity with respect to industry R & D. And the elasticity of innovative activity with respect to university research is about fifty percent greater for small enterprises than for large corporations.[8]

These results support the hypothesis that private corporation R & D plays a relatively more important role in generating innovative activity in large corporations than in small firms.[9] By contrast, spillovers from the research activities of universities play a more decisive role in the innovative activity of small firms.[10]

Geographic proximity between university and corporate laboratories within a state clearly serves as a catalyst to innovative activity for firms of all sizes. However, the impact is apparently greater on small firms than on large firms. The elasticity of innovative activity with respect to the geographic coincidence index multiplied by university research is nearly four times greater for small firms than for their larger counterparts.

A particular concern with the estimates in table 2 is multicollinearity between the log of industrial R&D expenditures and the log of university research expenditures. The simple correlation coefficient between the two variables is 0.61. When the log of industry R&D is omitted from the regressions, the coefficient of the log of university research rises somewhat to 0.592 (t-ratio of 8.733) in estimating large-firm innovative activity and to 0.828 (t-ratio of 8.036) in estimating small-firm innovative activity. Similarly, when the log of university research is omitted from the regressions, the coefficient of the log of industry R&D rises to 1.289 (t-ratio of 9.541) in estimating large-firm innovative activity and to 0.790 (t-ratio of 10.465) in estimating small-firm innovative activity. Thus, even when only one source of knowledge generating R & D is included in the regression equation, the results hold. The innovative activity of small enterprises responds more to university research, while the innovative activity of large corporations responds more to industry R & D.

[7] The logarithmic transformation of the left-hand side of equation (1) uses a linear transformation of $((I_{si} + 1) * 10)$. The Tobit censoring boundary is set to correspond to a zero observation.

[8] There may be greater measurement error for the small firm equation because large firms generate most of the state R & D expenditures. The size of any resulting bias is difficult to predict.

[9] There is a statistically discernible difference in the coefficients of industry R & D for small and large firms at the 95% level of confidence using a two-tailed test.

[10] There is a statistically discernible difference in the coefficients of university research expenditures for small and large firms at the 95% level of confidence using a two-tailed test.

V. Conclusions

The findings in this paper provide some insight into the puzzle posed by the recent wave of studies identifying vigorous innovative activity emanating from small firms in certain industries. How are these small, and frequently new, firms able to generate innovative output while undertaking generally negligible amounts of investment into knowledge generating inputs, such as R & D? At least one answer, implied by the findings in this paper, is that small firms innovate through exploiting knowledge created by expenditures on research in universities and on R & D in large corporations.

The empirical results suggest that the innovative output of all firms rises along with an increase in the amount of R & D inputs, both in private corporations as well as in university laboratories. However, R & D expenditures made by private companies play a particularly important role in providing inputs to the innovative activity of large firms, while expenditures on research made by universities serve as an especially key input for generating innovative activity in small enterprises. Apparently large firms are more adept at exploiting knowledge created in their own laboratories, while their smaller counterparts have a comparative advantage at exploiting spillovers from university laboratories.

REFERENCES

Acs, Zoltan J., and David B. Audretsch, *Innovation and Small Firms* (Cambridge, MA: MIT Press, 1990).

______, "Innovation in Large and Small Firms: An Empirical Analysis," *American Economic Review* 78 (Sept. 1988), 678–690.

— , "Innovation Market Structure and Firm Size," this REVIEW 69 (Nov. 1987), 567–575.

Acs, Zoltan J., David B. Audretsch, and Maryann P. Feldman, "Real Effects of Academic Research: Comment," *American Economic Review* 82 (Mar. 1992).

Dorfman, Nancy S., "Route 128: The Development of a Regional High Technology Economy," *Research Policy* 12 (Dec. 1983), 299–316.

Grilliches, Zvi, "Patent Statistics as Economic Indicators: A Survey," *Journal of Economic Literature* 28 (Dec. 1990), 1661–1707.

— , "Issues in Assessing the Contribution of R & D to Productivity Growth," *Bell Journal of Economics* 10 (Spring 1979), 92–116.

Jaffe, Adam B., "Real Effects of Academic Research," *American Economic Review* 79 (Dec. 1989), 957–970.

______, "Technological Opportunity and Spillovers of R & D: Evidence from Firms' Patents, Profits and Market Value," *American Economic Review* 76 (Dec. 1986), 984–1001.

Kleinknecht, Alfred, "Firm Size and Innovation: Observations in Dutch Manufacturing Industry," *Small Business Economics* 1 (4) (1989), 215–222.

______, "Measuring R & D in Small Firms: How Much Are We Missing?" *Journal of Industrial Economics* 36 (Dec. 1987), 253–256.

Link, Albert N., and John Rees, "Firm Size, University Based Research, and the Returns to R & D," *Small Business Economics* 2 (1), (1990), 25–32.

Mansfield, Edwin, "Comment on Using Linked Patent and R & D Data to Measure Interindustry Technology Flows," in Zvi Grilliches (ed.), *R & D, Patents, and Productivity* (Chicago, IL: University of Chicago Press, 1984), 462–464.

Pavitt, Keith, M. Robson, and J. Townsend, "The Size Distribution of Innovating Firms in the U.K.: 1945–1983," *The Journal of Industrial Economics* 55 (Mar. 1987), 291–316.

Scherer, F. M., "Changing Perspectives on the Firm Size Problem," in Z. J. Acs and D. B. Audretsch (eds.), *Innovation and Technological Change: An International Comparison* (Ann Arbor: University of Michigan Press, 1991), 24–38.

______, "The Propensity to Patent," *International Journal of Industrial Organization* 1 (Mar. 1983), 107–128.

[18]

JOURNAL OF URBAN ECONOMICS **42**, 422–448 (1997)
ARTICLE NO. UE972032

Local Geographic Spillovers between University Research and High Technology Innovations

Luc Anselin

Regional Research Institute and Department of Economics, West Virginia University, P.O. Box 6825, Morgantown, West Virginia 26506-6825

Attila Varga

Regional Research Institute, West Virginia University, P.O. Box 6825, Morgantown, West Virginia 26506-6825, and Department of Economics, Janus Pannonius University, Pecs, Hungary

and

Zoltan Acs

Office of Advocacy, U.S. Small Business Administration, Washington, DC 20416, and Merrick School of Business, University of Baltimore, Baltimore, Maryland 21201-5779

Received May 20, 1996; revised October 1996

This paper re-examines the empirical evidence on the degree of spatial spillover between university research and high technology innovations. The familiar Griliches–Jaffe knowledge production function is estimated at both the state and the metropolitan statistical areas (MSA) level and extended with more precise measures of spatial spillover. Alternatives based on the gravity potential and covering indices are formulated for Jaffe's "geographical coincidence index" and found to provide strong evidence of local spillovers at the state level. At the MSA level, a distinction is made between research and development activities and university research in the MSA and in the surrounding counties. Evidence is found of local spatial externalities between university research and high technology innovative activity, both directly and indirectly via private research and development. © 1997 Academic Press

INTRODUCTION

The accumulation of knowledge and its spillover into productive capacity through technological change is a central theme in the new theory of endogenous economic growth (e.g., Romer [66–68], Grossman and Helpman [38, 39]). An interesting aspect of this perspective has been the

0094-1190/97 $25.00

renewed attention to the geographic scope of the spillovers between knowledge creation and production, or, the extent of Marshallian spatial externalities (e.g., as exemplified in the new economic geography of David and Rosenbloom [21], Krugman [46], Glaeser *et al.* [31], and others). In urban and regional economics, a large number of empirical studies have focused on the creation of knowledge, on research and development activities (R & D), and on technological innovation as determinants of various aspects of local and regional economies. Examples are analyses of the location of R & D facilities and high technology production, and the effect of universities and technology parks on characteristics of the labor market and regional growth (e.g., Malecki [51–53], Markusen, Hall, and Glasmeier [57], Andersson, Anderstig, and Harsman [6], Goldstein and Luger [33], Luger and Goldstein [49], Glasmeier [32], Florax [28]).

An important aspect of studies of technological innovation at the regional scale is the role of spatial interaction and spatial structure, as expressed in the form of organizational networks of innovators, regional innovation complexes and regional knowledge infrastructure (e.g., Stohr [71], Von Hippel [74], Storper and Walker [72], DeBresson and Amesse [23], Feldman [26], Saxenian [69]). Universities play a central role in this process, not only as producers of basic research, but also by creating human capital in the form of higher skilled labor. Both of these aspects have received considerable attention in the literature, from a theoretical as well as from an empirical perspective. The importance of basic (university) research in the stimulation of technological innovation (and higher productivity) is derived from the public good nature of the research, and the resulting positive externalities to the private sector in the form of knowledge spillovers. The initial conceptualization of this process was provided by Arrow [12] and Nelson [59], and further refined by Griliches [35], Nelson [60], Von Hippel [74], Cohen and Levinthal [20], among others (for recent reviews, see, e.g., Dosi [24], Acs and Audretsch [1], Griliches [36, 37], Mansfield [55], Florax [28], Feldman [26]). The role of universities in the development of human capital that is more likely to create technological innovations is highlighted in the work of Bartel and Lichtenberg [16] and Lucas [48], among others. Both the knowledge spillovers and the human capital development constitute important locational attractors for private sector R & D and for high technology production (Malecki [53]).

While conceptually clear, the role of universities as "engines of growth" or "regional boosters" (Florax [28]), is not uniformly confirmed by empirical studies. A significant and positive effect of the presence of universities is found on the location of high technology production, new start ups and R & D facilities in, among others, Malecki [52], Nelson [61], Lund [50], Rees and Stafford [65], Harding [40], Bania, Calkins and Dalenberg [14],

Parker and Zilberman [62], and Sivitanidou and Sivitanides [70], but others have either qualified these findings or did not discover such a relationship. For example, Link and Rees [47], Rees [64], and Malecki and Bradbury [54] point out that the locational attraction may show little relationship to the research activities of universities, but instead may be determined by their teaching and training roles. Others, such as Howells [42], Markusen, Hall, and Glasmeier [57], Glasmeier [32], Florax and Folmer [29], Bania, Eberts, and Fogarty [15], and Beeson and Montgomery [17], find a weak, inconsistent (across sectors), or no relationship at all between the university research activities and high technology (or R & D) location, or other regional variables such as manufacturing investment, local labor market conditions, and high technology start ups. A number of explanations have been offered for this lack of uniform evidence, such as the difficulty of quantifying the inputs and outputs in a "knowledge production function" (Griliches [36]) and various sectoral and spatial aggregation problems (e.g., as argued in Markusen, Hall, and Glasmeier [57, p. 156]).

The goal of our paper is to shed further light on this issue by focusing on the spatial extent of the "local" geographic effects that university research (at both private and public universities) may have on the innovative capacity in a region, both directly and indirectly through its interaction with private sector R & D efforts. Our conceptual framework is based on the knowledge production function of Griliches [35] and is implemented at an aggregate level, as in other recent studies of this phenomenon by Jaffe [44], Acs, Audretsch, and Feldman [2–4], Audretsch and Feldman [13], and Feldman and Florida [27] (see Feldman [26] for a recent review). We extend the empirical evidence in two important respects:

(1) We broaden the cross-sectional basis for empirical analysis by utilizing data for 43 states (compared to 29 in the above mentioned studies) and for 125 metropolitan statistical areas (MSA). This is the first time MSA-level data are used in this type of analysis, which avoids many problems associated with the inappropriate spatial scale of a state as the areal unit of analysis. MSA-level results are obtained by using R & D laboratory employment as a proxy for R & D activity, based on a specially compiled data set.

(2) We focus on more precise measures of local geographic spillovers. At the state level, we introduce several alternatives to Jaffe's [44] "geographic coincidence index" that are more tightly integrated with the existing body of spatial interaction theory, and are able to significantly improve on his results. At the MSA scale, we formalize the spatial extent of the geographic spillovers by means of so-called spatial lag variables that capture the research activities in concentric rings around the MSA as well as in the MSA itself.

In the remainder of the paper, we first introduce the formal model of the knowledge production function and briefly review current empirical evidence on geographic knowledge spillovers of universities. We next elaborate on the data set used in the study. Subsequently, we present the results of our analysis at the state level and at the MSA level, focusing in particular on our alternative definitions of spatial interaction. We conclude with a summary and evaluation of our findings.

THE KNOWLEDGE PRODUCTION FUNCTION

Model

The conceptual framework for analyzing geographic spillovers of university research on regional innovative capacity is based on the knowledge production function of Griliches [35] (see also Jaffe [43, 44]). In essence, this is a two-factor Cobb–Douglas production function that relates an output measure for "knowledge" to two input measures: research and development performed by industry, and research performed by universities. Formally, this is expressed as

$$\log(K) = \beta_{K1} \log(R) + \beta_{K2} \log(U) + \varepsilon_K, \tag{1}$$

where K is a proxy for knowledge output (either patents or innovation counts), R is industry R & D, and U is university research, with ε_K as a stochastic error term. The analysis is typically carried out for aggregate cross-sectional units (e.g., states), possibly for several points in time and/or disaggregated by sector.

Following Jaffe [44], the potential interaction between university and industry research is captured by extending the model with two additional equations that allow for simultaneity between these two variables

$$\log(R) = \beta_{R1} \log(U) + \beta_{R2} Z_1 + \varepsilon_R \tag{2}$$

and

$$\log(U) = \beta_{U1}\, log(R) + \beta_{U2} Z_2 + \varepsilon_U \tag{3}$$

where U and R are as before, Z_1 and Z_2 are sets of exogenous local characteristics, and ε_U and ε_R are stochastic error terms.

Jaffe [44] introduced an additional term in Eq. (1) to compensate for the inappropriateness of using states as the units of observation. This acknowledges that states are too large to accurately capture the local spatial interaction between universities and firms. The additional term consists of the product of the logarithms of a "geographic coincidence index" (C) and university research. This term therefore adjusts the magnitude of university research in the state by a measure of the extent to which private and university research coincide in space (we return to a more detailed specification of this term below). The extended "geographic" knowledge production function then becomes

$$\log(K) = \beta_{K1} \log(R) + \beta_{K2} \log(U) + \beta_{K3}[\log(U) \cdot \log(C)] + \varepsilon_K \quad (4)$$

in the same notation as before.

The framework expressed in Eqs. (2)–(4) has become the basis of several empirical investigations, which we will briefly review in the next section. Before proceeding however, note that the geographic interaction term in Eq. (4) is an artifact of using states as the unit of analysis and is not needed when the spatial units of observation correspond more closely to the spatial scale of interaction between firms and universities. For example, this may be the case when the analysis is carried out at the MSA or county scale.

Previous Empirical Evidence

Almost all empirical investigations of geographic knowledge spillovers of universities in the U.S. have been aggregate in nature and based on the Griliches–Jaffe knowledge production function framework applied at the state level. A few notable exceptions are studies that use micro-data derived from surveys or patent information, e.g., Mansfield [56], Jaffe, Trajtenberg, and Henderson [45], and Almeida and Kogut [5]. These studies all provide strong evidence of the importance of spatial interaction at the *local* level. However, this does not find uniform confirmation in the aggregate studies. As noted above, this may be somewhat due to the fact that the unit of analysis—the state—only partially captures this interaction. We argue in this paper that it is also due to the formal specification of local spatial interaction in the form of a geographic coincidence index.

The aggregate empirical analyses in the literature are all based on data for 29 U.S. states. Knowledge output is proxied by patent counts in Jaffe [44] (also replicated in Acs, Audretsch, and Feldman [2]), while Acs, Audretsch and Feldman [2, 3] and Feldman and Florida [27] use a more direct measure of innovative activity, based on a 1982 data set of innovation counts compiled by the U.S. Small Business Administration (see below for further details).

The empirical studies in the literature vary somewhat in terms of research design, but they all find a strong and positive relationship between innovative activity and both industry R & D and university research at the state level. However, the situation is different in terms of the significance of a *local* geographic spillover effect. Overall, the evidence is non-existent, weak or mixed, only pertaining to a few individual sectors (e.g., in Jaffe [44] for only two out of four sectors considered).

We suggest that, in part, this lack in the evidence for local geographic spillovers (which contradicts the strong findings in micro-level studies) is due to the specification of the geographic effects in the model. All studies use the idea initially suggested in Jaffe [44, p. 959], where the product of the logarithm of state-level university expenditures with the logarithm of a geographic coincidence index is used. The latter is obtained as the uncentered correlation coefficient (i.e., without subtracting the mean) between university research and professional employees in R & D labs for the SMSAs in the state (i.e., R & D and university research outside SMSA are not included in the computation of the index). The resulting index is rescaled such that its mean is zero. While the coincidence index may make intuitive sense, it is unrelated to the existing large body of research on spatial interaction and measures of spatial accessibility (for a recent review, see, e.g., Frost and Spence [30], Talen and Anselin [73]). We will therefore consider the extent to which alternative indices provide a clarification of this issue.

DATA AND VARIABLE DEFINITIONS

We extend previous empirical evidence by using a detailed data set on innovation counts and R & D employment at both the state and MSA scale of geographic aggregation. Following Acs, Audretsch, and Feldman [2–4] and Feldman and Florida [27], we use a count of innovations as the dependent variable in the geographic knowledge production function [K in Eq. (4)], constructed from the U.S. Small Business Administration Innovation Database. This data set is a compilation of innovations that were introduced to the U.S. market in the year 1982, based on an extensive review of new product announcements in trade and technical publications. In the original data set, 4200 innovations are identified by the address of the innovating establishment, which allows for aggregation at different spatial scales, such as the state, MSA or county level (for further details on the data set and a discussion of their limitations, see Edwards and Gordon [25], Acs and Audretsch [1, Chapter 2], and Feldman [26]). In our study, we use aggregations at the state, county, and MSA level.

At the sectoral scale, the innovation data were aggregated to the two-digit SIC code level. Our interest focused on innovations in the "high technology" sector as an aggregate. Clearly, this is not an unambiguous determination and a number of different classifications have been suggested in the literature (e.g., Premus [63], Herzog, Schlottmann, and Johnson [41], Malecki [52], Glasmeier [32]). In general, the objective is to identify sectors dominated by the importance of non-routine functions, in contrast to standardized mass production. A number of criteria to identify such sectors have been suggested in the literature, such as the percentage of scientists and engineers employed, and the number of innovations per employee. At the two-digit SIC level, it is virtually impossible to designate sectors as "pure" high technology, although the SIC codes 28 and SIC 35–38 contain most of the commonly identified high technology sectors at the more appropriate 3 and 4 digit level (see, e.g., Herzog, Schlottman, and Johnson [41]). We consequently proceeded to use the aggregate of these five SIC codes as "high technology." To the extent that the sectoral mix in these sectors shows systematic variation over space in terms of its "pure" high tech content, our results on the relationship between innovation and research could be affected. However, we are confident that we will be able to detect such systematic variations by means of careful specification tests for spatial effects (Anselin [7], Anselin and Bera [10]).

Earlier studies of the aggregate knowledge production function were limited to data for 29 U.S. states. This was not due to the lack of data on innovations or patents (for which actual addresses are available), but to data limitations for the explanatory variables in the model, in particular for the variable on private R & D expenditures [R in Eqs. (2) and (4)]. In Jaffe [44], and also in Acs, Audretsch, and Feldman [2–4] and Feldman and Florida [27], this was computed from information on total industry R & D by state, compiled by the U.S. Bureau of the Census on the basis of survey data from the National Science Foundation (NSF). However, this data was only consistently reported for 29 states (for details, see, e.g., Jaffe [44, pp. 968–969]). Instead, we constructed a proxy for industrial R & D activity on the basis of data on professional employment in high technology research laboratories in the Bowker directories (Jaques Cattell Press [19]). While imperfect, this approach allowed us to construct a private R & D variable for 43 U.S. states and for 125 MSAs (see also Bania, Calkins, and Dalenberg [14, pp. 218–219], for a similar approach). As it turns out, our proxy variable is remarkably similar to the R & D expenditure variable used in Jaffe [44], yielding a correlation of 0.91 for the 29 states common to both studies (the data on innovations, private R & D employment, and university research expenditures for 43 U.S. states in 1982 are listed in Appendix A). Clearly, the use of lab employment as a proxy for expenditures assumes a constancy of the labor intensity and capital/labor ratio of R & D across the units of observation. Again, to the extent that this is not the case, it will tend to yield misspecifications in the form of heteroskedastic and/or spatially autocorrelated error terms, which will merit special attention in our analysis.

Our data for university research expenditures [U in Eqs. (3) and (4)] follow the common approach in the literature and are compiled from the NSF Survey of Scientific and Engineering Expenditures at Universities and Colleges for the year 1982 (National Science Foundation [58]). In addition, this data set also provides the source for exogenous variables used in the estimation of Eq. (3) at the MSA level: a proxy for the overall academic quality of high technology departments at universities, *RANK*; and a proxy for size, the total enrollment at universities, *ENRL*. We also consider data for total educational expenditures, *EDEXP*, from the County and City Data Book. We match the sectoral aggregation of the two-digit SIC industries to university departments using the same approach as in Feldman [26, p. 58].

In addition, we also included a number of variables compiled from County Business Pattern data for 1982: high technology employment, *HTEMP*; a location quotient for high technology employment, *LQ*; employment in business services (SIC 73), *BUS*; and the percent large firms (i.e., firms with employment exceeding 500), *LARGE*. Following general practice in the literature, the first three variables are included to capture agglomeration economies (see also Feldman and Florida [27]), the last one to assess the effect of firm scale (see also Acs, Audretsch, and Feldman [3]). A proxy for the coincidence of R & D and headquarter locations is a dummy variable for the presence of at least 10 headquarters of Fortune 500 companies in an MSA, *FORTUNE*, compiled from the May 2, 1982 listing in *Fortune Magazine*.

Our final data set only included those MSAs for which there were innovations in the high technology sector as well as both private industry R & D and university research expenditures (see Appendix B for a listing of the MSAs, innovation counts and research data). This results in 125 MSAs being considered that meet the three criteria, an almost complete match of university MSAs (125 out of 130), and a better than 70% portion of both the innovative MSAs (169) and the R & D MSAs (159). This selection leaves out a number of important non-MSA university locations, such as Ithaca, NY, and State College, PA, but it covers more than 80% of the relevant interaction (92% of all university research is in the 130 university MSAs). This limitation also excludes from consideration the joint determination of location and magnitude of high technology innovation and research. On the other hand, it avoids the problem of "zeros," and is motivated by a focus on the strength of interaction between the two forms of research and the generation of innovations where these are present. In other words, we address the strength and spatial extent of local geographic spillovers for metropolitan areas where an innovative complex is already in place. We leave the explanation of the more complex issue of the establishment and absence of these links for future research.

LOCAL GEOGRAPHIC SPILLOVERS AT THE STATE LEVEL

Alternative State-Level Geographic Coincidence Indicators

In addition to the familiar Jaffe indicator of geographic coincidence, we considered four alternative measures of local spatial interaction between university and industry research for states. One is constructed similar to Jaffe's approach, in that the log of a coincidence index is multiplied with the log of university research expenditures at the state level. However, rather than the uncentered correlation coefficient between university research and total professional employees in R & D laboratories for the MSAs in the state, we used the proportion of counties where industry research (proxied by R & D lab employment) and university research are co-located (out of all counties with university research). Formally, for state s,

$$CUR_s = \Sigma_i \delta R_i \cdot \delta U_i / \Sigma_i \delta U_i, \tag{5}$$

where the summation is over counties, and δR_i and δU_i are Kronecker deltas that take on a value of 1 when respectively $R_i > 0$ and $U_i > 0$, and a value of zero otherwise. In contrast to the Jaffe index, CUR_s also includes information on counties that do not belong to an MSA in the state.

The other three measures are more closely related to traditional accessibility indices from spatial interaction theory (see, e.g., Weibull [75]). The first is a gravity or potential measure with distance decay parameter 2 (see also Sitivanidou and Sivitanides [70] for a slightly different approach) for each industry R & D county in a state with respect to all university counties:

$$GRAV_i = \Sigma_j U_j / (d_{ij})^2, \tag{6}$$

where i is an index for each industry R & D county in the state, j an index for each university county, U is as before, and d_{ij} is the distance between the two counties. When an industry and university county coincide, no distance decay is applied to the U_i variable to avoid the familiar self-potential problem (e.g., Frost and Spence [30]). To obtain an index for each state, an average is taken over all counties in the state for which $GRAV_i > 0$.

The other two indices are based on the notion of "covering," i.e., a measure of how much university research is carried out in counties within a given distance band of a industry R & D county (note that the university counties do not need to be within the same state):

$$COV_i = \Sigma_j \delta_{ij} \cdot U_j \tag{7}$$

where δ_{ij} takes on a value of 1 when d_{ij} is less than or equal to a critical covering distance, respectively, 50 and 75 miles in our application. Again, to obtain an index for a state, an average is taken over all industry R & D counties in the state.

The computation of the indices was carried out by means of the access functions in the SpaceStat software (Anselin [9]) from data on boundary files for U.S. counties.

Empirical Results

In Table 1, we present the results of the estimation of a cross-sectional regression of the geographic knowledge production function (4) for 43 U.S. states in 1982. All variables are in logarithms. In addition to a replication of the standard Jaffe approach (but using innovation counts as the dependent variable, as in Acs, Audretsch, and Feldman [2, p. 365]), reported in the first column of the table, we also estimated the model with the four new indices of local geographic interaction (reported in columns 3–6).

All regressions yield highly significant and positive coefficients for both private R & D and university research (at $p < 0.01$), confirming the consensus result in the literature. The new gravity and covering indices are positive and significant as well (at $p < 0.05$), whereas the co-location index (5) is only marginally significant (at $p < 0.10$). Jaffe's geographic coincidence index does not turn out to be significant, confirming earlier findings of Acs, Audretsch, and Feldman [2, 3]. For all models, diagnostic tests were carried out for heteroskedasticity, using the White [76] test. In addition, specification tests for spatial autocorrelation were carried out, using the Lagrange multiplier tests for spatial error and spatial lag dependence implemented in the SpaceStat software (for technical details, see Anselin [7–9], Anselin and Hudak [11], Anselin and Bera [10]). The tests for spatial autocorrelation were computed for three different spatial weights matrices which reflect different prior notions on the spatial structure of dependence. The weights matrices have non-zero elements for each observation pair (an observation, row , and its potential neighbor, column) that is assumed to interact. The magnitude of the non-zero elements is either a function of contiguity (the presence of a common border), or some function of the distance between the two observations (for further details, see Anselin [7], Anselin and Bera [10]). The three weights used for the 43 states are based on simple contiguity (common border between states), on distance-based contiguity with a 500 mile cut-off (states are neighbors if there geographic centers are within 500 miles of each other); and on inverse distance squared (the magnitude of the interaction is proxied by the squared inverse distance between the geographic centers of states). Only the results for the most significant diagnostic are reported in Table 1. No evidence of heteroskedasticity was found, but the Jaffe model showed significant spatial error autocorrelation. Maximum likelihood estimation of the Jaffe model with spatial error autocorrelation (reported in column 2) did not alter the substantive interpretation of the model: both private R & D and university research remained strongly significant, but the

TABLE 1
Regression Results for Log(Innovations) at the State Level
($N = 43$, 1982)

Model	Jaffe (OLS)	Jaffe (ML-SER)	CUR (OLS)	GRAV (OLS)	COV(50) (OLS)	COV(75) (OLS)
Constant	−7.079	−7.084	−7.117	−10.181	−7.205	−8.386
	(0.100)	(0.093)	(1.032)	(1.574)	(1.001)	(1.103)
Log(R)	0.508	0.549	0.396	0.350	0.451	0.375
	(0.100)	(0.098)	(0.110)	(0.110)	(0.095)	(0.104)
Log(U)	0.574	0.544	0.676	0.633	0.402	0.549
	(0.142)	(0.129)	(0.144)	(0.133)	(0.153)	(0.133)
Log(local spillover)	0.0266	0.0164	0.0463	0.301	0.268	0.255
	(0.030)	(0.027)	(0.026)	(0.121)	(0.109)	(0.100)
λ		0.507				
		(0.218)				
R^2-adj	0.820		0.831	0.841	0.841	0.843
Log-Lik	−41.99	−39.91	−40.62	−39.28	−39.31	−39.08
White	6.29		9.09	5.42	6.91	5.38
LM-Err	5.40		1.01	0.84	0.60	0.52
	(IDIS2)		(D500)	(IDIS2)	(D500)	(D500)
LM-Lag	0.91	0.48	0.56	1.90	0.02	0.48
	(CONT)	(IDIS2)	(CONT)	(D500)	(D500)	(D500)

Notes: Estimated standard errors are in parentheses; critical value for the White statistic with 9 degrees of freedom is 16.92 ($p = 0.05$); critical value for LM-Err and LM-Lag statistics is 3.84 ($p = 0.05$); spatial weights matrices are row-standardized: CONT is first order contiguity; D500 is distance-based contiguity for 500 miles; and IDIS2 is inverse distance squared; only the highest values for a spatial diagnostic are reported.

geographic coincidence index was not (the spatial autoregressive coefficient was significant and positive; no evidence was found of any remaining higher order spatial dependence).

Interestingly, relative to the results of the Jaffe model, the effect of the introduction of the gravity and covering interaction indices seems to be to lower the coefficient of private R & D, whereas the coefficient of university research remains virtually unchanged. This may suggest that the original results overestimated the pure effect of private R & D on innovation, in the sense that some of it may be due to local geographic interaction with university research. The new indices seem to capture this interaction, while the original Jaffe index does not. Also, the introduction of the new indices removes any misspecification in the form of spatial autocorrelation, which is present in the original Jaffe formulation.

In sum, these results suggest that the spatial range of interaction between private R & D and university research reaches beyond the county (or MSA) where the R & D is carried out. Even the crude averages at the state level provide consistent evidence of this. Parenthetically, our results also demonstrate the usefulness of high technology lab employment as a proxy for private R & D activity, since no significantly different relationship was found compared to earlier results in the literature that used a purer measure of R & D expenditures.

LOCAL GEOGRAPHIC SPILLOVERS AT THE MSA LEVEL

Spatially Lagged Variables

The use of R & D lab employment as a proxy for private R & D activity allows us to carry out an analysis of the geographic knowledge production function at a smaller spatial scale than the state level. We constructed a data base for 125 MSAs in the U.S. for which some innovative and research activity was present (see Appendix B for a listing). Given the indication of a wider range of spatial interaction than purely within-county between university and private R & D from the results at the state level, we constructed two new variables that we refer to as *spatial lags*. These variables are designed to capture the effect of, respectively, university research and private R & D in counties surrounding the MSA, within a given distance band from the geographic center of the MSA. Specifically, for any MSA i, the spatial lags $U50_i$ and $R50_i$ are the sums of, respectively, the university research and private R & D in those counties surrounding the MSA (and not part of the MSA) whose geographic centers are within 50 miles of the geographic center of the core MSA county. Similar measures were computed for a 75 mile range as well (*U75*, *R75*). These measures correspond to our notion of covering used in the state level analysis.

Since the analysis is carried out at the scale at which we assume that the spatial interaction takes place (MSA and possibly its surrounding counties), there is no need to create an artificial index of geographic coincidence, as is the case at the state level. In fact, by explicitly including both the research magnitude at the MSA as well as for surrounding counties, we are able to get a much more precise insight into the spatial extent of local geographic spillovers.

Estimation Issues

Our model consists of three equations, the knowledge production function for K (1), an industry research equation, R (2), and a university research equation, U (3). The knowledge production function contains both R and U as explanatory variables and will be extended with the two spatial lag variables, *U50* (or *U75*) and *R50* (or *R75*). Both R and U equations contain the other as explanatory variable, as well as the spatial lags. The form of this system of equations raises a number of issues with respect to estimation and identification.

First, while the system is recursive between K and R and U, the exogeneity of the latter two in the knowledge production function should not be taken on faith. In fact, misspecifications (e.g., errors in variables) could easily lead to endogeneity and must be checked. We address this by means of the Durbin–Wu–Hausman test for exogeneity (e.g., Davidson and MacKinnon [22, pp. 237–242]). We also take this approach to test the extent to which R and U are endogenous to each other in Eqs. (2) and (3).

Second, even in a purely recursive system, ordinary least squares estimation of the knowledge production function would only be legitimate in the absence of interequation correlation, i.e., correlation between the error terms of the equations (e.g., Greene [34], p. 600). We check this by means of a Lagrange multiplier (LM) test on the diagonality of the error covariance matrix for the least squares residuals (Breusch and Pagan [18]).

Third, the spatially lagged variables *R50* (or *R75*) in the R equation, and *U50* (or *U75*) in the U equation could be endogenous as well. In spatial econometric models, such variables are often referred to as spatially lagged dependent variables and it is well known that their presence causes OLS to be biased and inconsistent (Anselin [7]). In our model however, the locations to which these lags pertain (counties surrounding the MSA, but not part of the MSA) are not part of the observation set (the MSAs), unless they belong to the MSA counties of a neighboring MSA.

There is some degree of overlap for 71 out of the 125 MSAs and we assess the extent to which this may cause endogeneity by means of a Durbin–Wu–Hausman (DWH) test.

Finally, the use of a cross-sectional sample potentially leads to spatial autocorrelation in the regression equations. We assess this by means of a Lagrange multiplier test for spatial error dependence using three spatial weights based on distance: the same 50 and 75 mile cut-offs as used in the construction of the lag variables, and a squared inverse distance weights matrix. These tests are only valid when the explanatory variables in the regression are exogenous and should be interpreted with caution when this is not the case. They are used here to assess the extent to which remaining unspecified spatial spillover may be present, even after the inclusion of the spatial lags (provided that the latter are exogenous, which turns out to be the case in our study).

Empirical Results

We first focus on the knowledge production function in Eq. (1), i.e., without a geographic coincidence index. The estimation results for a cross-sectional regression on 125 MSAs in 1982 are reported in Table 2. Three models are presented: a bare-bones Jaffe model, a spatial model which also includes the spatial lags for university and private R & D (only the most significant of, respectively, *U50* and *U75* and *R50* and *R75* are reported), and an extended spatial model, which includes local economic characteristics as explanatory variables as well.

The standard Jaffe model (in column 1 of Table 2) confirms the strong significance (at $p < 0.01$) of both private R & D and university research on the level of innovative activity in an MSA. However, compared to the analysis at the state level, there is a clear dominance of the coefficient of private R & D over university research, indicating an elasticity that is almost five times higher (compared to roughly equal elasticities at the state level in Table 1). There is no evidence of heteroskedasticity, but there is a strong indication of misspecification in the form of a spatial lag (at $p < 0.05$). When the spatial variables *U50* and *R75* are added to the Jaffe specification (instead of a coincidence index at the state level), the model fit improves slightly, with a positive and significant effect for *U50*, but an insignificant effect for the research spatial lag. Neither heteroskedasticity, nor spatial dependence remain. In the full model (in column 3), the spatial lag for university research, with a 50 mile range *(U50)* remains positive and significant ($p < 0.05$). However, there is no evidence that the effect of private R & D on MSA innovative activity spills over from outside the MSA (the coefficient of *R75* is negative, but not significant). The addition

TABLE 2
OLS Regression Results for Log(Innovations) at the MSA Level
(N-125, 1982)

Model	Jaffe	Spatial	Extended spatial
Constant	−1.045	−1.134	−1.407
	(0.146)	(0.172)	(0.212)
Log(R)	0.540	0.504	0.277
	(0.054)	(0.055)	(0.057)
Log($R75$)		0.001	−0.027
		(0.041)	(0.037)
Log(U)	0.112	0.132	0.093
	(0.036)	(0.036)	(0.034)
Log($U50$)		0.037	0.032
		(0.018)	(0.015)
Log(LQ)			0.652
			(0.163)
Log(BUS)			0.332
			(0.057)
Log($LARGE$)			−0.337
			(0.094)
$RANK$			0.202
			(0.101)
R^2-adj	0.599	0.611	0.725
White	1.18	9.02	37.85
LM-Err	2.69	2.18	0.10
	(D75)	(D75)	(D50)
LM-Lag	5.62	1.48	2.08
	(D50)	(D75)	(D75)

Notes: Estimated standard errors are in parentheses; critical values for the White statistic with, respectively 5, 20 and 35 degrees of freedom are 11.07, 31.41, and 49.52 ($p = 0.05$); critical values for LM-Err and LM-Lag statistics are 3.84 ($p = 0.05$) and 2.71 ($p = 0.10$); spatial weights matrices are row-standardized: D50 is distance-based contiguity for 50 miles; D75 is distance-based contiguity for 75 miles; and IDIS2 is inverse distance squared; only the highest values for a spatial diagnostic are reported.

of the local economic variables causes the elasticity with respect to private R & D to drop substantially, while the one for university research stays roughly the same. All three local economic variables are highly significant (with $p < 0.001$) and have the expected sign: positive for specialization in high technology (LQ) and the importance of business services (BUS), and negative for the presence of large firms ($LARGE$). Note that the negative sign for the presence of large companies confirms earlier evidence in Acs, Audretsch, and Feldman [3, p. 338] that smaller firms tend to be more innovative. In other words, *ceteris paribus*, MSAs dominated by the presence of larger firms tend to show less innovative activity. A fourth variable is added to correct for potential unmeasured quality effects that may cause inter-equation correlation. Following Jaffe [44], we use the rank of the universities as a proxy for quality. This variable turns out to be positive and significant ($p < 0.05$).

For the full specification, we further tested the exogeneity of each of the four variables *R*, *U*, *R75*, and *U50*, using the DWH test for a two stage least squares estimation with Log(*HTEMP*), Log(*ENRL*), Log(*EDEXP*), and *FORTUNE* as instruments. We failed to reject the null hypothesis for each (none achieved a p value less than 0.20). In other words, there was no evidence *against* exogeneity of these variables. In addition, the LM tests on diagonality of the error covariance between *K* and *R* and *K* and *U* were, respectively, 1.17 ($p = 0.28$) and 1.73 ($p = 0.19$), which did not provide evidence against the null hypothesis of no inter-equation error correlation.

The OLS estimation results in column 3 of Table 2 can thus be reliably interpreted to indicate the strong influence of university research on innovative activity in an MSA, not only of university research in the MSA itself, but also in the surrounding counties. By contrast, the effect of private R & D seems to be contained within the MSA itself. Of course, private and university R & D are not independent, and we turn to their interaction/ simultaneity in Tables 3 and 4.

Following Jaffe [44], we estimated two additional models that explicitly incorporate the potential simultaneity between private R & D and university research. The results in Table 3 show the estimation of the private R & D equation (2) in which, in addition to university research, the spatial lags of both university research and private R & D are included, as well as the log of the high technology employment (*HTEMP*), the *FORTUNE* dummy, and the *RANK* measure as exogenous variables (the latter to control for potential quality effects). Given the strong indication of heteroskedasticity (the White test in the OLS regression of the model yielded 42.54, which is significant at $p < 0.02$), both standard and robust results are reported. University enrollment (*ENRL*) and education expenditures (*EDEX*) were used as instruments in the 2SLS estimation reported in columns 3 and 4 of Table 3, with *U* as the endogenous variable.

The estimates and their significance vary considerably between the standard and robust results on the one hand and the OLS and 2SLS approaches on the other hand, indicating the strong effect of both heteroskedasticity and simultaneity bias. A Durbin–Wu–Hausman (DWH) test on the exogeneity of university research is clearly rejected at $p < 0.05$

TABLE 3
Regression Results for Log(Private R & D) at the MSA Level
($N = 125$, 1982)

Model	OLS	OLS-Robust	2SLS	2SLS-Robust
Constant	−0.952	−0.952	−1.036	−1.049
	(0.408)	(0.622)	(0.418)	(0.457)
Log(*R*50)	0.055	0.055	0.070	0.071
	(0.034)	(0.036)	(0.035)	(0.035)
Log(*U*)	0.155	0.155	0.259	0.257
	(0.047)	(0.065)	(0.067)	(0.078)
Log(*U*75)	0.044	0.044	0.040	0.040
	(0.041)	(0.040)	(0.042)	(0.038)
Log(*HTEMP*)	0.637	0.637	0.567	0.571
	(0.097)	(0.173)	(0.104)	(0.136)
FORTUNE	0.414	0.414	0.419	0.416
	(0.188)	(0.171)	(0.192)	(0.150)
RANK	0.122	0.122	−0.004	−0.007
	(0.163)	(0.154)	(0.176)	(0.154)
R^2	0.585	0.585	0.652	0.651

Notes: Estimated standard errors are in parentheses; Log(*U*) considered as endogenous; instruments in 2SLS estimation are Log(*ENRL*) and Log(*EDEX*). DWH test on exogeneity of Log(*U*), $F(1,117) = 5.27$, $p = 0.023$.

(the statistic was 5.27). The OLS results are therefore only reported for comparison purposes and should not be the basis of any substantive interpretation.

Focusing on the results in column 4 of Table 3, we find a strong positive and significant effect of university research on private R & D ($p < 0.001$), but only within the MSA itself (the coefficient of *U75* is not significant). On the other hand, there is an indication of spatial spill-over in private R & D, supported by a significant (at $p < 0.05$) and positive coefficient of *R50*. Both specialization in high technology and the presence of Fortune 500 companies also contribute significantly to the magnitude of private R & D. Note that the latter has a positive sign, in contrast to the effect of larger firms on innovations. A DWH test on the exogeneity of the spatial lag variable *R50* cannot be rejected [$F(1, 117) = 0.078$, $p = 0.78$], hence the results in the last column of Table 3 yield a valid interpretation.

The effect of private R & D on university research is quite different, as borne out by the results for Eq. (3) reported in Table 4, in the same format as in Table 3. There is no evidence of a significant influence of private

TABLE 4
Regression Results for Log(University Research) at the MSA Level
($N = 125$, 1982)

Model	OLS	OLS-Robust	2SLS	2SLS-Robust
Constant	−2.355	−2.355	−2.502	−2.244
	(1.715)	(1.846)	(1.749)	(1.702)
Log(*U*75)	0.096	0.096	0.102	0.107
	(0.057)	(0.075)	(0.058)	(0.062)
Log(*R*)	0.052	0.052	−0.025	0.051
	(0.109)	(0.152)	(0.202)	(0.160)
Log(*R*50)	−0.074	−0.074	−0.066	−0.066
	(0.046)	(0.044)	(0.049)	(0.039)
Log(*ENRL*)	1.777	1.777	1.846	1.760
	(0.179)	(0.230)	(0.234)	(0.227)
Log(*EDEXP*)	−1.187	−1.187	−1.176	−1.242
	(0.928)	(1.001)	(0.930)	(0.899)
RANK	0.510	0.510	0.539	0.469
	(0.208)	(0.159)	(0.218)	(0.157)
R^2	0.650	0.650	0.650	0.629

Notes: Estimated standard errors are in parentheses; Log(*R*) considered as endogenous; instruments in 2SLS estimation are Log(*HTEMP*) and *FORTUNE*. DWH test on exogeneity of Log(*R*), $F(1,117) = 0.210$, $p = 0.648$.

R & D in the MSA itself on university research, nor of spatial spillovers. The only strongly significant coefficients are for internal characteristics of the universities, i.e., their size (enrollment) and quality, both with the expected positive sign. In contrast to the R & D equation, there is no significant evidence of heteroskedasticity (White test yields 34.03, with 38.89 as the critical level for 26 degrees of freedom at $p = 0.05$), nor of spatial autocorrelation. More importantly, there is no indication of simultaneity bias (a DWH test on the exogeneity of R yields 0.21, which is clearly non-significant at $p = 0.65$). The correct interpretation should therefore be based on column 1 and the other results are only reported for completeness sake.

In sum, the estimates in Table 3 and 4 support the earlier finding of Jaffe [44, p. 968] about the direction of causality in the relation between private R & D and university research, but now at a much finer level of spatial detail and with an explicit measure of spatial spillovers. There is no evidence that private R & D in the MSA or its surrounding counties is endogenous to the university research equation. By contrast, there is a strong indication that university research in an MSA is endogenous in the private R & D equation.

CONCLUSIONS

In this paper, we have been able to shed additional light on the issue of *local* geographic spillovers between university research and high technology innovations. Our point of departure was Jaffe's [44, p. 968] often cited finding that "there is only weak evidence that spillovers are facilitated by geographic coincidence of universities and research labs within the state." We approached this issue from an explicit spatial econometric perspective and implemented the classic Griliches–Jaffe knowledge production framework for high technology innovations in 43 U.S. states, as well as in 125 MSAs. The latter became possible by using a specially compiled set of data on R & D laboratory employment. This yielded more precise insight into the range of spatial externalities between innovation and R & D in the MSA and university research, both within the MSA and in surrounding counties.

We also refined Jaffe's geographic coincidence index for analysis of local spillover based on data for states. We suggested four alternative measures, three of which were derived from established principles in spatial interaction theory, i.e., the gravity potential and covering index. We showed that the inclusion of the alternative indices in the geographic knowledge production function provided strong and unequivocal evidence of local spatial externalities. In addition, the evidence suggested that these externalities transcended the boundaries of counties.

Overall, we confirmed the positive and significant relationship between university research and innovative activity, both directly, as well as indirectly through its impact on private sector R & D. We found that the spillovers of university research on innovation extended over a range of 50 miles from the innovating MSA, but not with respect to private R & D. We also confirmed earlier findings on the direction causality between university and private research, the former being endogenous to the latter, but not vice versa.

Our findings are important in that they highlight the relevance of a precise consideration of the spatial range of interaction in the analysis of spatial externalities. However, some cautionary remarks are in order as well. Our analysis is limited by the use of a single cross-section. Unfortunately, there is currently no update of the 1982 U.S. SBA innovation data base for later points in time, precluding a more extensive analysis of the space-time dynamics. Also, we have elected to focus on studying the relations between research and innovations in those locations for which both were observed. This leaves aside the issue of why certain locations have research and innovative activity and others do not, especially when one of the two is present, but the other is not. We leave this aspect of the study for a separate paper.

APPENDIX A

Innovations, R & D Lab Employment and University Research for 43 U.S. States—1982

State	Innovations	R & D	University research
Alabama	4	1868	45293
Arizona	38	4086	62206
Arkansas	4	187	12303
California	851	48552	658487
Colorado	40	4244	87230
Connecticut	117	8799	136435
Delaware	11	6173	6962
Florida	49	2298	105177
Georgia	29	1585	102981
Illinois	185	20574	207669
Indiana	38	6970	92986
Iowa	13	1939	77435
Kansas	11	2029	21849
Kentucky	8	571	21224
Louisiana	4	701	48509
Maine	3	430	5527
Maryland	26	5244	131229
Massachusetts	316	24753	364871
Michigan	84	21816	149432
Minnesota	82	7478	103549
Mississippi	4	142	18220
Missouri	23	4863	100942
Nebraska	4	275	21893
Nevada	1	103	3966
New Hampshire	28	624	19368
New Jersey	352	42907	67887
New Mexico	3	339	57745
New York	347	27709	585312
North Carolina	23	5037	111994
Ohio	139	18886	161911
Oklahoma	14	2630	39175
Oregon	29	310	44788
Pennsylvania	197	22702	267614
Rhode Island	22	1083	22035
South Carolina	11	1131	27019
Tennessee	16	1397	55692
Texas	136	9896	296114
Utah	13	1926	62249
Vermont	3	67	14166
Virginia	27	8023	72208
Washington	41	1385	95530
West Virginia	2	1471	13589
Wisconsin	69	4504	127440

Sources: Compiled from U.S. SBA Innovations Data Base; compiled from R. R. Bowker Company Directories; compiled from NSF Survey of Scientific and Engineering Expenditures at Universities and Colleges.

APPENDIX B

Innovations, R & D Lab Employment and University Research for 125 U.S. MSAs—1982

MSA	Innovations	R & D	University research
Akron	7	2989	4504
Albany-Schenectady-Troy	1	2475	33600
Albuquerque	2	273	26064
Allentown-Bethlehem-Easton PA-NJ	7	946	9455
Anaheim-Santa Ana-Garden Grove	108	2904	31037
Ann Arbor	7	1428	84876
Atlanta	26	661	71321
Austin	12	1003	45079
Baltimore	12	2499	88697
Bellingham	1	78	255
Benton Harbor	1	388	19
Binghamton NY-PA	2	152	2988
Birmingham	1	646	35162
Bloomington-Normal	2	16	356
Boston	282	22391	325474
Bridgeport	67	3221	14
Bryan-College Station	2	37	40178
Buffalo	24	4404	25591
Burlington	3	41	13952
Charlotte-Gastonia	6	191	627
Chicago	164	15416	145814
Cincinnati OH-KY-IN	13	1247	26709
Cleveland	54	5195	42792
Colorado Springs	6	726	52
Columbia	1	135	23448
Columbus	20	2733	52252
Dallas-Fort Worth	77	1900	46974
Davenport-Rock Island-Moline Iowa-IL	5	4535	3
Dayton	11	2538	22608
Daytona Beach	1	97	31
Denver-Boulder	26	3386	65379
Detroit	51	11599	21511
El Paso	7	1160	834
Fort Collins	6	83	21717
Fort Lauderdale-Hollywood	9	285	99
Fresno	1	48	276
Gainesville	1	158	45324
Galveston-Texas	2	99	11882
Grand Rapids	4	359	21
Greensboro-Winston-Salem-High Point	5	1010	9159
Greenville-Spartanburg	10	506	9085
Hamilton-Middletown	4	39	709
Hartford	27	2087	52297
Houston	29	2545	120942
Huntsville	3	847	1843
Janesville-Beloit	2	207	49

APPENDIX B
(*Continued*)

MSA	Innovations	R & D	University research
Jersey City	11	709	4352
Johnson City-Kingsport-Bristol Tenn.-VA	2	737	736
Kalamazoo-Portage	5	950	146
Kansas City	12	1325	956
Knoxville	1	356	32526
Lafayette-West Lafayette	1	121	45048
Lancaster	4	671	69
Lansing-East	4	140	35951
Lincoln	2	127	13236
Lorain-Elyria	2	204	75
Los Angeles-Long Beach	161	20226	199028
Louisville	7	414	4992
Madison	4	741	107915
Melbourne-Titusville-Cocoa	11	81	779
Memphis TN.-AK.-MS.	3	122	799
Miami	4	177	32677
Milwaukee	34	2130	18832
Minneapolis-St. Paul	80	7181	103235
Nashville-Davidson	5	113	20604
Nassau-Suffolk	120	2534	31881
New Bedford	6	95	569
New Brunswick-Perth Amboy-Sayreville	30	2338	20408
New Haven-West Haven	19	1708	83974
New London-Norwich	1	1659	156
New Orleans	1	320	14394
New York NY-NJ	222	11136	285776
Essex County	143	26780	17500
Newburgh-Middletown	3	324	206
Newport News-Hampton	2	310	1676
Norfolk-Virginia Beach-Portsmouth	1	136	2819
Northeast	2	25	70
Oklahoma City	1	204	19703
Orlando	5	514	4704
Paterson-Clifton-Passaic	25	3098	226
Peoria	1	27	146
Philadelphia PA-NJ	139	16967	134499
Phoenix	29	3809	8037
Pittsburgh	39	4357	68160
Pittsfield	2	14	354
Portland	1	286	463
Portland OR-WA	22	285	20750
Portsmouth-Dover-Rochester	5	166	5468
Providence-Warwick-Pawtucket	15	480	22036
Provo-Orem	3	281	3419
Raleigh-Durham	8	2317	101136
Reading	1	423	123
Reno	1	26	3226
Riverside-San Bernardino-Ontario	13	193	9494

APPENDIX B
(*Continued*)

MSA	Innovations	R & D	University research
Rochester	32	4554	60999
Sacramento	7	311	32634
St. Louis	13	3312	69825
Salem	1	44	20
Salt Lake City	10	1585	40785
San Antonio	3	1674	19539
San Diego	59	4823	79857
San Francisco-Oakland	75	7856	154824
San Jose	374	9727	127897
Santa Barbara-Santa Maria-Lompoc	9	1945	10269
Santa Cruz	2	30	7168
Seattle-Everett	34	798	85514
South Bend	5	166	11192
Spokane	3	11	230
Springfield	3	131	19
Springfield-Chicopee-Holyoke	3	449	29455
Stockton	2	10	648
Syracuse	9	1020	31844
Tacoma	2	269	76
Tampa-St. Petersburg	12	668	4790
Toledo OH-MI	6	1129	6705
Trenton	29	2836	24680
Tucson	9	209	53599
Tulsa	12	89	1317
Waco	3	37	363
Washington, DC	21	8274	38311
Waterloo-Cedar Falls	1	14	5
Wichita	5	1490	909
Wilmington DE-NJ-MD	11	6706	6962
Worcester	17	861	4628
Youngstown-Warren	1	167	75

Sources: Compiled from U.S. SBA Innovations Data Base; compiled from R. R. Bowker Company Directories; compiled from NSF Survey of Scientific and Engineering Expenditures at Universities and Colleges.

REFERENCES

1. Z. Acs and D. Audretsch, "Innovation and Small Firms," MIT Press, Cambridge, MA (1990).
2. Z. Acs, D. Audretsch, and M. Feldman, Real effects of academic research: Comment, *American Economic Review*, **81**, 363–367 (1991).
3. Z. Acs, D. Audretsch, and M. Feldman, R & D spillovers and recipient firm size, *The Review of Economics and Statistics*, **76**, 336–340 (1994).
4. Z. Acs, D. Audretsch, and M. Feldman, R & D spillovers and innovative activity, *Managerial and Decision Economics*, **15**, 131–138 (1994).

5. P. Almeida and B. Kogut, "The geographic localization of ideas and the mobility of patent holders," Working Paper, Department of Management, The Warton School, University of Pennsylvania (1995).
6. A. E. Andersson, C. Anderstig, and B. Harsman, Knowledge and communications infrastructure and regional economic growth, *Regional Science and Urban Economics*, **20**, 359–376 (1990).
7. L. Anselin, "Spatial Econometrics, Methods and Models," Kluwer Academic, Boston (1988).
8. L. Anselin, "SpaceStat, a Program for the Analysis of Spatial Data," National Center for Geographic Information and Analysis, University of California, Santa Barbara, CA (1992).
9. L. Anselin, "SpaceStat Version 1.80 User's Guide," Regional Research Institute, West Virginia University, Morgantown, WV (1995).
10. L. Anselin and A. Bera, Spatial dependence in linear regression models with an introduction to spatial econometrics, *in* "Handbook of Economic Statistics" (D. Giles and A. Ullah, Eds.), Marcel Dekker, New York, (1997).
11. L. Anselin and S. Hudak, Spatial econometrics in practice: A review of software options, *Regional Science and Urban Economics*, **22**, 509–536 (1992).
12. K. Arrow, Economic welfare and the allocation of resources for invention, *in* "The Rate and Direction of Inventive Activity" (R. Nelson, ed.), Princeton University Press, Princeton, NJ (1962).
13. D. B. Audretsch and M. P. Feldman, Knowledge spillovers and the geography of innovation and production, *American Economic Review*, **86**, 630–640 (1996).
14. N. Bania, L. N. Calkins, and D. R. Dalenberg, The effects of regional science and technology policy on the geographic distribution of industrial R & D labs, *Journal of Regional Science*, **32**, 209–228 (1992).
15. N. Bania, R. Eberts, and M. S. Fogarty, Universities and the startup of new companies: Can we generalize from Route 128 and Silicon Valley? *The Review of Economics and Statistics*, **75**, 761–766 (1993).
16. A. Bartel and F. Lichtenberg, The comparative advantage of educated workers in implementing new technology, *The Review of Economics and Statistics*, **69**, 1–11 (1987).
17. P. Beeson and E. Montgomery, The effects of colleges and universities on local labor markets, *The Review of Economics and Statistics*, **75**, 753–761 (1993).
18. T. Breusch and A. Pagan, The LM test and its applications to model specification in econometrics, *Review of Economic Studies*, **47**, 239–254 (1980).
19. Jaques Cattell Press, Industrial Research Laboratories of the United States, 17th Edition, R. R. Bowker Company, New York and London (1982).
20. W. M. Cohen and D. A. Levinthal, Innovation and Learning, *The Economic Journal*, **99**, 569–596 (1989).
21. P. David and J. Rosenbloom, Marshallian factor market externalities and the dynamics of industrial localization, *Journal of Urban Economics*, **28**, 349–370 (1990).
22. R. Davidson and J. G. MacKinnon, "Estimation and Inference in Econometrics," Oxford University Press, New York (1993).
23. C. DeBresson and F. Amesse, Networks of innovators: A review and introduction to the issues, *Research Policy*, **20**, 363–380 (1992).
24. G. Dosi, Sources, procedures and microeconomic effects of innovation, *Journal of Economic Literature*, **26**, 1120–1171 (1988).
25. K. Edwards and T. J. Gordon, "Characterization of Innovations Introduced on the U.S. Market in 1982, Report Prepared for the U.S. Small Business Administration," The Futures Group, Washington, DC (1984).

26. M. Feldman, "The Geography of Innovation," Kluwer Academic, Boston (1994).
27. M. Feldman and R. Florida, The geographic sources of innovation: Technological infrastructure and product innovation in the United States, *Annals of the Association of American Geographers*, **84**, 210–229 (1994).
28. R. Florax, "The University: A Regional Booster? Economic Impacts of Academic Knowledge Infrastructure," Avebury, Aldershot (1992).
29. R. Florax and H. Folmer, Knowledge impacts of universities on industries: An aggregate simultaneous investment model, *Journal of Regional Science*, **32**, 437–466 (1992).
30. M. E. Frost and N. A. Spence, The rediscovery of accessibility and economic potential: The critical issue of self-potential, *Environment and Planning A*, **27**, 1833–1848 (1995).
31. E. L. Glaeser, H. D. Kallal, J. A. Scheinkman and A. Schleifer, Growth in cities, *Journal of Political Economy*, **100**, 1126–1152 (1992).
32. A. Glasmeier, "The High-Tech Potential. Economic Development in Rural America," Center for Urban Policy Research, New Brunswick, NJ (1991).
33. H. Goldstein and M. I. Luger, Science/technology parks and regional development theory, *Economic Development Quarterly*, **4**, 64–78 (1990).
34. W. H. Greene, "Econometric Analysis," Macmillan, New York (1993).
35. Z. Griliches, Issues in assessing the contribution of R & D to productivity growth, *Bell Journal of Economics*, **10**, 92–116 (1979).
36. Z. Griliches, Patent statistics as economic indicators: A survey, *Journal of Economic Literature*, **28**, 1661–1707 (1990).
37. Z. Griliches, The Search for R & D Spillovers, *Scandinavian Journal of Economics*, **94**, 29–47 (1992).
38. G. Grossman and E. Helpman, "Innovation and Growth in the Global Economy," MIT Press, Cambridge, MA (1991).
39. G. Grossman and E. Helpman, Endogenous innovation in the theory of growth, *Journal of Economic Perspectives*, **8**, 23–44 (1994).
40. C. F. Harding, Location choices for research labs: A case study approach, *Economic Development Quarterly*, **3**, 223–234 (1989).
41. H. W. Herzog, A. M. Schlottman and D. J. Johnson, High-technology jobs and worker mobility, *Journal of Regional Science*, **26**, 445–459 (1986).
42. J. R. Howells, The location of research and development: Some observations and evidence from Britain, *Regional Studies*, **18**, 13–29 (1984).
43. A. Jaffe, Technological opportunity and spillovers of R & D: Evidence from firms' patents, profits and market value, *American Economic Review*, **76**, 984–1001 (1986).
44. A. Jaffe, Real effects of academic research, *American Economic Review*, **79**, 957–970 (1989).
45. A. Jaffe, M. Trajtenberg and R. Henderson, Geographic localization of knowledge spillovers as evidenced by patent citations, *Quarterly Journal of Economics*, **108**, 577–598 (1993).
46. P. Krugman, Increasing returns and economic geography, *Journal of Political Economy*, **99**, 483–499 (1991).
47. A. Link and J. Rees, Firm size, university based research and returns to R & D, *Small Business Economics*, **2**, 25–32 (1990).
48. R. Lucas, On the mechanics of economic development, *Journal of Monetary Theory*, **22**, 3–39 (1988).
49. M. I. Luger and H. A. Goldstein, "Technology in the Garden: Research Parks in Regional Economic Development," University of North Carolina Press, Chapel Hill, NC (1991).

50. L. Lund, "Locating Corporate R & D Facilities," Conference Board Report No. 892, The Conference Board, New York (1986).
51. E. Malecki, Federal R & D spending in the United States of America: Some impacts on metropolitan economies, *Regional Studies*, **16**, 19–35 (1981).
52. E. Malecki, Research and development and the geography of high-technology complexes, *in* "Technology, Regions and Policy" (J. Rees, ed.), Rowman and Littlefield, Totowa, NJ (1986).
53. E. Malecki, "Technology and Economic Development," Longman Scientific and Technical, Essex (1991).
54. E. Malecki, and S. L. Bradbury, R & D facilities and professional labour: labour force dynamics in high technology, *Regional Studies*, **26**, 123–136 (1992).
55. E. Mansfield, Academic research and industrial innovation, *Research Policy*, **20**, 1–12 (1991).
56. E. Mansfield, Academic research underlying industrial innovations: Sources, characteristics and financing, *The Review of Economics and Statistics*, **77**, 55–65 (1995).
57. A. Markusen, P. Hall, and A. Glasmeier, "High Tech America: The What, How, Where and Why of Sunrise Industries," Allen and Unwin, Boston (1986).
58. National Science Foundation, "Academic Science and Engineering R & D Expenditures, Fiscal Year 1982," Data obtained from GASPAR data files (1982).
59. R. Nelson, The simple economics of basic scientific research, *Journal of Political Economy*, **67**, 297–306 (1959).
60. R. Nelson, The role of knowledge in R & D efficiency, *Quarterly Journal of Economics*, **97**, 453–470 (1982).
61. R. Nelson, Institutions supporting technical advance in industry, *American Economic Review*, **76**, 186–189 (1986).
62. D. D. Parker and D. Zilberman, University technology transfers: impacts on local and U.S. economies, *Contemporary Policy Issues*, **11**, 87–99 (1993).
63. R. Premus, "Location of high technology firms and regional economic development," U.S. Government Printing Office, Washington, DC (1982).
64. J. Rees, State technology programs and industry experience in the United States, *Review of Urban and Regional Development Studies*, **3**, 39–59 (1991).
65. J. Rees and H. Stafford, Theories of regional growth and industrial location: Their relevance for understanding high-tech complexes, *in* "Technology, Regions and Policy" (J. Rees, ed.), Rowman and Littlefield, Totowa, NJ (1986).
66. P. M. Romer, Increasing returns and long-run growth, *Journal of Political Economy*, **94**, 1002–1037 (1986).
67. P. M. Romer, Endogenous technological change, *Journal of Political Economy*, **98**, S72–102 (1990).
68. P. M. Romer, The origins of endogenous growth, *Journal of Economic Perspectives*, **8**, 3–22 (1994).
69. A. Saxenian, "Regional Advantage: Culture and Competition in Silicon Valley and Route 128," Harvard University Press, Cambridge, MA (1994).
70. R. Sivitanidou and P. Sivitanides, The intrametropolitan distribution of R & D activities: Theory and empirical evidence, *Journal of Regional Science*, **35**, 391–415 (1995).
71. W. Stohr, Regional innovation complexes, *Papers of the Regional Science Association*, **59**, 29–44 (1986).
72. M. Storper and R. Walker, "The Capitalist Imperative: Territory, Technology and Industrial Growth," Basil Blackwell, Oxford (1989).

73. E. Talen and L. Anselin, Assessing spatial equity: An evaluation of measures of accessibility to playgrounds, *Environment and Planning A*, (1997).
74. E. Von Hippel, "The Sources of Innovation," Oxford University Press, New York (1988).
75. J. Weibul, An axiomatic approach to the measurement of accessibility, *Regional Science and Urban Economics*, **6**, 357–379 (1976).
76. H. White, A heteroskedastic-consistent covariance matrix estimator and a direct test for heteroskedasticity, *Econometrica*, **48**, 817–838 (1980).

[19]

Patent Citations and the Geography of Knowledge Spillovers: A Reassessment

By PETER THOMPSON AND MELANIE FOX-KEAN*

Adam B. Jaffe, Manuel Trajtenberg, and Rebecca Henderson (1993) developed a matching method to study the geography of knowledge spillovers using patent citations, and found that knowledge spillovers are strongly localized. Their method matches each citing patent to a non-citing patent intended to control for the pre-existing geographic concentration of production. We show how the method of selecting the control group may induce spurious evidence of localized spillovers. This paper reassesses their findings using control patents selected under different criteria. Doing so eliminates evidence of strong intra-national localization effects at the state and metropolitan levels, but leaves largely unaffected evidence of international localization effects.

A revival of interest in economic geography during the last decade has renewed efforts at measuring location-specific externalities. These efforts have largely been guided by Alfred Marshall's (1920) three explanations for agglomeration economies: labor market pooling, scale economies in the provision of intermediate goods and services, and localization of knowledge spillovers. Perhaps because, as Paul R. Krugman (1991, p. 53) has argued, "knowledge flows. . . are invisible; they leave no paper trail by which they may be measured and tracked,"[1] the measurement of knowledge spillovers has proved the most challenging task.

The challenge was taken up most prominently by Jaffe et al. (1993, hereafter JTH), who pointed out that knowledge spillovers may well leave a paper trail in the citations to prior art recorded in patents. Moreover, because patents record the residence of the inventors, they are an invaluable resource for studying how knowledge flows are affected by geography. JTH undertook the considerable task of constructing a large dataset of patents and matching the locations of their inventors to the locations of inventors of all patents that subsequently cited them as prior art.

Of course, such an exercise would be futile unless one could also control for the existing geographic distribution of production. Patents linked by citation presumably not only share a technology, but are often developed by inventors working in a common industry. Patents linked by citation are therefore much more likely to share a geographic location than are a pair of patents drawn at random from the entire pool, but the observation tells us nothing about knowledge flows. JTH's important innovation was to construct a control group to mimic the existing geography of production.

JTH constructed three patent samples: a set of originating patents, a set of citing patents which referenced one of the originating patents, and a set of control patents matched to each citing patent. Each control patent shared the same technology class and (approximate) application date as its matched citing patent, but did not reference the matched originating patent. JTH's experiment compared the probabilities that the citing patent, and its matched control patent, were filed by inventors living in the same geographic location as the originating patent. The experiment yielded remarkable evidence that

* Thompson: Department of Economics, Florida International University, Miami, FL 33199 (e-mail: peter.thompson2@fiu.edu); Fox-Kean: Department of Economics, University of Houston, Houston, TX, 77204-5882 (e-mail: mefox@mail.uh.edu). We have benefited from insightful comments from three anonymous referees. This research was supported by the National Science Foundation under Grant No. SBR-0296192.

[1] Or perhaps this is because Marshall himself seemed less than convinced that knowledge spillovers would be localized: "Many of those economies in the use of specialized skill and machinery which are commonly regarded as within the reach of very large establishments do not depend on the size of individual factories. Some depend on the aggregate volume of production of the kind in the neighborhood; while others again, especially those connected with the growth of knowledge and the progress of the industrial arts, depend chiefly on the aggregate volume of production in the whole civilized world" (Marshall, 1920, Book IV, p. 220).

knowledge spillovers are localized. Citing patents were up to 1.2 times more likely than the control patents to come from the same country as the originating patent, up to two times more likely to come from the same state, and up to six times more likely to come from the same metropolitan area.

The number of citations a patent receives has also been used as a proxy for its value (see, e.g., Trajtenberg, 1990; Bronwyn H. Hall et al., 2000). By the same criterion, the citations to JTH's paper place it far in the upper tail of the value distribution. JTH's results have been used to motivate numerous theoretical models of growth and geography in which localized knowledge spillovers are simply assumed (e.g., Edward L. Glaeser, 1999), and their methodology has been applied with similar results in more specialized settings (e.g., Paul Almeida, 1996; Tony S. Frost, 2001; Diana Hicks et al., 2001; Jaffe and Trajtenberg, 1999). Now that a wealth of patent data has been released to the profession (Hall et al., 2001), we expect to see many more applications of their methods.

This paper argues that the evidence for localized knowledge spillovers generated by JTH's matched case-control methodology includes a significant spurious component. Controlling for unobservables using matching methods is invariably a dangerous exercise because one can rarely be confident that the controls are doing their job. In some applications imperfect matching may simply introduce noise and a corresponding loss of efficiency, but these are not the applications where matching is critical. In applications such as JTH's where matching is critical, imperfect matching induces systematic bias in the results. We show that JTH's selection of control patents using technology classifications cannot adequately control for the existing geographic distribution of production, and that failure to do so accounts for much of the evidence that knowledge spillovers are localized.

There are at least two reasons why JTH's method does not adequately control for existing patterns of industrial activity. First, control patents were selected using the broad, three-digit technological classification codes of the U.S. Patent and Trademark Office (USPTO), a level of aggregation that suppresses considerable within-class heterogeneity. Second, patents typically contain many distinct claims, to each of which a technological classification is assigned. The particular claim in a citing patent that can be associated with a citation to prior art may be quite distinct from the claim that generated the corresponding control patent. These two features of the control selection process mean there is no guarantee that the control patent has any industrial similarity either to the citing or to the originating patent.

We construct a new dataset in which control patents are selected under different criteria. First, we select control patents using the technology subclass, a much finer level of disaggregation than the three-digit classification scheme. Second, we restrict attention to originating-citing-control triads in which all three patents have at least one subclass in common. We then apply JTH's methods to the new dataset. While we continue to find evidence of international localization effects similar in magnitude to those found by JTH, there is no evidence of the remarkably strong intra-national localization reported in JTH.

The remainder of the paper is organized as follows. Section I explains further the problems induced by JTH's selection method. Section II describes the methods used to construct our dataset. Section III reports our results on geographic matching rates. Section IV concludes.

I. The Selection of Control Patents

This section describes in more detail our two principal concerns with the way in which control patents have been selected. The first is the level of aggregation in the technological classes used. The second is that the selection process does not ensure the existence of any industrial link between the originating and control patents.

A. *The Aggregation Problem*

The three-digit level in the U.S. classification system (USCS) for patents groups highly disparate technologies and industries. If a citing patent falls into the same three-digit technological class as the originating patent, it is also likely to fall into the same subclass. In contrast, the control patent is likely to be drawn from a different subclass, and consequently fails to control for pre-existing geographic patterns of production. It then follows that the control patent is less likely to be filed by an inventor in the same location as the originating patent for the simple reason that it is more likely to be relevant to a different industry.

As illustration, consider class "231-Whips and Whip Apparatus," which JTH cite as an example of the narrowness of the three-digit technological divisions. For reasons not relevant to the present paper, this is a technological class with which one of us happens to be quite familiar. And it turns out that there is a lot more to whips and whip apparatus than you might think. Class 231 contains seven distinct subclasses, which can usefully be grouped into three economically distinct activities. Subclass 231.1 consists of machines peculiar to whip manufacture, mostly rolling, pressing, and shaping machines. Subclasses 231.2 through 231.6 consist of whips, convertible whips and canes, lashes, lash fastenings, and joints, i.e., the pieces that one would expect to go into the average whip. The third subclass, "231.7-Electric Prods," is a rather different beast, and in the last 25 years has yielded patents for cattle prods, electric dog collars, personal security devices, and the rather more intriguing electrically chargeable trousers.[2] Class 231 has yielded 39 patents since January 1, 1976, one in 231.1, 17 in 231.2 through 231.6, and 21 in 231.7. These patents cite 305 domestic utility patents as prior art, of which 222 (73 percent) references are to patents sharing at least one three-digit classification with the citing patent. Once there is a match at the three-digit level, a match at the subclass level is likely: 179 (81 percent) of the pairs with a three-digit match also have at least one subclass match.[3]

These numbers are not atypical. In a sample of 7,627 citations to patents granted in January 1976 (the construction of which is explained in Section II), a somewhat lower 63 percent of the originating-citing pairs share at least one three-digit class, but an identical 81 percent of these also share at least one subclass.

In view of these numbers, how much spurious localization of knowledge might we expect to infer from selecting control patents at the three-digit level? Consider the following thought experiment. There is a patent class with k distinct subclasses, each of which is produced in the same number of geographic regions, and each of which generates the same number of patents. It then follows that the probability that a control patent selected at the three-digit level is taken from the same subclass is approximately $1/k$.[4] Assume also that $0.63 \times 0.81 = 51$ percent of originating-citing pairs share the same subclass. The extent to which inferences about localization of knowledge spillovers are biased depends on the fraction of output of each subclass that is produced in geographically distinct areas. It will be convenient to divide the class into two groups: group A consists only of subclass 1, while group B consists of the remaining $k - 1$ subclasses. Imagine that there are n geographic regions; that production of group B is uniformly distributed across all n regions; and production of group A is uniformly distributed across $m \leq n$ regions. It is then easy to calculate that, absent localization of knowledge spillovers, the probability that the citing patent is from the same region as the originating patent is given by $P_0 = 0.51/m + 0.49/n$, while the probability that the three-digit control patent is from the same region is

$$P_1 = 0.63\left(\frac{1}{km} + \frac{k-1}{kn}\right) + 0.37\left(\frac{1}{n}\right).$$

The ratio of the citing matching rate to the control matching rate, P_0/P_1, depends on the three parameters k, m, and n. At the metropolitan level, JTH used the 1981 definitions of the 17 consolidated metropolitan statistical areas (CMSAs) plus one artificial CMSA for foreign patents, and 50 artificial CMSAs, one for each state, containing all inventor locations not included in one of the 17 official CMSAs. Only 4.5 percent of patents were assigned to the artificial domestic CMSAs and, as the bias problem is rapidly increasing in n, we can set $n = 18$. k varies significantly across three-digit classes. Class 231, with seven subclasses, has an unusually small number, and many three-digit manufacturing classes have more than 100. Finally, we have no information on m. Figure 1 therefore plots the ratio P_0/P_1 for all

[2] Patent number 5,158,039, if you are interested.

[3] These numbers are essentially unchanged when one looks at the 214 citations made *to* the patents in class 231.

[4] The probability is slightly less than $1/k$ because all patents that cited the originating patent, which are more likely to come from the same subclass than from any other, must be removed from the pool of potential controls.

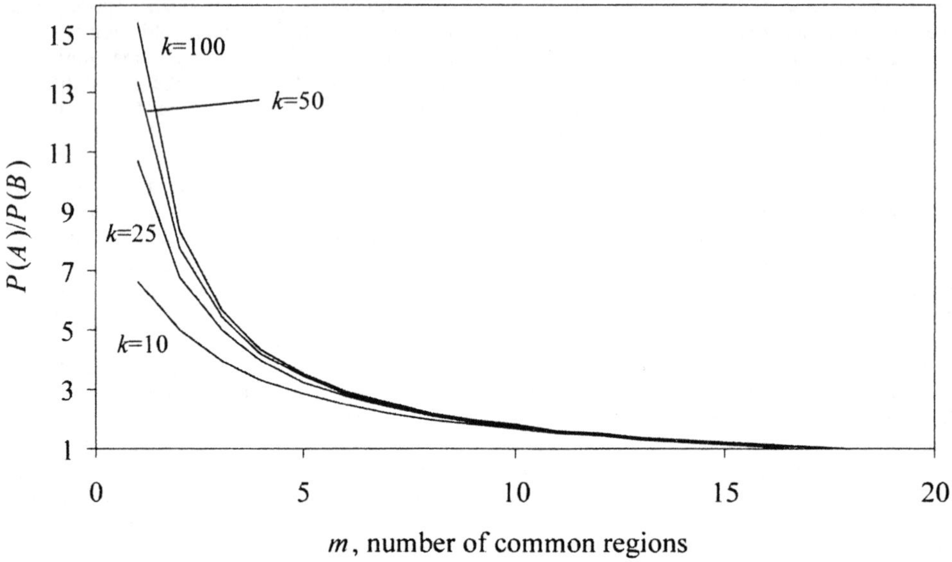

FIGURE 1. SPURIOUS LOCALIZATION EFFECTS

possible values of m when $n = 18$, and $k = \{10, 25, 50, 100\}$. A ratio of unity implies no spurious evidence for localization of knowledge spillovers, but is obtained only when the originating patent is produced in every single area. For $m < n$, in contrast, $P_0 > P_1$, and the ratio becomes especially large for large values of k/m. Note in particular that the ratio can handily exceed the values reported by JTH.

The importance of aggregation bias evidently turns on the number m of regions in which the average subclass is produced. But because production data coincide neither with CMSAs nor the USCS, we cannot address this question directly. What we can attempt to do is ameliorate the aggregation problem by constructing a version of the JTH dataset in which control patents are selected at the technological subclass level, rather than at the three-digit level.

B. *How Well Do Control Patents Do Their Job?*

Consider a citing patent that contains claims in two subclasses, A and B, where subclass A is the primary classification assigned to the patent. It cites prior art protected by a patent in subclass B, but a control patent is selected using the primary subclass A. It is possible under these circumstances that the control patent and the originating patent have no technological class in common. Sometimes, of course, this produces exactly what we want: the control and citing patents have an industry in common but only the citing patent exploits knowledge embodied in the originating patent. But in many other cases, the outcome is perverse: the originating and citing patents come from the same industry but the control and citing patents are essentially unrelated.

Figure 2 provides an illustration. An originating patent for a method to shape bacon bellies was assigned to packaged meat manufacturer Oscar Mayer Food Corporation. At a later date, Sara Lee Foods Corporation, also a major manufacturer of packaged meats, was assigned a patent for grading pork bellies. The latter patent required that the pork bellies be flattened prior to grading, and cited Oscar Mayer's prior art. Obtaining the matching control patent using the primary class is a deterministic process and in this case it leads, at the three-digit level, to a patent for a metal chip compactor, assigned to Prab, Inc., a manufacturer of metal scrap

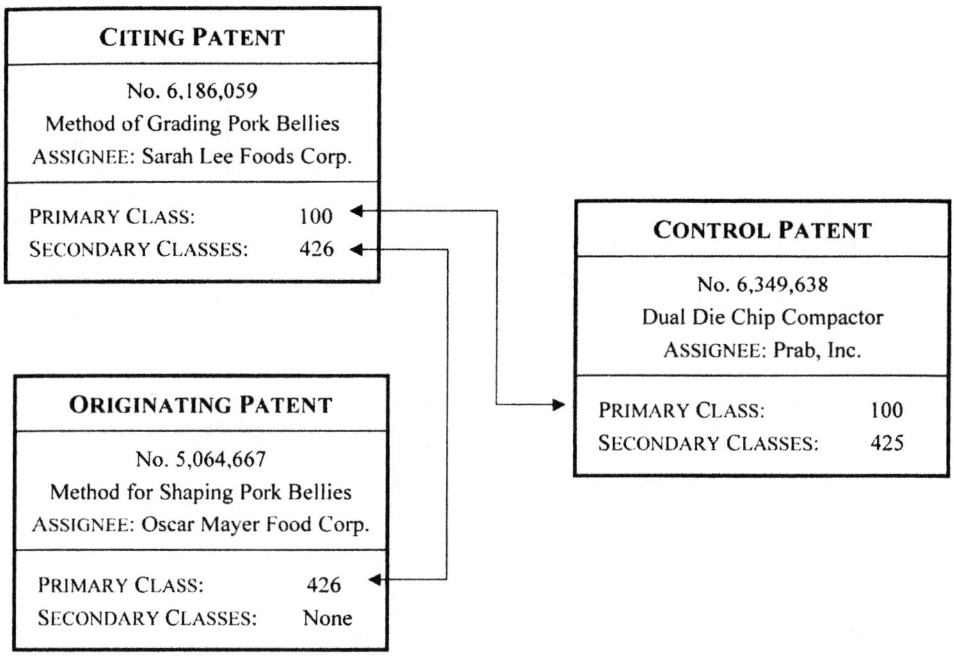

FIGURE 2. UNRELATED PATENTS

Notes: Although the citing and originating patents share a technological class, and the citing and control patents share a technological class, the originating and control patents are unrelated. In the example here, the citing and originating patents belong to firms operating in a common industry (meat processing), but the control patent does not.

reclamation systems. The technological classification that links the control and citing patents turns out to be a generic class covering any type of press, whose only commonality is that they are "apparatus for subjecting material to compressive force" (USPTO, 2002).[5]

The problems illustrated in Figure 2 do not simply introduce noise that can be overcome with a sufficiently large sample size. Sometimes the control patents do their job and sometimes they do not, and those that do not introduce systematic bias into the empirical analysis. Overcoming this problem is not straightforward. Selecting control patents using technological subclasses may reduce the bias, but it cannot eliminate it. For example, the primary subclass, 100/35, for the Sara Lee patent leads to a different control patent. In this case, however, we make little progress, because the new control patent is for a method to reduce variations in the gloss on certain types of paper. The patent is assigned to Stora Enso North America, a major paper manufacturer.

In an attempt to further reduce the bias, one might restrict the sample to observations in which the originating and control patents also share at least one subclass in common. This restriction will increase the likelihood that all three matched patents are drawn from firms engaged in similar activities. Doing so may go too far, because citing and originating patents

[5] The rules governing selection of the primary class are complex. The central rule (although subject to numerous detailed exceptions) is that the classification of the "broadest claim" in a patent is selected as the primary class (see USPTO, 2003a, 2003b). Unfortunately, the broadest claim in a patent need not represent the main thrust of technological or industrial activity that gave rise to the patent. These rules led to the assignment of primary class "426-Food or Edible Material: Processes, Compositions, and Product" to Oscar Mayer's patent, but class "100-Presses" to Sarah Lee's patent.

need not have a technological code in common. Nonetheless, if knowledge spillovers are localized, they should continue to be evident even in this restricted sample.

II. Data Construction

In this section we describe how we constructed a dataset to address the concerns raised above. We began with a sample of 2,724 originating patents consisting of all patents granted during January 1976 (the first month for which text search capabilities are offered), that had at least one inventor domiciled in the United States, and that were assigned to a company or institution. We then identified 18,551 patents granted between January 1976 and April 2001 that cite one or more of the originating patents. Finally, we constructed two control groups. First, for each citing patent we paired it with a control patent having a similar application date and that matched the primary classification of its paired citing patent at the three-digit level. The second group selected control patents that matched the primary classification of its paired citing patent at the level of the subclass.[6]

These datasets are analogs to those created by JTH, and our intention was to match their data construction procedure as closely as possible. Financial constraints, however, required us to extract our data from the USPTO Web site, with consequent limitations on access speed and search capabilities. These limitations imposed some minor differences between our procedure and those of JTH:

(a) JTH selected controls in which the primary class of the citing patent matched the primary class of its control patent. We matched the primary class of the citing patent to *any* class of the selected control patent;

(b) Among all admissible control patents (i.e., those with a technological match to the citing patent but that do not cite the corresponding originating patent), JTH deterministically selected the control patent that most closely matched the citing patent's application date. We randomly selected a control from all admissible patents with application dates within a one-month window on either side of the citing patent's application date. If no admissible patent was found within the plus-or-minus one-month window, the search was repeated with a plus-or-minus three-month window, and again, if necessary, with a plus-or-minus six-month window. If no control patent could be located within the widest window, a null observation was returned.[7]

The sample was then culled in three ways. First, following JTH we eliminated self-cites by removing observations in which the citing and originating patents had the same assignee. Second, we removed any observation for which we did not, as a result of our automated data extraction procedure, have readable data.[8] Third, we removed observations for which either the citing patent or either of the two control patents were not assigned to a corporation or institution.[9]

The final task was to assign geographic locations to each patent. Patents report the towns/cities and states of the inventors. These were first converted to counties and then to the 17 CMSAs as defined in 1981, using correlation files provided by the Office of Social and Economic Data Analysis (OSEDA) of the University of Missouri. Following JTH we also created one phantom CMSA for foreign patents and 50 phantom CMSAs, one for each state, containing all locations not included in one of the 17 CMSAs. For each patent we made a list of the locations of all domestic inventors and then selected at random a single inventor from this

[6] The three-digit and disaggregated controls could be the same patent.

[7] A null observation was returned about 40 percent of the time, reflecting the fact that there are over 100,000 subclasses in the USCS.

[8] The data were extracted from the USPTO Web site and coded using programs written in *perl*. Misplaced punctuation in the patent text readily defeats such methods and generates garbage.

[9] This culling was originally motivated by programming ease. If the inventor is also the assignee, the inventor's full name and address are given. If the inventor is not the assignee, only the inventor's hometown and state are given. This disparate treatment defeated our programming ability. In retrospect, however, we think the restriction creates a more interesting sample.

TABLE 1—SAMPLE SIZES

	Total	Rate of technological classification matching between controls and citing patents		Percentage of control patents with application date within a given time-window of citing patent application date		
	(N)	Primary USC (%)	Secondary USC (%)	+/−1 Month	+/−3 Months	+/−6 Months
Originating	1,913	—	—	—	—	—
Citing	7,627	—	—	—	—	—
3-digit control	7,627	58.1	41.9	>99.9	<0.1	0.0
Disaggregated controls	7,627	31.5	68.5	70.5	21.0	8.5

list to assign a unique location to the patent.[10]

The resulting dataset is composed of 7,627 observations generated from 1,913 originating patents, each of which has four elements: an originating patent, a citing patent, and two controls. Table 1 provides some details. At the three-digit level, 58 percent of the observations match the primary class of the citing patent to the primary class of the control. The primary matching rate at the subclass level is rather lower, at 31.5 percent. Unsurprisingly, in view of the modest number of three-digit classes and the very large number of subclasses defined by the USPTO, almost all three-digit controls have an application date within the narrowest window around the citing patent's application date, but this was the case for only 71.5 percent of the disaggregated controls.

III. Results

If, as we have argued, the disaggregation of technological classes matters for geographic location, the fraction of control patents that match the location of the *citing* patent will be higher the greater the disaggregation. Table 2 provides support for this prediction. The matching rates at the disaggregated level are 1.06, 1.35, and 1.38 times greater than at the aggregated level for the country, state, and CMSA, respectively, and all of these differences are statistically significant at the 1-percent level.

TABLE 2—GEOGRAPHIC MATCHES BETWEEN CITING AND CONTROL PATENTS

	Control patents	
	3-digit	Disaggregated
Percentage matched country	53.79	57.17 (4.20)
Percentage matched state	17.65	23.93 (9.58)
Percentage matched CMSA	16.14	22.23 (9.58)

Notes: Numbers in parentheses in Disaggregated column give the *t*-statistic for the test of equality between the indicated geographic matching rate and the corresponding matching rate for the 3-digit controls.

Table 3 presents the main results. Columns (1) and (2) summarize the sample-weighted geographic matching rates for the 1975 and 1980 cohorts of corporate originating patents presented in Table III of JTH. Their findings for this sample were that, excluding self-citations: (a) 68.0 [11.2, 7.3] percent of citing patents matched the country [state, CMSA] of the corresponding originating patent; and (b) 62.1 [6.4, 2.2] percent of control patents matched the country [state, CMSA] of the corresponding originating patent. That is, a citing patent is 1.1 times more likely than the control patent to match the originating patent country, 1.7 times more likely to match the state, and 3.3 times more likely to match the CMSA. These differ-

[10] While the main attraction of this method is that it is easy to program, it has other useful features. JTH assigned each patent to a "primary location," defined as the CMSA in which a plurality of inventors lived. An alternative is to accept locations of patents as matching if they share at least one inventor location in common. Consider matched pairs such that one patent in each pair has all inventors located in Los Angeles, while the other has two inventors located in Los Angeles and three in New York. JTH's plurality approach would fail to show a match between any such pairs; the alternative makes a match between every such pair. Because our interest is in the aggregate proportion of geographic matches, a methodology that weights the proportion of matches according to the degree of overlap seems appropriate. The methodology we use does this, and in the present example would produce a geographic match 40 percent of the time.

TABLE 3—GEOGRAPHIC MATCHING RATES
(*Self-citation removed*)

	JTH's corporate patents				Disaggregated controls		
					Citing-control code match		Control patent has any subclass code in common with originating patent
	Citing patents	3-digit controls	Citing patents	3-digit controls	Any code	Primary code	
	(1)	(2)	(3)	(4)	(5)	(6)	(7)
No. of citations	4,234	4,234	7,627	7,627	7,627	2,466	2,122
Percentage matched country	68.0	62.1	68.61	55.63	55.74	54.33	56.93
		(5.70)		(16.7)	(16.5)	(12.6)	(9.74)
Percentage matched state	11.2	6.4	7.75	5.00	6.03	6.69	7.73
		(7.53)		(6.96)	(4.20)	(1.80)	(0.03)
Percentage matched CMSA	7.3	2.2	5.22	3.47	4.10	4.37	5.18
		(11.1)		(5.31)	(3.28)	(1.76)	(0.07)

Notes: JTH numbers are a weighted average of their "top corporate" and "other corporate" samples for 1975 and 1980 originating cohorts. See their Table III. Sample sizes are adjusted from those reported by JTH to reflect self-citation rates reported in their Table I. Numbers in parentheses give the *t*-statistic for the test of equality between the indicated geographic matching rate and the corresponding matching rate for the citing patents excluding self-citations. Column (2) gives *t*-statistics for test of equality with the corresponding row entry in column (1). Columns (4) to (7) give *t*-statistics for test of equality with the corresponding row entry in column (3).

ences are statistically significant at a level much lower than the 1-percent level.

Results from our new sample are reported in columns (3) through (7). The geographic matching rates for citing patents [column (3)] at the state and CMSA levels are a little lower than those obtained by JTH. Our citing patents were awarded over a period of 25 years compared with periods of 9 and 14 years for JTH's two samples, so these lower matching rates are consistent with JTH's finding that apparent localization effects appear to fade with time. The numbers remain comparable, however, particularly in the rate of decline in matching rates as we move to finer geographic entities.[11]

While the country and state matching rates of our three-digit control patents are also lower, our CMSA matching rate for the three-digit controls, at 3.47 percent, is noticeably higher than the rate reported by JTH.[12] Nonetheless, our sample replicates JTH's central result that a citing patent is significantly more likely to match the location of the originating patent than is the three-digit control patent. As the *t*-statistics in column (4) show, the difference between the matching rates for the citing and for the three-digit controls are highly significant at each geographic level. Citing patents are 1.2 times more likely than the control patents to match the country, 1.6 times more likely to match the state, and 1.5 times more likely to match the CMSA. In conclusion, we also obtain strong evidence of international and intranational localization effects from three-digit controls. Columns (5) and (6) report the results of our attempts to resolve the bias created by the problem of aggregation. Column (5) reports the matching rates for the disaggregated controls in the full sample, while column (6) reports the rates when the sample is restricted to control patents whose primary subclass matches the primary subclass of the citing patent. At the state and CMSA levels, the introduction of the more disaggregated controls moves the matching rates in the direction expected. In column (5), the state and CMSA matching rates rise to 6.03 and 4.10 percent, respectively, an increase of about 20 percent in each case. Restricting the sample to observations in which the primary subclass of the citing and control patents coincide further increases the matching rates at the

[11] Moreover, the results reported below are very similar when we restrict the sample to citing patents with application dates prior to January 1986, corresponding approximately to JTH's shortest sample.

[12] We attribute this difference to our decision to restrict the sample, not only for originating patents but also for citing and control patents, to those patents with a corporate assignee.

TABLE 4—GEOGRAPHIC MATCHING RATES, REDUCED SAMPLE

	Citing patents		Control patents (from column (7) of Table 3)
	In sample	Not in sample	
	(1)	(2)	(3)
Sample size	2,122	5,505	2,122
Percentage matched country	66.11	69.57	56.93
		(−2.89)	(6.17)
Percentage matched state	8.77	7.36	7.73
		(2.00)	(1.23)
Percentage matched CMSA	6.07	4.89	5.18
		(1.98)	(1.26)

Note: Columns (2) and (3) give *t*-statistics for test of equality with the corresponding row entry in column (1).

state and CMSA levels. The new control group has a state matching rate of 6.7 percent and a CMSA matching rate of 4.4 percent. At the customary 5-percent standard, neither of these differs significantly from the citing-originating matching rates reported in column (3).

Column (7) further restricts the sample in an attempt to reduce the bias caused by a failure of the control selection procedure to ensure that there is some technological link across patents. The state and CMSA matching rates for the control sample, which now include only patents that have at least one technological subclass in common with the originating patent, show further changes in the expected direction. The state matching rate between control and originating patents is now 7.7 percent, while the CMSA matching rate rises to 5.18 percent, both of which are virtually identical to the citing-originating matching rates.

The results for state and CMSA matching rates stand in stark contrast to those obtained for country matches. Moving from three-digit to disaggregated controls, and further restricting the control group to patents sharing a primary code with the citing patent, induces essentially no change in the country-matching rate. Even with the restricted control groups, highly significant localization effects are apparent at the country level. The country-matching rate for citing patents is 68.6 percent, while for the controls it stays within the narrow range of 54.3 to 56.9 percent across all columns.

It might reasonably be argued that the comparisons made across different sample sizes are inappropriate. In moving from column (5) to column (7), the control group has declined in size by about two-thirds, and it is possible that the reduced sample of 2,122 citing patents that are matched to the control patents in column (7) is systematically different from the full sample. It is not obvious why the reduced sample might differ from the full sample but, somewhat surprisingly, it does. Columns (1) and (2) of Table 4 provide the matching rates for the 2,122 citing-originating pairs that correspond to the control patents in column (7) of Table 3, and the 5,505 that do not. The matching rate for the in-sample patents is lower at the country level but higher at the state and CMSA levels, and each difference is significant at the 5-percent level. Column (3) replicates the matching rates from Column (7) of Table 3, but now tests them against the reduced sample of citing patents. On the basis of the *t*-statistics, the changes in the matching rates for the citing patents do not alter our conclusions.

IV. Conclusions

We have evaluated the robustness of Jaffe, Trajtenberg, and Henderson's (1993) evidence on the localization of knowledge spillovers inferred from patent citations. We had been concerned with the level of aggregation employed in selected patents designed to control for the existing geography of production, and with the fact that the method used to select control patents did not ensure that the control and originating patents had any technological class in common. JTH selected control patents using the three-digit technological classification system of the patent office, but patent citations are commonly made within more detailed technological subclasses. We showed how, if there is geographic concentration of production within subclasses, the use of the three-digit classification system may generate spurious evidence of localized spillovers. Using a sample of corporate and university originating patents, we found that selecting control patents with a finer technological classification accounts for a significant fraction of the evidence for state and CMSA localization effects inferred from the three-digit controls.[13] The combination of using finer technological classifications and ensuring that the control and originating patents have at least one technological class in common eliminates any statistical support for intra-national localization effects. Remarkably, in view of these results, JTH's finding of significant localization effects at the country level easily survives our reassessment.

Reporting these results is easier than interpreting them. An optimistic interpretation is that the underlying methodology in JTH is in principle capable of identifying localization effects as long as the controls are carefully selected, that the present paper has done so, and it produced evidence that only national borders restrict knowledge flows. A natural corollary of this interpretation is that making financial resources available to merge subclass data into the Hall, Jaffe, and Trajtenberg (2000) database would be money well spent.

But this is just wishful thinking. The construction of controls using subclasses may enhance technological or industrial similarities between the citing and control patents, but subclasses are no panacea. It should not be forgotten that the USCS is a *library* classification system designed for the sole purpose of facilitating searches by patent examiners. Any correlation with industrial activity is purely incidental, and any application of the classification system to other ends is fraught with risk. Even disaggregated controls leave a lot of unobserved heterogeneity in industrial activity. Moreover, the use of subclasses introduces new problems, the full implications of which remain unexplored. Most important, for fully 40 percent of the time, we were unable to find any subclass control patent with an application date within one year of the application date of the citing patent, and further restricting the type of subclass match only exacerbates this problem. One can easily construct scenarios in which these failures induce selection bias, some of which would have caused us to overestimate the geographic matching rates, others to underestimate it.[14]

The fact that our new selection criteria for control patents induce significant movements in matching rates must lead us to conclude that imperfect matching at the three-digit level is quantitatively important. But there is no doubt that matching remains far from perfect with the disaggregated controls. Moreover, the disaggregated controls introduce additional concerns. Perhaps a more productive route in the search for localized knowledge spillovers is to move away from the USPTO classification scheme altogether, merge citation data with other measures of technological and industrial proximity, and supplement matching methods with regression methods.[15] But ultimately, *any* nonexperimental evidence for localization effects may always be attributed to unobserved within-technology heterogeneity. The implication is that we must devise identification strategies that do not rely on attempts to measure and classify technology. One recent approach (Thompson, 2004), which exploits within-patent variations in geographic matching for citations added by inventors and citations added by examiners, finds evidence of modest localization effects. It remains to be seen, however, whether alterna-

[13] Our findings are also consistent with Almeida and Bruce Kogut (1999). They analyze the geography of citations to major semiconductor patents, using controls selected with subclass data, and find evidence of localized spillovers in only three of 13 regions.

[14] One referee suggested the following scenario. Assume that older industries produce fewer patents and are less geographically concentrated. Then citing patents for which no satisfactory controls were found are more likely to come from an older industry and less likely to share the same location as the control patent than are observations included in the sample. This will induce an underestimate of the localization of knowledge spillovers. But consider an alternative scenario. Subclasses that are more narrowly defined are more likely to generate controls closely related to their corresponding citing patents, but they also generate fewer patents. Under this scenario, the omitted observations are *more* likely to share the same location as the citing patent than are observations included in the sample.

[15] See, for example, Jaffe and Trajtenberg (2002), and Jasjit Singh (2003a, 2003b).

tive identification schemes such as this can also withstand closer scrutiny.

REFERENCES

Almeida, Paul. "Knowledge Sourcing by Foreign Multinationals: Patent Citation Analysis in the U.S. Semiconductor Industry." *Strategic Management Journal*, 1996, *17*(Winter Special Issue), pp. 421–31.

Almeida, Paul and Kogut, Bruce. "Localization of Knowledge and the Mobility of Engineers in Regional Networks." *Management Science*, 1999, *45*(7), pp. 905–17.

Frost, Tony S. "The Geographic Sources of Foreign Subsidiaries' Innovations." *Strategic Management Journal*, 2001, *22*(2), pp. 101–23.

Glaeser, Edward L. "Learning in Cities." *Journal of Urban Economics*, 1999, *46*(2), pp. 254–77.

Hall, Bronwyn H.; Jaffe, Adam B. and Trajtenberg, Manuel. "Market Value and Patent Citations: A First Look." National Bureau of Economic Research, Inc., NBER Working Papers: No. 7741, 2000.

Hall, Bronwyn H.; Jaffe, Adam B. and Trajtenberg, Manuel. "The NBER Patent Citation Data File: Lessons, Insights and Methodological Tools." National Bureau of Economic Research, Inc., NBER Working Papers: No. 8498, 2001.

Hicks, Diana; Breitzman, Tony; Olivastro, Dominic and Hamilton, Kimberly. "The Changing Composition of Innovative Activity in the US—A Portrait Based on Patent Analysis." *Research Policy*, 2001, *30*(4), pp. 681–703.

Jaffe, Adam B. and Trajtenberg, Manuel. "International Knowledge Flows: Evidence from Patent Citations." *Economics of Innovation and New Technology*, 1999, *8*(1–2), pp. 105–36.

Jaffe, Adam B. and Trajtenberg, Manuel. *Patents, citations, and innovations: A window on the knowledge economy.* Cambridge, MA: MIT Press, 2002.

Jaffe, Adam B.; Trajtenberg, Manuel and Henderson, Rebecca. "Geographic Localization of Knowledge Spillovers as Evidenced by Patent Citations." *Quarterly Journal of Economics*, 1993, *108*(3), pp. 577–98.

Krugman, Paul. *Geography and trade. Gaston Eyskens Lecture Series.* Cambridge, MA: MIT Press, 1991.

Marshall, Alfred. *Principles of economics,* 8th ed. London: Macmillan, 1920.

Singh, Jasjit. "Multinational Firms and Knowledge Diffusion: Evidence Using Patent Citation Data." Unpublished Paper, 2003a.

Singh, Jasjit. "Social Networks as a Determinant of Knowledge Diffusion Patterns." Unpublished Paper, 2003b.

Thompson, Peter. "Patent Citations and the Geography of Knowledge Spillovers: What Do Patent Examiners Know?" Unpublished Paper, 2004.

Trajtenberg, Manuel. "A Penny for Your Quotes: Patent Citations and the Value of Innovations." *RAND Journal of Economics*, 1990, *21*(1), pp. 172–87.

United States Patent and Trademark Office. *U.S. manual of classification*, uspto.gov/web/patents/classification, Accessed 5 April 2002.

United States Patent and Trademark Office. *Examiner handbook to the U.S. patent classification system*, http://uspto.ustpo.gov/web/offoces/pac/dapp/sir/co/examhbk/index.html, accessed 25 July 2003a.

United States Patent and Trademark Office. *Manual of patent examining procedure,* 8th Ed. (Revised February 2003), http://www.ustpo.gov//web/offices/pac/mpep/index.html, accessed 25 July 2003b.

Part V
Knowledge Spillovers and New Firm Formation

[20]

Expropriation and Inventions: Appropriable Rents in the Absence of Property Rights

By JAMES J. ANTON AND DENNIS A. YAO*

We analyze the problem faced by a financially weak independent inventor when selling a valuable, but easily imitated, invention for which no property rights exist. The inventor can protect his or her intellectual property by negotiating a contingent contract (with a buyer) prior to revealing the invention or, alternatively, the inventor can reveal the invention and then negotiate with the newly informed buyer. Despite the risk of expropriation, we find that, in equilibrium, an inventor with little wealth can expect to appropriate a sizable share of the market value of the invention by adopting the latter approach. (*JEL* L13, O31, D23)

Most independent inventors cannot successfully create an organization to take commercial advantage of their invention and, therefore, must rely on another party. The ensuing relationship may involve a production contract, a licensing arrangement, or the outright sale of the invention. With all of these forms, the inventor's ability to capture rents depends on the marketplace value of the invention, the means by which the invention is taken to market, and the property rights and information of the inventor.

When an inventor can rely on patents or other forms of property rights to protect his intellectual property, theory suggests that the inventor can appropriate a substantial fraction of the invention's value. If property rights are weak or nonexistent, as is often the case, the inventor's ability to capture rents would seem to be substantially circumscribed. Under patent law, for example, many inventions are considered to be too closely related to "prior art" to justify a patent, and even if patents are granted "many...would be held invalid if ever litigated" (Phillip Areeda and Louis Kaplow, 1988 p. 177).[1] Reliance on trade-secret law (law pertaining to the use of confidentially disclosed, commercially useful information) to protect against the theft of ideas is often problematic because of the strategies that buyers employ to avoid legal challenges and

*Anton: Fuqua School of Business, Duke University, Durham, NC 27708; Yao: The Wharton School, University of Pennsylvania, Philadelphia, PA 19104, and Federal Trade Commission, Washington, DC 20580. We thank Beth Allen, Jean-Pierre Benoit, Ken Hendricks, Bob Nau, Bob Rosenthal, the referees, and workshop participants at Duke, Northwestern, Penn State, Rutgers, the University of Pennsylvania, the 1991 Econometric Society Summer meetings, the 1991 University of British Columbia Summer Conference, the 1992 AEA meetings, and the 1993 Stony Brook Game Theory Conference for comments and the Snider Entrepreneurial Center and Citibank for financial support. Additional support from the Fuqua Business Associates Fund is gratefully acknowledged.

[1]Consider, for example, the case of the ballpoint pen. Although L. J. Biro invented and patented the ballpoint pen, the first ballpoint pens were marketed in the United States by Milton Reynolds, who developed a pen (after seeing a Biro pen in Argentina) using "prior art" that did not infringe on Biro's patents (John Jewkes et al., 1969 pp. 234–35). Eric von Hippel's (1982) study of court records shows that, when litigation occurs, the likelihood of a patent being held valid and infringed is about 3 to 1 against the patent-holder.

the difficulty of establishing legal proof.[2] Finally, some forms of intellectual property, such as new product concepts and management-systems ideas, are just inherently difficult to protect.

In this paper we focus on the inventor's problem when no legal mechanism provides effective protection for the invention.[3] Our stylized version of the problem addresses the following concern. Suppose an inventor cannot individually bring an invention to the market and is concerned about disclosing key aspects of the invention to a potential buyer or partner. Sight unseen, a buyer may be hesitant to buy or contract for an unknown commodity, but if the inventor first reveals information regarding the invention, a lack of property rights makes it possible for the newly informed party to "steal" or "expropriate" the invention. Expecting that firms are likely to pay very little for an invention they have not seen, the inventor must decide if they would pay anything at all once they have a clear avenue to take the invention to market without the inventor.[4]

We model a setting with an inventor and two manufacturing firms that are currently competing in a duopoly market. The inventor may possess a valuable process invention that reduces the costs of production of the manufacturers. Initially, only the inventor knows whether the invention is valuable. A useless invention has no incremental value, while a valuable invention results in incremental profit increases to the manufacturers, the sizes of which depend on whether the other firm has the invention. The invention also has the property that, once revealed to a manufacturer by the inventor, it can be perfectly reproduced without cost or threat of legal action. Firms can also independently develop the invention, and a third party cannot verify the source of a valuable invention.[5]

Suppose that an inventor attempts to sell the invention while avoiding the risk of expropriation. One way to do this is to negotiate a contract before revealing the invention to any of the firms. When the inventor has large financial resources, it is straightforward to show that an inventor can credibly signal the value of the invention by agreeing to part with assets if the invention turns out to be useless. In this case, the inventor can obtain the full value of the invention to a "monopolist" firm. When, however, an inventor's wealth is small, this *ex ante* signal comes at the expense of inventor rents; when invention is difficult *ex ante* and the inventor's wealth goes to zero, the inventor's rents vanish.

Our primary result is that an inventor without property rights and with virtually no wealth can still appropriate a sizable share of the ultimate value of the invention. To obtain these rents the inventor chooses to reveal the invention prior to any agreement.

[2]Steven N. S. Cheung (1982), for example, notes that almost all U.S. firms require outside inventors to sign a waiver form acknowledging that no confidential relationship is established, thereby undermining a trade-secret remedy for idea theft. See also the opinions in *Burten* v. *Milton Bradley Co.* (592 F. Supp. 1021 [D.R.I. 1984], rev'd 763 F.2d 461 [1st Cir. 1985]) a case in which an electronic board game submitted to Milton Bradley from two inventors was allegedly "stolen" from the inventors. In addition, under trade-secret law (see J. H. A. Pooley, 1987) there is no violation if the invention was independently discovered or reverse-engineered. This is the basis of Kentucky Fried Chicken's defense against an inventor who claims that KFC stole his pressurized frying cooker (*Business Week*, 7 May 1990, p. 41).

[3]By analyzing the no-property-rights case, we adopt a different perspective from that of the R&D and innovation literature that assumes patent protection (see e.g., Partha Dasgupta and Joseph Stiglitz, 1980; Nancy T. Gallini, 1984; Gallini and Ralph A. Winter, 1985; Morton I. Kamien and Yair Tauman, 1986; Michael L. Katz and Carl Shapiro, 1986).

[4]David J. Teece (1986) argues, for example, that when imitation is easy the profits from innovation accrue to owners of complementary assets such as manufacturing capability rather than to the developers of the intellectual property. Gallini and Brian D. Wright (1990) identify when an inventor can capture rents by signaling the value of an innovation via licensing contracts.

[5]The amount of payment that the inventor can obtain from the firms is conditioned on the firms' beliefs about the usefulness of the invention. Thus, the model addresses adverse-selection problems but not moral-hazard issues.

This strategy allows the inventor to negotiate with the firm on the basis of full, rather than incomplete, information about the value of the invention. The manufacturer voluntarily negotiates a contract with the inventor because the manufacturer is concerned about the possible loss of its "monopoly" in the use of the invention. The equilibrium contract creates incentives that prevent the inventor from profiting by selling the invention to the other firm and undermining the first firm's monopoly. We also find that, when bargaining power lies with the firms, the inventor's share of the rents increases as the inventor's *ex ante* wealth decreases. This result derives from the fully informed firm's strategic use of the inventor's wealth (within the structure of the contract) to limit the potential contracts an inventor will accept from the other firm.

The results in our paper address the general issue of selling information, which has been emphasized by Kenneth J. Arrow and others (see Arrow, 1970). As the inventor chooses, in equilibrium, to reveal the invention prior to receiving a contract offer, our analysis is also related to the recent work on strategic information revelation. For example, see Masahiro Okuno-Fujiwara et al. (1990), which examines revelation via public disclosure. In contrast, we focus on a setting where information is privately revealed to only one of two firms in the market, and the only way the inventor can convince the firm that the invention is valuable involves transferring the necessary "how-to" knowledge to the firm. Thus, in terms of certifiable information, the inventor is unable to mitigate the adverse-selection problem without also opening the door for the firm to expropriate the invention via imitation.

The paper proceeds as follows. Section I describes the model and extensive form. Section II presents our main result. We characterize an equilibrium that exhibits *ex post* contracting on the equilibrium path: the inventor first reveals the invention to one of the firms and then receives and accepts an *ex post* contract offer for payment. Section III, which provides a benchmark for the main result, characterizes an equilibrium that exhibits *ex ante* contracting and derives a general bound on the payoff to an inventor in any *ex ante* allocation. We present our conclusions in Section IV. All proofs are in the Appendix.

I. The Model

We are interested in the extent to which an inventor can obtain rents from a useful invention when another party is needed to realize the value of the invention (i.e., put the invention to use), there are no property rights regarding the invention, and the inventor has small assets relative to the economic value of the invention. The essential features of this problem can be captured with a relatively simple model. Consider an industry in which there is the potential for a discovery that represents a discrete improvement over the status quo.[6] The industry consists of two firms or manufacturers, $i = 1, 2$, and there is a single inventor, I, who is outside of the industry. The inventor can make the discovery (a "good" invention) or fail to make the discovery (a "bad" invention). All parties are risk-neutral and maximize expected profits.

The model incorporates the following elements. *Ex ante*, neither manufacturer knows whether the inventor has a good invention; this is private information of the inventor. The inventor has the strategic option of "showing" or "revealing" a good invention to a manufacturer, which may be done before or after contracting. Thus, a good inventor can choose whether contracting is conducted on an *ex post* basis with an "informed" firm (the inventor reveals the invention prior to a contract offer) or on an *ex ante* basis with an "uninformed" firm (the inventor seeks a contract offer prior to revealing the invention). When the inventor reveals a good invention prior to a contract

[6]The discovery could, for instance, involve a process innovation that reduces production costs for an existing product. The model also applies if the discovery relates to a product innovation. To the extent that "reverse engineering" is possible, the benefit from innovating is the head start in bringing the product to market (see e.g., Jean-Pierre Benoit, 1985).

offer, the manufacturer is affected in two ways. In addition to learning that the inventor has a valuable invention, the manufacturer also gains the ability to reproduce and use the invention (imitate) without relying further on the inventor and without encountering any legal obstacles associated with expropriating the inventor's intellectual property.[7] We refer to this act of imitation as "expropriation."

We also allow for the possibility that a manufacturer may discover the invention independently of the inventor. Thus, the inventor or the firm may be the underlying source of a valuable invention, and this allows us to capture the lack of property rights as a structural feature of the analysis. We now turn to the formal specification of these elements and, with these in place, we present the extensive form of the model.

The profits of the manufacturers, gross of any payments to the inventor, are determined by which of the firms, if any, has a good invention. For simplicity, we specify these profits in reduced form.[8] There are four possible states of the world. When firm i has the (good) invention but firm j does not, then i earns a profit of π_M while j earns π_L, where M stands for "monopoly" and L for "low" profits. We use "monopoly" to refer to the case in which only one firm has the invention (the other firm is not necessarily driven from the market). If both firms have the invention, then each firm earns π_D where D stands for "duopoly" profits. Finally, profits are π_0 for each firm when neither has the invention. Assume that π_M exceeds π_D, and π_L and π_0 are each less than π_D. More importantly, we make the standard assumption that $\pi_M + \pi_L > 2\pi_D$ (industry profits are largest when one firm uses the invention).

The source of a good invention can be the inventor or a manufacturer. At the start of the game, I discovers a good or bad invention; neither is of direct value to I. The manufacturers do not observe this outcome. *Ex ante*, they believe I discovers a good invention with probability $q \in (0,1)$. Independently of each other and the inventor, each firm discovers the good invention with probability $\alpha \in (0,1)$; this occurs just before gross profits are realized.[9]

A good inventor can reveal the invention to a firm. In this event, the firm is able to make full use of the invention and earn a gross profit of π_M or π_D, depending on whether the other firm also possesses the invention. In our analysis, the ability of the inventor to appropriate rents is related to the rents available to the firms and the inventor collectively (the gross profit levels). These gross profits should be interpreted as including any downstream effects of imitation. If the invention involves process innovation, the invention may remain hidden from a competing firm when one firm adopts the invention in production. In contrast, for a product innovation in a market without a substantial first-mover advantage, the collective rents are likely to be smaller when a previously uninformed competitor can imitate the invention shortly after it is brought to market.

Now consider the specification of contracting within this framework. We assume that legally enforceable contracts take the form of payments that are contingent on the gross profit levels of the two firms. That is, we assume that *ex post* the gross profit levels are observable and verifiable. A contract between I and a firm is therefore a vector $\mathbf{R} = (R_0, R_i, R_j, R_D)$ where R_x is the payment by the firm to I in the event state x occurs and 0, i, j, and D correspond,

[7] One interpretation of this is that, once the invention is revealed, all relevant information regarding the invention (design, manufacture, etc.) becomes known to the firm, and the inventor cannot provide additional value with respect to the use of the invention by the firm.

[8] It is possible to derive the gross profits from an underlying model of duopoly competition. An example is provided by Cournot quantity competition, extended to include asymmetric cost information between the manufacturers.

[9] This allows us to explore expropriation without creating an added source of private information, as would be the case if firms had earlier discovery dates. The basic incentive for contracting is robust to this variation.

respectively, to gross profit outcomes when neither firm, only firm i, only firm j, and both firms have the invention.

Contracting on gross profits reflects the lack of property rights regarding the invention. In essence, we assume that a third party can observe only events that are consequences of using the invention. While a third party may conclude that the invention is being used, the third party cannot verify the source of the invention (as $\alpha > 0$ and $q > 0$, the source may be the inventor or the firms). For example, a payment of R_i in the event of a monopoly for firm i must be made whenever i earns π_M, independently of whether I revealed the invention to firm i or firm i discovered the invention.

Finally, we assume that the inventor has a limited liability equal to $L > 0$. We are particularly interested in the case of L approaching zero, which corresponds to a setting in which the inventor cannot pursue a self-use strategy (or a self-manufacture strategy in the case of product innovations), perhaps due to limited wealth or capital-market access. By assumption, then, a net payment by the inventor in excess of L in any state is not feasible. For simplicity, we assume that the value of L is common knowledge among the parties. Thus, if a contract **R** is agreed to and, subsequently, an additional contract **S** is agreed to, then we require

$$(1) \qquad R_x \geq -L \text{ and } R_x + S_x \geq -L \qquad x \in \{0, i, j, \mathrm{D}\}.$$

Note that (1) constrains each step in a sequence of contracts so that an initial contract payment is not allowed to violate (1) based on an expectation that a subsequent contract will resolve the violation.

The next step is to embed these elements in an extensive-form game of incomplete information. Figure 1 provides a schematic of the extensive form. It is a sequential process in which firms make contract offers and bargaining with one firm must be concluded prior to bargaining with the other firm.

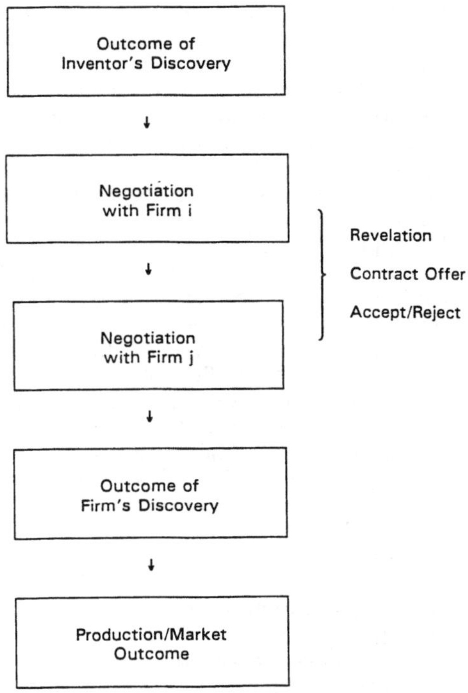

FIGURE 1. STRUCTURE OF THE GAME

The extensive form begins with I learning privately whether the invention is good or bad; we refer to this as a good I and a bad I, respectively. Next, I can choose to approach either firm for a contract offer or choose not to approach. I approaches a firm by (privately) sending one of two messages. If I reveals (only a good I can reveal), then the firm simultaneously learns that I is good and also acquires all necessary "how-to" knowledge (the ability to imitate the invention and, hence, to expropriate). Alternatively, I can choose not to reveal the invention. In response to the message, the firm chooses what contract, if any, to offer. Then, I chooses whether to accept the offer and whether to reveal (unless revelation has already occurred). This process is then repeated with the other firm.

There is only one bargaining round with each firm so that, at most, a firm makes one offer to the inventor. Thus, we employ a simple take-it-or-leave-it bargaining structure for contracting; this is conservative with respect to the payoffs of the inventor, as the firms will offer contracts that push the inventor to indifference with respect to accepting.

Consider now the information available to a firm when formulating a contract offer. If I approached firm k by revealing, then we refer to k as an "informed" firm and refer to an offer from k as an "*ex post*" contract offer. If I did not reveal, then k is "uninformed," and we refer to the offer as an "*ex ante*" contract offer. We assume that a signed contract between an inventor and a firm is observable to the other firm, but one firm cannot observe whether a good inventor reveals the invention to the other firm. Thus, an element of the set {informed, uninformed}$\times\{\mathbf{R}|R_x \geq -L,\ x=0,1,2,\mathrm{D}\}$ describes the information available to a firm, where an observation of $\mathbf{R}=(0,0,0,0)$ denotes no prior contract observed, and the contract offer, $\mathbf{S}$, must satisfy $S_x + R_x \geq -L$. A strategy for the firm maps observed information to a contract offer. Thus, a firm makes an offer based on whether a good invention has been revealed to it and on the observed contract, if any, between I and the other firm.

We now turn to the beliefs of the firms. Due to incomplete information, a firm cannot be sure of precisely where it is in the extensive form when it makes a contract offer. Rather, the firm has beliefs about its position which are based on available information. Suppose firm k observes that a signed contract is in place between I and firm $\sim k$. Then beliefs for k must specify probabilities for the events (a) I is bad, (b) I is good and has revealed the invention to the other firm, and (c) I is good and has not revealed to the other firm. If, in addition, k is informed, then the probability that I is bad must be zero. If k observes no signed contract between I and $\sim k$, then firm k is unsure whether it is the first or second firm to have been approached and, hence, whether $\sim k$ will have a subsequent opportunity to make an offer; in addition to (a)–(c) above, the beliefs must also specify probabilities for no approach, no offer, and a rejected offer as regards I and the other firm.

A strategy for the inventor is a plan for bargaining with the firms. An inventor has complete but imperfect information (as $\alpha > 0$) regarding position in the extensive form. First, immediately after discovering the invention, the inventor may choose one of the firms, denoted by i, to approach for a contract offer by revealing or not revealing the invention to firm i. If i makes an offer, then I chooses whether to accept or reject, and then whether to reveal to i or not. Next, I may approach the other firm, denoted by j, and choose whether to reveal. This is followed by the choice to accept or reject an offer from j and then a final decision to reveal to j and i.

We solve for a perfect Bayesian equilibrium. In this game, such an equilibrium is a triple of strategies and a pair of belief functions that satisfy three conditions: given the firm's available information and beliefs, and given the strategies of the inventor and the other firm, the firm's strategy specifies a contract offer that maximizes the firm's expected payoff; given an inventor type (good or bad), the strategies of the firms, and a decision node of the inventor, the inventor's strategy maximizes the inventor's expected payoff; and, whenever possible, beliefs are derived from Bayes' rule and the strategies of the other players.

II. The Structure of an *Ex Post* Contracting Equilibrium

In this section we develop our main result that an inventor who has made a difficult but valuable discovery can expect a significant payoff in equilibrium by freely revealing the invention prior to receiving a contract offer. This result emerges from the analysis of an equilibrium path with the following features: a good inventor "reveals" the invention to a randomly chosen firm, denoted by i, prior to any contractual relationship with either firm; firm i, which is now "informed," offers the "*ex post*" contract, $\mathbf{R}^*$; $\mathbf{R}^*$ is accepted by the inventor; and no other contracts are signed, and the good inventor does not reveal the invention to the other firm, j.

First, we develop necessary conditions for this "*ex post*" contracting equilibrium and pin down the payoffs and structure. Later, we present sufficient conditions and discuss how the equilibrium is supported.

In equilibrium, the expected payoff to a good I is $(1-\alpha)R_i^* + \alpha R_D^* \equiv \Pi_I^*$, since firm i has the invention and firm j independently discovers the invention with probability α. A bad I enters no contracts and thus earns zero. Upon becoming informed, $(1-\alpha)(\pi_M - R_i^*) + \alpha(\pi_D - R_D^*)$ is the expected payoff for firm i. Firm j, upon observing R^* between I and i, expects a payoff of $(1-\alpha)\pi_L + \alpha\pi_D$.

The first necessary condition for an *ex post* equilibrium derives from the need to structure $\mathbf{R}^*$ to remove any possible gains to trade between the inventor and firm j. That is, once a good I has accepted $\mathbf{R}^*$ it must not be possible for both I and j to benefit from an additional contract. Upon observing $\mathbf{R}^*$, firm j anticipates a monopoly outcome for firm i (with probability $1-\alpha$). Thus, to avoid receiving π_L, firm j can try to induce I to reveal to j as well, thus guaranteeing the duopoly outcome. To do so, j must find a contract, $\mathbf{S}$, that provides an incentive for I to reveal to j and is mutually profitable for I and j. Since the inventor has already revealed to firm i, this revelation incentive requires that $R_D^* + S_D > R_i^* + S_i$, so that the inventor earns more in state D than in state i.

The inventor will accept $\mathbf{S}$ if by doing so profits are increased, and this entails $R_D^* + S_D > (1-\alpha)R_i^* + \alpha R_D^*$. For j to benefit from offering $\mathbf{S}$, we need $\pi_D - S_D > (1-\alpha)\pi_L + \alpha\pi_D$. Mutual profitability thus requires

$$(2)\quad (1-\alpha)(\pi_D - \pi_L) > S_D > (1-\alpha)(R_i^* - R_D^*).$$

If this condition holds, then j can upset the equilibrium by offering a contract $\mathbf{S}$ such as $S_D = (1-\alpha)(\pi_D - \pi_L) - \varepsilon$ and $S_x = -(R_x^* + L)$ for states $x = 0, i, j$. This contract satisfies limited liability, induces a good I to reveal to firm j given that firm i is informed, and is mutually profitable for I and j given $\mathbf{R}^*$. To eliminate this possibility, condition (2) must fail, and thus, I must offer $\mathbf{R}^*$ such that

$$(3)\quad R_i^* - R_D^* \geq \pi_D - \pi_L.$$

That is, j's benefit from having the invention when j knows i has it (the most j is willing to pay) is less than I's loss under $\mathbf{R}^*$ if I reveals to j.

The second necessary condition concerns expropriation by firm i. Since I reveals to firm i without any prior contract, firm i has the option of expropriating the invention (a zero offer) rather than offering $\mathbf{R}^*$. The choice depends on what firm i expects I to do if no contract is offered. If i expects that I will end up revealing the invention to j, then expropriation is unprofitable for i when

$$(4)\quad (1-\alpha)(\pi_M - R_i^*) + \alpha(\pi_D - R_D^*) \geq \pi_D.$$

Intuitively, condition (4) reveals that firm i would always expropriate if the inventor were to receive too large a share of the rents from monopoly use of the invention. To see this, simply rearrange (4) by solving for the equilibrium payoff of the inventor to find $(1-\alpha)R_i^* + \alpha R_D^* \leq (1-\alpha)(\pi_M - \pi_D)$.

To build intuition for why expropriation leads to the duopoly outcome, note first that expropriation is a zero-payment contract offer. Given the above analysis of condition (2), from $R_i = R_D = 0$ we see that gains to trade exist between the inventor and firm j. This intuition underlies i's expectation that expropriation results in the duopoly outcome (we consider the equilibrium beliefs and contract offer of firm j later in this section).

We now examine the implications of these two necessary conditions and limited liability.[10] Begin by rearranging conditions (3)

[10]One potential interaction that we ignore is contracting between firms over the invention. In a duopoly context, contracting could be difficult because of antitrust considerations. It is easy, however, to expand the model to include this interaction, thus adding $\pi_M + \pi_L - 2\pi_D \geq R_i^* - R_D^*$ as an additional necessary condition.

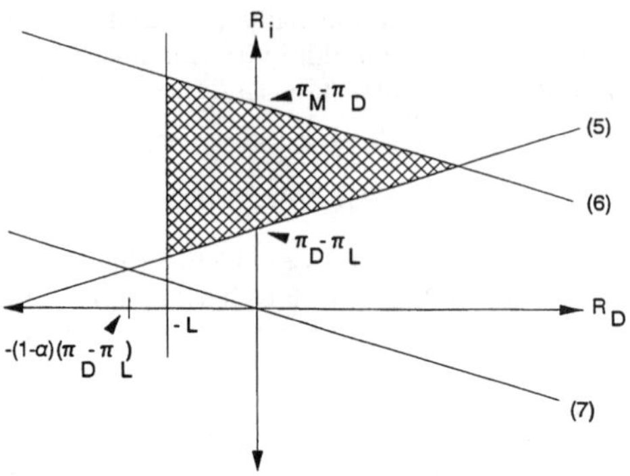

FIGURE 2. $L < (1-\alpha)(\pi_D - \pi_L)$

and (4) so that R_i is on the left side of the inequality, thus facilitating a comparison when R_i is graphed against R_D:

$$(5)\quad R_i^* \geq (\pi_D - \pi_L) + R_D^*$$

$$(6)\quad R_i^* \leq (\pi_M - \pi_D) + \alpha(1-\alpha)^{-1} R_D^*.$$

Figure 2 graphs these inequalities, including the lower bound set by limited liability, condition (1). By the direction of the inequalities, potential R_i^* and R_D^* equilibrium payments lie in the shaded area. Because the slopes in (5) and (6) are opposite in sign, the standard assumption that monopoly use of the invention leads to the greatest industry profits, $\pi_M + \pi_L > 2\pi_D$, implies that the shaded region exists. The inventor prefers contract payments to the northeast while firm i prefers those to the southwest. Because expected gross profits are constant at $(1-\alpha)\pi_M + \alpha\pi_D$ over the shaded region, movement from one contract to another has a zero-sum flavor: increases in profits to one party equal the decrease in the profits to the other party.

In equilibrium an inventor must earn nonnegative profits by accepting $\mathbf{R}^*$:

$$(7)\qquad \Pi_I^* \geq 0.$$

Define isoprofit lines for the inventor by $\Pi_I \equiv (1-\alpha)R_i + \alpha R_D$. In Figure 2, these lines are parallel to the boundary line of condition (6), and the zero isoprofit line passes through the origin. When L becomes small, the zero isoprofit line does not intersect the crosshatched region. As L approaches zero, the minimum expected payoff to a good inventor from accepting $\mathbf{R}^*$ approaches $(1-\alpha)(\pi_D - \pi_L)$, obtained from the contract at the intersection of the R_i axis and lower bound (5) in Figure 2.

This result reveals an important property of *ex post* equilibria involving low-probability discoveries: an inventor with little or no liability can expect a contract offer with a significant payoff. The reason is as follows. To prevent firm j from upsetting the equilibrium by offering a contract that leads to a duopoly outcome, $\mathbf{R}^*$ must establish a sufficiently large wedge between R_i^* and R_D^*, namely, $\pi_D - \pi_L$. For firm i, an attractive way of doing this would be to set R_D^* to a large negative value and thereby hold down the value of R_i^*. Since small liability prevents R_D^* from being a large negative amount, the wedge implies that the inventor must receive a large payment in the event of a monopoly for firm i. Thus, the positive expected payoff for a good inventor is a consequence of limited liability and incentives required to maintain a monopoly in the invention.

By the same logic, with a larger L firm i can maintain the wedge with a smaller monopoly state payment, R_i^*, by taking advantage of the larger liability level to reduce the duopoly state payment, R_D^*. Thus, the lower bound on the equilibrium payoff to I decreases as L rises. As shown in Figure 3, for $L \geq (1-\alpha)(\pi_D - \pi_L)$ the zero isoprofit line, $\Pi_I = 0$, intersects the crosshatched region, and the lower bound of $\Pi_I^* = 0$ is potentially binding.

In the extensive form, firms make take-it-or-leave-it offers, and in equilibrium, this allows firm i to push the payoff of the inventor to the lower bounds described above. Proposition 1 describes the equilibrium contract.

PROPOSITION 1: *Consider an equilibrium, for the extensive form, in which the equilibrium path exhibits ex post contracting, and let* $\mathbf{R}^*$ *denote an equilibrium contract. Then* $\mathbf{R}^*$

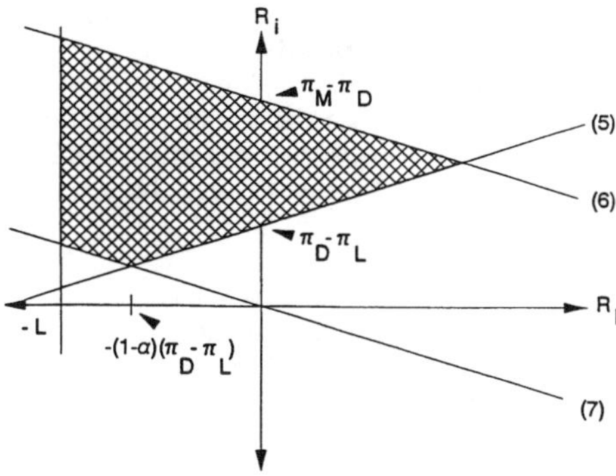

FIGURE 3. $L \geq (1-\alpha)(\pi_{\rm D} - \pi_{\rm L})$

satisfies:

(a) *when liability is small, that is,* $L < (1-\alpha)(\pi_{\rm D} - \pi_{\rm L})$, *we have*

$$R_i^* = \pi_{\rm D} - \pi_{\rm L} - L \quad \text{and} \quad R_{\rm D}^* = -L$$

(b) *when liability is large, that is,* $L \geq (1-\alpha)(\pi_{\rm D} - \pi_{\rm L})$, *we have*

$$R_i^* = \alpha(1-\alpha)^{-1} R_{\rm D}^*$$

$$\text{for} \quad -L \leq R_{\rm D}^* \leq -(1-\alpha)(\pi_{\rm D} - \pi_{\rm L}).$$

(All proofs are in the Appendix.) To develop an understanding of Proposition 1, begin with the case of large liability. Direct calculations reveal that the inventor receives an equilibrium payoff of $\Pi_I^* = 0$. The inventor would reject any such contract if the next best alternative, bargaining with firm *j*, yielded a positive expected payoff. With take-it-or-leave-it offers, however, firm *j* is able to capture all of the gains to trade that exist when the inventor has rejected the offer from firm *i* (this is demonstrated below in the proof of Proposition 2).

When L is sufficiently large, firm *i* can structure $\mathbf{R}^*$ so that gains to trade between the inventor and firm *j* are eliminated and the payoff to the inventor under $\mathbf{R}^*$ is forced to zero, the threshold level at which the inventor is just willing to accept the contract.

More important is the case of small liability in which the inventor receives

$$\Pi_I^* = (1-\alpha)(\pi_{\rm D} - \pi_{\rm L}) - L > 0.$$

Although the inventor can expect only a zero payoff from bargaining with *j*, this acceptance threshold is not binding. The reason, as discussed above, is that a small L makes it impossible for firm *i* to structure an $\mathbf{R}^*$ that eliminates gains to trade with firm *j* and also has a zero payoff for the inventor. To preserve a monopoly in the invention, firm *i* must respect the gains-to-trade condition, and so the inventor receives a positive expected payoff from $\mathbf{R}^*$.

The payoffs in Proposition 1 apply to equilibrium in our extensive form. Under this bargaining structure, where firms make take-it-or-leave-it offers, the payoff to the inventor declines as L rises, reflecting the role of L in setting the $(R_i^* - R_{\rm D}^*)$ wedge. Under alternative assumptions where bargaining power is less concentrated, this negative relationship between L and the equilibrium share of rents appropriated by I may not hold. Referring back to Figure 2, note that the lower bound on the payoff to I declines with L while the upper bound is independent of L. Under any bargaining structure, increases in L have this adverse effect on feasible payoffs and, in this sense, put the inventor at a strategic disadvantage.[11]

In each case, the firm structures $\mathbf{R}^*$ to take strategic advantage of the liability of the inventor. An interpretation of $\mathbf{R}^*$ which illustrates this point is the following. As a condition of the contract, firm *i* "requires" that the inventor "post a bond" equal to $R_{\rm D}^*$. This bond is forfeited if the duopoly state occurs, but it is returned with a bonus of $(R_i^* - R_{\rm D}^*)$ if the monopoly state for *i* occurs. By necessity, this bond is small when the liability of the inventor is small. The bonus, however, must remain sufficient to eliminate gains to trade with the other firm.

[11]Accounting for an initial wealth of L, we see that the net wealth to the inventor remains constant at $(1-\alpha)(\pi_{\rm D} - \pi_{\rm L})$ over the small L range. Thus, while there is no net advantage to reducing L, or burning money, the inventor could benefit by hiding assets. An interesting extension would involve allowing for private information regarding the size of L.

Thus, small liability reduces the strategic advantage of firm i, eliminating it as L approaches zero, and leads to a positive payoff for the inventor.

With the necessary structure of an *ex post* contracting equilibrium established in Proposition 1, we are ready to consider existence. Sufficient conditions for existence of this equilibrium are given in Proposition 2.

PROPOSITION 2: *Suppose that* $\pi_M + \pi_L > 2\pi_D$. *Then there exists an equilibrium with ex post contracting.*

As described above, on the equilibrium path a good I approaches one of the firms (randomly chosen) by revealing the invention, and then $\mathbf{R}^*$ is offered by this firm and accepted by I; a bad I chooses (optimally) not to approach either firm.[12] This path must be supported by the decisions of firms and the inventor "off the equilibrium path," and furthermore, these decisions must be optimal for a specified set of beliefs. The formal proof is deferred to the Appendix. For the present, we concentrate on developing the intuition behind the critical element in supporting the equilibrium, namely, the reason an informed firm finds it optimal to offer $\mathbf{R}^*$ to the inventor.

The reason is that if the informed firm does not offer $\mathbf{R}^*$, then the inventor will end up revealing the invention to the other firm. Thus, the monopoly in the invention will be lost, and the duopoly state will prevail. The exact way in which this occurs will depend on the nature of the deviation from $\mathbf{R}^*$. To illustrate the basic idea behind the supporting beliefs and strategies, we focus on the case of "expropriation" in which the informed firm deviates by making no contract offer.

Given expropriation, I has the option of approaching the other firm. If I were to reveal the invention before receiving an offer, then state D would occur. The monopoly in the invention for the expropriating firm would then be lost, and by the existence condition ($\pi_M + \pi_L > 2\pi_D$), offering $\mathbf{R}^*$ would yield higher profits.

If, on the other hand, I were to seek an *ex ante* contract, then the beliefs of an uninformed firm who observes no prior contract would come into play. These beliefs are specified as

$$(8)\quad \begin{cases} \rho = \text{probability } I \text{ is good and has revealed to the other firm} \\ 1-\rho = \text{probability } I \text{ is bad} \\ 0 = \text{probability } I \text{ is good and has not revealed to the other firm} \end{cases}$$

where ρ satisfies $0 < \rho < 1$, and where I approached the other firm with probability 1. Under these beliefs, an uninformed firm discounts completely the prospect of contracting for monopoly use of the invention and, instead, anticipates that *ex ante* contracting with a good I takes place only after the invention has been revealed to the other firm. In short, the belief is that a good I has been expropriated after revealing to the other firm.

These beliefs lead to a simple optimal contract offer. By offering a small positive payment in the duopoly state, $R_D = \varepsilon$, and a (negative) payment of $-L$ in all other states, the uninformed firm can screen a bad I while attracting a good I who then reveals.[13] As this results in the duopoly state, the monopoly in the invention for the expropriating firm would again be lost, and thus, the equilibrium-path offer of $\mathbf{R}^*$ is preferred to expropriation.

Of course, the complete analysis of the full range of deviations from equilibrium and the supporting beliefs and strategies involves a number of additional considerations. The essential idea for the overall supporting belief structure, however, is captured by this informal consideration of expropriation. Whenever an uninformed firm

[12] We avoid unnecessary complications off the equilibrium path with a random choice of firm by I, as a nonrandom choice would force one firm to treat revelation as a surprise. Specifying no approach also keeps matters simpler.

[13] Whenever contract payments push the inventor to indifference with respect to revealing the invention, the "tie" is broken in favor of the firm making the last offer, as this firm could always offer a slightly larger payment.

considers a contract offer and observes that $\mathbf{R}^*$ is not already in place, it believes that if I is good then I has previously revealed to the other firm. Consequently, as $\mathbf{R}^*$ is not in place, the uninformed firm and a good I can both benefit from a contract that leads a good I to reveal to the uninformed firm, thus resulting in the duopoly state. In turn, this supports the equilibrium-path offer of $\mathbf{R}^*$ by an informed firm.

III. *Ex Ante* Contracts and Limited Liability

Thus far we have focused on equilibrium behavior involving *ex post* contracting. We now consider equilibrium behavior when the inventor approaches a firm for a contract without revealing the invention. The fundamental difference between contracting on an *ex post* versus an *ex ante* basis is the information position of the firms. In the *ex post* case, the invention has been revealed to the firm, whereas in the *ex ante* case the firm remains unsure whether the inventor is good or bad.

We begin by examining equilibrium behavior (in the extensive form) involving *ex ante* contracting when the inventor has significant resources (large L). Afterwards, we turn to the general properties of *ex ante* contracting. We derive an upper bound on the size of the inventor's rents, which is valid for a wide range of extensive forms, and show that these rents must vanish as L and q go to zero. We conclude the section with a summary of *ex ante* and *ex post* contracting, focusing on the roles of L, α, q, and gross profit levels.

A. *Ex Ante Contracting in Equilibrium*

We will show that when L is sufficiently large, there exists an equilibrium involving *ex ante* contracting in which the inventor captures all of the rents. First, consider the maximum payoff an inventor could obtain when it is common knowledge that the inventor is good. The largest payoff possible for a firm is $(1-\alpha)\pi_M + \alpha\pi_D$, which occurs when a good inventor reveals only to this firm. Similarly, a firm can never be forced to a payoff below $(1-\alpha)\pi_L + \alpha\pi_D$. The difference between these extremes, $\Pi_I^* \equiv (1-\alpha)(\pi_M - \pi_L)$, is the largest payment an inventor could receive.

An equilibrium with *ex ante* contracting that supports a payoff of Π_I^* for I has a straightforward structure. The first feature is monopoly contracting: given that a good I accepts the equilibrium contract $\mathbf{R}^*$, the incentive structure of $\mathbf{R}^*$ ensures that the invention is revealed only to i, the contracting firm. As with the earlier analysis, this requires that, given $\mathbf{R}^*$, there are no remaining gains to trade between I and j; this reduces to setting sufficiently large payment wedges in $\mathbf{R}^*$, namely, $\pi_D - \pi_L < R_i^* - R_D^*$ and $\pi_M - \pi_L < R_i^* - R_j^*$.

The second feature is separation: a good I accepts and a bad I rejects the offer of $\mathbf{R}^*$. If

$$0 > E_{\alpha,\alpha}\{\mathbf{R}^*\}$$

$$\equiv (1-\alpha)^2 R_0^* + \alpha(1-\alpha)(R_i^* + R_j^*) + \alpha^2 R_D^*$$

so that the expected value of $\mathbf{R}^*$ (over independent discovery by the firms) is negative, then a bad I rejects $\mathbf{R}^*$. The intuitive structure of $\mathbf{R}^*$ is then simple. Set $R_0^* = R_j^* = R_D^* = -L$ and then set R_i^* to achieve the payoff of Π_I^* for I. This simultaneously makes the two wedges as large as possible and also minimizes the value of $\mathbf{R}^*$ to a bad I. The formal properties of this *ex ante* contract and the equilibrium are reported below.

PROPOSITION 3: *Necessary conditions for an equilibrium with ex ante contracting involving a monopoly contract and separation are*

(*a*) $$L > \alpha(\pi_M - \pi_L)$$

(*b*) $$\Pi_I^* = (1-\alpha)(\pi_M - \pi_L)$$

where Π_I^* *is the equilibrium payoff to a good inventor. This equilibrium exists if*

$$L > \alpha(\pi_M - \pi_L).$$

The equilibrium contract, $\mathbf{R}^*$, *is given by*

(*c*) $$R_i^* = \pi_M - \pi_L + \alpha(1-\alpha)^{-1}L$$

$$R_x^* = -L \qquad \text{for } x = 0, j, \text{D}.$$

On the equilibrium path, a good I approaches one of the two firms, denoted by i (randomly chosen), accepts the offer of $\mathbf{R}^*$ from i, and then reveals the invention to i; a bad I approaches neither firm. From (b) above, a good I captures all of the rents while the firms earn equal profits. The existence condition, $L > \alpha(\pi_M - \pi_L)$, which is also necessary for the equilibrium, is the pivotal part of Proposition 3. To develop intuition, note that separation is achieved by requiring a payment of L from I whenever the monopoly state does not occur. Calculation then reveals that $E_{\alpha,\alpha}\{\mathbf{R}^*\} = (1-\alpha)[\alpha(\pi_M - \pi_L) - L] < 0$. Thus, it is the existence condition which ensures that separation is possible. By contrast, sufficiently large payment wedges under $\mathbf{R}^*$ require only that $L > 0$. Consequently, the key to the equilibrium is that liability, L, is large relative to the increment from low to monopoly profits, $\pi_M - \pi_L$, weighted by the probability of independent discovery by a firm, α.[14]

Intuitively, then, the firms can screen a bad I in this case, and they are led to "compete away" their profits as they attempt to structure the *ex ante* contract offer, $\mathbf{R}^*$, so that a good I will choose one firm over the other.[15]

B. *The Limits of Ex Ante Contracting*

In contrast to the case when L is relatively large, separation is problematic when L is small. Consider, for example, the monopoly contract of Proposition 3 and let $L \to 0$. The payment R_i^* goes to $\pi_M - \pi_L$ while all other payments go to zero, and consequently, a bad I is no longer screened.[16] This example reflects a general limitation of *ex ante* contracts when the inventor has limited resources.

A bad I can mimic the actions of a good I with respect to entering into *ex ante* contracts since, by definition, any revelation of the invention does not occur until after the contract(s) are signed. Thus, any allocation involving *ex ante* contracting must present a good and bad I with the same options for entering into contracts. While the expected payoff for a bad I is generated by the firm discovery lottery of $\{(1-\alpha)^2, \alpha(1-\alpha), (1-\alpha)\alpha, \alpha^2\}$ over the states $\{0,1,2,D\}$, a good inventor accounts for the contractual benefits that derive from revealing the invention after contracts are signed.

The problem for a good inventor is that high payments in the monopoly or duopoly states are also attractive to a bad inventor. The value of q, the probability that I is good, then becomes important. Firms perceiving that attracting bad inventors will lead to negative expected profits will then be unwilling to enter into such contracts. As illustrated by the monopoly contract of Proposition 3, this problem can be avoided when L is large via contracts that require large payments from the inventor in the undesired states. As the resources of the inventor decline, however, the size of the payment an inventor can make is restricted, and consequently, contracts with large payments in a monopoly state will necessarily offer positive expected profits to bad inventors. In turn, negative expected profits for

[14]A simple benchmark may help to interpret what it means for L to be large relative to $\alpha(\pi_M - \pi_L)$. Ignore all strategic issues and suppose that a firm can pursue R&D which, if successful, will increase profits to π_M from a status quo of π_L. With α as the probability of success, the expected value to the firm of discovering the invention is just $\alpha(\pi_M - \pi_L)$, and the firm would be willing to invest in R&D an amount approaching this value. Relating back to Proposition 3, the equilibrium will not exist when the resources of the inventor are less than an amount that is of the same magnitude as such a firm's R&D budget (for this invention).

[15]This equilibrium-path behavior is analogous to the bidding incentives in an auction setting. A "monopoly" contract will emerge in equilibrium for a variety of contract models and extensive forms when L is large. One example is a sealed-bid auction where firms bid for a contract offered by the inventor (see e.g., Michael H. Riordan and David E. M. Sappington, 1987). Also, see Gallini and Wright (1990).

[16]There is also the possibility of an *ex ante* pooling equilibrium. One can show that such an equilibrium does not exist when $L > \alpha(\pi_M - \pi_L)$ or when L is below a (smaller) threshold value. Further, for L near the threshold, existence requires that q is bounded away from zero. This latter feature is troubling in that, if a bad inventor can earn positive expected profits, then an infinite number of "speculators" may be attracted to enter the industry.

firms create an incentive to reduce monopoly-state payments.

This intuition suggests that *ex ante* contracting does not allow a good inventor with small liability to obtain sizable rents. The formal statement of the result is given below.

PROPOSITION 4: *Suppose that (a) firms and the inventor enter into contracts voluntarily and (b) a good inventor makes an optimal choice regarding revelation of the invention only after entering into contracts. Then, as the liability of the inventor becomes small ($L \to 0$) and the adverse-selection problem becomes acute ($q \to 0$), the payoff to a good inventor goes to zero.*

Thus, contracting on an *ex ante* basis necessarily leads to a vanishingly small payoff for the inventor as L and q go to zero. Note that it is q, the probability that the inventor has a good invention, rather than α, the probability of independent invention by a manufacturer, that is important for adverse selection in *ex ante* contracting and limits the payoff to the inventor.

C. *The Structure of Ex Ante and Ex Post Contracts*

Propositions 3 and 4 provide useful benchmarks, and together with our earlier results, we can employ them to examine the structure of *ex ante* and *ex post* contracting in relation to the underlying environment. In particular, we are interested in how L, α, q, and gross profit levels relate to our primary question: can a financially weak independent inventor without property rights appropriate rents from a valuable invention? We begin with equilibrium payoffs and then turn to existence.

Payoffs to the inventor are $\Pi_I^* = (1-\alpha)(\pi_M - \pi_L)$ in the equilibrium with *ex ante* contracting (Proposition 3) and $\Pi_I^* = (1-\alpha)(\pi_D - \pi_L) - L$ in the equilibrium with *ex post* contracting (Proposition 1, small-L case; we are not focusing on the large-L case as it has $\Pi_I^* = 0$). Setting aside existence issues for the moment, it is clear that the *ex ante* payoff is always larger than the *ex post* payoff (as monopoly profits exceed duopoly, $\pi_M > \pi_D$). In both cases, decreases in α lead to larger payoffs for the inventor. Intuitively, as α declines and firms are less likely to discover the invention independently of the inventor, the firms are willing to pay more to acquire the invention. In contrast to the payoff in the *ex post* case (in Proposition 1), the payoff in the *ex ante* case (in Proposition 3) is not affected by L, as the inventor's resources are employed only to screen a bad inventor.

The crucial differences between *ex ante* and *ex post* contracting turn on the issue of existence. Recall that the existence condition in the *ex ante* case is

$$L > \alpha(\pi_M - \pi_L)$$

and for the *ex post* case it is

$$L < (1-\alpha)(\pi_D - \pi_L)$$

Each condition becomes easier to satisfy as α declines.[17] Now consider the role of liability (L) and adverse selection (q). For *ex ante* contracting, large L and small α function as substitutes with respect to existence: the less likely independent invention by the firms or the larger the liability of the inventor, the easier it is to screen a bad inventor. For a given $\alpha > 0$, however, as L declines the *ex ante* existence condition will eventually fail. Furthermore, from Proposition 4, we know that, in general, the payoff to the inventor in any *ex ante* contracting allocation will go to zero with L and q (screening

[17]The limiting case of $\alpha = 0$ highlights the importance of property rights. In this case the only possible gross profit outcome is π_0 when the inventor is bad and, consequently, screening becomes costless. Moreover, the outcomes π_M and π_D correspond one-to-one with the decision of a good inventor to reveal the invention to the firm(s). Consequently, the effect of $\alpha = 0$ is that the potential for contracting on gross profits essentially confers property rights for the invention on the inventor. In general, independent discovery is one of many factors that contribute to a stochastic relationship between gross profits and the revelation decision of a good inventor. In our analysis, the key assumption regarding the relationship is first-order stochastic dominance: revealing the invention to a firm makes higher profits more likely.

becomes more costly as q declines). In contrast, the existence condition for the *ex post* case necessarily holds as L declines. In addition, existence in the *ex post* case is independent of adverse selection (via q) since the inventor reveals prior to contracting with the firms.

Our primary focus is on a financially weak independent inventor who lacks property rights. For this case, our results imply that it will not be possible to appropriate significant rents through *ex ante* contracting. The *ex ante* equilibrium will not exist ($L < \alpha[\pi_M - \pi_L]$), and when combined with adverse-selection problems ($q \to 0$), rents via *ex ante* contracts are vanishingly small. In contrast, the *ex post* equilibrium will exist (when $L < (1-\alpha)[\pi_D - \pi_L]$), it is independent of adverse-selection problems, and the inventor captures a significant fraction of the available rents.

IV. Conclusion

In this paper we examined the problem confronting a financially weak, independent inventor whose only access to the market involves selling the invention to a firm. The fundamental problem facing an inventor with a good invention is how to convince the prospective buyer that the invention is good while also obtaining a significant share of rents. We focused on the polar case of perfect expropriability in which the inventor has no property rights with respect to the invention.

The inventor can choose an "*ex post* contracting" approach in which the inventor reveals the invention and then negotiates a contract or, alternatively, an "*ex ante* contracting" approach in which the parties try to mitigate adverse selection regarding the invention by employing an appropriately structured *ex ante* contract. In equilibrium, *ex post* contracting allows an inventor with limited resources to capture a sizable share of the market value of the invention.

This conclusion holds despite our conservative assumption regarding bargaining: firms make take-it-or-leave-it offers. Such an assumption is consistent with our view of the inventor as the "weaker" party in contract negotiations. An analysis of the other extreme, in which the inventor makes take-it-or-leave-it offers, leads to greater equilibrium expected profits for the inventor. For example, in Figure 2 the inventor would offer a contract in the extreme northeast corner and earn $(1-\alpha)(\pi_M - \pi_D)$.

With large financial resources, an inventor prefers the *ex ante* approach. This suggests the use of an intermediary, such as a venture capitalist (VC), to provide the resources a financially weak inventor needs in order to signal a valuable invention (see e.g., Raphael Amit et al., 1990). Such an arrangement could be structured so that the VC contracts with the inventor over the outcome states and then uses its financial resources (big L) to obtain the *ex ante* "monopoly" contract from a firm. Such an arrangement is attractive to the inventor as long as the VC market is competitive, verification of the value of the invention is inexpensive, and the VC cannot, itself, expropriate the invention.

Casual empiricism suggests that there are many problems that arise in the relationships between financial intermediaries and inventors.[18] Expropriation appears to be a problem that is endemic to all settings in which intellectual property can be stolen.[19] Thus, in our view, intermediaries do not undermine the importance of the expropriation issue.

Our analysis, which has focused on the case in which an inventor has no property

[18] There are many examples of inventors and entrepreneurs who have been unable to work deals with financial intermediaries or banks and have dealt directly with manufacturers. For example, Conner Peripherals, developers of a new 3.5-inch hard drive for laptop computers, "had trouble finding venture capitalists to finance the company, so went to Compaq Computer...(which) provided $12 million." In its first year of shipments Conner sold 90 percent to Compaq (*New York Times*, 27 May 1990, p. F-6). Similarly, the Gleasman Corporation, inventors of various automotive components, have had little luck with venture capitalists and have focused instead on selling their inventions to manufacturers (based on conversation with the CEO of Gleasman Corporation).

[19] For example, a VC could expropriate an invention by employing contracts with other inventors.

rights, may also be applicable for cases in which property rights, such as patents, are available. Patents have two problems: they are sometimes unenforceable, and they reveal underlying knowledge that may be useful for future inventions. For these reasons, firms often eschew patent protection in favor of relying on secrecy (Eli Lederman, 1989).

From the viewpoint of the inventor, this implies that the joint benefits associated with keeping an idea secret could exceed the benefits of a patented invention. Thus, because an inventor without property rights can appropriate a sizable share of the value of the invention, when joint benefits under secrecy are significantly greater than under patent protection, some inventors may choose to forgo patent protection.

Appendix

PROOF OF PROPOSITION 1:

Let $\mathbf{\Sigma} \equiv \{(R_i, R_D) |$ conditions (1), (5), and (6) hold$\}$. To begin, we find the payments in $\mathbf{\Sigma}$ with the lowest payoff to the inventor,

$$\begin{aligned}\Pi_I &\equiv (1-\alpha)R_i + \alpha R_D \\ &\geq (1-\alpha)[\pi_D - \pi_L + R_D] + \alpha R_D \\ &\geq (1-\alpha)(\pi_D - \pi_L) - L\end{aligned}$$

[by (5) and (1)], must hold for any $(R_i, R_D) \in \mathbf{\Sigma}$. When $L < (1-\alpha)(\pi_D - \pi_L)$, this lower bound is positive, and the payments $R_D = -L$ and $R_i = \pi_D - \pi_L - L$ are in $\mathbf{\Sigma}$ and achieve the lower bound. When $L > (1-\alpha)(\pi_D - \pi_L)$, this lower bound is negative. In this case the zero isoprofit line of $0 = \Pi_I = (1-\alpha)R_i + \alpha R_D$ intersects $\mathbf{\Sigma}$. By condition (7), $\Pi_I^* \geq 0$ is necessary, in equilibrium, for the inventor to accept $\mathbf{R}^*$. Payments $(R_i, R_D) \in \mathbf{\Sigma}$ with $0 = (1-\alpha)R_i + \alpha R_D$ then achieve $\Pi_I = 0$.

The following accounting identity relates payoffs to I and i at $(R_i, R_D) \in \mathbf{\Sigma}$:

$$\begin{aligned}\Pi_i &\equiv (1-\alpha)(\pi_M - R_i) + \alpha(\pi_D - R_D) \\ &= (1-\alpha)\pi_M + \alpha\pi_D - \Pi_I.\end{aligned}$$

Thus, over $\mathbf{\Sigma}, \Pi_i$ is at a maximum when Π_I is at a minimum.

The contracts in (a) and (b) of Proposition 1 imply that the payoff to I equals the lower bound developed above for each of the L cases. By the accounting identity, it is optimal for firm i to offer this $\mathbf{R}^*$ if I is expected to accept. In the proof of Proposition 2 we show that I expects a payoff of zero from bargaining with firm j following a rejection of a firm-i offer. Thus, when $L < (1-\alpha)(\pi_D - \pi_L)$, $\mathbf{R}^*$ is accepted by the inventor. When $L \geq (1-\alpha)(\pi_D - \pi_L)$, I necessarily accepts any offer from i with a positive payoff. The only possible equilibrium is for i to offer an $\mathbf{R}^*$ with $\Pi_I^* = 0$, thus leaving the inventor indifferent between acceptance and rejection.

PROOF OF PROPOSITION 2:

To establish Proposition 2, we must exhibit strategies and beliefs that support the equilibrium path described at the beginning of Section III. To begin, we construct the beliefs and solve for the strategy choice (contract offer) for the cases of an uninformed firm and an informed firm. Analysis of the inventor's strategy is then straightforward.

Case 1: Uninformed Firm.—Suppose that I has approached a firm for a contract without revealing the invention. Begin with the expropriation-branch case in which no prior contract is observed. Consider an optimal contract offer for the firm with beliefs as specified in (8) in the text. A bad I rejects an offer of $\mathbf{S}$ if

$$\begin{aligned}&E_{\alpha,\alpha}\{S_x\} \\ &\quad\equiv (1-\alpha)^2 S_0 + \alpha(1-\alpha)(S_i + S_j) + \alpha^2 S_D < 0.\end{aligned}$$

Thus, the maximum possible payoff for the firm is

$$\begin{aligned}E_{\alpha,\alpha}\{\pi_x\} &\equiv (1-\alpha)^2 \pi_0 \\ &\quad + \alpha(1-\alpha)(\pi_M + \pi_L) + \alpha^2 \pi_D\end{aligned}$$

if I is bad.

If I is good, the firm believes it is the second firm to be approached (firm j) and that the invention has been revealed to the other firm (firm i). Gross of contract payments, j expects a profit of π_D if I reveals to j and $(1-\alpha)\pi_L + \alpha\pi_D$ if I does not. Thus, for a good I who has revealed to i, the maximum possible expected profit for j is π_D, which occurs if I reveals to j without any payment from j. Then, $(1-\rho)E_{\alpha,\alpha}\{\pi_x\} + \rho\pi_D$ is an upper bound on the payoff to an uninformed firm when no prior contract is observed.

Suppose j offers I a contract, $\mathbf{S}$, where $S_x = -L$ for $x = 0, i, j$ and $S_D = \varepsilon$, where $0 < \varepsilon < \alpha^{-2}(1-\alpha^2)L$. Then, $E_{\alpha,\alpha}\{S_x\} < 0$, and a bad I rejects $\mathbf{S}$. A good I who revealed to the other firm would optimally accept $\mathbf{S}$ and reveal to j. The payoff to j from this offer is arbitrarily close to the upper bound, and letting $\varepsilon \to 0$ yields the optimal offer. The range of ε indexes a set of Pareto-improving contracts between a good I and an uninformed firm, and any contract in this set leads to the duopoly outcome.

When a prior contract, $\mathbf{R} \neq \mathbf{0}$, is observed, the uninformed firm can infer that it is firm j (the second firm). Beliefs are given by (8). Define a set of contracts $\mathfrak{R}$ by $\mathbf{R}' \in \mathfrak{R}$ if and only if $\mathbf{R}' \neq \mathbf{0}$, R' satisfies (1), and

$$\text{(A1)} \qquad 0 \le R_i' - R_D' < \pi_D - \pi_L .$$

The left-hand side of (A1) implies that, with i informed, a good I has no incentive under $\mathbf{R}'$ to reveal to j. By the right-hand side, j's incremental profit from becoming informed is greater than the incentive under $\mathbf{R}'$ for a good I not to reveal to j. The lemma below is used to find j's optimal offer.

LEMMA: *Suppose firm j is uninformed and observes* $\mathbf{R} \in \mathfrak{R}$. *Conditional on beliefs in (8), the expected payoff to j from any contract offer is bounded above by*

$$(1-\rho)E_{\alpha,\alpha}\{\pi_x\} + \rho[\pi_D - (1-\alpha)(R_i - R_D)].$$

PROOF:

Let $\mathbf{S}$ be a contract offer from j. With probability $1-\rho$, I is bad, and with $\mathbf{R}$ already in place, a bad I accepts $\mathbf{S}$ if $E_{\alpha,\alpha}\{S_x\} > 0$. Thus, if I is bad, the maximum possible expected profit for j is $E_{\alpha,\alpha}\{\pi_x\}$. With probability ρ, I is good and firm i is informed. We establish the upper bound for each possible case of contracting outcomes between firm j and a good I. (a) Firm j and a good I do not enter into a contract. Then j expects a payoff of $(1-\alpha)\pi_L + \alpha\pi_D$. Since $\mathbf{R} \in \mathfrak{R}$ we rearrange (A1) to obtain the upper bound of

$$(1-\alpha)\pi_L + \alpha\pi_D < \pi_D - (1-\alpha)(R_i - R_D).$$

(b) Firm j and a good I enter into a contract, $\mathbf{S}$, and I does not reveal to j. Then j expects a payoff of $(1-\alpha)\pi_L + \alpha\pi_D - [(1-\alpha)S_i + \alpha S_D]$. Without revelation, $(1-\alpha)S_i + \alpha S_D \ge 0$ for a good I to accept $\mathbf{S}$. Thus, j's payoff is bounded above by $(1-\alpha)\pi_L + \alpha\pi_D$, and applying (A1) yields the desired upper bound. (c) Firm j and a good I enter into $\mathbf{S}$, and I reveals to j. Then j expects $\pi_D - S_D$. With revelation, $R_D + S_D \ge (1-\alpha)R_i + \alpha R_D$ for a good I to accept $\mathbf{S}$. Hence, $S_D \ge (1-\alpha)(R_i - R_D)$ and $\pi_D - S_D \le \pi_D - (1-\alpha)(R_i - R_D)$. The lemma follows from the weights of $1-\rho$ and ρ in j's beliefs of a bad or a good I.

In order to achieve this payoff upper bound, firm j can employ an extension of the expropriation-branch contract which screens a bad I and provides incentives for a good I to reveal. For a given $\mathbf{R} \in \mathfrak{R}$, define a set $\mathbf{\Psi}(\mathbf{R})$ by $\mathbf{S}' \in \mathbf{\Psi}(\mathbf{R})$ if and only if (1) holds and

$$\text{(A2)} \qquad S_D' = -R_D + [(1-\alpha)R_i + \alpha R_D] + \varepsilon$$

$$\text{(A3)} \qquad S_x' + R_x = -L \qquad \text{for } x = 0, i, j$$

where

$$0 < \varepsilon < \alpha^{-1}(1-\alpha)[L + (1-\alpha)R_i + \alpha R_D].$$

With $\mathbf{R} \in \mathfrak{R}$ in place between I and i, firm j expects a bad I to reject $\mathbf{S} \in \mathbf{\Psi}(\mathbf{R})$, as $E_{\alpha,\alpha}\{S_x\} < 0$, and expects a good I to accept $\mathbf{S} \in \mathbf{\Psi}(\mathbf{R})$ and reveal to j. Firm j thus expects $(1-\rho)E_{\alpha,\alpha}\{\pi_x\} + \rho[\pi_D - S_D]$ as the payoff if it offers $\mathbf{S}$. Substituting for S_D from (A2), this payoff is within ε of the upper bound in the lemma. Letting $\varepsilon \to 0$ yields the optimal offer for firm j.

Finally, if the uninformed firm observes a prior contract, $\mathbf{R} \notin \mathfrak{R}$, so that (A1) fails, then no offer is optimal. If $R_i < R_D$, then a good I is expected to reveal to j without any additional incentives; j can earn π_{D} by making no offer. Thus, j expects $(1-\rho)E_{\alpha,\alpha}\{\pi_x\}+\rho\pi_{\mathrm{D}}$ in this case, and by the same logic as in the lemma, this is the maximum possible payoff for j. Next, if $R_i - R_{\mathrm{D}} \geq \pi_{\mathrm{D}} - \pi_{\mathrm{L}}$, then no Pareto-improving contracts exist between j and a good I. No offer yields $(1-\rho)\pi_{\mathrm{L}} + \alpha\pi_{\mathrm{D}}$ for j, and again, this is j's largest possible payoff.

Case 2: Informed Firm.—Suppose that I has approached a firm for a contract by revealing the invention. Where a prior contract between I and the other firm is observed, the informed firm need not fear any counteroffer from the other firm. We simplify matters by specifying that the informed firm believes I has revealed to the other firm with probability 1. Thus, no offer is optimal for the informed firm.

Now suppose that no prior contract is observed. This is consistent with the candidate equilibrium path; the updated beliefs are that I is good, that this firm is the first firm to be approached (denoted as firm i), and hence that the other firm is uninformed. We sort i's possible offers into four classes:

(a) Firm i makes no offer. Firm i expects an expropriation-branch response where I reveals to j, leading to a profit of π_{D} for i.

(b) Firm i offers $\mathbf{R} \in \mathfrak{R}$. If I refuses $\mathbf{R}$, then i expects π_{D} in profit as the expropriation branch ensues. If I accepts $\mathbf{R}$, then i expects j to counter with $\mathbf{S} \in \mathbf{\Psi}(\mathbf{R})$ and expects I to accept. Firm i's profits will be $\pi_{\mathrm{D}} - R_{\mathrm{D}}$. From the above analysis of $\mathbf{S} \in \mathbf{\Psi}(\mathbf{R})$, $(1-\alpha)R_i + \alpha R_{\mathrm{D}} \geq 0$ is necessary for I to accept $\mathbf{R}$. Combining this with the definition of $\mathfrak{R}$ then yields an upper bound on firm i's payoff of

$$\pi_{\mathrm{D}} - R_{\mathrm{D}} \leq \pi_{\mathrm{D}} + \min\{L,(1-\alpha)(\pi_{\mathrm{D}}-\pi_{\mathrm{L}})\}$$

from an offer $\mathbf{R} \in \mathfrak{R}$.

(c) Firm i offers $\mathbf{R} \notin \mathfrak{R}$, where $\mathbf{R}$ fails to satisfy (A1) due to $0 > R_i - R_{\mathrm{D}}$. If I rejects $\mathbf{R}$, then the expropriation branch ensues, and i expects a payoff of π_{D}. Otherwise, j is expected to make no offer, and the duopoly state results. Thus, I will accept $\mathbf{R}$ only if $R_{\mathrm{D}} \geq 0$, and at most, firm i earns π_{D}. An upper bound on i's payoff to an offer of $\mathbf{R}$ in these three classes is

$$\pi_{\mathrm{D}} + \min\{L,(1-\alpha)(\pi_{\mathrm{D}}-\pi_{\mathrm{L}})\}.$$

(d) The fourth class, which includes the candidate equilibrium-path contract ($\mathbf{R}^*$), involves an offer $\mathbf{R} \notin \mathfrak{R}$ with $R_i - R_{\mathrm{D}} \geq \pi_{\mathrm{D}} - \pi_{\mathrm{L}}$. Define $\mathbf{R}^*$ by $R_0^* = R_j^* = -L$, $R_i^* = \pi_{\mathrm{D}} - \pi_{\mathrm{L}} - L$, and $R_{\mathrm{D}}^* = -L$, when $L < (1-\alpha)(\pi_{\mathrm{D}} - \pi_{\mathrm{L}})$, and $(1-\alpha)R_i^* + \alpha R_{\mathrm{D}}^* = 0$ with $R_{\mathrm{D}}^* \in [-L, -(1-\alpha)(\pi_{\mathrm{D}} - \pi_{\mathrm{L}})]$, when $L \geq (1-\alpha)(\pi_{\mathrm{D}} - \pi_{\mathrm{L}})$. For large L, $\mathbf{R}^*$ is unique up to the payoff to a good I.

We now show that $\mathbf{R}^*$ is optimal for an informed firm when no prior contract is observed. Define a set of contracts by $\Gamma \equiv \{\mathbf{R} | R_i - R_{\mathrm{D}} \geq \pi_{\mathrm{D}} - \pi_{\mathrm{L}}\}$. We first show that $\mathbf{R}^*$ is optimal over $\mathbf{\Gamma}$. To begin, suppose i offers $\mathbf{R} \in \mathbf{\Gamma}$ to I. Now suppose I rejects $\mathbf{R}$. If I approaches the other firm but does not reveal, then we have the expropriation branch and an overall payoff of zero to I. If I reveals to the other firm, then this other firm must make the equilibrium-path offer of $\mathbf{R}^*$ as it is informed and observes no prior contract. In this case, the overall payoff is $R_{\mathrm{D}}^* < 0$ if I accepts and 0 if I rejects, since I has revealed to both firms. Thus, the largest possible payoff for I is zero if $\mathbf{R}$ is rejected.

Now, suppose I accepts the offer of $\mathbf{R} \in \mathbf{\Gamma}$. We know from the above analysis that I will not receive a second contract offer. As $\mathbf{R} \in \mathbf{\Gamma} \Rightarrow R_i > R_{\mathrm{D}}$, I has no incentive under $\mathbf{R}$ to reveal to the other firm. Thus, I will neither contract with nor reveal to the other firm if $\mathbf{R} \in \mathbf{\Gamma}$ is accepted.

We now show that $\mathbf{R}^*$ is optimal for i in $\mathbf{\Gamma}$. Begin with the case of L small:

$$L < (1-\alpha)(\pi_{\mathrm{D}} - \pi_{\mathrm{L}}).$$

Then any $\mathbf{R} \in \Gamma$ has

$$(1-\alpha)R_i + \alpha R_D \geq (1-\alpha)(\pi_D - \pi_L) - L > 0.$$

From above, it is optimal for I to accept an offer of $\mathbf{R} \in \Gamma$ from i, and as I will neither contract with nor reveal to the other firm, firm i expects $\mathbf{R} \in \Gamma$ to yield a payoff of $(1-\alpha)\pi_M + \alpha\pi_D - [(1-\alpha)R_i + \alpha R_D]$. The optimal offer for i over $\mathbf{R} \in \Gamma$ thus minimizes $(1-\alpha)R_i + \alpha R_D$ subject to $R_i - R_D \geq \pi_D - \pi_L$ and $R_x \geq -L$. This yields

$$R_D^* = -L$$

$$R_i^* = \pi_D - \pi_L - L$$

and $(1-\alpha)(\pi_M + \pi_L) - (1-2\alpha)\pi_D + L$ is the resulting payoff to i. Now suppose $L \geq (1-\alpha)(\pi_D - \pi_L)$. The only difference from the preceding case is that L is now large enough for firm i to capture all of the rents. Since there exists $\mathbf{R} \in \Gamma$ with $(1-\alpha)R_i + \alpha R_D > 0$ in this case, it is easy to verify that $\mathbf{R}^*$ as given in (b) of Proposition 1 is optimal over Γ.

Thus, $\mathbf{R}^*$ is optimal for i over Γ. It is then straightforward to show that the existence condition, $\pi_M + \pi_L > 2\pi_D$, implies that the payoff to i from $\mathbf{R}^*$ exceeds $\pi_D + \min\{L, (1-\alpha)(\pi_D - \pi_L)\}$ for each L case. Thus, $\mathbf{R}^*$ is optimal for i.

Case 3: Inventor.—The strategy for I involves specifying (a) which firms, if any, to approach and, if an approach is made, which message to send (reveal or not reveal), and (b) the accept/reject choices in response to contract offers, for the inventor's decision nodes in the extensive form. It is straightforward to calculate optimal approach and accept/reject decisions for I. As many of these calculations were done above, we omit some of the details and focus on two main points.

First, a bad I cannot reveal a good invention and always evaluates contracts according to the $E_{\alpha,\alpha}\{\cdot\}$ expectation. Thus, if $\mathbf{R}$ has been accepted, then an offer of $\mathbf{S}$ is accepted only if $E_{\alpha,\alpha}\{S_x\} \geq 0$. Let $\mathbf{S}$ denote the offer of an uninformed firm when $\mathbf{R}$ is observed and let $\mathbf{S}'$ denote the offer if no contract is observed. Then a bad I accepts $\mathbf{R}$ only if $E_{\alpha,\alpha}\{R_x\} + \max\{0, E_{\alpha,\alpha}\{S_x\}\} \geq \max\{0, E_{\alpha,\alpha}\{S_x'\}\}$. Note that $\mathbf{R}$ may be accepted even though $E_{\alpha,\alpha}\{R_x\} < 0$ if accepting $\mathbf{R}$ then allows a bad I to receive an offer $\mathbf{S}$ that offsets the initial loss. From the above analysis it is easy to check that a bad I will reject all contract offers.

Now consider a good I. We focus on verifying that a good I is at a best response to the equilibrium contract offers. A good I can follow one of three strategic approaches. One, on the equilibrium path, is to reveal to one firm before accepting any contracts. This results in an offer of $\mathbf{R}^*$ from the informed firm, and as shown above, it is optimal for I to accept $\mathbf{R}^*$. A second approach is to enter contracts only on an *ex ante* basis. When no prior contract is observed, an uninformed firm offers only the expropriation-branch contract. If I rejects, then the same offer would be made by the other uninformed firm; while if I accepts, then no offer would be made by the other firm. At best, a good I can net zero from this strategy. Finally, a good I can reveal at some intermediate point. Again, it is easy to show that I cannot benefit from this approach.

PROOF OF PROPOSITION 3:

To prove (b), consider the offer from an uninformed firm when no prior contract is observed. In equilibrium, the offer from each firm must provide an identical payoff of $(1-\alpha)(\pi_M - \pi_L) = (1-\alpha)R_i + \alpha R_D$ to a good inventor. Otherwise, one of the firms can profitably deviate by slightly raising the offered payment in the monopoly state. From (b), separation and (1) imply (a).

To demonstrate sufficiency, consider the contract $\mathbf{R}^*$ defined by

$$R_i^* = \pi_M - \pi_L + \alpha(1-\alpha)^{-1}L$$

$$R_x^* = -L \qquad \text{for } x = 0, j, D.$$

It is straightforward to verify that $E_{\alpha,\alpha}\{R_x^*\} < 0$, that a good inventor under $\mathbf{R}^*$ reveals only to firm i, and that no mutually profitable contracts exist between j and I given

$\mathbf{R}^*$ between i and I. Let $\mathbf{R}^*$ be the offer from each firm when a firm is uninformed and no prior contract is observed. A wide range of beliefs can support the equilibrium. Verification is trivial with the following set of beliefs. When a prior contract $\mathbf{R} \neq \mathbf{R}^*$ is observed, let an uninformed firm believe the inventor is bad with probability 1; thus, no offer is optimal. Let an informed firm believe that the other firm is informed with probability 1; thus, no offer is optimal for this firm.

PROOF OF PROPOSITION 4:

Allocations involving *ex ante* contracting are described with two pairs of contracts, $\{(\mathbf{R},\mathbf{S}),(\mathbf{R}',\mathbf{S}')\}$, where $\mathbf{R}$ and $\mathbf{S}$ denote a contract between a good I and firm i and firm j, respectively, while $\mathbf{R}'$ and $\mathbf{S}'$ denote the corresponding contracts for a bad I. Special cases include separation (the contracts are different for a good and bad I), pooling ($\mathbf{R}=\mathbf{R}'$ and $\mathbf{S}=\mathbf{S}'$), and no contracting ($\mathbf{R}=\mathbf{S}=\mathbf{R}'=\mathbf{S}'=\mathbf{0}$). We assume that *ex ante* contracting satisfies three basic conditions:

(a) A good I optimally reveals to firms i and j in response to the payoffs under $(\mathbf{R}+\mathbf{S})$. To describe this decision, let $\mathbf{T}=(T_0,T_1,T_2,T_D)$ be a payoff vector and define

$$\Pi^g(\mathbf{T}) \equiv \max\{E_{\alpha,\alpha}\{T_x\},(1-\alpha)T_1+\alpha T_D,$$

$$(1-\alpha)T_2+\alpha T_D, T_D\}.$$

Thus, $\Pi^g(\mathbf{T})$ is the maximum payoff for a good I under $\mathbf{T}$ when I makes the optimal choice among revealing nothing, revealing to firm 1 but not 2, revealing to 2 but not 1, or revealing to both firms.

(b) A good I and a bad I may enter into $(\mathbf{R},\mathbf{S})$ or $(\mathbf{R}',\mathbf{S}')$, or not enter into any contracts: revelation occurs after contracts are signed, so either type of I can mimic the behavior of the other. Hence, for a bad I we have $E_{\alpha,\alpha}\{R'_x+S'_x\} \geq \max\{0, E_{\alpha,\alpha}\{R_x+S_x\}\}$, and for a good I $\Pi^g(\mathbf{R}+\mathbf{S}) \geq \max\{0,\Pi^g(\mathbf{R}'+\mathbf{S}')\}$.

(c) Each firm prefers to enter into the contracts rather than refuse. If firm k refuses to enter any contracts, then k earns $E_{\alpha,\alpha}\{\pi_x\}$ when I is bad and, at a minimum, $(1-\alpha)\pi_L+\alpha\pi_D$ when I is good, as the lowest possible payoff for k occurs when a monopoly for $\sim k$ prevails. Thus, a firm can guarantee

$$\underline{\pi} \equiv (1-q)E_{\alpha,\alpha}\{\pi_x\} + q[(1-\alpha)\pi_L+\alpha\pi_D]$$

by refusing all contracts.

We employ (a)–(c) to derive an upper bound on a good I's payoff, $\Pi^g(\mathbf{R}+\mathbf{S})$. There are three cases, depending on a good I's revelation decision under $\mathbf{R}+\mathbf{S}$. For case 1, under $\mathbf{R}+\mathbf{S}$, a good I optimally reveals only to one firm, denoted by i. Then a good I's expected payoff is

$$\Pi^g(\mathbf{R}+\mathbf{S}) = (1-\alpha)(R_i+S_i)+\alpha(R_D+S_D) \equiv \pi^g.$$

Note that participation for a bad I, (b), implies that

$$E_{\alpha,\alpha}\{R'_x+S'_x\} \geq E_{\alpha,\alpha}\{R_x+S_x\} \geq -(1-\alpha)L+\alpha\pi^g$$

as follows from limited liability and the definition of π^g. By (c),

$$(1-q)E_{\alpha,\alpha}\{\pi_x-R'_x\} + q[(1-\alpha)(\pi_M-R_i)+\alpha(\pi_D-R_D)] \geq \underline{\pi}$$

and

$$(1-q)E_{\alpha,\alpha}\{\pi_x-S'_x\} + q[(1-\alpha)(\pi_L-S_i)+\alpha(\pi_D-S_D)] \geq \underline{\pi}$$

hold for i and j. Adding the two conditions above, we find

$$q(1-\alpha)(\pi_M-\pi_L) \geq (1-q)E_{\alpha,\alpha}\{R'_x+S'_x\}+q\pi^g$$

upon combining with the definition of $\underline{\pi}$. Then the desired upper bound is

$$\pi^g \leq (1-\alpha)\left[(1-q)L + q(\pi_M - \pi_L)\right] \times [q+\alpha(1-q)]^{-1}$$

as follows from the above participation constraint for a bad I.

By similar reasoning (for proofs, see the working-paper version of this article, available from the authors upon request), we arrive at the bounds

$$\pi^g \leq \{(1-q)(1-\alpha^2)L + 2q(1-\alpha)(\pi_D - \pi_L)\} \times [q+\alpha^2(1-q)]^{-1}$$

$$\pi^g \leq 2q(1-\alpha)[(1-\alpha)(\pi_0 - \pi_L) + \alpha(\pi_M - \pi_D)]$$

for the cases when a good I reveals to both firms and when a good I reveals to neither firm, respectively. For all three cases, $\pi^g \to 0$ as $L \to 0$ and $q \to 0$.

REFERENCES

Amit, Raphael; Glosten, Lawrence and Muller, Eitan. "Entrepreneurial Ability, Venture Investments, and Risk Sharing." *Management Science*, October 1990, *36*(10), pp. 1232–45.

Areeda, Phillip and Kaplow, Louis. *Antitrust analysis: Problems, text, cases*. Boston: Little, Brown, 1988.

Arrow, Kenneth J. *Essays in the theory of risk-bearing*. Amsterdam: North-Holland, 1970.

→ **Benoit, Jean-Pierre.** "Innovation and Imitation in a Duopoly." *Review of Economic Studies*, January 1985, *52*(1), pp. 99–106.

Cheung, Steven N. S. "Property Rights in Trade Secrets." *Economic Inquiry*, January 1982, *20*(1), pp. 40–53.

Dasgupta, Partha and Stiglitz, Joseph. "Uncertainty, Industrial Structure, and the Speed of R&D." *Bell Journal of Economics*, Spring 1980, *11*(1), pp. 1–28.

Gallini, Nancy T. "Deterrence by Market Sharing: A Strategic Incentive for Licensing." *American Economic Review*, December 1984, *74*(5), pp. 931–41.

Gallini, Nancy T. and Winter, Ralph A. "Licensing in the Theory of Innovation." *Rand Journal of Economics*, Summer 1985, *16*(2), pp. 237–52.

Gallini, Nancy T. and Wright, Brian D. "Technology Transfer under Asymmetric Information." *Rand Journal of Economics*, Spring 1990, *21*(1), pp. 147–60.

Jewkes, John; Sawyers, David and Stillerman, Richard. *The sources of invention*, 2nd Ed. New York: Norton, 1969.

Kamien, Morton I. and Tauman, Yair. "Fees versus Royalties and the Private Value of a Patent." *Quarterly Journal of Economics*, August 1986, *101*(3), pp. 471–91.

Katz, Michael L. and Shapiro, Carl. "How To License Intangible Property." *Quarterly Journal of Economics*, August 1986, *101*(3), pp. 567–89.

Lederman, Eli. "Criminal Liability for Breach of Confidential Commercial Information." *Emory Law Journal*, Fall 1989, *38*(4), pp. 921–1004.

Okuno-Fujiwara, Masahiro; Postlewaite, Andrew and Suzumura, Kotaro. "Strategic Information Revelation." *Review of Economic Studies*, January 1990, *57*(1), pp. 25–47.

Pooley, J. H. A. *Trade secrets: A guide to protecting proprietary business information*. New York: American Management Association, 1987.

Riordan, Michael H. and Sappington, David E. M. "Awarding Monopoly Franchises." *American Economic Review*, June 1987, *77*(3), pp. 375–87.

Teece, David J. "Profiting from Technological Innovation: Implications for Integration, Collaboration, Licensing and Public Policy." *Research Policy*, December 1986, *15*(6), pp. 285–305.

von Hippel, Eric. "Appropriability of Innovation Benefit as a Predictor of the Source of Innovation." *Research Policy*, April 1982, *11*(2), pp. 95–115.

[21]

Intellectual Human Capital and the Birth of U.S. Biotechnology Enterprises

By LYNNE G. ZUCKER, MICHAEL R. DARBY, AND MARILYNN B. BREWER*

The number of American firms actively using biotechnology grew rapidly from nonexistent to over 700 in less than two decades, transforming the nature of the pharmaceutical industry and significantly impacting food processing, brewing, and agriculture, as well as other industries. Here we demonstrate empirically that the commercialization of this technology is essentially intertwined with the development of the underlying science in a way which illustrates the significance in practice of the localized spillovers concept in the agglomeration literature and of the tacit knowledge concept in the information literature. Indeed we present here strong evidence that the timing and location of initial usage by both new dedicated biotechnology firms (*"entrants"*) and new biotech subunits of existing firms (*"incumbents"*) are primarily explained by the presence at a particular time and place of scientists who are actively contributing to the basic science as represented by publications reporting genetic-sequence discoveries in academic journals.

By quantifying separable effects of individual scientists, major universities, and federal research support we provide specific structure to the role of universities and their faculties in encouraging local economic development through what are conventionally described in the literature as geographically localized knowledge spillovers.[1] Such localized knowledge spillovers may play fundamental roles in both economic agglomeration and endogenous growth (Paul M. Romer, 1986, 1990; Gene M. Grossman and Elhanan Helpman, 1991). However, our evidence, like the other literature cited here, specifically indicates localized effects without demonstrating that they can be characterized as spillovers (or externalities).

Section I lays out our basic hypothesis. The data are described in Section II. Empirical results are reported and discussed in Section III. A summary and conclusions section (Section IV) and Data Appendix complete the article.

I. The Hypothesis

Innovations are generally treated in the growth literature as a nonrivalrous good—freely useable by an unlimited number of potential

* Zucker: Department of Sociology and Institute for Social Science Research, University of California, Box 951484, Los Angeles, CA 90095, and National Bureau of Economic Research; Darby: John E. Anderson Graduate School of Management, University of California, Box 951481, Los Angeles, CA 90095, UCLA Department of Economics, and National Bureau of Economic Research; Brewer: Department of Psychology, Ohio State University, 1885 Neil Avenue, Columbus, OH 43210. This research has been supported by grants from the National Science Foundation (SES 9012925), the University of California Systemwide Biotechnology Research and Education Program, the University of California's Pacific Rim Research Program, the UCLA Center for American Politics and Public Policy, and the UCLA Institute of Industrial Relations. We acknowledge very useful comments on earlier drafts from two anonymous referees, and from David Butz, Harold Demsetz, Robert Drazin, Martin Feldstein, Zvi Griliches, Keith Head, Adam Jaffe, Benjamin Klein, Josh Lerner, Gary Pisano, Jeff Rosensweig, L. G. Thomas, Ivo Welch, and others. We are indebted to a remarkably talented team of postdoctoral fellows Zhong Deng, Julia Liebeskind, and Yusheng Peng, and research assistants Paul J. Alapat, Jeff Armstrong, Lynda J. Kim, Amalya Oliver, Alan Paul, and Maximo Torero. Armstrong was principally responsible for conducting the analysis and cleaning the firm data set and Torero cleaned the scientist data set; comments from both substantially improved the paper. This paper is a part of the NBER's research program in Productivity. Any opinions expressed are those of the authors and not those of their employers or funders.

[1] Zvi Griliches (1992) has surveyed the importance of R&D spillovers as a major source of endogenous growth in recent "new growth theory" models and the difficult empirical search for their existence. Despite these difficulties, there have been a number of articles reporting evidence of geographic localization of knowledge spillovers, including Adam B. Jaffe (1989), Jaffe et al. (1993), and Edwin Mansfield (1995).

users at a zero marginal cost (Richard R. Nelson and Romer, 1996). A complementary literature recognizes that some information requires an investment of considerable time and effort to master. The human capital developed by this investment is seen as earning a normal return on the cost of the investment, both direct costs and foregone earnings. We believe that some innovations, particularly a breakthrough "invention of a method of inventing" (Griliches, 1957), may be better characterized as creating (rivalrous) human capital—intellectual human capital—characterized by natural excludability as opposed to a set of instructions for combining inputs and outputs which can be protected only by intellectual property rights. This natural excludability arises from the complexity or tacitness of the information required to practice the innovation (see Nelson [1959], Kenneth J. Arrow [1962], Nelson and Sidney G. Winter [1982], and Nathan Rosenberg [1982]).

Based on both extensive interviews and empirical work summarized in Zucker and Darby (1996), we believe that, at least for the first 10 or 15 years, the innovations which underlie biotechnology are properly analyzed in terms of naturally excludable knowledge held by a small initial group of discoverers, their coworkers, and others who learned the knowledge from working at the bench-science level with those possessing the requisite know-how. Ultimately the knowledge spread sufficiently widely to become part of routine science which could be learned at any major research university. After the initial 1973 discovery by Stanley Cohen and Herbert Boyer of the basic technique for recombinant DNA—the foundation of commercial biotechnology as well as of a burst of scientific innovation—the financial returns available to talented recombinant-DNA scientists first rose dramatically as the commercial implications became widely appreciated and then more gradually declined as more and more scientists learned the techniques, until knowledge of the new techniques per se earned only the normal return for the time required for a graduate student to master them. Further, mere knowledge of the techniques of recombinant DNA was not enough to earn these extraordinary returns; the knowledge was far more productive when embodied in a scientist with the genius and vision to continuously innovate and define the research frontier and apply the new research techniques in the most promising areas.

We hypothesize that entry of firms into biotechnology in a given year thus will be determined by the geographic distribution of stars and perhaps others then actively practicing the new science as well as by the geographic distribution of economic activity. Stars are properly viewed as locationally (semi-)fixed since few star scientists who knew how to do recombinant DNA were willing to abandon their university appointments and laboratory teams to pursue commercial applications of biotechnology. The primary pattern in the development of the industry involved one or more scientist-entrepreneurs who remained on the faculty while establishing a business on the side—businesses which, where successful, resulted in millions or even billions of dollars for the professors who acquired early ownership stakes. Thus, we see the university as bringing about local industrial benefits by permitting its professors to pursue private commercial interests while their faculty appointments tie them to the area. In preliminary work not reported here, we tried to develop measures of local economic activity for industries, like pharmaceuticals, specifically impacted by the new technology, but these attempts never added significantly to the measures of general activity used in the empirical work below. The *local* availability of venture capital is widely believed to play a significant role in the birth of new biotech entrants (Martin Kenney, 1986; Joshua Lerner, 1994, 1995); so we also include that variable in our regressions.

II. The Data

Data has been collected in panel form for 14 years (1976–1989) and 183 regions (functional economic areas as defined by the U.S. Department of Commerce, Bureau of Economic Analysis [BEA], 1992b). Frequently, the data are aggregates of data at the zip code or county level.[2] Lagged variables

[2] The BEA's functional economic areas divide all the counties in the United States into regions including one or more cities, their suburbs, and the rural counties most closely tied to the central city.

include data for 1975 in the unlagged form. See the Data Appendix for more details.

A. *Firms*

Our data set on firms was derived from a base purchased from the North Carolina Biotechnology Center (NCBC) (1992) which was cleaned and supplemented with information in *Bioscan* (1989–1993) and its precursor (Cetus Corp., 1988). We identified 751 distinct U.S. firms for which we could determine a zip code and a date of initial use of biotechnology. Of these 751 firms, 511 were entrants, 150 incumbents, and 90 (including 18 joint ventures) could not be definitively classified. By 1990, 52 of the 751 firms had died or merged into other firms.

We then calculated the number of births in each region by year of initial use of biotechnology for all 751 firms as well as for their identified subcomponents of entrants and incumbents. We also have the stocks of surviving firms, entrants, and incumbents by region and year.

B. *Scientists*

Early in our ongoing project studying the scientific development and diffusion of biotechnology, we identified a set of 327 star scientists based on their outstanding productivity through early 1990. The primary criterion for selections was the discovery of more than 40 genetic sequences as reported in *GenBank* (1990) through April 1990.[3] However, 22 scientists were included based on writing 20 or more articles, each reporting one or more genetic-sequence discoveries.[4] In the 1990's, sequence discovery has become routinized and is no longer such a useful measure of research success. These 327 stars were only three-quarters of one percent of the authors in *GenBank* (1990) but accounted for 17.3 percent of the published articles, almost 22 times as many articles as the average scientist.

We collected by hand the 4,061 articles authored by stars and listed in *GenBank* and recorded the institutional affiliation of the stars and their coauthors on each of these articles. These coauthors are called "*collaborators*" if they are not themselves a star. Some data on the stars and collaborators who ever published in the United States is given on the left side of Table 1, where the scientists are identified by the organization(s) with which they were affiliated on their first-such publication. The higher citation rate for firm-affiliated scientists is explored at length in Zucker and Darby (1996).

Figure 1 illustrates the time pattern of growth in the numbers of stars and collaborators who have ever published and the total number of firms using biotechnology in the United States. There was a handful of stars who published articles reporting genetic-sequence discoveries before the 1973 breakthrough, but even after 1973 their number increased gradually until taking off in 1980. The numbers of collaborators and firms lagged behind the growth in stars by some years.

To identify those scientists clearly working at the edge of the science in a given year, we term a star or collaborator as "*active*" if he or she has published three or more sequence-discovery articles in the three-year moving window ending with that year. As seen in the right side of Table 1, this stringent second screen provides an even more elite definition of star scientists as well as identifying some very significant collaborators. We count for each year the number of active stars and active collaborators who are affiliated with an organization in each region.

The locations of active stars and firms are both concentrated and highly correlated geographically, particularly early in the period. Figure 2 illustrates this pattern for the whole period by accumulating the number of stars who have ever been active in each region up to 1990 and plotting them together with the

[3] See Zucker et al. (1993). As will be obvious, much of the time between 1990 and the initial submission of this paper was spent in developing reasonable measures of intellectual human capital and in collecting and coding data necessary to locate the authors of the discoveries reported in the articles in question and to trace the diffusion process.

[4] Scientists advised that some sequence discoveries are more difficult than others and thus merit an article reporting only one sequence. Therefore we included scientists with 20 or more discovery articles to avoid excluding scientists who specialized in more difficult problems.

TABLE 1—DISTRIBUTION OF STAR SCIENTISTS AND COLLABORATORS WHO HAVE EVER PUBLISHED IN THE UNITED STATES

	Full data set		Ever active in U.S.[b]	
Organization type[a]	Number of scientists	Citations[c]/ scientist/years	Number of scientists	Citations[c]/ scientist/years
Stars:				
University	158	85.5	108	110.8
Institute	44	63.0	26	98.7
Firm	5	143.7	1	694.3
Dual	0	n/a	0	n/a
Total	207		135	
Collaborators:				
University	2901	10.4	369	30.6
Institute	776	13.7	88	35.8
Firm	324	29.2	43	99.1
Dual	3	7.2	0	n/a
Total	4004		500	

[a] The organization type refers to the affiliation listed on their **first** publication with a U.S. affiliation.

[b] Ever active in the U.S. means that in at least one three-year period beginning 1974 or later and ending 1989 or earlier, the scientist was listed on at least three articles appearing in our data set of 4,061 articles which reported genetic-sequence discoveries and were published in major journals and that the affiliation listed in the last of the three articles was located in the United States.

[c] Citation counts are for 1982, 1987, and 1992 for all articles in our data set (whenever published) for which the individual was listed as an author.

location of biotech-using firms as of early 1990.

C. *Other Measures of Intellectual Human Capital*

Active stars and collaborators may be incomplete measures of location of the scientific base because there are techniques other than recombinant DNA which have played an important role in commercial biotechnology. Some skeptical readers might also think that some simpler measures of regions' relevant academic resources would contain all the information which we have laboriously collected. We found two measures of regional scientific base which entered separately in regressions reported below, but none which were capable of eliminating the effects of the star scientists.

One measure is a count of the number of *"top-quality universities"* in a region where top quality is defined by having one or more *"biotech-relevant"* (biochemistry, cellular/molecular biology, and microbiology) departments with scholarly quality reputational ratings of 4.0 or higher in the 1982 National Research Council survey (Lyle V. Jones et al., 1982). There are 20 such universities in the United States.[5] Our second measure, *"federal*

[5] The 20 universities were: Brandeis University, California Institute of Technology, Columbia University, Cornell University, Duke University, Harvard University, Johns Hopkins University, Massachusetts Institute of Technology, Rockefeller University, Stanford University, University of California-Berkeley, University of California-Los Angeles, University of California-San Diego, University of California-San Francisco, University of

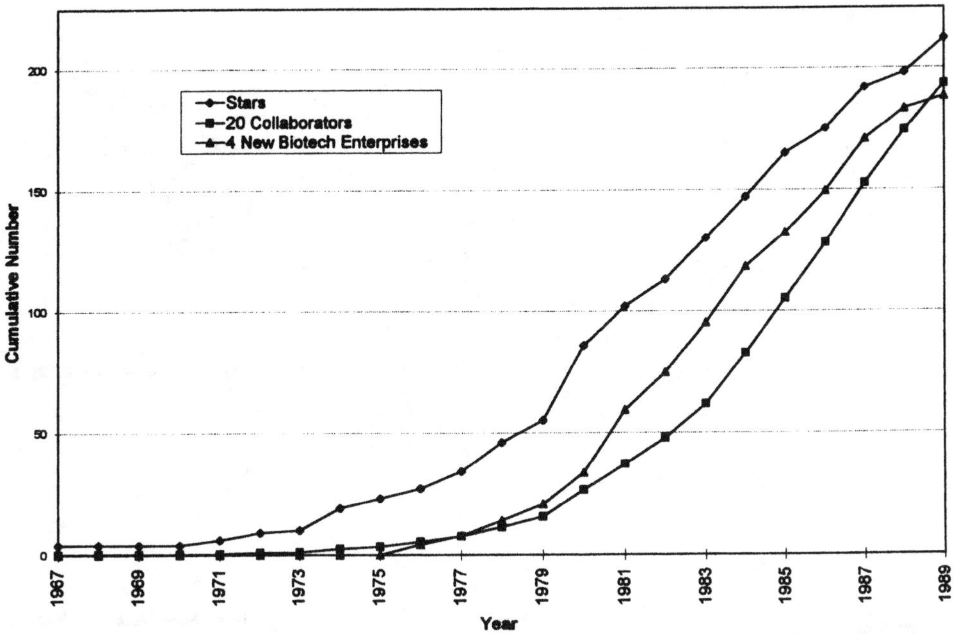

FIGURE 1. CUMULATIVE NUMBER OF U.S. STARS, COLLABORATORS, AND NEW BIOTECH ENTERPRISES, 1967–1989

support, " is the total number (in hundreds) of faculty supported by 1979–1980 federal grants to all universities in each region for biotech-relevant research.[6] These variables take on the same value for a given region in each year.

D. *Other Variables*

Using listings in Stanley E. Pratt (1982), we measure "*venture capital firms*" as the number of such firms in a region legally eligible to finance start-ups in each year up to 1981. For later years, the number of firms is fixed at the number in 1981 to avoid possible simultaneity problems once the major wave of biotech founding began.[7] (While great bookstores spring up around great universities, the former should not be counted as causing the latter.)

Since entry of biotech firms would be expected to occur where there is other economic activity, particularly involving a highly skilled labor force, we also include *total employment* in all industries (in millions of persons) and *average wages* (measured by deflated average earnings per job in thousands of 1987 dollars) for each region and year.

Finally, an increase in the (all-equity) cost of capital, as measured by the *earnings-price ratio* on the Standard & Poor's 500 Index would reduce the net present value of entry

Chicago, University of Colorado at Denver, University of Pennsylvania, University of Washington (Seattle), University of Wisconsin-Madison, and Yale University.

[6] We also tried a measure of biotech-relevant research expenditures as reported by the universities, but this variable was too collinear with the federal support variable to enter separately and appeared to be less consistently measured across universities.

[7] Instrumental variables would provide a more elegant approach to this problem if suitable instruments had been found.

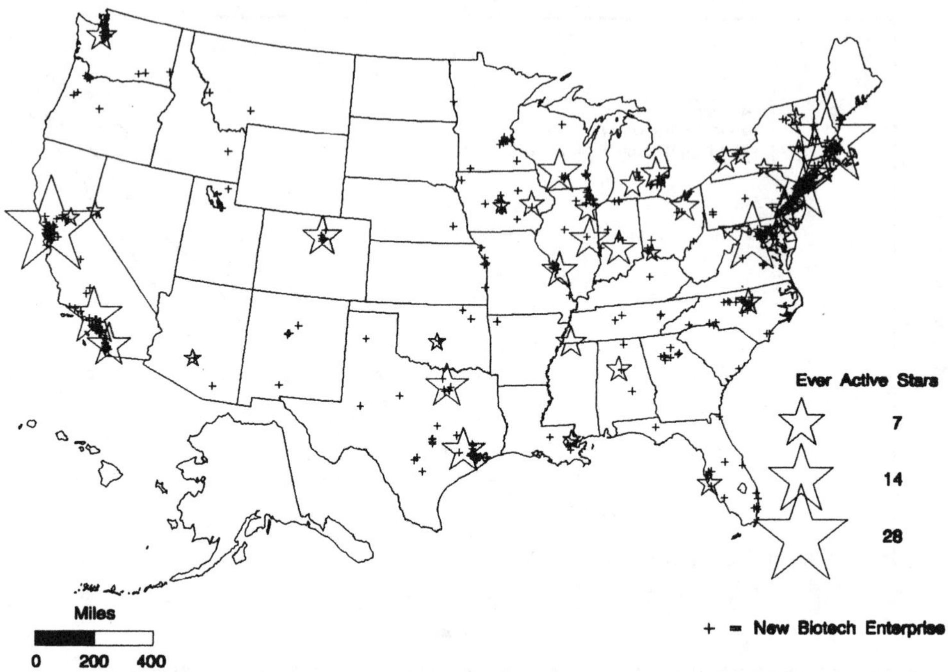

FIGURE 2. EVER ACTIVE STARS AND NEW BIOTECHNOLOGY ENTERPRISES AS OF 1990

and so should have a negative impact on birth of new firms, entrants or incumbents.

III. Empirical Results

We test our hypothesis using both the full panel data and by regressing the geographical distribution of the data in 1990 on values of the independent variables circa 1980. The former more fully exploit the available information while the latter avoid problems of possible simultaneity which might arise after 1980 when commercial biotechnology became a significant economic factor in some regions. All the regressions reported here, as well as an extensive sensitivity analysis noted below, were estimated in the poisson form appropriate for count variables with numerous zeroes using LIMDEP (William H. Greene, 1992 pp. 539–49), with the Wooldridge regression-based correction for the variance-covariance matrix estimates.[8] The poisson regressions estimate the logarithm of the expected number of firm births; so the signs and significance of coefficients have the usual interpretation. Although OLS regressions are inappropriate for our count dependent variables with most observations at zero and the rest tailing off through small positive integers, we

[8] As discussed in Jerry Hausman et al. (1984), the poisson process is the most appropriate statistical model for count data such as ours. In practice, overdispersion (possibly due to unobserved heterogeneity) frequently occurs. Given the problems with resort to the negative binomial (A. Colin Cameron and Pravin K. Trivedi, 1990), Jeffrey M. Wooldridge (1991) developed a flexible and consistent method for correcting the poisson variance-covariance matrix estimates regardless of the underlying relationship between the mean and variance. We are indebted to Wooldridge and Greene for advice in implementing the procedure in LIMDEP.

reported broadly consistent results using that technique in an earlier version of the paper (Zucker et al., 1994).

In our sensitivity analysis, we ran the same poisson regressions for entrants and incumbents defined both exclusive and inclusive of the arguable case of joint ventures. The results were generally very similar to the subcomponent regressions in Table 4. In other unreported poisson regressions, we found that eliminating those regions with no firms and no stars from the sample did not result in qualitatively different results.

A. *The Long-Run Model*

Because of concerns about possible simultaneity biases once the industry became a significant economic force, we begin our empirical discussion with models which relate the number of firms in each region at the beginning of 1990 to the distribution of intellectual human capital and other variables as of about 1980. These results provide something of an acid test of our approach.

In Table 2, we present cross-section poisson regressions across the 183 regions explaining the number of firms in each at the beginning of 1990 when our data set ends.[9] Column (a) restrains the analysis to only the numbers of stars and collaborators ever active in each region at any time up through 1980, while columns (b) and (c) add first their squares and then our other intellectual human capital variables. Column (d) considers alternatively other economic variables which might explain entry, and column (e) combines the variables in (c) and (d). Column (f) adds to this model the number of biotech firms existing in 1980.

Column (a) in Table 2 indicates that the number of stars and collaborators active through 1980 is a powerful predictor of the geographic distribution of biotech enterprises in 1990, since the log-likelihood increases to −871.9 compared to −1401.7 for a constant alone. It is the star scientists that contribute positively, with collaborators having a much smaller negative coefficient in this regression and most of the other long-run models discussed below. We had expected that the coefficient on collaborators would be much smaller than that on stars, but positive. We do obtain a positive coefficient on active collaborators when the squared terms are added in column (b), but that turns negative again upon addition of other variables in the remaining columns of Table 2.[10] (In the annual regressions discussed below, we generally estimate positive effects of active collaborators, but they are often statistically insignificant.)

We can offer two explanations for the generally negative sign on the number of active collaborators in the long-run regressions: (i) This coefficient reflects two partially offsetting influences; collaborators have a positive direct effect on the entry of firms but reduce the effect of stars who are devoting more of their time to training students and relatively less to starting their own firms. Training collaborators is surely a useful and rewarded activity—particularly for the academic stars—but it may take more of the stars' energy than it is worth if firm birth were the only criteria.[11] (ii) The sign and magnitude of the coefficient on collaborators may simply reflect significant multicollinearity among the intellectual human capital variables in the very early years. This is especially likely since when we examine the

[9] In an earlier version of this paper we included an alternative form of Table 2 (available from the authors upon request) in order to forestall interpretations that the results in Table 2 may reflect reverse causality. This alternative table reported regressions which explain the number of firms alive at the beginning of 1990 that were born after 1980. Nearly identical results were obtained, reflecting the fact that bulk of new biotechnology enterprises were founded after 1980.

[10] In column (b) of Table 2 (and Table 3), the negative coefficient on the squared term indicates that as the number of stars or collaborators increases, their marginal contribution diminishes eventually passing through zero. For collaborators, in columns (c)–(f) of these tables the sign pattern reverses so that the partial derivative of the log probability of birth with respect to collaborators starts out negative, and increases as their number increases, eventually becoming positive.

[11] In support of this explanation, we note that in our sensitivity analysis we tried regressions which substituted interaction terms multiplying the numbers of active stars and collaborators for the squared terms. In those regressions, we obtain significant positive coefficients on the numbers of stars and collaborators and a significant negative coefficient on their interaction.

TABLE 2—POISSON REGRESSIONS ON THE STOCK OF BIOTECH-USING FIRMS AT THE BEGINNING OF 1990 BY REGION

	(a)	(b)	(c)	(d)	(e)	(f)
Constant	0.911***	0.644***	0.468***	−2.595***	−2.718***	−2.607***
	(0.014)	(0.015)	(0.033)	(0.086)	(0.256)	(0.345)
Number stars active at any time during 1976–80	0.567***	0.587***	0.466***	—	0.877***	0.649***
	(0.029)	(0.072)	(0.090)	—	(0.076)	(0.084)
Number collaborators active at any time during 1976–80	−0.076***	0.175***	−0.183**	—	−0.333***	−0.261***
	(0.012)	(0.033)	(0.068)	—	(0.045)	(0.037)
(Number stars active at any time during 1976–80)^2	—	−0.028***	−0.019	—	−0.049***	−0.024*
	—	(0.007)	(0.014)	—	(0.012)	(0.012)
(Number collaborators active at any time during 1976–80)^2	—	−0.005***	0.002	—	0.007**	0.001
	—	(0.001)	(0.003)	—	(0.003)	(0.002)
Number top-quality universities in the region	—	—	1.388***	—	1.594***	0.442*
	—	—	(0.150)	—	(0.107)	(0.195)
Number faculty with federal support in the region	—	—	0.263	—	0.752***	0.711***
	—	—	(0.143)	—	(0.088)	(0.051)
Number venture capital firms in the region in 1980	—	—	—	0.017***	−0.045***	−0.013**
	—	—	—	(0.002)	(0.003)	(0.004)
Total employment (all industries) in the region in 1980	—	—	—	0.222***	−0.009	−0.213***
	—	—	—	(0.019)	(0.043)	(0.049)
Average wages per job in the region in 1980	—	—	—	0.166***	0.143***	0.139***
	—	—	—	(0.004)	(0.014)	(0.019)
Cumulative births of biotech firms during 1976–80	—	—	—	—	—	0.300***
	—	—	—	—	—	(0.025)
Log-likelihood	−871.9	−707.3	−543.2	−753.9	−416.0	−350.7
Restricted log-likelihood	−1401.7	−1401.7	−1401.7	−1401.7	−1401.7	−1401.7

Notes: N = 183. Standard errors (adjusted by Wooldridge, 1991 Procedure 2.1) are in parentheses below coefficients.
* Significantly different from 0 at the 5-percent level.
** Significantly different from 0 at the 1-percent level.
*** Significantly different from 0 at the 0.1-percent level.

full cross-section/time-series results just below we obtain (we think more reliable) zero or positive coefficients on collaborators, so the puzzle largely disappears.

The full "fundamentals" model (excepting the decade-lagged dependent variable) is presented in column (e) of Table 2, where all the coefficients are significant except that for total employment. Leaving aside the question of the negative collaborator coefficient, we note the strong, positive, separate effects of stars, top-quality universities, and federal research grants at universities on birth of firms in a given geographic region. The intellectual human capital variables alone increase the log-likelihood ratio from −1401.7 to −543.2 [see column (c)], with the final three variables bringing this quantity up to −416.0. As to the last three variables, the quality of the labor force, measured by average wages per job, seems much more relevant than its size. Surprisingly, to some observers, the number of venture capital firms in a region enters, but with a significantly negative sign. We interpret the negative sign as evidence that venture capitalists did play an active role in the formation of entrant firms, but they apparently resulted in fewer, larger firms being born in the areas in which they were more active.[12]

This sign of the coefficient of the number of venture capital firms in a region is robust in sensitivity experiments with other forms (not reported here) except for regressions which exclude the intellectual human capital variables such as in column (d). That regression looks good in terms of significance and expected sign pattern although it has a much lower explanatory power than the intellectual human capital variables alone [column (c)]. Just below, we report very similar results in a cross-section/time-series context. Thus, it is certainly easy to see why the evidence for an important positive impact of venture capital firms on the birth of biotech firms may have appeared stronger in previous work than seems warranted based on fuller models: Since venture capital firms have developed around a number of great universities, their presence proxies for intellectual human capital in the absence of more direct measures; if they are the only variable indicating presence of great universities and their faculties, they enter positively even if their packaging activities result in a negative direct effect on births.

The decade-lagged dependent variable is added to the full fundamentals model in column (f) of Table 2. Doing so primarily has the effect of weakening the significance of the top-quality universities variable (but, see the annual model below) due to significant multicollinearity between the variables.[13] One interpretation of this positive coefficient on the lagged dependent variable is that agglomeration effects strengthen the impact of fundamentals on regional development. However, the statistical properties of poisson regressions with lagged dependent variables are somewhat problematic so such regressions and their estimated standard errors should be viewed cautiously.

In conclusion, the intellectual human capital variables play a strong role in determining where the U.S. biotech industry developed during the 1980's. We have been able to identify particular star scientists who appear to play a crucial role in the process of spillover and geographic agglomeration over and above that which would be predicted based on university reputation and scientists supported by federal grants alone. The strong positive role of venture capital variable reported previously is not supported for firm births. Indeed, the data tell us that there were fewer firms founded, other things equal, where there more venture capital firms. It is left to future research to explore whether firms which are associated with particular star scientists or were midwifed by venture capitalists are more successful than other firms.[14]

B. *The Annual Model*

We next report analogous poisson regressions exploiting the panel nature of our data set with observations for the 183 regions for each of the years 1976 through 1989. Tables 3 and 4 report poisson regressions for this entire panel.

Column (a) of Table 3 reports the results using only the counts of stars and their collaborators active each year in each region. As with the long-run models in Table 2, examination of the data suggested that these effects—particularly for stars—were nonlinear so we add squared values in column (b). Again, as the number of stars increases, their marginal contribution diminishes eventually passing through zero.

These nonlinearities might reflect the declining value over time of the intellectual human capital as we have measured it. Basically,

[12] This hypothesis was derived from anecdotal evidence, but note that the top nine of Ernst & Young's list of top-ten companies by 1993 market valuation (G. Steven Burrill and Kenneth B. Lee, Jr., 1994 p. 54) were located and founded in regions richly endowed with venture capital firms: Boston (3), San Francisco (3), Los Angeles (1), San Diego (1), and Seattle (1).

[13] In the alternative version of Table 2 (see footnote 9 above), the coefficient on the lagged dependent variable was nearly as large as in Table 2, so the significant positive coefficient does not arise from firms born 1976–1980 appearing in both the current and lagged dependent variables.

[14] See Zucker et al. (1994) for our first effort to assess the determinants of success of firms after birth.

TABLE 3—ANNUAL POISSON REGRESSIONS ON THE BIRTH OF BIOTECH-USING FIRMS BY REGION AND YEAR, 1976–1989

	(a)	(b)	(c)	(d)	(e)	(f)
Constant	−1.591*** (0.032)	−1.918*** (0.041)	−2.148*** (0.057)	−4.447*** (0.226)	−4.491*** (0.349)	−4.687*** (0.565)
Number stars active in the region and year	0.157*** (0.020)	0.529*** (0.051)	0.270** (0.088)	— —	0.361*** (0.080)	0.282** (0.103)
Number collaborators active in the region and year	0.043*** (0.013)	0.083* (0.035)	0.047 (0.049)	— —	0.013 (0.047)	0.032 (0.052)
(Number stars active in the region and year)^2	— —	−0.022*** (0.002)	−0.014* (0.006)	— —	−0.015** (0.005)	−0.014 (0.008)
(Number collaborators active in the region and year)^2	— —	−0.001 (0.001)	0.000 (0.001)	— —	0.000 (0.001)	0.001 (0.002)
Number stars active in the region and year × DUMMY1986–89	— —	— —	−0.219 (0.113)	— —	−0.298** (0.102)	−0.245 (0.128)
Number collaborators active in the region and year × DUMMY1986–89	— —	— —	0.117 (0.067)	— —	0.115 (0.064)	0.027 (0.081)
(Number stars active in the region and year × DUMMY1986–89)^2	— —	— —	0.006 (0.007)	— —	0.009 (0.006)	0.007 (0.008)
(Number collaborators active in the region and year × DUMMY1986–89)^2	— —	— —	−0.001 (0.002)	— —	−0.001 (0.002)	0.001 (0.002)
Number top-quality universities in the region in 1981	— —	— —	0.444*** (0.125)	— —	0.472*** (0.095)	0.462*** (0.109)
Number faculty with federal support in the region in 1979–80	— —	— —	0.625*** (0.093)	— —	0.982*** (0.094)	0.930*** (0.093)
Number venture capital firms in the region and year[a]	— —	— —	— —	0.019** (0.007)	−0.028*** (0.006)	−0.024** (0.008)
Total employment (all industries) in the region and year	— —	— —	— —	0.173*** (0.051)	−0.081 (0.048)	−0.117* (0.055)
Average wages per job in the region and year	— —	— —	— —	0.153*** (0.010)	0.125*** (0.016)	0.132*** (0.017)
Earnings-price ratio (Standard & Poors 500) for year	— —	— —	— —	−0.024 (0.016)	−0.026 (0.026)	−0.017 (0.039)
Number firms active in the region at end of previous year	— —	— —	— —	— —	— —	0.020 (0.013)
Number firms active in all U.S. at end of previous year	— —	— —	— —	— —	— —	−0.000 (0.000)
Births of biotech firms in the region for previous year	— —	— —	— —	— —	— —	0.054 (0.034)
Log-likelihood	−1677.0	−1429.1	−1274.3	−1669.5	−1202.3	−1184.6
Restricted log-likelihood	−2238.5	−2238.5	−2238.5	−2238.5	−2238.5	−2238.5

Notes: N = 2562. Standard errors (adjusted by Wooldridge, 1991 Procedure 2.1) are in parentheses below coefficients.

[a] For years after 1981, the number of venture capital firms in a region is held constant at the 1981 level to avert simultaneity problems.

* Significantly different from 0 at the 5-percent level.

** Significantly different from 0 at the 1-percent level.

*** Significantly different from 0 at the 0.1-percent level.

as the knowledge diffuses we expect that more and more stars will result in less and less pay-off to any one of them if he or she were to start a firm, and indeed stars are less likely to result in birth of firms after 1985 than before. This is illustrated in column (c) of Table 3 where we add four interaction terms in which these counts and their squares have been multiplied by a dummy DUMMY1986–89 equal to 1 during 1986–1989 and 0 otherwise, as well as the other intellectual human capital terms. During 1986–1989 the positive effect of stars is sharply reduced while that of collaborators more than triples.[15] Nonetheless, we should view this inference cautiously since the significance values of the interaction terms for stars and collaborators with DUMMY1986–89 fall between 0.10 and 0.05, except for stars in the full fundamentals model in column (e) where the stars interaction term is significant at the 0.01 level.

Thus, we see that (at least during the first decade of this industry) localities with outstanding scientists having the tacit knowledge to practice recombinant DNA were much more likely to see new firms founded and preexisting firms begin to apply biotechnology. There is some evidence that as knowledge about gene splicing diffused and the tacit knowledge lost its scarcity and extraordinary value, the training function of universities became more important relative to the attraction of great scientists to an area. It is interesting that the quadratic term for stars is negative, suggesting diminishing returns (or possibly just proportionately fewer, larger firms) rather than the increasing returns suggested by standard views of knowledge spillovers which posit uninternalized, positive external effects from university scientists.[16] In the same regression in column (c), we see that, beyond the identified stars and collaborators, university quality and federal support are also significant measures of intellectual human capital relevant to firm founding.

Column (d) of Table 3 leads to the same conclusions with panel data as found for the same column in Table 2: The economic variables enter significantly with the expected sign if the intellectual human capital variables are omitted from the regression. However, unlike the previous long-run case, we can now enter the earnings-price ratio.[17] Here this variable enters with the correct sign, but does not even reach the 0.10 level of significance.

Column (e) of Table 3 presents the annual full fundamentals model incorporating the intellectual human capital and other variables. The results for the intellectual human capital measures are robust while the sign of the venture capital variable turns significantly negative as in the long-run model and the employment variable becomes insignificant (and negative).

Column (f) of Table 3, analogously to Table 2, adds a lagged dependent variable to the full fundamentals model. We also included the one-year lagged regional and national counts of firms using biotechnology as dynamic influences reflecting local agglomeration effects and market competition effects, respectively. None of the three dynamic variables enter significantly although their signs are consistent with some geographic agglomeration.

Thus, taken as a whole the results summarized in Table 3 support the strong role of intellectual human capital variables in determining the development of the American biotech industry.

The role of the economic variables, particularly the number of venture capital firms in the region, is explored further in Table 4. This table presents representative results for births in the entrant and incumbent subcomponents of firm entry into biotechnology. We see in columns (a) and (b) that if only the economic variables are introduced we get all the expected signs at appropriate significance [except for employment in (a) and the earnings-price ratio in both], including a result consistent with conventional wisdom that the number of venture capital firms has a signifi-

[15] To compute the effects of stars in the 1986–1989 period, we need to add the coefficients of the number of active stars and the coefficient of the same variable interacted with DUMMY1986–89 and then do the same for the two terms involving the squared values of these variables. An analogous approach yields the effect of collaborators during 1986–1989. We examined also interactions with dummy variables for 1976–1980 and with a time trend. Since the coefficients were very small and statistically insignificant for interaction terms involving 1976–1980 dummies, we believe the reported form more accurately reflects the time or diffusion dependence than a negative trend throughout the period.

[16] We are indebted to Jeff Armstrong for this point.

[17] The earnings-price ratio had to be dropped from these analyses because it is available only nationally over time.

TABLE 4—ANNUAL POISSON REGRESSIONS ON THE BIRTH OF BIOTECH-USING ENTRANTS AND INCUMBENTS BY REGION AND YEAR, 1976–1989

	(a) Entrants	(b) Incumbents	(c) Entrants	(d) Incumbents	(e) Entrants	(f) Incumbents
Constant	−4.726*** (0.284)	−5.798*** (0.563)	−4.843*** (0.409)	−5.673*** (0.902)	−4.928*** (0.669)	−5.228*** (1.285)
Number stars active in the region and year	— —	— —	0.414*** (0.095)	0.323 (0.165)	0.351** (0.124)	0.242 (0.169)
Number collaborators active in the region and year	— —	— —	−0.006 (0.053)	0.000 (0.105)	0.012 (0.059)	0.019 (0.101)
(Number stars active in the region and year)^2	— —	— —	−0.016** (0.006)	−0.016* (0.008)	−0.017 (0.009)	−0.015 (0.011)
(Number collaborators active in the region and year)^2	— —	— —	0.001 (0.002)	0.002 (0.003)	0.000 (0.002)	0.001 (0.003)
Number stars active in the region and year × DUMMY1986–89	— —	— —	−0.227* (0.113)	−0.519* (0.237)	−0.196 (0.147)	−0.456 (0.251)
Number collaborators active in the region and year × DUMMY1986–89	— —	— —	0.096 (0.071)	0.233 (0.141)	0.011 (0.090)	0.144 (0.153)
(Number stars active in the region and year × DUMMY1986–89)^2	— —	— —	0.007 (0.007)	0.018 (0.010)	0.006 (0.010)	0.015 (0.013)
(Number collaborators active in the region and year × DUMMY1986–89)^2	— —	— —	−0.001 (0.002)	−0.004 (0.003)	0.001 (0.003)	−0.002 (0.004)
Number top-quality universities in the region in 1981	— —	— —	0.440*** (0.110)	0.479* (0.205)	0.410** (0.126)	0.447 (0.238)
Number faculty with federal support in the region in 1979–80	— —	— —	0.973*** (0.112)	1.114*** (0.296)	0.932*** (0.107)	1.041*** (0.295)
Number venture capital firms in the region and year[a]	0.023** (0.009)	0.006 (0.013)	−0.029*** (0.007)	−0.027* (0.012)	−0.024** (0.009)	−0.024 (0.013)
Total employment (all industries) in the region and year	0.128 (0.067)	0.296** (0.098)	−0.110 (0.058)	−0.052 (0.098)	−0.149* (0.067)	−0.078 (0.103)
Average wages per job in the region and year	0.156*** (0.012)	0.139*** (0.024)	0.123*** (0.018)	0.113** (0.039)	0.127*** (0.020)	0.114** (0.040)
Earnings-price ratio (Standard & Poors 500) for year	−0.036 (0.021)	−0.033 (0.043)	−0.022 (0.031)	−0.056 (0.070)	−0.016 (0.046)	−0.082 (0.092)
Number firms active in the region at end of previous year	— —	— —	— —	— —	0.023 (0.015)	0.024 (0.025)
Number firms active in all U.S. at end of previous year	— —	— —	— —	— —	−0.000 (0.000)	−0.001 (0.001)
Births of biotech firms in the region for previous year	— —	— —	— —	— —	0.037 (0.041)	0.055 (0.061)
Log-likelihood	−1265.1	−486.3	−945.9	−386.8	−935.8	−382.9
Restricted log-likelihood	−1628.7	−607.9	−1628.7	−607.9	−1628.7	−607.9

Notes: $N = 2562$. Standard errors (adjusted by Wooldridge, 1991 Procedure 2.1) are in parentheses below coefficients.

[a] For years after 1981, the number of venture capital firms in a region is held constant at the 1981 level to avert simultaneity problems.

* Significantly different from 0 at the 5-percent level.

** Significantly different from 0 at the 1-percent level.

*** Significantly different from 0 at the 0.1-percent level.

cantly positive effect on the birth of new firms but an insignificant effect on the birth of subunits of existing firms which would not normally be financed by venture capital firms. The full fundamentals model is reported in columns (c) and (d) for births of entrants and incumbents, respectively, which is to be compared to column (e) for all firm births in Table 3. Again, in the presence of intellectual human capital the simple economic story does not hold up: the coefficients of venture capital firms and total employment turn negative, significantly so in the former case. Similar results are obtained in the dynamic versions of the full model reported in columns (e) and (f) of Table 4. The robustness of the negative venture capital coefficient remains a puzzle for future work, particularly in view of Yolanda K. Henderson's (1989) evidence that, despite some significant localization, most investments by venture capitalists cross regional boundaries.

IV. Summary and Conclusions

The American biotechnology industry which was essentially nonexistent in 1975 grew to 700 active firms over the next 15 years. In this paper, we show the tight connection between the intellectual human capital created by frontier research and the founding of firms in the industry. At least for this high-tech industry, the growth and location of intellectual human capital was the principal determinant of the growth and location of the industry itself. This industry is a testament to the value of basic scientific research. The number of local venture capital firms, which appears to be a positive determinant when intellectual human capital variables are excluded from the regressions, is found to depress the rate of firm birth in an area, perhaps due to the role of these venture capital firms in packaging a number of scientists into one larger firm which is likely to go public sooner.

We conclude that the growth and diffusion of intellectual human capital was the main determinant of where and when the American biotechnology industry developed. Intellectual human capital tended to flourish around great universities, but the existence of outstanding scientists measured in terms of research productivity played a key role over, above, and separate from the presence of those universities and government research funding to them. We believe that our results provide new insight into the role of research universities and their top scientists as central to the formation of new high-tech industries spawned by scientific breakthroughs. By being able to quantitatively identify individuals with the ability both to invent and to commercialize these breakthroughs, we have developed new specificity for the idea of spillovers and in particular raised the issue of whether spillovers are best viewed as resulting from the nonappropriability of scientific knowledge or from the maximizing behavior of scientists who have the ability to appropriate the commercial fruits of their academic discoveries.

Data Appendix

The data used here are generally in panel form for 14 years (1976–1989) and 183 regions (functional economic areas as defined by the BEA). Frequently, the data are aggregates of data at the zip code or county level. Lagged variables include data for 1975 in the unlagged form. These data sets, part of our ongoing project on "Intellectual Capital, Technology Transfer, and the Organization of Leading-Edge Industries: Biotechnology," will be archived upon completion of the project in the Data Archives at the UCLA Institute for Social Science Research. A full description of the data is available from the authors upon request.

Biotechnology Firms

The starting point for our firm data set covered the industry as of April 1990 and was purchased from NCBC (1991), a private firm which tracks the industry. This data set identified 1075 firms, some of which were duplicates or foreign and others of which had died or merged. Further, there were a significant number of firms missing which had exited prior to April 1990. For these reasons, an intensive effort was made to supplement the NCBC data with information from *Bioscan* (1989–1993) and an industry data set provided by a firm in the industry which was also the ancestor of the *Bioscan* data set (Cetus Corp., 1988).

We generally counted entry of firms by adding up for each year and region the number of entrants founded and incumbents first using biotechnology. A few special cases should be noted: Where a firm enters the data set due to the merger of a entrant and another firm, we count it for the purposes of this paper as a continuation of the original entrant and not a new birth (the older entrant if two are involved). If firms already in the data set merge and one continues with the other(s) absorbed, the enterprise is counted as the continuing enterprise and not a new birth.

Scientists

Star scientists and their collaborators were identified as described in the text. Individual scientists are linked to locations through the institutional affiliations reported in their publications in the article data set. The discovery of genetic sequences is recognized by *GenBank*'s assignment to an article of a "primary accession number" to identify each. The 22 additional stars added to the 315 with more than 40 primary accession numbers thus had 20 or more articles with at least one primary accession number and 20–40 primary accession numbers total.

Articles

Our article data set consists of all 4,061 articles in major journals listed in *GenBank* as reporting genetic-sequence discoveries for which one or more of our 327 stars were listed as authors. (A small number of unpublished papers and articles appearing in proceedings volumes and obscure journals were excluded to permit the hand coding detailed below.) All of these articles were assigned unique article ID numbers and collected by hand. For each article, scientist ID numbers are used to identify the order of authorship and the institutional affiliation and location for each author on each article. This hand coding was necessary because, under the authorship traditions for these fields, the head of the laboratory who is often the most prestigious author frequently appears last. Our stars, for example, were first authors on 18.3 percent of the articles and last authors on 69.1 percent of the 4,031 articles remaining after excluding the 30 sole-authored articles.[18] Unfortunately, only first- and/or corresponding-author affiliations are available in machine-readable sources.[19]

The resulting authorship data file contains 19,346 observations, approximately 4.8 authors for each of the 4,061 published articles. Each authorship observation gives the article ID number, the order of authorship, the scientist ID number of one of our stars and collaborators, and an institutional ID number for the author's affiliation which links him or her to a particular institution with a known zip code as of the publication date of the article.

Citations

We have collected data for 1982, 1987, and 1992 on the total number of citations to each of our 4,061 published articles listed in the Institute for Scientific Information's *Science Citation Index* (1982, 1987, 1992). These citation counts are linked to the article and authorship data set by the article ID number. The citations were collected for articles if and only if they appeared in the article data set; so scientists are credited with citations only insofar as they are to the 4,061 articles reporting genetic-sequence discoveries and published in major journals.

Universities

Our university data set consists of all U.S. institutions listed as granting the Ph.D. degree in any field in the Higher Education General Information Survey (HEGIS), Institutional Characteristics, 1983–1984 (U.S. Department of Education, National Center for Education

[18] This positional tradition holds across national boundaries: As a percentage of articles coauthored by their fellow nationals, American stars are 16.4 percent of first authors and 71.2 percent of last authors, compared to 21.2 percent and 63.1 percent, respectively, for Japanese, and 19.7 percent and 69.2 percent for other nationalities.

[19] The *Science Citation Index* lists up to six of the affiliations listed on the paper but only links the corresponding author to a particular affiliation.

Statistics, 1985). Each university is assigned an institutional ID number, a university flag, and located by zip code based on the HEGIS address file. Additional information described in the text was collected from Jones et al. (1982) for those universities granting the Ph.D. degree in biochemistry, cellular/molecular biology, and/or microbiology which we define as *''biotech-relevant''* fields.

Research Institutes and Hospitals

For those U.S. research institutions and hospitals listed as affiliations in the article data set, we assigned an institutional ID number and an institute/hospital flag, and obtained an address including a zip code as required for geocoding. No additional information has been collected on these institutions.

Venture Capital Firms

We created a venture capital firm data set by extracting from the Pratt (1982) directory the name, type, location, year of founding, and interest in funding biotech firms. This information was extracted for all venture capital which were legally permitted to finance start-ups. This latter requirement eliminated a number of firms which are chartered under government programs targeted at small and minority businesses. This approach accounts includes founding date of firms appearing in the 1982 Pratt directory, excluding those firms that may have either entered thereafter or existed in earlier years but exited before the directory was compiled.

Other Economic Variables

Total employment and average earnings per job by region and year are as reported by the Bureau of Economic Analysis based on county level data in U.S. Department of Commerce (1992b): Total employment is from Table K, line 010 (in millions of persons). Average earnings is from Table V, line 290 (wage & salary disbursements, other labor income, and proprietors income per job in thousands of current dollars), deflated by the implicit price deflator for personal consumption expenditures. The annual data for the implicit price deflator for personal consumption expenditures were taken from U.S. Department of Commerce (1992a p. 247, line 16) as updated in the July 1992 *Survey of Current Business*, (p. 92, line 16). The S&P 500 earnings-price ratio was taken from *CITIBASE* (1993), series FSEXP.

REFERENCES

Arrow, Kenneth J. ''Economic Welfare and the Allocation of Resources for Invention,'' in Richard R. Nelson, ed., *The rate and direction of inventive activity: Economic and social factors*. National Bureau of Economic Research Special Conference Series, Vol. 13. Princeton, NJ: Princeton University Press, 1962, pp. 609–25.

Bioscan, Vols. 3–7. Phoenix, AZ: Oryx Press, 1989–1993.

Burrill, G. Steven and Lee, Kenneth B., Jr. *Biotech 94: Long-term value, short-term hurdles*. Ernst & Young's Eighth Annual Report on the Biotechnology Industry. San Francisco: Ernst & Young, 1994.

Cameron, A. Colin and Trivedi, Pravin K. ''Regression-Based Tests for Overdispersion in the Poisson Model.'' *Journal of Econometrics*, December 1990, *46*(3), pp. 347–64.

Cetus Corp. ''Biotechnology Company Data Base,'' predecessor source for *Bioscan*. Computer printout, Cetus Corp., 1988.

CITIBASE: Citibank Economic Database. Machine-readable database, 1946–June 1993. New York: Citibank, 1993.

GenBank, Release 65.0. Machine-readable database. Palo Alto, CA: IntelliGentics, Inc., 1990.

Greene, William H. *LIMDEP: User's manual and reference guide, version 6.0*. Bellport, NY: Econometric Software, 1992.

Griliches, Zvi. ''Hybrid Corn: An Exploration in the Economics of Technological Change.'' *Econometrica*, October 1957, *25*(4), pp. 501–22.

______. ''The Search for R&D Spillovers.'' *Scandinavian Journal of Economics*, 1992 Supplement, *94*, pp. 29–47.

Grossman, Gene M. and Helpman, Elhanan. *Innovation and growth in the global economy*. Cambridge, MA: MIT Press, 1991.

Hausman, Jerry; Hall, Bronwyn H. and Griliches, Zvi. "Econometric Models for Count Data with an Application to the Patents-R&D Relationship." *Econometrica*, July 1984, *52*(4), pp. 909–38.

Henderson, Yolanda K. "Venture Capital and Economic Development." Paper presented to the New England Advisory Council, Federal Reserve Bank of Boston, Boston, MA, July 11, 1989.

Institute for Scientific Information. *Science Citation Index*, ISI compact disc editions, machine-readable database. Philadelphia: Institute for Scientific Information, 1982, 1987, 1992.

Jaffe, Adam B. "Real Effects of Academic Research." *American Economic Review*, December 1989, *79*(5), pp. 957–70.

Jaffe, Adam B.; Trajtenberg, Manuel and Henderson, Rebecca. "Geographic Localization of Knowledge Spillovers as Evidenced by Patent Citations." *Quarterly Journal of Economics*, August 1993, *63*(3), pp. 577–98.

Jones, Lyle V.; Lindzey, Gardner and Coggeshall, Porter E., eds. *An assessment of research-doctorate programs in the United States: Biological sciences*. Washington, DC: National Academy Press, 1982.

Kenney, Martin. *Biotechnology: The university-industrial complex*. New Haven, CT: Yale University Press, 1986.

Lerner, Joshua. "Venture Capitalists and the Decision to Go Public." *Journal of Financial Economics*, June 1994, *35*(3), pp. 293–316.

______. "Venture Capitalists and the Oversight of Private Firms." *Journal of Finance*, March 1995, *50*(1), pp. 301–18.

Mansfield, Edwin. "Academic Research Underlying Industrial Innovations: Sources, Characteristics, and Financing." *Review of Economics and Statistics*, February 1995, *77*(1), pp. 55–65.

Nelson, Richard R. "The Simple Economics of Basic Scientific Research." *Journal of Political Economy*, June 1959, *67*(3), pp. 297–306.

Nelson, Richard R. and Romer, Paul M. "Science, Economic Growth, and Public Policy," in Bruce L. R. Smith and Claude E. Barfield, eds., *Technology, R&D, and the economy*. Washington, DC: Brookings Institution and American Enterprise Institute, 1996, pp. 49–74.

Nelson, Richard R. and Winter, Sidney G. *An evolutionary theory of economic change*. Cambridge, MA: Harvard University Press, 1982.

North Carolina Biotechnology Center. *North Carolina Biotechnology Center U.S. companies database*, machine-readable database. Research Triangle Park, NC: North Carolina Biotechnology Center, April 16, 1992.

Pratt, Stanley E. *Guide to venture capital sources*, 6th Ed. Englewood Cliffs, NJ: Prentice-Hall, 1982.

Romer, Paul M. "Increasing Returns and Long-Run Growth." *Journal of Political Economy*, October 1986, *94*(5), pp. 1002–37.

______. "Endogenous Technological Change." *Journal of Political Economy*, October 1990, Part 2, *98*(5), pp. S71–S102.

Rosenberg, Nathan. *Inside the black box: Technology and economics*. Cambridge: Cambridge University Press, 1982.

U.S. Department of Commerce, Bureau of Economic Analysis, Economics and Statistics Administration. *National income and product accounts of the United States, volume 2, 1959–88*. Washington, DC: U.S. Government Printing Office, 1992a.

______. *Regional economic information system, version 1.3*, CD-ROM, machine-readable database. Washington, DC: Bureau of Economic Analysis, May 5, 1992b.

U.S. Department of Education, National Center for Education Statistics. *Higher education general information survey (HEGIS), institutional characteristics, 1983–84*, machine-readable database, ICPSR 8291. Ann Arbor, MI: Inter-University Consortium for Political and Social Research, circa 1985.

Wooldridge, Jeffrey M. "On the Application of Robust, Regression-Based Diagnostics to Models of Conditional Means and Conditional Variances." *Journal of Econometrics*, January 1991, *47*(1), pp. 5–46.

Zucker, Lynne G.; Brewer, Marilynn B.; Oliver, Amalya and Liebeskind, Julia. "Basic Science as Intellectual Capital in Firms: Information Dilemmas in rDNA Biotechnology

Research." Working paper, UCLA Institute for Social Science Research, 1993.

Zucker, Lynne G. and Darby, Michael R. "Star Scientists and Institutional Transformation: Patterns of Invention and Innovation in the Formation of the Biotechnology Industry." *Proceedings of the National Academy of Sciences*, November 12, 1996, *93*(23), pp. 12709–12716.

Zucker, Lynne G.; Darby, Michael R. and Armstrong, Jeff. "Intellectual Capital and the Firm: The Technology of Geographically Localized Knowledge Spillovers." National Bureau of Economic Research (Cambridge, MA) Working Paper No. 4946, December 1994.

Zucker, Lynne G.; Darby, Michael R. and Brewer, Marilynn B. "Intellectual Capital and the Birth of U.S. Biotechnology Enterprises." National Bureau of Economic Research (Cambridge, MA) Working Paper No. 4653, February 1994.

[22]

New Firms

3.1 Introduction

Professor Coase (1937) of the University of Chicago was awarded a Nobel Prize for explaining why a firm should exist. But why should more than one firm exist in an industry?[1] One answer is provided by the traditional economics literature focusing on industrial organization. An excess level of profitability induces entry into the industry. And this is why the entry of new firms is interesting and important—because the new firms provide an equilibrating function in the market, in that the levels of price and profit are restored to the competitive levels.

In this chapter a somewhat different lens through which to view the entry of new firms is suggested. A model is proposed that refocuses the unit of observation away from firms deciding whether to increase their output from a level of zero to some positive amount in a new industry, to individual agents in possession of new knowledge that, due to uncertainty, may or may not have some positive economic value. In fact, it is the uncertainty inherent in new economic knowledge, combined with asymmetries across agents with respect to its expected value, that potentially leads to a gap between the valuation of that knowledge.

To appropriate the value of that new knowledge, an economic agent can either effectively sell it to an incumbent firm—typically through an employment contract—or use it to launch a new firm. Of course, an important insight of Arrow (1962 and 1983), Stigler (1961), Williamson (1975), and more recently Milgrom and Roberts (1988, 1990, and 1991) is that a firm's decisionmaking process is made within the context of that firm's organization, that is, within a hierarchical bureaucracy.

1. Coase (1937, p. 23) himself asked, "A pertinent question to ask would appear to be (quite apart from the monopoly considerations raised by Professor Knight), why, if by organizing one can eliminate certain costs and in fact reduce the cost of production, are there any market transactions at all? Why is not all production carried on by one big firm?"

That new economic knowledge is inherently asymmetric is an insight that dates back at least to Schumpeter (1911) and Knight (1921). For example, Schumpeter (1911) emphasized that an innovation is the result of new economic knowledge that is embodied in an individual. This led him to observe that, "The function of entrepreneurs is to reform or revolutionize the pattern of production by exploiting an invention, or more generally, an untried technological possibility for producing a new commodity or producing an old one in a new way ... To undertake such new things is difficult and constitutes a distinct economic function, first because they lie outside of the routine tasks which everybody understands, and secondly, because the environment resists in many ways" Schumpeter (1942, p. 13).

Because new economic knowledge is inherently asymmetric, a divergence in beliefs between an incumbent firm, or more exactly stated, the decisionmakers embedded in the hierarchical organization of an incumbent firm, and an individual agent can emerge with respect to the potential economic value of that new knowledge. This leads to the prediction that the greater the extent to which knowledge asymmetries exist, the more likely new economic knowledge will result in the agent deciding that she or he must start a new enterprise in order to appropriate the expected economic value of that new knowledge.

Therefore, new firms are expected to be more pervasive in industries characterized by what is termed as the *entrepreneurial technological regime*, or where the underlying knowledge conditions tend to be particularly asymmetric. By contrast, under the *routinized technological regime*, less knowledge asymmetry tends to result in innovative activity occurring within the boundaries of the incumbent enterprises. There is less incentive for an agent to resort to starting a new firm to appropriate the potential economic value of her or his new economic knowledge, and therefore presumably less startup activity under the rountinized technological regime.

Based on a large panel data base enabling the examination of the startup of new firms within both a cross section and time series context, there is considerable evidence consistent with the hypothesis that new-firm startups are shaped by the underlying knowledge conditions. There tends to be less startup activity in industries better characterized by the routinized technological regime. By contrast, in those industries more closely conforming to the entrepreneurial technological regime, new-firm startups clearly play a more important role. These findings provide at least some support for the view that the entry of new firms is not only shaped by the decisionmaking process of individuals, but that new entrants represent, at least in some cases, not merely smaller replica of the existing incumbent enterprises but also agents of change.

3.2 Why Firms Exist

Before tackling the questions of why new firms would come into existence and how this relates to technological change and the evolution of industries, it is appropriate first to understand why firms should exist at all. In fact, what has become known as the theory of the firm is one of the major topics of modern microeconomics.[2] But to say that the topic draws a significant portion of the attention and energy of the economics profession is not to say that there is anything approaching unanimity about the subject. Thus, in reflecting, "What does economics have to say about the role of the business firm in a market economy?" Sidney Winter (1991, p. 179) concludes that, "I think we must acknowledge that the present state is one of incoherence."

At least part of the reason for this state of incoherence is attributable to what Edith Penrose (1959, p. 10) observed some three decades ago: "A 'firm' is by no means an unambiguous clear-cut entity; it is not an observable object physically separable from other objects, and it is difficult to define except with reference to what it does or what is done within it. Hence each analyst is free to choose any characteristics of firms that he is interested in, to define firms in terms of those characteristics, and to proceed thereafter to call the construction so defined a 'firm'. Herein lies a potential source of confusion." Still, a number of schools of thought, or focal points, have emerged in economic discourse on the theory of the firm.

There are three major categories, or schools of thought, with respect to the theory of the firm that are relevant here.[3] The first views the firm in the context of a production function—that is, by combining inputs to produce outputs.[4] The second view focuses on the firm in terms of the information and transactions it processes and is able to process. And the third view is evolutionary in that it considers the firm as a repository of knowledge.

This first school of thought, or view of the firm, considers firms as possessing a specific capability for transforming inputs into outputs generally through a process characterized by a production function. Firms are assumed to be solitary actors which reach decisions rationally. In particular, the goal of the firm is to maximize profit or, in terms of invested resources,

2. See, for examples, Kreps (1991) and Milgrom and Roberts (1987 and 1988).
3. Others can be found in Holmstrom and Tirole (1989).
4. Nelson and Winter (1982, pp. 6–11) refer to the production-function view of the firm as the "orthodox" viewpoint in economic theory. According to Winter (1991, p. 180), "Orthodox economic theory is the theoretical view that dominates the leading textbooks of intermediate microeconomics, together with the extensions and elaborations of that basic viewpoint found in more advanced work."

the present value of a stream of profits over time. Of particular importance here is the assumption that firms produce in markets for homogeneous commodities. Both inputs and outputs are typically purchased and sold on a spot market. And technology is considered to be exogenous. Beyond anything else, this view of the firm focuses on how an exogenous technology and production process transforms inputs into outputs.

The second view of the firm dates back at least to Coase (1937), who redirected the focus of attention away from the production process and toward transactions as the unit of analysis. Coase's answer to the question, "What is a firm?" was that a firm is a form of coordination of economic activity by authority, that is by administration, rather than by contract on an exchange market. Coase recognized that there are different ways of organizing transactions, or what has been more recently termed as "modes of governance" for transactions. These different organizational structures or modes of governance incur differing costs, and it is the costs of transactions that will ultimately dictate the organizational form. Thus Coase (1937, p. 38) concluded that, "The main reason why it is profitable to establish a firm would seem to be that there is a cost of using the price mechanism." That is, firms come into existence because the costs of coordinating economic activities through the market exceed the costs of administering those activities within the boundaries of the firm. What sets the limits of the firm's boundaries? Here, too, Coase (1937, p. 30) was unequivocal, "The question always is, will it pay to bring an extra exchange transaction under the organizing authority? At the margin, the costs of organizing within the firm will be equal either to the costs of organizing in another firm or to the costs involved in leaving the transaction to be 'organized' by the price mechanism."

For Williamson (1975), as well, the unit of observation is the transaction.[5] Williamson makes two key behavioral assumptions regarding human action, both of which follow from the notion that it is people or economic agents who engage in transactions. That is, goods or services are transferred a a result of human action: "Transaction cost economics characterizes human nature as we know it by reference to bounded rationality and opportunism. The first acknowledges limits on cognitive competence. The second substitutes subtle for simple self-interest seeking" (Williamson, 1985, p. 44). Because rationality is bounded and economic agents engage in opportunistic behavior additional costs are incurred in making transactions.

5. According to Williamson (1985, p. 1), "A transaction occurs when a good or service is transferred across a technologically separable interface. One stage of activity terminates and another begins."

As for Coase, Williamson (1985, p.46) concludes that the motivation to minimize costs of transaction shapes the emerging organizational structures that ultimately govern the costs of transactions, or what has become known as the governance structures:

> Economizing on bounded rationality takes two forms. One concerns decision processes, and the other involves governance structures. The use of heuristic problem-solving ... is a decision process response. Transaction cost economics is principally concerned, however, with the economizing consequence of assigning transactions to governance structures in a discriminating way. Confronted with the realities of bounded rationality, the costs of planning, adapting, and monitoring transactions need expressly to be considered. Which governance structures are more efficacious for which types of transactions? *Ceteris paribus*, modes that make large demands against cognitive competence are relatively disfavored.

Thus "the modern corporation is mainly to be understood as the product of a series of organizational innovations that have had the purpose and effect of economizing on transaction costs" (Williamson, 1985, p. 273). This leads Williamson (1985, p. 13) to conclude that, "the firm is more usefully regarded as a governance structure."

The third school of thought on the theory of the firm considers the repository of knowledge to be the most distinguishing feature of a firm. In fact, this view dates back at least to Alfred Marshall (1920, p. 115), who defined capital largely in terms of knowledge and organization, "Capital consists in a great part of knowledge and organization: and of this some part is private property and the other part is not. Knowledge is our most powerful engine of production; it enables us to subdue Nature and force her to satisfy our wants. Organisation aids knowledge."

Edith Penrose (1959, pp. 31–32) similarly attributed the distinguishing characteristic of an enterprise to the knowledge embedded in the firm, "The productive activities of a firm are governed by what we shall call its 'productive opportunity', which comprises all of the productive possibilities that its entrepreneurs see and can take advantage of ... It is clear that this opportunity will be restricted to the extent to which a firm does not see opportunities for expansion, is unwilling to act upon them, or is unable to respond to them." According to Penrose, the source of these ideas is the experience and knowledge generated within the firm.

Penrose's view focusing on the characteristic of firm-specific knowledge as the distinguishing feature of a firm was a predecessor to the evolutionary theory developed by Nelson and Winter. At the heart of their theory are three concepts underlying the evolution of industries—routine, search, and, selection. The organizational knowledge of each firm is stored in its routines. A routine refers to the "regular and predictable" (Nelson and Winter, 1982, p. 14) aspects regulating the behavior of the firm. Thus Nelson and Winter (1982, p. 99) argue that, "It is easy enough to suggest that a plausible answer to the question, 'Where does the knowledge reside?' is 'In the organization's memory.' But where and what is the memory of an organization? We propose that the routinization of activity in an organization constitutes the most important form of storage of the organization's specific operational knowledge."

The notion of a routine emerges as an organizational mechanism for dealing with Simon's (1984) bounded rationality. That is, confronted with incomplete information, and ill-equipped for digesting an abundance of information—much of it certain—organizations resort to their routines in guiding not only actions but, beyond everything else, the decision-making process. As the agents employed by a firm accumulate experience and knowledge, they incorporate them into the routines characterizing the firm. Nelson and Winter (1982, p. 36) assume that, although these routines may change in the long run, they are stable over the short and medium run. Most importantly, the routines are stable enough to yield predictions as to how a given firm will cope with a change in its external environment.

According to Nelson and Winter (1982), the firm can engage in actively searching for new economic knowledge. This function is distinct from that embedded in routines: "Organizations that are involved in the production or management of economic change as their principal function—organizations such as R & D laboratories and consulting firms—do not fit neatly into the routine operation mold" (Nelson and Winter, 1982, p. 97). The two key distinguishing features of search are irreversibility and uncertainty. Still, Nelson and Winter (1982, p. 247) incorporate a routine-like structure in the search mechanism: "The decision maker is viewed as having a set of decision rules ... these rules determine the direction of 'search'."

The third key element in Nelson and Winter's (1982) evolutionary theory involves selection, which is determined partially by the external environment. Nelson and Winter (1982, p. 19) characterize the selection mechanism as, "Search and selection are simultaneous, interacting aspects of the evolutionary process: the same prices that provide selection feedback also influence the directions of search. Through the joint action of search and selection, the firms evolve over time, with the condition of the industry in each period bearing the seeds of its condition in the following period."

3.3 The Traditional View of Entry

After summarizing the state of knowledge with respect to entry, Mueller (1991) concludes that at least five different forms of entry can be distinguished, only one of which involves a newly created firm.[6] Still, a new-firm startup not only represents the entry of a firm, but as Dunne, Roberts, and Samuelson (1988 and 1989) show, it is by far the most prevalent and common form of entry. Which is to say that if the standard theories about why firms enter industries—and why the propensity to enter may vary from industry to industry—are about anything, they are about the startup of new firms.

Why do firms enter an industry? The traditional view to be found in the literature of industrial organization[7] is that a level of profitability in excess of long-run equilibrium induces entry into the industry. And this is why the entry of new firms is interesting and important in the traditional theory—the new firms provide an equilibrating function in the market, in that the levels of price and profit are restored to their long-run competitive levels.

In this traditional theory, outputs and inputs in an industry are assumed to be homogeneous. That is, the entry of new firms is about business as usual—it is just that with the new entrant there is more of it. For example, Geroski (1991a, p. 65) assumes that,

> If we think of entry as an error-correction mechanism which is attracted by and serves to bid away excess profits, it is natural to suppose that entry will occur whenever profits differ from their long-run levels. Given this maintained hypothesis, observations of actual entry rates and current (or expected post-entry) profits can be used to make inferences about the unobservable of interest—long-run profits. In particular, entry in an industry is hypothesized to occur whenever expected post-entry profits exceed the level of profits protected in the long run.

This leads to the traditional model of entry:

$$E_{jt} = \lambda(P_{jt} - b_j) + u_{jt}, \tag{3.1}$$

where E_{jt} represents entry in industry j at time t, P_{jt} represents expected post-entry profits, and b_j represents the level of profits protected in the

6. The other four types are (1) entry by an existing firm that builds a new plant in the industry; (2) entry by an existing firm that purchases a plant or firm already existing in the industry; (3) entry by an existing firm that alters the product mix in an existing plant; and (4) entry by a foreign-owned firm in one of the above ways.

7. See, for examples, Geroski (1989b, 1990, 1991a, 1991c, and 1992).

long run by entry barriers b_j, and u_{jt} represents stochastic disturbance. In this standard model of entry, which theoretically dates back to Bain (1956) and empirically to Orr (1974), λ measures the speed with which entrants respond to excess profits, and has the dimension of a flow per unit of time. The level of profits which can be sustained in perpetuity without attracting entry is b_j, and serves as the limit to what Geroski (1991a, pp. 65–66) terms as "limit profits" and is a natural measure of the height of barriers to entry.

The point to be emphasized here is that this traditional model assumes that the impact that the new entrant has on the market, and on equilibrium price and industry profits, is through the additional amount of output that is contributed. And the fundamental motivation for entering an industry, either through a new firm or through diversified entry (Schwalbach, 1987), as reflected in equation (3.1), is that profits exceed their long-run equilibrium level, even after accounting for structural barriers to entry. In fact, little consensus has emerged in the growing plethora of studies trying to link industry profitability, growth, and structural barriers to entry rates, as implied by equation (3.1).

The level of industry profitability was found to have only a weak positive impact on entry in studies by Orr (1974) and Duetsch (1984), whereas Khemani and Shapiro (1986) found past profits to be a strong determinant of entry.[8] And five of the six country studies contained in Geroski and Schwalbach (1991) found a positive link between industry profitability and entry.

As equation (3.1) implies, a positive relationship has generally been found to exist between industry growth rates and entry rates. Presumably higher rates of growth enable incumbent enterprises to raise prices (Bradburd and Caves, 1982), thereby inducing more entry, or else raise expectations about future profits. For example, in five of the six country studies (Germany, Norway, Portugal, Belgium, and Korea) contained in Geroski and Schwalbach (1991) entry rates were found to be positively influenced by industry growth rates. Only for the United Kingdom did a negative relationship emerge between entry rates and industry growth.

One of the most startling results that has emerged in empirical studies is that entry by firms into an industry is apparently not substantially deterred or even deterred at all in capital-intensive industries in which scale economies play an important role (Austin and Rosenbaum, 1990; Siegfried and

8. While Orr (1974) used net entry as the dependent variable, the dependent variable in Khemani and Shapiro (1986) was de novo gross plant entry, net of exit, plant openings by incumbents, mergers, and diversification.

Evans, 1992). For example, Acs and Audretsch (1990, chapter 5; 1989a and 1989b) found that even small firms are not significantly deterred from entering industries that are relatively capitalintensive.

Empirical evidence in support of the model represented by equation (3.1) is ambiguous at best, leading Geroski (1991c, p.282) to conclude, "Right from the start, scholars have had some trouble in reconciling the stories told about entry in standard textbooks with the substance of what they found in their data. Very few have emerged from their work feeling that they have answered half as many questions as they have raised, much less that they have answered most of the interesting ones."

Perhaps one reason for this trouble is the inherently static model used to capture an inherently dynamic process. Manfred Neumann (1993, pp. 593–594) has criticized this traditional model of entry, as found in the individual country studies contained in Geroski and Schwalbach (1991), because they

> are predicated on the adoption of a basically static framework. It is assumed that startups enter a given market where they are facing incumbents which naturally try to fend off entry. Since the impact of entry on the performance of incumbents seems to be only slight, the question arises whether the costs of entry are worthwhile, given the high rate of exit associated with entry. Geroski appears to be rather sceptical about that. I submit that adopting a static framework is misleading ... In fact, generally, an entrant can only hope to succeed if he employs either a new technology or offers a new product, or both. Just imitating incumbents is almost certainly doomed to failure. If the process of entry is looked upon from this perspective the high correlation between gross entry and exit reflects the inherent risks of innovating activities ... Obviously it is rather difficult to break loose from the inherited mode of reasoning within the static framework. It is not without merit, to be sure, but it needs to be enlarged by putting it into a dynamic setting.

3.4 Asymmetric Information, Transaction Costs, and the Principal-Agent Relationship

The starting point for most theories of innovation is the firm. The relative innovative advantages of large and small firms are then analyzed by comparing factors such as the cost of and access to finance, scale economies in production and marketing, and access to a supply of new technological knowledge that has the potential for being transformed—through innovative activity—into new economic knowledge. In such theories the firms are exogenous and their performance in generating technological change is endogenous.[9] For example, in the most prevalent model found in the litera-

9. See, for example, Williamson (1975), Scherer (1984 and 1991), Cohen and Klepper (1991, 1992a, and 1992b), and Arrow (1962 and 1983).

ture of technological change, the model of the knowledge production function introduced by Zvi Griliches (1979), firms exogenously exist and then engage in the pursuit of new economic knowledge as an input into the process of generating innovative activity.[10]

Here I propose shifting the unit of observation away from exogenously assumed firms to individuals—agents confronted with new knowledge and the decision whether and how to act upon that new knowledge.[11] Of course, agents can and do work for firms, and even if they do not, they can potentially be employed by an incumbent firm. In fact, in a model of perfect information with no agency costs, any positive economies of scale or scope will ensure that, in the absence of monopoly rents being earned by the incumbent firm, no incentive will exist for an agent to start a firm. If an agent has an idea for doing something different than is currently being practiced by the incumbent enterprises—both in terms of a new product or process and in terms of organization—the idea, which we will term here as an innovation, will be presented to an incumbent enterprise. Because of the assumption of perfect knowledge, both the firm and the agent would agree upon the expected value of the innovation. But to the degree that any economies of scale or scope exist, the expected value of implementing the innovation within the incumbent enterprise will exceed that of taking the innovation outside of the incumbent firm to start a new enterprise. Thus the incumbent firm and the inventor of the idea would be expected to reach a bargain splitting the value added to the firm contributed by the innovation. The payment to the inventor—either in terms of a higher wage or some other means of remuneration—would be bounded between the expected value of the innovation if it is implemented by the incumbent

10. See also Jaffe (1986 and 1989), Acs, Audretsch, and Feldman (1992 and 1994), and Feldman (1994a and 1994b).

11. Winter (1991) points out that J. de V. Graaf (1957, p. 16) characterized the "clash" between focusing on the firm and on the individual as, "When we try to construct a transformation function for society as a whole from those facing the individual firms comprising it, a fundamental difficulty confronts us. There is, from a welfare point of view, nothing special about the firms actually existing in an economy at a given moment of time. The firm is in no sense a 'natural unit'. Only the individual members of the economy can lay claim to that distinction. All are potential entrepreneurs. It seems, therefore, that the natural thing to do is to build up from the transformation function of men, rather than the firms, constituting an economy. If we are interested in eventual empirical determination, this is extremely inconvenient. But it has conceptual advantages. The ultimate repositories of technological knowledge in any society are the men comprising it, and it is just this knowledge which is effectively summarized in the form of a transformation function. In itself a firm possesses no knowledge. That which is available to it belongs to the men associated with it. Its production function is really built up in exactly the same way, and from the same basic ingredients, as society's."

enterprise on the upper end, and by the return that the agent could expect to earn if she used it to launch a new enterprise on the lower end. Or, as Frank Knight (1921, p. 273) observed more than seventy years ago, "The laborer asks what he thinks the entrepreneur will be able to pay, and in any case will not accept less than he can get from some other entrepreneur, or by turning entrepreneur himself. In the same way the entrepreneur offers to any laborer what he thinks he must in order to secure his services, and in any case not more than he thinks the laborer will actually be worth to him, keeping in mind what he can get by turning laborer himself."

Kihlstrom and Laffont (1979), Evans and Leighton (1989, 1990a, and 1990b), Holmes and Schmitz (1990), Evans and Jovanovic (1989), Jovanovic (1994), and Blanchflower and Meyer (1994) have all reinterpreted Knight's theory of the decision confronting each economic agent to become either an employee in an incumbent enterprise or an entrepreneur, that is start a new firm, in terms of a model of income choice:[12]

$$Pr(e) = f(P_i - w), \tag{3.2}$$

where the probability of an agent starting a new firm is $Pr(e)$, the expected profits accruing from such a new firm are represented by P_i, and w is the wage that the agent would earn if she or he chooses to be employed by an incumbent enterprise. In fact, the model of income choice has served as the underlying framework for estimating those individual characteristics shaping the choice individuals make between starting a new firm or being an employee in an incumbent firm (Evans and Leighton, 1989, 1990a, and 1990b; Evans and Jovanovic, 1989; Blanchflower and Meyer, 1994).

But, of course, as Knight (1921), and later Arrow (1962) and Stigler (1961) were among the first to emphasize, new economic knowledge is anything but certain. Not only is new economic knowledge inherently risky, but substantial asymmetries exist across agents both between and within firms (Milgrom and Roberts, 1987). Which is to say that the expected value of a new idea—or what has been termed loosely here as a potential innovation—is likely to be anything but unanimous between the inventor of that idea and the decisionmaker, or group of decisionmakers,[13] of the firm confronted with evaluating proposed changes or innovations. In fact, it is because information is not only imperfect but also asymmetric that Knight (1921, p. 268) argued that the primary task of the firm is to process

12. See also Vivarelli (1991), Santarelli and Sterlacchini (1994), and Foti and Vivarelli (1994).
13. For example, as of 1993 a proposal for simply modifying an existing product at IBM had to pass through 250 layers of decisionmaking to gain approval ("Überfördert und Unregierbar," *Der Spiegel*, No. 14, 1993, p. 127).

imperfect information in order to reach a decision: "With the introduction of uncertainty—the fact of ignorance and the necessity of acting upon opinion rather than knowledge—into this Eden-like situation (that is a world of perfect information), its character is entirely changed ... With uncertainty present doing things, the actual execution of activity, becomes in a real sense a secondary part of life; the primary problem or function is deciding what to do and how to do it."[14]

Alchian (1950) pointed out that the existence of knowledge asymmetries would result in the inevitability of mistaken decisions in an uncertain world. Later, Alchian and Demsetz (1972) attributed the existence of asymmetric information across the employees in a firm as resulting in a problem of monitoring the contribution accruing from each employee and seeing the rewards correspondingly. This led them to conclude that, "The problem of economic organization is the economical means of metering productivity and rewards" (Alchian and Demsetz, 1972, p. 783).

Combined with the bureaucratic organization of incumbent firms to make a decision, the asymmetry of knowledge leads to a host of agency problems, spanning incentive structures, monitoring, and transaction costs. It is the existence of such agency costs, combined with asymmetric information that not only provides an incentive for agents with new ideas to start their own firms, also at a rate that varies from industry to industry, depending on the underlying knowledge conditions of the industry.

Coase (1937) and later Williamson (1975) argued that the size of an (incumbent) enterprise will be determined by answering what Coase (1937, p. 30) articulated as, "The question always is, will it pay to bring an extra exchange transaction under the organizing authority?" In fact, Coase (1937, p. 24) pointed out that, "Other things being equal, a firm will tend to be larger the less likely the (firm) is to make mistakes and the smaller the increase in mistakes with an increase in the transactions organized." And both Coase (1937) and Williamson (1975) were quick to note that an

14. Knight (1921, pp. 268 and 295) emphasizes that the existence of imperfect information leads to the two most important characteristics of firms, "In the first place, goods are produced for a market, on the basis of entirely impersonal prediction of wants, not for the satisfaction of the wants of the producers themselves. The producer takes the responsibility of forecasting the consumers' wants. In the second place, the work of forecasting and at the same time a large part of the technological direction and control of production are still further concentrated upon a very narrow class of the producers, and we meet with a new economic functionary, the entrepreneur ... When uncertainty is present and the task of deciding what to do and how to do it takes the ascendancy over that of execution the internal organisation of the productive groups is no longer a matter of indifference or a mechanical detail. Centralisation of this deciding and controlling function is imperative, a process of "cephalization" is inevitable."

obvious source causing an increase in transaction costs is uncertainty and imperfect information. According to Coase (1937, p. 27), "The fact of uncertainty means that people have to forecast future wants. Therefore, you get a special class springing up who direct the activities of others to whom they give guaranteed wages. It acts because good judgement is generally associated with confidence in one's judgement."

The basic agency problem (Alchian and Demsetz, 1972; Jensen and Meckling, 1976; Holmstrom and Milgrom, 1987; Holmstrom and Tirole, 1979; and Milgrom, 1988) arises in the context of an organization responding to an agent who possesses potentially new economic knowledge. In fact, it may even be the task of that agent within the context of a bureaucratic organization to search out, either through production or acquisition, such economic knowledge. Because the principal is not able to directly observe either the efforts or the outcome of the agent, both monitoring and incentive problems, as well as possible hostage problems, emerge. Notice that if either information asymmetries or a divergence in risk preferences along the lines described in Kihlstrom and Laffont (1979) exists then the principal and the agent will tend to value any given proposed project differently. Further, the agent will have a clear incentive to exaggerate the expected value of a potential innovation along with the amount of effort required by her to develop and implement it.

If the principal introduces some mechanism for monitoring, it can induce something about the agent's efforts from a particular signal. Alchian and Demsetz (1972) pointed out that a principal needs a monitor to meter inputs, particularly effort. However, the riskier any project might be, the less able the principal is to infer from monitoring. That is, the costs of monitoring rise and the expected value of the project falls as risk increases. This, in turn, reduces the incentive for the agent to develop risky projects, or at least their likelihood of being approved and implemented by an incumbent firm.

Holmstrom (1989) and Milgrom (1988) have pointed out the existence of a "bureaucratization dilemma," where, "To say that increased size brings increased bureaucracy is a safe generalization. To note that bureaucracy is viewed as an organizational disease is equally accurate" (Holmstrom, 1989, p. 320). Milgrom (1988) has argued that bureaucratic rules are a rational way for organizations to limit investments by agents in influence activities; and Tirole (1986) and Milgrom and Roberts (1991) have pointed out that bureaucratic decisionmaking is a mechanism to avoid collusion among coalitions of subordinates and bosses, because the integrity of the evaluation of subordinates depends on the incentives facing the monitor. Bureaucratic decisionmaking favoring nondiscretionary over discretionary rules reduces such collusions.

To minimize agency problems and the cost of monitoring, bureaucratic hierarchies develop objective rules. In addition, Kreps (1991) has argued that such bureaucratic rules promote internal uniformity and that a uniform corporate culture, in turn, promotes the reputation of the firm. These bureaucratic rules, however, make it more difficult to evaluate the efforts and activities of agents involved in activities that do not conform to such bureaucratic rules. As Holmstrom (1989, p. 323) points out, "Monitoring limitations suggest that the firm seeks out activities which are more easily and objectively evaluated. Assignments will be chosen in a fashion that are conducive to more effective control. Authority and command systems work better in environments which are more predictable and can be directed with less investment in information. Routine tasks are the comparative advantage of a bureaucracy and its activities can be expected to reflect that."

Williamson (1975, p. 201) has also emphasized the inherent tension between hierarchical bureaucratic organizations and entrepreneurial activity, "Were it that large firms could compensate internal entrepreneurial activity in ways approximating that of the market, the large firm need experience no disadvantage in entrepreneurial respects. Violating the congruency between hierarchical position and compensation appears to generate bureaucratic strains, however, and is greatly complicated by the problem of accurately imputing causality." This leads Williamson (1975, pp. 205–206) to conclude that,

> I am inclined to regard the early stage innovative disabilities of large size as serious and propose the following hypothesis: An efficient procedure by which to introduce new products is for the initial development and market testing to be performed by independent inventors and small firms (perhaps new entrants) in an industry, the successful developments then to be acquired, possibly through licensing or merger, for subsequent marketing by a large multidivision enterprise ... Put differently, a division of effort between the new product innovation process on the one hand, and the management of proven resources on the other may well be efficient.

The degree to which agents and incumbent firms are confronted with knowledge asymmetries and agency problems with respect to seeking out new economic knowledge and (potential) innovative activity would not be expected to be constant across industries. This is because the underlying knowledge conditions vary from industry to industry. In some industries new economic knowledge generating innovative activity tends to be relatively routine and can be processed within the context of incumbent hierarchical bureaucracies. In other industries, however, innovations tend to come from knowledge that is not of a routine nature and therefore tends to be rejected by the hierarchical bureaucracies of incumbent corporations. Nelson and Winter (1974, 1978, and 1982) described these different underlying knowledge conditions as reflecting two distinct technological regimes—the entrepreneurial and routinized technological regimes: "An entrepreneurial regime is one that is favorable to innovative entry and unfavourable to innovative activity by established firms; a routinized regime is one in which the conditions are the other way around" (Winter, 1984, p. 297). At least some empirical evidence was provided by Acs and Audretsch (1987, 1988, and 1990) supporting the existence of these two distinct technological regimes.

Gort and Klepper (1982) posited, and found evidence, that the relative innovative advantage between newly established enterprises and incumbent firms depends on the source of information generating innovative activity. If information based on nontransferable experience in the market is an important input in generating innovative activity, then incumbent firms will tend to have the innovative advantage over new firms. This is consistent with Winter's (1984) notion of the routinized regime, where the accumulated stock of nontransferable information is the product of experience within the market, which firms outside of the industry, by definition, cannot possess.

By contrast, when information outside of the routines practiced by the incumbent firms is a relatively important input in generating innovative activity, newly established firms will tend to have the innovative advantage over the incumbent firms. Arrow (1962), Mueller (1976), and Williamson (1975) have all emphasized that when such information created outside of the incumbent firms cannot be easily transferred to those incumbent firms—presumably due to the type of agency and bureaucracy problems analyzed above—the holder of such knowledge must enter the industry to exploit the market value of her or his knowledge.

Thus, when the underlying knowledge conditions are better characterized by the routinized technological regime, there is likely to be relatively little divergence in the evaluation of the expected value of a (potential) innovation between the inventor and the decisionmaking bureaucracy of the firm. Under the routinized regime a great incentive for agents to start their own firms will not exist, at least not for the reason of doing something differently. When the underlying knowledge conditions more closely adhere to the entrepreneurial technological regime, however, a divergence in beliefs between the agent and the principal regarding the expected value of a (potential) innovation is more likely to emerge. Therefore, it is under the entrepreneurial regime where the startup of new firms is likely to play a more important role, presumably as a result of the motivation to appropriate the value of economic knowledge; due to agency problems, this knowledge cannot be easily and costlessly transferred to the incumbent enterprise.

The analytical framework supporting the decision of how best to appropriate the value of new economic knowledge confronting an individual economic agent seems useful when considering the actual decision to start a new firm taken by entrepreneurs. For example, Chester Carlsson started Xerox after his proposal to produce a (new) copy machine was rejected by Kodak. Kodak based its decision on the premise that the new copy machine would not earn very much money, and, in any case, Kodak was in a different line of business—photography. It is perhaps no small irony that this same entrepreneurial startup, Xerox, decades later turned down a proposal from Steven Jobs to produce and market a personal computer, because they did not think that a personal computer would sell, and, in any case, they were in a different line of business—copy machines (Carrol, 1993). After seventeen other companies turned down Jobs for virtually identical reasons, including IBM and Hewlett Packard, Jobs resorted to starting his own company, Apple Computer (Rose, 1989).

Similarly, IBM turned down an offer from Bill Gates, "the chance to buy ten percent of Microsoft for a song in 1986, a missed opportunity that would cost $3 billion today."[15] IBM reached its decision on the grounds that "neither Gates nor any of his band of thirty some employees had anything approaching the credentials or personal characteristics required to work at IBM."[16] In fact, even as recently as 1991, IBM clung to the belief, reflected in an executive's memo, that "the technical superiority of IBM's OS/2 and Presentation Manager over DOS and Windows 3.0 is universally unquestioned" (Ichbiah and Knepper, 1993, p. 244).

Divergences in beliefs with respect to the value of a new idea need not be restricted to what is formally known as a product or even a process innovation. Rather, the fact that economic agents choose to start a new firm due to divergences in the expected value of an idea applies to the sphere of managerial style and organization as well. One of the most vivid examples involves Bob Noyce, who founded Intel. Noyce had been employed by Fairchild Semiconductor, which is credited with being the pio-

15. "System Error," *The Economist*, 18 September 1993, p. 99.
16. Paul Carrol, "Die Offene Schlacht," *Die Zeit*, No. 39, 24 September 1993, p. 18.

neering semiconductor firm. In 1957, Noyce and seven other engineers quit en masse from Schockley Semiconductor to form Fairchild Semiconductor, an enterprise that in turn is considered the start of what is today known as Silicon Valley (Cringley, 1993, p. 36). Although Fairchild Semiconductor had "possibly the most potent management and technical team ever assembled" (Gilder, 1989, p. 89), "Noyce couldn't get Fairchild's eastern owners to accept the idea that stock options should be part of compensation for all employees, not just for management. He wanted to tie everyone, from janitors to bosses, into the overall success of the company ... This management style still sets the standard for every computer, software, and semiconductor company in the Valley today ... Every CEO still wants to think that the place is being run the way Bob Noyce would have run it" (Cringley, 1993, p. 39). That is, Noyce's vision of a firm excluded the dress codes, reserved parking places, closed offices, and executive dining rooms, along with the other trappings of status that were standard in virtually every hierarchical and bureaucratic U.S. corporation. But when he tried to impress this vision upon the owners of Fairchild Semiconductor,[17] he was flatly rejected. The formation of Intel in 1968 was the ultimate result of the divergence in beliefs about how to organize and manage the firm.

The key development at Intel was the microprocessor, which has been described as a computer on a chip, in that it has a central processing unit (CPU), is programmable, and could readily be connected to memory chips and input-output devices. When Ted Hoff approached IBM and DEC with his new microprocessor in the late 1960s, "IBM and DEC decided there was no market. They could not imagine why anyone would need or want a small computer; if people wanted to use a computer, they could hook into time-sharing systems" (Palfreman and Swade, 1991, p. 108).

3.5 New-Firm Startups Over Time and Across Industries

Studies examining the determinants of entry generally suffer from two well-known limitations. First, while several notable exceptions exist (Dunne, Roberts, and Samuelson, 1988 and 1989), the most common measure of entry used in studies attempting to empirically identify the determinants of entry has been the change in the number of firms over a given period, or what has become referred to as "net entry."[18] Measuring the change in the

17. Fairchild Semiconductor was actually controlled by an eastern firm, Fairchild Camera and Instrument of Syosset, New York.

18. For examples of this literature, see Orr (1974), Duetsch (1984), and Baldwin and Gorecki (1985, 1987, and 1989).

number of firms does not account for enterprises that exited from the industry during the relevant time period. That is, given an amount of gross entry, the measure of net entry will increase as the number of exits from the industry decreases. Thus it is quite conceivable that an industry could have a negative amount of net entry, if many firms actually entered the industry (i.e., if gross entry was positive), but even more firms exited from the industry. Because the pattern of industry exits varies across industries, the extent to which net entry deviates from actual gross entry will also vary substantially from industry to industry.

The second limitation is that entry has typically been measured over a single time period. Although it has been possible to measure the number of new-firm startups at the aggregate macroeconomic level (Highfield and Smiley, 1987), it has not been systematically done at the disaggregated industry level.[19]

These two limitations have made it virtually impossible to disentangle the macroeconomic influences on new-firm startups from the microeconomic influences. All that can be concluded from the existing literature is that both are probably important. To overcome these traditional data limitations, I rely upon the U.S. Small Business Administration's Small Business Data Base (SBDB), which was documented in chapter 2. The annual number of new-firm startups is aggregated to major manufacturing sectors and shown for alternate years between 1976 and 1986 in table 3.1. The share of the total number of enterprises and in the sector accounted for by new-firm startups is listed in parentheses. Three major points from table 3.1 should be emphasized. First, the number of new-firm startups and their share of the total number of enterprises vary considerably across manufacturing sectors.

Second, the number of startups varies substantially from year to year. That is, in 1976 there were 11,154 new-firm startups in all of U.S. manufacturing; this fell by nearly one-quarter to 8,525 startups in 1980, and by nearly two-thirds to 4,239 in 1982. By 1984 the number of manufacturing startups had more than doubled to 10,055, which was nearly again at the 1976 and 1978 levels. This volatility in the number of new-firm startups is attributable, at least to some extent, to macroeconomic fluctuations. This is reflected by the fluctuations in annual growth rates of real gross national product (GNP) of 4.9 percent in 1976, 5.3 percent in 1978, −0.2 percent

19. Yamawaki (1991) examines the determinants of net entry into 135 three-digit Japanese manufacturing industries for five one-year periods between 1980 and 1984. However, he was not able to identify new-firm startups from his measure of net entry.

in 1980, −2.5 percent in 1982, 6.8 percent in 1984, and 2.8 percent in 1986.[20] The extent of startup activity for manufacturing as a whole corresponds quite closely to these macroeconomic fluctuations. In addition, there is also a clear tendency for the number of startups within each manufacturing sector to reflect the phase of the business cycle.

The third major point from table 3.1 is that, although no industrial sector is immune from the influences of macroeconomic fluctuations, the impact varies considerably from sector to sector. New-firm startups in certain sectors, such as petroleum, textiles and apparel, and communications are apparently quite susceptible to the prevailing phase of the business cycle, at least over this period of time. By contrast, in the computer and food sectors, the number of startups seems to be less vulnerable to macroeconomic fluctuations. Just as the strong intertemporal tendency toward fewer startups in the transportation (other) sector probably reflects a longer-term decline, the pronounced tendency toward an increase in the number of startups in computers seems to suggest long-term sectorial expansion.

3.6 The Model

The dependent variable in the model to be estimated is based on the panel of data described in section 3.5, in which the unit of observation is the number of new-firm startups in a given four-digit SIC industry for a given year over the period 1976–1986, for alternate years. The theory from section 3.4 of this chapter suggests that the propensity for new firms to be started will be shaped by the underlying technological regime. But that is not to say that the technological regime is the only factor influencing new-firm startups. A rich and expansive literature has developed linking the entry of firms to the existence of scale economies and capital intensity.[21] This literature points to an immediate application to the model of an agent and bureaucratic hierarchy deciding how to deal with new economic information. The greater the extent of scale economies and capital requirements for establishing a new firm, the higher the cost will be to the agent for starting a new firm. That is, for any given gap in the expected value of a (potential) innovation arising between the agent and the decisionmaking bureaucracy of the incumbent firm, the agent will be less likely to decide to start a new firm as the startup costs rise.

20. The annual growth rates of real gross national product are from the U.S. Department of Commerce, Bureau of Economic Analysis.

21. For recent examples, see the individual country studies contained in Geroski and Schwalbach (1991) as well as Siegfried and Evans (1992).

Table 3.1
New-firm startups by industrial sector

	1976	1978	1980	1982	1984	1986
Food	474	481	374	209	480	535
	(2.53)	(2.67)	(2.15)	(1.22)	(2.79)	(3.09)
Textiles and apparel	1,172	1,254	854	491	1,026	992
	(4.03)	(4.20)	(2.95)	(1.69)	(3.41)	(3.34)
Lumber and furniture	1,325	1,375	868	425	1,060	1,106
	(3.71)	(3.70)	(2.28)	(1.12)	(2.75)	(2.79)
Paper	126	191	101	50	149	152
	(2.97)	(4.24)	(2.20)	(1.11)	(3.19)	(3.15)
Chemicals	322	390	384	164	332	335
	(2.95)	(3.52)	(2.54)	(1.45)	(2.85)	(2.83)
Industrial	91	99	98	41	85	84
	(3.62)	(3.76)	(3.69)	(1.55)	(3.12)	(3.09)
Drugs and medicinals	34	54	22	26	48	49
	(2.75)	(4.47)	(1.82)	(2.11)	(3.60)	(3.50)
Other	123	154	116	65	130	138
	(2.39)	(3.00)	(2.24)	(1.23)	(2.42)	(2.52)
Petroleum	41	42	57	11	43	46
	(3.21)	(3.16)	(4.01)	(0.76)	(3.02)	(3.17)
Rubber	430	469	312	158	382	385
	(4.72)	(4.78)	(2.97)	(1.44)	(3.26)	(3.18)
Stone, clay, and glass	545	493	292	133	337	358
	(3.86)	(3.41)	(2.00)	(0.93)	(2.41)	(2.58)
Primary metals	168	179	141	79	195	201
	(2.97)	(3.07)	(2.36)	(1.32)	(3.22)	(3.25)
Ferrous metals	85	90	61	45	102	110
	(3.21)	(3.28)	(2.24)	(1.67)	(3.70)	(3.80)
Nonferrous metals	83	89	80	34	93	91
	(2.76)	(2.89)	(2.47)	(1.04)	(2.82)	(2.76)
Fabricated metal products	962	1,042	782	362	913	877
	(3.19)	(3.30)	(2.37)	(1.07)	(2.65)	(2.52)
Machinery	1,519	1,731	1,407	586	1,433	1,314
	(3.14)	(3.38)	(2.60)	(1.02)	(2.43)	(2.22)
Office and computers	50	66	62	43	118	100
	(4.73)	(5.27)	(4.14)	(2.21)	(4.58)	(3.64)
Other machinery, non-electrical	1,469	1,665	1,345	543	1,315	1,214
	(3.10)	(3.34)	(2.56)	(0.98)	(2.33)	(2.15)
Electrical equipment	635	620	461	274	665	606
	(4.41)	(3.98)	(2.88)	(1.62)	(3.63)	(3.21)
Radio and TV equipment	79	82	53	21	43	49
	(5.02)	(4.63)	(3.00)	(1.15)	(2.38)	(2.73)
Communications equipment	128	94	74	48	168	119
	(5.27)	(3.59)	(2.77)	(1.62)	(5.02)	(3.40)
Electronic components	193	204	163	110	233	211
	(4.83)	(4.64)	(3.40)	(2.10)	(3.98)	(3.46)
Other	235	240	171	95	221	227
	(3.67)	(3.54)	(2.52)	(1.39)	(3.03)	(3.04)

Table 3.1 (cont.)

	1976	1978	1980	1982	1984	1986
Motor vehicles	147	149	116	57	147	148
	(4.55)	(4.31)	(3.11)	(1.49)	(3.62)	(3.33)
Other transport equipment	247	250	142	76	191	127
	(5.31)	(5.23)	(3.17)	(1.76)	(4.30)	(2.99)
Aircraft and missiles	26	36	26	13	33	42
	(2.50)	(3.24)	(2.09)	(0.94)	(2.19)	(2.68)
Instruments	312	323	226	160	308	353
	(3.94)	(3.72)	(2.41)	(1.60)	(2.79)	(3.01)
Scientific and measuring	130	120	104	66	145	141
	(4.56)	(3.63)	(2.76)	(1.55)	(2.99)	(2.72)
Optical, surgical, and photographic	182	203	122	94	163	212
	(3.59)	(3.78)	(2.18)	(1.63)	(2.63)	(3.24)
Other manufacturing	2,703	2,703	2,082	1,081	2,361	2,435
	(3.62)	(3.48)	(2.67)	(1.35)	(2.81)	(2.82)
Total manufacturing	11,154	11,728	8,525	4,329	10,055	10,012
	(3.56)	(3.60)	(2.56)	(1.27)	(2.86)	(2.80)

Source: SBIDB.
Note: The share (percentage) of the total number of firms accounted for by new-firm startups is indicated in the parentheses.

To measure the extent to which scale economies exist in an industry (Caves and Porter, 1977), we rely on a common proxy measure of the minimum efficient scale (MES), developed and applied by Comanor and Wilson (1967), who measured the MES as the mean size of the largest plants accounting for one-half of the industry value-of-shipments. Along with the 1977 proxy for scale economies, we include a measure of the 1977 capital intensity, defined as gross assets divided by the labor force.[22]

As mentioned in section 3.4, a second traditional explanation for why entry varies across industries has been variations in expected profitability. Although expected profitability is virtually impossible to measure, it has been generally positively associated with industries experiencing higher growth rates (Bradburd and Caves, 1982). Thus we measure industry growth as the annual percentage change in value-of-shipments. Presumably the extent of new-firm startups will be greater in industries exhibiting greater rates of growth.

22. White (1982) points out that higher capital-labor ratios tend to be associated with greater scale economies. This is partially because capital equipment tends to be "lumpy" in nature. Also, by enabling firms to take advantage of increased specialization and greater rates of utilization, the use of larger machines tends to reduce costs per unit of output.

Although the above measures have been the main focus of most prior studies focusing on the entry behavior of firms, it was argued in section 3.4 that the underlying knowledge conditions that shape the decisionmaking process at the level of the individual economic agent to choose between pursuing an innovation within an incumbent firm or starting a new firm results in a propensity for people to start firms in a manner that systematically reflects the underlying technological regime. While the concept of technological regimes does not lend itself to precise measurement, the major conclusion of Acs and Audretsch (1987, 1988, and 1990) was that the existence of these distinct regimes can be inferred by the extent to which small firms are able to innovate relative to the total amount of innovative activity in an industry. That is, when the small-firm innovation rate is high relative to the total innovation rate, the technological and knowledge conditions are more likely to reflect the entrepreneurial regime. The routinized regime is more likely to exhibit a low small-firm innovation rate relative to the total innovation rate.

Using the innovation data base described in chapter 2, the total innovation rate is defined as the total number of innovations recorded in 1982 divided by industry employment. The small-firm innovation rate is defined as the number of innovations contributed by firms with fewer than 500 employees divided by small-firm employment. The rates are used to standardize the amount of innovative and small-firm innovative activity in an industry for the size of that industry (as Acs and Audretsch do in their 1987 and 1990 studies). Since high small-firm innovation rates, given a total innovation rate, presumably reflect the entrepreneurial regime, the small-firm innovation rate is expected to have a positive influence on new-firm startups. By contrast, a high total innovation rate, given a small-firm innovation rate, which reflects the routinized regime, would be expected to lead to a lower number of new-firm startups.

Studies by Yamawaki (1991) and Highfield and Smiley (1987) have shown that the extent of entry is influenced by the business cycle. And certainly table 3.1 indicates that new-firm startups tend to exhibit procyclical tendencies. To capture these influences for a data set that contains a time series element, we include three measures that typically reflect the stage of the business cycle—macroeconomic growth, the interest rate, and the unemployment rate. To measure macroeconomic growth, the annual percentage change in real GNP is used. The average three-month interest rate paid on U.S. treasury bills should reflect the cost of capital. These three measures are taken from the 1989 *Economic Report of the President*. Both the growth rate of real GNP and the unemployment rate are expected to exert

Table 3.2
Regression results for new-firm startups

	(1)	(2)
Scale economies$_{(i)}$	0.043	0.004
	(5.27)**	(0.39)
Capital intensity$_{(i)}$	0.245	0.243
	(5.83)**	(5.63)**
Total innovation rate$_{(i)}$	−0.094	−0.098
	(−2.27)**	(−1.86)*
Small-firm innovation rate$_{(i)}$	6.002	4.263
	(2.45)**	(1.86)*
Industry growth$_{(it)}$	0.496	0.750
	(0.46)	(0.70)
Macroeconomic growth$_{(t)}$	0.576	0.573
	(5.59)**	(5.43)*
Interest rate$_{(t)}$	−0.437	−0.432
	(−2.88)**	(−2.77)**
Unemployment$_{(t)}$	−1.626	−1.655
	(−5.24)**	(−5.20)**
Industry sales$_{(it)}$	0.147	—
	(10.57)	
R^2	0.121	0.081
F	37.027**	26.702**
Sample size	2442	2442

Note: *t*-statistics listed in parentheses.
*Statistically significant at the 90 percent level of confidence for a two-tailed test.
**Statistically significant at the 95 percent level of confidence for a two-tailed test.

a positive influence on new-firm startups, and the interest rate is expected to be negatively related to the number of startups.

It should be emphasized that although these macroeconomic variables vary over time—but not across industries for any given year—most of the industry-specific variables are measured only at one point in time. It is implicitly assumed that variables such as the total innovation rate and the small-firm innovation rate, along with the measures of capital intensity and scale economies in an industry are invariant over a relatively short time period. Only the measure of industry growth varies over time and across industries.

3.7 Results

The regression results based on pooled cross-section estimation using the number of new-firm startups in each industry for each year as the dependent variable are reported in table 3.2. It should be emphasized that the subscript i denotes variables that are measured only at one point in time within a cross-sectional context, while the subscript t denotes those variables that are constant across industries but vary over time. Those variables with a subscript it have both industry- and time-specific components. Because the dependent variable, the number of new-firm startups for a particular industry in a given year, does not control for variations in market size, an additional variable—industry value-of-shipments—is included in the first equation to account for variations in size across industries.

There is no evidence that the existence of scale economies or capital intensity serves to deter the startup of new firms. Although this result is somewhat startling, it is certainly consistent with the same findings in a number of the individual country studies included in Geroski and Schwalbach (1991), as well as in Acs and Audretsch (1989a, 1989b, and 1990, chapter 5), Austin and Rosenbaum (1990), and Siegfried and Evans (1992).

The regression results are consistent with the central hypothesis of this chapter—that new-firm startups are shaped by the underlying knowledge conditions in an industry. As the negative and statistically significant coefficient of the total innovation rate and the positive and statistically significant coefficient of the small-firm innovation rate indicate, new-firm startups tend to be lower in industries that are best characterized by the routinized regime—that is, where the overall innovation rate of the industry is high but the small-firm innovation rate is low. On the other hand, new-firm startups tend to be greater in industries that are best characterized by the entrepreneurial regime, where for any given level of innovative output, the small-firm innovation rate is particularly high.

Contrary to the findings in most of the traditional studies linking entry to market structure (Orr, 1974 and Geroski and Schwalbach, 1991), new-firm startups are not significantly influenced by the growth rate at the industry level. However, macroeconomic growth clearly influences startup activity. In fact, all of the variables reflecting the business cycle influence new-firm startups. The startup of new firms is found to be greater during periods of strong macroeconomic growth and low unemployment. At the same time, high interest rates, presumably reflecting the cost of capital, are found to exert a negative influence on startup activity.

3.8 Conclusions

Neither the theory nor the results of this chapter contradict previous studies linking the entry behavior of new firms to industry specific characteristics. Indeed, our findings that growth is an important determinant of entry (albeit at the aggregated macroeconomic level and not at the disaggregated industry level) are certainly consistent with the conclusions of virtually every study in the literature. We have argued here, however, that there is another industry-specific characteristic that shapes entry behavior that has generally been overlooked in the literature—the knowledge conditions underlying the industry. It is the existence of asymmetric information in the context of agency costs and bureaucratic decisionmaking that leads agents to value an anticipated innovation differently from the decisionmaking bureaucracy of an incumbent organization. But the extent to which asymmetric information between agents and principals exists is not at all constant across industries. Rather, these underlying knowledge conditions vary from industry to industry and can be characterized alternatively by the routinized technological regime and the entrepreneurial technological regime.

The evidence from a large panel data set used throughout this book lends support to the hypothesis that the startup of new firms is influenced by the underlying knowledge conditions or what has been termed as the technological regime. Because in industries better characterized by the routinized regime, new-firm startups tend to be less important. Presumably in these industries there is less divergence in beliefs about the expected value of an anticipated innovation. Under the routinized regime innovative activity tends to occur within the incumbent enterprises, and there is less of an incentive for an agent to start a new firm to exploit the expected value of her or his economic knowledge.

By contrast, under the entrepreneurial regime asymmetric information leads to a divergence in beliefs regarding the expected value of anticipated innovations. Thus it is under the entrepreneurial regime that agents resort to starting a new firm in order to appropriate the expected value of their knowledge. The findings of this chapter are consistent with this view because new-firm startups are significantly greater in industries characterized by the entrepreneurial regime than in industries characterized by the routinized regime.

The industrial organization literature has generally considered entry to be an important economic phenomenon because it is the mechanism by which market equilibrium is restored. The focus has been on firms which, by entering, restore prices and profits back to their equilibrium levels of output. Under this view, entry is about business as usual, it is just that the new entrant supplies more of it. By contrast, in this chapter I have shifted the focus away from firm to individual economic agents confronted by maximizing the value of new economic knowledge. Because new economic knowledge is not only imperfect but also asymmetric, agents must decide whether to pursue their anticipated innovation within the boundaries of an incumbent enterprise or start a new firm.

Focusing on the individual possessing uncertain knowledge leads to an additional economic role for the new entrant—not just to equilibrate the market by increasing the supply of the product already produced by the incumbent firms—but rather by doing something different and thereby serving as an agent of change. This role is reminiscent of the concepts of "voice" and "exit" introduced and popularized by Albert O. Hirschman (1970). An agent in possession of new economic knowledge may be unable to exercise "voice" in the context of the decisionmaking hierarchy of an incumbent firm, that is, to actualize her potential innovation within the boundaries of an incumbent enterprise. This drives her to choose instead to exercise "exit"—out of the incumbent enterprise, and ultimately to start a new business entity.

An important insight is that the implicit model of technological change currently dominating the literature, the model of the knowledge production function introduced by Griliches (1979)—that firms exogenously exist and then engage in the pursuit of new knowledge—is perhaps somewhat misguided, or in any case not complete. It is because new knowledge held by individual agents, which here is treated as exogenous, is not being utilized within the context of the incumbent enterprises, that endogenously leads economic agents to the decision to start a new firm.

References

Acs, Zoltan J., and David B. Audretsch. 1990. *Innovation and Small Firms*. Cambridge, MA: MIT Press.

Acs, Zoltan J., and David B. Audretsch. 1989a. "Small-Firm Entry in U.S. Manufacturing." *Economica* 56(2): 255–265.

Acs, Zoltan J., and David B. Audretsch. 1989b. "Births and Firm Size." *Southern Economic Journal* 56(2): 467–475.

Acs, Zoltan J., and David B. Audretsch. 1988. "Innovation in Large and Small Firms: An Empirical Analysis." *American Economic Review* 78(4): 678–690.

Acs, Zoltan J., and David B. Audretsch. 1987. "Innovation, Market Structure and Firm Size." *Review of Economics and Statistics* 69(4): 567–575.

Acs, Zoltan J., David B. Audretsch, and Maryann P. Feldman. 1994. R & D Spillovers and Recipient Firm Size." *Review of Economics and Statistics* 100(2): 336–340.

Acs, Zoltan J., David B. Audretsch, and Maryann P. Feldman. 1992. "Real Effects of Academic Research." *American Economic Review* 82(1): 363–367.

Alchian, Almerin. 1950. "Uncertainty, Evolution, and Economic Theory." *Journal of Political Economy* 58:211–21.

Alchian, Almerin, and H. Demsetz. 1972. "Production, Information Costs, and Economic Organization." *American Economic Review* 62: 777–795.

Arrow, Kenneth J. 1983. "Innovation in Large and Small Firms." In J. Ronen, ed. *Entrepreneurship*. Lexington, MA: Lexington Books, pp. 15–28.

Arrow, Kenneth J. 1962. "Economic Welfare and the Allocation of Resources for Invention." In R. R. Nelson, ed. *The Rate and Direction of Inventive Activity*. Princeton, NJ: Princeton University Press, pp. 609–626.

Austin, John S., and David I. Rosenbaum. 1990. "The Determinants of Entry and Exit Rates into U.S. Manufacturing Industries." *Review of Industrial Organization* 5(2): 211–223.

Baldwin, John R., and Paul K. Gorecki. 1989. "Firm Entry and Exit in the Canadian Manufacturing Sector." Mimeo.

Baldwin, John R., and Paul K. Gorecki. 1987. "Plant Creation versus Plant Acquisition: The Entry Process in Canadian Manufacturing." *International Journal of Industrial Organization* 5(1): 27–42.

Baldwin, John R., and Paul K. Gorecki. 1985. "The Determinants of Small Plant Market Share in Canadian Manufacturing Industries in the 1970s." *Review of Economics and Statistics* 67(1): 156–161.

Blanchflower, D., and B. Meyer. 1994. "A Longitudinal Analysis of Young Entrepreneurs in Australia and the United States." *Small Business Economics* 6(1): 1–20.

Bradburd, Ralph, and Richard E. Caves. 1982. "A Closer Look at the Effect of Market Growth on Industries' Profits." *Review of Economics and Statistics* 64(4): 635–645.

Carrol, Paul. 1993. *Big Blues: The Unmaking of IBM*. New York: Crown Publishers.

Caves, Richard E., and Michael E. Porter. 1977. "From Entry to Mobility Barriers." *Quarterly Journal of Economics* 91: 241–261.

Coase, R. H. 1937. "The Nature of the Firm." *Economica* 4(4): 386–405. Reprinted in Oliver E. Williamson and Sidney G. Winter, eds. *The Nature of the Firm: Origins, Evolution, and Development*. New York: Oxford University Press, pp. 18–33.

Cohen, Wesley M., and Steven Klepper. 1992a. "The Tradeoff between Firm Size and Diversity in the Pursuit of Technological Progress." *Small Business Economics* 4(1): 1–14.

Cohen, Wesley M., and Steven Klepper. 1992b. "The Anatomy of Industry R & D Intensity Distributions." *American Economic Review* 82(4): 773–799.

Cohen, Wesley M., and Steven Klepper. 1991. "Firm Size Versus Diversity in the Achievement of Technological Advance." In Z. J. Acs and D. B. Audretsch, eds. *Innovation and Technological Change: An International Comparison*. Ann Arbor: University of Michigan Press, pp. 183–203.

Comanor, William S., and Thomas A. Wilson. 1967. "Advertising, Market Structure, and Performance." *Review of Economics and Statistics* 49(4): 423–440.

Cringley, Robert X. 1993. *Accidental Empires: How the Boys of Silicon Valley Make Their Millions, battle Foreign Competition, and Still Can't Get a Date*. New York: Harper Business.

Duetsch, Larry L. 1984. "Entry and the Extent of Multiplant Operations." *Journal of Industrial Economics* 32(June): 477–487.

Dunne, Timothy, Mark J. Roberts, and Larry Samuelson. 1989. "The Growth and Failure of U.S. Manufacturing Plants." *Quarterly Journal of Economics* 104(4): 671–698.

Dunne, Timothy, Mark J. Roberts, and Larry Samuelson. 1988. "Patterns of Firm Entry and Exit in U.S. Manufacturing Industries." *Rand Journal of Economics* 19(4): 495–515.

Evans, David, and Boyan Jovanovic. 1989. "Estimates of a Model of Entrepreneurial Choice under Liquidity Constraints." *Journal of Political Economy* 97(3): 808–827.

Evans, David, and Linda S. Leighton. 1990a. "Small Business Formation by Unemployed and Employed Workers." *Small Business Economics* 2(4): 319–330.

Evans, David S., and Linda S. Leighton. 1990b. "Some Empirical Aspects of Entrepreneurship." In Z. J. Acs and D. B. Audretsch, eds. *The Economics of Small Firms: A European Challenge*. Boston: Kluwer Academic Publishers, pp. 79–97.

Evans, David S., and Linda S. Leighton. 1989. "The Determinants of Changes in U.S. Self-Employment. 1968–1987." *Small Business Economics* 1(2): 111–120.

Feldman, Maryann P. 1994a. "Knowledge Complementarity and Innovation." *Small Business Economics* 6(5).

Feldman, Maryann P. 1994b. *The Geography of Innovation*. Boston: Kluwer Academic Publishers.

Foti, Alessandro, and Marco Vivarelli 1994. "An Econometric Test of the Self-Employment Model: The Case of Italy." *Small Business Economics* 6(2): 81–94.

Geroski, Paul A. 1992. *Market Dynamics and Entry*. Oxford: Basil Blackwell.

Geroski, Paul A. 1991a. "Domestic and Foreign Entry in the United Kingdom: 1983–1984." In P. Geroski and J. Schwalbach, eds. *Entry and Market Contestability: An International Comparison*. Oxford: Basil Blackwell, pp. 63–88.

Geroski, Paul A., 1991c. "Some Data-Driven Reflections on the Entry Process." In P. Geroski and J. Schwalbach, eds. *Entry and Market Contestability: An International Comparison*. Oxford: Basil Blackwell, pp. 282–286.

Geroski, Paul A. 1990. "Innovation and the Evolution of Market Structure." *Journal of Industrial Economics* 38(3): 299–314.

Geroski, Paul A. 1989b. "The Interaction Between Domestic and Foreign Based Entrants." In D. B. Audretsch, L. Sleuwaegen, and H. Yamawaki, eds. *The Convergence of International and Domestic Markets*. Amsterdam: North-Holland, pp. 59–83.

Geroski, Paul, and Joachim Schwalbach, eds. 1991. *Entry and Market Contestability: An International Comparison*. Oxford: Basil Blackwell.

Gilder, George. 1989. *Microcosm*. New York: Touchstone.

Gort, Michael, and Steven Klepper. 1982. "Time Paths in the Diffusion of Product Innovations." *Economic Journal* 92(3): 630–653.

Graf, J. de V. 1957. *Theoretical Welfare Economics*. Cambridge: Cambridge University Press.

Griliches, Zvi. 1979. "Issues in Assessing the Contribution of R & D to Productivity Growth." *Bell Journal of Economics* 10(Spring): 92–116.

Highfield, Richard, and Robert Smiley. 1987. "New Business Starts and Economic Activity: An Empirical Investigation." *International Journal of Industrial Organization* 5(1): 51–66.

Hirschman, Albert O. 1970. *Exit, Voice, and Loyalty*. Cambridge, MA: Harvard University Press.

Holmes, Thomas J., and James A. Schmitz Jr. 1990. "A Theory of Entrepreneurship and its Application to the Study of Business Transfers." *Journal of Political Economy* 98(4): 265–294.

Holmstrom, Bengt. 1989. "Agency Costs and Innovation." *Journal of Economic Behavior and Organization* 12: 305–327.

Holmstrom, Bengt, and Paul Milgrom. 1987. "Aggregation and Linearity in the Provision of Intertemporal Incentives." *Econometrica* 55(2): 303–328.

Holmstrom, Bengt, and Jean Tirole. 1989. "The Theory of the Firm." In R. Schmalensee and R. Willig, eds. *Handbook for Industrial Organization*. Amsterdam: Elsevier.

Ichbiah, Daniel, and Susan I. Knepper. 1993. *The Making of Microsoft: How Bill Gates and His Team Created the World's Most Successful Software Company*. Rocklin, CA: Prima.

Jaffe, Adam B. 1989. "Real Effects of Academic Research." *American Economic Review* 79(5): 957–970.

Jaffe, Adam B. 1986. "Technological Opportunity and Spillovers of R & D: Evidence from Firms' Patents, Profits and Market Value." *American Economic Review* 76(5): 984–1001.

Jensen, Michael C., and William H. Meckling. 1976. "Theory of the Firm: Management Behavior, Agency Costs and Ownership Structure." *Journal of Financial Economics* 3: 305–360.

Jovanovic, Boyan. 1994. "Entrepreneurial Choice When People Differ in Their Management and Labor Skills." *Small Business Economics* 63(3): 185–192.

Khemani, R. S., and Daniel M. Shapiro. 196. "The Determinants of New Plant Entry in Canada." *Applied Economics* 18 (November): 1243–1257.

Kihlstrom, Richard E., and Jean-Jacques Laffont. 1979. "A General Equilibrium Entrepreneurial Theory of Firm Formation Based on Risk Aversion." *Journal of Political Economy* 87(4): 719–748.

Knight, Frank H. 1921. *Risk, Uncertainty and Profit*. New York: Houghton Mifflin.

Kreps, David. 1991. "Corporate Culture and Economic Theory." In J. Alt and K. Shepsle, eds., *Positive Perspectives on Political Economy*. Cambridge: Cambridge University Press.

Marshall, Alfred. 1920. *Principles of Economics*, 8th ed. London: Macmillan.

Milgrom, Paul. 1988. "Employment Contracts, Influence Activities and Organization Design." *Journal of Political Economy* 96(1): 42–60.

Milgrom, Paul, and John Roberts. 1991. "Bargaining and Influence Costs and the Organization of Economic Activity." In J. Alt and K. Shepsle, eds. *Positive Perspectives on Political Economy*. Cambridge: Cambridge University Press.

Milgrom, Paul, and John Roberts. 1990. "The Economics of Modern Manufacturing: Technology, Strategy, and Organization." *American Economic Review* 80(3): 511–528.

Milgrom, Paul, and John Roberts. 1988. "An Economic Approach to Influence Activities in Organizations." *American Journal of Sociology* 94: 154–179.

Milgrom, Paul, and John Roberts. 1987. "Information Asymmetries, Strategic Behavior, and Industrial Organization." *American Economic Review* 77(2): 184–193.

Mueller, Dennis C. 1991. "Entry, Exit, and the Competitive Process." In P. Geroski and J. Schwalbach, eds. *Entry and Market Contestability: An International Comparison*. Oxford: Basil Blackwell, pp. 1–22.

Mueller, Dennis C. 1976. "Information, Mobility, and Profit." *Kyklos* 29(3): 419–448.

Nelson, Richard R., and Sidney G. Winter. 1982. *An Evolutionary Theory of Economic Change*. Cambridge, MA: Harvard University Press.

Nelson, Richard R., and Sidney G. Winter. 1978. "Forces Generating and Limiting Concentration under Schumpeterian Competition." *Bell Journal of Economics* 9(Autumn): 524–548.

Nelson, Richard R., and Sidney G. Winter. 1974. "Neoclassical vs. Evolutionary Theories of Economic Growth: Critique and Prospectus." *Economic Journal* 84(December): 886–905.

Neumann, Manfred. 1993. Review of *Entry and Market Contestability: An International Comparison*. P. A. Geroski and J. Schwalbach, eds. *International Journal of Industrial Organization* 11(4): 593–594.

Orr, Dale. 1974. "The Determinants of Entry: A Study of the Canadian Manufacturing Industries." *Review of Economics and Statistics* 56(1): 58–66.

Palfreman, Jon, and Doron Swade. 1991. *The Dream Machine: Exploring the Computer Age*. London: BBC Books.

Penrose, Edith T. 1959. *The Theory of the Growth of the Firm*. Oxford: Basil Blackwell.

Rose, Frank. 1989. *West of Eden: The End of Innocence at Apple Computer*. New York: Viking Press.

Santarelli, Enrico, and Alessandro Sterlacchini. 1994. "New Firm Formation in Italian Industry: 1985–89." *Small Business Economics* 6(2): 95–106.

Scherer, F. M. 1991. "Changing Perspectives on the Firm Size Problem." In Z. J. Acs and D. B. Audretsch, eds. *Innovation and Technological Change*. An International Comparison, Ann Arbor: University of Michigan Press, pp. 24–38.

Scherer, F. M. 1984. *Innovation and Growth: Schumpeterian Perspectives*. Cambridge, MA: MIT Press.

Schumpeter, Joseph A. 1911. *Theorie der wirtschaftlichen Entwicklung. Eine Untersuchung über Unternehmergewinn, Kapital, Kredit, Zins und den Konjunkturzyklus*. Berlin: Duncker und Humblot.

Schumpeter, Joseph A. 1942. *Capitalism, Socialism and Democracy*. New York: Harper and Row.

Schwalbach, Joachim. 1987. "Entry by Diversified Firms into German Industries." *International Journal of Industrial Organization* 5(March): 43–50.

Siegfried, John J., and Laurie B. Evans. 1992. "Entry and Exit in U.S. Manufacturing Industries from 1977 to 1982." In D. B. Audretsch and J. J. Siegfried, eds., *Empirical Studies in Industrial Organization: Essays in Honor of Leonard W. Weiss*. Ann Arbor: University of Michigan Press, pp. 253–275.

Simon, Herbert A. 1984. "On the Behavioral and Rational Foundations of Economic Dynamics." *Journal of Economic Behavior and Organization* 5(1): 35–36.

Stigler, George J. 1961. "The Economics of Information." *Journal of Political Economy* 59(2): 185–193.

Tirole, Jean. 1986. "Hierarchies and Bureaucracies." *Journal of Law, Economics and Organization* 2(3): 181–214.

Vivarelli, Marco. 1991. "The Birth of New Enterprises." *Small Business Economics* 3(3): 215–223.

White, Lawrence J. 1982. "The Determinants of the Relative Importance of Small Business." *Review of Economics and Statistics* 64(1): 42–49.

Williamson, Oliver E. 1985. *The Economic Institutions of Capitalism*. New York: The Free Press.

Williamson, Oliver E. 1975. *Markets and Hierarchies: Antitrust Analysis and Implications*. New York: The Free Press.

Winter, Sidney G. 1991. "On Coase, Competence, and the Corporation." In O. E. Williamson and S. G. Winter, eds. *The Nature of the Firm: Origins, Evolution, and Development*. Oxford: Oxford University Press, pp. 179–195.

Winter, Sidney G. 1984. "Schumpeterian Competition in Alternative Technological Regimes." *Journal of Economic Behavior and Organization* 5(Sept.–Dec.): 287–320.

Yamawaki, Hideki. 1991. "The Effects of Business Conditions on Net Entry: Evidence from Japan." In P. Geroski and J. Schwalbach, eds. *Entry and Market Contestability: An International Comparison*. Oxford: Basil Blackwell, pp. 168–186.

[23]

Technological Opportunities and New Firm Creation

Scott Shane
3355 Van Munching Hall, R.H. Smith School of Business, University of Maryland, College Park, Maryland 20742
sshane@rhsmith.umd.edu

Research on the creation of new high-technology companies has typically focused either on industry-level factors such as market structure and technology regime or on individual-level factors such as the work experience of entrepreneurs. This study complements these approaches by examining the effect of technological opportunities on firm formation. In particular, the study shows that the probability that an invention will be commercialized through firm formation is influenced by its importance, radicalness, and patent scope.
(*Entrepreneurship; Patents; Technology Opportunities*)

The creation of new firms is an important mechanism through which entrepreneurs use technology to bring new products, processes, and ways of organizing into existence (Schumpeter 1934). Moreover, recent research has shown that the rate at which new firms are created is increasing in the U.S. economy (Gartner and Shane 1995), and that the pace of technological change influences the rate of firm formation (Shane 1996). However, inventors do not always start firms to exploit technological opportunities. Sometimes they sell these opportunities (Audretsch 1995); other times they simply abandon them prior to exploitation (Roberts 1991). The fact that opportunity discovery does not correlate perfectly with firm creation raises the question of when new firms are created to commercialize new technological opportunities.

Previous research has provided two categories of explanations for the creation of new technology firms. The first is that firm formation depends on industry-level factors, such as market structure (Audretsch 1995) or technology regime (Winter 1984). Industry-level arguments hold that when industries are young, unconcentrated, composed of small firms, have limited requirements for complementary assets, have access to capital, and are not R&D intensive, people tend to form new firms to exploit opportunities (Audretsch 1995). The second explanation is that firm formation depends on individual-level factors, such as the psychology of entrepreneurs (Roberts 1991) or their career experience (Carroll and Mosakowski 1987). Individual-level arguments hold that when the individuals who discover opportunities are more experienced in firm creation (Carroll and Mosakowski 1987), more creative (Schumpeter 1934), more imaginative (Shackle 1979), more risk tolerant (Khilstrom and Laffont 1979), higher in need for achievement (Roberts 1991) or more tolerant of ambiguity (Begley and Boyd 1987), they tend to form new firms to exploit opportunities.

Unfortunately, empirical tests to predict the creation of new firms on the basis of the characteristics of individuals (see Low and MacMillan 1988, Busenitz and Barney 1997) or industry (see Geroski 1995, Cohen and Levin 1989) have provided limited explanatory power. One reason for these weak empirical results is that industry and individual-based explanations fail to consider the nature of the opportunity discovered. Technological opportunities differ from each other on several dimensions, which influence the decision to found a new firm. Using a unique dataset of 1,397 inventions patented by the Massachusetts Institute of Technology over the 1980–1996 period,

0025-1909/01/4702/0205$5.00
1526-5501 electronic ISSN

I examine the effect of three dimensions of technological opportunities—importance, radicalness, and patent scope—on the probability that they will be commercialized through firm formation.

This paper makes three contributions to our understanding of high-technology entrepreneurship. First, it provides an empirical test of the effect of technological opportunity on the decision to establish a new firm. Such a test complements existing explanations of new firm formation that examine only industry or individual-level factors. Second, the study disentangles the behaviour of established firms entering new markets from the behaviour of independent entrepreneurs. Although previous research has shown that new entrants are more likely to adopt certain technologies than are incumbent firms, this research has confounded established firms entering new markets with independent entrepreneurs. This study shows that independent entrepreneurs make similar exploitation decisions to those of new entrants in general. Third, this study controls for individual-level factors in the examination of firm formation. Most studies of firm formation in high-technology settings assume away learning curves and individual-level variation in preferences for firm creation, and treat all potential entrants as equally willing and able to exploit an opportunity by founding a company (Audretsch 1995, Caves 1998). However, extensive research shows that entrepreneurship involves a significant component of learning-by-doing (Carroll and Mosakowski 1987). This finding means that whoever obtains decision rights over a new technology can influence the mode of commercialization. This study improves upon prior examination of the firm-formation decision in technology settings by controlling for the attributes of the individuals making the commercialization decision.

I have divided the paper into three sections. In the next section, I explain why dimensions of technology opportunity should influence the decision to found a firm. In the third section, I describe the dataset and the methods used for analysis. In the final section, I summarize the findings and discuss their implications for future research and the practice of entrepreneurship.

Theoretical Development

The linkage between technological change and firm formation is one of the oldest relationships in the study of business organization. Schumpeter (1934) argued that the creation of new technology firms that displace incumbent firms through a process of creative destruction is a major source of innovation in a capitalist system. In particular, Schumpeter argued that the process by which independent entrepreneurs use exogenously created inventions to produce new goods, services, raw materials, and organizing methods is central to understanding business organization, the process of technical change, and economic growth.

Subsequent research in the Schumpeterian tradition has expanded the complexity of this framework by incorporating the possibility that entrepreneurial opportunities will be sold in markets (Casson 1982) or abandoned prior to exploitation (Roberts 1991). Consideration of multiple phases in the invention-to-firm creation process has led researchers to propose three categories of factors that influence the decision to exploit an invention through firm creation: the nature of the individual making the decision (Roberts 1991), the nature of the industry in which the opportunity would be exploited (Audretsch 1995), and the nature of the opportunity itself (Henderson 1993). In the decades since Schumpeter's seminal work, economists of technical change have tested the effects of numerous industry characteristics on firm formation (Audretsch 1995). Similarly, applied entrepreneurship researchers have tested the effects of a variety of individual attributes on firm formation (Roberts 1991). To date, however, researchers have not directly examined the effect of the attributes of new technologies themselves on firm formation.

Nevertheless, researchers working in the related area of technological innovation and industry entry (e.g., Christiansen and Bower 1996, Lerner 1994, Audretsch 1995) have proposed several attributes of technological opportunities that could influence the probability that new firms will be created to exploit them. This paper explores three of these attributes: importance, radicalness, and patent scope.[1]

[1] Data limitations preclude the measurement of other constructs. However, this limitation should not severely handicap the

Importance measures the magnitude of the economic value of an invention. Radicalness measures the degree to which an invention, however large or small in economic value, differs from previous inventions in the field. Patent scope measures the breadth of intellectual property protection for the invention, however large or small its economic value, and however similar or different the invention is to previous inventions in the field.

Both prior research and field interviews conducted with MIT inventors indicate that each of these attributes captures a different dimension of technological opportunity, that entrepreneurs see these dimensions as conceptually distinct, and that these dimensions increase the probability of firm formation. In the subsections below, I argue why these dimensions of opportunity should influence the decision to create a new firm.

Importance. The importance of an invention should increase the likelihood that a new firm will be founded to commercialize it because more important inventions have higher economic value (Harhoff et al. 1999)[2] and, therefore, provide a larger potential payoff to firm formation. In explaining the decision of a person to start a firm, Knight (1921, p. 273) argued that "the laborer asks what he thinks the entrepreneur will be able to pay, and in any case, will not accept less than he can get from some other entrepreneur, or by turning entrepreneur himself." For a person to "turn entrepreneur," he or she must expect an entrepreneurial profit which exceeds (a) the opportunity cost of not engaging in other activities (Amit et al. 1995), (b) a premium for the illiquidity that results from the investment of financial resources in a form that cannot be easily turned into cash (Venkataraman 1997), and (c) a premium for bearing uncertainty (Khilstrom and Laffont 1979).

Technological change is an uncertain process, in which developments are sometimes small and other times large. As a result, the potential economic value created by technological advances is highly varied (Trajtenberg 1990). Empirical evidence of the skewness of the distribution of patent value (Pakes 1986) indicates that most patents have no commercial value and a few have a large value (Trajtenberg et al. 1997). Consequently, "a great majority of patents are never exploited commercially, and only a very few are associated with major technological improvements" (Cohen and Levin 1989, p. 1063).

Given the skewness in their commercial value, most inventions do not provide sufficient potential to justify the investment of time and money in the uncertain process of creating a new firm. Field interviews with MIT inventors confirmed the effect of this process on inventors' decision making. When individuals make what they believe to be important technological discoveries, they anticipate greater economic value than when they make minor discoveries. Moreover, inventors point out that the importance of technological discoveries increases their willingness to incur the opportunity cost of not undertaking other activities, making illiquid investments in new firms, and bearing the technical, market, competitive, and financial uncertainty of establishing a new company. Consequently, both the prior theoretical literature and fine-grained case information from inventors suggests that the more important the inventors' technical discoveries, the greater their impetus to create new firms. This argument leads to the first hypothesis:

Hypothesis 1. *The more important the invention, the higher the probability that it will be commercialized through the establishment of a new firm.*

Radicalness. One of the central concepts in the literature on technological change is that refining and improving an existing technology (an incremental improvement) and introducing a new approach to technical practice (a radical improvement) are fundamentally different things (Reinganum 1983). In particular, an incremental improvement reinforces the activities of established firms, while a radical improvement may undermine those activities (Tushman and Anderson 1986).

Researchers have argued that independent entrepreneurs, rather than managers in established

investigation described here because the paper does not posit a fully specified model of the relationship between technological opportunity and firm formation.

[2] Some inventions may be more important than others because they have great social value, even if they have little economic value. This study does not consider the social dimension of importance.

firms, will be the ones most likely to introduce radical technological developments. Three separate arguments have been put forth for why this should be the case. First, radical technologies destroy the capabilities of existing firms because they draw on new technical skills. Since organizational capabilities are difficult and costly to create (Nelson and Winter 1982, Hannan and Freeman 1984), established firms are organized to exploit established technologies. Firms find it difficult to change their activities to exploit technologies based on different technical skills. Therefore, established firms often choose not to pursue radical opportunities, leaving them to independent entrepreneurs. Second, established firms have less incentive to invest in the development of technologies that undermine the value of the assets that they already possess. The desire not to cannibalize existing assets leads established firms to underinvest in the development of new technologies (Arrow 1962). Independent entrepreneurs, with no existing assets to protect, do not face this resistance to invest, and are more likely to be the ones to pursue these opportunities. Third, firms develop routines for filtering information based on what is likely to be valuable for what they are doing currently (Henderson 1993). These routines limit the search process to those technologies which are conceptually close to the technologies with which firms are currently working (Podolny and Stuart 1995). Knowledge transfer from outside is impeded when new information is not the logical extension of existing knowledge possessed by the organization. Since radical inventions are often based on things that are not the logical extension of the firm's internal information, they are difficult for established firms to understand and evaluate (Rosenbloom and Christiansen 1994). Consequently, established firms often filter out information about radical technologies in situations in which independent entrepreneurs do not.

Field interviews with fifty MIT inventors support the arguments of the previous theoretical literature. These inventors explained that the technical skills upon which an invention draws influence their decisions to start companies. In particular, these inventors identified the problem of market failure in finding established companies interested in their technological developments when those developments draw on technologies unfamiliar to established companies. Moreover, these inventors also pointed out the relative ease of obtaining venture capital funding for technologies that did not fit with the activities of established companies, particularly those that had the potential for generating new industries. Consequently, both the prior theoretical literature and fine-grained case information from inventors suggests that the more radical the inventors' technical discoveries, the greater their impetus to create new firms. This argument leads to the second hypothesis:

Hypothesis 2. *The more radical the invention, the higher the probability that it will be commercialized through the establishment of a new firm.*

Patent Scope. Technological opportunities with broader intellectual property protection are more likely to be commercialized through firm creation. When an entrepreneur decides to start a new firm in response to the invention of a new technology, he or she typically does not yet possess complementary assets, like a distribution system, which might provide a competitive advantage in the industry in which the technology would be exploited (Teece 1986). The more effectively an opportunity can be protected against appropriation by competitors, the more likely the entrepreneur will be to obtain necessary complementary assets before the competitive advantage of its new technology has dissipated.

Broad patents enhance the probability that the entrepreneur can protect the technology against appropriation. Patents provide a legal right to prevent others from imitating a particular technological development in areas delineated by the patent claims. The scope of the patent is important because "the broader the scope, the larger number of competing products and processes that will infringe the patent" (Merges and Nelson 1990, p. 839). When a patent is narrow in scope, the holder of the patent will have less incentive to develop the technology through the creation of a new firm because it will have a smaller space of technology that is protected against imitation by other firms (Merges and Nelson 1990).

Although established firms might also be more likely to commercialize broad patents, they are disproportionately important to independent entrepreneurs who lack complementary assets. "For small, start-up ventures, patents may be a relatively effective means of appropriating R&D returns, in part because some other means, such as investment in complementary sales and service efforts may not be feasible. The patents held by a small, technologically oriented firm may be its most marketable asset" (Levin et al. 1987, p. 797).

Potential entrepreneurs and their investors appear to respond to the influence of patent scope in the firm-formation decision process. Field interviews with fifty MIT inventors revealed that many of them ask patent attorneys for a judgement as to the scope of patent protection before they decide to start new firms to exploit their inventions. Moreover, they report that their investors are concerned with the breadth of the protection of the technology against appropriability and prefer broader patents in the decision of whether or not to fund a new venture. Lerner (1994) also provides large-sample statistical evidence for this investor effect. He found that patent scope has a substantive and significant impact on the valuation of venture capital financed biotechnology start-ups. Consequently, both the prior theoretical literature and fine-grained case information from inventors suggest that the broader the scope of patent protection, the greater the impetus to create new firms to exploit inventions. This argument leads to the third hypothesis:

HYPOTHESIS 3. *The broader the scope of the patent, the higher the probability that the invention will be commercialized through the establishment of a new firm.*

Methodology

Sample

This study examines the population of 1,397 patents assigned to the Massachusetts Institute of Technology (MIT) between 1980 and 1996. This population includes all inventions made by faculty, staff, or students of the university that made material use of MIT property during the course of their development.

I focus on patented inventions rather than other entrepreneurial opportunities in this study for several reasons. First, by looking at patents, I can examine the decision to exploit entrepreneurial opportunities in a setting in which I have information both about who discovered the opportunity and the discovery itself. Second, the population of patented inventions is recorded, mitigating selection biases that plague much of the survey and case study work on the commercialization of new technologies, particularly in the context of firm formation. Third, patents have been studied in many contexts other than firm creation, providing comparability of the results to findings about other aspects of technological change.

I examine MIT's inventions from the 1980–1996 period for several reasons: First, universities are not in the business of manufacturing products or providing services (other than education). They do not refrain from patenting to protect new knowledge through trade secrets, refrain from licensing for competitive purposes, collect patents, or cross-license (Henderson et al. 1998). Therefore, one can examine the process of firm creation in response to technical discoveries by universities absent the biases that would occur from the examination of patents belonging to companies. Second, university research is an important source of new knowledge (Winter 1984) that is particularly relevant to economic growth (Caballero and Jaffe 1993, Romer 1990). Therefore, the study of firm creation to exploit university inventions is important in its own right, even if results cannot be generalized to other settings. Third, the post-1980 period follows a change in federal law granting universities the property rights to federally funded inventions. Since federally funded inventions account for roughly 70 percent of all university inventions, this change provided a strong incentive to universities to commercialize their inventions (Henderson et al. 1998). Therefore, the post-1980 period is an important one in which to understand university firm-formation activity.[3]

Analysis

I use event history analysis to examine the probability that each patent results in the formation of a

[3] The selection of MIT over other universities was a matter of convenience. I had access to data on MIT, but not other universities.

new firm. When a patent is issued, a spell begins.[4] The spell continues until a firm is formed. If a patent does not result in firm formation, it is treated as censored. Because each patent is observed until a firm is formed or until the observation period ends, the 1,397 patents generate 9,002 case-year observations. I use Cox regression to analyze the data because I make no claims about the functional form of time dependence. The analysis incorporates both time-varying and time-invariant covariates, which are specified below.

Dependent Variable: Firm Formation

I used the MIT Technology Licensing Office records of its licensees to construct the dependent variable. The MIT records indicate the legal status of the licensee, and therefore the ability to identify its year of incorporation. The dependent variable was coded 1 in the year that the patent was first licensed if the patent was licensed to an (unincorporated) individual or was licensed to a firm that was incorporated after the patent application was submitted. If these conditions were not met in a given year, the patent was coded as 0 in that year.[5]

This coding scheme allowed two types of licensees to be coded as new firms. The first type was an (unincorporated) individual who licensed the patent. The second type was a new firm incorporated to commercialize the MIT invention. Under this coding arrangement, two types of patents are treated as censored: patents granted to MIT, but never licensed, and patents licensed to an already incorporated company (e.g., Sony). Twenty-six percent of the patents were licensed to "new firms" during the period of investigation.

Predictor Variables

Radicalness. Following Rosenkopf and Nerkar (1999), I measure the radicalness of the patent as a time-invariant count of the number of three-digit patent classes in which *previous* patents cited by the given patent are found, but the patent itself is not classified. Citations influence the legal boundaries of the property rights to an invention (Jaffe et al. 1993). Patents are divided into approximately 100,000 nine-digit patent classes, which aggregate to approximately 600 three-digit classes, and which represent distinct technical areas (Jaffe et al. 1997). The assignment of a patent to a particular patent class represents the U.S. Patent and Trademark Office's (USPTO's) assessment that the patent belongs in a particular technical field. Patent examiners also determine what previous inventions must be cited in a patent by searching prior patents. Because patents belong to technical classes and because they cite previous patents, citations to patents in particular technical fields represent the USPTO's assessment that a particular invention builds upon (cites) knowledge in that technical field. When a patent cites previous patents in classes other than the ones it is in, that pattern suggests that the invention builds upon different technical paradigms from the one in which it is applied.

Importance. Following Henderson et al. (1998), I measure the importance of a patent as a time-invariant measure of the total number of citations that the given patent received from *subsequent* patents from patent issue until 1997. The use of a time-invariant measure of citations provides both advantages and disadvantages in the examination of the economic value of patents. A time-invariant measure of citations captures the idea that inventions have a distribution of underlying economic value at the time of invention. It is this underlying economic value to which people respond in making the decision whether or not to found a firm. A time-varying measure of citations would fail to capture this underlying economic value because it would consider all patents as having the same importance (zero) at the time of invention.

However, a time-invariant measure of citations does not treat all patents equally. A patient issued in 1996 has had only one year to be cited, whereas a patent granted in 1980 has had seventeen years to be cited. Although controlling for the technical field and the year of patent filing (or issue) incorporates the effect of time and technology into the measure of

[4] Since MIT has no records of firm formation for patents that were never issued, starting the spells with the filing date would create selection bias.

[5] Patents licensed to nonprofit organizations were treated as non-firm formation patents.

importance, these controls do not allow the researcher to directly measure the relative importance of one patent as compared to another. Therefore, I also examine an alternative specification of importance: the total number of citations that the given patent received from *subsequent* patents from patent issue until 1997, divided by the number of citations received by an average university patent of that age and in that technology field.

Research confirms the construct validity of the importance measures. Patent citations are significantly correlated with the economic value of inventions. Harhoff et al. (1999) found that the higher the estimated private value of an invention, the more the patent was cited by later patents. Hall et al. (1998) found that companies with highly cited patents have higher stock market values, all other things being equal.

Researchers familiar with the journal review process might question the use of citations on the grounds that inventors might cite friends to help them or enemies to describe their work as inadequate or trivial. However, this problem is unlikely to occur with patent citations. Jaffe et al. (1993) explain that gratuitous citations in patent applications are costly to inventors. Patent citations determine the scope of the inventor's monopoly and unnecessary citations limit what the inventor can claim as his or her monopoly right. Moreover, even if the inventor were to include these citations, the patent examiner's job requires the correct citation of previous patents. Hence, the examiner (unlike the editor of an academic journal) has a sworn duty to remove gratuitous citations from the patent before it issues.

Scope. Following Lerner (1994), I measure patent scope as a time-invariant count of the number of international patent classes into which the USPTO assigns a patent. Lerner (1994) explained that because international patent classes are nested, the more classes to which a patent is assigned, the broader is its intellectual property protection. Lerner (1994) provided construct validity for this measure by showing that venture-capital-backed-biotechnology firms with broader patents, as measured by a count of four-digit international patent classes, received higher valuations.[6]

Control Variables

Firm-Founding Experience. I measure firm-founding experience as a time-invariant estimate of the log of the average number of prior MIT firm-founding inventions across the set of inventors on the patent. I control for founding experience because several researchers have demonstrated that the probability of starting a firm increases with entrepreneurial experience (e.g., Evans and Leighton 1989, Carroll and Mosakowski 1987). Prior experience provides knowledge about resources that help to start new firms, entrepreneurial skills, and reputations that help to influence the reallocation of resources to the new venture (Shane and Khurana 1999). This knowledge influences the propensity to found firms because it increases the expectation of the value that would be gained from firm formation.

Patenting Experience. To control for patenting propensity in the examination of the effect of prior firm-founding experience, I also include a time-invariant measure of the log of the average number of prior MIT inventions across the set of inventors on the patent.

Age of the Technology Class. I measure the age of the technology class as the number of years since the three-digit patent class was established by the U.S. Patent and Trademark Office. Research has shown that people are more likely to start new firms to exploit a new technology when the technical field is young (Utterback and Abernathy 1975) because the markets for new technologies are often too small to interest established firms, because established firms possess learning-curve advantages in mature technologies (Nelson 1995), and because the maturation of technology makes capital-intensity, the reduction of production costs, and scale economies important (Pavitt and Wald 1971).

[6] I construct the radicalness measure on the basis of three-digit U.S. patent classes and the patent scope measure on the basis of international patent classes to make the measures comparable to the measures developed by previous authors (e.g., Rosenkopf and Nerkar 1999, Lerner 1994).

R&D Intensity. I measured R&D intensity in an industry as a time-varying estimate of research and development expenditures as a percentage of the value-added of industry shipments in the three-digit SIC code, using the research and development expenditures obtained from various years of Science and Engineering Indicators.[7] Previous researchers have argued that entrepreneurs will be less likely to establish new companies to exploit inventions in R&D intensive industries (Audretsch 1995) because small firms face a scale disadvantage in research and development (Mansfield 1981, Kamien and Schwartz 1975).

Manufacturing Value-Added. I measured the importance of complementary assets in manufacturing by calculating a time-varying estimate of the value-added by manufacture in the industry year, using data from various years of the Census of Manufacturers. Teece (1987, p. 191) explained that "the successful commercialization of an innovation requires that the know-how in question be utilized in conjunction with other capabilities or assets ... such as marketing, competitive manufacturing, and after-sales support." Where complementary assets are important in an industry, entrepreneurs will be less likely to establish firms to commercialize technology.

Importance of Sales or Service. Another measure of the importance of complementary assets in an industry is the importance of superior sales or service as a competitive advantage. I measured the importance of sales or service in the industry as a time-invariant industry average score for the Levin et al. (1987)[8] survey item: "In this line of business superior sales or service efforts are very effective measures of capturing and protecting the competitive advantages of new or improved production processes."

[7] The USPTO does not assign patents to SIC codes. However, the USPTO has developed a concordance which allows the assignment of patents to three-digit SIC codes by reference to the patent's six-digit technology classification. To assign the correct industry variables to the patents, I used the USPTO concordance to identify the SIC code.

[8] Levin et al. (1987) generated line-of-business measures of innovation and appropriability from a survey of R&D managers.

Incumbent Innovation. Some industries have lower rates of new firm formation than others because established firms are the major innovators of new products and processes in those industries (Acs and Audretsch 1990). I measured the rate of incumbent innovation as a time-invariant industry average score for the Levin et al. (1987) survey item: "Firms within this business have been a very important source of contribution to technological progress in this line of business since 1970."

Market Segmentation. New firms often successfully enter markets with new technologies by targeting the niche customers, because large firms allocate resources for innovation to satisfy the demands of their major customers (Christiansen and Bower 1996). Therefore, the ability of new firms to enter industries depends, in part, upon the degree to which markets in those industries are segmented. Using data from the Levin et al. (1987) survey of research and development managers, I measured market segmentation as a time-invariant industry average score for the Levin et al. (1987) survey item: "designing markets for specific market segments" is a very important technological activity for the line of business.

Capital Availability. An important source of capital to start new technology companies is venture capital. To measure the availability of venture capital across industries, I included a time-varying estimate of the amount of venture capital funding as a percentage of industry shipments in the three-digit SIC code, using data obtained from Securities Data Corporation's venture capital database and the Commerce Department's Survey of Manufacturers. I used Securities Data Corporation's concordance between its proprietary industry scheme and SIC codes to map the data by SIC code. I include this control because entrepreneurship is less likely to take the form of new firms when capital market imperfections make it difficult for independent entrepreneurs to secure financing (Cohen and Levin 1989).

Concentration. I measured industry concentration as a time-varying estimate of the market share of the four largest companies in the industry, using data from the Census of Manufacturers: Concentration Ratios. Since values were available only for 2-digit and 4-digit SIC codes, values for three-digit SIC codes were estimated by taking an average of the relevant 4-digit SIC codes. A straight-line interpolation was used to determine the values for noncensus years. High levels of concentration in an industry discourage people from creating new firms because concentration enhances the power of incumbents to attack new entrants and to collude (Geroski 1995).

Firm Size. I measured the size of the firms in the industry as a time-varying estimate of the average number of employees per firm, using data from the Census of Manufacturers. High levels of minimum efficient scale discourage entrepreneurs from creating firms because they raise the cost of entry (Audretsch 1995). Because new firms that enter a market at below minimum efficient scale must operate at a cost disadvantage vis-a-vis established firms, the greater the amount of production that is necessary before a firm reaches minimum efficient scale, the less likely entrepreneurs should be to enter that market (Weiss 1976).

Technical Fields. The patents examined in this study had their primary U.S. patent classification in 188 different three-digit patent classes. Since attempts to control for such a large number of classes creates incidental parameter bias, I control for technical field using the modified version of Scherer's (1965) classification scheme common in many patent studies (Cohen and Levin 1989). I include dummy variables for mechanical inventions, electrical inventions, chemical inventions, and drugs (the other is the omitted field). I include these variables because citation patterns vary across technologies, necessitating the controls to measure importance correctly (Henderson et al. 1998). Moreover, firm-formation rates vary substantially by type of technology.

Time. I control for the year of patent application. Since citation patents vary over time, this control is necessary to measure importance accurately (Henderson et al. 1998). Moreover, changes in federal law and MIT policy have changed the incentives for firm formation over time.

Results

The descriptive statistics are reported in Table 1. Table 2 provides the distribution of firm-formation patents by year of patent issue. This table demonstrates a clear trend towards an increase in firm-formation patents over time.

Table 3 provides the correlation matrix. It shows that the highest correlation between any two of the independent variables is $r = 0.56$ between previous patents and previous start-ups. This level of correlation indicates that problems of multicollinearity are unlikely to be manifest in the data. Moreover, the table also indicates a low correlation between the radicalness, importance, and patent scope measures. This low level of correlation provides support for the argument that these three measures capture independent dimensions of technological opportunity.

Table 4 reports the effects of technology attributes on the likelihood of firm formation. I provide six models. Model 1 predicts the likelihood of firm formation on the basis of technical field and year of patent filing. Overall, this model is significant (Chi-square $= 112.04, p < 0.0001$). As Model 1 shows, the technical field and year of issue have strong effects on the probability of firm formation. Relative to other technologies, chemical technologies ($\text{Exp}(B) = 0.59, p < 0.001$), electrical technologies ($\text{Exp}(B) = 0.40, p < 0.0001$), and mechanical technologies ($\text{Exp}(B) = 0.56, p < 0.05$) are less likely to lead to firm formation, and drug technologies ($\text{Exp}(B) = 1.38, p < 0.05$) are more likely to lead to firm formation. (However, the effects for mechanical technologies and drug technologies are not robust to the inclusion of other variables in the regression equations). Moreover, the year of patent filing ($\text{Exp}(B) = 1.08, p < 0.0001$) shows a significant and robust pattern of increasing probability of firm formation over time.

Model 2 adds the industry-level control variables. Overall, this model is significant (Chi-square $= 135.40, p < 0.0001$). As Model 2 shows, two dimensions of industry have a significant effect on the probability of firm formation. First, the greater the importance of designing for specific market segments, the greater the probability of new firm formation ($\text{Exp}(B) = 1.60, p < 0.01$). This result is consistent with

Table 1 Descriptive Statistics ($N = 9002$)

Variable	Mean	S.D.	Min	Max
Technology Opportunity				
Radicalness	3.56	3.83	0.00	41.00
Importance	6.82	9.82	0.00	117.00
Patent Scope	1.30	0.62	1.00	6.00
Industry				
Venture Capital (percent of market value)	0.02	0.02	0.00	0.04
Age of Technology Class (years)	22.37	22.87	0.00	96.00
Manufacturing (percent of value added)	51.00	13.00	12.00	66.00
Importance of Sales or Service Efforts	4.98	0.48	3.94	6.25
R&D Intensity (percent of sales)	5.38	3.22	0.30	14.90
Four Firm Concentration Ratio	36.47	9.77	9.00	72.00
Average Firm Size	90.84	29.61	34.16	198.44
Importance of Market Segmentation	5.25	0.50	4.00	6.50
Incumbent Innovation	6.17	0.50	4.00	7.00
Inventor Attributes				
Log Previous Patents	0.66	0.37	0.00	1.60
Log Founding Experience	0.09	0.22	0.00	1.34
Technical Field				
Chemical Invention	0.35	0.48	0.00	1.00
Electrical Invention	0.36	0.48	0.00	1.00
Mechanical Invention	0.05	0.22	0.00	1.00
Drug Invention	0.14	0.35	0.00	1.00
Other Invention	0.09	0.29	0.00	1.00
Time				
Year Filed	1983.67	5.02	1974.00	1995.00

prior research which argues that new entrants displace incumbent firms by entering market segments with radical technologies and then expanding to the mainstream once they have established a foothold (Christiansen and Bower 1996). (However, the effect for market segmentation is not robust to the inclusion of other variables in the regression equations).[9] Second, the greater the average firm size in the industry, the greater the probability of firm formation ($\text{Exp}(B) = 0.99, p < 0.05$). This result is counter to expectations from prior research, which suggests that new firm formation should be more likely in industries in which the average size of firms is smaller.

Model 3 adds the individual-level control variables. Overall, this model is significant (Chi-square $= 299.70, p < 0.0001$). As Model 3 shows, the log of the number of previous new firm patents belonging to the inventors at the time of founding increases the probability of new firm formation ($\text{Exp}(B) = 6.19, p < 0.0001$), even after controlling for the log of the total number of patents belonging to the inventors at the time of founding. These results suggest that career experience of the inventors influences the decision to exploit a technological opportunity through the formation of a firm. Consequently, accurate measurement of the effect of technological

[9] The lack of robust effects for several industry variables should be interpreted with caution. First, to create industry variables, the USPTO patent concordance was applied. Since the concordance is inexact, the lack of effects could result from inexact classification of patents to industry. Second, the construction of the industry variables required some averaging of data across fields and years which might have mitigated the industry effects. Third, the industry variables are measured at a relatively high level of aggregation. Fourth, the technology field dummy variables may be capturing industry effects.

Table 2 The Distribution of Start-up Patents By Year of Patent Issue

Year	Number of Start-up Patents
1980	13
1981	14
1982	13
1983	14
1984	15
1985	9
1986	12
1987	13
1988	17
1989	31
1990	29
1991	26
1992	35
1993	32
1994	28
1995	25
1996	38

attributes on firm formation requires controlling for inventor attributes.

Model 4 adds the three predictor variables to the regression equation. Overall, this model is significant (Chi-square $= 330.16, p < 0.0001$). Specifically, the results show strong support for the three hypotheses. More important inventions ($\text{Exp}(B) = 1.02, p < 0.0001$) are more likely to be commercialized through the creation of new firms (Hypothesis 1). More radical patents ($\text{Exp}(B) = 1.03, p < 0.01$) are more likely to be commercialized through the creation of new firms (Hypothesis 2). Patents with broader scope ($\text{Exp}(B) = 1.18, p < 0.05$) are more likely to be commercialized through the creation of new firms (Hypothesis 3).

Model 5 tests the alternative specification of the importance measure. It examines the relative importance of the patent—the deviation from the average for university patents in the same technological field and of the same age.[10] Overall, this model is significant (Chi-square $= 251.91, p < 0.0001$). Moreover, the effect of the relative importance measure is substantively the same as that of the original importance measure ($\text{Exp}(B) = 1.03, p < 0.0001$), indicating that the original importance measure is robust. Finally, the use of the relative importance measure does not change the significance or magnitude of the coefficients for radicalness ($\text{Exp}(B) = 1.04, p < 0.01$) or patent scope ($\text{Exp}(B) = 1.26, p < 0.05$) in any substantive way.

Model 6 predicts firm formation on the basis of technological opportunity alone, without controlling for the time-varying covariates that measure industry characteristics. Since this model predicts firm formation on the basis of time-invariant measures only, I examine a time-invariant Cox regression, with 1,397 spells—one for each of the patents. Overall, this model is significant (Chi-square $= 436.17, p < 0.0001$). Moreover, the results are substantively the same as the results for the time-varying Cox regressions that control for industry attributes. Patent importance ($\text{Exp}(B) = 1.02, p < 0.0001$), patent scope ($\text{Exp}(B) = 1.16, p < 0.05$), and radicalness ($\text{Exp}(B) = 1.03, p < 0.01$) all increase the likelihood of firm formation.

Some researchers might argue that the radicalness of an invention should have a curvilinear relationship with the probability of firm formation. Initially, the argument goes, the probability that an invention will be commercialized through firm formation increases as radicalness rises, but beyond a certain level, it decreases. In unreported regressions, I tested for this curvilinear relationship. However, the results supported the linear relationship reported here.

Researchers might also argue that people will be more likely to forms firms to exploit radical patents only if the inventions have significant economic value. In unreported regressions, I also tested the interaction between radicalness and importance. However, the results did not support the interaction hypothesis.

Finally, researchers might argue that the correct measure of time to predict firm formation is the year of patent issue rather than the year of patent filing because patents are granted to new companies only after issue. I tested this alternative specification of time. I found substantively the same results as I found for the date of patent filing, indicating that the time measure is robust.

[10] The number of case-years for this regression is 8,420 because the relative importance measure cannot be calculated for all of the patents.

Table 3 Bivariate Correlations For The Independent Variables ($N = 9002$)

Variable	1	2	3	4	5	6	7	8	9	10	11	12	13	14	15	16	17	18	19
1. Radicalness	1.00																		
2. Importance	0.04	1.00																	
3. Patent Scope	0.03	0.05	1.00																
4. Year Filed	0.32	−0.23	−0.00	1.00															
5. Previous Patents	0.06	0.05	−0.04	0.17	1.00														
6. Founding Experience	0.02	−0.05	−0.05	0.19	0.56	1.00													
7. Manufacturing	0.04	−0.01	0.02	0.09	0.03	0.07	1.00												
8. Sales/Service	−0.04	−0.03	−0.09	0.01	−0.02	−0.05	−0.25	1.00											
9. Class Age	−0.08	−0.10	−0.06	−0.03	−0.14	−0.12	−0.05	0.14	1.00										
10. Venture Capital	0.13	0.02	0.01	0.02	0.04	−0.01	0.00	−0.08	0.04	1.00									
11. R&D Intensity	0.06	0.01	0.15	0.03	0.05	0.13	0.20	−0.45	−0.21	0.16	1.00								
12. Concentration	−0.02	−0.00	−0.06	0.03	−0.10	−0.07	−0.05	0.11	−0.03	0.02	0.26	1.00							
13. Firm Size	−0.15	−0.02	−0.00	−0.11	0.10	0.03	−0.16	0.28	−0.03	0.02	0.05	0.28	1.00						
14. Segmentation	0.15	0.14	−0.01	0.08	0.05	0.07	0.10	0.19	−0.07	0.29	0.24	−0.00	0.02	1.00					
15. Incumbent Innov.	−0.05	−0.01	0.12	0.03	0.20	0.15	0.01	−0.17	−0.25	−0.09	0.47	0.07	0.32	−0.00	1.00				
16. Chemical	0.11	0.06	0.13	0.01	0.10	−0.08	−0.03	−0.02	−0.03	0.00	−0.28	−0.28	−0.03	0.02	−0.18	1.00			
17. Electrical	−0.01	0.03	−0.14	0.10	−0.13	−0.07	−0.07	0.22	0.11	0.09	0.13	0.45	0.09	0.11	−0.02	−0.56	1.00		
18. Mechanical	−0.02	−0.05	−0.03	−0.10	−0.14	−0.06	0.02	−0.01	0.17	−0.06	−0.20	−0.10	−0.27	−0.15	−0.24	−0.18	−0.18	1.00	
19. Drugs	−0.15	−0.09	0.13	−0.08	0.16	0.21	0.02	−0.25	−0.18	−0.12	0.37	−0.17	0.13	−0.14	0.53	−0.30	−0.30	−0.10	1.00

Discussion

This article examined the effect of technological opportunities on the probability that they would be commercialized through the creation of new firms. Using a unique dataset which examines the population of inventions assigned to the Massachusetts Institute of Technology during the 1980–1996 period, I have shown that three dimensions of technological opportunity influence the probability of firm formation. Controlling for who invented the technology, the characteristics of the industry, the time period when the technological development took place, and the nature of the technology, more important inventions, more radical inventions, and inventions with a broader scope of patent protection were more likely to be commercialized through the creation of new firms.

Limitations

This study is not without limitations. First, it uses proxy variables to measure several of the constructs. Although previous research and field interviews supported the construct validity of the measures used, future research could corroborate the findings of this study by demonstrating similar results through the use of alternative measures.

Second, the results of this study may not be generalizable outside the context of university inventions. While I have no a priori reason to expect that the results here would fail to apply to other settings, the empirical examination looked only at one setting. Consequently, future research is necessary to demonstrate the generalizability of the results.

Third, sampling solely from the Massachusetts Institute of Technology during the 1980–1996 period might generate results that cannot be generalized to other universities at other points in time. Location advantages, ties to the venture capital community, and intellectual leadership in many technical fields might enhance the rate at which MIT inventions lead to firm formation.[11] Therefore, the results of this study

[11] However, some rough estimates suggest that the Massachusetts Institute of Technology is representative of the population of universities in terms of start-up activity. The USPTO calculates that between 1980 and 1996, U.S. colleges and universities were issued 17,647 patents, of which MIT received 1,397, or 7.92 percent. The Association of University Technology Managers (AUTM 1996) estimates that between 1980 and 1996, 1,444 new companies were founded to exploit patented inventions to which U.S. universities and colleges were assignees. During that same period, MIT reported to AUTM that 105 new companies were founded to exploit patented inventions to which MIT was the assignee. Thus, MIT

Table 4 Cox Regressions Predicting New Firm Formation: 1980–1996

Variable	Model 1 Hazard Ratio	Model 2 Hazard Ratio	Model 3 Hazard Ratio	Model 4 Hazard Ratio	Model 5 Hazard Ratio	Model 6 Hazard Ratio
Technical Field						
Chemical	0.59***	0.59**	0.70*	0.66*	0.65*	0.67*
Electrical	0.40****	0.40****	0.46****	0.50****	0.55*	0.55***
Mechanical	0.56*	0.57†	0.71	0.73	0.51	0.91
Drugs	1.38*	1.47†	1.00	1.15	1.15	1.09
Time						
Year Filed	1.08****	1.08****	1.06****	1.06****	1.05**	1.07****
Industry						
Age of Technology Class (years)	#	0.99	1.00	1.00	1.00	#
Average Firm Size	#	0.99*	0.99*	0.99†	0.99*	#
4-Firm Concentration Ratio (%)	#	1.01	1.01†	1.01†	1.01	#
Venture Capital (% of market)	#	0.96	0.97	0.96	0.95	#
R&D Intensity (% of sales)	#	1.02	1.01	1.00	1.01	#
Manufacturing Share of Value-added	#	1.03	0.65	0.69	0.64	#
Importance of Sales or Service	#	1.08	0.97	0.95	1.05	#
Incumbent Innovation	#	1.09	1.03	1.02	0.94	#
Market Segmentation	#	1.60***	1.37*	1.26	1.17	#
Inventor Attributes						
Log Number of Previous Patents	#	#	1.24	1.03	1.25	1.00
Log Founding Experience	#	#	6.19****	9.19****	7.76****	8.02****
Technology Opportunity						
Radicalness	#	#	#	1.03**	1.04**	1.03**
Importance	#	#	#	1.02****	1.03****	1.02****
Patent Scope	#	#	#	1.18*	1.26*	1.16*
N	9002	9002	9002	9002	8420	1397
Log Likelihood	−2497.33	−2479.14	−2396.99	−2381.47	−1718.44	4782.92
Chi-square	112.04****	135.40****	299.70****	330.16****	251.91****	436.17****

Key: **** = $p < 0.0001$; *** = $p < 0.001$; ** = $p < 0.01$; * = $p < 0.05$; † = $p < 0.10$; # = not included; two-tailed tests. 9,002 case-years and 364 events.

should not be generalized to other universities until future research shows their validity in other university settings.

Fourth, this study cannot determine if the inventions were developed to allow inventors to found firms or the founding of firms resulted from the development of the inventions. Therefore, the findings presented here cannot resolve the debate over whether the existence of technological opportunities leads to firm creation or the need to create firms leads to the creation of technological opportunities.

represents 7.29 percent of the new companies founded to exploit inventions to which U.S. universities and colleges were assignees, in line with its proportion of university inventions.

Implications for Research

The major implication of this study is straightforward. The probability that an invention will be commercialized through the creation of a new firm varies with the nature of the technological opportunity discovered. This finding enhances the literature on firm formation by adding the nature of the technological discovery to previous findings about the effect of industry and individual attributes.

The specific findings also have implications for related and future research. This study provides empirical evidence that technologically more important inventions are more likely to be commercialized through the creation of new firms. This finding is important because previous research has shown that patent importance influences several aspects of technology commercialization and economic growth, but has not considered the possibility that patent importance would also influence the decision to establish a new company. In particular, previous research has explored the relationship between patent importance and knowledge spillovers (Jaffe et al. 1993) and between knowledge spillovers and economic growth (Romer 1990). The fact that more important technological discoveries are more likely to be exploited through the creation of new firms suggests that students of technological change should examine the relationship between firm creation, knowledge spillovers, and economic growth. While this relationship is precisely the core of Schumpeter's (1934) explanation for a capitalist system, the subsequent empirical and theoretical literature is resoundingly silent on this central question.

In addition, inventions are more likely to be commercialized through the creation of new organizations when the scope of patent rights is broader, enabling the inventor to protect the development against imitation from a broader range of competing technologies. This result demonstrates that inventors are influenced by the breadth of intellectual property protection in making the decision of whether or not to start a firm to commercialize their inventions. This finding is important because it extends research about the effect of patent scope (Lerner 1994) to a broader context than biotechnology and to broader questions about firm formation than just venture capital financing.

Finally, the finding that inventions are more likely to be commercialized through the creation of new firms when the inventions are technologically more radical extends the work of strategic management researchers (Tushman and Anderson 1986) and students of technological change (Henderson 1993). These researchers have argued that new entrants are more likely than incumbents to commercialize radical technologies. This study shows that the radicalness of technology also influences the decisions of independent entrepreneurs to create new companies. While previous research suggested the plausibility of this argument, this study provides the first empirical evidence to support it.

The results also provide support for the position of applied entrepreneurship researchers, who have argued that firm formation from technlogical change cannot be accurately explained without consideration of the individuals who possess decision rights over inventions (Roberts 1991). These researchers have argued that one of the reasons for poor results for theories of new firm entry has been the failure to incorporate individual-level heterogenity into theories of technological change and industry entry. The results of this study provide support for a recent strand of individual-level research on entrepreneurship. Career experience influences the propensity to use firm formation as a mechanism to commercialize a new technology (Shane and Khurana 1999, Carroll and Mosakowski 1987). These results suggest that future empirical research on technical change should incorporate the role of the entrepreneur.

Normative Implications

This study examined firm formation in the context of university technologies and provides some useful implications for the management of that process. First, universities earn revenues from the commercialization of technology and therefore adopt policies to enhance it. Many university policies regarding start-up organizations differ from those regarding established organizations. For example, some universities are willing to take equity positions in start-up organizations to commercialize university inventions; and licensing officers play an important role in brokering relationships between faculty entrepreneurs and the venture capital community (Shane and Cable 1998). Consequently, knowledge of which inventions are more likely to be commercialized through the creation of new organizations may prove useful in determining university policies toward firm formation.

Second, from a management science perspective, an understanding of which inventions are commercialized through the creation of new firms is also important to potential entrepreneurs. The creation of a new firm to commercialize a university invention is a costly exercise that requires significant investment of time and capital. Consequently, an understanding of which technology opportunities are most often commercialized through firm creation is useful to university researchers who have to make a decision about how to commercialize their inventions.

Conclusion

This paper showed that three attributes of technology—importance, radicalness, and patent scope—influence the probability that an invention will be exploited through the creation of a new firm. This finding generates several important implications for development of theory about, and the practice of, entrepreneurship. Hopefully, future researchers will consider the attributes of technological opportunities presented here to generate a robust explanation for technology entrepreneurship.

Acknowledgments

The author would like to thank Don Kaiser, Lita Nelsen, and Lori Pressman at the MIT Technology Licensing Office for access to the data on MIT patents and for answering many questions about the data and TLO policies and procedures. He would also like to thank Adam Jaffe for providing the calculations of importance for the patents and Alvin Klevorick and Richard Nelson for the Yale data on appropriability. Pierre Azoulay, David Hsu, Adam Jaffe, Josh Lerner, Lita Nelsen, Richard Nelson, Lori Pressman, Ken Smith, Scott Stern, and Arvids Ziedonis provided helpful comments.

References

Acs, Z., D. Audretsch. 1990. *Innovation and Small Firms*. MIT Press, Cambridge, MA.

Amit, R., E. Mueller, I. Cockburn. 1995. Opportunity costs and entrepreneurial activity. *J. Bus. Venturing* **10**(2) 95–106.

Arrow, K. 1962. Economic welfare and the allocation of resources for inventions. R. Nelson, ed. *The Rate and Direction of Inventive Activity*. Princeton University Press, Princeton, NJ.

Audretsch, D. 1995. *Innovation and Industry Evolution*. MIT Press, Cambridge, MA.

AUTM. 1996. *The AUTM Licensing Survey*. Association of University Technology Managers, Norwalk, CT.

Begley, T., D. Boyd. 1987. Psychological characteristics associated with performance in entrepreneurial firms and smaller businesses. *J. Bus. Venturing* **2** 79–93.

Busenitz, L., J. Barney. 1997. Differences between entrepreneurs and managers in large organizations: Biases and heuristics in strategic decision-making. *J. Bus. Venturing* **12**(1) 9–30.

Caballero, R., A. Jafffe. 1993. How high are the giant's shoulders?: An empirical assessment of knowledge spillovers and creative destruction in a model of economic growth. O. Blanchard, S. Fischer, eds., *NBER Macroeconomics Annual*, Vol. 8. MIT Press, Cambridge, MA.

Carroll, G., E. Mosakowski. 1987. The career dynamics of self-employment. *Admin. Sci. Quart.* **32** 570–589.

Casson, M. 1982. *The Entrepreneur*. Barnes and Noble Books, Totowa, NJ.

Caves, R. 1998. Industrial organization and new findings on the turnover and mobility of firms. *J. Econ. Literature* **36** 1947–1982.

Chen, C., P. Greene, A. Crick. 1998. Does entrepreneurial self-efficacy distinguish entrepreneurs from managers? *J. Bus. Venturing* **13** 295–316.

Christiansen, C., J. Bower. 1996. Customer power, strategic investment, and the failure of leading firms. *Strategic Management J.* **17** 197–218.

Cohen, W., R. Levin. 1989. Empirical studies of innovation and market structure. R. Schmalensee, R. Willig, eds. *Handbook of Industrial Organization*, II. Elsevier, New York.

Evans, D., L. Leighton. 1989. Some empirical aspects of entrepreneurship. *Amer. Econom. Rev.* **79**(3) 519–535.

Gartner, W., S. Shane. 1995. Measuring entrepreneurship over time. *J. Bus. Venturing* **10** 283–301.

Geroski, P. 1995. What do we know about entry? *Internat. J. Indust. Organ.* **13** 421–440.

Hall, B., A. Jaffe, M. Trajtenberg. 1998. Patent citations and market value: A first look. Paper presented at the National Bureau of Economic Research.

Hannan, M., J. Freeman. 1984. Structural inertia and organizational change. *Amer. Soc. Rev.* **49** 149–164.

Harhoff, D., F. Narin, F. Scherer, K. Vopel. 1999. Citation frequency and the value of patented inventions. *Rev. Econom. and Statist.* **81**(3) 511–515.

Henderson, R. 1993. Underinvestment and incompetence as responses to radical innovation: Evidence from the photolithographic alignment equipment industry. *RAND J. Econom.* **24**(2) 248–270.

———, A. Jaffe, M. Trajtenberg. 1998. Universities as a source of commercial technology: A detailed analysis of university patenting, 1965–1988. *Rev. Econom. and Statist.* **80** 119–127.

Jaffe, A., M. Fogarty, B. Banks. 1997. Evidence from patents and patent citations on the impact of NASA and other federal labs on commercial innovation. NBER Working paper 6044.

———, M. Trajtenberg. 1998. International knowledge flows: Evidence from patent citations. NBER Working paper 6507.

———, ———, R. Henderson. 1993. Geographic localization of knowledge spillovers as evidenced by patent citations. *Quart. J. Econom.* **108** 557–586.

Kaish, S., B. Gilad. 1991. Characteristics of opportunities search of entrepreneurs versus executives: Sources, interests, and general alertness. *J. Bus. Venturing* **6** 45–61.

Kamien, M., N. Schwartz. 1975. Market structure and innovation: A survey. *J. Econom. Literature* **13** 1–37.

Khilstrom, R., J. Laffont 1979. A general equilibrium entrepreneurial theory of firm formation based on risk aversion. *J. Political Economy* **87**(4) 719–748.

Knight, F. 1921. *Risk, Uncertainty and Profit*. Augustus Kelley, New York.

Lerner, J. 1994. The importance of patent scope: An empirical analysis. *RAND J. Econom.* **25**(2) 319–333.

Levin, R., A. Klevorick, R. Nelson, S. Winter. 1987. Appropriating the returns from industrial R&D. *Brookings Papers on Econom. Activity* **3** 783–831.

Low, M., I. MacMillan. 1988. Entrepreneurship: Past research and future challenges. *J. Management* **14**(2) 139–161.

Mansfield, E. 1981. Composition of R and D expenditures: The relationship to size of firm, concentration, and innovative output. *Rev. Econom. and Statist.* **63** 610–615.

Merges, R., R. Nelson. 1990. On the complex economics of patent scope. *Columbia Law Rev.* **90**(4) 839–916.

Nelson, R. 1995. Recent evolutionary theorizing about economic change. *J. Econom. Literature* **33** 48–90.

———, S. Winter. 1982. *An Evolutionary Theory of Economic Change.* Belknap Press, Cambridge, MA.

Pakes, A. 1986. Patents as options: Some estimates of the value of holding European patent stocks. *Econometrica* **54**(4) 755–784.

Pavitt, K., S. Wald. 1971. *The Conditions For Sucess In Technological Innovation.* OECD, Paris, France.

Podolny, J., A. Shephard. 1997. When are technological spillovers local?: Patent citation patterns in the semiconductur industry. Working paper, Stanford University, Palo Alto, CA.

———, T. Stuart. 1995. A role-based ecology of technological change. *Amer. J. Soc.* **100**(5) 1224–1260.

Reinganum, J. 1983. Uncertain innovation in the persistence of monopoly. *Amer. Econom. Rev.* **73** 741–748.

Roberts, E. 1991. *Entrepreneurs in High Technology: Lessons from MIT and Beyond.* Oxford University Press, New York.

Romer, P. 1990. Endogenous technological change. *J. Political Economy* **98**(5) S71–S101.

Rosenbloom, R., C. Christiansen. 1994. Technological discontinuities, organizational capabilities, and strategic commitments. *Indust. and Corporate Change* **3** 655–685.

Rosenkopf, L., A. Nerkar. 1999. Beyond local search: Boundary spanning, exploration and impact in the optical disk industry. Working paper, University of Pennsylvania, Philadelphia, PA.

Scherer, F. 1965. Firm size, market structure, opportunity, and the output of patented inventions. *Amer. Econom. Rev.* **55** 1097–1125.

———, 1988. Testimony before the Subcommittee on Monopolies and Commercial Law, Committee of the Judiciary, U.S. House of Representatives (February 24).

Schumpeter, J. 1934. *Theory of Economic Development.* Harper and Row, New York.

Shackle, G.S. 1979. *Imagination and the Nature of Choice.* Edinburgh University Press, Edinburgh, UK.

Shane, S. 1996. Explaining variation in rates of entrepreneurship in the United States: 1899–1988. *J. Management.* **22**(5) 747–781.

———, D. Cable. 1998. Social relationships and the financing of new ventures. Working paper, Massachusetts Institute of Technology, Cambridge, MA.

———, R. Khurana. 1999. Career experience and firm founding. Working paper, Massachusetts Institute of Technology, Cambridge, MA.

Teece, D. 1987. Profiting from technological innovation: Implications for integration, collaboration, licensing, and public policy. D. Teele, ed. *The Competitive Challenge.* Ballinger, Cambridge, MA.

Trajtenberg, M. 1990. A penny for your quotes: Patent citations and the value of innovations. *Rand. J. Econom.* **21** 172–187.

———, R. Henderson, A. Jaffe. 1997. University versus corporate patents. A window on the basicness of invention. *Econom. Innovation and New Tech.* **5** 19–50.

Tushman, M., P. Anderson. 1986. Technological discontinuities and organizational environments. *Admin. Sci. Quart.* **31** 439–465.

Utterback, J., W. Abernathy. 1975. A dynamic model of product and process innovation. *Omega* **3** 639–656.

Venkataraman, S. 1997. The distinctive domain of entrepreneurship research: An editor's perspective. J. Katz, R. Brockhaus, eds. *Advances in Entrepreneurship, Firm Emergence, and Growth,* Vol. 3. JAI Press, Greenwich, CT, 119–138.

Weiss, L. 1976. Optimal plant scale and the extent of suboptimal capacity. R. Masson, P. Qualls, eds. *Essays on Industrial Organization in Honor of Joe S. Bain.* Ballinger, Cambridge, MA, 126–134.

Winter, S. 1984. Schumpeterian comeptition in alternative technological regimes. *J. Econom. Behaviour and Organ.* **4** 287–320.

Accepted by Ralph Katz; received September, 1999. This paper was with the author for 3 months for 2 revisions.

[24]

New Technology and The Small Firm*

Boyan Jovanovic

ABSTRACT. This paper examines how and why the role of small business has become more important over time. A special focus is placed on the role of small business in the New Economy.

1. Wealth and the small firm

How much of our wealth is created by small firms? The question is somewhat misleading because many of today's giant firms did not exist thirty years ago. Consider Table I, which reports the twenty largest companies by market capitalization, as of the end of August 1999, and the dates when they were incorporated. Some like Lucent are spun off from an old company, while others like Citigroup comprise companies that were founded at much earlier dates than the column reports. But four of these giants – Microsoft, Cisco Systems, MCI and Dell – are clearly less than twenty years of age, and their value adds up to 12 or 13 percent of our gross domestic product.

The broader picture emerges when we look at a larger sample of companies. Most stock-market incumbents are identifiable from the Center for Research in Security Prices (CRSP) data. These data can identify who, precisely created the additional wealth? To a large measure, the once-small firms in the NASDAQ replaced the 1968 incumbents. These firms represent a particularly high-tech subset of the post-1968 entrants. To get a broader representation of the fate of incumbents, we now present plots of market shares of various generations of incumbents, which we endearingly call "dinosaurs" because of their tendency to atrophy with the passage of time. One way to see the process of decay is to look at the share of market value, in December1996, of each vintage of CRSP entrant, starting in 1928 when the observations start.

The vertical axis measures the share that each vintage holds in total stock-market capitalization in 1996. The large spike in 1928 that we see in Figure 1 is the value of companies that were started before that date. From Table I we see that GE, IBM, Exxon, AT&T, etc., are all in this group of firms. The spikes reflect the intensity of investment in the year in which a particular "vintage" of firms entered the stock market, as well the survivorship of that vintage of capital. We see, for instance, that not much value from the World War II vintages survives today, but that is probably because very few firms entered during the wartime years. Similar remarks apply to the years of the great depression. Moreover, two of the spikes are artificial because the 1961 spike reflects largely the entry of the AMEX firms into the CRSP, and 1971 spike reflects the entry of the NASDAQ firms. Once these anomalies are recognized, we see a definite drop-off in the contribution of the vintage of a firm as the age of the capital rises.

2. Small firms versus large over the past seventy years

One barometer of how well small business does is to compare the stock market values of small firms with those of large firms. What I do here is report the relative performance of a broad-based small company index against that of the S&P 500 index over the seventy-year period 1926–1996.[1] The figure shows two striking things. The first of these is well known – small companies grow faster

Final version accepted on September 15, 2000

University of Chicago
1126 East 59th St.
Chicago, IL 60637
U.S.A.
E-mail: bjovanov@uchicago.edu

Small Business Economics **16**: 53–55, 2001.

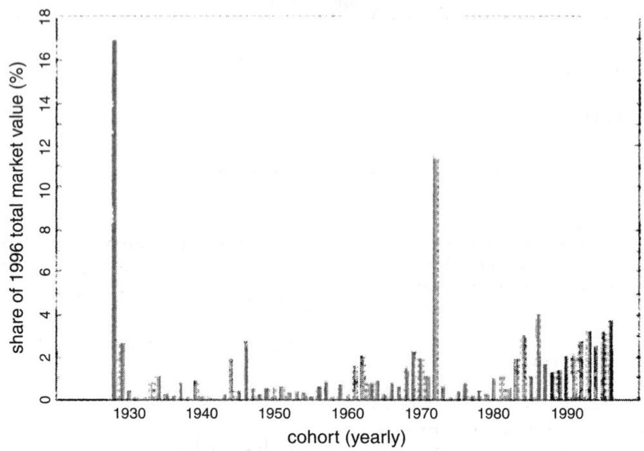

Figure 1. 1996 capitalization shares by cohort.

in value than large companies. The difference in trend is about 1% per year. This is a risk premium, but it is nowhere near as large as the premium that both indexes would command over bonds and treasury bills.

The second striking thing we see is rather larger waves or departures from trend. The arrival of the microprocessor in the early 1970s seemed, for about ten years or so, to favor the small firm. This is what one would expect, in as much as the arrival of any major new technology should favor young, small firms – firms that tend to have a nonhierarchical structure, fewer unionized or tenured workers, and fewer outdated management practices.

Figure 2 plots the ratio of the small cap index to the S&P 500 index. During the period 1974–1982, small cap stocks outperformed the S&P 500 by a factor of nearly 4. Since then, the S&P 500 has done better than the small caps, probably because, by the early 1980s and the advent of the junk-bond, inefficient large firms began to feel stronger hostile-takeover pressures, and responded by becoming more efficient.

3. What is new in the "New Economy"

The new economy is one in which technologies and products become obsolete at a much faster rate than a few decades ago. Recently, this is the result of the information age, but it is merely a continuation of a trend that we have been witnessing for at least the past century. Shrinking imitation lags: A startling finding by Agarwal and Gort is that imitation lags have shrunk. Response times are shrinking. That is, new products are produced faster than before, and new technologies are

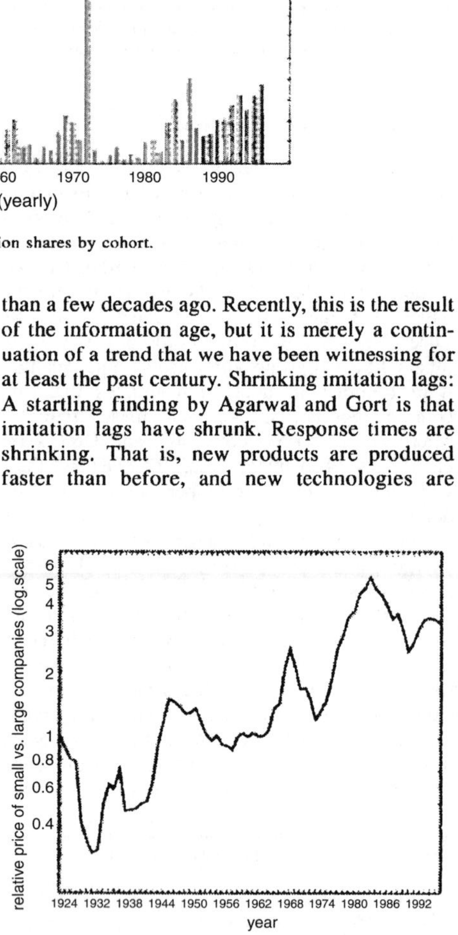

Figure 2. Performance of small companies vs. large companies.

TABLE I
The twenty largest companies by market capitalization, August 29, 1999, and the dates when they were incorporated

Largest Dow companies	Market value* $ billion	Year incorporated
General Electric	389	1892
IBM	223	1911
Wal-Mart	211	1969
Exxon	195	1882
Merck	162	1934
Citigroup	161	1968**
AT&T	159	1885
Coca-Cola	152	1919
Johnson & Johnson	140	1887
Procter & Gamble	133	1905

Largest non-Dow companies	Market value* $ billion	Year incorporated
Microsoft	483	1981
Intel	269	1968
Cisco	220	1984
Lucent	202	1995**
Pfizer	154	1942
AIG	153	1967**
MCI	146	1983
Bristol Myers Squibb	145	1933
Dell Computer	122	1987
Bank of America	111	1968**

Notes:
* In billions.
** Predecessor companies were founded at significantly earlier dates.
Source: Wall Street Journal August 30, 1999.

adopted faster. Companies are now going public on the internet – a practice unheard of before 1995. That is, the "speed of diffusion" has risen sharply over the last century. But the same is true about the IT revolution. The early companies – Microsoft, for instance – took a long time to IPO, but now the process is faster, partly because the technology is mature.

It is clear that we are entering the era of the young firm. The average age of all companies in the stock market – and the average age of the giants – is shrinking as well. Since the death rate of old capital is higher, this means that the "birth rate" of new capital must rise if we are to maintain the stock of existing capital at existing levels. The small firm will thus resume a role that, in its importance, is greater than it has been at any time in the last seventy years or so.

Notes

* Paper presented at the Conference on Research Issues in Industrial Organization and Small Business, January 21, 2000, Washington DC Office of Advocacy Small Business Administration.

[1] The data come from Ibbotson et al. (1997). More precisely, the data underlying the plot in Figure 1 are as follows: 1. Large company stocks: S&P 500 index with dividends reinvested (S&P 500, 1957-present, S&P 90, 1926–1956). 2. Small Company stocks: Firth capitalization quintile of stock on the NYSE for 1926–1981. Performance of the Dimensional Fund Advisors (DFA) Small Company Fund, 1982-present. [This is also a total return index, i.e. dividends (which are probably insignificant anyway for small firms) are also assumed to be reinvested.]

References

Agarwal, Rajshree and Michael Gort, 1998, 'First Mover Advantage and the Speed of Competitive Entry 1887–1983', Mimeo, University of Central Florida.

Greenwood, Jeremy and Boyan Jovanovic, May 1999, 'The Information – Technology Revolution and the Stock Market', *American Economic Association* (Papers and Proceedings) **89**(2), 116–122.

Hobijn, Bart and Boyan Jovanovic, August 1999, 'The Information – Technology Revolution and the Stock Market: Evidence', unpublished manuscript.

Ibbotson, Roger et al., 1999, *Stocks, Bonds, Bills, and Inflation*, 1997 Yearbook. Chicago.

[25]

INTERNATIONAL REGIONAL SCIENCE REVIEW 25, 1: 132–148 (January 2002)

GEOGRAPHY, ENDOGENOUS GROWTH, AND INNOVATION

ZOLTAN J. ACS
University of Baltimore, zacs@ubmail.ubalt.edu

ATTILA VARGA
University of Pécs, Hungary, vargaa@ktk.pte.hu

If one is to understand why some regions grow and others stagnate, there are three fundamental questions that need to be answered. First, Why and when does economic activity become concentrated in a few regions, leaving others relatively underdeveloped? Second, What role does technological change play in regional economic growth? Third, How does technological advance occur, and what are the key processes and institutions involved? To answer these three questions, the authors surveyed three separate and distinct literatures that have a long and distinguished history, and all three have been recently reexamined. They include the new economic geography (Krugman), the new growth theory (Romer), and the new economics of innovation (Nelson).

1. INTRODUCTION

For the past two decades, scientists, policy makers, and the general public have been fascinated by developments in the Silicon Valley. The developments we have in mind are the ubiquitous creation of new product innovations in this part of the world. The Valley has had an unprecedented record of success in the introduction of computers, software, semiconductors, biotechnology, and a host of other innovations that have come to dominate industry after industry throughout the world. Of course, this is not the only region of the world that has experienced an outflow of innovations. But the truth of the matter is that regions that are innovative appear to grow faster than regions that are not (see, e.g., Suarez-Villa 2000).

In other words, there appears to be a growing consensus that innovation is the key driving force behind economic growth, standards of living, international competitiveness, and regional development. Recently, several books have appeared that try to identify the underlying processes and interconnections that govern regional innovation (Braczyk, Cooke, Heidenreich 1998; de la Mothe and Pacquet 1998;

The idea of this article originated out of a series of discussions we had during the fall of 1998. The article was outlined while Zoltan J. Acs visited at the Vienna University of Economics and Business Administration in the summer of 1999. We would like to thank seminar participants at the International Regional Science Association meetings in Montreal, Canada, in November 1999 and two anonymous referees for valuable suggestions on how to revise the article.

Ratti, Bramanti, and Gordon 1997; De Bresson 1996; Acs 2000). While these books take different approaches, rely on different methodologies, use different data, and define the unit of analysis differently, they all suggest that there is something fundamental at work at the regional level. While these works are all interesting, illuminating pieces of the regional innovation puzzle, neither singularly nor in concert answer the bigger question of why some regions are more innovative than others.[1]

If we are to understand why some regions grow and others stagnate, there are three fundamental questions that need to be answered. First, Why and when does economic activity become concentrated in a few regions, leaving others relatively underdeveloped? Second, What role does technological change play in regional economic growth? Third, How does technological advance occur, and what are the key processes and institutions involved? To answer these three questions, we surveyed three separate and distinct literatures that have a long and distinguished history, and all three have been recently reexamined. They include the new economic geography (Krugman 1991a), the new growth theory (Romer 1990), and the new economics of innovation (Nelson 1993).

The purpose of this article is to survey these three literatures to help us better understand the relationship between economic geography, economic growth, and innovation. Why these three literatures, you might ask? The reason for choosing these three literatures is that while each one sheds some light on the relevant question, neither completely answers the larger question of divergent regional growth. The new economic geography answers the question why economic activity concentrates in certain regions but not others but leaves out innovation and economic growth. The new growth theory explains the causes of economic growth but leaves out regional considerations and ignores the key processes and institutions involved in innovation. The new economics of innovation, while explaining the institutional arrangements in the innovation process, leaves out regional issues and economic growth.

What we are looking for from this new and evolving literature are insights that will help us develop a clear analytical framework that integrates economic growth, spatial interdependencies, and the creation of new technology as an explicit production process to formulate production-oriented regional policies (Nijkamp and Poot 1997). In the next section, we examine the new economic geography; in the third section, we survey the endogenous growth theory, while the fourth section examines systems of innovation. We finish with an outline for a new model of technology-led regional economic development.

2. Geography and Economic Theory

Despite that space-related economic problems such as firm location or externalities resulting from agglomeration are of prominent theoretical and practical importance, economic geography has been a neglected field of mainstream economics for a long time. Ignorance of spatial issues is even more astonishing if one considers that major accomplishments in the history of spatial economics have approached geographical issues within generally accepted frameworks of economics (Blaug 1979; Greenhut and Norman 1995; Krugman 1991a, 1995). Johann von Thünen, the father of location theory, made constant use of algebra and differential calculus long before mathematics had become the primary language of economic theory (Blaug 1979, 1992). Alfred Weber (1929) created a location theory within a comparative static microeconomics environment. Likewise, recent efforts in regional science (Isard 1956) and the new urban economics (Beckmann 1969) to bring spatial issues into the framework of mainstream economics have also failed despite their purely neoclassical framework (Krugman 1991a).

One possible reason for the negligence of geographic issues could be that from the very beginning, neoclassical economics has a nonspatial framework for analyzing economic problems. Introducing space into microeconomics can result in substantial changes in well-established theorems (Ohta 1988). Of course, reconstruction of such a huge intellectual framework is always very expensive, and those costs are rarely incurred. Krugman (1991a, 1995) provided an alternative explanation. According to this, a major analytical problem with previous work in economic geography is that issues related to geography were treated within the perfectly competitive neoclassical framework. As a consequence, spatial economic problems have not attracted the attention of mainstream economists until the recent development of the analytics of imperfect competition.

Instead of an in-depth survey of the extensive works of Paul Krugman (for such surveys, see Isserman 1996; Martin and Sunley 1996; and Martin 1999), or the dynamically evolving field of the new economic geography, this section provides a concise description of Krugman's basic theory on economic concentration to assess its potential for modeling technology-led economic growth.

Krugman, a prominent international trade theorist, had turned to economic geography issues after recognizing that regions and not countries are the real units of economic analysis (Krugman 1991a, 1993c). Economic activities are not evenly distributed across space. An issue having central importance in economics is the explanation of why economic activities concentrate in certain areas while others remain relatively underdeveloped. In his basic theory (Krugman 1991a, 1991b) and its subsequent extensions to specific issues (Krugman 1993a, 1993b, 1996; Fujita, Krugman, and Venables 1999), a general equilibrium model was developed that explains spatial concentration of economic activities with the interrelations of three parameters: increasing returns, transport cost, and demand for manufacturing goods. According to the model, each of these parameters should pass certain threshold values before any kind of geographic concentration emerges. Similar to models in the new trade theory and the new (endogenous) growth theory, it builds extensively on the monopolistic competition model of Dixit and Stiglitz (1977).

An important prerequisite for any kind of spatial concentration to occur is the presence of some kind of increasing returns to scale in manufacturing production.

However, without an appropriately low level of transport costs, economies of scale cannot facilitate concentration: production remains dispersed in the proximity of their markets. When suitable levels of transport costs and increasing returns are attained, manufacturing producers start concentrating in areas where the initial demand for their goods is relatively high to save at least on a portion of transportation expenditures. Demand and supply externalities ("backward and forward linkages" in Hirschman 1958) have important roles in facilitating further concentration: manufacturers locate where demand is high and the supply of their inputs is available. However, demand and supply externalities induce a "cumulative causation" (Myrdal 1957) process: location of new firms reinforces these externalities and will attract further manufacturers to the region.

The following briefly summarize the basic two-region model in Krugman (1991a, 1991b).

Individuals share the Cobb-Douglas type utility function of

$$U = C_M^{\pi} C_A^{1-\pi}, \tag{2.1}$$

where C_M is consumption of aggregate manufacturing goods, C_A is agricultural consumption, and π is the share of expenditures spent on manufacturing goods.

The aggregate of manufacturing consumption is determined by the constant elasticity of substitution function of

$$C_M = \left[\sum_i C_i^{(\sigma-1)/\sigma}\right]^{\sigma/(\sigma-1)}, \tag{2.2}$$

where $\sigma > 1$ is the elasticity of substitution among manufacturing products. When i is sufficiently large, σ represents demand elasticity for manufacturing good i (Krugman 1980).

Agricultural production is assumed to have constant returns to scale technology, whereas manufacturing follows increasing returns. Costs in labor are determined by

$$L_{Mi} = \alpha + \beta x_{Mi}, \tag{2.3}$$

where i stands for individual firms, α is the fixed cost parameter, β is the parameter of marginal costs, and x_i is the good's output.

In equilibrium, price of good i is determined in monopolistic competition as

$$P_i = [\sigma/(\sigma - 1)]\, \beta w, \tag{2.4}$$

which implies that (since $P_i = AC_i$ and βw is marginal cost) AC/MC (as a measure of firm size) is represented by $\sigma/(\sigma - 1)$. It follows that σ is an inverse index of the importance of increasing returns in the model.

After defining two of the crucial parameters (i.e., π and σ) of the model, transport cost is defined in the manner of Samuelson's iceberg cost formulation: as distance increases, the portion of the good delivered gets smaller. The fraction that arrives is denoted τ.

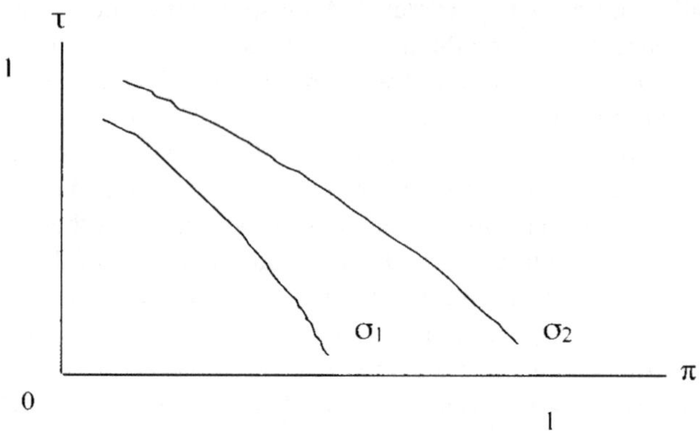

FIGURE 1. The Combination of Parameters Leading to Geographical Concentration
Source: Krugman (1991b, 498).

The basic model assumes that there are two regions: East, dominated by manufacturing, and West, a perfectly agricultural area. π is the portion of manufacturing workers in East, whereas agricultural workers are evenly distributed in space with $(1 - \pi)/2$ in each region.

Instead of modeling the emergence of the two regions within some complicated dynamic settings, prerequisites for a representative manufacturing firm to relocate from East to West are searched for. This approach is based on the assumption that parameter conditions for leaving an established manufacturing region are the same as the conditions for locating there. K stands for the ratio of shares from a West location of a representative firm to that from an East location. When $K > 1$, the firm will relocate. It is represented in the model as follows in (2.5).

$$K = S_w \tau^{-\pi} / S_E\, 1/2\, \tau^{\pi\sigma}\left[(1+\pi)\tau^{\sigma-1} + (1-\pi)\tau^{-(\sigma-1)}\right], \tag{2.5}$$

with S_W and S_E representing the value of sales in West and East and $\tau^{-\pi}$ being the price index. Partial derivatives of K evaluated in the vicinity of $K = 1$ in (2.5) indicates the effects of parameter values on the representative firm's decisions. $\partial K / \partial \pi < 0$ shows that higher shares of expenditures on manufacturing goods urge firms to concentrate in space, whereas $\partial K / \partial \sigma > 0$ indicates that economies of scale has a positive effect on geographical concentration (i.e., the less important increasing returns are, the less strong is the incentive of firms to agglomerate). Also, $\partial K / \partial \tau < 0$ at the boundary between concentration and nonconcentration.

Evaluated at $K = 1$, equation 2.5 makes simulations of any combination of the three parameters possible as illustrated in Figure 1. The two curves associated with σ_1 and $\sigma_2 (\sigma_1 < \sigma_2)$ illustrate two boundaries between spatial concentration and dispersion of manufacturing activities. Any specific combination of π and τ represented by a point above a curve will induce concentration of manufacturing production in space. It is shown in the figure that agglomeration emerges at a certain combination of large economies of scale, low transportation cost, and a large share of manufacturing in expenditure.

The model described above can be extended to cases of more than two locations as exemplified in Krugman (1993a, 1993b) to model the emergence of city hierarchies. Novelties of Krugman's theory on economic concentration are not in its elements but in the way the system was put together. It has already been well known in economic geography and regional economics that decreasing transportation costs, economies of scale, or increasing demand favor agglomeration. Also, demand and supply externalities (Hirschman 1958) and economic growth resulting from "circular causation" (Myrdal 1957) have been standard tools in spatial sciences for a long time. However, the way Krugman put these elements together in a general equilibrium model is novel. The model provides a case for the treatment of spatial issues in a way economists are accustomed to (Fujita, Krugman, and Venables 1999): not only in that it makes simulations to study the interrelations of parameters possible but also in that it opens the way for empirical testing of the theory.

The model provides a technique to analyze geographical concentration of economic activities as being induced by some initial combinations of basic parameters. However, the model in its current form does not seem to be suitable for modeling technology-led regional economic growth at least for two reasons. First, Krugman's definite insistence of avoiding modeling the role of technological externalities in regional economic growth prevents the model from being applicable in modeling innovation-led regional development since spillovers and innovation networks are in the core of this type of development, as exemplified by the literature of innovation systems.

Second, while the model is very strong in working out the characterization of specific initial combinations of the three parameters favoring geographical concentration, it is weak in actually modeling the growth process. Cumulative causation effects of Myrdal (1957) induced by forward and backward linkages of Hirschman (1958) seem not to be sufficient enough to trace economic growth of individual regions.

3. Endogenous Technical Change

This section is not considered a detailed survey of the endogenous economic growth literature (for such surveys see, for example, Grossman and Helpman 1991; Helpman 1992; Romer 1994; Barro and Sala-i-Martin 1995; Nijkamp and Poot 1997; Aghion and Howitt 1998); it focuses instead on certain aspects having crucial importance for the subject of this article. The distinguishing feature of endogenous economic growth theory as compared to the neoclassical growth model is in its modeling of technological change as a result of profit-motivated investments in knowledge creation by private economic agents. The novel formulation of technological knowledge in economic theory in Romer (1990) is the key in establishing this new and fast-evolving field of economic growth theory. According to this formulation, technological knowledge is a nonrival, partially excludable good. Such formulation of technological knowledge as a key factor in the production function results in a departure from the constant returns to scale, perfectly competitive world of the neoclassical growth theory.

Central to the neoclassical theory of economic growth as formulated in Solow (1956) is the production function. Assuming that capital does not depreciate, labor force does not grow, and technology does not change over time (Helpman 1992), the production function has the form of

$$Y = F\,(K, L), \tag{3.1}$$

where Y represents aggregate production, K the capital stock, and L the labor force. $F(\bullet)$ is the constant returns to scale production function. It is assumed that the capital stock grows without bounds. However, the growth rate of per capita income is bounded. Growth rate of per capita income is

$$g = s\,F_K\,(K, L), \tag{3.2}$$

where g is the growth rate of per capita income, s is the savings rate, and F_K is the marginal product of capital. Equation 3.2 says that per capita income grows as long as the marginal product of capital exceeds zero. However, assuming constant growth in the capital stock, per capita income approaches zero. Relaxing the assumptions of stable labor force and no depreciation of capital does not change essentially the main point of the model. The condition for a sustained per capita income growth in the long run is that, resulting from continuous capital accumulation, the marginal product of capital should not decrease below a positive lower bound.

Development in the state of technology is an essential force to offset the effect of capital accumulation on per capita income to decline in the neoclassical model of economic growth. Introducing technological progress in the production function, it takes the form

$$Y = F\,(A, K, L), \tag{3.3}$$

where A stands for the state of technology. Assuming that A increases, it will increase the marginal product of capital, which will lead to a higher per capita income. As a result, in steady state the rate of technical development equals the rate of capital accumulation.

The essential role of technological progress in economic growth has been emphasized above. However, technological development remains unexplained in the neoclassical theory of economic growth. As a public good, it is considered exogenously determined, although (as data show in Solow 1957 and Maddison 1987) the major portion of economic growth can be attributed to technological change whereas capital accumulation (the main concern in the neoclassical model) explains only a fraction of it.

Primary attempts in the literature to endogenize technological progress include Arrow (1962) by introducing "learning by doing" in technological development, Lucas (1988) by modeling human capital as the determinant factor in technical change and Romer (1986) by explicitly including research in the production function. In Arrow's (1962) formulation,

$$Y_i = A(K)F(K_i, L_i), \quad (3.4)$$

the state of technology depends on the aggregate capital stock in the economy. Subscript i denotes individual firms. According to the Lucas' (1988) model of endogenous technological change, spillovers resulting from human capital accumulation instead of the accumulation of physical capital increase the technological level in the economy:

$$Y_i = A(H)F(K_i, L_i), \quad (3.5)$$

where H stands for the general level of human capital in the economy. In Romer (1986), it is assumed that spillovers from private research efforts lead to the development in the public stock of knowledge. It could be written as

$$Y_i = A(R)F(R_i, K_i, L_i), \quad (3.6)$$

where R_i stands for the results of private research and development efforts by firm i and R denotes the aggregate stock of research results in the economy.

As summarized in Romer (1990), the major conceptual problem with the formulation of endogenous growth in equations 3.4 through 3.6 is that in those models the entire stock of technological knowledge is considered to be public good. However, as daily evidence suggests, new technological knowledge can become partially excludable (at least for a limited length of time) by means of patenting. Consequently, each firm developing a new technological knowledge has some market power and earns monopoly profits on its discoveries. The new theory of economic growth, which follows from this first in Romer (1990), builds on a more suitable view of the available stock of technological knowledge as well as formulates the economy within the framework of imperfect competition.

Not until the formulation of monopolistic competition in Dixit and Stiglitz (1997), applied in the dynamic context by Judd (1985), did modeling economic growth within an imperfectly competitive market structure become attainable. In Romer (1990), the approach by Judd (1985) was combined with learning by doing in innovation to create the first model of endogenously determined technical change with imperfectly competing firms.

In the core of the new economic growth theory there is a concept of technological knowledge as being a nonrival and partially excludable good as opposed to the neoclassical view of knowledge as an entirely public good. According to the concept of Romer (1990), knowledge is a nonrival good because it can be used by one agent without limiting its use by others. However, at the same time it is also a partially excludable good given the possibility to prevent its use by others by means of patenting.

Knowledge enters into production in two ways. For the first one, newly developed technological knowledge is used in production by the firm invested in the development of this new set of technological knowledge to produce output. In this role, knowledge can be protected from being used by others in producing the same type of output. However, this new set of knowledge increases the total stock of publicly available knowledge as being spilled over to other researchers by way of studying its patent documentation (Romer 1990). As such, it increases the productivity of creating further inventions in the research sector. This second role of knowledge in production can be formalized as

$$dA = G(H, A), \tag{3.7}$$

where H stands for human capital used in research and development, A is the total stock of technological knowledge available at a certain point in time, and dA is the change in technological knowledge resulting in private efforts to invest in research and development. Human capital creates new knowledge, whereas at the same time, the productivity of human capital depends on the total stock of already available knowledge (A). The larger A, the higher the productivity of H, and the less expensive it is to create new technological knowledge.

A principal assumption in the theory of endogenous growth is that for creating new sets of technological knowledge, the total stock of knowledge (A in equation 3.7) is freely accessible for anyone engaged in research. However, this assumption is not verified in the growing literature of geographic knowledge spillovers (e.g., Jaffe 1989; Acs, Audretsch, and Feldman 1991,1994; Glaeser et al. 1992; Anselin, Varga, and Acs 1997, 2000; Varga 1998, 2000) and innovation systems (e.g., Saxenian 1994; Braczyk, Cooke, and Heidenreich 1998; Fischer and Varga 2001; Oinas and Malecki 1999; Sternberg 1999; Acs 2000). New technological knowledge (the most valuable type of knowledge in innovation) is usually in such a tacit form that its accessibility is bounded by geographic proximity and/or by the nature and extent of the interactions among actors of an innovation system.

Similar to the case of relaxing the neoclassical assumption of equal availability of technological opportunities in all countries of the world (Romer 1994), a relaxation of the assumption that the term A in equation 3.7 is evenly distributed across space within countries seems also to be necessary. The nonexcludable part of the total stock of knowledge seems rather to be correctly classified if it is assumed to have two portions: a perfectly accessible part consisting of already established knowledge elements (obtainable via scientific publications, patent applications, etc.) and a novel, tacit element, accessible by interactions among actors in the innovation system. While the first part is available without restrictions, accessibility of the second one is bounded by the nature of interactions among actors in a system of innovation.

4. Systems of Innovation

Models and formal theories characterize the previous two sections of this article. Systems of innovation are a new approach to the study of innovations in the economy that has emerged during the past decade. The systems approach is certainly not a formal theory. It does not provide convincing propositions as regards established and stable relations between variables. The most it does in this direction is to provide a basis for the formulation of conjectures that various factors are important for technological innovation.

The systems approach, like many other institutionally oriented approaches, is a conceptual framework that many scholars and policy makers consider useful for the analysis of innovation (Edquist 1997). Although the systems of innovation approach is not considered a formal and established theory, its development has been influenced by different theories of innovation such as interactive learning theories (Arrow 1962) and evolutionary theories (Nelson and Winter 1982).

Much work has been done in the tradition of neo-Schumpeterian evolutionary economics to advance our understanding of the micro foundations of innovation. Perhaps the touchstone volume of this tradition is Nelson and Winter's (1982) *An Evolutionary Theory of Economic Change*. In recent years, efforts have been made to percolate general theoretical and empirical observations from this literature into a conceptual framework capable of guiding policy and loosely organized around the idea of a national system.

Despite their different interpretations of innovation, all versions of the systems of innovation approach place innovation at the very center of focus. Technological innovation is a matter of producing new knowledge or combining existing knowledge in new ways and of transforming this into economically significant products and processes. Many different kinds of actors and agents in the system of innovation are involved in these processes. Such activities involve learning by doing, increasing the efficiency of production operations, learning by using, increasing the efficiency of the use of complex systems, and learning by interacting, involving users and producers in a interaction resulting in product innovations (Lundvall 1992, 9).

In the systems approach, innovation is viewed as a ubiquitous phenomenon. In all parts of the economy, and at all times, we expect to find ongoing processes of learning, searching, and exploring, which result in new products, new techniques, new forms of organization, and new markets. The first step in recognizing innovation as a ubiquitous phenomenon is to focus on its gradual and cumulative aspects. Such a perspective gives rise to simple hypotheses about the dependence of future innovation on the past. In this context, an innovation may be regarded as a new use of preexisting possibilities and components. Almost all innovations reflect existing knowledge combined in new ways.

The systems of innovation approaches can be characterized as holistic in the sense that they have the ambition to encompass a wide array of the determinants of innovation that are important. The systems approach allows for the inclusion not only of economic factors influencing innovation but also of institutional, organizational, social, and political factors. In this sense, it is an interdisciplinary approach. Elements of the systems approach, such as firms, behave differently concerning innovation in different contexts. To understand such phenomena, it is important to have a structural concept. The systems of innovation approach can fruitfully serve as such, because it can be considered to have a structural and actor-orientated approach.

The institutional setup is the second important dimension of the systems of innovation approach. One of the most striking characteristics the systems of innovation approaches have in common is their emphasis on the role of institutions. Institutions are of crucial importance for innovation processes. However, what is meant by institutions varies from author to author. The concept of institutions can be quite heterogeneous and very complex. It includes normative structures, regimes, and organizations of various kinds. For others, it includes firms and industrial research laboratories involved in innovation. It can also include supporting institutions that include research universities, government laboratories, and technology policies.

The systems of innovation of various countries can be quite different. In addition, the organizations and institutions constituting elements of the systems of innovation may be different in various countries, regions, or sectors. In the systems of innovation approaches, the differences between the various systems are stressed and focused on rather than abstracted from. This makes it not only natural but also vital to compare various systems. Without comparisons between existing systems, it is impossible to argue that one national system is specialized in one way or the other. We cannot define an optimal system of innovation because evolutionary learning processes are important in such systems, and they are thus subject to continuous change. The system never reaches equilibrium.

In the real world, the state and the public sector are rooted in national states, and national borders define their geographical sphere of influence. The focus on national systems reflects the fact that national economies differ regarding the structure of the production system and regarding the general institutional setup. Specifically, it is assumed that basic differences in historical experience, language, and culture will reflect in national idiosyncrasies in internal organization of firms, interfirm relationships, role of the public sector, institutional set up of the financial sector, and research and development organizations. The concept of a national system of innovation was proposed virtually simultaneously by Lundvall (1988), Freeman (1988), and Nelson (1988) and examined in two volumes of Lundvall (1992) and Nelson (1993). These two approaches were different. Lundvall was influenced by theories of interactive learning, and Nelson approached innovation systems from an evolutionary perspective.

One of the central concerns of Nelson's (1993) fifteen-country study was to establish "whether, and if so in what ways, the concept of a national system made any sense today" (p. 1). For Nelson there were two issues. First, unless one draws the analysis of innovation very narrowly, it is inevitable that analysis of innovation in a country gets drawn into discussions of labor markets, financial systems, monetary policy, and trade policy. Second, the national level may be at the same time too broad since policies that support industry$_j$ might not support industry$_k$, and in some industries a number of the institutions may act internationally, making the concept of national systems too narrow. The concept of a national innovation system may be problematic. Lundvall (1992) recognized that both globalization and regionalization weaken the nation-state. However, Lundvall suggested that we focus on the national level precisely because of the weakened position of the nation-state.

We are therefore left with the question, "What is the proper unit of analysis for innovation systems?" In other words, at what level should innovation systems be defined: nation, industrial sector, technology, region, or global? Of course, this depends in part on the size of the nation-state. For small states, the system might very well be larger than the nation. For large countries, the nation-state might be too large. Nevertheless, there are many reasons for balkanization to proceed as globalization sets in. For example, the disfunctionality of the nation-state has triggered the emergence of a genuine shared community of economic interests at the regional level (Acs, de la Mothe, and Paquet 1996).

Krugman (1995) suggested that as economies become less constrained by national frontiers (as globalization spreads), they become more geographically specialized. Important elements of the process of innovation tend to become regional rather than national. The trends are more important in the science-based and high-technology industries. Some of the largest corporations are weakening their ties to their home country and are spreading their innovation activities to source different regional systems of innovation. Regional networks of firms are creating new forms of learning and production. These changes are important and challenge the traditional role of national systems of innovation.

EPILOGUE: TOWARD A NEW MODEL OF REGIONAL ECONOMIC DEVELOPMENT?

As emphasized in the introduction, a spatialized theory of technology-led regional economic growth needs to reflect three fundamental issues. First, it should provide an explanation of why knowledge-related economic activities start concentrating in certain regions leaving others relatively underdeveloped; second, it needs to answer the questions of how technological advances occur and what the key processes and institutions involved are; and third, it has to present an analytical framework in which the role of technological change in regional economic growth is clearly explained. To answer these three questions, we surveyed three separate and distinct literatures: the new economic geography, the new growth theory, and the new economics of innovation. We suggest that each one of the above three approaches has its strengths and weaknesses that can be integrated to develop an appropriate model of technology-led regional economic development.

The new economic geography answers the question of why economic activity concentrates in certain regions but not in others, but it leaves out innovation and economic growth. The contribution of Krugman's theory on economic concentration is not in its elements but in the way the system was put together. It has already been well known in economic geography and regional economics that decreasing transportation costs, economies of scale, or increasing demand favor agglomeration. However, the way Krugman put these elements together in a general equilibrium model is novel. The model provides a case for the treatment of spatial issues in the way economists are accustomed to. The model provides a technique to analyze geographical concentration of economic activities as being induced by some initial combinations of basic parameters. However, the model in its current form does not seem to be suitable for modeling technology-led regional economic growth at least for two reasons. First, Krugman's definite insistence of avoiding modeling the role of technological externalities in regional economic growth prevents the model from being applicable in innovation-led regional development since spillovers and innovation networks are in the core of this type of development, as exemplified in the literature of innovation systems. Second, while the model is very strong in working out the characterization of specific initial combinations of parameters favoring geographical concentration, it is weak in actually modeling the growth process.

The new growth theory explains the causes of economic growth but leaves out regional considerations and ignores the key processes and institutions involved in innovation. A principal assumption in the theory of endogenous growth is that for creating new sets of technological knowledge, the total stock of knowledge is freely accessible for anyone engaged in research. However, this assumption is not verified in the growing literature of geographic knowledge spillovers. New knowledge (potentially leading to either product or process innovations) is usually in such a tacit form that its accessibility is bounded by geographic proximity and/or by the nature and extent of the interactions among actors of an innovation system (see, e.g., Anselin, Varga, and Acs 1997). Similar to the case of relaxing the neoclassical assumption of equal availability of technological opportunities in all countries of the world, a relaxation of the assumption that knowledge is evenly distributed across space within countries seems also to be necessary. The nonexcludable part of the total stock of knowledge seems rather to be correctly classified if it is assumed to have two portions: a perfectly accessible part consisting of already established knowledge elements (obtainable via scientific publications, patent applications, etc.) and a novel, tacit element, accessible by interactions among actors in the innovation system. While the first part is available without restrictions, accessibility of the second one is bounded by the nature of interactions among actors in a system of innovation.

The new economics of innovation, while explaining the institutional arrangements in the innovation process, leaves out regional issues and economic growth.

The systems approach is a conceptual framework that many scholars and policy makers consider useful for the analysis of innovation. Although the systems of innovation approach is not considered a formal and established theory, its development has been influenced by different theories of innovation such as interactive learning theories and evolutionary theories. In recent years, efforts have been made to percolate general theoretical and empirical observations from this literature into a conceptual framework capable of guiding policy and loosely organized around the idea of a national system of innovation. The concept of a national innovation system may be problematic. Krugman has suggested that as economies become less constrained by national frontiers (as globalization spreads), they become more geographically specialized. Important elements of the process of innovation tend to become regional rather than national. Some of the largest corporations are weakening their ties to their home country and are spreading their innovation activities to source different regional systems of innovation. Regional networks of firms are creating new forms of learning and production. These changes are important and challenge the traditional role of national systems of innovation.

We suggest that a specific combination of the Krugmanian theory of initial conditions for spatial concentration of economic activities with the Romerian theory of endogenous economic growth complemented with a systematic representation of interactions among the actors of Nelson's innovation system could be a way of developing an appropriate model of technology-led regional economic development. It is not the purpose of this article to present an explicit solution for this problem; however, a brief outline of a possible synthesis may be provided here. The central element of the model could be the "regional knowledge production equation" distilled from the predominantly empirical literature of innovation networks as presented in the literature of the new economics of innovation. In the traditional model of the knowledge production function, firms exogenously exist and then engage in the pursuit of new economic knowledge as an input into the process of generating innovative activity. As suggested by Audretsch (1995), we "propose shifting the unit of observation away from exogenously assumed firms to individuals—agents confronted with new knowledge and the decision whether and how to act upon that new knowledge" (p. 48). This equation would facilitate the presence of knowledge in the Krugmanian economic geography model. Here the analytical technique for deriving initial conditions of spatial concentration can be adapted to come up with the preconditions for the emergence of knowledge-induced agglomerations. Together with other parameters of the model, threshold values of knowledge may be calculated following the technique developed by Krugman. Finally, to actually model the equilibrium path of regional economic growth induced by the threshold values of knowledge and other regional parameters, the combined framework of the new economic geography and the new economics of innovation can be complemented with the Romerian analytics of economic growth.

Note

1. For an alternative view of the economic development, see Webber (1996).

References

Acs, Z., ed. 2000. *Regional innovation, knowledge and global change*. London: Pinter.

Acs, Z., D. Audretsch, and M. Feldman. 1991. Real effects of academic research: Comment. *American Economic Review* 81: 363-67.

———. 1994. R&D spillovers and recipient firm size. *The Review of Economics and Statistics* 76: 336-40.

Acs, Z., J. de la Mothe, and G. Paquet. 1996. Regional innovation: In search of an enabling strategy. In *The implications of knowledge-based growth for micro-economic policies*, edited by P. Howitt. Calgary, Alberta, Canada: Calgary Press, 339-58.

Aghion, P., and P. Howitt. 1998. *Endogenous growth theory*. Cambridge, MA: MIT Press.

Anselin, L., A. Varga, and Z. Acs. 1997. Local geographic spillovers between university research and high technology innovations. *Journal of Urban Economics* 42: 422-48.

———. 2000. Geographic spillovers and university research: A spatial econometric approach. *Growth and Change* 31: 501-15.

Arrow, K. 1962. The economic implications of learning by doing. *Review of Economic Studies* 29: 155-73.

Audretsch, D. B. 1995. Innovation and industry evolution. Cambridge, MA: MIT Press.

Barro, R., and Sala-i-Martin. 1995. *Economic growth*. New York: McGraw-Hill.

Blaug, M. 1979. The German hegemony of location theory: A puzzle in the history of economic thought. *History of Political Economy* 11: 3-11.

———, ed. 1992. Johann von Thünen (1783-1850). In *Pioneers in economics* (an Elgar reference collection). Aldershot, UK: Elgar.

Beckmann, M. 1969. On the distribution of urban rent and residential density. *Journal of Economic Theory* 1: 60-67.

Braczyk, H.-J., P. Cooke, and M. Heidenreich, eds. 1998. *Regional innovation systems. The role of governances in a globalized world*. London: UCL Press.

De Bresson, C. 1996. *Economic interdependence and innovative activity*. Cheltenham, UK: Edward Elgar.

de la Mothe, J., and G. Paquet, eds. 1998. *Local and regional systems of innovation*. Boston: Kluwer.

Dixit, A., and J. Stiglitz. 1997. Monopolistic competition and optimum product diversity. *American Economic Review* 67: 297-308.

Edquist, C. 1997. *Systems of innovation*. London: Cassel.

Fischer, M., and A. Varga. 2001. Technological innovation and interfirm cooperation. An exploratory analysis using survey data from manufacturing firms in the metropolitan region of Vienna. *International Journal of Technology Management*. Forthcoming.

Freeman, C. 1988. Japan, a new system of innovation. In *Technical change and economic theory*, edited by G. Dosi, C. Freeman, R. Nelson, G. Silverberg, and L. Soete. London: Pinter.

Fujita, M., P. Krugman, and A. Venables. 1999. *The spatial economy*. Cambridge, MA: MIT Press.

Glaeser, E., H. Kallal, J. Sheinkman, and A. Shleifer. 1992. Growth in cities. *Journal of Political Economy* 100: 1126-52.

Greenhut, M., and G. Norman. 1995. Introduction. In *The economics of location*. Aldershot, UK: Ashgate.

Grossman, M., and E. Helpman. 1991. *Innovation and growth in the global economy*. Cambridge, MA: MIT Press.

Helpman, E. 1992. Endogenous macroeconomic growth theory. *European Economic Review* 36: 237-67.

Hirschman, A. 1958. *The strategy for economic development*. New Haven, CT: Yale University Press.

Isard, W. 1956. *Location and space-economy*. Cambridge, MA: MIT Press.

Isserman, A. 1996. "It's obvious, it's wrong, and anyway they said it years ago"? Paul Krugman on large cities. *International Regional Science Review* 19: 37-48.

Jaffe, A. 1989. Real effects of academic research. *American Economic Review* 79: 957-70.

Judd, K. 1985. On the performance of patents. *Econometrica* 53: 567-86.

Krugman, P. 1980. Scale economies, product differentiation, and the pattern of trade. *American Economic Review* 70: 950-59.

———. 1991a. *Geography and trade*. Cambridge, MA: MIT Press.

———. 1991b. Increasing returns and economic geography. *Journal of Political Economy* 99: 483-99.

———. 1993a. First nature, second nature, and metropolitan location. *Journal of Regional Science* 33: 129-44.

———. 1993b. On the number and location of cities. *European Economic Review* 37: 293-98.

———. 1993c. On the relationship between trade theory and location theory. *Review of International Economics* 1: 10-122.

———. 1995. *Development, geography and economic theory*. Cambridge, MA: MIT Press.

———. 1996. Urban concentration: The role of increasing returns and transport costs. *International Regional Science Review* 19: 5-30.

Lucas, R. 1988. On the mechanics of economic development. *Journal of Monetary Economics* 22: 3-42.

Lundvall, B. 1988. Innovation as an interactive process: From user-producer interaction to the national system of innovation. In *Technical change and economic theory*, edited by G. Dosi, C. Freman, R. Nelson, G. Silverberg, and L. Soete. London: Pinter.

———, ed. 1992. *National systems of innovation*. London: Pinter.

Maddison, A. 1987. Growth and slowdown in advanced capitalist economies. *Journal of Economic Literature* 25: 649-98.

Martin, R. 1999. The new "geographical turn" in economics: Some critical reflections. *Cambridge Journal of Economics* 23: 65-91.

Martin, R., and P. Sunley. 1996. Paul Krugman's geographical economics and its implications for regional development theory: A critical assessment. *Economic Geography* 72: 259-92.

Myrdal, G. 1957. *Economic theory and under-developed nations*. London: Duckworth.

Nelson, R. 1988. Institutions supporting technical change in the United States. In *Technical change and economic theory*, edited by G. Dosi, C. Freeman, R. Nelson, G. Silverberg, and L. Soete. London: Pinter.

———, ed. 1993. *National innovation systems*. New York: Oxford.

Nelson, R., and S. Winter. 1982. *An evolutionary theory of economic change*. Cambridge, MA: Harvard University Press.

Nijkamp, P., and J. Poot. 1997. Endogenous technological change, long run growth and spatial interdependence: A survey. In *Innovative Behavior in Time and Space*, edited by C. Bertuglia, S. Lombardo, and P. Nijamp. Heidelberg, Germany: Springer.

Ohta, H. 1988. *Spatial price theory of imperfect competition*. College Station: Texas A&M University Press.

Oinas, P., and E. J. Malecki. 1999. Spatial innovation systems. In *Making connections. Technological learning and regional economic change*, edited by E. J. Malecki and P. Oinas, 261-75. Aldershot, UK: Ashgate.

Ratti, R., A. Bramanti, and R. Gordon. 1997. *The dynamics of innovative regions*. Aldershot, UK: Ashgate.

Romer, P. 1986. Increasing returns and long-run growth. *Journal of Political Economy* 94: 1002-37

———. 1990. Endogenous technological change. *Journal of Political Economy* 98: S71-S102.

———. 1994. The origins of endogenous growth. *Journal of Economic Perspectives* 8: 3-22.

Saxenian, A. 1994. *Regional advantage: Culture and competition in Silicon Valley and Route 128*. Cambridge, MA: Harvard University Press.

Solow, R. 1956. A contribution to the theory of economic growth. *Quarterly Journal of Economics* 70: 65-94.

———. 1957. Technical change in an aggregative model of economic growth. *International Economic Review* 6: 18-31.

Sternberg, R. 1999. Innovative linkages and proximity: Empirical results from recent surveys of small and medium sized firms in German regions. *Regional Studies* 33: 529-40.

Suarez-Villa, L. 2000. *Invention and the rise of technocapitalism*. New York: Rowman & Littlefield.

Varga, A. 1998. *University research and regional innovation: A spatial econometric analysis of academic knowledge transfer.* Boston: Kluwer.

———. 2000. Local academic knowledge spillovers and the concentration of economic activity. *Journal of Regional Science* 40: 289-309.

Weber, A. 1929. *The theory of the location of industries*. Chicago: University of Chicago Press.

Webber, W. J. 1996. *The golden age illusion*. London: Guilford.

[26]

The Economic Journal, **113** (*January*), 207–225. © Royal Economic Society 2003. Published by Blackwell Publishing, 9600 Garsington Road, Oxford OX4 2DQ, UK and 350 Main Street, Malden, MA 02148, USA.

LOW RETURNS IN R&D DUE TO THE LACK OF ENTREPRENEURIAL SKILLS*

Claudio Michelacci

This paper proposes a model of endogenous growth where innovating requires both researchers, who produce inventions, and entrepreneurs who implement them. As research and entrepreneurship compete in the allocation of aggregate resources, the relation between growth and research effort is hump-shaped. When entrepreneurs appropriate too little rents from innovation, too few resources are allocated to entrepreneurship and returns to R&D are low because of this lack of entrepreneurial skills. When so, innovation should be promoted by encouraging entrepreneurship rather than research.

Since the Malthusian (1986 [1798]) and Ricardian (1951 [1817]) prophecies of the eventual coming of a stationary state, the spectre of diminishing returns has hovered over economics. Arguably, such diminishing returns may arise because of the progressive exhaustion of investment opportunities which makes the returns to R&D effort decrease over time.[1] According to Schumpeter (1943, 1946) this is, however, just one of two competing explanations of the 'state of decay' of the 'capitalist society'. A 'more plausible' one is that 'the individual leadership of the entrepreneur tends to lose importance'.

According to this view, entrepreneurial ability is as important as research ability in determining the growth rate of an economy. Indeed, Schumpeter (1947) stresses the distinction between the inventor and the entrepreneur: 'The inventor produces ideas, the entrepreneur "gets things done" [...] an idea or scientific principle is not, by itself, of any importance for economic practice'. Hence, in order to grow, an economy requires both researchers who produce inventions and entrepreneurs who implement them.[2] Scientific knowledge has no economic impact unless some effort is made to spread it.[3]

* I would like to thank Philippe Aghion, Samuel Bentolila, Chris Pissarides and Javier Suarez for very helpful discussions and suggestions at different stages of the project. I am also particularly indebted to Christopher Bliss, another Editor and two anonymous referees for valuable and thoughtful comments on previous versions of the paper. The usual disclaimer applies.

[1] Recent empirical evidence seems to support some form of diminishing returns in research over time. For example, Griliches (1990) and Kortum (1993) note that, in the US, the ratio of the number of patent applications to the scientists and engineers involved in R&D has fallen over time in the post-war period, while Jones (1995*a*, *b*) points out that the economy growth rates have remained constant and even declined despite an increase in the amount of R&D effort. Indeed, Griliches (1979) concludes that 'the exhaustion of inventive and technological opportunities remains a major suspect for the productivity slow-down in the 70's'.

[2] The history of innovation is full of examples of successful inventor-entrepreneur partnerships. For example, Mokyr (1990) notices that even James Watt would have hardly succeeded in the implementation of the steam engine without the help of his partner, Matthew Boulton.

[3] Economists are well aware of the importance of allocating resources to both research and entrepreneurial activities. They must allocate their time between the studying required to write new papers and teaching students, talking with colleagues, presenting work at the NBER meetings, etc.

In order to model the aggregate trade off between allocating resources to research and entrepreneurship, I consider an endogenous growth model where an innovation requires the matching of an entrepreneur with a successful invention. In the model, individuals face the choice of whether to become researchers, who produce inventions, or entrepreneurs who implement them.[4] Hence an economy that allocates too many individuals to the research sector produces too many inventions which will be wasted because there will not be enough entrepreneurs to implement them. Thus the relationship between productivity growth and research effort is hump-shaped.

The existence of search frictions in the process whereby entrepreneurs and researchers are matched produces a natural incompleteness of contracts. Hence, the share of the rents from innovation appropriated by either researchers or entrepreneurs does not necessarily reflect their contribution to innovation. In general equilibrium, however, such rents determine whether individuals choose to become researchers or entrepreneurs. Consequently, the allocation of skill sustained in equilibrium might depart substantially from the socially efficient one. Interestingly, any point on the growth-research effort technological frontier can be sustained as a decentralised equilibrium of the economy.

One of the key results of the paper is that the innovation process can suffer because of the lack of either research effort or entrepreneurial skills. Economic history may provide examples of either case. Islamic civilisation at the end of the twelfth century may be an example of the former. Indeed Mokyr (1990) argues that it declined because it 'was not capable of adding much new to the existing stock of ideas it retrieved and applied so brilliantly'. Mokyr also discusses notorious examples of societies, such as the Greeks and the Romans, that suffered because of the lack of entrepreneurial skills. Such a deficit may also well explain the economic decline of the late Victorian Britain usually attributed to the inability of British entrepreneurship in promoting the diffusion of new and advanced production methods.[5]

Some relation to the literature. By positing that innovation requires the random matching of entrepreneurs with valuable inventions, my model borrows elements from the standard search model, e.g. Pissarides (1990). This setting is, however, extended to capture two important features of the innovation process. First, I introduce a technological externality generated by firm creation that leads to endogenous growth. Second, I allow the skill allocation to be endogenous by

[4] Schumpeter (1949) was concerned that his model resembled a model of 'exogenous' technological growth: 'a stumbling block' of the theory 'may be expressed by saying that the entrepreneur simply does nothing but take advantage of technological progress, which therefore appears, implicitly or explicitly as something that goes along entirely independently of entrepreneurial activity [...] It is perhaps not difficult to understand that technological progress, so obvious in some societies and so nearly absent in others, is a phenomenon that needs to be explained'. My model stresses the role of the entrepreneurial function in an endogenous growth framework.

[5] The 1998 White Paper by the British Government argues that the UK is still lacking in entrepreneurial skills. The recent experience of 'Silicon Fen' – the cluster of over 1,600 high tech firms that have grown up around Cambridge University – is sometimes perceived as a successful attempt to reverse this secular deficit in entrepreneurship.

permitting agents to decide on whether to become an entrepreneur or a researcher.[6] In this context I generalise earlier results, originally due to Diamond (1982) and Hosios (1990), on the efficiency of economies subject to search frictions. In particular I show that restoring efficiency under the presence of technological externalities requires research to have more bargaining power relative to the Hosios's benchmark.

The paper also relates to those strands of the growth literature that have analysed the relation between growth and R&D effort. It shares with Helpman and Trajtenberg (1994) the premise that inventions have widespread economic consequences only if implemented. The finding that research effort can be excessive arises also in Tirole (1988, p.399), in Aghion and Howitt (1992), and in Aghion and Howitt (1998, ch. 6), while the emphasis on the allocation of talent is shared with Baumol (1990) and Murphy *et al.* (1991). It is the interaction between these strands of the literature that is new here.

The model has also some implications for the recent literature which has analysed the relation between growth and the scale of the economy.[7] Jones (1995*b*), Kortum (1997), Young (1998) and Howitt (1999) propose a theoretical solution to the observed absence of relation between the scale of R&D effort and growth. Their models have in common the assumption that the output cost of an innovation is increasing with the level of development. Due to this form of exhaustion of investment opportunities, the amount of research effort increases over time to keep growth constant. Importantly such decreasing returns to R&D arise as the efficient response of a competitive economy to the increase in its scale.[8]

The paper might propose a complementary reading of the same empirical facts. I show theoretically that the returns to R&D might be decreasing if they are the result of an inefficient equilibrium shift that increases the amount of research effort to the detriment of more socially useful entrepreneurial skills. Furthermore, I show that the data are consistent with the existence of a constant-returns-to-scale matching function that links a measure of innovation (number of patent applications) to a measure of research effort (number of scientists involved in R&D) and entrepreneurship (population of self-employed). This implies that, at the optimal allocation, the returns to R&D never decrease since well-balanced increases in the amount of research and entrepreneurial effort always exhibit a constant marginal effect on the innovation rate.

Section 1 expounds the general set-up while Section 2 solves for the steady state equilibrium. Section 3 deals with efficiency. Section 4 carries out some exercises of comparative statics. Section 5 discusses some empirical implications. The conclusions are in Section 6. The Appendix contains some technical derivations.

[6] See Acemoglu (1996, 1997) for an analysis of workers' decisions to accumulate human capital in a labour market subject to search frictions.

[7] See Jones (1999) for a survey.

[8] For example, both in Jones (1995*b*) and Kortum (1997) the economy growth rate is the socially optimal one, while in Young (1998) the total amount of R&D effort is optimal even if its allocation among different dimensions of innovation might not be.

1. The Model

The economy is populated by a continuum of agents of size C. Each agent is infinitely lived, risk-neutral and maximises expected returns in output units discounted at rate $r > 0$. At each point in time agents can choose either to become researchers or entrepreneurs. I indicate with f_t the time-t fraction of researchers in the population and I refer to it as the level of *research effort* in the economy.

A researcher discovers *inventions* according to a Poisson process with intensity λ. When he makes a discovery, he starts searching for entrepreneurs able to implement it. Keeping the invention up-to-date requires a flow cost of χx_t per unit of time, where x_t represents the leading technology in the economy at time t and $\chi \geq 0$. I assume that the opportunity of transforming an invention into an innovation vanishes according to a Poisson process with rate of arrival v. This implies that at each point in time the stock of scientific knowledge, k_t, measured by the number of inventions suitable for economic exploitation, increases in response to the discovery of new inventions, while it tends to fall as the old ones become obsolete. This allows one to capture some key characteristics of the process of accumulating scientific knowledge; see Adams (1990). First, the stock of scientific knowledge is *fundamental* in that its applicability is not immediate. Secondly, it recognises that the stock of knowledge is made up of *heterogenous* pieces of information since the implementation of an invention requires a costly search for a suitable entrepreneurs. Thirdly, the use of the stock of knowledge is *repetitive*, as an invention can give rise to a 'cluster' of innovations concentrated 'in certain sectors and their surroundings' (Schumpeter, 1939, pp. 100–1). Finally it recognises its *time specificity*, as scientific knowledge becomes obsolete as time goes by.[9]

An *innovation* requires an invention discovered by a researcher and an entrepreneur. Once they match, a new firm run by the entrepreneur is created. A firm created at time t accesses the leading technology of that date x_t. A firm can produce, at each point in time t, a flow of goods equal to Px_t and is shut down according to a Poisson process with rate of arrival δ.[10]

The entrepreneur can be either running a firm or working at home; see Benhabib *et al.* (1991). When the entrepreneur runs a firm, he receives a profit flow π_t. Profits of the entrepreneur are chosen, so as to share with the researcher the gains from running the firm, by using a generalised Nash bargaining solution. The bargaining powers of the entrepreneur and the researcher are β and $1 - \beta$, respectively. When the entrepreneur works at home, he produces a flow of goods equal to hx_t where $h \geq 0$ measures the level of human capital of the entrepreneur who uses it according to the technology available at time t. In equilibrium an entrepreneur running a firm has no incentive to search. Hence only entrepreneurs

[9] The model could be extended by assuming that the rate v at which inventions become obsolete, depends on the rate of growth of the leading technology of the economy x_t so that technological progress causes the obsolescence of scientific knowledge – an effect reminiscent of Schumpeter's idea of creative destruction. The main implications of the model would remain however unchanged.

[10] I model firms in this way for analytical convenience. In fact the results of the paper would remain qualitatively unchanged if either a firm's death probability or output were assumed to be a function of its age, as they arguably are in reality.

working at home are 'available' for innovating. I will denote their number at time t by s_t and I will refer to it as the amount of *entrepreneurial slackness* available in the economy.

The rate at which free entrepreneurs find suitable inventions is determined by the homogeneous-of-degree-one matching function $m(k_t, s_t)$; see Pissarides (1990). This function is increasing and concave in each of its arguments. The matching function allows one to represent in a parsimonious fashion two key characteristics of the innovation process: the fact that both entrepreneurial skills and inventions are heterogenous, so that the search for viable partnerships is time consuming, and the fact that entrepreneurial skills are a scarce resource for which different researchers compete. The probability that an invention matches with an entrepreneur is given by $q(\theta_t) = m(k_t, s_t)/k_t$ where $\theta_t = k_t/s_t$ will be referred to as the level of *knowledge intensity* in the economy. By analogous considerations it follows that $p(\theta_t) = \theta_t q(\theta_t)$ is the instantaneous probability that a free entrepreneur finds a valuable invention. I also assume that

$$p(0) = q(\infty) = m(0, s_t) = m(k_t, 0) = 0, \quad \text{and} \quad p(\infty) = q(0) = \infty \tag{A1}$$

so as to avoid dealing with uninteresting corner solutions.

Vertical innovations are the only source of growth in this economy. I follow Caballero and Jaffe (1993) and Aghion and Howitt (1998), among others, in assuming that the rate of growth of the technological parameter x_t is given by the product of the size of the innovation, σ, and the frequency of innovation. With the above assumptions, this frequency coincides with the number of innovations introduced in the economy at time t, $\theta_t q(\theta_t) s_t$. Hence the growth rate of the technological parameter x_t is given by $g_t = \sigma\theta_t q(\theta_t) s_t$.[11]

A sufficient condition to assure finite present values is $\sigma C\delta < r$ since it will imply that in steady state equilibrium $g_t < r$. Whereas obtaining a positive equilibrium level of research requires that $(\lambda/v)(1 - \beta)(P - h) > 0$. As in Romer (1990), in order to sustain growth we must assign a strictly positive capacity of rent appropriation to researchers, $\beta < 1$, since future rents are the only compensation for the sunk investment in research necessary to produce an invention.

The endogenous variables of the model are the level of knowledge intensity θ, the level of research effort f and the steady state growth rate g.

2. Equilibrium

In this section, I first write down value functions and characterise individual behaviour. I then analyse the level of research effort and the growth rate that arise in a steady state equilibrium.

[11] A natural interpretation is that there is a continuum of increasingly productive techniques indexed by a real number which corresponds to the log of its productivity parameter. Creation of a new firm means that the entrepreneur has succeeded in discovering a new technique and all entrepreneurs can thereafter search for the next technique in the continuum. So, at any point in time, the rate of technological progress is proportional to the flow of newly created firms.

2.1. *The Value of Research and Entrepreneurship*

The value of doing research at time t, R_t, solves

$$R_t = E_{\tau \geq 0}\{[(L_{t+\tau} + \sup(R_{t+\tau}, H_{t+\tau})]e^{-r\tau}\}, \tag{1}$$

where L_t and H_t are the time-t value of an invention and of being an entrepreneur working at home, respectively. $t + \tau$ is the arrival date of the first invention whose arrival rate is λ, while the second term in the right-hand side captures the option a researcher has of becoming an entrepreneur at a later date.

Since an invention becomes obsolete at rate v and requires a cost of χx_t to remain up-to-date, it has a value L_t to the researcher which follows the asset-type equation

$$(r + v)L_t = -\chi x_t + q(\theta)I_t + \dot{L}_t. \tag{2}$$

I_t measures the time-t value to a researcher of an innovation, $\dot{L}_t$ is the time derivative of L_t while $q(\theta)$ is the probability that an invention matches with an entrepreneur. As a firm produces a flow of goods equal to Px_t and entrepreneurs' profits are equal to π_t, I_t can be obtained by solving

$$(r + \delta)I_t = Px_t - \pi_t + \dot{I}_t, \tag{3}$$

where $\dot{I}_t$ represents the time derivative of I_t while δ is the firm's death probability.

Analogously, the value to an entrepreneur of working at home at time t, H_t, solves the asset equation

$$rH_t = hx_t + \theta q(\theta)(E_t - H_t) + \dot{H}_t, \tag{4}$$

where hx_t, E_t and $\dot{H}_t$ are respectively the real flow of goods produced by an entrepreneur working at home, the entrepreneur's value of running a firm and the time derivative of H_t. $\theta q(\theta)$ is the instantaneous probability that a free entrepreneur finds an invention that he can implement. Analogously E_t solves

$$rE_t = \pi_t + \delta[\sup(R_t, H_t) - E_t] + \dot{E}_t, \tag{5}$$

where the above equation embodies the option entrepreneurs have of becoming researchers.

Profits are the outcome of bilateral bargaining between the researcher with the invention and the entrepreneur. Hence they maximise the weighted product of the researcher's and entrepreneur's net return from the creation of a new firm equal to I_t and $E_t - H_t$, respectively. Hence

$$\pi_t = \arg\max(I_t)^{1-\beta}(E_t - H_t)^{\beta},$$

where H_t is the entrepreneur's outside option which from the point of view of the maximisation is taken as constant. As a result, profits are such that

$$I_t = (1 - \beta)(I_t + E_t - H_t) = (1 - \beta)S_t, \tag{6}$$

where $S_t = I_t + E_t - H_t$ is the private net surplus generated by the creation of a new firm. Of this surplus, researchers and entrepreneurs appropriate fractions $1 - \beta$ and β respectively.

2.2. *Free Entry Condition*

In steady state, each variable is growing at the same rate g as the economy so

$$R_t = Rx_t, \quad I_t = Ix_t, \quad L_t = Lx_t, \quad H_t = Hx_t, \quad E_t = Ex_t. \tag{7}$$

From making use of (7) to rewrite (2), (3), (4) and (5) it follows that

$$L = \frac{1}{r - g + v}[-\chi + q(\theta)(1 - \beta)S], \tag{2'}$$

$$I = (1 - \beta)S, \tag{3'}$$

$$H = \frac{1}{r - g}[h + \theta q(\theta)\beta S], \tag{4'}$$

$$E = H + \beta S \tag{5'}$$

where $S = I + E - H$ measures (once divided by x_t) the private net surplus associated with the creation of a new firm and is equal to

$$S = \frac{P - h}{r - g + \delta + \theta q(\theta)\beta}. \tag{8}$$

In equilibrium, agents must be indifferent between becoming a researcher or an entrepreneur, so that the *free entry condition* $R_t = H_t$ must hold at each point in time. But then, from using this condition, the fact that $R_t = Rx_t$ and then substituting into (1) it follows that

$$R = \frac{\lambda L}{r - g}.$$

Together with (2′) and (4′) this implies that the condition $R_t = H_t$ reads as

$$h + \theta q(\theta)\beta S = \frac{\lambda[-\chi + q(\theta)(1 - \beta)S]}{r - g + v}. \tag{9}$$

The left-hand side of (9) represents the relative profitability of being an entrepreneur as an increasing function of the amount of knowledge intensity θ: the larger the stock of scientific knowledge available per entrepreneur, the more profitable is entrepreneurship. Analogously, an increase in θ reduces the profitability of being a researcher because the implementation of inventions becomes progressively more difficult. As a result the right-hand side of (9), which measures the relative profitability of being a researcher, is decreasing in θ. Hence, for any given level of the steady state growth rate g, there exists a unique value of knowledge intensity θ such that the right-hand side equals the left-hand side. At such value individuals are indifferent between becoming researchers or entrepreneurs.[12]

[12] One can also check that steady state profits π_t are equal to πx_t, where, from substituting (3), (4), (5) into (6), it follows that $\pi = \beta P + (1 - \beta)h + \theta q(\theta)\beta I$.

2.3. *Technological Frontier*

Since firms are closed at rate δ, entrepreneurial slackness, s_t, evolves according to the differential equation

$$\dot{s}_t = \delta[C(l-f) - s_t] - \theta q(\theta)s_t, \tag{10}$$

so that in steady state it is equal to

$$s_t = s = \frac{\delta C(1-f)}{\delta + \theta q(\theta)}. \tag{11}$$

Analogously, the dynamics of the stock of knowledge, k_t, is governed by

$$\dot{k}_t = \lambda Cf - vk_t. \tag{12}$$

so that in steady state $k_t = k = (\lambda/v)Cf$.

From the assumption that the frequency of innovation is equal to the number of successful matches between k_t and s_t, it follows that the steady state growth rate g of the technological parameter x_t is equal to

$$g = \sigma\theta q(\theta)s = \sigma Cm\left[\frac{\lambda}{v}f, \frac{\delta(1-f)}{\delta + \theta q(\theta)}\right], \tag{13}$$

where $\theta = k/s$ solves the simple non-linear equation

$$\theta = \theta(f) = \frac{\lambda}{v}\frac{[\delta + \theta q(\theta)]f}{\delta(1-f)}. \tag{14}$$

Equation (14) merely expresses knowledge intensity as the ratio of the steady state amount of scientific knowledge to the amount of slackness given by (11). Notice that (14) defines an function $\theta(f)$ that associates to each f the corresponding value of θ that solves the equation. Such function is increasing since a larger f associates with a larger k and a lower s. But then, after substituting $\theta(f)$ for θ in (13), one can define a function $g = g(f)$ which relates the level of research effort, f, to the steady state growth rate of the economy. Since the function $g(f)$ determines the frontier of technological possibilities of the economy, I refer to it as the *TE-schedule*. Such a schedule is hump-shaped and it satisfies the property that no growth can be the outcome of either too much or too little research effort, $g(0) - g(1) = 0$ (see Propositions 1 and 2 in the Appendix for a formal proof).[13] As both research and entrepreneurial skills are required to innovate, an over-allocation to either factor harms technological process.

2.4. *Equilibrium*

Equations (9) and (13), given the constraint imposed by (14), completely solve the model in the research effort, f, growth rate, g, space. Equation (9) determines, for

[13] Denote by $0 < \eta < 1$ the elasticity of the matching function with respect to the level of entrepreneurial slackness. Then one can show that $g(f)$ is globally strictly concave if, for example, η is weakly increasing in θ. One can also show that $g(f)$ has a unique maximum under the milder condition: $-d\eta/d\theta < \eta^2$. For example, both conditions would obviously be satisfied in the case of a Cobb–Douglas matching function.

any given level of the growth rate, the relative amount of resources that the economy will end up devoting to research. Given (14) and the associated function $\theta(f)$, (9) allows to define the function $g = e(f)$ which relates the level of research effort that solves the free-entry condition to the given level of the growth rate. Hereafter I will refer to the function $e(f)$ as the *FE-schedule*. Such function maps the zero one interval onto the whole real line (see Proposition 3 in the Appendix). Furthermore $e(f)$ is increasing. To understand why it is so, notice that research requires a longer time than entrepreneurship to become productive and yield some (positive) revenue. Also notice that in the model an increase in the growth rate, g, affects individual decisions through a capitalisation effect that reduces the effective discount rate of agents. But a change in the discount factor always tends to have a greater effect on the investment with the longer time horizon so, as a result of the increase in g, the value of both research and entrepreneurship increase but the value of research increases relatively more so f must rise to restore the free entry condition.

The steady-state equilibrium is defined by the point at which the *FE*-schedule, $e(f)$, crosses the *TE*-schedule, $g(f)$, like at point A in Figure 1. At such a point no researcher has an incentive to become an entrepreneur (or *vice-versa*) and the economy grows at the constant steady state growth rate determined by the technological frontier.

3. Welfare Analysis

In this section, I compare the equilibrium to the constrained social optimum. I define social welfare $\tilde{W}$ as the net present discounted value of the aggregate income flows produced by entrepreneurs and researchers. The social planner

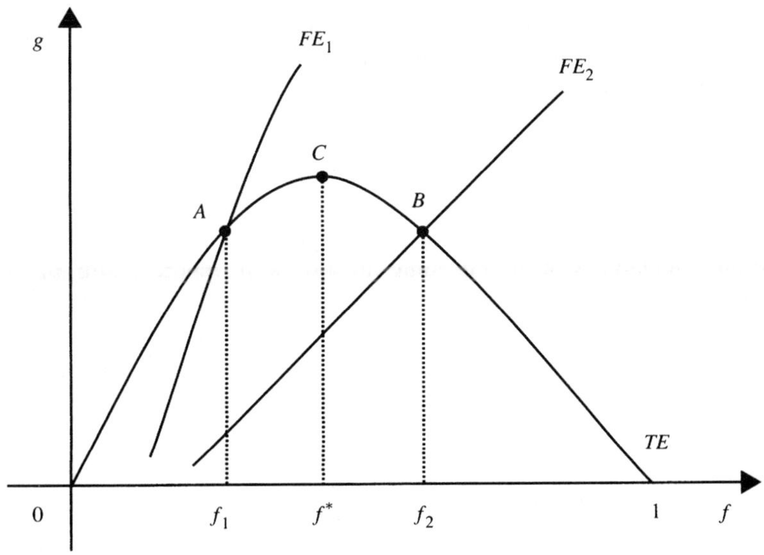

Fig. 1. *Steady State Equilibrium*

takes as given the search frictions involved in the innovation process. At any point in time, the state of the economy is fully summarised by the quantities k_t and n_t that represent the time-t stock of scientific knowledge and total number of firms, respectively. Given the focus on steady state equilibria, I consider time invariant allocations described by a fixed amount of research effort f.

3.1. *The Social Planner Problem*

Once multiplied by x_t,

$$y(k_t, n_t; f) = Pn_t + h[C(1-f) - n_t] - \chi k_t$$

denotes the net income flow produced by the economy at time t. Also notice that the evolution of the number of firms in the economy is governed by

$$\dot{n}_t = m[k_t, C(1-f) - n_t] - \delta n_t \tag{15}$$

while the stock of scientific knowledge grows at a rate $\dot{k}_t$ given by (12). Hence one can implicitly define social welfare as the product of x_t and the function $W(k, n, f)$ which solves

$$rW(k_t, n_t; f) = y(k_t, n_t; f) + W_k \dot{k}_t + W_n \dot{n}_t + g_t W(k_t, n_t; f) \tag{16}$$

where $g_t = \sigma m(k_t, s_t)$ while $W_k = \partial W/\partial k_t$ and $W_n = \partial W/\partial n_t$ measure the social shadow value of an invention and of an additional firm, respectively. Three components account for social welfare in (16): the first is current income, the second is the capital gain associated with changes in the state of the system – the second and third term in (16) – and last the capitalisation effect associated with technological progress, the fourth term in (16).

To characterise the constrained socially optimum allocation, denote by g and n the economy's steady state value of the growth rate and number of firms, respectively. Furthermore let

$$\bar{W} = \frac{Pn + h[C(1-f) - n] - \chi k}{r - g} \tag{17}$$

denote (once divided by x_t) the steady state level of social welfare which is equal to the present discounted value of the level of output net of research costs. Finally let η indicate the elasticity of the matching function with respect to entrepreneurial slackness.[14] Then differentiating (16) with respect to f and after some algebra, the Appendix shows that

$$\frac{\mathrm{d}W}{\mathrm{d}f} = -\frac{C}{r-g}\left\{h + \theta q(\theta)\eta S^E - \frac{\lambda[-\chi + q(\theta)(1-\eta)S^E]}{r - g + v}\right\} \tag{18}$$

where

$$S^E = W_n + \sigma\bar{W} = \frac{P - h + \sigma(r - g + \delta)\bar{W}}{r - g + \delta + \theta q(\theta)\eta} \tag{19}$$

[14] Notice that η is in general a function of θ.

is the net social surplus associated with the creation of a new firm. Such surplus is equal to the sum of the shadow value of a new firm W_n plus the gains associated with the technological externality generated by innovation, which have value $\sigma\bar{W}$.

3.2. *Results*

I can now state my first result on efficiency: the social optimum always lies on the positively sloped arm of the *TE*-schedule, $g(f)$. To prove this result, the Appendix first shows that the steady state growth rate g is maximised at the 'golden rule' level of research effort f^* such that the level of knowledge intensity $\theta^* = \theta(f^*)$ solves

$$\theta^* = \frac{\lambda}{\nu}\frac{1-\eta}{\eta}. \tag{20}$$

Then the Appendix shows that the derivative of W with respect to f evaluated at f^* is negative which indeed implies that at the growth maximising allocation the level of research effort in the economy is too large.

The economic intuition of the result is pretty simple. Given the objective of maximising the present discounted value of output, the central planner must equate the marginal benefit of increasing research effort to its marginal cost. Notice that (13) and (15) imply that the value of f that maximises the steady state number of operating firms, n, also maximises the steady state growth rate, g. So the sign of the social marginal benefits of research coincides with that of the derivative of the *TE*-schedule, $g(f)$. But then, since increasing research always involves a strictly positive marginal cost (relative to increasing entrepreneurship), the social optimum must be at a point where the marginal benefits from increasing f are also strictly positive and this is so only on the strictly increasing arm of *TE*-schedule.

It is interesting to compare the level of research effort sustained by the decentralised economy to the social optimal one. This can be done by comparing (9) with (18) and checking when the decentralised economy implements a level of research effort which induces $\mathrm{d}W/\mathrm{d}f = 0$. One can immediately see that the two conditions associate with the same level of research effort only if the private and social net-surplus from the creation of a new firm coincide, $S = S^E$. In the absence of technological externalities, $\sigma = 0$, this occurs when entrepreneurs' bargaining power is such that $\beta = \eta$ so that (8) and (19) are identical. Hosios (1990) first proved that, in an economy with search frictions, a constrained efficient allocation of resources requires bargaining power to reflect each side's contribution to the creation of net surplus as measured by the elasticity of the matching function. Here I have generalised this result to an environment where the allocation of skill is endogenous.

When technological externalities are present, $\sigma > 0$, however, the Appendix shows that entrepreneurs' bargaining power (β must be smaller than η for the decentralised economy to implement the socially efficient allocation. To understand the result, notice that search frictions simultaneously induce a negative externality to agents on the same side of the market (congestion externalities) and a positive one to agents on the other side of the market (thin market externalities). At ($\beta = \eta$ the congestion and thin-market externalities balance exactly. Yet, at this point, technological externalities are still operating and tend to discourage activities, namely research, characterised by a longer time horizon and larger initial investment. Hence restoring efficiency requires $\beta < \eta$. There is only one case where the Hosios's rule $\beta = \eta$ still applies even in the presence of technological externalities. This occurs when the type of investment required to specialise in different activities is the same, that is when the instantaneous return from becoming a researcher or an entrepreneur are equal $h = \chi = 0$, (see Appendix).

4. Comparative Statics

Proposition 4 in the Appendix states that, depending on the parameters of the model, any point on the *TE*-schedule, $g(f)$, can be sustained as an equilibrium of the economy, including those situated on the negatively sloped arm of the *TE*-schedule, like point B in Figure 1. A point like B is associated with a Pareto dominated equilibrium since a reduction in research effort would simultaneously lead to an increase in the output level and in the growth rate (both over the transition path and in the new steady state). These types of equilibria can be sustained for reasons similar to those analysed by Hosios (1990) and Caballero and Hammour (1996). An economy characterised by *ex-ante* competitive relationships but *ex-post* bilateral monopolies has no compelling tendency to coordinate itself towards the social optimum, since there is no reason why the amount of rents appropriated by each party should reflect his contribution to the creation of social surplus.

In equilibria like point B in Figure 1, research effort crowds out more socially useful entrepreneurial skills and the growth process stagnates precisely because of this *lack of entrepreneurial skills.* Indeed, as the amount of research effort rises, the stock of scientific knowledge increases, but it also becomes progressively more difficult to implement inventions both because more researchers are competing for the same resources (congestion externalities) and because an increase in research effort crowds out useful entrepreneurial skills (thin market externalities). When the stock of scientific knowledge is already large while the amount of entrepreneurial skills is low, an increase in research effort reduces the growth rate of the economy, because it misallocates socially useful resources. For example, an increase in research effort can be translated into stagnant or declining growth rates, if the equilibrium of the economy shifts from point A to point B in Figure 1.

Such shifts in the equilibrium can occur, for example, in response to a fall in the bargaining power of entrepreneurs β. A fall in β makes research relatively more profitable for any given level of the growth rate so that the *FE*-schedule shifts to the right while it has no effect on the *TE*-schedule. Such a fall in β might relate to the Schumpeterian (1943, 1946) claim that cultural and sociological shifts tend to destroy the 'protective strata' able to sustain the entrepreneurial function.[15] Since β measures the outcome of a bilateral bargaining problem here specified

[15] Culture, understood as prevailing values and beliefs, is often perceived as an important determinant of the level of entrepreneurship in the economy, see Davidson and Wilklund (1997). In terms of the model, a cultural environment more favourable to entrepreneurship may translate into a larger either β or h. This last case may correspond to societies that reward more (in terms of social status) entreprenerial 'success' and 'leadership'.

parsimoniously, one can also argue that, in the presence of some asymmetric information that favours researchers, β might be affected by the technological content of the innovation: innovations that are more technological advanced might associate with a greater ability of the researcher to appropriate rents and therefore with a lower β.

Particularly interesting are the consequences of a change in the firms' destruction rate δ. A change in δ has generally ambiguous effects on the *FE*-schedule. To see this notice that, a fall in δ requires an increase in the level of knowledge intensity, θ, to restore the free entry condition (9). However, (14) implies that for given f, a lower δ associate with a larger value of θ, since the amount of entrepreneurial slack is reduced. If this last effect is sufficiently strong the *FE*-schedule, $e(f)$, shifts to the left, otherwise f must rise to restore the free entry condition and $e(f)$ shifts to the right.

What is not ambiguous is the impact of a change in δ on the technological frontier of the economy. A fall in δ implies a downward shift of the *TE*-schedule, with the hump of the schedule moving leftward. To prove the downward movement, notice that for given f a fall in δ reduces the steady state level of entrepreneurial slack given by (11) both directly through δ and indirectly through the induced positive effect on θ, see (14). The leftward movement of the hump follows from the fact that entrepreneurial slack falls for any given level of f, so the constraint imposed by the number of free entrepreneurs on the innovation process becomes progressively more binding as f increases. In other words, an economy that liquidates more often (i.e. with a larger δ) has more frequent opportunity to rebuild its productive stock at a higher technological level and grows faster thanks to the intertemporal spill-overs typical of standard endogenous growth models. Thus the model suggests a channel whereby an increase in the pace of reallocation promotes growth by increasing the amount of resources available for innovating.[16]

5. Empirical Implications

I next discuss some empirical implications of the model. The analysis is simply meant to be illustrative of some of the theoretical properties of the model.

If one assumes that the matching function is Cobb–Douglas,

$$m(k_t, s_t) = A(k_t)^{\alpha}(s_t)^{\gamma}, 0 < \alpha, \gamma < 1, \tag{21}$$

then (13) suggests running the following regression

$$\ln(g_t) = \alpha \ln(f_t C_t) + \gamma \ln(s_t) + \alpha \ln\left(\frac{\lambda}{\nu}\right) + \ln \sigma + \ln(A). \tag{22}$$

The model has two important testable implications. First, the parameters α and γ must both be positive numbers strictly between zero and one. Indeed this would imply that both entrepreneurial and research skills affect the innovation rate of

[16] See Mortensen and Pissarides (1998) for a model where the causality between reallocation and growth reverses. In their model an increase in the rate of exogenous technological progress raises the job reallocation rate in the economy.

the economy. Second, the matching function (21) exhibits constant returns to scale, that is $\alpha + \gamma = 1$. This would imply that, at the optimal allocation, the returns to R&D can never be decreasing since well-balanced increases in the amount of research and entrepreneurial effort have constant marginal effects on the innovation rate.

To illustrate the methodology, I estimate (22) for the US over the period 1950–90.[17] I follow Griliches (1979, 1990) in taking the total number of patent applications as an index of innovative activity in order to proxy for g_t,[18] and Jones (1995*b*) in considering the ratio of scientists and engineers involved in R&D over the population if working age as a measure of (relative) research effort f_t. As a proxy for the scale of the economy C_t in (22) I take the population if working age. Identifying entrepreneurship is more problematic. According to Schumpeter (1949) the entrepreneurial function can be identified only *a posteriori.*[19] In fact, Evans and Jovanovic (1989) note how, in their sample, 20% of the individuals who switched into self-employment formed incorporated businesses. That is why I follow Evans and Jovanovic (1989), Evans and Leighton (1989) and Blanchflower and Oswald (1998) in using as a proxy of the amount of entrepreneurship, s_t, the population of self-employed.

To control for cyclical disturbances, I also consider some additional specifications with the unemployment rate, the growth rate of *per capita* GDP and an index of the number of vacancies as independent variables.[20] The unemployment rate together with the other cyclical indicators could correct for that part of the population of self employed that does not represent real entrepreneurship but simply people that moved into self-employment from unemployment to find a job.[21]

Table 1 reports the estimates obtained by running the regression (22), allowing for different lags in the relation between research effort, entrepreneurship and innovation. The table shows that, in these data, both implications of the model tends to receive empirical support. In fact, α and γ are both positive and significantly different from zero with α ranging between 0.47 and 0.54 while γ ranges

[17] Data before 1970 are taken from *Historical Statistics of the United States: Colonial Times to 1970*; after 1970 they are from various issues of *The Statistical Abstract of the United States.*

[18] Whether a patent corresponds more to the notion of invention or innovation, is a topic beyond the scope of this paper. The Patents Laws-United State Code, U.S.C. 101, states that 'whoever invents or discovers any new and useful process, machine, manufacture, or composition of matter, or any new and useful improvement thereof, may obtain a patent'. The UK Patents Act 1977 states that the invention to be patented must be 'capable of industrial application'. More generally Cornish (1989) states that 'a patent cannot be granted for a thing or process which however interesting or suggestive it might be to scientists, has no known practical application at the priority date'. The requisite of 'industrial applicability' is not contained in the US law. It seems however, that the requisite of 'usefulness' roughly coincides with that of 'industrial applicability', so that even if a patent still represents something between an invention and an innovation as defined in this paper, it seems to be closer to the latter.

[19] Indeed, he writes: 'our definitions of entrepreneur, entrepreneurial function and so on can only grow out of it *a posteriori'* and again 'when we speak of the entrepreneur we do not mean so much a physical person as we do a function, but even if we look at individuals who at least at some juncture in their lives fill the entrepreneurial function it should be added that these individuals do not form a social class. They hail from all the corners of the social universe'.

[20] Data on the unemployment rate, growth rate of GDP *per capita* and the number of vacancies are taken from the OECD-CEP data set, see Bagliano *et al.* (1991).

[21] Blanchflower and Oswald (1998) document that flows from unemployment into self-employment are large.

Table 1

Results from Running Regression (22) in the Text

Indep. variable:	Lagged $\ln(f_t C_t)$	Lagged $\ln(s_t)$	Unempl. rate	GDP per cap. growth rate	Vacancy index	R^2	Wald test (p-value) $\alpha + \gamma = 1$
1lag	0.47 (0.03)	0.24 (0.10)	–	–	–	0.84	0.005
2lags	0.49 (0.03)	0.29 (0.10)	–	–	–	0.86	0.03
3lags	0.50 (0.04)	0.37 (0.10)	–	–	–	0.86	0.24
4lags	0.52 (0.04)	0.48 (0.10)	–	–	–	0.86	0.94
1lag	0.50 (0.04)	0.30 (0.10)	−0.02 (0.008)	–	–	0.86	0.06
2lags	0.53 (0.04)	0.37 (0.10)	−0.02 (0.008)	–	–	0.88	0.36
3lags	0.54 (0.04)	0.45 (0.10)	−0.02 (0.009)	–	–	0.88	0.97
4lags	0.54 (0.05)	0.53 (0.11)	−0.01 (0.009)	–	–	0.87	0.62
1lag	0.51 (0.04)	0.33 (0.10)	−0.02 (0.009)	−0.53 (0.45)	–	0.88	0.15
2lags	0.54 (0.04)	0.39 (0.10)	−0.02 (0.009)	−0.33 (0.42)	–	0.89	0.56
3lags	0.54 (0.04)	0.44 (0.11)	−0.02 (0.009)	0.10 (0.43)	–	0.88	0.92
1lag	0.46 (0.07)	0.32 (0.10)	−0.02 (0.009)	−0.54 (0.45)	0.08 (0.08)	0.88	0.10
2lags	0.47 (0.07)	0.39 (0.10)	−0.02 (0.009)	−0.34 (0.41)	0.10 (0.08)	0.89	0.25
3lags	0.46 (0.07)	0.44 (0.11)	−0.01 (0.01)	0.09 (0.42)	0.10 (0.07)	0.89	0.48

The regression was run using OLS, the dependent variable being the (logged) number of patent applications. Similar results are obtained when allowing for first-order auto-correlation in the residuals and a non-scalar auto-covariance matrix.

from a value of 0.24 to 0.48. This implies that the relation between the number of innovations and research effort $f_t C_t$ turns out to be concave and hump-shaped. Furthermore, Table 1 documents that, if we consider reasonable a lag of over two years before research effort shows up in a patent application, the hypothesis that the matching function (21) exhibits constant returns to scale cannot be rejected even at a 10% level of significance. Moreover, in the specification with the unemployment rate as independent variable, this hypothesis cannot be rejected whatever the number of lags considered.

6. Conclusions

This paper has shown how an increase in the amount of resources devoted to research does not necessarily increase the growth rate of the economy. In a world with rent sharing and inter-temporal spill-overs, an increase in research effort can crowd out more socially useful entrepreneurial skills, reduce the growth rate and ultimately lead to an allocation of skills which is Pareto dominated. Thus, the model might provide some foundations to the Schumpeterian warning that the 'state of decay of the capitalist society' can ultimately be driven by the lack of entrepreneurial skills.

The question on whether the economy's skill allocation is optimal for growth is likely to be particularly relevant in advanced economies where innovation is so *complex* that a single agent fails to possess both the scientific knowledge and the entrepreneurial ability required to innovate. When this is so, individuals must specialise and innovation requires the contribution of different, highly specialised agents. Consequently, the allocation of skills to different tasks changes from being internal to the *individual* to being internal to the *market* and the amount of rents from innovating appropriated by each task may become a crucial determinant of the economy's skill allocation. For instance, the division of rents could affect the choice between allocating resources to accumulating either education or experience as well as any occupational choice, especially among those occupations that are likely to participate to the sharing of the rents from innovation – say, whether to become a manager, a politician, a financier or an entrepreneur.

Investigating whether the actual skill allocation tends to depart from the optimal one as the economy develops and gets more specialised, remains however a question for further research that requires more careful micro-based empirical investigation. Looking at the skill composition either by occupations or along the experience-schooling dimension could be a promising avenue to address this question.

CEMFI, Madrid

Date of receipt of first submission: August 2000
Date of receipt of final typescript: January 2002

Dataset is available for this paper: www.res.org.uk

Appendix

In this Appendix I derive some technical results discussed in the text. Before proceeding, notice that the derivatives of $m(k,s)$ with respect to k and s can be computed as

$$\frac{\mathrm{d}m}{\mathrm{d}k} = (1-\eta)q(\theta) \quad \text{and} \quad \frac{\mathrm{d}m}{\mathrm{d}s} = \eta\theta q(\theta),$$

respectively.

Properties of the function $\theta(f)$

PROPOSITION 1: *Equation* (14) *defines the level of knowledge intensity θ as an increasing function of f, $\theta = \theta(f)$, with the property that $\theta(0) = 0$ and $\theta(1) = \infty$.*

Proof: By using $\mathrm{d}\theta q(\theta)/\mathrm{d}\theta = (1-\eta)q(\theta)$, total differentiating (14) with respect to θ and f yields

$$\frac{\mathrm{d}\theta}{\mathrm{d}f} = \frac{\delta + \theta q(\theta)}{\delta + \eta\theta q(\theta)} \frac{(\lambda/v)[\delta + \theta q(\theta)] + \delta\theta}{\delta(1-f)} \tag{23}$$

which proves monotonicity. The properties of the function $\theta(f)$ at the boundary follow directly from (14) and (Al) together with the fact that the elasticity of the right hand side of (14) with respect to θ is less than one.

Properties of the TE-schedule

PROPOSITION 2: *Given* (14), (13) *determines the growth rate g, as a function of f, $g = g(f)$ with the property that* $g(0) = g(1) = 0$. *Steady state growth is maximised at an allocation such that* (20) *is satisfied.*

Proof: Note that (11) together with proposition 1 allow to define entrepreneurial slackness as a decreasing function of f, $s = s(f)$. After substituting in (13), one obtains that

$$g = g(f) = \sigma m\left[\frac{\lambda}{v}Cf, s(f)\right],$$

which explicitly defines the function $g(f)$.

To conclude the proof, notice that from (23), it follows that

$$\frac{ds(f)}{df} = \frac{d}{df}\left[\frac{\delta(1-f)}{\delta+\theta q(\theta)}\right] = -\frac{\delta+(\lambda/v)(1-\eta)q(\theta)}{\delta+\eta\theta q(\theta)},$$

thus the derivative of $g(f)$ reads as

$$\frac{dg(f)}{df} = \frac{\sigma C\delta q(\theta)[(\lambda/v)(1-\eta)-\eta\theta]}{\delta+\eta\theta q(\theta)},$$

which proves (20).

Properties of the FE-schedule

PROPOSITION 3: *Given* (14) *and the associated function* $\theta = \theta(f)$, (9) *defines the growth rate of the economy* g *as an increasing function of* f, $g = e(f)$, *with the properties that* $e(0) = -\infty$ *and* $e(1) = \infty$.

Proof: Multiply the left and right hand side of (9) by $[r - g + \delta + \theta q(\theta)\beta]$. The resulting left hand side of the equation is strictly increasing in θ and equal to $h(r + \delta - g)$ when $\theta = 0$, while the resulting right hand side is strictly decreasing in θ and it approaches infinity when θ goes to zero since $(\lambda/v)(1-\beta)(P-h) > 0$ by assumption. This implies that, for given g, a θ that solves (9) always exists. Furthermore, an increase in g causes a downward (upward) shift in the left (right) hand side so that θ unequivocally increases. As $\theta = \theta(f)$ is increasing, $e(f)$ is also increasing. The properties of the function $e(f)$ at the boundary follow immediately from condition (Al) and the properties of $\theta(f)$ established in Proposition 1.

PROPOSITION 4: *There exist values of the parameters such that any point on the TE-schedule,* $g(f)$, *can be sustained as an equilibrium.*

Proof: Note that, for a given level of g, changes in P, h, χ and β modify the value of θ that solves (9) without affecting the function $\theta(f)$. Hence, one can find a combination of parameters such that arbitrary given values of f and g are on the FE-schedule. Since changes in these parameters do not affect the TE-schedule, any point along $g(f)$ can possibly be sustained as an equilibrium.

Efficiency Results

The derivative of W with respect to f. Partially differentiating (16) with respect to k_t and n_t and after using the fact that in steady state $\dot{k}_t = \dot{n}_t = 0$ yields

$$W_k = \frac{-\chi + (1-\eta)q(\theta)(W_n + \sigma\bar{W})}{r - g + v} \tag{24}$$

and

$$W_n = \frac{P - h - \sigma\eta\theta q(\theta)\bar{W}}{r - g + \delta + \eta\theta q(\theta)}. \tag{25}$$

Differentiating (16) with respect to f and evaluating this derivative in steady state yields

$$\frac{dW}{df} = \frac{C}{r-g}[-h - \eta\theta q(\theta)(W_n + \sigma\bar{W}) + \lambda W_k]$$

which after using (24) immediately leads to expression (18) in the main text.

Relation between θ^* and the social optimum. To prove that the social optimum always lies on the positively sloped arm of the *TE*-schedule, evaluate (18) at $\theta^* = [\lambda(l - \eta)]/v\eta$. This immediately yields

$$\frac{\mathrm{d}W}{\mathrm{d}f} = -\frac{C}{r-g}\left[h + \frac{\lambda\chi}{r-g+v} + \frac{(r-g)\eta\theta^* q(\theta^*)S^E}{r-g+v}\right]$$

which is negative.

The efficient sharing rule. I now prove that it exists $\beta < \eta$ such that the decentralised economy implements the optimal allocation. From (9) it follows that

$$h + \frac{\lambda\chi}{r-g+v} = \left[\frac{\lambda(1-\beta)}{r-g+v} - \beta\theta\right] q(\theta)S, \tag{26}$$

which substituted into (18) and after using the definition of S^E in (19) yields

$$\frac{\mathrm{d}W}{\mathrm{d}f} = \frac{C(\beta-\eta)q(\theta)\{\theta(r-g+\delta) - \lambda[r-g+\delta+\theta q(\theta)]/(r-g+v)\}S}{(r-g)[r-g+\delta+\eta\theta q(\theta)]} + \frac{\sigma C q(\theta)(r-g+\delta)[\lambda(1-\eta)/(r-g+v) - \eta\theta]\bar{W}}{(r-g)[r-g+\delta+\eta\theta q(\theta)]}.$$

Evaluating this derivatives at $\beta = \eta$, one can see that the first term is equal to zero while (26) implies that the second term is strictly positive, so that, at this level of β, research effort is too low. Now note that (9) implies that f is negatively related to β, so that β must fall below η for the economy to reach efficiency. There are two exceptions where at $\beta = \eta$, $\mathrm{d}W/\mathrm{d}f = 0$. The former arises when $\sigma = 0$, the latter when $h + \lambda\chi/(r - g + v) = 0$ so that, by (26), $[\lambda(1 - \eta)]/(r - g + v) - \eta\theta = 0$.

References

Acemoglu, D. (1996). 'A micro-foundation for social increasing returns in human capital accumulation', *Quarterly Journal of Economics*, vol. 111(3), pp. 779–804.

Acemoglu, D. (1997). 'Training and innovation in an imperfect labour market', *Review of Economic Studies*, vol. 64, pp. 445–64.

Adams, J. (1990). 'Fundamental stocks of knowledge and productivity growth', *Journal of Political Economy*, vol. 98, pp. 673–702.

Aghion, P. and Howitt, P. (1992). 'A model of growth through creative destruction', *Econometrica*, vol. 60, pp. 323–51.

Aghion, P. and Howitt, P. (1998). *Endogenous Growth Theory*, Cambridge, MA: MIT Press.

Bagliano, F., Brandolini A. and Dalmazzo, A. (1991). 'The OECD-CEP data set', Centre for Economic Performance Working Paper No. 118, June.

Baumol, W. (1990). 'Entrepreneurship: productive, unproductive, and destructive', *Journal of Political Economy*, vol. 98(5), pp. 893–921.

Benhabib, J., Rogerson, R. and Wright, R. (1991). 'Homework in macroeconomics: household production and aggregate fluctuations', *Journal of Political Economy*, vol. 99(6), pp. 1166–87.

Blanchflower, D. G. and Oswald, A. (1998). 'What makes an entrepreneur?', *Journal of Labour Economics*, vol. 16, pp. 26–60.

Caballero, R. and Hammour, M. (1996). 'The fundamental transformation in macroeconomics', *American Economic Review, Papers and Proceedings*, vol. 86(2), pp. 181–6.

Caballero, R. and Jaffe, A. (1993). 'How high are the giants shoulders: an empirical assessment of knowledge spillovers and creative destruction in a model of economic growth', in (O. Blanchard, and S. Fischer, eds), *NBER Macroeconomics Annual 1993*, Cambridge MA: MIT Press, pp. 15–74.

Cornish, W. R. (1989). *Intellectual Property: Patents, Copyright, Trade Marks and Allied Rights*, 2nd edn, London: Sweet & Maxwell.

Davidson, P. and Wilklund, J. (1997). 'Values, beliefs and regional variations in new firm formation rates', *Journal of Economic Psychology*, vol. 18, pp. 179–99.

Diamond, P. (1982). 'Wage determination and efficiency in search equilibrium', *Review of Economic Studies*, vol. 49, pp. 217–27.

Evans, D. and Jovanovic, B. (1989). 'An estimated model of entrepreneurial choice under liquidity constraints', *Journal of Political Economy*, vol. 97(4), pp. 808–27.

Evans, D. and Leighton, L. (1989). 'Some empirical aspects of entrepreneurship', *American Economic Review*, vol. 79(3) (June), pp. 519–35.

Griliches, Z. (1979). 'Patents: recent trends and puzzles', *Brookings Papers on Economic Activity, Microeconomics*, pp. 291–330.

Griliches, Z. (1990). 'Patent statistics as economic indicators: a survey', *Journal of Economic Literature*, vol. 28 (December), pp. 1661–707.

Helpman, E. and Trajtenberg, M. (1994). 'A time to sow and a time to reap: growth based on general purpose technologies', NBER Working Paper Series, No 4854, September.

Hosios, A. (1990). 'On the efficiency of matching and related models of search and unemployment', *Review of Economic Studies*, vol. 57, pp. 279–98.

Howitt, P. (1999). 'Steady endogenous growth with population and R&D inputs growing', *Journal of Political Economy*, vol. 107(4), pp. 715–30.

Jones, C. (1995*a*). 'Time series tests of endogenous growth models', *Quarterly Journal of Economics*, vol. 110, pp. 495–525.

Jones, C. (1995*b*). 'R&D based models of economic growth', *Journal of Political Economy*, vol. 103(4), pp. 759–83.

Jones, C. (1999). 'Growth: with or without scale effects', *American Economic Review Papers and Proceedings*, vol. 89(2), pp. 139–44.

Kortum, S. (1993). 'Equilibrium R&D and the decline in the patent-R&D ratio', *American Economic Review: Papers and Proceedings*, vol. 83(2), pp. 450–7.

Kortum, S. (1997). 'A model of research, patenting and productivity growth', *Econometrica*, vol. 65(6), pp. 1389–419.

Malthus, T. R. (1986 [1798]). *An Essay on the Principle of Population*, London: W.Pickering.

Mokyr, J. (1990). *The Lever of Riches: Technological Creativity and Economic Progress*, Oxford: Oxford University Press.

Mortensen, D. T. and Pissarides, C. (1998). 'Technological progress, job creation, and job destruction', *Review of Economic Dynamics*, vol. 1(4), pp. 733–53.

Murphy, K., Shleifer, A. and Vishny, R. (1991). 'The allocation of talent: implications for growth', *Quarterly Journal of Economics*, vol. 56, pp. 503–30.

Pissarides, C. A. (1990). *Equilibrium Unemployment Theory*, Oxford: Basil Blackwell.

Ricardo, D. (1951 [1817]). *On the Principles of Political Economy and Taxation*, Cambridge: Cambridge University Press.

Romer, P. (1990). 'Endogenous technological change', *Journal of Political Economy*, vol. 98(5), pp. 71–102.

Schumpeter, J. A. (1939). *Business Cycles: a Theoretical, Historical and Statistical Analysis of the Capitalist Process*, New York: McGraw-Hill.

Schumpeter, J. A. (1943). 'Capitalism in the postwar world', in (S. E. Harris, ed.), *Postwar Economic Problems*, New York: McGraw-Hill, pp. 113–26.

Schumpeter, J. A. (1946). 'Capitalism', in *Encyclopaedia Britannica*, vol. IV, Encyclopaedia Britannica Inc., pp. 801–7.

Schumpeter, J. A. (1947). 'The creative response in economic history', *Journal of Economic History*, vol. 7 (November), pp. 149–59.

Schumpeter, J. A. (1949). 'Economic theory and entrepreneurial history', in (Research Center in Entrepreneurial History, Harvard University) *Change and the Entrepreneur*, Cambridge, MA: Harvard University Press, pp. 63–84.

Tirole, J. (1988). *The Theory of Industrial Organization*, Cambridge, MA: MIT Press.

Young, A. (1998). 'Growth without scale effects', *Journal of Political Economy*, vol. 106(1), pp. 41–63.

Part VI
Theoretical Extensions and Evidence

[27]

Industrial and Corporate Change, Volume 18, Number 6, pp. 1193–1229
doi:10.1093/icc/dtp043
Advance Access published October 13, 2009

Knowledge creation, entrepreneurship, and economic growth: a historical review

Bo Carlsson, Zoltan J. Acs, David B. Audretsch and Pontus Braunerhjelm

This article explores the relationship between knowledge creation, entrepreneurship, and economic growth in the United States over the last 150 years. Distinguishing between general knowledge and economically useful knowledge, we examine the changes over time in the locus and content of new knowledge creation: the role of universities, particularly engineering schools and land-grant universities, industrial laboratories, and corporate research and development (R&D) laboratories prior to World War II. The practical orientation of US academic R&D and the close research interaction between academia and industry are noted. We study the unprecedented increase in R&D spending in the United States during and after World War II and how it was converted into economic activity via incumbent firms in the early postwar period and increasingly via new ventures in the last few decades.

1. Introduction

It is generally accepted that the creation of new knowledge is an important driver of economic growth. But the mechanisms through which new knowledge gives rise to economic growth are not well understood. For example, why is it that large investments in research and development (R&D; such as in Japan and Sweden) have not resulted in high economic growth in the last two decades? Why does entrepreneurial activity play a more important role in some countries (e.g. the United States) than in others (Europe and Japan)? And how and why do these relationships shift over time? The underlying question is: why doesn't new knowledge always result in (new) economic activity?

In a series of papers (Acs *et al.*, 2004, 2005, 2009) we have developed a model that distinguishes between knowledge and economically useful knowledge (following Arrow, 1962) by introducing the notion of a knowledge filter that prevents knowledge from becoming economically useful. We have also identified entrepreneurship as a mechanism (in addition to incumbent firms) that converts economic knowledge into economic growth.

The purpose of this article is to explore the locus and content of knowledge creation over time and to explain how it influences innovation and economic growth. We focus primarily on the United States, occasionally comparing the US experience with that elsewhere.

The article is organized as follows. We begin with a conceptual overview of the knowledge creation system and the role of filters or obstacles that prevent knowledge from resulting in economic activity. We then present a historical overview of the organization of academic and industrial research in the United States and the links between knowledge creation and economic growth. We start with a discussion of how the industrial revolution was based in part on turning knowledge into economically useful knowledge and how university education and research in the United States became practically and vocationally oriented (in comparison with European universities), partly through the land-grant universities established in the mid- to late 19th century. In the early part of the 20th century, corporate R&D labs began to emerge as major vehicles of basic industrial research. Virtually all of the funded research prior to World War II was conducted in corporate or federal labs. In conjunction with a rapidly increasing share of the population with a college education, this made for high absorptive capacity on the part of industry and, as a result, strong links between knowledge creation and economic growth. In subsequent sections we discuss the emergence of the research university, the dramatic increase in R&D spending, and the shift of basic research toward the universities, especially during and following World War II. During the 1960s and 1970s, knowledge creation became less focused on economically useful knowledge, leading to an increasing need to "translate" basic (academic) research into economic activity. New firms have increasingly become the vehicle to translate research into growth; this can be seen in the greater role of small business and entrepreneurship from the 1970s onward. Meanwhile, the R&D of large incumbent firms has become more applied and oriented toward building absorptive capacity rather than pushing out the knowledge frontier. We conclude with a discussion of the implications for innovation, entrepreneurship, and economic growth.

2. The knowledge creation system

A schematic picture of the knowledge creation system is presented in Figure 1.[1] There are two main types of research: academic and industrial. The former is primarily basic and is carried out in universities or research institutes, while industrial R&D primarily involves applied research or development work and is carried out in

[1]Even though Figure 1 may give the impression of a linear process from academic research to commercialization, there are numerous cases of loops and feedbacks, as indicated later in the text.

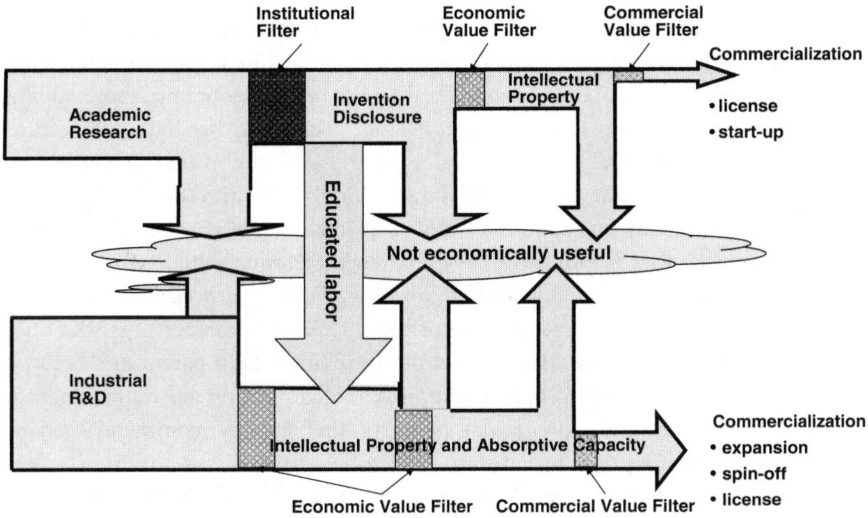

Figure 1 The knowledge filter in the knowledge creation system.

industrial firms or government laboratories.[2] In the United States, basic R&D (about 60% of which is performed in universities) currently makes up slightly less than 20% of the total R&D, while applied research makes up a little more than 20%. The remaining 60% is development R&D (NSF, 2006a). A large part of academic research and a substantial part of industrial R&D have no direct commercial value (i.e. is not economically useful).

The part of industrial R&D (in the lower part of Figure 1) that is judged to have potential economic value (i.e. passes the economic value filter) can be turned into intellectual property to be subsequently commercialized (if it turns out to have sufficient commercial value), or it can be indirectly commercialized by increasing the absorptive capacity or knowledge base of the company. Commercialization can be accomplished via expansion of the activities (new or improved goods and services) in existing firms—the main vehicle—or via spin-off to new entities, or via licensing to other firms.

[2]The Frascati Manual defines *basic research* as "experimental or theoretical work undertaken primarily to acquire new knowledge of the underlying foundation of phenomena and observable facts, without any particular use in view. *Applied research* is also original investigation undertaken in order to acquire new knowledge. It is, however, directed primarily toward a specific practical aim or objective. Experimental development is systematic work, drawing on existing knowledge gained from research and/or practical experience, which is directed to producing new materials, products, or devices, to installing new processes, systems and services, or to improving substantially those already produced or installed" (OECD, 2002: 30).

Most academic research (in the upper part of Figure 1) is basic and has no immediate economic value. Basic sciences such as chemistry, physics, and mathematics have no immediate economic value but are essential in more applied fields whose output does have economic value.[3] There are two types of research output that has potential economic value. The most important is in the form of knowledge that graduating students take with them into the labor market, i.e. educated labor (human capital, including skilled researchers). Another important output is research that is judged to have potential economic value, i.e. it passes the institutional filter. The first step in the commercialization process is an invention disclosure. The researcher/inventor and/or the technology transfer organization of the employer (university) decides whether or not to apply for a patent. If this results in a patent application and if a patent is approved, the invention passes the economic value filter and becomes intellectual property that can be commercialized via licensing to an existing firm or via start-up of a new firm.[4]

The basic knowledge produced in academia may be in areas (such as the humanities) that have little or no economic value (even though it may have significant human value). The research may not be advanced enough to be on the knowledge frontier, or it may not be sufficiently broad, deep, or systematic to be useful in applied research or development that can lead to commercialization. In practical terms, the allocation of funding of academic research in the United States may be viewed as reflecting the perceived economic value of research in various disciplines. In 2004, the life sciences received nearly 60% of all R&D funding at US universities and colleges, all other sciences (including social sciences) about 25%, and engineering the remaining 15% (NSF, 2006a).

The barriers to converting research into commercialized knowledge may be referred to as the "knowledge filter". The first component of the knowledge filter as far as academic research is concerned (besides the "roundaboutness" of converting basic science into applied knowledge) may be referred to as the institutional filter. It consists of organizational barriers, university policies, attitudes among faculty and university administrators against commercialization of research, and lack of incentives to pursue commercialization. The main output of academic research is often highly skilled labor. The higher the barriers to commercialization of research, the greater is the share of research dissemination via skilled labor.

---

[3]The relationship between the science of chemistry and the body of knowledge concerning certain manufacturing processes referred to as chemical engineering is a case in point (Rosenberg, 2000: 84).

[4]Only about half of the invention disclosures in US universities result in patent applications; half of the applications result in patents; only one-third of patents are licensed; and only 10–20% of licenses yield significant income (Carlsson and Fridh, 2002: 231). In other words, only 1% or 2% of the inventions are successful in reaching the market and yielding income.

The second and third components of the academic knowledge filter are the economic and commercial value filters which reflect the capability to convert invention disclosures into intellectual property (primarily in the form of patents) and then to commercialize the intellectual property via licenses and start-ups.

There are similar filters for industrial R&D reflecting the difficulty in business organizations to convert research into intellectual property and to commercialize new products. The greater are the obstacles to commercialization of research, the thicker is the knowledge filter.

The research efforts of academic institutions have increased enormously over time, with varying economic impact. On the whole, US universities have been much more successful in commercializing research than their foreign counterparts. But external funding of academic research is a relatively recent phenomenon; the research university as we know it today did not emerge in the United States until around World War II. It is only in the last few decades that basic academic research has begun to play an important role in the economy. We turn now to a brief historical overview.

3. Historical overview of academic and industrial research and the knowledge filter

3.1 Universities and their role in knowledge creation

Universities began to arise in Europe during the Middle Ages.[5] They developed out of monastery and cathedral schools, attended by adolescents and taught by monks and priests. One of the firsts was in Bologna, established in 1088, followed by Paris (1160), Modena (1175), and Oxford (1190). The curriculum consisted of the art of reading and writing, focusing on the Bible and the Latin and Greek classics, rhetoric, and logic. Knowledge meant logic, grammar, and rhetoric; it did not mean ability to do, or utility. The role of the universities was to collect, codify, and teach general knowledge. Utility or economic knowledge was not thought of as knowledge but rather as skill—the Greek word *téchne*. The only way to learn *téchne* or skill was through apprenticeship and experience.

The establishment of the first engineering schools in the mid-18th century represents the beginning of codification of economically useful knowledge. The French *École des Ponts et Chaussées* was the first engineering school, founded in 1747. It was followed around 1770 in Germany by the first school of agriculture, and in 1776 by the first school of mining. The first technical university, *École Polytechnique* in France, was established in 1794. Meanwhile in Britain the application of patents shifted from establishing monopolies to enrich royal favorites

[5]There were earlier precedents in China, India, and the Middle East.

(first granted in the mid-15th century) to patents being granted to encourage the application of knowledge to tools, products, and processes, and to reward inventors, provided that they published their inventions (Drucker, 1998, pp. 18–21).

> None of the technical schools of the eighteenth century aimed at producing *new* knowledge... None even talked of the application of science to tools, processes and products, that is, to technology. This idea had to wait for another hundred years until 1849 or so, when a German chemist, Justus Liebig (1803–1873), applied science to invent first, artificial fertilizers and then a way to preserve animal protein, the meat extract. What the early technical schools... did was, however, more important perhaps. They brought together, codified and published the *téchne*, the craft mystery, as it had been developed over millennia. They converted experience into knowledge, apprenticeship into textbook, secrecy into methodology, doing into applied knowledge. These are the essentials of what we have come to call the 'Industrial Revolution,' i.e. the transformation by technology of society and civilization worldwide (Drucker, 1998: 21).

The first university in the United States was Harvard College, founded in 1636 and established in the tradition of European universities. Seven of the nine colleges founded in the colonial era (all of which were private)[6] were oriented to a classical curriculum in preparation for civic leadership or ministry. After independence, private and public universities were established in parallel. The University of Georgia (1785) was the first state university. The US Military Academy at West Point, founded in 1802, was the first engineering school, followed by Rensselaer Polytechnic Institute in 1824 (Rosenberg and Nelson, 1994: 327).

After the Civil War, the Morrill Land-Grant College Act (1862) led to the establishment of universities in every state. Unlike most private universities, these state universities and colleges were charged with public service obligations in agricultural experimentation and extension services,[7] industrial training, teacher education, home economics, public health, and veterinary medicine (Graham and Diamond, 1997: 18). Some of the land-grant universities were private (e.g. MIT). One of their main features was a strong practical/vocational orientation in both education and research. While the emphasis was on teaching branches of learning related to agriculture and the mechanical arts in addition to the liberal arts, there was also research. The agricultural experiment stations at the land-grant universities

[6]Two of these, the College of Willliam and Mary and Queen's College (now Rutgers) were later converted to public universities.

[7]After the Hatch Act of 1887 which allowed government support of the experiment station program (Nelson and Wright, 1992: 1942).

played a particularly important role not only in advancing knowledge in fields of practical and economic relevance but also in making the practical application of research acceptable if not required in US academic institutions. The Morrill Act also stimulated engineering education. The number of engineering schools increased from 6 to 70 within a decade, and the number of engineering graduates grew from 100 in 1870 to 4300 in 1914 (Nelson and Wright, 1992: 1942). Since the research agenda was driven primarily by the needs of the rapidly expanding economy and largely involved applied research, this made for a thin knowledge filter. The new knowledge was easily converted into economic activity.

Meanwhile in Europe, the tradition of higher education continued to focus on teaching, but the advent of Humboldt University in Berlin in 1809 represented a new approach, focused on research. This was the first research university; adherence to scientific and scholarly discipline emerged as preeminent, not worldly application of knowledge.

As a result of the different origins of US colleges and universities with respect to both ownership and mission, the American system of higher education has been decentralized and pluralistic as well as competitive from the very beginning, in clear contrast to Europe:

> By the late nineteenth and early twentieth centuries, higher education systems in Europe were typically centralized under a national ministry of education. As a consequence, higher education policy was essentially government policy. Like most public bureaucracies, Europe's state-dominated systems of higher education were organized hierarchically by function. Competition was minimized by bureaucratic boundaries, much as it was in ministries of justice, war, or public health. Teaching faculty were typically civil servants. The European university was the training ground for the middle and professional classes, and for this reason, attendance was confined to a small, closely screened cadre of academically talented students who sought advanced professional and vocational training rather than general education in the liberal arts, the goal of American college students... Even though chair professors dominated their institutes, the basic decisions about university budgets, student admissions, and academic programs were made by central ministry officials, not by campus academic officers. The centralized system of European higher education achieved organizational rationality and bureaucratic efficiency at the expense of competition and innovation (Graham and Diamond, 1997, p. 12–13).

In European countries, ministries of education have typically paid the costs and set the agenda for the university. As a result, European universities have had limited authority to manage their own size and shape, their entry or exit requirements, and their broad character and function (*ibid.*: 23). By contrast, American private universities are typically governed by a board of nonresident, nonacademic trustees, led by a powerful president, and independent of control or support by the state.[8]

The decentralized and pluralistic American system allowed for much faster expansion than that in Europe without requiring structural change. In 1910, about 330,000 students were studying at almost 1000 colleges and universities in the United States (whose population was 92 million), while there were only 14,000 students in sixteen universities in France (with a population of 39 million). This represented about 4% of the college-age population in the United States vs. about 0.5% in France.[9] In most European countries (including Britain), college enrollments did not exceed 5% of the college-age group until after World War II (*op. cit.*: 15). By 1940, the percentage of college-age Americans attending institutions of higher education was three times higher than the European average (roughly 12% in the United States vs. 4% in Europe) (*op. cit.*: 24).

The expansion of the US system of higher education allowed it to cater not only to a rapidly growing population but also to increasing percentages of each cohort demanding higher education. This set the US system apart from its European competitors. It contributed importantly to the creation of a relatively highly educated industrial labor force, i.e. a relatively high capacity to absorb new technology. And this contributed significantly to the rising economic power and competitiveness of the United States.

---

8 Almost without exception, American universities were built around a large, core college of arts and sciences, organized into discipline-based departments in which faculty appointments and tenure were based. This common organizational form traced its origin to the colonial colleges and, by the nineteenth century, to the peculiar need for American undergraduate education to fill the void left by a democratic system of public secondary schools that valued high graduation rates over demanding standards. The raison d'être of "the college" was thus to provide baccalaureate education in the liberal arts and sciences for residential undergraduates, while the graduate school offered masters and doctoral degrees, and separate schools offered professional degrees...

When the German model of the research university was imported to the United States by way of Johns Hopkins late in the nineteenth century, it was admired and emulated. But it was also quickly Americanized. The research and graduate orientation of the German model took a prominent and permanent place in the hierarchy of American academic prestige. But the American graduate school was superimposed on the colleges of arts and sciences, with their undergraduate-centered departmental organization. By the early twentieth century even Johns Hopkins looked more like Yale, or like the University of North Carolina, than like von Humboldt's model in Berlin" (Graham and Diamond, 1997: 19).

9 The average years of formal higher educational experience of the population aged 15–64 years in 1913 was 0.20 years in the United States vs. 0.10 in France, 0.09 in Germany, 0.11 in the Netherlands, and 0.08 in the UK—Rosenberg and Nelson, 1994: 325, citing Maddison (1987).

The rapid growth of the US economy between the Civil War and World War I was founded on rising productivity in agriculture and the emergence of new engineering-based industries: mechanical, electrical, and chemical engineering, telecommunications, and instrumentation. To a large extent this was the result of the "hands-on" practical problem-solving nature of academic research in the United States. As early as 1830, Alexis de Tocqueville commented on the practical orientation of science and attitudes toward science in America, in contrast to the more theoretical and abstract orientation in Europe (Rosenberg and Nelson: 324). "Whereas in Great Britain, France and Germany, engineering subjects tended to be taught at separate institutions, in the United States such subjects were introduced at an early date into the elite institutions. Yale introduced courses in mechanical engineering in 1863, and Columbia University opened its School of Mines in 1864" (*ibid.*: 327). This interdependence between the needs of the growing economy and the rise of university education is the reason Rosenberg has referred to American universities as "endogenous institutions" (Rosenberg, 2000).

After the breakthroughs in electricity research around 1880, American universities responded almost instantly to the need for electrical engineers. In the same year (1882) in which Edison's Pearl Street Station in New York City went into operation, MIT (founded in 1865) introduced its first course in electrical engineering. Cornell followed in 1883 and awarded the first doctorate in the subject in 1885. By the 1890s schools like MIT had become the chief suppliers of electrical engineers. Throughout the entire twentieth century, American schools of engineering have provided the leadership in engineering and applied science research upon which the electrical industries have been based (*ibid.*, p. 327–328).

The story is similar in chemical engineering. Even though Britain was the "workshop of the world" and had the largest chemical industry in 1850, this industry was based on its role as supplier to the textile manufacturers of such essential inputs as bleaches and detergents and of soda and sulfuric acid to the glass industry, not on professional engineering competence. In fact, there were no departments of chemical engineering in Britain or anywhere else outside the United States until the 1930s,[10] and the Institution of Chemical Engineers was not founded until 1922. By contrast, MIT offered the first course in chemical engineering in 1888 and established the School of Chemical Engineering Practice in 1915 (Rosenberg, 2000: 88). Several American universities established chemical engineering departments in the first decade of the 20th century, and the American Institute of Chemical Engineers was founded in 1908 (Rosenberg, 1998, p. 193–200). The preeminence of the United States in chemical engineering was based on the insight that industrial applications of chemistry involve not only a scaling up of scientific discoveries but also integration

[10]In Germany, chemical engineering did not emerge as a distinct subject until after World War II (Rosenberg, 2000: 103).

with skills from a wide variety of engineering fields. It was based also on close collaboration between academic scientists and industrial scientists in corporate R&D laboratories that were being established during the same time period. The knowledge that was being transferred was practical and informal rather than formalized as intellectual property. And the knowledge flows were not necessarily vertical (from upstream to downstream); for example, fluidized bed catalytic cracking was largely a joint achievement of MIT and the Esso laboratories. Also, chemical engineering as an academic discipline became a vehicle (general-purpose technology) for "horizontal" diffusion of knowledge into the petrochemical industry (Rosenberg, 2000, p. 98–103).

But while the US performance was strong in the practical application of scientific discoveries, basic science itself was still relatively weak. Even as late as during the first half of the 1920s Americans often traveled to Europe to learn, and when European scientists traveled to the United States, they did so primarily to teach. An aspiring student seeking the best available academic education in basic scientific disciplines such as physics or chemistry would have been well advised to study in Germany, Britain, or France (Nelson and Wright, 1992: 1941). But this situation was changing by the end of the decade as American scientists began to reach the scientific frontier. Leading US scientists attained parity with the leading Europeans in some fields before events in Europe forced the intellectual migration of the 1930s (Geiger, 1986, p. 233–234). But by and large, basic science in the United States remained behind that in the leading European countries, and external research funding was still quite small.

During the interwar period, US academic research was funded primarily by philanthropic foundations (such as the Rockefeller and Carnegie Foundations) and large corporations (e.g. Du Pont, General Electric, Borden, and Lilly). The federal government was not involved in funding academic research at this time. The total value of foundation grants to academic institutions was only on the order of $50 million in 1931 and then fell dramatically as the depression deepened. It rose again in the late 1930s but attained only $40 million (about $450 million in 2006 dollars) in 1940.[11] The externally funded academic research was also concentrated in just a handful of institutions. Among sixteen preeminent universities[12] only six spent more than $2 million annually (from all sources, including internal) in the late

[11]The fact that external research funding was so limited does not mean that no research was conducted, only that it was focused on areas that did not require expensive equipment or large laboratories. "Big science" projects in science, engineering, and medicine were new phenomena that did not come about until during and after World War II.

[12]These sixteen were UC Berkeley, Chicago, the California Institute of Technology (Caltech), Columbia, Cornell, Harvard, Illinois, Johns Hopkins, MIT, Michigan, Minnesota, Pennsylvania, Princeton, Stanford, Wisconsin, and Yale.

1930s, and four spent less than $1 million. Eleven of these were private institutions (Graham and Diamond, 1997: 28). Most research grants were small and made to individuals rather than to institutes or schools; there were few if any institutional grants.

But even though external funding of academic research was small, the issue of commercialization of academic research arose. As early as in 1912, a new entity, the Research Corporation, was set up to develop and strengthen the patents covering electrostatic precipitation technologies developed by Professor Frederick Cottrell at the University of California. In the 1930s, the Research Corporation began to manage the patenting and licensing activities of other research universities who were reluctant to handle these activities themselves (Mowery and Sampat, 2001a, p. 318–319). Also, in 1924 the Wisconsin Alumni Research Foundation (WARF) was set up as a legally independent organization to receive and license patents to University of Wisconsin faculty (Shane, 2004). Several other state universities also set up similarly affiliated but legally separate foundations to manage patenting and licensing. And in 1932, MIT instituted a policy that asserted its rights to all faculty inventions. But rather than taking on a direct role in managing patents and establishing its own affiliated foundation, MIT signed a contract with the Research Corporation to manage its patent portfolio. Other universities also signed similar agreements with the Research Corporation. However, most US universities did not have any patent policies prior to World War II, and several leading universities (including Harvard, Penn, Chicago, and Johns Hopkins) had adopted policies that discouraged patenting or prohibited faculty from patenting (Mowery and Sampat, 2001b, p. 788–789).

Thus, although both American universities and American industry had increased their capacities for scientific research during the interwar period, the best among them having reached parity with Europe's best in some areas by the early 1930s, the total research effort was still small compared to what was to come during and after World War II. The contribution of university research to economic growth was quite modest. Most of the limited basic research that was done was carried out in industrial and government labs. Industrial expenditures for basic and applied research had reached $200 million in 1939, up from just over $100 million a decade earlier and $30 million at the start of the 1920s (Geiger, 1993: 4). Expenditures for research in government and industry, overwhelmingly applied in character, were ten times university expenditures for basic research in 1940 (*op. cit.*: 14). As a result, a large part of the new knowledge created (though limited in scope, particularly in the universities) was directly economically useful and therefore relatively easily transformed into economic activity.

By 1945 there were 641 public institutions in 48 state systems and about 1100 private institutions with a combined total enrollment of about 1.6 million students evenly split between public and private institutions. Both the public and the private institutions varied widely in purpose, size, and quality. Even though the number of public institutions more than doubled between 1950 and 1988 (from 641 to 1548), the private institutions were still in a majority. The strongest growth was in public two-year community colleges (Graham and Diamond, 1997, p. 15–16; Snyder, 1993).

3.1.1 Early 20th century: the emergence of corporate R&D labs

As mentioned already, the research conducted in American and European universities in the earlier part of the 20th century was quite limited. But as the number of universities grew in the United States—particularly in the first decade after the Civil War and in the 1890s when many land-grant universities were established—and with it the number of graduates, the skills of the industrial labor force increased. New science-based industries emerged that constituted the core of the "Second Industrial Revolution"—those relying on chemical engineering, electricity, and the internal combustion engine. But "relatively little of the American performance during this era was based in science, nor even on advanced technical education. American technology was practical, shop-floor oriented, built on experience. The level of advanced training in German industry was substantially higher" (Nelson and Wright, 1992: 1938). The new industries needed new knowledge, but the universities did not possess the specialized knowledge, equipment, and organization that was required. Instead, a new mechanism emerged in the form of industrial laboratories.

During the latter half of the 19th century a number of industrial labs were established in the United States. There were at least 139 by the turn of the century (Mowery, 1981, cited in Rosenberg, 1985: 51). The earliest industrial labs did not perform activities that could be regarded as research; they were set up to apply existing knowledge, not to make new discoveries. They were organized to engage in a variety of routine and elementary tasks such as testing and measuring in the production process, assuring quality control, standardizing both product and process, and meeting the precise specifications of customers (Chandler, 1985: 53). But as Rosenberg points out, this development was linked to the expansion of higher education in the United States:

> The growing utilization of scientific knowledge and methodology in industry was vastly accelerated by an expanding pool of technically trained personnel—especially engineers. Associated with this expansion was the growth in the number of engineering schools, engineering programs, and the engineering subspecialties in the second half of the nineteenth century.... But it is essential... to realize that it was the larger body of scientific knowledge, and not merely frontier science, that was relevant to the needs of an expanding industrial establishment (Rosenberg, 1985: 24).

Thus, the pioneering efforts by an increasing number of US universities to establish not only individual courses but also whole departments in chemical and electrical engineering ahead of their European counterparts built a foundation for rapid economic growth. They also established close collaboration between academia and engineering-based industries. The links between *science* and industry—the emergence of new bodies of scientific knowledge that were subsequently applied to industry—were established somewhat later in a second stage of development in the form of corporate R&D laboratories. The expansion of higher education coincided with the establishment of such laboratories.

US universities played an important role in the creation of corporate R&D laboratories, especially in chemical engineering, via collaborative research and consulting, and in developing expanded research capabilities over time, in addition to serving as the launching pad for the careers of individuals who found employment in private firm laboratories. There is also evidence of influence in the opposite direction, from firms to universities. By providing both financial support for university research laboratories and a market for future trained labor, firms supported the growth of scientific capabilities at local universities (MacGarvie and Furman, 2005, p. 4–5).[13]

> Critical to this second stage [the corporate R&D labs] was a separation of the testing, standardizing, and quality control functions from those of product and process development. This separation involved, first, the creation of a laboratory physically separate and usually geographically distant from the factory or factories of the enterprise. Of even more importance, it called for the creation of a separate, specialized organization to exploit the laboratory's activities. This organization usually took the form of a department separate from those responsible for production, distribution, and purchasing activities of the enterprise. The new department's objective was to define programs for the laboratory by monitoring both market and technological opportunities. Its most critical task was to integrate the activities of the research personnel with those of university professors working closer to the sources of scientific knowledge and with those of managers in the company's design, manufacturing, and marketing offices (Chandler, 1985: 54).

The primary function of this new type of department—the *corporate research and development* laboratory—was not basic research but rather the commercial development of products and processes. Shortly after the turn of the century, a handful of relatively new—but still big—businesses made a marriage between science and business by creating such laboratories. The first corporate R&D lab was established

[13]The Du Pont company funded graduate fellowships at 25 universities during the 1920s and expanded its program during the 1930s to include support for postdoctoral researchers. Other companies had similar programs (Mowery and Rosenberg, 1988: 24).

by Du Pont in 1902. It was soon followed by the leaders in the electrical, chemical, photographic, and telecommunications industries. These firms had one thing in common: They were based on technologies that had emerged in one way or another from scientific discoveries or developments and were particularly susceptible to significant, further improvement through a scientific approach to problem solving. As a result, science became a part of corporate strategy (Hounshell and Smith, 1988: 1). Another result was that a much larger fraction of business-supported research was conducted *within* firms in the United States compared with Europe where industry-wide associations or other arrangements played an important role (Rosenberg, 1985: 24).

But the modern corporate R&D laboratory did not emerge full blown from the minds of executives at such firms as Du Pont, General Electric, Eastman Kodak, and American Telephone and Telegraph. The first organized industrial laboratories had appeared in Germany in the 1870s, in firms that sought to commercialize inventions based on new breakthroughs in organic chemistry (MacGarvie and Furman, 2005: 9).What distinguished these German corporate chemical laboratories and set them apart from other approaches to innovation was that modern corporate chemical research called for massive scientific teamwork rather than the efforts of individual chemists. It also set the international pattern for the conduct of research in the chemical-related industries: most of it has been in-house at the firms, although some has been outsourced to universities. Also, once researchers had uncovered the mechanisms of reactions, they could then pursue the invention of new dyes in a highly regular, systematic way.

> The number of routine experiments that had to be conducted to find a single promising color was large. When such a color was discovered, it was sent to the dye-testing division, where it was subjected to a battery of tests to indicate whether and under what conditions it would tint anyone of the common fibers, or such other items as wood, paper, leather, fur, or straw. Then each item successfully tinted was subjected to several agents of destruction to determine fastness. Of 2378 colors produced and tested [by Bayer] in the year 1896, only 37 reached the market. This tedious, meticulous experimentation, in which a thousand little facts were wrenched from nature through coordinated massed assault, admirably illustrates the method and spirit introduced into scientific inquiry by the rising industrial laboratory of the late nineteenth century (John J. Beer, quoted in Hounshell and Smith, 1988: 5).

As a result of these developments, while an increasing share of American industrial production was based on new knowledge in the form of scientific discoveries, the supporting research activity took place primarily in corporate R&D labs, not in universities. The employment of scientists and engineers in corporate laboratories grew about tenfold, from less than 3000 in 1921 to about 28,000 in 1940 (Mowery and Rosenberg, 1998, p. 20–23). Up until the 1920s and 30s, research at universities had little to do with industrial innovation and economic growth, with the exception of organic chemicals and to a limited extent electrical equipment. The main contribution of universities was the education of highly skilled labor.

3.2 World war II and the emergence of the research university

The research university in the United States is largely a postwar phenomenon, although as indicated above, its foundations were laid long before. It was patterned after the German model (von Humboldt's University of Berlin), but in contrast to its European predecessor it emerged into a "decentralized, pluralistic, and intensely competitive academic marketplace fueled by federal research dollars" (Graham and Diamond, 1997: 2). Competitive pressures forced the American research universities to change much more quickly than did the sheltered public universities elsewhere.

The war effort led to an enormous scaling up of US research to an unprecedented level. Total federal R&D expenditures grew from $83 million in 1940 to $1314 million in 1945, in 1930 dollars (Mowery and Rosenberg, 1998: 28), while total industrial R&D expenditures were on the order of $200 million annually just before World War II. During the war, the Army Corps of Engineers alone spent $2 billion on the atomic bomb and the Radiation Laboratory at MIT expended $1.5 billion for radar systems (Geiger, 1993: 9). The increased research was guided by military needs and involved both basic research and its immediate application and development in the form of military goods and services (which translated directly into economic activity as these goods had to be paid for by the federal government).

> During World War II the US government harnessed the talent of the top scientists and engineers at these [top] universities, not—as had been done in World War I—by inducting them into uniformed service in war-related bureaucracies but by developing a new contract and grant system under the leadership of civilian scientific elites. The success story of the wartime Office of Scientific Research and Development (OSRD) is well known. To lead the mobilization of scientific manpower, President Roosevelt in 1940 summoned Vannevar Bush, former dean of engineering at MIT and, since 1938, president of the Carnegie Institution of Washington. Bush in turn brought in a powerful trio of senior associates: Karl T. Compton, president of MIT; James B. Conant, president of Harvard; and Frank B. Jewett, board chairman of Bell Telephone Laboratories and president of the National Academy of Sciences. A formidable group, they represented the major sectors of science and technology outside of government. The OSRD developed an intense

> and intimate model of collaboration between Washington and the nation's leading universities. The wartime development of radar, the proximity fuse, penicillin, DDT, the computer, jet propulsion, and the climactic trump card, the atomic bomb, brought enormous prestige to the scientific community. From the development of radar at MIT, through the control of nuclear fission at the University of Chicago, to the University of California's secret operations at Los Alamos, scientific brilliance in the national interest was associated with the great universities (Graham and Diamond, 1997: 28).

Thus, the magnitude, nature, and locus of US R&D changed dramatically in conjunction with World War II, necessitated by the war effort and funded by the federal government. Prior to the war, the federal government had almost no role in funding academic research. Its research funding went to government (intramural) labs and to defense contractors. Only government and industrial labs had the necessary resources to carry out large-scale, systematic, programmatic research. The sheer scale of the wartime research effort required the engagement of academic researchers as well. This, in turn, required massive investments in building up the research infrastructure of the universities.

Impressive as the federal effort was in mobilizing the nation's scientific resources to develop the means necessary to win the war, it is a story not only about research funding, brilliant scientists, and organization at the federal level. It is also a story about the internal culture and organization of the universities and about the relations between the universities and their external environment, particularly industry.[14] Granted that academic research in the United States was highly concentrated to a few elite institutions prior to World War II, had these elite universities not been ready to take up the challenge, it is unlikely that their response would have been as rapid and strong as it was. As a result of changes in its internal organization and incentive system during the 1930s, MIT was in a leading position; it is no accident that Vannevar Bush was picked by President Roosevelt to head up the wartime research effort.

As a private land-grant university, MIT had stood virtually alone as a university that embraced rather than shunned industry.

> From its start MIT developed close ties with technology-based industrialists, like Edison and Alexander Graham Bell, then later with

[14]During the interwar period the output of trained professionals was clearly more important than the research carried out in academic institutions: "In 1919... MIT launched its Cooperative Course in electrical engineering, a program that divided the students' time between courses at the Institute and at General Electric, which hired one-half of the students after graduation. The program was later joined by AT&T, Bell Labs, Western Electric, and other firms." (Nelson and Wright: 1949, quoting Noble, 1977: 192).

> its illustrious alumnus Alfred P. Sloan, during his pioneering years at General Motors, also with close ties to the growing petroleum industry. In the 1930s, MIT generated The Technology Plan, to link industry with MIT in what became the first and is still the largest university–industry collaborative, the MIT Industrial Liaison Program (Roberts, 1991: 33).

MIT's involvement with industry was not confined to established companies, however:

> The traditions at MIT of involvement with industry had long since legitimatized active consulting by faculty about one day per week, and more impressive for its time had approved faculty part-time efforts in forming and building their own companies, a practice still questioned at many universities. Faculty entrepreneurship, carried out over the years with continuing and occasionally heightened reservations about potential conflict of interest, was generally extended to the research staff as well, who were thereby enabled to 'moonlight' while being 'full-time' employees of MIT labs and departments. The result is that approximately half of all MIT spin-off enterprises, including essentially all faculty-initiated companies and many staff-founded firms, are started on a part-time basis, smoothing the way for many entrepreneurs to 'test the waters' of high-technology entrepreneurship before making a full plunge (Roberts, 1991: 34).

In his role as head of the OSRD, Bush not only spearheaded the mobilization of the nation's science research capabilities; he also revolutionized the relationship between science and government by channeling the funding to universities rather than to government labs, despite the fact that it involved military research (Roberts, 1991, p. 13–14). In World War I, scientists had been recruited to government labs. The new way of organizing the war effort made it possible to tap into existing research facilities and capabilities and thus scale up the research much more quickly than would have been the case otherwise. It also suddenly established the federal government as *the* major source of funding for basic research—something that had not been the case until then.

Another important step was an institutional change in the organization of academic research. In order to accommodate both the academic requirement of education and research for the common good (i.e. via publication) on one hand and the need to protect the confidentiality of military research on the other, it was necessary to find a new way to organize the research. The solution was to set up independent laboratories such as the Draper and Lincoln labs, both organized by MIT faculty who could then divide their time between the normal academic activities and the defense-related research.

Yet another innovation associated with MIT was that in the immediate postwar years MIT president Compton pioneered efforts toward commercialization of new technology, including military developments. Among other things he helped to create the first institutionalized venture capital fund, American Research and Development (ARD), set up in 1946. It was largely Compton's brainchild. He became a board member, along with three MIT department heads. ARD's first several investments were in MIT developments, and some of the emerging companies were initially housed at MIT (Roberts, 1991, p. 33–34). Thus was established another crucial link connecting knowledge creation and commercialization, keeping the knowledge filter as clean as possible. Other universities eventually followed suit. However, in the early postwar years it is clear that the vast majority of research funding continued to go to defense contractors, not universities, and that the bulk of commercialization of new technology took place via incumbent firms rather than via establishment of new firms.

3.3 Postwar developments: 1945–1965

The impact of World War II on knowledge creation in the United States was fourfold: (i) There was a tremendous increase in R&D spending. (ii) The federal government (particularly the Department of Defense, DoD) became by far the dominant provider of research funding. (iii) The primary thrust was toward systematic, programmatic research. (iv) The enrollment in higher education rose dramatically as a result of the so-called G.I. Bill.

The aftermath of World War II involved a huge increase of the role of the federal government not only in research but also in funding of higher education. The G.I. Bill, signed into law by President Roosevelt in 1944, was designed to provide educational opportunities to returning war veterans. The US system of higher education expanded rapidly and contributed significantly to increasing capacity in the society to absorb new technology, thereby contributing indirectly to economic growth. The number of colleges increased dramatically. Whereas there were 18 new colleges started annually between 1861 and 1943, the number rose to 25 in 1944–1959 and 50 between 1960 and 1979 (Adams, 2000). The total enrollment in higher education went from 1.5 million (9.1% of the 18–24-year-old cohort) in 1939–1940 to 2.4 million (15%) in 1949 (Snyder, 1993), more than a 50% increase. By the early 1950s, approximately 8 million veterans had received educational benefits. The GI Bill also brought to the university campuses a new kind of student—older, more oriented to work than to traditional college-age pursuits, and ultimately more entrepreneurial—different from the pre-war campus population.

The R&D initiated during the war was continued after the war. There was a huge increase in total R&D spending from the early 1950s to the mid-1960s, mostly due to a sharp increase in federal R&D spending. R&D expenditures as a share of gross domestic product (GDP) more than doubled, rising from 1.4% in the early 1950s to

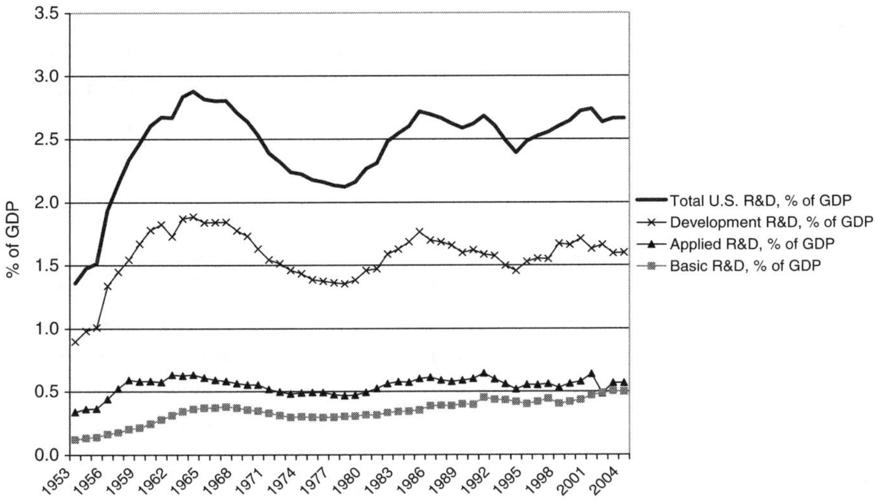

Figure 2 US R&D as percentage of GDP, 1953–2004. *Source*: NSF, http://www.nsf.gov/statistics/. U.S. Bureau of the Census, Statistical Abstract of the USA, various issues.

2.9% in 1964 (Figure 2). Nearly 80% of federal R&D funding came from the DoD and NASA. More than half of this defense-related R&D spending was intramural (i.e. carried out within federal agencies), and most of the rest was carried out by industry; less than 5% was carried out by academic and nonprofit institutions. While modest as a fraction of total US R&D, the funding for academic research still represented a major increase in R&D funding for universities. It was the beginning of a long-term shift toward more academic research in the national R&D system. In these early postwar years, less than 2% of the DoD R&D spending involved basic research, while about one-third of NASA's R&D spending was for basic research.[15] By far the largest component of the overall R&D spending involved applied work that was converted very rapidly into economic activity.

When demobilization after the war closed down the OSRD, national political leaders agreed that the government–university collaboration should continue. The question now became how to organize and institutionalize the nation's research effort. A deep split developed, however, over its structure and control.

The debate over the continuation of federal funding for scientific research started immediately after the war. The main issues were whether a single central agency should be set up to shape and coordinate scientific research, or whether research funding should be handled through existing departments of the federal government.

[15]The small fraction of DoD-funded research classified as "basic" may be due in part to the fact that much of the funding went to research in engineering fields viewed as "applied" rather than "basic".

Another issue was on what basis research contracts should be allocated. The result of the debate was a trade-off: the creation of a new agency, the National Science Foundation (NSF), working in parallel with the traditional agency structure and thus politically accountable to elected officials, while the science establishment and the university community won a commitment to peer-reviewed merit competition for basic science research funding.

The debate took 5 years: the NSF was finally established in 1950, but with much smaller funding than originally envisioned by the scientific community. In its first year of funding (1951), the NSF was allocated only $150,000, and its funding did not exceed $10 million annually until 1955 (source: NSF). The NSF thus became much less dominant as a source of federal funding for academic research than the academic community had originally expected. Meanwhile, the total federal government R&D spending continued to increase and soon exceeded $1.5 billion annually with $1.1 billion allocated to the DoD alone. Also, at this time several other agencies emerged. The navy was eager to crack the wartime monopoly of the army and Army Air Corps on developing the atomic bomb; this led to the establishment of the Office of Naval Research in 1946. At the same time, scientists who opposed military control of atomic technology won support for the creation of the Atomic Energy Commission (AEC). "Like the wartime OSRD, the AEC undertook much of its research via contractual agreements with universities and industry, and it used the contract model to shift from government to university management most of the Manhattan Project's secret government-owned laboratories, including those at Los Alamos, Lawrence, Argonne, Ames, and Brookhaven, and parts of Oak Ridge" (Graham and Diamond, 1997: 31).

In biomedical research, the demise of the OSRD left a vacuum that was filled by the National Institute of Health (NIH), the research branch of the Public Health Service.[16]

> The NIH, which had operated its own in-house or intramural research laboratories since 1930, seized the moment of opportunity in 1945. Taking over fifty wartime research grant projects from the expiring OSRD, the NIH became the chief supporter of extramural research in the nation's expanding network of medical schools. Congress... encouraged the NIH initiative and between 1945 and 1950 expanded its budget from $3 million to $52 million.
>
> By 1950 the pluralistic nature of the federal research system was well established, and the new NSF, despite its distinctive primary research mission, was disadvantaged by its tardy entry. The NSF's research funds would modestly enlarge the federal aid pot, but the foundation

[16]The NIH became plural—the National Institutes of Health—in 1948 when Congress added a separate heart institute to the cancer institute it had created in 1937.

> would not significantly reshape or coordinate national science policy. During the 1950s, federal mission agencies expanded their programs of R&D support under the broad rubric of 'mission-related basic research'. Under this umbrella, agencies stretched their traditionally applied R&D programs to include basic research, thereby offering universities a growing cafeteria of funding opportunities (Graham and Diamond, 1997: 31).

By 1954, federal agencies still accounted for as much as 69% of the funding for the now vastly expanded total university research budgets, while the universities' own funds contributed only 8.5%. Meanwhile, the share of private foundations was reduced to only 11% and that of industry to 9%—both shares sharply reduced from the peaks they had reached in the interwar years (*ibid.*: 32).

Thus, in conjunction with World War II, a comprehensive array of new federal support programs provided both vastly increased research funding opportunities and considerable competitive pressure. As a result, federal science policy was shaped not by a new ministry of higher education or federal science agency but rather by a whole set of new agencies and programs in already established departments of the federal government. This led to overlapping and often duplicative sponsored research programs which on the one hand invited individual researchers to shop for funding and on the other allowed a broad, experimental research agenda (*ibid.*: 26).

However, the new, federally subsidized research economy of the postwar years by no means provided a blank check to subsidize the research agenda of university scientists and scholars. It is important to note that federal R&D expenditures were allocated primarily to development, not to research. In the early 1950s only 6–7% of federal R&D expenditures were for basic R&D (and less than 2% of the R&D funding by the DoD). Furthermore, most of the federal money (over 80%) was spent either by intramural federal agencies or by industry, particularly defense contractors, not universities. In fact, during the 1950s roughly 75% of all federal R&D funding was provided by the DoD. In addition, most federal research funds were provided to support applied, programmatic research, not basic, "pure," or "disinterested" research. Even the federally funded basic research program was dominated by "big science" projects with military applications (*ibid.*: 32). For example, in 1956 the budget for basic science research was $78 million at the DoD and $45 million at the AEC, while at the NIH, the major supporter of "little science" projects in fundamental research, the basic research budget was only $26 million and that of the NSF $15 million. Even by 1960, after a whole decade of rapid funding growth, the NSF still provided only 12% of federal R&D funding for basic research (source: NSF).

It is noteworthy that while most research funding went to industry, not universities, and was provided by the DoD and other defense-related agencies, the military was also the largest provider of research funding to universities. In 1963, the DoD provided 26.4% federal R&D funding to academic institutions, the AEC 8.4%, and NASA 5.8%. Thus, together these three agencies were responsible for 40.6%, while the NIH provided 36.6%, USDA 5.0%, and other federal sources (including the NSF) 4.9%. Altogether, the federal government funded 70% of the total funds for academic research. Philanthropic organizations, state governments, and internal sources provided the remaining 30% of funding (Graham and Diamond, 1997, p. 34–35).

Under the terms of the federal research contracts, the funding agencies owned the intellectual property rights to the research results and were responsible for their application and implementation in their own domains—a form of commercialization. However, as early as the late 1960s, federal funding agencies including the DoD, the NIH, and the NSF began requiring the universities to develop formal patent policies. About 40 universities had done so during the war, and about 45 announced patent policies during the first decade after the war ended. But even though most major universities had developed patent policies, many of them avoided direct involvement in patent administration, preferring instead to let the Research Corporation do that. Thus, the number of institutions that had invention administration agreements with the Research Corporation grew from 6 in 1946 to 89 in 1956 and continued to rise to 278 in 1979 (Mowery and Sampat, 2001b, p. 790–792). Federal research sponsors started to allow universities with approved patent policies to patent and license the results of their federally funded research under the terms of Institutional Patent Agreements (IPAs) negotiated by individual universities with each federal funding agency (Mowery *et al.*, 2001: 102).

The Sputnik crisis provided the leaders of academic science with a window of opportunity during the Eisenhower and Kennedy presidencies. During the period 1958–1963 two historic decisions were made: (i) the federal government assumed primary responsibility for supporting basic research in the United States; and (ii), the research was to be carried out primarily by the universities as an integral component of graduate education (Graham and Diamond, 1997: 33–34).

As a result of the post-Sputnik surge in research expenditures, federal funding for basic science raised the "pure science" component of university research expenditures on American campuses from 52% in 1953 to 76% in 1963. The total federal budget for R&D grew by 250% in constant dollars between 1953 and 1963, and federal funding of university research grew by 455% (*ibid.*: 34).

Military considerations clearly dominated the national research agenda during the first two decades after the war. However, it took more than a decade before new research started to generate significant start-up of new firms. The fact that the bulk of the R&D continued to involve applied research or development work and was carried out either intramurally or by large defense contractors meant that the research was commercialized rapidly. Even though the "big science" projects were targeted primarily for the military, the postwar work was mainly applied, and much of it resulted in products that found increasing civilian use (and largely via existing companies, not start-ups). Several of the major technologies developed during the war originated in Britain and Germany and came to the United States via Britain. For example, the Radiation Lab at MIT was initially set up in 1940 as a joint Anglo-American project to further develop British radar technology and produce radar equipment. Penicillin was mass produced in the United States during the war, organized by the War Production Board on the basis of discoveries made in the UK and Australia. The first American jet engine was built by General Electric in 1943 after a British prototype, and the first American commercial jetliner was the Boeing 707, built in 1957 as an adaptation of a military tanker plane (KC-135), following the British de Havilland Comet jetliner first in commercial jet service in 1952. The civilian commercialization of these technologies was carried out primarily by existing companies.

While the conversion from military to civilian production was based in part on imported technology and largely benefitted existing companies, the development of the computer industry represents a different story, one in which the technology originated in the United States and in which US universities played a prominent role.

> The first digital electronic computer, the ENIAC, was brought to the full stage of a working prototype at the Moore School of Electrical Engineering at the University of Pennsylvania, in the fall of 1945 . . . In the case of the computer, moreover, American universities not only designed and assembled the initial hardware of the computer industry; they created an entirely new discipline, of huge economic importance, along with the research infrastructure that had to be built in order to exploit the vast potential of the new hardware (Rosenberg, 2000: 49).

The first computer company was Eckert–Mauchly Computer Corporation, formed in 1946. The company was sold in 1950 to Remington Rand (later Sperry Rand), which was an established maker of office equipment and electric shavers. Many other companies entered the emerging computer business during the early postwar years, most of them responding to burgeoning government needs . . . Some companies were founded *de novo* to pursue computer development opportunities, but in contrast to the history of semiconductors, most start-ups sold out to become the nucleus of computer operations of established companies (Scherer, 1996: 240).

The development of the computer and the computer industry is the most important example of the increasingly prominent role of US universities in research and in creating new technologies for economic exploitation following World War II. In conjunction with the war and its aftermath, US universities not only closed the gap with leading European universities in basic science but also, following in the paths established earlier, pushed ahead in research in applied fields such as engineering and medicine. As a result, American universities have played a much more prominent role than their European counterparts in generating new technologies as a basis for economic growth. And the role has often been one of collaboration and co-development with private partners than one of unilateral transfer of technology.

Another example of such postwar high-tech industries (beyond the computer) is scientific instruments: new instrumentation has often been an unintended byproduct of basic science research at universities and has subsequently been taken up, produced, and sold by private firms. At the same time, new scientific instruments have played a major role in shaping the research agenda at universities (Rosenberg, 1992). Yet another example is the development of numerically controlled machine tools. Numerically controlled machine tools represent the application of computers to solving machining problems; the technology was first developed in collaboration between the Propeller Lab at the Wright-Patterson Air Force base, MIT, IBM, and a small company in Michigan (Carlsson, 1984: 103). Another example is semiconductors: the transistor was first developed by William Shockley and colleagues at Bell Labs and then shared with some 30 university professors (in part through a 6-day course at Bell Labs). This was clearly a case of knowledge transfer from industry to university (Rosenberg, 1992: 34). Similarly, laser and optical fibers are examples of the result of interaction between academic researchers and industry.

Several leading research universities, most prominently MIT, which had received significant research funding during the war began to spin off new companies to commercialize the new technologies in precision machinery, electronic components, acoustics, and noise control, in addition to computer hardware and software and machine tools (Shane, 2004: 46). As mentioned earlier, the ARD Corporation, the first modern venture capital company, was set up at the end of the war to commercialize the military technologies invented at MIT. But even though this spinoff activity represented a major increase in academic entrepreneurship, its overall impact on new company formation was modest.

In terms of generating economic growth, the conversion of products and production facilities from military to commercial applications clearly dominated. Huge investments had been made during the war to produce tanks, trucks, jeeps, airplanes, warships, and ammunition. These investments and technologies as well as the knowledge about how to organize and manage mass production were converted to civilian use. The effects lingered for several decades as mass production thinking became dominant in US manufacturing (Carlsson, 1984). The trade liberalization and economic integration that began at the end of World War II (manifested in the establishment of the United Nations, The World Bank, IMF, GATT, OECD, etc.) provided opportunities for the industrial giants to expand as they converted from military to civilian production.

As a result, the economic growth in the early postwar years was largely concentrated in large existing companies, not startups. The total number of concerns in business was about 2.1 million in 1946 (about the same as in 1930) and rose to 2.7 million by 1949 and then stayed at that level until the end of the 1950s. Few new companies were formed; the number of new business incorporations was around

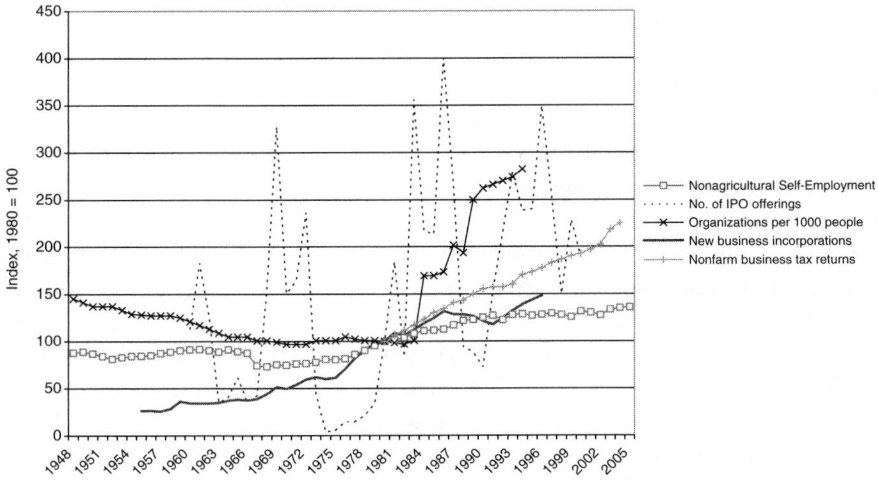

Figure 3 Indicators of entrepreneurial activity, 1948–2005 index, 1980 = 100.
Source: Number of IPOs: Ibbotson et al. (2001); Ritter (2006). New Business Incorporations, Nonfarm business tax returns, and nonagricultural self-employment: Statistical Abstract of the USA, various issues. Organizations per 1000 people: Gartner and Shane (19945), p. 295.

100,000 per year in the latter half of the 1940s, rose gradually in the 1950s and 1960s but did not really take off until the late 1970s. Nonagricultural self-employment stayed constant from the late 1940s until the mid-1960s, even though the labor force grew (i.e. the self-employment rate declined). There were few Initial Public Offerings (IPOs), and those that did occur were quite modest in size (in terms of proceeds per IPO) and often involved companies that had been started many years earlier (Ibbotson *et al.*, 2001). The number of organizations per person declined continuously from 1948 to about 1970 and then leveled off (Figure 3).

During the 1950s and early 1960s the knowledge filter was thin and easily penetrated. While knowledge creation in traditional fields went on as before, with little desire to commercialize the results, there was a large increase in knowledge creation in science and engineering. Most of the new knowledge was applied and quickly converted into new products, mostly military in the beginning but increasingly civilian. As owners of the intellectual property, the federal agencies were responsible for commercialization, and incumbent firms were the main vehicles to do that.

3.4 1965–1980: a period of transition

After the first two decades following World War II the economic landscape shifted. Concerns were being raised about the rise of the military–industrial complex against whose consequences President Eisenhower had warned. After the onset of the Cold War, the Korean War, and the Sputnik challenge, military expenditures began to decline, at least in relative terms. Although the DoD was still the major source of R&D funding, federal R&D expenditures grew more slowly than GDP. As a result, R&D as a share of GDP fell from 2.9% in 1964 to 2.1% in 1979 (Figure 2). Meanwhile, nonfederal (mostly industrial) R&D spending stayed constant in relation to GDP during the 1960s and 1970s and began to increase in the late 1970s, surpassing federal R&D expenditures in 1978. By the end of the 1970s the federal government was no longer the primary source of research funding.

While the share of military R&D declined, NIH funding increased.[17] As a result, there was a shift toward more basic R&D in the national research portfolio. R&D funding for basic research as a share of total federal R&D funding increased from 7% in 1959 to 13% by 1972. This share then increased steadily from the end of the 1970s to reach 25% in the early 2000s.

In addition to this shift from applied to basic R&D there was also a change in the locus of the R&D effort. The fact that the research funding was still highly concentrated to only the top universities, most of them private, became a political issue. In 1963, out of a total of 2139 institutions of higher education, only 492 were awarded federal R&D funds. Of the $830 million total, the top 100 recipients won 90% (Graham and Diamond, 1997: 34).

Predictably, this allocation of resources was defended by the representatives of the top universities. Former Harvard president Conant wrote: "In the advance of science and its application to many practical problems, there is no substitute for first-class men. Ten second-rate scientists or engineers cannot do the work of one who is in the first rank" (*ibid.*: 36). Nevertheless, this concentration of taxpayer support to a few elite universities was challenged in the 1960s.

As egalitarian, populist pressures grew in the 1960s, the Kennedy and Johnson administrations responded by retaining the core formula of agency funding through peer-review competition, but added new policies and programs in three areas. First, agencies were directed to widen the geographic distribution of federal research support, emphasizing physical facilities and attempting to double the number of strong research universities. Second, federal research support was extended to include the social sciences, the humanities, and the visual and performing arts. Third, federal support was significantly expanded, extending beyond the roughly 100 doctorate-granting universities, to provide funding for construction and non-scientific programs, including student financial aid, to more than 3000 institutions. These included community colleges, private liberal arts colleges, state colleges

[17]Defense funding of R&D went mostly to engineering schools and was classified as "applied", while NIH funding went to the life sciences and was classified as "basic".

and regional universities, historically black institutions, and vocational and proprietary schools. Thus, in the Great Society agenda, federal science policy expanded and blurred into higher education and social policy (*ibid.*: 27).

As a result of these changes, the national R&D effort became less focused and more dispersed. The rate of new knowledge creation declined, as indicated by the falling share of R&D in GDP. Also, while the share of basic R&D increased sharply (as that of applied R&D declined), the potential of commercialization diminished as the beneficial economic effects would occur only with greater delay. The economic impact of the war-related innovations was wearing off. In addition, a significant share of the basic R&D funding went to educational purposes and building of research infrastructure rather than into "pure" research. As research projects were increasingly built around clusters of large-scale equipment, often requiring special facilities and support staff, they became enmeshed in the difficult process of institutional-level priority setting, since accepting these project assignments often meant making major capital investments. In this sense, the knowledge that was created provided fewer stimuli to economic growth. The institutional barriers to commercialization that had always existed were still there; the research effort was less focused; and the federal agencies were becoming less successful in converting research into economic activity.

At the same time, the industrial landscape was also shifting. Antitrust regulation and its enforcement prevented both horizontal and vertical integration, leaving conglomeration as the only option beyond organic growth. As time went on and companies engaged in conglomeration in order to expand, they started to lose touch with their knowledge base (their absorptive capacity was reduced). In this sense, too, the knowledge filter started to clog up. The civilian spin-offs from military products began to decline. Many large firms found it difficult to grow during this period. The share of small firms in manufacturing began to grow, reversing the trend toward larger firm size during the previous 100 years (Carlsson, 1992). Some indicators of entrepreneurial activity began to change. The number of self-employed persons outside of agriculture plummeted in 1966–1968 and then recovered slowly, reaching its previous level only in 1978, rising slowly until about 1990 and remaining roughly constant since then. The number of organizations per 1000 people, which had declined steadily since 1948, leveled off around 1970. New business incorporations started to increase in the late 1960s. The number of IPOs was low in the 1960s but rose temporarily to a much higher level in the early 1970s, only to fall to extreme lows in the latter half of the 1970s.[18] These developments are illustrated in Figure 3. With the arrival of the microprocessor and recombinant DNA, as well as higher oil

[18]The rise in IPOs was probably due mainly to institutional reforms that lowered taxes on capital gains and allowed institutional investors such as pension funds to invest in venture capital firms. This made it possible for old but privately held firms to go public.

prices, in the early 1970s, additional forces were set in motion. The industrial giants were no longer able to absorb the new technologies; new mechanisms and new institutions were necessary.

3.5 1980–2006: new patterns of knowledge creation and economic growth

It was against this background of diminished dynamism in the economy (but with a few signs of a new era of entrepreneurial activity emerging) that a number of institutional reforms were made in the early 1980s. There was a broad shift toward stronger intellectual property rights. In 1980, the US Supreme Court affirmed the validity of patents on life forms. Protection was extended to cover new types of artifacts such as research tools (Mowery *et al.*, 2004: 5). The Court of Appeals for the Federal Circuit was established in 1982 to serve as the court of final appeal for patent cases throughout the federal judiciary, making it easier for people to protect their intellectual property (Shane, 2004: 60).

The Bayh–Dole Act, passed by the US Congress in 1980, was another important reform. The purpose of the legislation was to facilitate commercialization of the results of federally funded research by transferring the intellectual property rights from the funding agencies to the universities in which the research was carried out. This provided incentives for the universities to commercialize the results of their research. The Bayh–Dole Act extended the Institutional Patent Agreements (IPAs) that federal funding agencies had made with a number of universities to all performers of federally funded research. In the subsequent period, from the early 1980s onward, the number of licenses issued by universities and the number of startups began to increase, due in part to the Bayh–Dole Act but also due to other reforms and structural changes.[19] For example, there was a shift in universities toward a policy of taking equity in licensees of university technology. Also, the experience

[19]Several authors have questioned the efficacy of the Bayh–Dole Act in bringing about increased commercialization of academic research. For a discussion, see Mowery *et al.* (2004). Concerns have also been raised about growing participation of universities in commercial relationships. But there is not much evidence that the orientation and nature of academic research have shifted (see e.g. Thursby and Thursby, 2002) or that academic publishing has diminished (Mowery *et al.*, 2004) as a result of Bayh–Dole. Nor has industrial funding of academic R&D changed much as a percentage of total funding. It is difficult to isolate the impact of the Bayh–Dole Act as it was only one part of a broad set of institutional changes that occurred in the early 1980s. As important as Bayh–Dole was, it was essentially an attempt to "reverse engineer" the technology transfer process that had worked so effectively in prior years at a few very special institutions such as MIT and CalTech. As mentioned earlier, the Research Corporation had been set up in 1912 and played an important role in commercializing research results at several leading academic institutions. In the face of the incentives offered by Bayh–Dole, a wide range of universities adapted to the new set of rules and began promoting technology transfer, but the vast majority of them never developed the kind of permissive, entrepreneurial culture that marked the early models.

of university spin-offs in the early postwar years gave increased momentum to academic entrepreneurship (Shane, 2004, p. 61–63).

About this time, reforms were also made that permitted pension funds to invest in venture capital partnerships, giving rise to institutionalized venture capital. This helped unleash the growth of a new form of entrepreneurial finance. Between 1946 and 1977 the creation of new venture funds amounted to less than a few hundred million dollars annually. Starting in the late 1970s, and culminating in the late 1990s, fundraising in the venture capital industry increased sharply (Kortum and Lerner, 2000). While most of this money went to IPOs and new business startups, some of it was used to fund industrial innovation. It is estimated that venture capital accounted for 8% of industrial innovations during 1983–1992 and 14% by 1998 (Kortum and Lerner, 2000: 691).

Another step in unclogging the knowledge filter was the establishment in 1982 of the Small Business Innovation Research (SBIR) program. The program requires a wide range of federal agencies with extramural research budgets to set aside a portion (2.5%) of their research funding for small businesses (less than 500 employees). Over $10 billion has been awarded by the SBIR program to various small businesses since 1982. Also, deregulation, especially in telecommunications and transportation, gave rise to new entrepreneurial opportunities. Further, the 1980s and 1990s witnessed a fundamental re-organization of US business. In the biggest merger and acquisition wave in US history, most conglomerates were broken up and business units were re-bundled. Companies became more focused on their core competencies. One result of this restructuring was a re-organization of corporate R&D. Many central R&D labs were closed, and others were reduced in size while also focusing more narrowly on applied research with a clear commercial objective; basic research (with no foreseeable commercial application) was reduced in industrial firms and shifted instead to the universities which have picked up the role as major performers of basic R&D.

Meanwhile, R&D spending as a share of GDP rose again, from 2.1% in 1979 to 2.7% in 1984 and has since remained at about that level (Figure 2). The increase was due to a modest increase in federal R&D spending and a sustained increase in nonfederal (primarily industrial) R&D that continued until 2000. After 1987 there has been a decline in federal R&D spending as a percentage of GDP, but the continued growth in nonfederal R&D has kept R&D/GDP constant. Meanwhile, federal R&D funding has continued to shift toward basic R&D which now represents about 25% of total R&D funding.[20] Applied and development R&D is now funded overwhelmingly by industrial firms. A larger share than ever before of the creation of applied knowledge is now both funded and performed by the entities closest

[20]As noted earlier, the shift toward basic R&D is largely a consequence of increased funding by the NIH both in absolute terms and as a share of total federal R&D funding.

to commercialization, the industrial firms themselves.[21] This makes for a thin knowledge filter.

The structure of R&D at the national level has changed also in another way as the life sciences have absorbed an increasing share of R&D expenditures. Particularly in the biotech arena, this has meant a shift of basic research toward academic institutions. The new discoveries have typically been made in universities, as the absorptive capacity of existing companies (mostly pharmaceutical firms based on chemistry and chemical engineering know-how) was quite limited in the new technology arena. The time lag between scientific discovery and commercial application has turned out to be much longer than in the electronics field—clearly, part of this is due to the approval process requirements of the US Food and Drug Administration and other key agencies. And the transfer mechanism has turned out to be not the large pharmaceutical firms but rather new firms dedicated to biotechnology research who can translate the basic research into commercial knowledge. But in most cases (Genentech being a notable exception), the commercialization in the form of production, marketing, and distribution has been carried out by established pharmaceutical firms, creating a new division of labor and symbiosis between these two groups of firms.

New entrepreneurial opportunities have arisen through scientific and technical advances and changes in government policies. Changes have occurred in the entrepreneurial infrastructure, e.g. via greater availability of professional services and stronger intellectual property protection. There have also been shifts in values, preferences, and attitudes toward entrepreneurship. For example, a study of MIT alumni shows that the entrepreneurial activity of MIT alumni increased steadily (as measured by first firm founding per 1000 alumni) during the 1950s, 1960s, and 1970s and has then stayed at a high level (Hsu, Roberts, and Eesley, 2006). The impact of universities on industry occurs not only directly through transfer of technical knowledge but also through the entrepreneurial activities of alumni.

As a result of all these institutional and structural changes, the last two decades have witnessed a period of a new kind of dynamism in the form of growth of new industries and renewal of old ones as the digital economy has penetrated into all parts of society. Whereas in the early postwar period innovation tended to be carried out by large firms in capital-intensive, concentrated industries characterized by highly differentiated goods, the last two decades have been characterized by a different technological regime in which innovation is carried out primarily by small firms in industries that are highly innovative, skilled-labor intensive, and have a large share of large firms (Winter, 1984; Acs and Audretsch, 1987;

[21]It remains true, at least in fields other than biotechnology/biomedicine, that firms source most of their technology from internal R&D. Joint ventures, alliances, and acquisitions of companies are the most common external sources. Universities are seldom listed as primary sources of technology but serve rather as places for recruiting employees (Carlsson *et al.*, 2008).

Plummer and Acs, 2005). Jovanovic has shown that the performance of small companies vs. large ones (as measured by the price of small capitalization stocks relative to the S&P 500) was about equal from the end of World War II to the late 1960s and then rose dramatically to a 4:1 ratio by the mid-1980s (Jovanovic, 2001: 54). He attributes this rise in the relative performance of small firms to the arrival of the microprocessor. He also notes that among the 20 largest US companies by market capitalization in 1999, 10 were incorporated in or after 1967.

There are two main reasons why small firms have become more important in recent decades. One is that small firms simply do certain things (such as certain types of innovation) better than large firms. As a result, through division of labor between small and large firms, the efficiency of the economy is increased. The other reason is that small firms provide the entrepreneurship and variety required for macroeconomic growth and stability (Carlsson, 1999).

Some of the results of the increased dynamism of the US economy in the last two decades are captured in Figure 3.[22] As knowledge creation (as measured by the R&D/GDP ratio) picked up after 1980, and as various institutional reforms were made, there was a dramatic increase in entrepreneurial activity in the United States. The number and gross proceeds of IPOs rose markedly in the early 1980s. New business incorporations started to rise in the late 1970s and continued to rise to the mid-1990s. Nonagricultural self-employment also rose from the late 1970s to the mid-1990s.[23] Nonfarm business tax returns have been on a steady rise since 1980 (the earliest year for which data are available). The number of organizations per person made a dramatic jump in 1984 and has continued to rise rapidly.

Thus, the early 1980s appears to have marked a turning point in US economic growth, representing a transition to a new technological regime in which new business formation plays an increasing role in converting new knowledge into economic growth, thereby reducing the knowledge filter.

4. Summary and conclusions

This article has examined the sources of knowledge creation in the United States over the last century and a half and the mechanisms through which economic knowledge has been converted into economic activity. Prior to the establishment of land-grant universities after the Civil War, there was not much academic research that resulted directly in economic growth. But the rapid growth of universities (particularly

[22]This is consistent with Phelps's view that "dynamism requires that innovation be ample, aimed in good directions and finding managers and consumers open to novelty" (Phelps, 2006).

[23]Blau attributes the rise in self-employment to technological change in the form of personal computers, "government safety and health regulations that favor the self-employed over businesses with employees, and demand-induced changes in industrial structure favoring industries in which self-employment is common." (Blau, 1987: 464)

land-grant universities and their agricultural extension programs) and enrollment in higher education in the latter half of the 19th century created a much larger college-educated labor force in the United States than in Europe. The resulting increased capacity to absorb new technology was strengthened by the practical orientation of US higher education. Economically useful knowledge creation in the late 19th century was largely informal and took place in industrial laboratories whose main tasks were testing, measuring, and standardizing industrial products and processes.

As science-based industries such as the chemical, electrical machinery, and telecommunications industries began to grow in the late 19th century, new scientific knowledge was needed. The creation of such knowledge required research facilities utilizing specialized equipment and skills on a scale beyond the capabilities of the universities. Industrial firms such as Du Pont and General Electric therefore built their own corporate R&D labs. But in contrast to Europe, universities in the United States made important contributions by establishing new disciplines such as electrical and chemical engineering—thereby building up a significant and highly skilled labor force—and by working closely with industrial scientists. Many leading scientists in the new disciplines worked as consultants and had much of their research funded by industrial firms. It is in such corporate laboratories, as well as in federal government laboratories, that most economically useful R&D, including basic research, was conducted until after World War II. The vast bulk of industrial R&D has always been commercially oriented, resulting in new and improved products and processes. This is true even today.

Before World War II, basic research in the universities was quite small. Total external funding of academic research did not exceed $50 million annually prior to World War II, and all of this came from foundations and a few industrial companies. The federal government played no role in funding academic research. But World War II led to fundamental changes in the knowledge creation system. The military effort led to a massive increase in R&D spending. Most of the funding came from the DoD and was conducted largely in government labs and by defense contractors. But the war-related research was organized in such a way that it also involved universities. For the first time, and quite suddenly, the federal government became the major source of funding for academic research which now included "big science" projects (systematic, programmatic research) on a scale never seen before. Basic research started to shift toward the universities. As the war ended, the G.I. bill generated a large increase in college enrollment. The war-related products (such as computers, jet engines, and radar) that resulted from the R&D were commercialized almost immediately through the military and soon after the war were converted into civilian products. The commercialization took place mainly through incumbent firms. Even though leading research universities such as MIT began spinning off new companies based on the military technologies they had developed during the war, these new start-ups were too few to affect the total number of firms materially. Overall, there were few new firms created; entrepreneurial activity declined or stagnated between 1950 and 1965.

The federal funding of defense-related R&D continued to grow until total R&D expenditures reached a peak in the mid-1960s. The total R&D spending then fell as the federal expenditures declined. Also, the federal funding of academic research now became less focused, dispersed to a much larger number of institutions of higher education, and used for building research infrastructure. As a result, much less of the knowledge created was economically useful; the knowledge filter was clogging up. The civilian spin-offs from military products began to decline, and the economic growth rate fell.

The year 1980 represents something of a turning point. A number of institutional reforms (including strengthening of intellectual property rights, the enactment of the Bayh–Dole Act, changes in tax laws, and deregulation of financial institutions that created new financial instruments) mark a transition to a new technological regime in which new business formation plays an increasing role in converting new knowledge into economic growth. The breakthrough in DNA research and the microprocessor revolution also played a role. Entrepreneurial activity (as measured by nonfarm business tax returns, new business incorporations, the number of organizations per capita, the number of IPO offerings, and nonagricultural self-employment) began to pick up as the dynamism of the economy increased.

The "big picture" that emerges is one which shows the economy becoming increasingly dependent on economically useful knowledge and on the effectiveness with which that knowledge is converted into economic activity. In the late 19th century the knowledge creation was linked to a rise in the share of college-educated people in the population and codification and standardization of economically useful knowledge in industrial labs. At the turn of the 19th century, economic growth was driven primarily by science and engineering-based industries such as chemicals, electrical equipment, and telecommunications. The leading companies in these industries built their own corporate R&D labs. These, as well as several federal laboratories, working closely with academic scientists, were the main producers of economically useful knowledge, and they were quick to reap the economic benefits.

World War II led to a massive scaling up of R&D in the United States. While most of the funding went to federal and corporate laboratories, the federal government now also began to fund academic research. At first the defense-related R&D was immediately converted into economic activity via incumbent firms producing such products as jet engines, radar, and computers. In the mid-1960s, the total R&D expenditures started falling in relation to GDP, and the economic impact of the war-related products was diminishing. The decline in R&D spending was reversed in the early 1980s, carried largely by biotechnology and the microprocessor.

The main conclusions of this article are (i) that in order to explain economic growth it is necessary to distinguish between general knowledge and economically useful knowledge; (ii) that the effectiveness with which such knowledge is converted into economic activity varies over time and depends on institutional arrangements and on the nature of the knowledge created; and (iii) that the mechanism converting economically useful knowledge into economic growth also varies over time. Until the late 1970s, incumbent firms were the main vehicles. Subsequently, new ventures have been the main mechanism, at least in the United States.

Addresses for correspondence

Bo Carlsson, Department of Economics, Case Western Reserve University, 111 19 Bellflower Road, Cleveland, Ohio 44106-7235, USA. e-mail: bo.carlsson@case.edu
Zoltan J. Acs, School of Public Policy, George Mason University, Fairfax, VA 22030-4444, USA. e-mail: zacs@gmu.edu
David B. Audretsch, Institute for Development Strategies, Indiana University, Bloomington, IN 47405, USA. e-mail: daudrets@indiana.edu
Pontus Braunerhjelm, Department of Transport and Economics, Royal Institute of Technology, 100 44 Stockholm, Sweden. e-mail: pontusb@infra.kth.se

References

Acs, Z. J. and D. B. Audretsch (1987), 'Innovation, market structure, and firm size,' *Review of Economics and Statistics*, **69**(4), 567–574.

Acs, Z. J., D. B. Audretsch, P. Braunerhjelm and B. Carlsson (2004), 'The missing link: the knowledge filter and entrepreneurship in economic growth,' CEPR working paper no. 4358.

Acs, Z. J., D. B. Audretsch, P. Braunerhjelm and B. Carlsson (2005), 'Growth and entrepreneurship – An Empirical Assessment,' working paper, April.

Acs, Z. J., D. B. Audretsch, P. Braunerhjelm and B. Carlsson (2009), 'The knowledge spillover theory of entrepreneurship,' *Small Business Economics*, **32**(1), 15–30.

Acs, Z. J. and L. A. Plummer (2005), 'Penetrating the "Knowledge Filter" in regional economies,' *The Annals of Regional Science*, **39**, 439–456.

Adams, J. A. (2000), 'The G.I. Bill and the Changing Place of U.S. Higher Education after World War II,' paper presented at the Association for the Study of Higher Education (ASHE) annual meeting, Sacramento, November.

Arrow, K. (1962), 'The economic implication of learning by doing,' *Review of Economics and Statistics*, **80**, 155–173.

Blau, D. M. (1987), 'A time-series analysis of self-employment in the United States,' *Journal of Political Economy*, **95**(3), 445–467.

Carlsson, B. (1984), 'The development and use of machine tools in historical perspective,' *Journal of Economic Behavior and Organization*, **5**, 91–114.

Carlsson, B. (1992), 'The rise of small business: causes and consequences,' in W. J. Adams (ed.), ''*Singular Europe: Economy and Polity of the European Community after 1992.* University of Michigan Press: Ann Arbor, MI, pp. 145–169.

Carlsson, B. (1999), 'Small business, entrepreneurship, and industrial dynamics,' in Z. J. Acs (ed.), ''*Are Small Firms Important? Their Role and Impact.* Kluwer Academic Publishers: Boston, pp. 99–110.

Carlsson, B., M. Dumitriu, J. T. Glass, C. A. Nard and R. Barrett (2008), 'Intellectual property (IP) management: organizational processes and structures, and the role of IP donations,' *Journal of Technology Transfer*, **33**(6), 549–559.

Carlsson, B. and A-C. Fridh (2002), 'Technology transfer in United States universities: a survey and statistical analysis,' *Journal of Evolutionary Economics*, **12**(1–2), 199–232.

Chandler, A. D. Jr. (1985), 'From industrial laboratories to departments of research and development,' in B. C. Kim, H. H. Robert and L. Christopher (eds), ''*The Uneasy Alliance: Managing the Productivity-Technology Dilemma.* Harvard Business School Press: Boston, MA, pp. 53–61.

Drucker, P. F. (1998), 'From capitalism to knowledge society,' in N. Dale (ed.), ''*The Knowledge Economy.* Butterworth-Heinemann: Boston, pp. 15–34.

Gartner, W. B. and S. A. Shane (1995), 'Measuring entrepreneurship over time,' *Journal of Business Venturing*, **10**, 283–301.

Graham, H. D. and N. Diamond (1997), *The Rise of American Research Universities: Elites and Challenges in the Postwar Era.* The Johns Hopkins University Press: Baltimore.

Geiger, R. L. (1986), *To Advance Knowledge: The Growth of American Research Universities, 1900–1940.* Oxford University Press: Oxford.

Geiger, R. L. (1993), *Research and Relevant Knowledge: American Research Universities Since World War II.* Oxford University Press: Oxford.

Hounshell, D. A. and J. K. Smith, Jr. (1988), *Science and Corporate Strategy: Du Pont R&D, 1902–1980.* Cambridge University Press: Cambridge.

Hsu, D. H., E. R. Roberts and C. E. Eesley (2006), 'Entrepreneurs from technology-based universities: evidence from MIT,' Working paper, M.I.T., August.

Ibbotson, R. G., J. L. Sindelar and J. R. Ritter (2001), 'The market's problems with the pricing of initial public offerings,' *Journal of Applied Corporate Finance*, 66–74, http://bear.cba.ufl.edu/ritter.

Inklaar, R., van-A. B. McGuckin and S. M. Dougherty (2005), 'The structure of business R&D: recent trends and measurement implications,' *Paper for the Conference on,* The Measurement and Analysis of Innovation and Productivity Growth,' November, University of Groningen.

Jovanovic, B. (2001), 'New technology and the small firm,' *Small Business Economics*, **16**(1), 53–55.

Kortum, S. and J. Lerner (2000), 'Assessing the contribution of venture capital to innovation,' *RAND Journal of Economics*, **31**(4), 674–692.

Lucas, R. (1988), 'On the mechanics of economic development,' *Journal of Monetary Economics*, **22**, 3–39.

Maddison, A. (1987), 'Growth and slowdown in advanced capitalist economies: techniques of quantitative assessment,' *Journal of Economic Literature*, **25**(2), 649–698.

MacGarvie, M. and J. L. Furman (2005), 'Early academic science and the birth of industrial research laboratories in the U.S. pharmaceutical industry,' NBER Working paper 11470, June.

Mowery, D. C. (1981), 'The emergence and growth of industrial research in American manufacturing, 1899–1945.' PhD dissertation. Stanford University, Stanford.

Mowery, D. C., R. R. Nelson, B. N. Sampat and A. A. Ziedonis (2001), 'The growth of patenting and licensing by U.S. universities: an assessment of the effects of the Bayh-Dole Act of 1980,' *Research Policy*, **30**(1), 99–119.

Mowery, D. C., R. R. Nelson, B. N. Sampat and A. A. Ziedonis (2004), *Ivory Tower and Industrial Innovation: University-Industry Technology Transfer Before and After the Bayh-Dole Act in the United States.* Stanford University Press: Stanford.

Mowery, D. C. and N. Rosenberg (1998), *Paths of Innovation: Technological Change in 20th-Century America.* Cambridge University Press: Cambridge.

Mowery, D. C. and B. N. Sampat (2001a), 'Patenting and licensing university inventions: lessons from the history of the research corporation,' *Industrial and Corporate Change*, **10**(2), 317–355.

Mowery, D. C. and B. N. Sampat (2001b), 'University patents and patent policy debates in the USA, 1925–1980,' *Industrial and Corporate Change*, **10**(3), 781–814.

National Science Foundation, Division of Science Resources Statistics (2006a), *Academic Research and Development Expenditures: Fiscal Year 2004.* NSF 06-323. NSF: Arlington, VA.

National Science Foundation, Division of Science Resources Statistics (2006b), *National Patterns of R&D Resources: 2004 Date Update.* NSF 06-327. NSF: Arlington, VA.

Nelson, R. R. and G. Wright (1992), 'The rise and fall of American technological leadership: the postwar era in historical perspective,' *Journal of Economic Literature*, **30**(4), 1931–1964.

Noble, D. (1977), *America by Design.* Oxford University Press: Oxford.

Organization of Economic Cooperation and Development (2002), *Frascati Manual. Proposed Standard Practice for Surveys on Research and Experimental Development.* OECD: Paris.

Phelps, E. S. (2006), 'Understanding the great changes in the world: gaining ground and losing ground since World War II,' *Capitalism and Society*, **1**(2), 1010.

Ritter, J. R. (2006), 'Some factoids about the 2005 IPO Market.' http://bear.cba.ufl.edu/ritter.

Roberts, E. B. (1991), *Entrepreneurs in High Technology: Lessons from MIT and Beyond.* Oxford University Press: Oxford.

Romer, P. (1986), 'Increasing returns and economic growth,' *American Economic Review*, **94**, 1002–1037.

Romer, P. (1990), 'Endogenous technical change,' *Journal of Political Economy*, **98**, 71–102.

Rosenberg, N. (1985), 'The commercial exploitation of science by American industry,' in B. C. Kim, H. H. Robert and L. Christopher (eds), ''*The Uneasy Alliance: Managing the Productivity-Technology Dilemma*. Harvard Business School Press: Boston, MA, pp. 19–51.

Rosenberg, N. (1992), 'Scientific instrumentation and university research,' *Research Policy*, **21**(3), 381–390.

Rosenberg, N. (1998), 'Technological change in chemicals: the role of university–industry relations,' in A. Arora, R. Landau and N. Rosenberg (eds), ''*Chemicals and Long-Term Economic Growth: Insights from the Chemical Industry*. New York: Wiley, pp. 193–230.

Rosenberg, N. (2000), *Schumpeter and the Endogeneity of Technology. Some American Perspectives*. Routledge: New York.

Rosenberg, N. and R. R. Nelson (1994), 'American universities and technical advance in industry,' *Research Policy*, **23**(3), 323–348.

Scherer, F. M. (1996), *Industry Structure, Strategy, and Public Policy*. HarperCollins College Publishers: New York.

Shane, S. (2004), *Academic Entrepreneurship: University Spinoffs and Wealth Creation*. Edward Elgar: Cheltenham, UK.

Snyder, T. D. (1993), '120 years of American education: a statistical portrait,' *Center for Education Statistics*. http://nces.ed.gov/pubs93/93442.pdf.

Thursby, J. G. and M. C. Thursby (2002), 'Who Is selling the ivorytower? Sources of growth in university licensing,' *Management Science*, **48**(1), 90–104.

Winter, S. G. (1984), 'Schumpeterian competition in alternative technological regimes,' *Journal of Economic Behavior and Organization*, **5**(304), 287–320.

[28]

Small Bus Econ (2009) 32:15–30
DOI 10.1007/s11187-008-9157-3

The knowledge spillover theory of entrepreneurship

**Zoltan J. Acs · Pontus Braunerhjelm ·
David B. Audretsch · Bo Carlsson**

Accepted: 5 November 2008 / Published online: 13 December 2008

Abstract Contemporary theories of entrepreneurship generally focus on the recognition of opportunities and the decision to exploit them. Although the entrepreneurship literature treats opportunities as exogenous, the prevailing theory of economic growth suggests they are endogenous. This paper advances the microeconomic foundations of endogenous growth theory by developing a knowledge spillover theory of entrepreneurship. Knowledge created endogenously results in knowledge spillovers, which allow entrepreneurs to identify and exploit opportunities.

Z. J. Acs (✉)
School of Public Policy, George Mason University, Fairfax, VA, USA
e-mail: zacs@gmu.edu

P. Braunerhjelm
Department of Transport and Economics, The Royal Institute of Technology, Stockholm, Sweden
e-mail: pontusb@infra.kts.se

D. B. Audretsch
School of Environmental and Public Affairs, Indiana University, Bloomington, IL, USA
e-mail: daudrets@indiana.edu

B. Carlsson
Weatherhead School of Management, Department of Economics, Case Western Reserve University, Cleveland, OH, USA
e-mail: Bo.Carlsson@case.edu

Keywords Opportunity · Knowledge · Entrepreneurship · Endogenous growth · Start-ups · New product innovation

JEL Classifications J24 · L26 · M13 · O3

> Any course of action must expose the chooser to numberless different sequels, rival hypotheses, some desired and some counter-desired…The entrepreneur is a maker of history, but his guide in making it is his judgment of possibilities and not a calculation of certainties.
>
> G.L.S. Shackle (1982, vii)

1 Introduction

Where do entrepreneurial opportunities come from? Endogenous growth models suggest R&D activities are "purposeful investment in new knowledge" undertaken by profit-maximizing firms, where knowledge is an input in the process of generating endogenous growth (Romer 1990; Segerstrom et al. 1990; Aghion and Howitt 1992).[1] The set of technological opportunity is endogenously created by

[1] Griliches (1979) formalized the knowledge production function. In this model, firms are also exogenous and pursue new economic knowledge as an input into future innovative activity.

investments in new knowledge. Technological change is central in explaining economic growth: The rate of per capita GDP growth equals the rate of technological change on the steady state growth path. This explanation assumes that efficiency of knowledge production is enhanced by the historically developed stock of scientific-technological knowledge.[2]

In addition to facilitating technological change, knowledge also generates opportunities for third-party firms (Jaffe et al. 1993; Thompson and Fox-Kean 2005), which are often entrepreneurial start-ups (Shane 2001). This occurs through intra-temporal knowledge spillovers. Therefore, entrepreneurial activity involves both arbitrage of opportunities (Kirzner 1973) and exploitation of new opportunities created, but not appropriated by incumbent firms (Schumpeter 1934).

Endogenous growth models are based on strong assumptions for the technical ease and analysis. However, these advantages impose drawbacks of deviations from real-world behavior.

The endogenous growth framework offers no insight into what role, if any, entrepreneurial activity plays in the intra-temporal spillover of tacit knowledge. While the new growth theory enhances our understanding of the growth process, the essence of the Schumpeterian (1934) entrepreneur is missed. As a result, endogenous growth models fail to incorporate a crucial element in the process of economic growth: Transmission of knowledge spillovers through entrepreneurship (Audretsch 1995). This implies that knowledge by itself is only a necessary condition for the exercise of successful enterprise in a growth model. An interesting approach recently focuses on the allocation of societal resources spent on R&D and entrepreneurship. Michelacci (2003) concludes that low rates of return to R&D may be due to lack of entrepreneurial skills. Hence, the ability to transform new knowledge into economic opportunities involves a set of skills, aptitudes, insights and circumstances that is neither uniformly nor widely distributed in the population.

This paper develops a knowledge spillover theory of entrepreneurship to improve the microeconomic foundations of endogenous growth models, in which the creation of knowledge expands technological opportunity. The theory shifts the unit of analysis from exogenously assumed firms to individual agents with new knowledge endowments. Agents with new economic knowledge endogenously pursue the exploitation of such knowledge, implying that the existing stock of knowledge yields spillovers. This further suggests a strong relationship between such knowledge spillovers and entrepreneurial activity. The theory provides an explanation for the role of the individual and the firm in an economy. According to Romer (1996, 204), such an approach "...removes the dead end in neoclassical theory and links microeconomic observations on routines, machine designs, and the like with macroeconomic discussions of technology."

The model is one where new product innovations can come both from either incumbent firms or start-ups (Acs and Audretsch 1988).[3] We can think of incumbent firms as reliant on incremental innovation from the *flow* of knowledge, such as product improvements. Start-ups with access to entrepreneurial talent and intra-temporal spillovers from the *stock* of knowledge are more likely to engage in radical innovation leading to new industries or replacing existing products.[4] According to Baumol (2004, 9): "...the revolutionary breakthroughs continue to come predominantly from small entrepreneurial enterprises, with large industry providing streams of incremental improvements that also add up to major contributions." Entry by start-ups has played a major role in radical innovations, such as software, semiconductors, biotechnology (Zucker et al. 1998) and the information and communications technologies (Jorgenson 2001). Start-ups are especially important at early stages of the life cycle, when technology is still fluid. Therefore, this paper makes the strong assumption that radical innovation comes from new firm start-ups.

[2] This is not a fixed stock of knowledge. For example, change introduced by an entrepreneur can make part of the existing stock of knowledge obsolete.

[3] Acs and Audretsch (1988) find that, *ceteris paribus*, the greater extent to which an industry comprises large firms, the greater will be innovative activity, but increased innovative activity will tend to come from small firms rather than large firms.

[4] A large amount of literature exists that shows how entrepreneurial start-ups use networks to access the stock of knowledge.

The main predictions of the model are:

1. An increase in the stock of knowledge has a positive effect on the level of entrepreneurship.
2. The more efficiently incumbents exploit knowledge flows, the smaller the effect of new knowledge on entrepreneurship.
3. Entrepreneurial activities decrease under greater regulation, administrative burden and market intervention by government.

Thus, entrepreneurship contributes to economic growth by acting as a conduit[5] through which knowledge created by incumbent firms spills over to agents who endogenously create new firms. Opportunities are created when incumbent firms invest in, but do not commercialize, new knowledge. In this theory, entrepreneurship is a response to these opportunities. We suggest that, *ceteris paribus*, entrepreneurial activity will be greater where investments in new knowledge are relatively high, since start-ups will exploit spillovers from the source of knowledge production (the incumbents). In an environment with relatively low investments in new knowledge, there will be fewer entrepreneurial opportunities based on potential spillovers.

Our theoretical model explains entrepreneurship as a function of the following factors: knowledge stock, R&D exploitation by incumbents and barriers to entrepreneurship. It considers factors such as risk aversion, legal restrictions, bureaucratic constraints, labor market rigidities, taxes, lack of social acceptance, etc. This explains why economic agents might decide against starting up, even when in possession of knowledge that promises potential profit opportunity. In addition, culture, traditions and institutions are more difficult factors to identify than strictly economic factors, but they also play an important role in entrepreneurship. To capture such country-specific differences, we estimate a reduced form equation with a fixed-effect panel-regression technique.

The paper is organized as follows. The next section examines knowledge spillovers as a source of entrepreneurial opportunity in the endogenous growth framework. We present a formal model in Sect. 2. We test and discuss results in Sects. 3 and 4, using data for the period 1981–2002 for 19 OECD countries. Our results show that entrepreneurial activity is strongly influenced by knowledge created but not exploited by incumbent firms. We conclude in the final section.

[5] It is not, of course, the sole conduit.

2 Knowledge spillovers as source of entrepreneurial opportunity

In order to enable more realistic applicability, the theory relaxes two central assumptions of the endogenous growth model. The first is that all knowledge is economic knowledge. Arrow (1962) emphasized knowledge as inherently different from traditional factors of production, resulting in a gap between new knowledge (K) and what he termed economic knowledge (K^c).[6] The second assumption is the assumed spillover of knowledge. In endogenous growth models, the existence of the factor of knowledge is equated with *inter-temporal spillover*, which yields endogenous growth. In our model we assume *intra-temporal knowledge spillovers* from incumbent organizations to start-ups. Moreover, *institutions* impose a gap between knowledge and economic knowledge ($0 < K^c/K < 1$), yielding a lower volume of intra-temporal knowledge spillovers (Acs et al. 2004).

Romer (1990) separates economically useful scientific-technological knowledge into two parts: The total set of knowledge consists of non-rival, partially excludable knowledge elements, and the rival, excludable elements of knowledge. Codified knowledge published in books, scientific papers or patent documentations belongs to the first set. This can be only partially excludable: The right of applying a technology for production of a particular good is guaranteed by patenting, but the same technology can be used in other applications as others learn from the patent documentation. Rival, excludable knowledge elements comprise personalized (tacit) knowledge of

[6] New knowledge leads to opportunities that can be exploited commercially. However, harnessing new ideas for economic growth requires converting new knowledge (K) into "economic knowledge" that holds commercial opportunity. For example, only about half of the invention disclosures in US universities lead to patent applications. Of these applications, about half result in patents issued, of which only 1/3 are actually licensed. Between 10–20% of licenses actually yield significant income (Carlsson and Fridh 2002, 231). In other words, only 1 or 2% of the inventions are successful in reaching the market and yielding income.

individuals and groups, including experiences and insights of researchers and business people. This does not go far enough. In the model proposed by Romer (1990), the movement of knowledge from firms producing it to other firms is exogenous. That model explains the effect of knowledge spillovers on technological change without elaborating *why* or *how* these spillovers occur.[7]

Why should entrepreneurship play an important role in the intra-temporal spillover of knowledge? New knowledge is characterized by greater uncertainty and asymmetry than other economic goods. Therefore, both the mean expected value of any new idea and its *variance* will differ across economic agents. If an incumbent firm decides the expected economic value of a new idea is not sufficiently high to warrant its development and commercialization, other economic agents may (or may not) assign a higher expected value to the idea. These agents can operate within or outside of the incumbent firm. This divergence in expected valuation can lead to market entry by economic agents to appropriate new knowledge. The knowledge that induces the decision to start new firms is generated by investments made by an incumbent firm. Thus, the start-up serves as the mechanism through which knowledge spills over from sources that produced it (such as a university or research laboratory in an incumbent firm) to a new organizational form where it is actually commercialized.

One way to reconcile the difference in the role of opportunities in models of entrepreneurship and endogenous growth models is the unit of analysis. Most models of entrepreneurship focus on the individual as the decision-making unit of analysis, whereas the literature on endogenous growth focuses on the firm as the decision-making unit of analysis. In such theories, the firm is exogenous, but its role in generating technological change is endogenous. Therefore, our theory focuses not on exogenously assumed firms, but rather, on the individual agent endowed with new economic knowledge. With this new focus, the issue of appropriability remains, but the central question is: How can economic agents with a given endowment of new knowledge best appropriate its returns (Audretsch 1995)?

In the knowledge spillover theory of entrepreneurship, the knowledge production function is reversed. The agent decides to start a new firm based on expected net return from a new product. Accordingly, the inventor would expect compensation for the future value of the potential innovation. In both cases, the employee in the incumbent firm will weigh the alternative of starting a new firm. If expected return from commercialization is sufficiently different for the inventor and for the incumbent decision-maker, and if the cost of starting a new firm is sufficiently low, the employee may choose to leave the incumbent firm to start a new firm. These start-ups typically do not have direct access to a large R&D laboratory. Rather, they rely on knowledge and experience gained in R&D laboratories of previous employers, i.e., the incumbents.

This type of labor mobility is likely to be an important source of intra-temporal knowledge spillovers (Pakes and Nitzan 1983). Essentially, R&D capital is knowledge that can earn a monopoly rent, and this potential rent motivates investment in R&D. To a large extent, R&D capital is embodied in R&D employees (L_R).[8]

Hellmann (2007), Lazear (2005), Hvide (2006) and Anton and Yao (1994) proposed models to examine conditions under which agents pursue entrepreneurial activity by starting rival enterprises. Most of this literature addressed incentives where potential externalities could be internalized (Moen 2005), while some have examined circumstances under which employees take advantage of intellectual human capital (L_E) through start-ups (Bhide

[7] Knowledge spillovers operate more strongly in some parts of the economy than in others. Particular characteristics tend to be associated with locations with high density of opportunities, such as those hosting high-tech industries. Most innovations occur in high-technology opportunity industries and not low-technology opportunity industries (Scherer 1965). The extent to which the results of innovation can be appropriated by incumbent firms also varies among industries.

[8] Such *intellectual human capital* is human capital that is neither publicly available nor perfectly protected. This distinguishes it from *ordinary human capital* that is widely diffused knowledge and can be acquired at a cost and earns a normal rate of return (Zucker et al. 1998). This distinction between intellectual human capital and ordinary human capital is almost identical to the distinction between rival and non-rival knowledge found in Romer (1990) where rival, excludable knowledge elements are primarily the personalized (tacit) knowledge of individual agents. We assume that the firm owns the intellectual property that results in a start-up.

1994). If an innovation makes the incumbent firm a true monopolist, it will not be profitable for the agent to pursue a start-up, since the sum of rents in a duopolistic market will be less than monopoly rent. This paper does not model the incentive structure under which individual agents become mobile. Instead, it focuses on institutions that prevent the creation of a monopoly in the first place.

3 Theoretical framework

To model entry, we partly draw on previous contributions in the endogenous growth literature. In particular, those models introduce a mechanism that fosters innovative entry through investments in R&D. We will retain that channel of entry, but also demonstrate how knowledge investment by incumbents can spur entry by entrepreneurs that do not engage in R&D themselves. Since our emphasis is on entry and not growth per se, our aim is not to derive a full-fledged growth model, but rather to focus on the mechanism for entry.[9]

Consider an economy with demand, supply and a financial market.[10] There are two types of firms: Incumbents undertaking R&D to *improve existing products* and start-ups exploiting knowledge spillovers and the existing stock of knowledge to *innovate new products*. Firms that develop an improved or new variety demanded by consumers are rewarded by temporary monopoly profits until new products out compete old ones.

3.1 Demand side

Starting with the demand side, consumers maximize standard linear intertemporal utility,

$$U = \int_0^{\infty} e^{-\rho t} \ln[h(x)]dt, \tag{1}$$

where $\rho t > 0$ equals consumer rate of time preference (discount rate) and h is the sub-utility function. Assume that different varieties of the x-goods are perfect substitutes and that v^Irefers to the most recent innovated product or variety, with improved quality or novel features. If $vp_t < p_{t-1}$, then all consumers will prefer the new product,

$$h(x_o, x_1, x_2 \ldots\ldots) = \sum_{I=0}^{\infty} v^I x_I, \quad v > 1. \tag{2}$$

The novel products/qualities demanded by consumers may range from research-intensive varieties to products characterized by combining existing knowledge. Hence, high R&D intensity by itself does not guarantee successful introduction of a new product.

3.2 Supply side

On the production side, new products/qualities can be invented either by incumbent firms investing in R&D or by entrepreneurial start-ups.[11] Successful entry means a temporary monopoly, where the price of the new product/quality equals the improved property of the product, $v = p_I$. The only factor of production is labor, which is allocated among three different activities: R&D production (L_R), self-employment through start-ups (L_E) or a residual sector employing R&D findings and producing final goods (L_F),

$$L_R + L_E + L_F = \overline{L}. \tag{3}$$

Perfect mobility across sectors assures that wages are equalized.[12] Initial profit conditions for firms/products that successfully enter the market are

$$\pi = (p_I - 1)Y/p_I = (v - 1)Y/v \tag{4}$$

where p_I represents the price of the new good, corresponding to the quality improvement (v), and wage is set equal to one. Total consumption expenditure is captured by **Y**, that is, demand for a new variety. In the long run, free entry implies zero profits. Hence, in the period preceding the introduction of a new product/firm, prices equal wage costs,

---

[9] See Braunerhjelm et al. (2006) for a growth model.

[10] For details, see Intriligator (1971), Aghion and Howitt (1992) and Dinopoulos (1996).

[11] The general production function is $x = AL^{\gamma}$, $0 < \gamma \leq 1$.

[12] The final good sector is not modeled in order to enhance transparency. It can be viewed as a sector with constant returns to scale, where labor embodied with findings in the R&D sector at each given time "t" is employed (i.e., labor does not possess skills related to ongoing R&D).

which are set to one. The first-order condition implies $v = p_I \geq 1$.

The introduction of new product innovations occurs either through R&D outlays by incumbents or through start-ups where existing knowledge is combined in innovative ways. The latter does not require any investment in R&D.[13] Instead, individuals combine their given entrepreneurial ability ($\overline{e}_j$) with the overall knowledge stock (K) in an economy to discover commercial opportunities. The societal knowledge stock is a composite of previous knowledge stemming from activities by incumbents and start-ups, i.e., knowledge refers not only to scientific discoveries, but also to novel ways of production and distribution in traditional businesses, changing business models, new marketing strategies, etc. Both types of entry are assumed to occur through a Poisson process.

Hence, the first type of product improvement is related to R&D expenditures, i.e., it is a *flow* variable, taking previous scientific knowledge as the departure point. The second type of innovative new product draws instead on the overall *stock* of knowledge and applies it in a novel way. All innovation implies that some fixed costs are incurred, such as for R&D or marketing. Innovation is thus modeled consistently with real world behavior.

Starting with incumbents, the aggregate probability of a successful product improvement increases in an economy's R&D outlays, measured as R&D employees.[14] As shown above, labor is the only input. The production technology is characterized by decreasing returns to scale ($0 < \gamma < 1$). At the firm level, each firm i's probability (μ) of successfully launching a new product increases with higher R&D investments. Thus, innovation by incumbents can be modeled as

$$\mu \sum_{i=1}^{n} (\text{R\&D})\mathrm{d}t = \mu \left(\sum_{R=1}^{\overline{L}} l_R \right) \mathrm{d}t \equiv (1/\sigma) L_R^{\gamma} \mathrm{d}t \tag{5}$$

where dt denotes an infinite increment of time and (σ)refers to an efficiency parameter that reflects how smoothly a new discovery is introduced to the market. The second type of innovative new product occurs through start-ups, where the probability of a successful start-up (η) is related to the given knowledge stock $\overline{K}$ (at each point in time) times the average entrepreneurial ability ($\hat{e}$) in the economy,

$$\eta \left(\sum_{j=1}^{\overline{L}} \overline{e}_j \right) \overline{K} \mathrm{d}t = \eta \, (\hat{e}\overline{K}) \mathrm{d}t \equiv (1/\sigma) \overline{K}^{\gamma} \mathrm{d}t. \tag{6}$$

At the individual level, the probability of success depends on each individual j's given endowment of entrepreneurial talent, which is unevenly distributed across the population of $\overline{L}$ individuals in an economy (Lucas 1978).[15] Also here, decreasing returns to scale ($0 < \gamma < 1$) prevail since an increase in entrepreneurial ability will not translate into a proportional increase in start-ups.[16]

The total rate of innovative entry in an economy can be calculated by employing the additive property of Poisson distributions,

$$\kappa \, \mathrm{d}t = \mu \, \mathrm{d}t + \eta \, \mathrm{d}t = (1/\sigma)(\overline{K}^{\gamma} + L_R^{\gamma}) \mathrm{d}t. \tag{7}$$

[13] To some extent, this parallels the classification of horizontal and vertical innovations (Howitt 1999; Gancia and Zilibotti 2005). The former refers to new products, whereas the latter implies quality improvements in existing products. In the model specification used by Howitt (1999), an increase in R&D directed towards horizontal innovations may decrease the profit flow accruing to vertical R&D, thereby undermining incentives to undertake vertical innovations. This impairs growth. However, there are considerable differences between the models. The previous literature refers to the discussion on "scale effect" and growth, initiated by Jones (1995). Moreover, in Howitt's model, different production technologies but identical inputs (R&D staff) are assumed for the two types of innovation, whereas in the present context, identical technologies are assumed, but different factors of production are employed. Finally, the line between an improved quality of an existing variety or a new product that replaces the former product is thin and not necessary for our argument.

[14] The assumed Poisson entry process means that the time frequency with which entry will occur is a random variable whose distribution is exponential with parameter μ, i.e., μ is the probability per unit of time.

[15] We follow Lucas (1978), who assumes that managerial talent is distinct from labor talent. Lazear (2005) assumes that workers and managers have the same two skills in different combinations. Those with more balanced skills are more likely to become entrepreneurs. Those with varied work and educational backgrounds are much more likely to start their own businesses than those who have focused on one role at work or concentrated in one subject at school. The implications for the size distribution of firms are similar in both models.

[16] Moreover, it would not be optimal for all economic activities to be undertaken by entrepreneurs.

Hence, incumbents may now be replaced by rival firms in an R&D race or by start-ups.[17]

3.3 The financial market

To cover investment costs in R&D or other costs, such as marketing, both incumbents and Schumpeterian entrepreneurs must turn to the financial market.[18] Investors take a risk since start-ups may replace firms, or entrepreneurs may fail. Start-ups are included in investor portfolios prior to entering the market.

Assume that investors can buy shares in all firms in order to minimize risk, implying an average return of $r(t)$. Investors calculate expected returns on their investments over time in the following way. First, a firm's instantaneous profits ($\pi = (p_I - 1)(Y/p_I)$) and the discounted return (V)—or the value of the firm—are linked through the financial market. The (expected) discounted profit is simply the value of the firm at a given time, times the probability it will succeed in inventing (μ_j) or innovating (η_j) new varieties, minus incurred labor costs.

In each period of time (dt) the shareholder receives a dividend, which is related to profits, and the firm appreciates in value $\dot{V}(t)dt = (dV/dt)dt$. However, incumbents (whether R&D-based or the entrepreneurial type) run the risk of being replaced by the introduction of new qualities (κ), thereby risking a loss of $V(t)$,

$$[\pi(t)/V(t)]dt + \dot{V}(t)/V(t)[1 - \kappa]dt + \left[(\dot{V}(t) - V(t))/V(t)\right]\kappa dt = r(t)dt \tag{8}$$

where $(1 - \kappa)$ is the probability that the firm survives and κ represents the probability that the firm will be forced out of business. Consequently, investors will incur losses on previous investments. From Eq. 8, as dt goes to zero,

$$\dot{V}(t)/V(t) + \pi/V(t) = r(t) + \kappa, \quad \dot{V}(t) = 0 \tag{9}$$

$$\pi/V(t) = r(t) + \kappa \equiv \tilde{r} \tag{10}$$

i.e., the higher risks associated with an investment in incumbents (because they may become replaced and a capital loss may be incurred) require a higher return in steady state.

To close the model intertemporal consumption must match intertemporal production—i.e., entry of new goods and start-ups reliant upon access to capital. Such an exercise implies solving a dynamic consumption (growth) model subject to a budget constraint (returns on savings and wages). This will not be undertaken here since our predominant interest is entry, not growth. However, in equilibrium the standard dynamic equilibrium condition will apply, implying that the consumer rate of time-preferences (ρ) must equal the rate of return ($\tilde{r}$) of investments over time. Then capital flows to the financial market (savings that is invested in new ventures) correspond exactly to demand for new products of intertemporally utility-maximizing consumers. Note that this does not imply a continuous flow of innovation for each period of time.

Embarking from a traditional consumer utility function, where utility is increasing in new and high quality goods, it was shown how either incumbents or start-ups supply such goods. The production technology only requires labor. Incumbents will employ labor in R&D, whereas new entrepreneurs will engage in production by drawing upon the existing stock of knowledge. Both types of firms depend on capital injections to finance innovation. This is supplied by the financial market through savings by households. Since firms may be threatened by innovation, investors require a risk-adjusted rate of return to invest in either incumbents or start-ups. Equilibrium in the labor market is assured by the assumption of free mobility across sectors, while free entry in the long-run drives profits to zero.

4 Empirical analysis

According to this model, expected profits from entrepreneurship are enhanced by the magnitude of new knowledge, but constrained by the commercialization capabilities of incumbent firms.

[17] The same factors that are identified by the theory developed here, as influencing entrepreneurship is also likely to influence imitation in the same direction. This paper does not explore the implications of replictive entrepreneurship.

[18] To improve tractability, we do not consider self-financing as a viable possibility. This connects with Schumpeter (1911 [1934]), who was adamant that the entrepreneur is not a risk-bearer. Risk bearing is the function of the capitalist who lends his funds to the entrepreneur.

4.1 The hypotheses

Given that entrepreneurial activity exceeds zero, the following testable hypotheses are derived from our model (see Appendix A):

Hypothesis 1 An increase in the stock of knowledge is expected to positively impact the degree of entrepreneurship.

Hypothesis 2 The more efficient incumbents are at exploiting R&D, the smaller the effect of a given knowledge stock on entrepreneurship.[19]

Hypothesis 3 Entrepreneurial activities can be expected to decrease under higher regulations, administrative barriers and governmental intervention in the market.

Our model explains entrepreneurship as a function of the following factors: knowledge stock (KSTOCK), R&D exploitation by incumbents (INC) and barriers to entrepreneurship captured by σ (BARR). It considers factors such as risk aversion, legal restrictions, bureaucratic constraints, labor market rigidities, taxes, lack of social acceptance, etc. (Parker 2004). The existence of such barriers is reflected by a low value of σ. This explains why economic agents might decide against starting up, even when in possession of knowledge that promises potential profit opportunity. In addition, culture, traditions and institutions are more difficult factors to identify than strictly economic factors, but they also play an important role in entrepreneurship. To capture such country-specific differences, we estimate a reduced form equation with a fixed-effect panel regression technique,[20]

$$\mathrm{ENT}_{j,t} = \lambda_j + \alpha \mathrm{KSTOCK}_{j,t} + \mathrm{BARR}'_{j,t}\beta + \mathrm{INC}'_{j,t}\gamma + \alpha_4 Z'_{j,t}\delta + \varepsilon_{j,t}$$

where j denotes country, t represents time and Z is a vector of control variables. The error term is expected to exhibit standard properties: That is, $\varepsilon_{j,t}$ is assumed to have an independent and identical distribution with a zero mean and variance σ^2 for all j and t.

4.2 The variables

The dependent variable, entrepreneurship (ENT), proxies country share of self-employed as a percentage of the labor force.[21] There are several reasons to expect the self-employment rate to decrease as economies become more developed. Blau (1987) argues this is a fundamental economic change. He shows that the time series of self-employment is correlated with a measure of the extent to which technological change has been biased towards industries in which self-employment is important. Acs et al. (1994a) document the diversity in self-employment across countries and in time-series by examining variations in self-employment rates across OECD countries. However, the convergence of several factors, notably the decline in heavy manufacturing, growth of services and possibly the bias towards technological change in the 1990s tended to stem the decline in self-employment for most OECD countries. The self-employment rate is the best available measure across-countries and over time, and it serves as an acceptable proxy for high-impact entrepreneurship.[22]

With respect to explanatory variables, our main focus is the endowment of knowledge within an economy. We first elaborate a stock measure composed of accumulated annual R&D flows, assuming an annual depreciation rate of ten percent (KSTOCK). For the time period we are investigating (1981–2002), this implies a rapid accumulation of knowledge stock in the first 10 years (up to 1991) followed by more stable development where change in the knowledge stock is determined by annual outlays on R&D. Obviously, this can insert biases into the estimations. We have therefore chosen to approximate knowledge stocks over the entire period with annual R&D flows,

[19] The efficiency with which incumbents exploit knowledge is, in part, related to the incentive structure and intellectual property rights of the firm (Moen 2005; Hellmann 2007; Hvide 2006).

[20] The choice of empirical model is based on an F-test to check the validity of using a fixed-effect regression technique as compared to OLS. The test clearly rejects the null hypotheses of all fixed effects jointly being zero. We also estimated the model without fixed effects, which yields unstable results. A dynamic panel estimation is also a possible approach, but is more appropriate for limited time series data with many panels. The current paper has rather long time series with few panels.

[21] The agricultural sector has been excluded.

[22] Start-up data are available from the World Bank, but for a different set of countries across a considerably shorter period (1997–2004). See Klapper et al. (2007) for more. The data are shown to be positively (but weakly) correlated with the self-employment data used in the current study.

which actually constitutes variation in the stock variable once the stock has been built up. Considering that the correlation between the knowledge stock variable and R&D flows was very strong, annual R&D outlays are an acceptable approximation for knowledge stocks.[23] In line with our model, we expect an increase in relative knowledge endowment to increase the profitability of entrepreneurial activity by facilitating the recognition of entrepreneurial opportunities. The knowledge variable is normalized by GDP.

The most difficult variable to model empirically is incumbent intra-temporal exploitation of knowledge. We use two variables that are important indicators of the extent to which incumbents draw upon knowledge flows. The first is the number of patents (PATENTS) in relation to population, where a higher proportion implies that incumbents use more of the existing knowledge flows (Griliches 1986).[24] The second approximation is the gap between actual and potential GDP (GAP). The argument is that full employment of the economy resources, a small percentage difference between actual and potential GDP, diminishes possibilities for start-ups through exploitation of knowledge flows. Both variables are assumed to influence entrepreneurship negatively.

We use two variables to capture the extent of barriers to entrepreneurship in an economy. First, we incorporate public expenditure in relation to GDP (GEXP) as an approximation of total tax pressure and the extent of regulatory interventions in the economy (Nicoletti et al. 2000). As an alternative, we also include tax share in GDP for both individual (TAXPERS) and corporate firms (TAXCORP). Start-ups are less likely to occur if incentive structures are distorted through high taxes (Henrekson 2005). We expect these variables to be negatively associated with entrepreneurship.[25]

In addition to the above variables, we also include a number of control variables where previous research indicates influence on entrepreneurship. Numerous studies claim urban environments are particularly conducive to entrepreneurial activities, innovation and growth because of agglomeration economies (Jacobs 1969; Krugman 1991). Information flows are denser in cities, where different competencies and financial resources are more accessible, and market proximity is obvious (Acs et al. 1994b). We therefore include a variable to capture the share of the population living in urban areas (URBAN). We expect greater urbanization to be reflected in higher entrepreneurial activities. Studies on demographic variables conclude that individuals in the age cohort 30–44 are most likely to undertake entrepreneurial activities. A large share of population in this cohort (AGE) is expected to relate positively to the share of entrepreneurs. In order to smooth out business fluctuations, we control for economic growth, defined as a 5-year moving average (GROWTH). Finally, we control for time-specific effects by implementing either period dummies, annual dummies or a time trend.[26]

All regressions are based on data for 19 OECD countries for the period 1981 to 2002. Data come predominantly from the OECD, but other sources are also used (see Appendix B).[27] Summary statistics and correlations are listed in Tables 1 and 2, respectively.

5 Regression results

Regression results estimating the entrepreneurship rate (ENT) for 1981–2002 are presented in Tables 3 and 4. These results are consistent with the predictions of our model. Entrepreneurial opportunities do not appear to be exogenous but rather systematically created by a high presence of knowledge spillovers. As the positive and statistically significant coefficients of the knowledge stock suggest, entrepreneurial activity tends to be greater where knowledge is more prevalent.

[23] For the full sample the correlation coefficient between knowledge stocks and R&D flows varied between 0.95 and 0.98 (5-year averages), depending on the lag-structure. There are some variations across countries. R&D data are not available prior to 1981; hence, it was not possible to construct knowledge stocks using data from the 1970s.

[24] Giuri et al. (2007) and Braunerhjelm and Svensson (2008) provide evidence that large companies hold the dominating part of patents.

[25] We could also use the World Bank cost of doing business as an alternative measure barrier to entry.

[26] For instance, Jorgenson (2001) argues that increased technological change enhanced entrepreneurial activity, particularly in the 1990s.

[27] The following countries are included: Australia, Austria, Belgium, Canada, Denmark, Finland, France, Germany, Ireland, Italy, Japan, The Netherlands, New Zealand, Norway, Portugal, Spain, Sweden, the UK and US. For some variables with missing values, we have used the closest year available.

Table 1 Descriptive statistics

Variable	Minimum	Mean	Maximum	SD	Observations
ENT	2.83	4.90	8.41	1.37	418
KSTOCK	.28	1.77	4.63	.75	411
GEXP	30.2	47.87	72.93	8.43	412
TAXPERS	3.83	11.54	26.22	4.74	410
TAXCORP	.27	2.87	8.92	1.31	410
PATENT	$9.8E^{-4}$	.28	1.38	.26	414
GAP	−10.90	−.65	6.83	2.65	418
GROWTH	−3.02	2.64	9.77	1.51	418
URBAN	43.3	74.13	97.10	11.66	418
AGE	16.87	21.88	25.60	1.72	418

Table 3 reports results where no lags are implemented on the knowledge stock variable. In the first three regressions, variables are defined as 5-year averages, while the remaining three implement annual data.[28] Different versions of the time variables are used in the estimations, albeit the results are quite robust regardless of variable specification. When data are defined as 5-year averages in the estimations, the knowledge stock is shown to be positive in all regressions and significant in two of three. With respect to the remaining variables, only demographic structure (AGE) has a consistent and significant impact on entrepreneurship. Incumbent exploitation of knowledge (PATENT and GAP) and barriers to entrepreneurship (GEXP and taxes) have the expected negative signs but are insignificant. Only personal income taxes have a weak negative and significant impact on entrepreneurship.

One reason for the relatively low explanatory value of these regressions may be the limited number of observations from using 5-year averages for the estimations. When the regressions are repeated using annual data, the significance of the knowledge stock variable is strengthened considerably. This is also the case for variables representing entrepreneurial barriers and the impact of incumbents. Note that the patent variable is significant in all regressions, suggesting that extensive knowledge exploitation by incumbents is negatively related to entrepreneurial activity. The lower the ability of incumbents to appropriate new knowledge, the more knowledge will spill over to third parties, as predicted by the theory. Hence, to the degree that incumbent firms take advantage of opportunities, there will be less entrepreneurial activity. The GAP variable, admittedly defined at an aggregate level, fares less well. Public expenditure (GEXP), which indicates a wide set of barriers to entrepreneurship, has a negative and significant effect on entrepreneurship. As we substitute government expenditure for the two tax variables, the negative and statistically significant coefficient of the personal tax rate indicates that personal taxes are a barrier to entrepreneurship. The positive and weak significance of the corporate tax rate may actually indicate that a higher rate of corporate taxes reduces the propensity for incumbent firms to appropriate returns from opportunities. This can generate more entrepreneurial opportunities.

A lag structure ranging from 1 to 3 years is imposed on the knowledge stock variable, as shown in Table 4. One regression is shown with variables defined as 5-year averages; remaining regressions use annual data. We only present results for estimations using annual dummies.[29] In general, the estimations conform with the results in Table 3, particularly for variables related to knowledge stock, barriers to entrepreneurship and incumbents exploitation of knowledge. The patent variable always presents a negative sign (as does the GAP variable with one exception), and is significant in four of the five regressions using annual data.

Table 2 reports a relatively high correlation coefficient (.74) between the knowledge stock variable (KSTOCK) and the patent variable (PATENT). A similar but weaker coefficient (.47) is

[28] Averages are calculated for the periods 1981–85, 1986–90, 1991–95 and 1996–2002. The average for the last period is based on 7 years.

[29] Independent of the time variable used (trend, period dummy or year dummy), the results are quite robust. Regression results using alternative time specifications are available from the authors on request.

Table 2 Correlation matrix, independent variables, 5-year averages, period dummies

	KSTOCK	KSTOCK-1	GEXP	TAX-PERS	TAX-CORP	PATENT	GAP	GROWTH	URBAN	AGE
KSTOCK	1									
KSTOCK-1	.95	1								
GEXP	.11	.02	1							
TAXPERS	.15	.02	.45	1						
TAXCORP	.09	.10	−.44	−.17	1					
PATENT	.74	.76	.35	.05	−.19	1				
GAP	−.04	−.00	−.26	.00	.20	.02	1			
GROWTH	−.20	−.22	−.46	−.16	.29	−.32	.47	1		
URBAN	.19	.24	.11	.44	−.00	.08	.02	−.18	1	
AGE	.09	.11	−.06	.14	.01	.04	−.01	−.16	.19	1

Note: Period dummies not shown. The correlation matrix for annual data is highly similar and is not shown, but of course available on request

Table 3 Regression results, country fixed-effect panels, 5-year averages and annual data

Dep. var: ENT	5-year averages, trend or period dummies			Annual data, trend or year dummies		
KSTOCK	.54**	.51*	.19	.35***	.44***	.19**
	(2.09)	(1.97)	(.67)	(4.02)	(4.93)	(1.95)
GEXP	−.03	−.03	–	−.02***	−.02***	–
	(−1.60)	(−1.58)		(−2.59)	(−3.62)	
TAXPERS	–	–	−.11*	–		−.08***
			(−1.98)			(−4.67)
TAXCORP	–	–	.04	–		.05*
			(.53)			(1.81)
PATENT	−.89	−.56	−.50	−.30*	−.74***	−.66***
	(−1.27)	(−.91)	(−.82)	(−1.68)	(−3.39)	(−3.04)
GAP	.01	−.01	.00	−.01	.01	−.01
	(.14)	(−.25)	(.01)	(−.71)	(.75)	(−.59)
GROWTH[a]	−.04	−.04	.04	.03	−.02	.06***
	(−.45)	(−.66)	(.58)	(1.54)	(.87)	(2.79)
URBAN	.07	.07	.01	.09***	.08***	.03
	(1.46)	(1.56)	(.20)	(5.21)	(4.70)	(1.49)
AGE	.21***	.23***	.22***	.23***	.21***	.19***
	(3.55)	(4.37)	(4.19)	(12.79)	(10.29)	(10.12)
TREND[b]	–	.01	.10	−.00	–	–
		(.09)	(.86)	(−.33)		
YEAR/PERIOD DUMMIES	YES	NO	NO	NO	YES	YES
CONSTANT	−3.99	−4.94	−.08	−6.69***	−6.16***	−1.19
	(−.92)	(−1.19)	(−.02)	(−4.57)	(−3.51)	(−.76)
R^2	.64	.63	.64	.60	.63	.61
F	8.04	9.88	8.64	71.87	22.13	19.15
No. of obs.	73	73	72	408	408	403

[a] Growth is defined as a 5-year moving average when annual data are implemented

[b] Linear time trend

Table 4 Regression results, country fixed-effect panels, 5-year averages (first column) and annual data, lagged knowledge stock variable

Dep. var: ENT	5-year average	Annual data				
KSTOCK-1	.85*	.23**	–	–	–	–
	(1.70)	(2.26)				
KSTOCK-2	–	–	.47***	.26**	–	–
			(4.52)	(2.38)		
KSTOCK-3	–	–		–	.46***	.27**
					(3.95)	(2.28)
GEXP	−.06*	–	−.03***	–	−.03***	–
	(−1.77)		(−3.44)		(−3.49)	
TAXPERS	–	−.09***	–	−.09***	–	−.09***
		(−5.10)		(−5.26)		(−5.34)
TAXCORP	–	.04	–	.02	–	.02
		(1.44)		(1.01)		(.87)
PATENT	−.26	−.59***	−.57***	−.50**	−.45*	−.40
	(−.25)	(−2.63)	(−2.32)	(−2.09)	(−1.74)	(−1.62)
GAP	−.06	−.01	.00	−.01	−.00	−.02
	(−1.56)	(−.61)	(.10)	(−.92)	(−.28)	(−1.18)
GROWTH[a]	−.05	.06***	.03	.05***	.03	.06***
	(−.60)	(2.82)	(1.07)	(3.00)	(1.07)	(2.97)
URBAN	.13**	.03*	.10***	.04**	.10***	.05***
	(2.18)	(1.80)	(5.52)	(2.20)	(5.44)	(2.48)
AGE	.23***	.21***	.22***	.21***	.21***	.21***
	(3.22)	(10.30)	(10.19)	(10.40)	(9.83)	(10.32)
TREND[b]	−.17	–	–	–	–	–
	(−1.01)					
YEAR/PERIOD DUMMIES	NO	YES	YES	YES	YES	YES
CONSTANT	7.66	−1.71	−6.47***	−2.30	−6.50***	−2.72*
	(−1.39)	(−1.07)	(−4.19)	(−1.42)	(−4.01)	(−1.66)
R^2	.60	.61	.62	.60	.60	.59
F	5.28	18.88	20.82	17.80	18.74	16.44
No. of obs.	55	385	372	367	353	349

[a] Growth is defined as a 5-year moving average when annual data are implemented

[b] Linear time trend

obtained for the variables GROWTH and GAP, i.e., the difference between potential and actual GDP. Finally, the variables GROWTH and governmental expenditure (GEXP) display a similar degree of correlation (.46). This may introduce multicollinearity into the regressions, making the estimators less efficient, albeit still unbiased. We therefore rerun the regressions where we have excluded GAP (all regressions) and control for the impact of the exclusion in the regressions of either PATENT or GEXP on the other estimates.

As shown in Tables 5 and 6, removing the patent variable from the regressions does not influence the estimates of the knowledge stock variable. However, when we omit GEXP the control variable GROWTH turns significantly positive, but the remaining variables are basically unaffected (if the growth variable is omitted, the significance of the GEXP variable is strengthened). Hence, these correlations have only a minor influence on the results.

Thus, the empirical findings that entrepreneurship tends to be systematically greater in the presence of

Table 5 Regression results, country fixed-effect panels, 5-year averages and annual data

Dep. var: ENT	5-year averages, trend			Annual data, trend		
KSTOCK	.42*	.51**	.51**	.31***	.35***	.36***
	(1.71)	(1.96)	(2.03)	(3.63)	(4.12)	(4.13)
GEXP	–	−.02	−.03	–	−.02***	−.02***
		(−1.11)	(−1.60)		(−2.62)	(−2.57)
PATENT	–	–	−.58	–	–	−.30*
			(−.96)			(−1.67)
GROWTH[a]	.03	−.01	−.05	.06***	.03	.02
	(.75)	(−.11)	(−.77)	(4.10)	(1.60)	(1.43)
URBAN	.11***	.11***	.07	.09***	.10***	.09***
	(2.54)	(2.60)	(1.56)	(5.89)	(6.09)	(5.17)
AGE	.26***	.25***	.23***	.25***	.25***	.23***
	(45.55)	(3.55)	(4.41)	(14.07)	(14.26)	(12.78)
TREND[b]	−.08	−.09	.01	−.01	−.01*	−.01
	(−.88)	(.1.05)	(.10)	(−1.42)	(−1.83)	(−.34)
CONSTANT	−9.38***	−8.74***	−4.81	−8.08***	−7.57***	−6.50
	(−2.72)	(−2.50)	(−1.18)	(−6.04)	(−5.63)	(−4.52)
R^2	.64	.65	.63	.60	.61	.60
F	17.88	15.17	11.52	117.53	100.57	182.17
No. of obs.	74	74	73	411	411	408

[a] Growth is defined as a 5-year moving average when annual data are implemented

[b] Linear time trend

Table 6 Regression results, country fixed-effect panels, annual data, lagged knowledge stock variable

Dep. var: ENT						
KSTOCK-1	.46***	.41***	–	–	–	–
	(3.87)	(4.39)				
KSTOCK-2	–	–	.39***	.45***	–	–
			(3.96)	(4.44)		
KSTOCK-3	–	–	–	–	.43***	.48***
					(3.99)	(4.32)
GEXP	–	−.02***	–	−.02***	–	−.02***
		(−2.64)		(−2.65)		(−2.63)
PATENT	–	−.22	–	−.12	–	−.02
		(−1.17)		(−.58)		(−.07)
GROWTH[a]	.06***	.02	.05***	.02	.05***	.02
	(3.88)	(1.20)	(3.73)	(1.05)	(3.57)	(.88)
URBAN	.10***	.09***	.11***	.10***	.11***	.11***
	(6.23)	(5.69)	(6.48)	(6.19)	(6.52)	(6.18)
AGE	.25***	.24***	.25***	.24***	.25***	.24***
	(13.98)	(12.91)	(13.71)	(12.91)	(13.32)	(12.67)
TREND[b]	−.01*	−.01	−.01*	−.01	−.01*	−.01*
	(−1.71)	(−.97)	(−1.86)	(−1.54)	(1.86)	(−1.86)

Table 6 continued

Dep. var: ENT						
CONSTANT	−8.79***	−7.32***	−9.30***	−8.07***	−9.58***	−8.25***
	(−6.40)	(−5.07)	(−6–62)	(−5.58)	(−6.66)	(−5.54)
R^2	.60	.61	.60	.60	.58	.58
F	11.73	78.80	102.43	73.52	91.17	64.60
No. of obs.	392	390	373	372	354	353

[a] Growth is defined as a 5-year moving average when annual data are implemented

[b] Linear time trend

knowledge spillovers are strikingly robust. Although the significance and even sign of some control variables are more sensitive to time period and variable specification, entrepreneurial activity responds positively to economic knowledge, regardless of time and variable specification.

6 Conclusions

This paper has developed a Knowledge Spillover Theory of Entrepreneurship in which the creation of new knowledge expands the set of technological opportunity. Therefore, entrepreneurial activity does not involve simply the arbitrage of opportunities, but also the exploitation of intra-temporal knowledge spillovers not appropriated by incumbent firms. The theory focuses on individual agents with endowments of new economic knowledge as the unit of analysis in a model of economic growth, rather than exogenously assumed firms. Agents with new knowledge endogenously pursue the exploitation of knowledge. This suggests that knowledge spillovers come from the stock of knowledge, and there is a strong relationship between such spillovers and entrepreneurial activity. If incumbent firms appropriated all the rents of R&D, there would be no intra-temporal knowledge spillovers.

There are several implications of these findings for future research. First, theories of entrepreneurship need to explain where opportunities come from, how intra-temporal knowledge spillovers occur, and the dynamics of occupational choice leading to new firm formation. Prevailing theories of entrepreneurship do not address these questions. Second, the theory helps better understand the contradictions in Smith's *Wealth of Nations* between increasing returns (the pin factory) and how the market economy can harness self-interest to the common good, leading individuals to unintentional ends (the invisible hand). The real challenge in endogenous growth theory is not that the firm will under-invest in new knowledge, but how to balance increasing returns with competition. The theory provides an explanation of the role of the individual and the firm in the economy. If Romer inspired a new economics of knowledge, The Knowledge Spillover Theory of Entrepreneurship brings us a step closer to understanding the essential role of the entrepreneur in the market economy. Finally, the role of intellectual property rights protection needs to be revaluated in light of the theory. If intellectual property protection becomes too strong, and all rents accrue to the producer of knowledge, it will reduce intra-temporal knowledge spillovers, and ultimately innovation and growth (Acs and Sanders 2008).

Acknowledgements We thank discussants at the 2007 DRUID conference in Copenhagen, Sameeksha Desai, Larry Plummer, Mark Sanders, Scott Shane, Mariagrazia Squicciarini, Siri Terjesen and two anonymous referees for valuable comments. We would also like to thank Per Thulin and Benny Borgman for excellent research assistance and the Marianne and Marcus Wallenberg Foundation for generous financial support.

Appendix A

From Eqs. 5, 6 and 7, the impact of a change in the knowledge stock (K), R&D efforts by incumbents (L_R) and in the efficiency parameter (σ, $0 < \sigma < 1$), where values close to zero in the efficiency variable implies fewer obstacles in introducing new products, can be derived as

Table 7 Definition of variables and data source

Variable	Definition	Sources
ENT	Dependent variable. Non-agricultural self-employed, as percentage of total non-agricultural employment	EIM, The COMPENDIA Database
KSTOCK	Gross domestic expenditure on R&D as percentage of GDP. KSTOCK-t refers to different lag structures	OECD, Main Science and Technology Indicators
GEXP	Total government disbursements as percentage of GDP	OECD, Economic Outlook
TAXPERS	Taxes on personal income, as percentage of GDP	OECD, Revenue Statistics
TAXCORP	Taxes on corporate income, as percentage of GDP	OECD, Revenue Statistics
PATENT	Number of patents applications to the EPO by residence of the inventor (date of grant) divided by total population expressed in 10,000	OECD, Patent Database
GAP	Difference between actual annual GDP and potential GDP, divided by potential GDP	OECD, Economic Outlook
GROWTH	Average growth over 5 years. Alternatively (see notes in the regression tables), 5-year moving average of gross domestic product growth (at the price levels and PPPs of 1995)	OECD, National Accounts vol. 1, Main Aggregates; OECD, Economic Outlook for Germany
URBAN	The share of the total population living in urban areas	World Bank, World Development Indicators
AGE	Share of population between 30 and 44 years of age. Data published every 5th year. Intermediate values estimated by cubic splines	United Nations; OECD, Economic Outlook
TREND	Linear time trend	

$$\kappa_K = (1/\sigma)\gamma K^{\gamma-1} > 0, \quad \kappa_{KK} = (1/\sigma)(\gamma-1)\gamma K^{\gamma-2} < 0 \quad (\gamma<1), \tag{A.1a}$$

$$\kappa_{L_R} = (1/\sigma)\gamma L_R^{\gamma-1} > 0, \quad \kappa_{L_R L_R} = -(1/\sigma)(\gamma-1)\gamma L_R^{\gamma-2} < 0 \quad (\gamma<1) \tag{A.1b}$$

$$\kappa_\sigma = -[(K^\gamma)+(L)^\gamma]/\sigma^2 < 0, \quad \kappa_{\sigma\sigma} = 2[(K^\gamma)+(L)^\gamma]/\sigma^3 > 0, \tag{A.1c}$$

implying that the probability of entrepreneurial start-ups is increasing in the knowledge stock (weighted by average entrepreneurial ability), but at a decreasing rate (A.1a). By increasing R&D staff, incumbent firms increase the probability of launching a new quality, albeit at a decreasing rate (A.1b). Both types of innovation are positively affected by a more efficient economy, but at a decreasing rate (A.1c).

Appendix B

See Table 7.

References

Acs, Z. J., & Sanders, M. (2008). *Intellectual property rights and the knowledge spillover theory of entrepreneurship*. Working paper. Jena: Max Planck Institute of Economics.

Acs, Z. J., Audretsch, D. B., Braunerhjelm, P., & Carlsson, B. (2004). *The missing link: The knowledge filter and entrepreneurship in endogenous growth*. Discussion Paper no. 4783. London, UK: Center for Economic Policy Research.

Acs, Z. J., Audretsch, D. B., & Evans, D. S. (1994a). *The determinants of variations in self-employment rates across countries and over time*. Discussion Paper no. 871. London, UK: Centre for Economic and Policy Research.

Acs, Z. J., Audretsch, D. B., & Feldman, M. P. (1994b). R&D spillovers and recipient firm size. *The Review of Economics and Statistics, 76*, 336–340.

Acs, Z. J., & Audretsch, D. B. (1988). Innovation in large and small firms: An empirical analysis. *The American Economic Review, 78*, 678–690.

Aghion, P., & Howitt, P. (1992). A model of growth through creative destruction. *Econometrica, 60*, 323–351.

Anton, J. J., & Yao, D. A. (1994). Expropriation and inventions: Appropriable rents in the absence of property rights. *The American Economic Review, 84*, 190–209.

Arrow, K. (1962). Economic welfare and the allocation of resources for invention. In R. Nelson (Ed.), *The rate and direction of inventive activity*. NJ: Princeton University Press and NBER.

Audretsch, D. B. (1995). *Innovation and industry evolution*. Cambridge, USA: MIT Press.

Baumol, W. J. (2004). Entrepreneurial enterprises, large established firms and other components of the free-market growth machine. *Small Business Economics, 23*, 9–21.

Bhide, A. (1994). How entrepreneurs craft strategies that work. *Harvard Business Review, 72*, 150–161.

Blau, D. M. (1987). A time-series analysis of self-employment in the United States. *The Journal of Political Economy, 95*, 445–467.

Braunerhjelm, P., & Svensson, R. (2008). The inventor's role: Was Schumpeter right? *Journal of Evolutionary Economics*, forthcoming.

Carlsson, B., & Fridh, A.-C. (2002). Technology transfer in United States universities: A survey and statistical analysis. *Journal of Evolutionary Economics, 12*, 199–232.

Dinopoulos, E. (1996). Schumpeterian growth theory: An overview. In E. Helmstädter & M. Perlman (Eds.), *Behavioral norms, technological progress and economic dynamics*. Ann Arbor: The University of Michigan Press.

Gancia, G., & Zilibotti, F. (2005). Horizontal innovation in the theory of growth and development. In P. Aghion & S. Durlauf (Eds.), *Handbook of economic growth*. North Holland, Amsterdam: New York Elsevier.

Giuri, P., Mariani, M., Brusoni, S., Crespi, G., Francoz, D., Gambardella, A., et al. (2007). Inventors and invention processes in Europe: Results from the PatVal-EU survey. *Research Policy, 36*, 1107–1127.

Griliches, Z. (1979). Issues in assessing the contribution of R&D to productivity growth. *Bell Journal of Economics, 10*, 92–116.

Griliches, Z. (1986). Productivity, R&D, and basic research at the firm level in the 1970s. *The American Economic Review, 76*, 141–154.

Hébert, R. F., & Albert, N. (1982). *The entrepreneur: Mainstream views and radical critiques*. New York: Praeger.

Hellmann, T. (2007). When do employees become entrepreneurs? *Management Science, 53*, 919–933.

Henrekson, M. (2005). Entrepreneurship: A weak link in the welfare state? *Industrial and Corporate Change, 14*, 437–467.

Howitt, P. (1999). Steady endogenous growth with population and R&D inputs growing. *Journal of Political Economy, 107*, 715.

Hvide, H. K. (2006). Firm size and the quality of entrepreneurs. *Economic Journal*, in press.

Intriligator, M. D. (1971). *Mathematical optimization and economic theory*. NJ: Prentice-Hall Englewood Cliffs.

Jacobs, J. (1969). *The economy of cities*. NY, USA: Random House.

Jaffe, A. B., Trajtenberg, M., & Henderson, R. (1993). Geographic localization of knowledge spillovers as evidenced by patent citations. *Quarterly Journal of Economics, 108*, 577–598.

Jones, C. I. (1995). R&D-based models of economic growth. *Journal of Political Economy, 103*, 759–784.

Jorgenson, D. W. (2001). Information technology and the US economy. *The American Economic Review, 91*, 1–32.

Kirzner, I. M. (1973). *Entrepreneurship and competition*. Chicago: University of Chicago Press.

Klapper, L., Amit, R., Guillén, M. F., & Quesada, J. M. (2007). *Entrepreneurship and firm formation across countries. Development Research Group Working Paper*. Washington DC: World Bank.

Krugman, P. (1991). Increasing returns and economic geography. *The Journal of Political Economy, 99*, 483–499.

Lazear, E. P. (2005). Entrepreneurship. *Journal of Labor Economics, 23*, 649–680.

Lucas, R. E. J. R. (1978). On the size distribution of business firms. *The Bell Journal of Economics, 9*, 508–523.

Michelacci, C. (2003). Low returns in R&D due to the lack of entrepreneurial skills. *The Economic Journal, 113*, 207–225.

Moen, J. (2005). Is mobility of technical personnel a source of R&D spillovers? *Journal of Labor Economics, 23*, 81–114.

Nicoletti, G., Scarpetta, S., & Boylaud, O. (2000). *Summary indicators of product market regulation with an extension to employment protection legislation. Concepts and measurement of labour markets flexibility/adaptability indicators*. Paris: OECD Economics Department.

Pakes, A., & Nitzan, S. (1983). Optimum contracts for research personnel, research employment, and the establishment of "Rival" Enterprises. *Journal of Labor Economics, 1*, 345–365.

Parker, S. C. (2004). *The economics of self-employment and entrepreneurship*. New York: Cambridge University Press.

Romer, P. M. (1990). Endogenous technological change. *The Journal of Political Economy, 98*, 71–102.

Romer, P. M. (1996). Why, indeed, in America? Theory, history, and the origins of modern economic growth. *The American Economic Review, 86*, 202–206.

Scherer, F. M. (1965). Firm size, market structure, opportunity, and the output of patented inventions. *The American Economic Review, 55*, 1097–1125.

Schumpeter, J. (1911 [1934]). *The theory of economic development*, New York: Oxford University Press.

Segerstrom, P. S., Anant, T. C. A., & Dinopoulos, E. (1990). A Schumpeterian model of the product life cycle. *The American Economic Review, 80*, 1077–1091.

Shackle, G. L. S. (1982). Foreword. In R. F. Hebert & A. Link (Eds.), *The entrepreneur: Mainstream views and radical critiques*. New York: Praeger Publishers.

Shane, S. (2001). Technological opportunities and new firm creation. *Management Science, 47*, 205–220.

Thompson, P., & Fox-Kean, M. (2005). Patent citations and the geography of knowledge spillovers: A reassessment. *American Economic Review, 95*, 450–460.

von Hayek, F. A. (1937). Economics and knowledge. *Economica, 4*, 33–54.

Zucker, L. G., Darby, M. R., & Brewer, M. B. (1998). Intellectual human capital and the birth of us biotechnology enterprises. *The American Economic Review, 88*, 290–306.

[29]

Ann Reg Sci (2009) 43:989–1012
DOI 10.1007/s00168-008-0234-y

ORIGINAL PAPER

Penetrating the knowledge filter in "rust belt" economies

Zoltan Acs · Lawrence A. Plummer · Ryan Sutter

Received: 13 August 2007 / Accepted: 6 April 2008 / Published online: 20 May 2008

Abstract A new model of economic growth introduces the knowledge filter between new knowledge and economically useful knowledge. It identifies both new ventures and incumbent firms as the mechanisms that penetrate the knowledge filter. Recent empirical work has shown that new firms are more proficient at penetrating the knowledge filter than are incumbent firms; however, the analysis has only examined expanding economies and has relied on purely cross-sectional regression methodologies. This study explores the role of new and incumbent firms in penetrating the knowledge filter utilizing recent developments in spatial panel estimation techniques to provide a more robust set of findings. The results suggest those new firms are more proficient at penetrating the knowledge filter in declining and growing regions alike.

JEL Classification L26 · O1 · O18 · O3 · R1

1 Introduction

The production and application of new knowledge is often seen as pivotal to economic growth and prosperity. This idea, strongly voiced by endogenous growth theorists (Romer 1990), forms the basis for several policies intended to rev the engine of economic progress. Included in these is the Bayh-Dole Act, enacted in 1980, which transfers the intellectual property rights of federally funded research to the research performing university as well as the small business innovation research (SBIR) program

Z. Acs (✉) · R. Sutter
School of Public Policy, George Mason University, Fairfax, VI 22030, USA
e-mail: zacs@gmu.edu

L. A. Plummer
Clemson University, 165 Sirrine Hall, Clemson, SC 29634, USA

established in 1982.[1] These developments—coupled with deregulation, the biggest wave of merger and acquisition in the US history, pension fund reforms that gave rise to institutionalized venture capital, and the reorganization of most corporate R&D activities—have been followed by two decades of unprecedented economic dynamism with the emergence of new industries and the renewal of old ones (Acs and Armington 2006).

Indeed, whereas in the early postwar period innovation tended to be carried out by large firms in capital-intensive, concentrated industries characterized by highly differentiated goods, the last two decades have been characterized by a different technological regime in which innovation is carried out primarily by new firms in highly knowledge and skilled-labor intensive industries having a large share of big firms (Winter 1984; Acs and Audretsch 1987; Acs and Plummer 2005; Holtz-Eakin and Kao 2003). Jovanovic, for example, finds that the performance of small companies versus large ones (as measured by the price of small capitalization stocks relative to the S&P 500) is about equal from the end of World War II to the late 1960s and then rises dramatically to a 4:1 ratio by the mid-1980s (Jovanovic 2001, p. 54).[2]

This structural transformation of the United States economy emphasizing the contribution of new firms suggests that the production and application of knowledge, although a necessary condition, is not sufficient alone for economic growth in local economies. Instead, it seems that any general knowledge available in the economy must be actively "converted" into economically useful knowledge and that this conversion is a particular specialty of new firms.[3] Moreover, the conversion of knowledge seems a highly localized process given evidence that the flow and diffusion of knowledge is spatially constrained (Anselin et al. 1997). As a result, the link between the production of knowledge and economic growth appears most evident at a regional level of analysis.

Acs, Audretsch, Braunerhjelm, and Carlsson conceive of the conversion of available knowledge into economically useful knowledge as the "penetration" of a "knowledge filter" by the actions of both new and existing firms. The knowledge filter is the sum of all the barriers inhibiting the conversion of knowledge produced by research into commercialized knowledge (Carlsson et al. 2007). By characterizing the knowledge filter as being "semi-permeable", Acs et al. (2004) contended that the conversion of knowledge in regional economies occurs only through the concerted *actions* and the bearing of relevant costs by new and existing firms. Thus, the knowledge filter

[1] As important as Bayh–Dole was, it was essentially an attempt to "reverse engineer" the technology transfer process that had worked so effectively in prior years at a few very special institutions such as MIT and CalTech. In the face of the incentives offered by Bayh-Dole, a wide range of universities adapted to a new landscape and began promoting technology transfer, but the vast majority of them never developed the kind of permissive, entrepreneurial culture that marked the early models.

[2] Jovanovic attributes this rise in the relative performance of small firms to the application of the microprocessor. He also notes that among the 20 largest US companies by market capitalization in 1999, 10 were incorporated in or after 1967.

[3] There are two reasons why this may be true: One, new firms may simply do certain things (such as certain types of innovation) better than large firms. As a result, through division of labor between small and large firms, the efficiency and growth of the economy is increased. Two, new firms are indicative of the entrepreneurship and variety required for particularly meaningful economic growth and stability (Carlsson 1999).

conjecture suggests that the contribution of knowledge to regional economic growth *depends* on the absorptive capacity of existing firms as well as the creation of new firms by individual entrepreneurs; Acs and Plummer (2005) found support for the conjecture using economic data for the state of Colorado.

The purpose of this paper is to test the knowledge filter model of endogenous growth in the context of *declining* regional economies. For the past 30 years, the performance of Colorado's economy has been exceptionally strong with a gross state product that has increasingly outpaced the national average. Over the same period, Ohio is a declining "rustbelt" region once dominated by large firms and heavy manufacturing with gross state products increasingly falling behind the national average.[4] The focus of this paper, then, is to ask, does the knowledge filter conjecture hold in declining local economies? Far from being a mere replication, this study carries important theoretical implications for the Acs et al. (2004) knowledge filter model of endogenous growth and the Acs et al. (2005) knowledge spillover theory of entrepreneurship model by serving to assess the validity and generalizability of the theoretical models.

The paper is organized as follows. Section 2 discusses the existing economic growth theory and outlines the basic assumptions regarding endogenous growth models. Section 3 lays out the basic elements of the model and develops the hypotheses to be tested. Section 4 describes the research design and Sect. 5 reports the results of our analysis. Finally, Sect. 6 will provide some conclusions.

2 Endogenous growth

The pivotal contributions of Romer (1986, 1990), Lucas (1988), and their followers to the theory of economic growth are celebrated. Their efforts theoretically endogenize the production of knowledge within an economy and thereby disconnect growth from investment in physical capital or increases in the supply of labor. The model has the following basic structure: at the firm-level, knowledge is produced by profit-maximizing firms, while at the macro-level, the production of knowledge has important implications for growth. In Romer's original formulation, knowledge enhances growth in two ways. First, the knowledge-producing firm runs its operations more efficiently, and, second, the produced knowledge spills over to other firms, acting as a shift factor in their production functions. In subsequent variants, referred to as Schumpeterian growth models, economic growth is propelled by the combination of competition and temporary monopoly profits stemming from knowledge-based innovations.

The endogenous growth models provide little micro-economic foundation for explaining the mechanisms that promote growth at the macro-level. In other words, the focus is chiefly on growth at the national level. As applied, however, the emphasis in these models is often on the macro-economic consequences of innovation and knowledge. As Acs et al. (2004) contended, the simplistic firm-level formulation of endogenous growth models misguides policy-makers and makes empirical testing and

[4] The manufacturing "rust belt" of the United States covering many of the Great Lakes states including Ohio experienced a dramatic decline from the early 1970s well into the 1980s characterized chiefly by "deindustrialization" (High 2003). Among the monumental shifts in these regional economies were large reductions in manufacturing employment, plant closures, rising crime rates, and net population losses.

validation of the models much more difficult. In this vein, Acs et al. (2004) explored the underlying assumptions of the basic endogenous growth model intended to better capture how and why knowledge contributes precipitously to economic growth.

2.1 Assumptions on firms and technology

Endogenous growth models rely on several assumptions regarding the nature of the firms themselves and the production technology they employ. In the case of the former, for example, the model builds on the assumption of a "representative" firm intended to capture firm-level behavior at a macroeconomic level of analysis.[5] As for production technology, it is generally assumed that the *production of goods* is characterized by increasing returns to scale as a function of the increasing marginal productivity of knowledge, but the *production of knowledge* is subject to diminishing returns to scale.[6] Given the these assumptions, there is an optimal level of knowledge for the firm to produce and thus, all things equal, an optimum rate of growth.

2.2 Assumptions on knowledge

A particularly important, and often problematic, set of assumptions concerns the nature of knowledge. In particular, it is typically assumed that firms employ *firm-specific knowledge* in the production of goods. The knowledge produced exists forever in a non-depreciating stock implying that zero research by a firm means that the firm's stock of knowledge is constant. The assumption of firm-specific knowledge serves an important theoretical purpose, but is somewhat inconsistent with assumptions previously mentioned. Indeed, if "representative" firms are symmetric—i.e., the same size and producing the same goods, etc.—why then is firm-specific knowledge necessary? The answer is that the assumption of firm specificity is necessary to justify that only a *portion* of the knowledge produced by a firm spills over to another. This assumption is necessary for the dynamics of the model, but seems inconsistent with other firm-level assumptions.[7]

2.3 Assumptions on the spatial distributions of knowledge

Perhaps, the most crucial assumption in the theory of endogenous growth is that the total stock of knowledge produced by firms is evenly distributed across geographic space. This assumption, however, is not supported empirically in the literature on

[5] In particular, the scale and number of firms are indeterminate and all are assumed to be price-takers implying that many firms are operating in a competitive market and are earning zero profits. In addition, the number of firms is given, all firms operate at the same output level, and either no start-up of new firms occurs (in the Romer model) or new products are introduced through R&D races (in the neo-Schumpeterian models).

[6] On the firm level, empirical evidence demonstrates a concave relation prevails between firm performance and knowledge investment (Braunerhjelm 1999).

[7] As Acs et al. (2004) pointed out, if knowledge at the firm level was identical any subsequent spillovers would be direct and involve 100% of the produced knowledge. If this were the case, other firms would have no incentive to invest in the production of knowledge resulting in no, or at least less, growth.

geographic knowledge spillovers. Complex technological knowledge (seemingly the most valuable type of knowledge) usually contains a strong element of tacitness meaning that its flow and diffusion is constrained by the geographic proximity and extent of interaction among individuals within whom the tacit component resides. A host of recent empirical studies have confirmed that knowledge spillovers are geographically bounded (Jaffe 1989; Jaffe et al. 1993; Audretsch and Feldman 1996; Anselin et al. 1997; Keller 2002).

2.4 The "missing link"

Endogenous growth models do not adequately explain knowledge spillovers accruing from aggregate knowledge investment. Even in the Schumpeterian models, entry is restricted to existing firms investing in R&D that comply with the behaviors assumed of incumbents. In essence, at the firm-level, knowledge spillovers occur automatically without regard to the absorptive capacity of firms or the entrepreneur's ability and actions. The condition imposed by the discussed assumptions lacks both theoretical and intuitive appeal as well as empirical backing. Indeed, it is one thing for technological opportunities to exist, but an entirely different matter for them to be discovered, exploited and commercialized (Acs and Varga 2002).[8]

3 The knowledge filter model

The term "Schumpeterian" growth model already implies some of the mechanisms deemed missing from the basic endogenous growth model: innovative entry, the reorganization and rationalization of existing firms, and firm exits as the result of "creative destruction" (Schumpeter 1911; Hayek 1945). Although these factors are implied, they must be better and explicitly integrated theoretically into the endogenous growth process in order to capture the interdependency among knowledge, opportunity, and commercialization. In particular, newly produced knowledge—embodied in patents, products, processes, organizations and the like—defines opportunities that can be exploited commercially. With that said, for new ideas to translate into economic growth, new knowledge must be converted into what Arrow (1962) identified as economic knowledge.

3.1 The knowledge filter

The most fundamental argument made by Acs et al. (2004) is that knowledge by itself is a necessary, but not sufficient, condition for economic growth. Michelacci (2003), for example, focused on the allocation of societal resources spent on R&D and entrepreneurship and concludes that low rates of return to R&D may be due to

[8] Acs and Varga (2002) suggested that if one is to understand endogenous economic growth one needs to answer the question of how technological advance occurs, and what are the key processes and institutions involved.

lack of entrepreneurial skills. Thus, the ability to transform new knowledge from economic opportunities to growth-improving products and processes involves a set of skills, aptitudes, insights and circumstances that is neither uniformly nor widely distributed in the population. This suggests that the conversion of new knowledge into economic knowledge occurs with the expenditures both tangible and otherwise (e.g., effort) of relevant economic agents.

Complicating the knowledge conversion process are the uncertainty, asymmetries, and high transactions cost making it difficult to evaluate the expected value of new ideas; indivisibilities in the production of knowledge; and limits to the appropriation of any expected returns (Arrow 1962). Acs et al. (2004) conceived the combination of barriers to converting new knowledge (produced by research activities) into economic knowledge as the "knowledge filter." This knowledge filter is conceptualized as being "semi-permeable" in the sense that the collection of obstacles to the knowledge conversion process can be overcome with the effort and actions of firms and individuals.

3.2 "Arrowian" conversion of knowledge

As Romer (1990) assumed that new knowledge is a non-rivalrous and partially excludable good. Such new knowledge, however, passes through an "Arrowian" conversion process that determines the rate at which the stock of knowledge (K) is converted into economically useful firm-specific knowledge (K^{c}), $0 \leq K^{c}/K < 1$. In addition, knowledge spillovers are spatially (regionally) bounded and access to any localized stock of knowledge is assumed to be equal to all local entities. There are two mechanisms by which new knowledge (K) is converted into economically useful knowledge (K^{c}). The first involves incumbent firms, $K^{cI,}$ and the second involves the entrepreneurial startup of new (Schumpeterian) firms, K^{cSch},

$$K^{c} = K^{cI} + K^{cSch}. \tag{1}$$

As a result, the conversion of economic knowledge from new knowledge is based on the combination of the absorptive capacity of incumbent firms (θ) and the propensity for entrepreneurship in the local economy (λ). Policy and previous history (path dependence) in the form of regulations, attitudes, networks, and technology transfer mechanisms determine the absorptive capacity (θ) of incumbents and the region's propensity for entrepreneurship (λ),

$$K^{c} = (\theta + \lambda)K, \quad 0 \leq \lambda + \theta < 1. \tag{2}$$

3.3 Incumbent firms

Incumbent firms transform knowledge as a function of their absorptive capacity (Cohen and Levinthal 1990). In particular, a firm converts new knowledge into economically useful knowledge, K^{cI}, by a combination of investing in R&D and learning-by-doing; these activities add to the firm's firm-specific knowledge. The firm's absorptive capacity to exploit spillovers, which we denote θ, depends at each given point in time on previous accumulation of firm-specific knowledge $k^I_{i,t}$,

$$k^I_{i,t} = f\left(\int_{t=0}^{t} k^I_{i,t}\mathrm{d}t, K\right), \quad \sum_i^n k^I_{i,t} = K^I_t, \quad K^{cI}_t = \theta K, \; \theta < 1 \tag{3}$$

Given this perspective, we propose,
Hypothesis 1. The contribution of newly created knowledge in a region to economic growth depends on the absorptive capacity of incumbent firms in a region.

3.4 New firms

A set of individuals S can either be employees in the production of goods (L_{M}) or knowledge (L_{R}), or become entrepreneurs (L_{E}). Entrepreneurial ability is distributed unevenly across individuals; these individuals deploy their endowments of entrepreneurial capabilities to evaluate the new knowledge available to them and decide how best to appropriate the returns from that knowledge. Individuals make profit-maximizing inter-temporal choices whether to remain an employee or become entrepreneurs (Knight 1921).

Entrepreneurial start-ups are the manifestation of the knowledge transformation process. In short, each start-up represents a new idea (innovation), which represents any kind of new combination of new or existing knowledge, where individuals draw on their entrepreneurial ability ($\overline{e}_i$) and the aggregate stock of knowledge (K).[9] Start-ups occur through a Poisson process, which leads to the successful entry of a share λ of new firms,

$$K^{\mathrm{cSch}} = \lambda K, \quad \lambda < 1. \tag{4}$$

Thus, we contend,
Hypothesis 2. The contribution of newly created knowledge in a region to economic growth depends on the propensity of a region to create new business ventures.

3.5 Incumbent versus new firms

Aspects of the knowledge filter—especially those concerning the evaluation and assessment of the future expected values—have particularly perverse effects within established firms. In particular, there are strong disincentives for incumbent firms to invest in the production of new knowledge at socially optimal levels and/or deploy truly novel knowledge. There is, for example, a concern that new products will "cannibalize" revenue streams of existing ones or that the minimum required investment

9 Schumpeter (1911).

in R&D is, due to indivisibilities in the production of knowledge, too great (Bernard et al. 2006). Thus, there is the possibility that the knowledge passed over by existing firms may be deemed “too risky” or “too revolutionary” to merit investment. As a result, the decision-making process within incumbent firms can induce agents to start new firms as a mechanism to appropriate the (expected) value of new knowledge.

Indeed, empirical findings suggest that entrepreneurial startups are important links between knowledge creation and the commercialization of such knowledge, particularly at the early stage of the firm or innovation lifecycle when knowledge is still fluid (Utterback and Abernathy 1975). Thus, by serving as a conduit for the spillover of knowledge that might not otherwise be commercialized by incumbent firms, entrepreneurship is the mechanism most likely constituting the strongest link between knowledge and economic growth (Acs et al. 2005). Thus,

Hypothesis 3. The contribution of newly created knowledge to economic growth in a region depends more strongly on newly created business ventures than on the absorptive capacity of existing incumbent firms.

3.6 Booming versus declining economies

As mentioned, Acs and Plummer (2005) found support for the knowledge filter conjecture using cross-sectional county-level data covering 1990 to 2000 gathered for the state of Colorado. The pressing issue that this study addresses is that these received findings constitute support for the model in a particular context. Assessing the generalizability of the model, while primarily an empirical exercise, is critical to validating the knowledge filter model as conceived. Economic growth is itself somewhat self-fulfilling in that the expansion of incomes and increases in standards of living carry forward from year to year in a way vital to an economy’s future prosperity (Henderson 2003).[10] Given this perspective, it is essential to theoretically validating the knowledge filter model to test the central conjecture in the context of diminishing economic conditions. This is the theoretical basis for our focus on the state of Ohio.

4 Research design

4.1 Sample and data collection

Aside from the theoretical basis discussed earlier, the sample of Ohio counties is based on criteria suggested by Acs and Plummer (2005) and the data are collected to afford comparability with their analysis. In particular, the sample for this study contains adequate variance in the variables in the model and encompasses counties large enough to represent statistically workable regions of knowledge spillovers. The period of the current study is 1990 to 1999 comparable to Acs and Plummers’ 1990 to 2000 study period. Data for the year 2000 was not included here because county-level patent data for Ohio was not available for that year. As detailed in the next section,

[10] Henderson (2003) found that these effects carry forward between 5 and 20 years depending on the industry structure of the local economy.

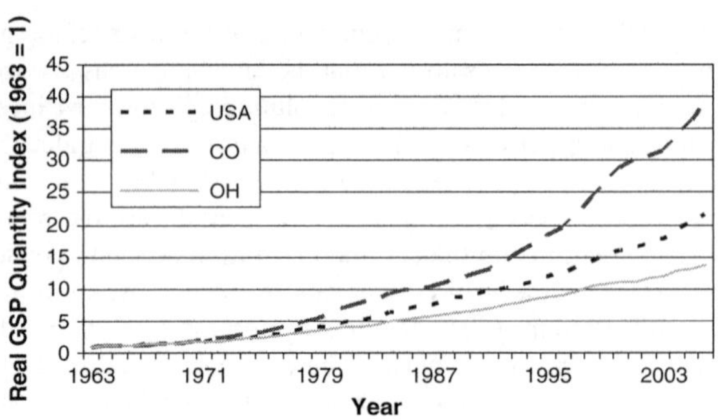

Fig. 1 Annual change in real gross state product 1963–2006. Source: Bureau of Economic Analysis, US Department of Commerce

the data came from the US Census Bureau, the US Patent and Trademark Office, the Bureau of Economic Analysis (BEA), the Small Business Administration (SBA), and the National Science Foundation (NSF).

Ohio, comprised 88 counties, is a mid-western state with a rich history of industrial dominance and, in contrast to Colorado, has experienced economic decline in recent decades. As shown in Fig. 1, Ohio's gross state product kept pace with the United States average until 1979 and began to lag considerably thereafter. Colorado, by comparison, began to surge ahead in 1975 with a slight retraction toward the national average during the recession of the early 1990s. After that period, Colorado surged strongly ahead of the US average with a minor retraction after 2000. Over the period 1963 – 2006, Colorado's real GSP increased by a factor of over 40 while Ohio's real GSP increase was far more modest. In addition, personal income growth in Ohio after 1982 lagged behind the United States in general and the state of Colorado in particular. Likewise, manufacturing output declined in Ohio by 20% from 1978 to 1983 and never again exceeded its 1978 level of output until the early 1990s. This makes the state of Ohio an appropriate context in which to examine the knowledge filter model.

4.2 Variables

The variables for this study are defined in a manner consistent with Acs and Plummer (2005). This facilitates a comparison of the current studies' results to those found in their study.

Personal income growth. The dependent variable is calculated from data obtained from the Bureau of Economic Affairs regional economic accounts. Personal income growth is the annual change in personal income from one year to the next.

Knowledge. Knowledge is notoriously difficult to measure and little data beyond patent counts exists as a county-level measure. As a result, in this paper, we measure the county's stock of knowledge as the number of patents granted in a given county using data obtained from the US Patent and Trade Mark Office. For standardization purposes,

the number of patents granted is divided by the total number of establishments in the given county.
Research and development. Since patents capture the output from knowledge production activities in the county, we include an indicator of research and development activities in a given county. Information on research and development *expenditures* is not available at the county-level. Thus, as an alternative, we assigned a dummy code equal to 1 to the counties with universities, federally funded R&D centers, and non-profits receiving federal R&D funding sometime between 1990 and 1999. The dummy code equals 0 in those counties receiving no federal funding in the period. The data for this variable came from the National Science Foundation.
New ventures. We define new ventures as number of "high technology" single-establishment births in the county divided by the number of existing establishments. The Census defines a single establishment as a single physical location where business is conducted or where services or operations are carried out.[11] The "high technology" sectors are defined using Varga's (1998) three criteria: industries with (1) an above average research and development to industry sales ratio at the 3-digit SIC level, (2) an above average percentage of mathematicians, scientists, engineers and engineering technicians when compared to total industry occupations, and (3) the total number of innovations per 1,000 employees. The single-establishment birth and existing establishment data were obtained from the US Bureau of the Census. The establishment birth tabulations are broken out by year, by Standard Industrial Classification code (SIC) for 1990–1997 and by North American Industrial Classification System (NAICS) for 1998–1999 at the four and five digit levels, respectively.
Incumbents. The number of establishments with more than 100 employees divided by the total number of establishments is our measure for incumbent firms. This proxy measure is used due to the lack of information in the Census data regarding the establishment's age. Age and size are generally correlated and it seems unlikely that a single establishment firm would start out with 100 or more employees.
Density. Density is defined as the total number of establishments in a given county divided by the county's total area in square miles. It is included to capture the relevant effects of the geographic concentration of economic activity, resources, and people. The data for the numerator and denominator in this variable were obtained from the US Census Bureau.
Log total personal income. Using the BEA regional accounts, we include the log of total personal income in the county to account for the possibility that subsequent growth is likely a function of previous wealth. Furthermore, the inclusion of the logged level of total personal income facilitates inferences regarding the notion that richer economies grow more slowly (Barro et al. 1991). Stated another way, the logged value of total personal income is included as an explanatory variable to account for variation in the dependent variable resulting from higher initial income levels. The convergence hypothesis predicts a negative coefficient estimate for this variable.
Interaction terms. The last two variables included in this analysis embody the interactions of knowledge with new and incumbent firms. These variables constitute the

[11] A single-establishment birth for a given year is defined as an establishment having no payroll anytime the prior year and positive payroll in the first quarter of the current year.

primary variables of interest in this study as all of the proposed hypotheses employ the crucial proposition that the production of knowledge is a necessary yet insufficient condition for achieving economic growth. The inclusions of these two variables provide a means for testing this important concept; therefore, specific emphasis will be placed on the inferences supplied by these variables.

4.3 Estimation issues

There are a number of relevant regression issues that must be diagnosed when carrying out regression modeling of economic growth. These issues are spatial dependence, heteroscedasticity and outliers, as well as collinearity. These statistical problems are particularly important because the existence of spatial dependence has been shown to be a source of both bias and inefficiency under traditional regression methodologies such as OLS (LeSage 1997). When the dependent variable vector exhibits spatial autocorrelation, the resulting parameter estimates are known to be biased; where as when the spatial dependence is contained simply in the residuals, the problems involve only efficiency (LeSage 1997). Therefore, the particular structure of the spatial dependence existing in the sample data governs which type of spatial regression model ought to be employed in one's empirical pursuits.

Heteroscedasticity is a concern because its existence is known to cause inefficiency whereas outliers have been shown to cause bias in the resulting parameter estimates. Lastly, collinearity is a common problem in regression modeling and is associated with an over estimation of standard errors and, hence, is a problem with regard to efficiency. As well, a lack of consistency in the magnitudes and significance levels of the parameter estimates is commonly associated with datasets suffering from collinearity.

4.3.1 Spatial dependence

We expect that the dependent variable, personal income growth, exhibit spatial dependence. Spatial dependence becomes an issue when observations at one location, y_i, depend on neighboring observations, y_j, where j denotes the set of neighboring observations to any observation, y_i. The existence of spatial dependence invalidates the use of ordinary least-squares (OLS) regression methods (LeSage 1997) and requires that we apply an alternative estimation procedure.

To assess and account for this statistical problem, we employ diagnostic test statistics, such as Moran's I, as well as spatial regression methodologies that investigate the particular nature of the dependence relationship and explicitly deal with its existence. To carry out this analysis, we began by testing for the existence of spatial dependence using the test statistic commonly referred to as Moran's I. Moran's I provides a statistical procedure for testing for the existence of spatial autocorrelation; however, the test provides no means of correcting for this problem. Estimating a spatial autoregressive regression model provides both a test for the existence of spatial dependence among the dependent variable observations as well as it provides a means of accounting for this problem, should it exist.

If the estimation of this type of regression model reveals that the spatial dependence parameter, ρ, is statistically significantly different from zero, then one can conclude that spatial dependence exists amid the $n \times 1$ vector of dependent variable observations. Upon estimation of this type of regression model, one can then analyze the resulting residuals to examine whether or not spatial autocorrelation remains in the residuals. This can be carried out via either Moran's I or the Wald statistic or by estimating a spatial error regression model (SEM). These procedures provide a simple way of investigating whether or not the spatial autoregressive regression model adequately deals with the spatial dependence inherent in this particular application or whether a spatial error model or even a general spatial regression model[12] ought to be utilized.

4.3.2 Heteroscedasticity and collinearity

To take into account possible heteroscedasticity in the data, we rely on the estimation of a Bayesian heteroscedastic linear variant of the spatial autoregressive model that is robust to both outliers and heteroscedasticity (LeSage 1997). By comparing robust and standard SAR results, we can determine whether heteroscedasticity and outliers are a concern in the data. Supposing that heteroscedasticity exists, we would observe an increase in the t statistics associated with the coefficient estimates resulting from the Bayesian heteroscedastic linear variant of the SAR model when compared to the traditional SAR model's coefficient estimates. The existence of outliers would cause the two estimation routines to produce coefficient estimates that differ in magnitude on a scale dependent on the extent and size of the outliers. Therefore, if the estimation of these two regression models produce coefficient estimates that are statistically equivalent to each other, along with similar t statistics, then one can conclude that neither heteroscedasticity or outliers are a problem with regard to the given sample data.

White or Breusch–Pagan tests are commonly used for determining these matters. However, these test statistics are considerably less valuable than is the above strategy for a number of reasons. For one, these test statistics commonly fail to provide the proper inferences. As a result, the use of these methods commonly provides incorrect information. For example, suppose one is presented with a situation where the White test statistic is used and is statistically significant at only the 85% level. The general approach would be to imply that since the White test-statistic is not statistically significant at least at the 90% level, then one cannot reject the null hypothesis that there exists no heteroscedasticity. However, how is one to know whether any bias or inefficiency exists in subsequent least-squares parameter estimates? The short answer is they simply do not know. The common approach is to simply imply that, since the test statistic is not statistically significantly different from zero, the impacts associated with heteroscedasticity are not influential. However, there is little basis for that

[12] The use of the term general spatial model pertains to a particular spatial regression model that incorporates spatial autocorrelation amongst both the dependent variable observations and residuals. It is described as a general spatial model due to its generalization of both the spatial autoregressive regression model and the spatial error regression model (LeSage 1998). It should be relied on when the spatial dependence structure is more complex than assumed by both the SAR and SEM models, and as a result, when spatial correlation exists in the residuals associated with the estimation of a SAR model.

conclusion other than the rules of thumb associated with this testing procedure. As well, test-statistics only indicate whether or not the problem exists, thus, providing nothing for correcting for these problems.

The strategy relied on in this paper is substantially better than relying on information provided by test statistics. The reasons for this are that this procedure is:

1. Capable of diagnosing whether a systematic heteroscedastic trend exists across the observations in the sample or whether the particular heteroscedasticity takes on the form of a few significant outliers.
2. This procedure provides a means of identifying the extent of the influence of these problems on the resulting parameter estimates.
3. This procedure provides a means of correction should heteroscedasticity (in either form) exist in the sample data. The use of a testing procedure completely fails in this regard.

To test for the presence of collinearity problems we subject the sample data to the Belsley, Kuh, and Welsch collinearity diagnostic, which lends a variance-decomposition proportions matrix (Belsley et al. 1980). Secondly, we implement a method of estimating ten alternative specifications of each regression model, where the alternative specifications are based on different specifications of the matrix X. Stability in the parameter estimates and significance levels across the different permutations mitigates problems associated with lack of precision, as one can draw inferences from commonalities across the alternative specifications of the regression models. Furthermore, the spatial panel estimation procedures, laid out in Sect. 4.5, further mitigate any existing collinearity that may linger in the sample data (Elhorst 2003).

4.4 Spatial autoregressive model

To test for the influence of spatial dependence across the observations in our sample, we estimate a spatial autoregressive (SAR) model of the following form:[13]

$$\begin{aligned} y &= \rho W y + X\beta + \varepsilon \\ \varepsilon &\sim N(0, \sigma^2 I_n) \end{aligned} \tag{5}$$

W denotes an 88 × 88 spatial weight matrix that defines the set of neighboring counties to each observation. ρ denotes a scalar parameter measuring the strength of the relationship between the dependent variable, y_i, and the spatially lagged variable vector, Wy_i. It is important to note here that inferences regarding the existence of spatial dependence are, then, provided by the coefficient estimate of ρ. If ρ is non-zero and is statistically significant, then the existence of spatial dependence is confirmed to exist

[13] This particular spatial regression model was initially utilized because previous empirical work (Acs and Plummer 2005) demonstrated that spatial dependence existed amongst the vector of dependent variable observations. Therefore, it was determined that bias in the resulting parameter estimates was a bigger problem than was inefficiency. However, the residuals stemming from estimating this regression model revealed that no significant spatial correlation remained in the residuals and, hence, the estimation of a spatial error model was unnecessary at that point.

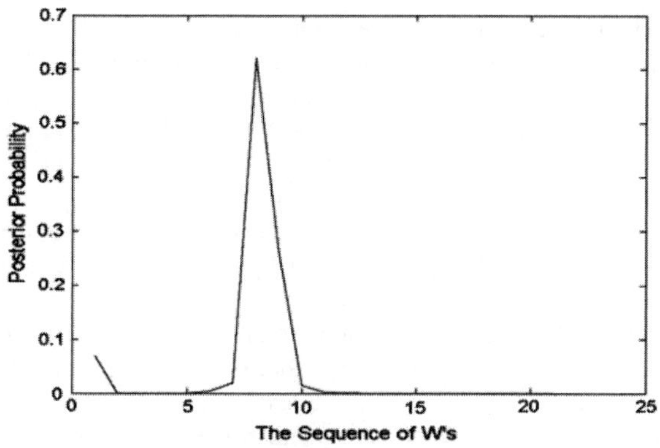

Fig. 2 Posterior model probabilities for the sequence of W's

in the sample data. In this situation SAR and OLS parameter estimates should differ as OLS parameter estimates are biased in the face of spatial dependence.

A variety of approaches have been used to define W with the most common being a first-order contiguity-based specification. To help validate our definition of the spatial weights matrix, we specify a set of 21 row-standardized spatial weights matrices. The set of alternative weight matrices are based on (1) a first-order contiguity based specification and (2) on a sequence of 20 weight matrices that are specified to select one through 20 of the nearest neighboring counties, respectively. We select the weights matrix associated with the largest posterior model probability, thereby, selecting the weights matrix that "best" fits the sample data.[14] It should also be noted here that the selection of this sequence of weight matrices was initially quite ad hoc, for a one could specify many more such matrices; all the way up to specifying one that selects every observation that being one that selects the 88 nearest neighbors. However, inspection of the posterior model probabilities associated with these weight matrices revealed that specifying this matrix to extract a larger number of neighboring observations than this tended to decrease the posterior model probability, indicating that our sequence was more than adequate. This fact is demonstrated in Fig. 2, where the first specification is one based on first order contiguity while the others are the 1 through 20 nearest neighboring specifications.

4.5 Spatial panel data models

Spatial panel regression models provide several important advantages when compared to their cross sectional equivalents. For one, the reliance on our panel of data significantly increases the number of observations underlying the inquiry (going from 88 observations in the cross-sectional analysis to 880 observations when relying on our 10 year panel), thereby providing a considerable increase in precision and robustness.

[14] This was accomplished using LeSage's "sar_g" and "model_probs" Matlab functions.

Second, the spatial panel routines are known to help extinguish the impacts of any undetected collinearity that may exist in the sample data (Elhorst 2003). As a result, we exploit the advantages of our panel data by estimating two panel data equations. The first (Eq. 6) is a pooled model with the inclusion of a spatially lagged dependent variable. The second (Eq. 7) is a pooled model that includes a spatially lagged dependent variable as well as time fixed effects.

The specification of Eq. 7 is based on three factors. First, random effects are inappropriate when observations are based on irregular spatial units such as counties (Elhorst 2003; Anselin and Bera 1988). Second, the assumption of zero correlation between μ and X in the random effects model is restrictive and unlikely to hold. Lastly, we exclude spatial (i.e., county) fixed effects because spatial fixed effects cannot be consistently estimated (Elhorst 2003).[15] In contrast, time fixed effects can be consistently estimated. The two equations—with W the row standardized first-order contiguity weighs matrix described above—take the form:

$$\begin{gathered} y_t = \rho W y_t + X_t \beta + \varepsilon_t \\ \varepsilon_t \sim N(0, \sigma^2 I_n) \end{gathered} \tag{6}$$

and

$$\begin{gathered} y_t = \rho W y_t + X_t \beta + \mu + \varepsilon_t \\ \varepsilon_t \sim N(0, \sigma^2 I_n) \end{gathered} \tag{7}$$

These equations are estimated by first demeaning the Y and X variables such that the Y and X variables for each spatial unit are expressed in deviation from their average over time (Elhorst 2003). Given the inclusion of the spatial lag and the resulting statistical complications, we use a two-stage procedure, with the intercept estimated as $\beta_1 + \mu_i$ (Anselin and Bera (1988), pp. 181–182), to maximize the log-likelihood function (Elhorst 2003, p. 250) resulting in maximum likelihood estimates (MLE) of the relevant parameters.[16]

5 Results

5.1 Descriptive statistics and correlations

Table 1 contains descriptive statistics associated with the 10-year annual averaged variables. Row 1 of Table 1 contains the dependent variable, the average annual growth rate of total personal income. As can be seen from the table, the average annual growth rate of county-level total personal income was approximately 5%. The minimum value

[15] In our panel, T = 10 and N =88. In this spatial panel context, T can be viewed as fixed while N tends toward infinity (in other words, N is considerably larger than T). Thus, only time fixed effects can be consistently estimated.

[16] The models were estimated using the "sar_panel" Matlab function (Elhorst 2003).

Table 1 Descriptive statistics

Variable	Mean	STD	Min	Max
Total Personal Inc.	0.047	0.011	0.028	0.108
Density	6.545	12.524	0.362	79.13
LogTotInc	14.071	1.103	12.029	17.42
Knowledge	0.008	0.007	0.001	0.036
RDdummy	0.125	0.333	0.000	1.000
NewVentures	0.002	0.001	0.000	0.006
Incumbents	0.023	0.006	0.011	0.039

was around 3%, occurring in Noble County in South East Ohio, while the maximum was approximately 11% occurring just North of Columbus, Ohio in Delaware County.

As for the independent variables, the mean density was 6.55 establishments per square mile with a standard deviation of 12.52, indicating considerable variability in the density of Ohio's counties. The minimum density occurred in Vinton County Ohio, located in South East Ohio, while the maximum density occurred in Cuyahoga County, the county containing Cleveland, Ohio. The mean value of the log of total income was 14.07 with a standard deviation of 1.10. Cuyahoga County was associated with the maximum value while Vinton County was associated with the minimum value. There is on average eight patents per 1,000 establishments with a range of 1 patent per 1,000 establishments, occurring in Belmont County Ohio along the South East border, to 36 patents per 1,000 establishments, occurring in Delaware County Ohio, North of Columbus, Ohio. On average, there were 2 new high technology ventures per 1,000 establishments, within a range of 0 to 6 per 1,000 establishments. The maximum value, again, occurred in Delaware County Ohio. In contrast, there were 23 incumbent firms per 1,000, with a range of 1.1 to 39 incumbents per 1,000 establishments. The minimum value was again located in the South East region of the state, in Vinton County, while the maximum value occurred in Shelby County Ohio, located in West Central Ohio.

Table 2 reports the correlations among the variables. All of the explanatory variables except density, log total income, and incumbent firms were positively correlated with total personal income. Knowledge was most correlated with total personal income growth (0.542) while the log of total personal income was the least correlated with this variable (−0.033). Interestingly, new ventures showed the strongest correlation with knowledge (0.768) while density (0.227) and incumbents (0.268) showed the weakest correlations with this variable.

5.2 Heteroscedasticity, outliers, collinearity and spatial dependence

We find that heteroscedasticity and outliers are not influential factors in the data. This is due to the fact that the coefficient estimates produced by the SAR and Bayesian linear heteroscedastic SAR model are statistically equivalent to each other (based on a t test between the two sets of coefficient estimates). If outliers were of statistical importance then the coefficient estimates produced by the estimation of a SAR model

Table 2 Correlation matrix

Variable	1	2	3	4	5	6	7
1. Total Personal Inc.	1.000						
2. Density	−0.169	1.000					
3. LogTotInc	−0.033	0.786	1.000				
4. Knowledge	0.542	0.227	0.470	1.000			
5. RDdummy	0.106	0.558	0.472	0.336	1.000		
6. NewVentures	0.484	0.439	0.632	0.768	0.386	1.000	
7. Incumbents	−0.043	0.349	0.420	0.268	0.316	0.303	1.000

Table 3 Belsley Kuh and Welsch collinearity diagnostic results

Condition index	Knowledge	Rddummy	NewVentures	Incumbents	Density	LogTotInc	K*NV	K*Incumb
1	0.00	0.00	0.00	0.00	0.13	0.01	0.00	0.00
2	0.00	0.00	0.00	0.00	0.41	0.02	0.00	0.00
63	0.00	0.88	0.00	0.00	0.22	0.00	0.00	0.00
2426	0.03	0.07	0.00	0.01	0.00	0.05	0.00	0.00
3216	0.00	0.01	0.00	0.43	0.00	0.34	0.00	0.00
14567	0.11	0.01	0.01	0.04	0.00	0.01	0.45	0.00
44307	0.35	0.01	0.55	0.08	0.03	0.40	0.49	0.31
49397	0.51	0.01	0.45	0.45	0.22	0.17	0.07	0.69

would be statistically significantly different from the coefficient estimates produced by the estimation of the robust Bayesian linear heteroscedastic SAR model (LeSage 1997). Furthermore, the fact that these two models produce asymptotic t statistics of similar magnitude, yielding the exact same inferences, renders a situation where one can confidently infer that systematic heteroscedasticity across the sample observations is not a significant problem in our sample. Therefore, we only present the SAR results as no additional information was provided by estimating the Bayesian linear heteroscedastic SAR model.

The presence of collinearity was investigated in two ways. First, the data were subjected to the Belsley, Kuh, and Welsch (BKW) collinearity diagnostic. This technique is based on the singular value decomposition, where this decomposition is applied to the variance–covariance matrix of our OLS estimates and is rearranged to create a variance-decomposition proportions matrix, which is shown in Table 3 (Belsley et al. 1980).

The first column contains the condition indices while the other columns contain variance-decomposition proportions. The joint condition of a condition index greater than 30 along with a variance-decomposition proportion in excess of 0.5 indicates the possibility of collinear relations among the associated variables (Belsley et al. 1980).

As is evident by examination of Table 3, there is only one instance where this joint condition is satisfied. The respective instance is located in the bottom row of this table and pertains to the knowledge variable and the knowledge-incumbent interaction term. Here, one can observe that the variance-decomposition proportion associated with knowledge just barely exceeds the BKW threshold, while the proportion associated with the knowledge-incumbent interaction term definitely exceeds it. As a result, this evidence suggests that collinearity is not a significant problem with regard to our sample data; however, it does suggest that the knowledge and knowledge-incumbent interaction terms may suffer from this problem to a minor degree. Although the BKW evidence suggests that the influence of collinearity is slight, we relied on a method of estimating a set of alternative specifications of the regression model, that are based on alternative permutations of the explanatory variable matrix, to be on the safe side. Furthermore, our usage of spatial panel techniques, below, will work to overcome any lingering collinearity, should its existence produce any significant effects on our coefficient estimates in the first place (Elhorst 2003).

Evidence supporting the existence of spatial dependence in the sample data comes from the computation of Moran's I, which is approximately equal to a value of 0.35 in this application. The t statistic associated with this value is 6.12, which is clearly large than the 99% critical value, yielding a marginal probability level of 0.000. This evidence provides a clear indication of statistically significant spatial dependence.

As a result, OLS results are not presented as the existence of spatial dependence has been established by Moran's I, as well as below in Table 3, where the coefficient on the spatial dependence parameter, ρ, is positive and statistically significant at the 99% level in all specifications of the model. In light of this finding, OLS estimates are not to be relied upon as spatial dependence has been shown to exist in the sample data (LeSage 1997).

Lastly, the model comparison techniques utilized to ascertain the most appropriate specification of the spatial weights matrix indicated that a weights matrix selecting the seven nearest neighboring observations was associated with the largest posterior model probability and, hence, this specification of the spatial weight matrix was utilized in every instance where W was required (see Fig. 2).

5.3 Spatial autoregressive results

Three sets of results, obtained from estimation of the maximum likelihood SAR regression equation (Eq. 5) are reported in Table 4. Column 2 of Table 4 contains Model 1 including all of the variables, except the interaction terms, discussed in Sect. 4.1. Column 3 contains Model 2 that includes the interaction terms, while omitting new and incumbent firm variables. Column 4 contains Model 3 that includes all the variables discussed in Sect. 4.1.

Examination of the results obtained from estimation of Model 1 indicates that knowledge and new high technology business ventures have positive and statistically significant impacts on the average annual growth rate of total personal income. Density and the log of total personal income are associated with negative and statistically significant impacts on the growth rate of total personal income. Incumbent firms are

Table 4 Maximum likelihood spatial autoregressive results

1990–2000 (average) Ohio Counties (N = 88)	Model 1 ($r^2 = 0.53$)	Model 2 ($r^2 = 0.60$)	Model 3 ($r^2 = 0.60$)
Constant	0.064*** (3.940)	0.053*** (3.540)	0.057*** (3.718)
Knowledge	0.525*** (3.291)	0.317 (0.753)	0.327 (0.641)
Rddummy	0.002 (0.934)	0.000 (0.133)	0.001 (0.257)
NewVentures	6.226*** (4.635)		1.873 (1.123)
Incumbents	−0.276*** (−2.168)		−0.024 (−0.137)
Density	−0.000** (−2.473)	−0.000* (−1.835)	−0.000** (−1.948)
LogTotInc	−0.004*** (−3.037)	−0.003** (−2.490)	−0.003*** (−2.615)
K*NV		3.090*** (5.576)	2.578*** (3.621)
K*Incumb		−1.678 (−1.431)	−1.579 (−0.946)
Rho	0.622*** (6.736)	0.608*** (6.543)	0.612*** (6.670)

Significance: * at the 90% level, ** at the 95% level, and *** at the 99% level
t statistics are shown in parentheses

associated with a negative and statistically significant parameter estimate. The spatial dependence parameter, ρ, is quite large (it ranges from 0 to 1) and is statistically significant. This last result provides clear indication that total personal income growth in one county is heavily dependent on total personal income growth in neighboring counties.

The results of Model 2 indicate that knowledge becomes statistically insignificant when the interaction of knowledge and new business ventures is included as an explanatory variable in the regression model. The implication here is that it is the interaction of these two variables that impacts growth in total personal income, rather than knowledge alone. This result supports the assumption of the knowledge filter model that the production of knowledge is a necessary yet insufficient provision for yielding economic growth. It also appears that the contribution of newly created knowledge in a region to economic growth depends on the propensity of a region to create new business ventures that are adept at commercializing this newly created knowledge. The interaction of knowledge with incumbent firms suggests that the knowledge commercialized by incumbent firms has little to no impact on the growth rate of total personal income. All other variables provide the same inferences as Model 1.

The estimation and presentation of Model 3 (which includes all of the variables) provides the last important piece of additional information presented in this section of this analysis. Examination of the results obtained from the estimation of Model 3 indicate that the interaction of knowledge with new business ventures have positive and statistically significant impacts on the growth rate of total personal income. Both density and the log of total personal income have negative and statistically significant impacts on total personal income growth while knowledge, the R&D dummy, new ventures, incumbents, and the interaction of knowledge with incumbents have effects that are not statistically significantly different from zero. This set of results also provides evidence that incumbents are not a proficient mechanism for converting newly created knowledge into commercialized, economically useful knowledge. In

Table 5 Pooled model with spatially lagged dependent variable and no fixed effects

1990–2000 (panel), Ohio Counties (N = 880)	Model 4 ($r^2 = 0.53$)	Model 5 ($r^2 = 0.54$)	Model 6 ($r^2 = 0.54$)
Constant	0.045*** (3.813)	0.044*** (3.835)	0.046*** (3.932)
Knowledge	0.734*** (8.672)	1.242*** (5.760)	1.127*** (4.149)
Rddummy	0.003 (1.578)	0.002 (1.090)	0.002 (1.262)
NewVentures	1.601*** (4.002)		0.783 (1.470)
Incumbents	−0.320*** (−3.610)		−0.118 (−.0971)
Density	−0.001*** (−3.266)	−0.001*** (−2.504)	−0.001*** (−2.635)
LogTotInc	−0.002*** (−2.476)	−0.003*** (−2.984)	−0.003*** (−2.800)
K*NV		0.965*** (4.472)	0.667** (2.300)
K*Incumb		−2.960*** (−3.878)	−2.281** (−2.179)
Rho	0.720*** (26.039)	0.722*** (26.353)	0.720*** (25.920)

Significance: * at the 90% level, ** at the 95% level, and *** at the 99% level
t statistics are shown in parentheses

summary, the results obtained from the estimation of a maximum likelihood SAR model provides evidence in support of both Hypotheses 2 and 3 while the evidence does not support Hypothesis 1.

5.4 Spatial panel results

Table 5 contains the results obtained from estimating the pooled model (Eq. 6) with the inclusion of a spatially lagged dependent variable for same three equations presented in Table 4. The results associated with Model 4 show that knowledge and new business ventures are positively related to the growth rate of total personal income at the 1% level. Incumbents, density, and the log of total personal income are associated with negative and statistically significant impacts on total personal income growth, while the R&D dummy is associated with an impact that is not statistically significantly different from zero.

Model 5 again demonstrates that the interaction of knowledge with new business ventures has an impact on economic growth. Under this estimation procedure, however, knowledge remains positive and statistically significant, whereas, it became insignificant in Model 2 of Table 4 when the new venture and knowledge interaction term was introduced. Both density and the log of total personal income remain negative and statistically significant; again indicating convergence in total personal income across the period 1990–1999 and that density has a negative relationship to total personal income growth in our sample declining economy.

In Model 6, knowledge and the interaction of knowledge with new business ventures have positive and statistically significant impacts on the growth rate of total personal income, while new and incumbent firms, by themselves, have influences that are not statistically significantly different from zero. Density and the log of total personal income, again, have negative and statistically significant impacts. In addition, the interaction of knowledge with incumbent firms is associated with a negative and

Table 6 Pooled model with a spatially lagged dependent variable and time period fixed effects

1990–2000 (panel), Ohio Counties (N = 880)	Model 7 ($r^2 = 0.55$)	Model 8 ($r^2 = 0.56$)	Model 9 ($r^2 = 0.56$)
Knowledge	0.772*** (9.125)	1.213*** (5.579)	1.152*** (4.294)
Rddummy	0.004* (1.904)	0.003 (1.454)	0.003 (1.531)
NewVentures	1.575*** (3.804)		0.580 (1.081)
Incumbents	−0.276*** (−3.153)		−0.066 (−0.551)
Density	−0.001*** (−2.947)	−0.001*** (−2.148)	−0.001** (−2.274)
LogTotInc	−0.002** (−2.522)	−0.003*** (−2.981)	−0.003*** (−2.852)
K*NV		1.008*** (4.640)	0.797*** (2.802)
K*Incumb		−2.736*** (−3.619)	−2.356** (−2.297)
Rho	0.425*** (10.014)	0.417*** (9.730)	0.434*** (10.292)

Significance: * at the 90% level, ** at the 95% level, and *** at the 99% level
t statistics are shown in parentheses

statistically significant impact on the growth rate of total personal income. This last finding suggests that the commercialization of knowledge by incumbent firms may actually have a negative influence on total personal income growth when a more robust estimation technique is utilized.

Table 6 reports the set of estimation results of Eq. 7. This set of results extends the spatial panel modeling framework to include both a spatially lagged dependent variable as well as time period effects when estimating the three equations presented above. The results associated with the estimation of Model 7 indicate that the knowledge, R&D and the new business venture variables have positive and statistically significant influences on total personal income growth. Incumbents, density, and the log level of total personal income are associated with negative and statistically significant impacts on economic growth.

Model 8 again replaces new and incumbent firms with their respective interactions with knowledge. The results of the estimation of this relationship suggest that the interaction of new firms with knowledge is positive while the interaction of knowledge with incumbent firms is negative; both of these findings are statistically significant. As well, knowledge remains positive and significant while the R&D dummy is no longer statistically significantly different from zero. Density and the log of total personal income remain negative and statistically significant, providing evidence of convergence in total personal incomes and that density in this sample has a negative relationship to total personal income growth.

Results obtained from the estimation of Model 9 indicate that knowledge and the interaction of knowledge with new business ventures have positive and statistically significant impacts on total personal income growth while density, the log level of total personal income, and the interaction of knowledge with incumbent firms have negative and statistically significant impacts. Once again, new and incumbent business ventures become insignificant when the interaction of knowledge with these variables were included as an explanatory variable.

Taken together, the spatial panel regression models provide considerable evidence in support of Hypotheses 2 and 3, but no support Hypothesis 1. Density appears to have a negative relationship with regard to our declining economy while the negative sign on the log level of total personal income suggests convergence has occurred in total personal incomes during the 1990s. These results provide considerable evidence in favor of the knowledge filter growth model, as they are provided by one of the most robust applicable regression models currently available.

6 Discussion

Given the similarity of approach, data, and results to the previous paper by Acs and Plummer (2005) it would be useful and instructive to expand the comparison between the results for Ohio with the previous results for Colorado in a more explicit and somewhat more detailed manner. The findings in Acs and Plummer (2005) suggested that new firms were more important in the commercialization of new knowledge than incumbents. This made some sense in an expanding high technology economy where new knowledge played an important role in the economy. In a declining state, where knowledge in large industrial research laboratories dominated the economy, we would have thought that incumbents would play a more, not less, important role in contributing to economic growth.

The coefficient on the interaction term for knowledge by high technology births was 609.23 in Colorado, significant at the 95% level, while in Ohio it was 2.578 and significant at the 99% level. While the coefficient on the interaction term for knowledge by incumbents in Colorado was negative and tiny (−0.09 at the 95% level) in Ohio it was −1.579 and statistically insignificant. The results seem to suggest that the role of high tech births, if anything, were at least as important in Ohio, not less. Moreover, the results on the knowledge by incumbent's interaction were also similar in Ohio, although not statistically significant. These results were robust with respect to both time and alternative model specifications. How do we interpret these results given the history of incumbents and the research laboratory in Ohio?

One answer to this anomaly can be explained by the fact that Ohio has restructured away from a "rustbelt" state to a more progressive state based on knowledge creation, government services, universities and entrepreneurship. In this environment high tech startups play a more, not less important role in the economy. This restructuring in Ohio is evidenced by the fact that Columbus, Ohio is now a larger and more important city than Cleveland, Ohio in terms of economic activity. In other words, Cleveland was never able to recover from the industrial decline in heavy manufacturing. Therefore, the coefficient on the interaction term between knowledge and incumbents is both larger and more significant in Ohio than in Colorado, a state that did not have to restructure to the extent of Ohio. Ohio exhibits the same results as Colorado, in part, because the declining industry and knowledge base has never recovered yet a new knowledge base has been created where new knowledge-based high tech startups play a more important role than incumbents.

7 Conclusions

The contributions made by endogenous growth theorists have done much to improve the understanding of the complex process of economic growth to be sure. However, the basic model does not sufficiently explain the transition of newly created knowledge to commercialized, or rather, economically useful knowledge at the micro-level. Furthermore, the explanation of the diffusion of outputs from aggregate knowledge investments, in the form of "knowledge spillovers," is inadequate as the assumptions associated with these types of growth models lack both theoretical and intuitive appeal. As well, and perhaps more importantly, they lack empirical backing. It is one thing for technological opportunities to exist, but an entirely different thing for them to be discovered, exploited and commercialized (Shane and Eckhardt 2003).

The purpose of this paper was to test the validity and generalizability of the theoretical knowledge filter model by applying the model to previously untested regions of the US economy, i.e. the regions in economic decline. To make this assessment, a dataset reflecting a "typical" declining economy was identified and analyzed by the utilization of recent developments in the spatial econometric literature, most notably the extension of spatial econometric models to spatial panel datasets. The estimation procedures facilitate inferences regarding the specific hypotheses that are specifically derived from the theoretical model. The emphasis of this paper was placed on determining the validity and generalizability of the knowledge filter model to economies suffering from recent economic decline.

The results of our analysis have provided evidence in support of the theoretical model with regard to two of our three specific hypotheses. Specifically, the contribution of knowledge in a region to economic growth depends on the propensity of a region to create new business ventures and that the contribution of newly created knowledge to economic growth in a region depends more strongly on newly created business ventures than on the absorptive capacity of existing incumbent firms. The evidence has also shown that knowledge is, indeed, a necessary condition, yet, by itself an insufficient explanation of economic growth.

We did not, however, find evidence to support our hypothesis that the contribution of newly created knowledge in a region to economic growth depends on the absorptive capacity of incumbent firms in a region. In fact, we found that the interaction of knowledge with incumbent firms has a negative impact on economic growth, as measured by the growth rate of total personal income. This finding is consistent with that found in Acs and Plummer (2005) and, as they indicate, may be a function of the specification of the incumbent firms variable. Furthermore, this result may reflect the fact that when local operations of corporations do absorb knowledge spillovers, the contribution to growth may occur in other regions, such as the location of the corporate headquarters.

References

Acs ZJ, Armington C (2006) Entrepreneurship, geography and american economic growth. Cambridge University Press, Cambridge

Acs ZJ, Audretsch DB (1987) Innovation, market structure, and firm size. Rev Econ Stat 69(4):567–574

Acs Z, Plummer L (2005) Penetrating the "knowledge filter" in regional economics. Ann Reg Sci 39:439–456

Acs Z, Varga A (2002) Geography, endogenous growth and innovation. Int Reg Sci Rev 25(1):132–148

Acs Z, Audretsch D, Braunerhjelm P, Carlsson B (2004) The missing link: the knowledge filter and endogenous growth. Center for Economic Policy Research, London, UK, December, no. 4783

Acs ZJ, Audretsch DB, Braunerhjelm P, Carlsson B (2005) The knowledge spillover theory of entrepreneurship, Center for Economic Policy Research. London, UK, December, no. 5326

Anselin L, Bera A (1988) Spatial dependence in linear regression models with an introduction to spatial econometrics. In: Cummings LL (eds) The handbook of applied economic statistics. Marcel Dekker, New York, pp 237–289

Anselin A, Varga A, Acs Z (1997) Local geographic spillovers between university research and high-technology innovation. J Urban Econ 42:422–448

Arrow K (1962) The economic implication of learning by doing. Rev Econ Stat 80:155–173

Audretsch D, Feldman M (1996) R&D spillovers and the geography of innovation and production. Am Econ Rev 86(3):630–640

Barro RJ, Sala-I-Martin X, Blanchard OJ, Hall RE (1991) Convergence across states and regions. Brookings Pap Econ Act 1:107–182

Belsley D, Kuh E, Welsch R (1980) Regression diagnostics. Wiley, New York

Bernard AB, Redding SJ, Schott PK (2006) Multi-product firms and product switching. NBER working papers, no. 12293

Braunerhjelm P (1999) Sunk costs, firm Size, and internationalization. Weltwirtsch Archiv 135(4):657–674

Carlsson B (1999) Small business, entrepreneurship, and industrial dynamics. In: Acs ZJ (ed) Are small firms important? Their role and impact. Kluwer, Boston pp 99–110

Carlsson B, Fridh A-C (2002) Technology transfer in United States Universities: a survey and statistical analysis. J Evol Econ 12(1–2):199–232

Carlsson B, Acs ZJ, Audretsch DB, Braunerhjelm P (2007) The knowledge filter, entrepreneurship and economic growth. Case Western Reserve University

Cohen W, Levinthal D (1990) Absorptive capacity: a new perspective on learning and innovation. Adm Sci Q 35(1):128–152

Elhorst JP (2003) Specification and estimation of spatial panel data models. Int Reg Sci Rev 26:244–268

Hayek J (1945) The use of knowledge in society. Am Econ Rev 35:520–530

Henderson VJ (2003) Marshall's scale economies. J Urban Econ 53(1):1–28

High S (2003) Industrial sunset: the making of North America's rust belt, 1969–1984. University of Toronto Press, Toronto

Holtz-Eakin D, Kao C (2003) Entrepreneurship and economic growth: the proof is in the productivity, Center for Policy Research – Working paper no. 50, available at http://www.cpr.maxwell.syr.edu

Jaffe A (1989) The real effects of academic research. Am Econ Rev 79(5):957–970

Jaffe A, Tratjenberg M, Henderson R (1993) Geographic localization of knowledge spillovers as evidenced by patent citations. Q J Econ 108:483–499

Jovanovic B (2001) New technology and the small firm. Small Bus Econ 16(1):53–55

Keller W (2002) Geographic localization of international technology diffusion. Am Econ Rev 92(1):120–142

Knight F (1921) Risk, uncertainty and profit. Houghton Mifflin, Boston

LeSage JP (1997) Bayesian estimation of spatial autoregressive models. Int Reg Sci Rev 20:113–129

LeSage JP (1998) Spatial econometrics. http://www.spatial-econometrics.com/html/wbook.pdf

Lucas R (1988) On the mechanics of economic development. J Monet Econ 22:3–39

Michelacci C (2003) Low returns in R&D due to the lack of entrepreneurial skills. Econ J 113:207–225

Romer P (1986) Increasing returns and economic growth. Am Econ Rev 94:1002–1037

Romer P (1990) Endogenous technical change. J Pol Econ 98:71–102

Schumpeter J (1911) Theorie der wirtschaftlichen entwicklung. English translation: The theory of economic development, harvard University Press, Cambridge

Shane S, Eckhardt J (2003) The individual-opportunity Nexus. In: Acs ZJ, Audretsch D. Handbook of entrepreneurship research. Boston, Kluwer, pp 161–194

Utterback JM, Abernathy W (1975) A dynamic model of process and product innovation. Omega 33:639–656

Varga A (1998) University research and regional innovation. Kluwer, Boston

Available online at www.sciencedirect.com
SCIENCE @ DIRECT®
ELSEVIER
Research Policy 34 (2005) 1191–1202
research policy
www.elsevier.com/locate/econbase

Does the Knowledge Spillover Theory of Entrepreneurship hold for regions?

David B. Audretsch [a,*], Erik E. Lehmann [b]

[a] *Department Entrepreneurship, Growth, and Public Policy, Max-Planck Institute of Economics, Jena, Kahlaische Strasse 10, 07745 Jena, Germany*
[b] *University of Augsburg and Max-Planck Institute of Economics, Jena, Universitätsstrasse 16, 86159 Augsburg, Germany*

Received 21 November 2004; received in revised form 12 March 2005; accepted 15 March 2005
Available online 19 July 2005

Abstract

The purpose of this paper is to test whether the Knowledge Spillover Theory of Entrepreneurship holds for regions. We do this by linking investments in knowledge by universities and regions to the amount of entrepreneurial activity associated with each university. Using binomial regressions we estimate how the number of young and high-tech firms located around universities depends on regional factors and the output of universities. The results clearly show that the number of firms located close to a university is positively influenced by the knowledge capacity of this region and the knowledge output of a university. Thus, there is considerable evidence suggesting that the Knowledge Spillover Theory of Entrepreneurship holds for regions as well as for industries.

Keywords: University spillover; Firm location

1. Introduction

As described by Audretsch and Thurik (2001), globalization is shifting the comparative advantage in the OECD countries away from being based on traditional inputs of production toward knowledge. As the comparative advantage has become increasingly based on new knowledge, public policy has responded enabling the creation and commercialization of knowledge. Furthermore, new policy approaches are emerging shifting the focus from national and international focuses towards regions and regional clusters. Examples of such policies include encouraging R&D spillovers, venture capital and new firm startups. In this new entrepreneurship based policy, universities play a key role in providing spillovers by academic research and human capital in the form of well trained and educated students. The success of a number of different technology clusters is the

* Corresponding author. Fax: +49 3641 68 6900.
E-mail address: audretsch@mpiew-jena.mpg.de
(D.B. Audretsch).

0048-7333/$ – see front matter
doi:10.1016/j.respol.2005.03.012

direct result of enabling policies, such as the provision of research support by universities.

The traditional view of knowledge and innovation is that the firm exists exogenously and then invests in research and development or the augmentation of human capital through training and education of workers to endogenously create new knowledge and ideas. This view was formalized by Griliches (1979) in what he termed as the Model of the Firm Knowledge Production Function.

While Griliches' model of the knowledge production function was statistically confirmed through a plethora of econometric studies linking knowledge inputs to innovative outputs at the levels of the country and the industry, the relationship proved to be considerably more ambiguous at the level of the firm, particularly when new and small enterprises were included in the sample. Small firms were found to contribute more to innovative output than would have been expected from their rather meager investments in R&D and other knowledge inputs.

This paradox of the high innovative output of small enterprises given their low level of knowledge inputs that seemingly contradicts the Griliches model of the firm knowledge production function was resolved by Audretsch (1995), who introduced the Knowledge Spillover Theory of Entrepreneurship, "The findings in this book challenge an assumption implicit to the knowledge production function—that firms exist exogenously and then endogenously seek out and apply knowledge inputs to generate innovative output. It is the knowledge in the possession of economic agents that is exogenous, and in an effort to appropriate the returns from that knowledge, the spillover of knowledge from its producing entity involves endogenously creating a new firm" (pp. 179–180).

What is the source of this entrepreneurial knowledge that endogenously generated the startup of new firms? The answer seemed to be through the spillover of knowledge from the source creating to commercialization via the startup of a new firm, "How are these small and frequently new firms able to generate innovative output when undertaking a generally negligible amount of investment into knowledge-generating inputs, such as R&D? One answer is apparently through exploiting knowledge created by expenditures on research in universities and on R&D in large corporations" (p. 179). While policy makers have no direct influence on large corporations as a source of spillovers, they can, more or less, influence universities as a promoter for entrepreneurship and important source of spillovers. As an example for Germany, the Federal Ministry for Education and Research established regions where startups from universities and government research laboratories are encouraged (Federal Ministry of Education and Research, 2004). Strongly influenced by federal and state policies, universities can alter their curricula, initiate research, and participate in collaborations that serve these ventures and thus allowing the startups to be more competitive with larger and better financed competitors.

The empirical evidence supporting the Knowledge Spillover Theory of Entrepreneurship was provided from analyzing variations in startup rates across different industries reflecting different underlying knowledge contexts. In particular, those industries with a greater investment in new knowledge also exhibited higher startup rates while those industries with less investment in new knowledge exhibited lower startup rates, which was interpreted as the mechanism by which knowledge spillovers are transmitted.

Thus, compelling evidence was provided suggesting that entrepreneurship is an endogenous response to the potential for commercializing knowledge that has not been adequately commercialized by the incumbent firms. This involved an organizational dimension involving the mechanism transmitting knowledge spillovers—the startup of new firms. In addition, Jaffe (1989), Audretsch and Feldman (1996) and Audretsch and Stephan (1996) provided evidence concerning the spatial dimension of knowledge spillovers. In particular their findings suggested that knowledge spillovers are geographically bounded and localized within spatial proximity to the knowledge source. None of these studies, however, identified the actual mechanisms which actually transmit the knowledge spillover; rather, the spillovers were implicitly assumed to automatically exist (or fall like Manna from heaven), but only within a geographically bounded spatial area.

The purpose of this paper is to bring these two literatures together by asking whether the Knowledge Spillover Theory of Entrepreneurship also has a spatial component in that the startups tend to cluster within geographic proximity to knowledge sources. The second section explains why the knowledge spillover theory should also have a spatial component in that the knowledge startups cluster geographically around the knowledge source. The third and fourth sections provide an empirical test within a spatial context—the region around a university. Finally, in the last section a summary and conclusion are provided. In particular, the empirical evidence suggests that those universities with a greater investment in knowledge and where the regional investment in knowledge is higher tend to generate more technology startups. This supports the view that the spillover theory of knowledge holds for regional contexts as well as for industries.

2. Localizing the Knowledge Spillover Theory of Entrepreneurship

That entrepreneurial activity varies across geographic space has long been observed. Efforts to systematically link spatial variations in entrepreneurship with locational specific characteristics showed that such spatial activity is not at all random but rather shaped by factors associated with particular regions (Reynolds et al., 1994). A series of studies, dating back at least to Carlton (1983) and Bartik (1985) and more recently Reynolds et al. (1994), have tried to identify characteristics specific to particular regions that account for inter-spatial variations in entrepreneurship. However, while a large literature exists linking new-firm startup activity to region-specific characteristics and attributes (Fritsch, 1997; Reynolds et al., 1994; Carlton, 1983; Bartik, 1985; Audretsch and Fritsch, 1994), virtually none of these studies provided a theory linking knowledge spillovers to new-firm startup activity, nor did any of these studies provide a measure of knowledge spillovers.

For example, Audretsch and Fritsch (1994), examine the impact that location plays on entrepreneurial activity in (West) Germany. Using a database derived from the social insurance statistics, which covers about 90% of employment, they identify the birth rates of new startups for each of 75 distinct economic regions. These regions are distinguished on the basis of planning regions, or *Raumordungsregionen*. They find that, for the late 1980s, the birth rates of new firms was higher in regions experiencing low unemployment, which have a dense population, a high growth rate of population, a high share of skilled workers, and a strong presence of small businesses.

The Knowledge Spillover Theory of Entrepreneurship provides a focus on the generation of entrepreneurial opportunities emanating from knowledge investments by incumbent firms and public research organizations which are not fully appropriated by those incumbent enterprises. It is a virtual consensus that entrepreneurship revolves around the recognition of opportunities and the pursuit of those opportunities (Shane and Eckhardt, 2003). Much of the more contemporary thinking about entrepreneurship has focused on the cognitive process by which individuals reach the decision to start a new firm. According to Sarasvathy et al. (2003, p. 142), "An entrepreurial opportunity consists of a set of ideas, beliefs and actions that enable the creation of future goods and services in the absence of current markets for them". Sarasvathy, Dew, Velamuri and Venkataraman provide a typology of entrepreneurial opportunities as consisting of opportunity recognition, opportunity discovery and opportunity creation.

While much has been made about the key role played by the recognition of opportunities in the cognitive process underlying the decision to become an entrepreneur, relatively little has been written about the actual source of such entrepreneurial opportunities (see Audretsch and Stephan, 1999). The Knowledge Spillover Theory of Entrepreneurship identifies one source of entrepreneurial opportunities—new knowledge and ideas. In particular, the Knowledge Spillover Theory of Entrepreneurship posits that it is new knowledge and ideas created in one context but left uncommercialized or not vigorously pursued by the source actually creating those ideas, such as a research laboratory in a large corporation or research undertaken by a university, that serves as the source of knowledge generating entrepreneurial opportunities. Thus, in this view, one mechanism for recognizing new opportunities and actually implementing them by starting a new firm involves the spillover of knowledge. The source of the knowledge and ideas, and the organization actually making (at least some of) the investments to produce that knowledge, is not the same as the organization actually attempting to commercialize and appropriate the value of that knowledge—the new firm. If the use of that knowledge by the entrepreneur does not involve full payment to the firm making the investment that originally produced that knowledge, such as a license or royalty, then the entrepreneurial act of starting a new firm serves as a mechanism for knowledge spillovers. However, theories of localization suggest that just because universities are the sources of knowledge spillovers does not mean that knowledge transmits costlessly across geographic space. In particular, these theories argue that geographic proximity reduces the cost of accessing and absorbing knowledge spillovers. Thus, a basic tenet in the literature is that university spillovers lower the costs of firms to accessing and absorbing knowledge spillovers. If an entrepreneur decides to locate nearby a university, the benefits must outweigh the costs. Locating close to universities, mostly in the center of a city, is associated with high costs of living, housing, and others. Though, firms also have to pay higher wages to their employees since their costs of living. If the basic resources gathered from a university are not essential to bear those costs, it is more advantageous to locate outside such a metropolitan area.

There are at least two principle mechanisms facilitating the knowledge spillovers from universities to firms. The first one involves scientific research published in scholarly journals. Such published research is codified knowledge. This is because knowledge provided by articles can be transferred and transmitted with low cost, or with costs which are independent from the location. Academic papers can be downloaded from the Internet, obtained from publishers or found in libraries. However, an important qualification is that not all university knowledge is the same. In fact, the knowledge output of a university is heterogeneous. One useful distinction differentiates natural and social science knowledge. Social science knowledge is not based on a unified and established scientific methodology, but it rather is idiosyncratic to very specific disciplines, subdisciplines and even research approaches. Compared to the natural sciences, research in the social sciences is considerably less codified. Thus, geographic proximity to high output universities may be more important for accessing social science research than for accessing natural science research.

This suggests that an important testable hypothesis is that the amount of scientific articles published by a university has no effect on firm location, since accessing (codified) knowledge is more or less invariant to locational distance from the university producing that knowledge. Strict adherence to the scientific method assures that academic research embodies a high component of codified and specific knowledge in the natural sciences (Stephan, 1996). By contrast, the more limited applicability of the scientific method implies that research in the social sciences will embody less codified knowledge (Stephan, 1996). In contrast, academic research in the natural sciences is more codified. Because of its high codification, scientific knowledge can be largely accessed by (competently) reading scientific journals. Thus, we assume that knowledge in the social sciences is more tacit and less codified.

The second type of spillover mechanism involves human capital embodied in students graduating from the university. As Saxenien (1994) points out, one of the important mechanisms facilitating knowledge spillovers involves the mobility of human capital, embodied in graduating students, as they move from the university to a firm. Spatial proximity to universities can therefore generate positive externalities that can be accessed by the firm through the spillover mechanism of human capital. As Varga (2000) shows, university graduates may be one of the most important channels for disseminating knowledge from academia to the local high technology industry. In addition, other related externalities may result from close geographic proximity. For example, local proximity lowers the search costs for both firms and students. This may lead to some competitive advantage over similar firms which are not located close to universities, especially when high skilled labor is a scarce resource and there is intense competition about high potentials.

If a universities graduate's educational background correspondents to the needs of a firm, the probability of employment is higher which lowers the costs of matching for both students and firms. This contains to educate students and prepare them for later employment in the economy and academic research activities and thus provides spillovers. The business sector receives inputs from universities in the form of highly educated human capital. Although these individuals may require further training, university education provides the foundation for subsequent specialized industrial training. Universities may also provide further training of employees. Thus, it seems that the most frequent types of interactions between firms and universities are the employment of university graduates (Schartinger et al., 2001, p. 259).

Why should entrepreneurship play an important role in the spillover of new knowledge and ideas? And why should new knowledge play an important role in creating entrepreneurial opportunities? In the Romer (1986) model of endogenous growth new technological knowledge is assumed to automatically spill over. Investment in new technological knowledge is automatically accessed by third-party firms and economic agents, resulting in the automatic spill over of knowledge.[1] The assumption that knowledge automatically spills over is, of course, consistent with the important insight by Arrow (1962) that knowledge differs from the traditional factors of production – physical capital and (unskilled) labor – in that it is non-excludable and non-exhaustive. When the firm or economic agent uses the knowledge, it is neither exhausted nor can it be, in the absence of legal protection, precluded from use by third-party firms or other economic agents. Thus, in the spirit of the Romer model, drawing on the earlier insights about knowledge from Arrow, a large and vigorous literature has emerged obsessed with the links between intellectual property protection and the incentives for firms to invest in the creation of new knowledge through R&D and investments in human capital.

However, the preoccupation with the non-excludability and non-exhaustability of knowledge first identified by Arrow and later carried forward and assumed in the Romer model, neglects another key insight in the original Arrow (1962) article. Arrow also identified another dimension by which knowledge differs from the traditional factors of production. This other dimension involves the greater degree of uncertainty, higher extent of asymmetries, and greater cost of transacting new ideas. The expected value of any new idea is highly uncertain, and as Arrow pointed out, has a much greater variance than would be associated with the deployment of traditional factors of production. After all, there is relative certainty about what a standard piece of capital equipment can do, or what an (unskilled) worker can contribute to a mass-production assembly line. By contrast, Arrow emphasized that when it comes to innovation, there is uncertainty about whether the new product can be produced, how it can be produced, and whether sufficient demand for that visualized new product might actually materialize.

In addition, new ideas are typically associated with considerable asymmetries. In order to evaluate a proposed new idea concerning a new biotechnology product, the decision maker might not only need to have a Ph.D. in biotechnology, but also a specialization in the exact scientific area. Such divergences in education, background and experience can result in a divergence in the expected value of a new project or the variance in outcomes anticipated from pursuing that new idea, both of which can lead to divergences in the recognition and evaluation of opportunities across economic agents and decision-making hierarchies. Such divergences in the valuation of new ideas will become greater if the new idea is not consistent with the core competence and technological trajectory of the incumbent firm.

Thus, because of the conditions inherent in knowledge – high uncertainty, asymmetries and transactions cost – decision making hierarchies can reach the decision not to pursue and try to commercialize new ideas that individual economic agents, or groups or teams of economic agents think are potentially valuable and should be pursued. The basic conditions characterizing new knowledge, combined with a broad spectrum of institutions, rules and regulations impose what Acs et al. (2003) term *the knowledge filter*. The knowledge filter is the gap between new knowledge and what Arrow (1962) referred to as economic knowledge or commercialized knowledge. The greater is the knowledge filter, the more pronounced is this gap between new knowledge and new economic, or commercialized, knowledge.

The knowledge filter is a consequence of the basic conditions inherent in new knowledge. Similarly, it is the knowledge filter that creates the opportunity for entrepreneurship in the Knowledge Spillover Theory of Entrepreneurship. According to this theory, opportunities for entrepreneurship are the duality of the knowledge filter. The higher is the knowledge filter, the greater are the divergences in the valuation of new ideas across economic agents and the decision-making hierarchies of incumbent firms. Entrepreneurial opportunities are generated not just by investments in new knowledge and ideas, but in the propensity for only a distinct subset of those opportunities to be fully pursued by incumbent firms.

Thus, the knowledge theory of entrepreneurship shifts the fundamental decision making unit of observation in the model of the knowledge production function

[1] For more recent models of endogenous growth combining both new economic geography models and R&D investments see Fujita and Thisse (2002, Chapter 11) or Baldwin and Forslid (2000).

away from exogenously assumed firms to individuals, such as scientists, engineers or other knowledge workers—agents with endowments of new economic knowledge. As Audretsch (1995) pointed out, when the lens is shifted away from the firm to the individual as the relevant unit of observation, the appropriability issue remains, but the question becomes, *How can economic agents with a given endowment of new knowledge best appropriate the returns from that knowledge?* If the scientist or engineer can pursue the new idea within the organizational structure of the firm developing the knowledge and appropriate roughly the expected value of that knowledge, he has no reason to leave the firm. On the other hand, if she places a greater value on his ideas than do the decision-making bureaucracy of the incumbent firm, he may choose to start a new firm to appropriate the value of his knowledge.

In the Knowledge Spillover Theory of Entrepreneurship the knowledge production function is actually reversed. The knowledge is exogenous and embodied in a worker. The firm is created endogenously in the worker's effort to appropriate the value of his knowledge through innovative activity. Typically an employee from an established large corporation, often a scientist or engineer working in a research laboratory, will have an idea for an invention and ultimately for an innovation. Accompanying this potential innovation is an expected net return from the new product. The inventor would expect to be compensated for his/her potential innovation accordingly. If the company has a different, presumably lower, valuation of the potential innovation, it may decide either not to pursue its development, or that it merits a lower level of compensation than that expected by the employee.

In either case, the employee will weigh the alternative of starting his/her own firm. If the gap in the expected return accruing from the potential innovation between the inventor and the corporate decision maker is sufficiently large, and if the cost of starting a new firm is sufficiently low, the employee may decide to leave the large corporation and establish a new enterprise. Since the knowledge was generated in the established corporation, the new start-up is considered to be a spin-off from the existing firm. Such startups typically do not have direct access to a large R&D laboratory. Rather, the entrepreneurial opportunity emanates from the knowledge and experience accrued from the R&D laboratories with their previous employers. Thus, the knowledge spillover view of entrepreneurship is actually a theory of endogenous entrepreneurship, where entrepreneurship is an endogenous response to opportunities created by investments in new knowledge that are not commercialized because of the knowledge filter.

As investments in new knowledge increase, entrepreneurial opportunities will also increase. Contexts where new knowledge plays an important role are associated with a greater degree of uncertainty and asymmetries across economic agents evaluating the potential value of new ideas. Thus, a context involving more new knowledge will also impose a greater divergence in the evaluation of that knowledge across economic agents, resulting in a greater variance in the outcome expected from commercializing those ideas (see Audretsch et al., 2006). It is this gap in the valuation of new ideas across economic agents, or between economic agents and decision-making hierarchies of incumbent enterprises, that creates the entrepreneurial opportunity.

As already discussed, a vigorous literature has already identified that knowledge spillovers are greater in the presence of knowledge investments. Just as Jaffe (1989) and Audretsch and Feldman (1996) show, those regions with high knowledge investments experience a high level of knowledge spillovers, and those regions with a low amount of knowledge investments experience a low level of knowledge spillovers, since there is less knowledge to be spilled over. Bode (2004) shows that in general interregional spillovers contribute significantly to regional knowledge production. Also Anselin et al. (1997) conclude that the geographical scope of knowledge spillovers is restricted to a limited number of neighboring regions or to regions within a given maximum distance from the region of interest.

The Knowledge Spillover Theory of Entrepreneurship analogously suggests that, ceteris paribus, entrepreneurial activity will tend to be greater in contexts where investments in new knowledge are relatively high, since the new firm will be started from knowledge that has spilled over from the source actually producing that new knowledge. A paucity of new ideas in an impoverished knowledge context will generate only limited entrepreneurial opportunities. By contrast, in a high knowledge context, new ideas will generate entrepreneurial opportunities by exploiting (potential) spillovers of that knowledge. Thus, the knowledge spillover view of entrepreneurship provides a clear link, or prediction that entrepreneurial activity will result from investments in new knowledge and that entrepreneurial activity will be spatially localized within close geographic proximity to the knowledge source.

3. Dataset, methodology, and descriptive statistics

To test the hypothesis that knowledge-based startups are located within close proximity to knowledge sources, we examine the spatial relation between knowledge-based startups and their proximity to universities. We do this by analyzing a unique dataset consisting of 281 firms which made an initial public offering (IPO) in Germany between March 1997 and March 2002. The dataset was collected combining individual data from IPO prospectuses, along with publicly available information from on-line data sources including the *Deutsche Boerse AG* (www.deutsche-boerse.com). We pooled this dataset by adding university-specific variables, which are individually collected from the 73 public universities in Germany (see Warning, 2004). The dependent variable is the number of firms located closest to a university (Firms). We do this by measuring the distance (in kilometers) between a firm and the surrounding universities. The distance is measured in kilometers using the online database of the *German Automobile Club* (www.adac.de). All firms located within a radius of 1.5 km are classified as belonging in the distance category of 1 km. Thus, the variable "Firms" contains all firms for which the distance towards this respective university is the closest one. No firm is allocated to more then one university.

To analyze the amount of knowledge investment that has been made by the university and elsewhere in the region, we include a measure that should reflect the potential supply of knowledge spillovers. This variable (regional capacity) is an index constructed by using spending in R&D and technological innovations in a specific region (see Sternberg and Litzenberger, 2003). This variable serves also as a measure for the "knowledge filter", which is significantly influenced by the structure of the innovation system within a regional context.

As university output we measure the number of students in the natural sciences (StudentsSCI) and in the social sciences as well as the articles published in the natural sciences (SCI) and the social sciences (SSCI). Information about the number of students is provided by the Federal Statistical Office. Since those data are provided sporadically we take 1997 as the base year where we have information of all universities. Publication data are hand-collected from the research database ISI (Information Sciences Institutes). We included the number of listed papers for each university published from 1993 until 2000.[2]

These measures are shown for the top 20 universities in Germany in Table 1. Using four distinct measures of university output enables us to examine any potential heterogeneous impact of university output.

We further include several control variables which may influence the degree of technology entrepreneurship located around a university. First, we include a dummy variable indicating a technical university (TU) (see Audretsch et al., in press). Technical universities are assumed to play a special role in the technology transfer, since their focus is especially on engineering and natural sciences (Audretsch and Lehmann, 2005). Those universities get more funds compared to other universities to foster and promote spillovers from new technologies to firms. Thus, technical universities may attract more young and high-tech firms. Secondly, we include the age (Age) of a university as measured in years to control for reputation effects. Thirdly, we use the number of inhabitants as the size of the city (City) of the respective university as well as the number of universities of applied sciences (universities). Those universities differ from the included public universities since their main goal is to educate students and are not engaged in academic research. They also do not have the status of a "university" in Germany. However, they also provide human capital and are thus included for this effect. Most of the included cities have a couple of such universities of applied sciences. Finally, a dummy variable is included to control for re-unification effects (West).

We employ the negative binomial regression model as the analytical technique for estimating the impact of the technology index and university research output on the role of geographic proximity. The underlying assumption is that the number of knowledge

[2] The publications in social science and natural science did not vary across the universities during time.

Table 1
List of the top 20 universities in Germany

University	Firms[a]	ϕ (km)[b]	SSCI Students[c]	SCI Students[d]
LMU München	51	17.9	43,633	8,119
Uni Frankfurt	26	19.5	26,324	5,715
Uni Hamburg	24	13.25	1,361	329
Uni Stuttgart	16	16.2	4,779	12,104
HU Berlin	14	7.78	20,769	4,936
Uni Köln	12	5.9	47,112	9,395
TU München	10	8.8	1,619	14,976
TU Karlsruhe	9	36.7	4,102	11,818
Uni Düsseldorf	7	18.57	14,697	4,762
Uni Erlangen-Nbg.	6	14	12,861	7,144
Uni Freiburg	6	40.33	12,334	4,942
FU Berlin	5	5.5	30,290	6,260
TU Aachen (RWTA)	5	3.4	7,884	20,570
Uni Jena	5	15.8	7,615	2,864
U-GH Paderborn	4	17	6,993	8,676
Uni Bielefeld	4	15.5	15,831	4,400
Uni Bremen	4	25.5	11,749	4,800
UdB München	4	14.25	1,054	1,104
Uni Kiel	4	40.75	13,000	6,513
Uni Regensburg	4	40	11,192	3,696

[a] Measured by the number of firms located closest to this university.
[b] ϕ (km) is the average distance of the firms located closest to this university.
[c] Number of students in the social sciences.
[d] Number of students in the natural sciences.

spillover startups located within geographic proximity to a university could be interpreted as count data. Since ordinary least squares regression is inappropriate for the count dependent variables that have large numbers of the smallest observation and remaining observations taking the form of small positive numbers, Poisson-regression seems to be more appropriate. However, the assumption for a Poisson-regression, the equality of mean and variance of the exogenous variable, is rejected by several tests. Thus, we apply the negative binomial regression model to overcome this problem of over-dispersion (Greene, 2003, pp. 740–752). Also, this statistical technique is designed for maximum likelihood estimation of the number of occurrence of nonnegative counts like the event of location.

In the first model (Model (1)), we estimate the following basic regression with the technology indices:

$$\begin{aligned}\text{Number of firms} &\\ &= \text{const.} + \beta_1 \text{regional technology capacity}\\ &\quad + \text{Controlvariables} + \varepsilon \qquad (1)\end{aligned}$$

Secondly, we test for the impact of university spillovers on the number of firms located to a university:

$$\begin{aligned}\text{Number of firms} &\\ &= \text{const.} + \beta_1 \text{StudentsSCI} + \beta_2 \text{StudentsSSCI}\\ &\quad + \beta_3 \text{SSCI} + \beta_4 \text{SCI} + \text{Controlvariables} + \varepsilon \qquad (2)\end{aligned}$$

then we test the combination from (1) and (2):

$$\begin{aligned}\text{Number of firms} &\\ &= \text{const.} + \beta_1 \text{regional capacity} + \beta_2 \text{StudentsSCI}\\ &\quad + \beta_4 \text{StudentsSSCI} + \beta_5 \text{SSCI}\\ &\quad + \beta_6 \text{Controlvariables} + \varepsilon \qquad (3)\end{aligned}$$

Finally, we estimate model (3) for selected industries. Our base hypotheses is that in models (1) and (3) $\beta_1 > 0$, that the numbers of firms clustered around a respective university depends on the regional capacity.

From the 73 public universities in Germany, only 54 universities are chosen as the closest university for the included firms. The descriptive statistics, as presented in Table 1, show the highly skewed number of firms located around universities. From the included 281 firms, 51 of them are located closest to the LMU Munich. The median distance of all firms located around the closest university is 7 km and the average distance is about 16 km. It could also be shown that the universities differ highly in their number of students in the natural sciences and the social sciences (see Warning, 2004, for more details).

4. Empirical evidence

In this section, we provide the results of the negative binomial regressions with the number of firms clustered around a university as the dependent variable. The empirical results for the negative binomial regression equations estimating the number of knowledgebased-startups located within geographic proximity to each university are presented in Table 2. The dependent variable is the number of firms located within geographic proximity to a university. As explained in the previous section, the first model includes only the technology index, the second model includes only the measures of university outputs, and the third model includes both types of measures.

In the first model, as the positive and statistically significant coefficient of the measure of technology capacity of the region in which the university is located suggests, the greater the technological capacity of the region, the greater are the number of firms locating in that region.

The negative coefficient of university age does not support the hypothesis that university reputation, at least measured by age, has a positive influence on startups. Rather, the younger a university is, the more firms tend to locate with close geographic proximity. This finding could be explained by the fact that old and traditional universities are more focused on the social sciences, resources which are not really necessary for young and high-tech firms. The same holds for technical oriented universities which are mainly focused on the traditional research in engineering and machineries. While the size of the city is insignificant, the number of firms is also explained by the number of universities.

The results of the second estimation model show that university output influences the location decision of firms. As found by others (Schartinger et al., 2001; Stephan, 2001) the amount of university educated human capital is one of the major factors influencing firm location. Proximity offers the possibility of linking students to industry more efficiently, by providing industry and students a pre-employment look at each other. Thus, universities with a high output of students tend to generate more knowledge-based startups.

New technology oriented firms are particularly dependent on technological innovations and scientific progress and are therefore more than others

Table 2
Negative binomial estimating regional technology startups

	Model I	Model II	Model III
Regional knowledge capacity	0.03753 (10.19)***		0.4279 (16.95)***
SCI students		0.0004 (3.77)***	0.0001 (1.36)
SSCI students		0.00003 (10.03)***	0.0003 (14.36)***
SSCI		−0.0009 (8.43)***	−0.0009 (11.06)***
SCI		0.1100 (3.13)***	0.1149 (3.99)***
Age	−0.0004 (2.27)**		
	0.0005 (2.30)**	−0.00057 (2.90)**	
TU	−1.4786 (10.20)***	−1.1976 (6.65)***	−1.389 (9.53)***
City	−1.17E07 (1.10)	−0.856E07 (10.09)***	−3.06E07 (3.96)***
Universities and universities of applied sciences	0.4047 (6.24)***	1.1935 (20.90)***	0.681 (12.90)***
West	−0.2029 (0.93)	−0.6351 (3.32)***	−0.5320 (3.61)***
Const.	0.4892 (1.65)*	1.7352 (6.65)***	0.6887 (3.07)***
Pseudo R2	0.1792	0.2067	0.3412
LL-ratio	−897.76	−850.11	−705.998

The endogenous variable is the number of technology startups located within geographic proximity to a university. Z-values are in brackets. The asterisks, *, **, and *** indicate significance at the 10%, 5%, and 1% level, respectively. The dataset includes 281 firms and 54 universities.

inclined to engage in interactions with research intense universities. Thus, research intense universities especially in the natural sciences are more attractive for high-tech firms. This could be confirmed by the positive significance of the coefficient indicating the number of articles published in the natural sciences. However, research intensity in the social sciences seems to be less important for attracting high-tech firms.

The third regression model includes both the regional innovation capacity measures and the university output measures and is consistent with the previous results, indicating that the results are robust. The number of new knowledge-based firms located within close geographic proximity to a university is positively influenced by the knowledge output of the respective university and the innovative capacity of the region. In both regressions, the dummy variable controlling for the reunification effect is significant negative for West Germany. This could be explained by the entrepreneurial policy which is concentrated around universities in East Germany, especially Dresden in Saxony and Thüringen (see Acs, 2002, p. 190).

This third model is also estimated for major technology industries or groups. As the results in Table 3 suggest, the evidence suggests that not all of the knowledge measures have a homogeneous impact on entrepreneurship across the particular industries. The results show that the demand for high skilled labor, as expressed by the number of students, differs across the industries. Especially in the high-technology and knowledge intense industries, the amount of students in the natural sciences enters the regressions significantly. While research activities in the social sciences seem to have no impact on the number of firms clustered around a university, the results differ for research in the natural sciences. The number of articles published in this field is significant and positive in the Hardware and Technology sector as well as in the Biotechnology and Medicine Technique sector. This strongly supports the thesis that university research in those fields can provide knowledge spillovers in high-technology and knowledge based industries.

Thus, there is no reason to think that knowledge spillover entrepreneurship is invariant to the type of knowledge or the industry context. Still, these results confirm that there is compelling evidence suggesting that knowledge-based startups tend to cluster within geographic proximity to the knowledge source.

Table 3
Negative binomial estimating regional technololgy startups

	Software	Service	Media and Entertainment	Hardware/technology	Biotec/medtec
Regional capacity	0.0437 (7.59)***	0.0402 (7.81)***	0.04377 (4.50)***	0.06128 (8.08)***	0.0706 (2.36)**
SCI students	−1.50E06 (0.09)	0.0001 (0.98)	0.0001 (2.30)***	0.0002 (2.59)***	0.0003 (3.04)***
SSCI students	0.00002 (3.17)***	0.0003 (8.84)***	0.0004 (3.35)***	0.0004 (4.49)***	0.0006 (3.69)***
SSCI	−0.001 (1.13)	−0.012 (1.39)	−0.001 (1.13)	−0.0005 (1.52)	−0.0018 (1.35)
SCI	−0.1971 (0.92)	0.2076 (0.71)	−0.0043 (0.08)	0.2356 (2.05)**	0.0103 (1.67)*
Age	0.00002 (0.05)	−0.0004 (1.16)	−0.006 (1.12)	−0.003 (2.96)***	0.0025 (0.85)
TU	−1.0577 (3.46)***	−0.9135 (3.15)***	−2.284 (3.01)***	−0.5618 (1.59)	1.468 (1.67)*
City	−0.282E07 (5.34)***	−4.74E07 (2.85)***	−3.45E07 (1.48)	−1.17E07 (0.24)	−1.18E06 (1.47)
West	−0.4105 (1.01)	−0.5606 (1.67)*	−0.4876 (1.12)	−0.7883 (0.88)	−3.130 (1.45)
Number of universities	0.8234 (5.34)***	0.9007 (7.66)***	0.7666 (3.42)***	0.4493 (1.26)	0.7164 (1.72)*
Const.	0.3309 (0.62)	0.7129 (1.41)	0.3122 (0.23)	0.2576 (0.22)	3.5109 (2.12)*
Pseudo R2	0.3722	0.3560	0.4042	0.3999	0.4143
LL-ratio test (χ^2)	−130.485	−164.059	−93.357	−110.326	−51.769
Number of firms in each industry	55	67	37	47	25

The endogenous variable is the number of technology startups located within geographic proximity to a university. *Z*-values are in brackets. The asterisks, *, **, and *** indicate significance at the 10%, 5%, and 1% level, respectively. *N* = 54 universities. We dropped all industries with less than 20 firms and matched Hardware and Technology as well as Biotech and Medtec to increase the number of firms in the regressions.

5. Conclusion

The Knowledge Spillover Theory of Entrepreneurship suggests that investment in the creation of new knowledge will generate opportunities for entrepreneurship as a mechanism for knowledge spillovers. While empirical evidence has already been provided supporting the Knowledge Spillover Theory of Entrepreneurship for the different contexts provided by industry specific knowledge investments, little was known about the spatial dimension. This paper has found that those universities in regions with a higher knowledge capacity and greater knowledge output also generate a higher number of technology startups. Thus, at least on the basis of data from Germany, there is considerable evidence that the Knowledge Spillover Theory of Entrepreneurship holds for regions as well as for industries.

As comparative advantage has become increasingly based on new knowledge, the results show that public policy can respond in two fundamental ways. First, by providing an infrastructure that enables young firms to absorb necessary resources. This is shown by the positive effect of the regional knowledge capacity on firm location. Secondly, by influencing universities to increase their research activities, especially in the natural sciences and in providing well educated students.

Acknowledgements

Financial support by the German Research Foundation (DFG) through the research group #FOR454 "Heterogeneous Labor" at the University of Konstanz and the ZEW, Mannheim, is gratefully acknowledged. We also would like to thank Adam Lederer, two anonymous referees and Michael Fritsch for helpful comments and suggestions.

References

Acs, Z., 2002. Innovation and the Growth of Cities. Edward Elgar, Cheltenham.

Acs, Z., Audretsch, D.B., Braunerjhelm, P., Carlsson, B., 2003. The Missing Link: The Knowledge Filter and Endodenous Growth. Center for Business and Policy Studies, Stockholm, Sweden.

Anselin, L., Varga, A., Acs, Z., 1997. Local geographic spillovers between university research and high technology innovations. Journal of Urban Economics 42, 422–448.

Arrow, K., 1962. Economic welfare and the allocation of resources for invention. In: Nelson, R. (Ed.), The Rate and Direction of Inventive Activity. Princeton University Press, Princeton.

Audretsch, D.B., 1995. Innovation and Industry Evolution. MIT Press, Cambridge/Mass.

Audretsch, D.B., Lehmann, E.E., 2005. Mansfields's missing link: the impact of knowledge spillovers on firm growth. Journal of Technology Transfer 30 (1/2), 207–210.

Audretsch, D.B., Keilbach, M., Lehmann, E.E., 2006. Entrepreneurship and Growth. Oxford University Press, in press.

Audretsch, D.B., Feldman, M.P., 1996. R&D spillovers and the geography of innovation and production. American Economic Review 86, 630–640.

Audretsch, D.B., Fritsch, M., 1994. The geography of firm births in Germany. Regional Studies 28 (4), 359–365.

Audretsch, D.B., Stephan, P.E., 1996. Company-scientist locational links: the case of biotechnology. American Economic Review 86 (3), 641–652.

Audretsch, D.B., Stephan, P.E., 1999. How and why does knowledge spill overs in biotechnology? In: Audretsch, D.B., Thurik, R. (Eds.), Innovation, Industry Evolution, and Employment. Cambridge University press, Cambridge, pp. 216–229.

Audretsch, D.B., Thurik, R., 2001. What is new about the new economy: sources of growth in the managed and entrepreneurial economies. Industrial and Corporate Change 19, 795–821.

Baldwin, R.E., Forslid, R., 2000. The core-periphery model and endogenous growth: stabilizing and destabilazing integration. Economics 67, 307–324.

Bartik, T., 1985. Business location decisions in the United States: estimates of the effects of unionization, taxes, and other characteristics of the states. Journal of Business and Economic Statistics 3 (January), 16–22.

Bode, E., 2004. The spatial pattern of localized R&D spillovers: an empirical investigation for Germany. Journal of Economic Geography 4, 43–64.

Carlton, D.W., 1983. The location and employment choices of new firms: an econometric model with discrete and continuous endogenous variables. Review of Economics and Statistics 65 (3), 440–449.

Federal Ministry of Education and Research, 2004. Facts and Figures 2002 (http://www.bmbf.de).

Fritsch, M., 1997. New firms and regional employment change. Small Business Economics 9 (5), 437–448.

Fujita, M., Thisse, J.-F., 2002. Economics of Agglomeration. Cities, Industrial Location, and Regional Growth. Cambridge University Press, Cambridge.

Greene, W.H., 2003. Econometric Analysis, fifth ed. Prentice Hall, New York.

Griliches, Z., 1979. Issues in assessing the contribution of research and development to productivity growth. Bell Journal of Economics 10, 92–116.

Jaffe, A.B., 1989. Real effects of academic research. American Economic Review 79, 957–970.

Reynolds, P., Storey, D.J., Westhead, P., 1994. Cross-national comparisons of the variation in new firm formation rates. Regional Studies 28 (4), 443–456, July.

Romer, P.M., 1986. Increasing returns and long-run growth. Journal of Political Economy 94, 1002–1037.

Sarasvathy, S.D., Dew, N., Ramakrishna Velamuri, S., Venkataraman, S., 2003. Three views of entrepreneurial opportunity. In: Acs, Z., Audretsch, D.B. (Eds.), Handbook of Entrepreneurship Research. Kluwer Academic Publishers, Dordrecht, pp. 141–160.

Saxenien, A., 1994. Regional Advantage: Culture and Competition in Silicon Valley and Route 128. Harvard University Press, Cambridge, MA.

Schartinger, D., Schibany, A., Gassler, H., 2001. Interactive relations between universities and firms: empirical evidence for Austria. Journal of Technology Transfer 26 (3), 255–268.

Shane, S., Eckhardt, J., 2003. The individual-opportunity nexus. In: Acs, Zoltan J., Audretsch, D.B. (Eds.), Handbook of Entrepreneurship Research. Kluwer Academic Publishers, Dordrecht, pp. 161–194.

Stephan, P., 1996. The economics of science. Journal of Economic Literature 34 (3), 1199–1235.

Stephan, P., 2001. Educational implications of university-industry technology transfer. Journal of Technology Transfer 26 (3), 199–205.

Sternberg, V.R., Litzenberger, T., 2003. Regional clusters—operationalisation and consequences for entrepreneurship. Working paper, University of Cologne.

Varga, A., 2000. Local academic knowledge transfers and the concentration of economic activity. Journal of Regional Science 40, 289–309.

Warning, S., 2004. Performance differences in the German Higher Education System: empirical analysis of strategic groups. Review of Industrial Organization 24, 393–408.

[31]

Small Business Economics (2005) 24: 323–334
DOI 10.1007/s11187-005-1998-4

Entrepreneurship, Agglomeration and Technological Change

Zoltán J. Ács
Attila Varga

ABSTRACT. A growing body of literature suggests that variations across countries, in entrepreneurial activity and the spatial structure of economies could potentially be the source of different efficiencies in knowledge spillovers, and ultimately in economic growth. We develop an empirical model that endogenizes both entrepreneurial activity and agglomeration effects on knowledge spillovers within a Romerian framework. The model is tested using the GEM cross-national data to measure the level of entrepreneurship in each particular economy. We find that after controlling for the stock of knowledge and research and development expenditures, both entrepreneurial activity and agglomeration have a positive and statistically significant effect on technological change in the European Union.

KEY WORDS: agglomeration, economic geography, economic growth; entrepreneurship, knowledge spillovers, technological change

JEL CLASSIFICATION: O3, R1, J24, M13

1. Introduction

The story of the entrepreneurial process is one of the entrepreneur recognizing and acting on unexploited opportunity. Opportunity frequently exists in a crowded space of knowledge creating institutions, networks and venture capitalists. This 'Silicon Valley' story has been told in the context of other high technology agglomerations including Seattle, WA, Austin, TX, Boston, MA, and Washington, D.C. It also has its international counter parts in Bangalore, India, London, UK and Baden-Würtenberg, Germany. While this suggests that entrepreneurship and agglomerations play an important role in economic growth this has not been worked out theoretically (Bresnahan, Gambardella and Saxenian, 2001).

The seminal contribution of Romer (1986, 1990) to the literature on economic growth was to endogenize technological change within an economy in a general equilibrium model with well-specified market forms, selfish agents, perfect foresight and clearing markets. This provided a more realistic explanation of economic growth than the neoclassical theory that focuses on the role of investment in physical capital, increases in the supply of labor, and an exogenous change in technology. According to Romer (1996, 204):

> New growth theory started on the technology-as-public good path and worried about where technology came from, but it soon backed up and reconsidered the initial split that economists make in the physical world. New growth theorists now start by dividing the world into two fundamentally different types of productive inputs that can be called "ideas" and "things". Ideas are nonrival goods that could be stored in a bit string. Things are rival goods with mass (or energy). With ideas and things, one can explain how economic growth works. Nonrival ideas can be used to rearrange things, for example, when one follows a recipe that transforms noxious olives into tasty and healthful olive oil. Economic growth arises from the discovery of new recipes and the transformation of things from low to high value configurations.

The above view leads to insights that do not follow from the neoclassical model. It emphasizes

Final version accepted on February 9, 2005

Zoltán J. Ács
Max-Plank-Institute for Research into Economic Systems
Jena
Germany
and Merrick School of Business
University of Baltimore
Baltimore, MD 21201, USA
E-mail: zacs@ubalt,edu

Attila Varga
Center for Research in Economic Policy (CREP)
and Department of Economics
Faculty of Business and Economics
University of Pécs
Pécs, Rákóczi 80, H-7622, Hungary
E-mail: vargaa@ktk.pte.hu

that ideas are goods that are produced and distributed just as other goods are. If economic growth arises from the creation of new recipes to rearrange things, might not entrepreneurship and geography play some role in this transformation? However, the theory offers no insight into what role, if any, entrepreneurship and agglomeration might play in economic growth. In other words, it does not answer the question, "What is the role of entrepreneurship and agglomeration in technological change?"

An answer to this question can be pursued through the lens of the new economic geography and the modern theory of entrepreneurship. The distinguishing characteristic of the new economic geography is that it studies the economy within a framework that integrates space into general equilibrium theory (Fujita, et al., 1999; Krugman, 1991). One aspect of economic geography is the agglomeration of knowledge. Over the past decade, the new economic geography literature has tried to explain the development, and the economic role of geographic structures, and one of the important questions is related to the role of agglomeration in technological change and ultimately in macroeconomic growth (Fujita and Thisse, 2002).

The recent literature on entrepreneurship has shifted the emphasis in entrepreneurship from cultural and psychological traits to the discovery and exploitation of new knowledge by profit-seeking agents (Shane and Venkataraman, 2000). If entrepreneurs play an important role in the exploitation of technological opportunity the impact of entrepreneurship on growth becomes an important research question (Carree and Thurik, 2003). This relationship is especially important at the spatial level (Acs and Armington, 2004).

Both the relationship between geography and technological change, and between entrepreneurship and technological change, is interesting because these lines of research may prove fruitful in better explaining economic growth through knowledge spillovers. However, both approaches have severe limitations. For example, while there have been several attempts to model new growth theory with endogenously generated spatial structures this work is still in its infancy, until the very recent attempts by Fujita and Thisse (2002) and Baldwin et al. (2003). The purpose of this paper is to develop an empirical framework that endogenizes both entrepreneurial activity and agglomeration effects on knowledge spillovers. We present the first empirical test of the impact of entrepreneurial activity and agglomeration effects on the spillover of new knowledge, based on the Romerian (1990) model of endogenous technological change, as modified by Jones (1995). The model allows us to directly test the relationship between technological change and knowledge creation, conditioned by entrepreneurship and agglomeration effects, while controlling for knowledge spillovers.

We use a new and novel data set from the Global Entrepreneurship Monitor (GEM) project to test the effects of entrepreneurship on knowledge spillovers, and an index is created to measure the effect of agglomeration on knowledge spillovers in the European Union. Section 2 examines the relationship between agglomeration, knowledge spillovers and economic growth and the relationship between entrepreneurship, knowledge spillovers and economic growth. Section 3 extends the basic Romer framework, develops the empirical specification and presents the data, Section 4 has the results and the final section presents the conclusions. We find significant empirical support for the Romer model, where the coefficient on the stock of knowledge is significant but less then one. We also find support for the hypothesis that both agglomeration effects and entrepreneurship facilitate the knowledge spillover mechanism of new knowledge in economic growth.

2. Entrepreneurship and agglomeration

Technological change is the most important factor in long-run macroeconomic growth (Solow 1957). In new growth theory the technological element of the growth process is directly modeled within the economic system as a result of profit motivated choices of economic agents. Recently published findings in entrepreneurship, the geography of innovation and the new economic geography suggest that the extent to which a country is 'entrepreneurial' and its economic system is 'agglomerated' could be a factor that explains technological change. In this section we outline these literatures from an economic growth perspective.

2.1. *Entrepreneurship and technological change*

The origins of the discussion about the existence of opportunity can be traced to Joseph Schumpeter (1934). Schumpeter believed that the existence of opportunity required the introduction of new knowledge not just differentiated access to existing knowledge (Kirzner, 1973). One source of new knowledge came from changes in technology. These technological opportunities are innovative and break away from existing knowledge.

Opportunity therefore comes in part from the research and development (R&D) process that takes place in society. Technological change is an important source of entrepreneurial opportunity because it makes it possible for people to allocate resources in different and potentially more productive ways (Casson, 1995).

However, as was pointed out by Arrow (1974) the link between knowledge and economic knowledge is not well understood. The central problem is a gap in our understanding between technological change and the market that come into existence based on that innovation – this gap in our understanding is filled by the concept of entrepreneurial opportunity. An entrepreneurial opportunity consists of a set of ideas, beliefs and actions that enable the creation of future goods and services in the absence of current markets for them.

If technological opportunity is in part created by the production of new knowledge how is this opportunity discovered? One way in which people discover technological opportunity is through knowledge spillovers. Entrepreneurial discovery is in fact a process of knowledge spillover where knowledge is a non-rival good. Once entrepreneurs discover new opportunities, which are only partially excludable, they have the chance to exploit the opportunity. While most R&D is carried out in large firms and universities it does not mean that the same individuals that discover the opportunity will carry out the exploitation. In fact, because knowledge spills over, one person may discover an opportunity and another may exploit it.

The uncertainty inherent in new economic knowledge, combined with asymmetries between the agent possessing that knowledge and the decision making of the incumbent organization with respect to its expected value potentially leads to a gap between the valuations of the knowledge. This initial condition of not just uncertainty but greater degree of uncertainty vis-à-vis incumbent enterprises in the industry is captured in the theory of firm selection and industry evolution proposed by Jovanovic (1982). An implication of the theory of firm selection is that new firms may begin at a small scale of output, and then if merited by subsequent performance expand. What emerges from the new evolutionary theories and empirical evidence on the role of new firms is that markets are in motion, with a lot of new firms entering the industry and lots of firms leaving (Audretsch, 1995). The empirical evidence supports such an evolutionary view of entrepreneurship (Caves, 1998; Sutton, 1997).

The empirical evidence also supports the argument that technological change is a source of entrepreneurial opportunity. The evidence is indirect since we cannot measure the existence of opportunity. Acs and Audretsch (1989) found that young entrepreneurial firms play a key role in generating technological innovations, at least in some industries. Blau (1987) examined self-employment rates in the United States over a two-decade period and found that an increase in the rate of technological change led to an increase in the self-employment rate, Shane (1996) looked at the number of organizations per capita from 1899 to 1988 found that the rate of technological change, measured as the annual number of new patents issued, had a positive effect on the number of organizations per capita in the economy in the subsequent year.

While the relationship between technological change and opportunity cannot be measured directly, several authors have tried to measure the relationship between entrepreneurship and employment growth. Acs and Armington (2004) found that differences in the level of entrepreneurial activity, and the extent of human capital are positively associated with variation in growth rates. Holtz-Eakin and Kao (2003) using a rich panel of state-level data to quantify the relationship between productivity growth and entrepreneurship found that the effect of new firm formation on productivity is quite persistent.

These results are consistent with Audretsch and Keilbach (2004) who estimate a production function model for German regions based on start-up data from the 1990s. The empirical evidence suggests that entrepreneurship plays an important role in the discovery and exploitation of technological opportunity through knowledge spillovers, lead to higher economic growth.

Theories of entrepreneurship and economic growth are still relatively new even though the entrepreneurship literature does recognize that R&D is an important source of technological opportunity. The process by which knowledge spills over from the firm producing it for use by a third-party firm is exogenous in the model proposed by Romer (1990). The emphasis was on the influence of knowledge spillovers on technological change without specifying *why* and *how* new knowledge spills over. Yet, the critical issue in modeling knowledge-based growth rests on the spillover of knowledge. This was to some extent remedied by the neo-Schumpeterian models of endogenous growth (Aghion and Howitt, 1992, Cheng and Dinopoulos, 1992, Schmitz, 1989, Segerstrom,1991, Segerstrom, et al.,1990). However, these neo-Schumpeterian models design entrepreneurship as an R&D race where a fraction of R&D will turn into successful innovations.

While this implies a step forward, the essence of the Schumpeterian entrepreneur is missed. The innovation process stretches far beyond R&D races that predominantly involve large incumbent firms and concern quality improvements of existing goods. As pointed out by Schumpeter (1947) "the inventor produces ideas, the entrepreneur 'gets things done' ... an idea or scientific principle is not, by itself, of any importance for economic practice." Indeed, the Schumpeterian entrepreneur, by and large, remains absent in those models (Acs et al., 2004).

2.2. *Agglomeration and technological change*

As long as the knowledge necessary for technological change is codified (i.e., it can be studied in written forms either in professional journals and books or in patent documentations) the access to it is essentially not constrained by spatial distance: among other means libraries or the Internet can facilitate the flow of that knowledge to the interested user no matter where the user actually locates.

However, in case knowledge is not codified, because it is not yet completely developed, or it is so practical that it can only be transmitted while knowledge is actually being applied, the flow of knowledge can only be facilitated by personal interactions. Thus, for the transmission of tacit knowledge spatial proximity of knowledge owners and potential users appears to be critical (Polanyi, 1967). For example, several firms move their research facilities to geographic areas where significant amounts of related knowledge has already been accumulated in order to get easier access to that knowledge. Knowledge from other (industrial or academic) research facilities can be channeled via different means, such as, a web of social connections, the local labor market for scientists and engineers or by different types of consultancy relations between universities and private firms.

A large body of literature exists on the spatial extent of knowledge spillovers. At different levels of spatial aggregation (such as states, metropolitan areas, countries) in different countries (e.g. the US, France, Germany, Italy, Austria) and with the application of different econometric methodologies (e.g., various spatial or a-spatial methods) many of these studies conclude that geographical proximity to the knowledge source significantly amplifies spillovers between research and innovating firms. Strong evidence is provided both for the US (Acs et al., 2002; Jaffe et al., 1993; Varga 1998) and for Europe (e.g., Autant-Bernard, 2001, Fischer and Varga, 2003) that knowledge flows are bounded within a relatively narrow geographical range. Although certain industrial differences exists (such as for innovation in the microelectronics, instruments or biotechnology sectors proximity is more significant than for new technology development in the chemicals or the machinery industries) the hypothesis that spatial proximity is an important factor in innovation is strongly supported in the literature.

Varga (2000, 2001) provides empirical evidence that the spillover impact in knowledge production is positively related to the size of the region. Different types of agglomeration effects are at work to explain this phenomenon. Larger regions inhabit more firms connected by richer network linkages and as such the same knowledge generated by research in the area spills over to potentially more applications. Larger regions also offer a wider selection of producer services essential in technological innovation (e.g., information technology, legal, marketing services) contributing to a larger number of new technologies developed from the same knowledge base generated by (public and private) research in the area.

The new economic geography literature provides a general equilibrium framework where spatial economic structure is endogenously determined simultaneously with equilibrium in goods and factor markets (Fujita et al., 1999; Krugman,1991). This is a real breakthrough in economics given that before the appearance of the new economic geography no any school of economics since von Thünen' *Der Isolirte Staat* in the early nineteenth century had been able to build an economic model where the development of spatial structure is treated endogenously within a general equilibrium framework (Samuelson, 1983). The most recent models in the new economic geography incorporate the effects of knowledge spillovers on the formation of spatial economic structure as well as provide the first attempts to explicitly integrate the two 'new' schools of economics: the new growth theory and the new economic geography (Baldwin et al., 2003 Fujita and Thisse 2002). The need for the integration of the two schools is clear if one takes into account that agglomeration facilitates knowledge spillovers (according to the new economic geography) and knowledge spillovers determine per-capita GDP growth (according to the new growth theory) then it is not an unrealistic assumption that spatial economic structure affects macroeconomic growth.

Unfortunately, empirical investigations in the area of agglomeration and technological change are still relatively uncommon in the literature. The very few exceptions include Ciccone and Hall (1996), Ciccone (2002) and Varga and Schalk (2004). The following section presents the empirical modeling framework to integrate entrepreneurship and agglomeration into the explanation of technological change (and implicitly into the explanation of macroeconomic growth).

3.The empirical modeling framework

Our framework for empirical investigation is based on the Romer (1990) model of aggregate knowledge production as extended by Jones (1995). One of the most original contributions of Romer (1990) is the separation of economically useful scientific technological knowledge into two parts. The total set of knowledge consists of the subsets of non-rival, partially excludable knowledge elements that can practically be considered as public goods and the rival, excludable elements of knowledge. Codified knowledge published in books, scientific papers or in patent documentations belongs to the first group. This knowledge is non-rival since eventually it can be used by several actors at the same time and many times historically. On the other hand it is only partially excludable since only the right of applying a technology for the production of a particular good can be guaranteed by patenting while the same technology can spill over to further potential economic applications as others can study the patent documentation. Rival, excludable knowledge elements include the personalized (tacit) knowledge including particular experiences, insights developed and owned by the researchers themselves.

Equation (1) presents the manner the two types of knowledge interact in the production of economically useful new technological knowledge.

$$\mathring{A} = \delta H_A \lambda A^{\varphi} \tag{1}$$

where H_A stands for the number of researchers working on knowledge production in the business sector, A is the total stock of technological knowledge available at a certain point in time whereas $\mathring{A}$ is the change in technological knowledge resulting from private efforts to invest in research and development: δ, λ and φ are parameters. Equation (1) plays a central role in economic growth explanation since on the steady state growth path the rate of per capita GDP growth equals the rate of technological change ($\mathring{A}/A$).

Technological change is generated by research and it depends on the number of researchers involved in knowledge creation (H_A). However, their efficiency is directly related to the total stock of already available knowledge (A). Knowledge spillovers are central to the growth process: the higher A, the larger the change in technology produced by the same number of researchers. The same number of researchers with a similar value of A can raise the level of already existing technological knowledge with significant differences depending on the size of the parameters. Consider $\delta > 0$, which is the research productivity parameter. The larger is δ, the more efficient is H_A in producing economically useful new knowledge.

The size of φ reflects the extent to which the total stock of already established knowledge impacts knowledge production. Given that A stands for the level of codified knowledge (available in the academic literature or patent documentation), φ is called the parameter of codified knowledge spillovers. The size of φ reflects the portion of A that spills over and as such its value largely influences the effectiveness of research in generating new technologies. Hence the value of the aggregate codified knowledge spillovers parameter φ should be between 0 and 1.

However, not only codified but also non-codified, tacit knowledge can spill over as detailed in the previous section. The value of λ, in (1) reflects the extent to which tacit knowledge spills over within the research sector. The larger is λ in equation (1), the stronger the impact the same number of researchers plays in technological change. In contrast to φ and δ, which are determined primarily in the research sector and as such their values are exogenous to the economy, λ is endogenous because its value depends on the spatial pattern of economic activities and the level of entrepreneurial activity in the country. The size of λ in equation (1) relates to spatial economic structure in two ways. First, the higher the spatial concentration of research, the higher λ because spatial proximity promotes spillovers to a large extent. Second, as recent empirical studies suggest, the magnitude of localized knowledge spillovers from research is also influenced by agglomeration.

The value to λ is also influenced by the amount of entrepreneurial activity because the value of new economic knowledge is uncertain. While most R&D is carried out in knowledge creating institutions (large firms and universities) it does not mean that the same individuals that discover the opportunity will carry out the exploitation. An implication of the theory of firm selection is that new firms may enter an industry in large numbers to exploit knowledge spillovers. The higher the rate of new firm entry the greater should be the value of λ because of knowledge spillovers.

Based on the literature we assume that these spillovers are influenced largely by agglomeration as well as by the level of entrepreneurial activity in the country.To empirically investigate the extent to which entrepreneurship and agglomeration affect knowledge spillovers we develop an empirical model in which we endogenize the parameter λ in equation (1).

$$\log(\text{NK}) = \delta + \lambda \log(H) + \varphi \log(A) + \varepsilon \qquad (2)$$

$$\lambda = (\beta_1 + \beta_2 \log(\text{ENTR}) + \beta_3 \log(\text{AGGL})) \qquad (3)$$

where NK. stands for new knowledge (i.e., the change in A), ENTR is entrepreneurship, AGGL is agglomeration, A is the set of publicly available scientific-technological knowledge and ε is stochastic error term, Implementation of (3) into (2) results in the following estimated equation:

$$\begin{aligned}\log(\text{NK}) = \delta &+ \beta_1 \log(\text{H}) \\ &+ \beta_2 \log(\text{ENTR}) \log(H) \\ &+ \beta_3 \log(\text{AGGL}) \log(H) \\ &+ \varphi \log(A) + \varepsilon \end{aligned} \qquad (4)$$

In (4) the estimated values of the parameters β_2 and β_3 measure the extent to which research interacted with entrepreneurship and agglomeration contributes to knowledge creation.

In the estimation of (4) the units of observation are selected industrial sectors in European countries for the year 2001.[1] The selection of countries and sectors is determined by data availability. The number of patent applications operationalizes NK. Although patents are good indicators of new technology creation, they do not measure the economic value of these technologies (Hall et al., 2001). According to Griliches (1979) and Pakes and Griliches (1980, 378) "patents are a flawed measure (of innovative output) particularly since not all new innovations are patented and since patents differ greatly in their economic impact." However, a recent comparative study of direct innovation measures and patent applications evidence that patent applications are good proxy for innovation in econometric models (Acs et al, 2002).

R&D expenditures in Euro measure H. Data accessibility explains this choice to measure H in (1). An earlier comparison of R&D expenditures and employment data conclude that the two measures are highly correlated with each other and the results of regression analyses are not significantly different (Varga, 1998). This makes us confident to apply R&D expenditures in our study.

A is operationalized by the total number of available patents in all the sectors in the country (i.e., the total number of patents granted by inventors of the country in the last 20 years). The source of patent data is the OECD patent database. International patent classification (IPC) classes are assigned to industrial classes (ISIC Rev. 2) by the application of the MERIT concordance table developed by Verspagen, Moergastel and Slabbers (1994), Eurostat provides R&D expenditures data.

AGGL is measured by the agglomeration index that is calculated as the share of employment of city regions where the number of employees exceeds 500,000 in country total employment.[2] Eurostat provides regional employment data. ENTR is empirically measured by the total entrepreneurial activity (TEA) index developed within the framework of the GEM project.

The intent of GEM is to systematically assess two things: the level of start-up activity or the prevalence of nascent firms and the prevalence of new or young firms that have survived the start-up phase. First, start-up activity is measured by the proportion of the adult population (18–64 years of age) in each country that is currently engaged in the process of creating a nascent business. Second the proportion of adults in each country who are involved in operating a business that is less than 42 months old measures the presence of new firms. The distinction between nascent and new firms is made in order to determine the relationship of each to national economic growth. For both measures, the research focus is on entrepreneurial activity in which the individual involved have a direct but not necessarily full, ownership interest in the business. There are numerous ways to measure entrepreneurial activity. One important distinction is between opportunity-based entrepreneurial activity and necessity-based entrepreneurial activity. Opportunity entrepreneurship represents the voluntary nature of participation and necessity reflecting the individual's perception that such actions presented the best option available for employment but not necessarily the preferred option. Opportunity entrepreneurship differs from necessity by sector of industry and with respect to growth aspirations. Opportunity entrepreneurs expect their ventures to produce more high growth firms and provide more new jobs. The 16 European Union countries in 2001 had an average prevalence rate of about 8 percent (Reynolds et al., 2001).

To further explore this question of high potential entrepreneurs the 2002 GEM researchers added several new questions to the GEM protocol in order to isolate those ventures widely believed to have the greatest possibility for having a substantial impact on the economy. These new items were utilized to locate those ventures with potential to create new markets. Two additional criteria were added to further distinguish those new ventures with the potential to make a major contribution to the national economy: (1) the expectation of 20 or more jobs created within five years and (2) the intention to export goods or services. Of the 9615 star-ups and new firms identified in the 37 countries, only 926 met all of these criteria, about 9.6 percent (Reynolds et al., 2002).

4. Empirical results

In empirically estimating equation (4) two issues should get particular attention: multicollinearity (because H appears three times in the equation) and heteroskedasticity (since the expected heterogeneity of the country-industry dataset).

Table I presents empirical estimation results for equation (4). The equation is estimated by OLS. Standard errors are based on the White heteroskedasticity consistent covariance matrix estimator that provides correct estimates of the coefficient covariances in the presence of heteroskedasticity of unknown form (White 1980).

In Model 1 the estimated parameter of log(H) is highly significant. The logarithm of R&D expenditures explains 75 percent of the variations in the logarithm of patent applications at the country-industry level. The additional effect of log(A) where A is measured by the aggregate stock of available patents is considerable: it improves regression fit by 24 percent. Both coefficients are highly significant indicating that the original Romer (1990) equation captures well the main factors in technological change.

In Model 2 ENTR is measured by the high potential index. Extending Model 1 by the interaction terms Log(H)*Log(AGGL) and Log(H)*-Log(ENTR) improves regression fit only slightly.[3] The estimated parameter for Log(H)*log(ENTR) is significant at ($p < 0.05$) while the agglomeration effect is not significant in Model 2. This latter result might be the outcome of a shortcoming of the measurement applied (i.e., the agglomeration index is not sensitive to differences in the relative geographical positions of city regions within a country). Technical constraints could not make it possible for us to improve on this agglomeration measure. However, we also assumed that perhaps the United Kingdom as an outlier observation of AGGL might cause the unexpected result.[4]

To account for this effect in agglomeration a dummy variable (DUMUK) is included in Model 2a. The parameter of DUMUK is marginally significant ($p < 0.10$) whereas the interaction term's parameter becomes significant ($p < 0.05$). This supports the hypothesis that in Model 2a the high potential index enters the estimated equation with a highly significant parameter value ($p < 0.01$). Regression fit increases only slightly as compared to the original Romer equation in

TABLE I

Ols regression results for Log(Patent applications) for selected industries in selected European countries (N = 63, 2001)

	Model 1 Romer	Model 2 Extended Romer with TEA high potential	Model 2a Extended Romer with TEA high potential	Model 3 Extended Romer with TEA Opportunity	Model 4 Extended Romer with TEA Necessity	Model 5 Extended Romer with TEA
Constant	−3.175[a]	−2.843[a]	−2.966[a]	−3.711[a]	−3.506[a]	−3.513[a]
	(0.346)	(0.360)	(0.368)	(0.467)	(0.468)	(0.482)
Log (H)	0.298[a]	0.355[a]	0.442[a]	0.204[c]	0.396[a]	0.257[c]
	(0.061)	(0.074)	(0.080)	(0.102)	(0.079)	(0.134)
Log(H) *Log(AGGL)		0.024	0.086[b]	0.112[b]	0.100[c]	0.089[c]
		(0.032)	(0.042)	(0.046)	(0.054)	(0.047)
Log(H)*Log(ENTR)		0.069[b]	0.079[a]	0.143[b]	0.038	0.073
		(0.028)	(0.030)	(0.056)	(0.048)	(0.062)
Log(A)	0.723[a]	0.679[a]	0.698[a]	0.794[a]	0.775[a]	0.773[a]
	(0.063)	(0.055)	(0.056)	(0.072)	(0.077)	(0.075)
DUMUK			−0.740[c]	−0.852[b]	−0.716[c]	−0.711[c]
			(0.378)	(0.327)	(0.407)	(0.407)
R^2-adj	0.92	0.93	0.93	0.93	0.93	0.93
F-statistic	357[a]	189[a]	160[a]	162[a]	145[a]	147[a]

Note: TEA high potential index is available only for year 2002; White heteroscedasticity–consistent estimated standard errors are in parentheses.

[a] Denotes significance at least at 0.01.

[b] Denotes significance at least at 0.05.

[c] Denotes significance at least 0.10; variables are introduced in the main text.

Model 1. Estimated parameters of the Romer equation as well as their significance patterns are stable across models 1 to 2 These observations suggest that multicollinearity is not a serious issue in the model.

Table I also shows estimates for the different additional measures of entrepreneurship provided by the GEM project. Model 3 shows results for opportunity entrepreneurship, Model 4 gives the results for necessity entrepreneurship and Model 5 is the output with TEA. While in Model 3 the coefficient for opportunity entrepreneurship is positive and statistically significant, for both Model 4 and Model 5, the interaction term for entrepreneurial activity is not significant. This result is consistent with the fact that necessity entrepreneurship, having to become an entrepreneur because you have no better option, will not lead to technological change, although it may lead to employment growth. The same is true of entrepreneurial activity in general.

The estimated elasticity of technological change with respect to available codified knowledge is less than 1 (0.7) that corresponds to what is suggested by Jones (1995). The estimated research spillover effect in Model 2a is related to high potential entrepreneurship and agglomeration according to the following equation:

$$\lambda = 0.44 + 0.086^{*}\text{Log(AGGL)} + 0.079^{*}\text{Log(ENTR)} \quad (5)$$

What do these results suggest for entrepreneurship and agglomeration? Table II shows the coefficients for knowledge spillovers for the nine countries with and without the effect of entrepreneurship and agglomeration. As shown in column two agglomeration varies considerably from country to country with a high of 0.59 in the UK to a low of 0.14 in Poland. The TEA high potential index also varies from a low of 0.19 percent in Belgium to a high of 1.52 in Ireland. The elasticity of R&D spillovers with respect to new knowledge is 0.30 (Model 1 in Table I), This number is relatively small with respect to the 0.70 elasticity found for the total stock of knowledge.

One would suspect that entrepreneurship and agglomeration would be able to significantly raise the effect of R&D spillovers from new knowledge. This would especially be true if entrepreneurs played an important role in knowledge spillovers as suggested by the entrepreneurship literature. In Table II, column four shows the value of the coefficient of λ extended by entrepreneurship and agglomeration effects. The coefficient varies between 0.31 and 0.42 and is greater than 0.30 in all the countries in the sample. How should we interpret these results? Compared to what? The only other study we know of that measures the interaction between entrepreneurship and research and development is Michelacci (2003) who found that the value of λ for the United States varied between 0.24 and 0.48 for the post-war period. These findings are broadly consistent with those results. The last column shows the ratio of the extended coefficient divided by the non-extended coefficient. The ratio varies from 1.04 to 1.41.

TABLE II
Country coefficients with and without extension for entrepreneurship and agglomeration

	AGGL	TEA HIGH	Coefficient (Model 2a)	Coefficient (Model 1)	Coefficient ratio
Belgium	0.31	0.64	0.38	0.30	1.28
France	0.34	0.57	0.38	0.30	1.28
Germany	0.18	1.47	0.39	0.30	1.31
Hungary	0.24	0.76	0.38	0.30	1.27
Ireland	0.35	1.52	0.42	0.30	1.40
Italy	0.18	0.71	0.37	0.30	1.23
Poland	0.14	0.19	0.31	0.30	1.04
Spain	0.43	0.64	0.40	0.30	1.33
United Kingdom	0.59	0.99	0.42	0.30	1.41

These results suggest that increasing research and development expenditures, without increasing entrepreneurial activity, and or agglomeration effects, may not achieve the same result as if it was accompanied by entrepreneurial activity and agglomeration effects (Acs et al., 2004). After taking into account the effect of the stock of knowledge and research and development expenditures, both agglomeration and entrepreneurship have a weak positive effect on technological change. It is unlikely that a substantial high technological entrepreneurial sector will develop in the absence of broad national participation in entrepreneurship. If one wanted to increase technological change in the European Union more agglomeration of economic activity and more entrepreneurship may increase the amount of knowledge spillovers.

5. Conclusion

This paper developed an empirical model that endogenizes both entrepreneurial activity and agglomeration effects on knowledge spillovers within a Romerian framework. We tested a modified model of technological change to ascertain the impact of agglomeration effects and entrepreneurial activity on endogenous technological change. Specifically, we examine the impact of entrepreneurial activity and agglomeration effects on the spillover of new knowledge. The effect of agglomeration on technological change is positive and statistically significant. The effect of entrepreneurship on technological change is positive and highly significant. When the interactive terms are taken into account the regression fit increases only slightly. Consequently we found significant, but not too strong agglomeration and entrepreneurship effects on technological change for selected European countries. These results are broadly consistent with a growing body of literature suggesting that entrepreneurial activity and agglomeration play a positive role in knowledge spillovers and therefore in technological change and economic growth.

We also find that the endogenous growth model developed by Romer (1990) does a good job of modeling economic growth. We found support for Jones (1995) that the spillover effects from codified knowledge are less then one. There are several caveats that should be kept in mind. First, the GEM data might not be a good measure of either opportunity entrepreneurship in general or high technology entrepreneurship in particular. However, strong co-occurrence among these diverse measures indicated that the indices are reasonable measures of the overall level of different types of entrepreneurial activity. Second, we only have one year of data at the start of a recession. A longer time period may reveal different results. Finally, the inter relationship between agglomeration effects and entrepreneurial activity may be important. However, we did not model this in our paper and is an interesting topic for further research.

Notes

1 Industrial sectors include Chemistry and Pharmaceuticals, Computers and Office Machines, Electrical Machinery, Electronics, Instruments, Other Machinery, Transportation Vehicles. The following European countries are included; Belgium, Germany, France, Hungary, Ireland, Italy, Poland, Spain and the United Kingdom.

2 With the exception of Belgium and the UK where the highest level of data aggregation is NUTS 2 for the rest of the countries the agglomeration index is calculated using NUTS 3 level aggregated employment data.

3 Equation (4) forms the basis of our empirical investigations. This explains why we do not account for the direct effect of either ENTR or AGGL on technological changer.

4 A closer investigation of the agglomeration index suggests that economic activities exhibit an exceptionally high level of spatial concentration in the UK (relative to the rest of the countries in the sample). While the agglomeration index for the UK is 0.59 the corresponding average value for the rest of the sample countries is 0.27. According to this the UK economy appears to be more than two times concentrated geographically than the rest of Europe. This observation might well be the sign of aggregation bias as the lowest level of data aggregation for the UK is NUTS 2 while for the rest of the countries (with the exception of Belgium) data are aggregated at NUTS 3 level. While this does not seem to distort the value of the index for Belgium (i.e., the size of NUTS 2 regions are compatible with the average size of NUTS 3 regions in our sample) it might not be the case for the UK. Running separate regressions with and without the UK further supports the presence of regional data aggregation bias: with the exception of the 'agglomeration effect' the UK follow very similar patterns to the rest of Europe: parameter estimates and their significances are similar for $\log(H)$, $\log(A)$ and $\log(H)$*log (ENTR) This is not repeated for the geography effect: the number of patents in the UK exceeds the rest of the sample

average only slightly as it is 632 in the UK and 478 otherwise while AGGL is more than twice as high in the UK

Acknowledgements

We wish to thank Edward Malecki, participants at the first GEM conference in Berlin, Germany and at the international workshop on "Regions in Action: The Nexus of Innovation, Entrepreneurship and Public Policy" at the Tinbergen Institute in Amsterdam, the Netherlands, two anonymous referees and the editors of this special issue for valuable comments. The usual caveat applies.

References

Acs, Zoltan J., Luc Anselin, and Attila Varga, 2002, 'Patents and Innovation Counts as Measures of Regional Production of New Knowledge', *Research Policy* **31**, 1069–1085.

Acs, Zoltan J. and Catherine Armington, 2004, 'Employment Growth and Entrepreneurial Activity in Cities', *Regional Studies* **38**(8), 911–927.

Acs, Zoltan J. and David B. Audretsch, 2003, *Handbook of Entrepreneurship Research,* Boston: Kluwer Academic Publishers.

Acs, Zoltan J., David B, Audretsch, Pontus Braunerhjelm and Bo Carlsson, 2004, *The Missing Link: The Knowledge Filter and Entrepreneurship in Endogenous Growth,* Discussion Paper, No.4783, December, Center for Economic Policy Research, London, UK.

Aghion, Philip and Peter Howitt, 1992, 'A Model of Growth through Creative Destruction', *Econometrica* **60**, 323–351.

Arrow, Keneth, 1974, 'Limited Knowledge and Economic Analysis', *Amercian Economic Review* **64**(1), 1–10.

Audretsch, David, 1995, *Innovation and Industry Evolution,* Cambridge: The MIT Press.

Audretsch, David and Max Keilbach, 2004, 'Entrepreneurship Capital and Economic Performance', *Regional Studies* **38**(8), 949–959.

Autant-Bernard, Corinne, 2001, 'Science and Knowledge Flows: Evidence from the French Case', *Research Policy* **30**, 1069–1078.

Baldwin, Richard, Rikard Forslid, Philippe Martin, Gianmarco Ottaviano, and Frederic Robert-Nicoud, 2003, *Economic Geography and Public Policy* , Princeton: Princeton University Press.

Blau, David, 1987, 'A Time-series Analysis of Self-employment in the United States', *Journal of Political Economy* **95**, 445–467.

Bresnahan, Timothy, Alfonso Gambardella and Annalee Saxenian, 2001, 'Old Economy' Inputs for 'New Economy' Outcomes: Cluster Formation in the New Silicon Valleys', *Industrial and Corporate Change* **10**, 835–860.

Carree, Martin and Roy Thurik, 2003, 'The Impact of Entrepreneurship on Economic *Growth', Handbook of Entrepreneurship Research,* Boston: Kluwer Academic Publishers.

Casson, Mark, 1995, *Entrepreneurship and Business Culture,* Brookfield: Aldershot, US: Edward Elgar.

Caves, Richard, 1998, 'Industrial Organization and New Findings on the Turnover and Mobility of Firm', *Journal of Economic Literature* **36**, 1947–1982.

Cheng, Leonard K. and Elias Dinopoulos, 1992, 'Schumpeterian Growth and International Business Cycles', *American Economic Review* **82**, 409–414.

Ciccone, Antonio, 2002, 'Agglomeration Effects in Europe', *European Economic Review* **46**, 213–227.

Ciccone, Antonio, and Robert Hall, 1996, 'Productivity and the Density of Economic Activity', *American Economic Review* **86**, 54–70.

Fischer, Manfred and Attila Varga, 2003, 'Spatial Knowledge Spillovers and University Research: Evidence from Austria', *Annals of Regional Science* **37**, 303–322

Fujita, Masahisa and Jacques-François Thisse, 2002, *Economics of Agglomeration Cities, Industrial Location, and Regional Growth, Cambridge:* Cambridge University Press.

Fujita, Masahisa, Paul Krugman, and Anthony J. Venables, 1999, *The Spatial Economy,* Cambridge: MIT Press.

Griliches, Zvi, 1979, ' Issues in Assessing the Contribution of R&D to Productivity Growth', *Bell Journal of Economics* **10**, 92–116.

Hall, Bronwyn, Adam Jaffe, Manuel Trajtenberg, 2001, *The NBER Patent Citations Data File: Lessons, Insights and Methodological Tools,* WP 8498, National Bureau of Economic Research.

Holtz-Eakin, Doug and Chihwa Kao, 2003, *Entrepreneruship and Economic Growth: The Proof is in the Productivity,* Center for Policy Research, Syracuse University mineo.

Jaffe, Adam, Manuel Trajtenberg, Rebecca Henderson, 1993, 'Geographic Localization of Knowledge Spillovers as Evidenced by Patent Citations', *Quarterly Journal of Economics* **108**, 577–598.

Jones, Charles, 1995, 'R&D Based Models of Economic Growth', *Journal of Political Economy* **103**, 759–84.

Jovanovic, Boyan, 1982, 'Selection and Evolution of Industry', *Econometrica* **50**, 649–670.

Kirzner, Israel M., 1973, *Competition and Entrepreneurship,* Chicago: University of Chicago Press.

Krugman, Paul, 1991, 'Increasing Returns and Economic Geography', *Journal of Political Economy* **99**(3), 483–499.

Michelacci, Claudio, 2003, 'Low Returns in R&D due to the Lack of Entrepreneurial Skills', *The Economic Journal* **113**, 207–225.

Pakes, Ariel and Zvi Griliches, 1980, 'Patents and R&D at the Firm Level: A First Report', *Economics Letters* **5**, 377–381.

Polanyi, Michael, 1967, *The Tacit Dimension.* New York: Doubleday Anchor.

Reynolds, Paul D., S. Michael Camp, William D. Bygrave, Erkko Autio, and Michael Hay, 2001, *Global Entrepreneurship Monitor, 2001 Executive Report.*

Reynolds, Paul. D., William D. Bygrave, Erkko Autio, L. W. Cox, Michael Hay, 2002, *Global Entrepreneurship Monitor, 2002 Executive Report.*

Romer, Paul, 1986, 'Increasing Returns and Long-run *Growth* ', *Journal of Political Economy* **94**, 1002–1037.

Romer, Paul, 1990, 'Endogenous Technological Change', *Journal of Political Economy* **98**, S71–S102.

Romer, Paul, 1996, 'Why, Indeed, in America? Theory, History, and the Origins of Modern Economic Growth', *American Economic Review* **86**, 202–207.

Samuelson, Paul, 1983, 'Thunen at Two Hundred', *Journal of Economic Literature* **21**, 1468–1488.

Schmitz, James, 1989, 'Imitation, Entrepreneurship, and Long-Run Growth', *Journal of Political Economy* **97**, 721–739.

Schumpeter, Joseph, 1934, *The Theory of Economic Development,* Cambridge, MA: Harvard University Press.

Schumpeter, Joseph, 1947, 'The Creative Response in Economic History', *Journal of Economic History* **7**, 149–159.

Segerstrom, Paul, 1991, 'Innovation, Imitation and Economic Growth', *Journal of Political Economy* **99**, 190–207.

Segerstrom, Paul, T. C. A. Anant and Elias Dinopoulos, 1990, 'A Schumpeterian Model of the Product Life Cycle', *American Economic Review* **80**, 1077–1091.

Shane, Scott, 1996, 'Explaining Variation in Rates of Entrepreneurship in the United States 1899–1988', *Journal of Management* **22**, 747–781.

Shane, Scott and Sankaran Venkataraman, 2000, 'The Promise of Entrepreneurship as a Field of Research', *Academy of Management Review* **25**, 217–221.

Solow, Robert, 1957 'Technical Change in an Aggregative Model of Economic Growth', *International Economic Review* **6**, 18–31.

Sutton, John, 1997, 'Gibrat's Legacy', *Journal of Economic Literature* **35**, 40–59.

Varga, Attila, 1998, *University Research and Regional Innovation: A Spatial Econometric Analysis of Academic Technology Transfers, Boston* : Kluwer Academic Publishers.

Varga, Attila, 2000, 'Local Academic Knowledge Spillovers and the Concentration of Economic Activity', *Journal of Regional Science* **40**, 289–309.

Varga, Attila, 2001, 'Universities and Regional Economic Development: Does Agglomeration Matter?', in Johansson Bent, Karlsson Charlie and Stough Roger (eds.), *Theories of Endogenous Regional Growth –Lessons for Regional Policies,* Berlin: Springer, pp. 345–367.

Varga, Attila and Hans J. Schalk, 2004, 'Knowledge Spillovers, Agglomeration and Macroeconomic Growth: An Empirical Approach', *Regional Studies* **38**, 977– 989.

Verspagen, Bart, T. Moergastel and M. Slabbers, 1994, 'MERIT Concordance Table: IPC ISIC (rev. 2)', *MERIT Research Memorandum 2/94/004,* Maastricht Economic Research Institute on Innovation and Technology, University of Limburg.

White, Halbert L., 1980, 'A Heteroskedasticity-Consistent Covariance Matrix and a Direct Test for Heteroskedasticity', *Econometrica* **48**, 817–838.

Part VII
The Entrepreneur

[32]

Entrepreneurship

Edward P. Lazear, *Hoover Institution and Stanford University*

The theory below is that entrepreneurs must be jacks-of-all-trades who need not excel in any one skill but are competent in many. A model of the choice to become an entrepreneur is presented. The primary implication is that individuals with balanced skills are more likely than others to become entrepreneurs. Using data on Stanford alumni, the predictions are tested and found to hold. Those who have varied work and educational backgrounds are much more likely to start their own businesses than those who have focused on one role at work or concentrated in one subject at school.

What is entrepreneurship? Economic growth may be related to the formation of new businesses, but what is it exactly that entrepreneurs do?[1] The view of entrepreneurship taken here is that it is the process of assembling necessary factors of production consisting of human, physical, and information resources and doing so in an efficient manner. Entrepreneurs put people together in particular ways and combine them with physical capital and ideas to create a new product or to produce an existing

I thank first Boyan Jovanovic for a useful derivation that is used in this article. Comments from Anat Admati, Uschi Backes-Gellner, George Baker, Gary Becker, Lanier Benkard, Jennifer Hunt, Andrea Ichino, Stephen Jones, Kenneth Judd, David Kreps, Robert Lucas, John McMillan, Paul Oyer, Canice Prendergast, Korok Ray, John Roberts, Kathryn Shaw, Steven Tadelis, Robert Wilson, Justin Wolfers, and Jeffery Zwiebel are gratefully acknowledged. Able research assistance was provided by Yuliy Sannikov on theoretical issues and Eugene Kwok and Zeynep Emre on data compilation.

[1] Lazear (1995) found that those Eastern European economies that grew the fastest during the transition from communism to market economies were those for which new business formation was most rapid.

[*Journal of Labor Economics,* 2005, vol. 23, no. 4]

0734-306X/2005/2304-0001$10.00

one at a lower or competitive cost. Because the entrepreneur must bring together many different resources, he or she must have knowledge, at least at a basic level, of a large number of business areas. An entrepreneur must possess the ability to combine talents and manage those of others. Why do some choose to become entrepreneurs, and what characteristics create successful ones? Most of the past work on entrepreneurship has been empirical,[2] but it is useful to have theory to guide the empirics and to assist in interpretation of the results.[3]

It is tempting to argue that the most talented people become entrepreneurs because they have the skills required to engage in creative activity. Perhaps so, but this flies in the face of some facts. The man who opens up a small dry-cleaning shop with two employees might be termed an entrepreneur, whereas the half-million-dollar-per-year executive whose suit he cleans is someone else's employee. It is unlikely that the shop owner is more able than the typical executive.

The reverse might be true. As necessity is the mother of invention, perhaps entrepreneurs are created when a worker has no alternatives. Rather than coming from the top of the ability distribution, they are what is left over.[4] This argument also flies in the face of some facts. Any ability measure that classifies John D. Rockefeller, Andrew Carnegie, or, more recently, Bill Gates near the bottom of the distribution needs to be questioned.

The idea explored below is that entrepreneurs differ from specialists in that entrepreneurs have a comparative disadvantage in a single skill but have more balanced talents that span a number of different skills. Specialists can work for others who have the talent to spot and combine a variety of skills, but an entrepreneur must possess that talent. Although entrepreneurs can hire others, the entrepreneur must be sufficiently well versed in a variety of fields to judge the quality of applicants.

How shall we define *entrepreneur*? There are a number of possible definitions. In keeping with the empirical analysis to be performed below, an entrepreneur is defined for this study as someone who responds af-

[2] See, e.g., the early paper by Evans and Leighton (1989) and the more recent one by Hamilton (2000), who examines the returns to self-employment.

[3] The theoretical papers on the subject rarely speak to the issue that is central to this article. For example, Otani (1996) examines the theoretical relation of firm size to entrepreneurial ability. Perhaps the closest to this study in terms of discussing specialization (although from a very different point of view) is Holmes and Schmitz (1990), where it is argued that certain agents specialize in entrepreneurial skills. This differs from the approach here, where entrepreneurial skills are implicitly defined to be a cross section of all possible skills. De Meza and Southey (1996) build a model where new entrants are excessively optimistic.

[4] Landier (2002) argues that the part of the ability distribution from which entrepreneurs are drawn may differ across countries and provides a multiple equilibrium approach in an information framework to discuss the differences.

firmatively to the question "I am among those who initially established the business." Such individuals, even if they leave the business early, are usually responsible for the conception of the basic product, hiring the initial team, and obtaining at least some early financing. Other definitions are possible. For example, CEOs who "reinvent" a company might also consider themselves entrepreneurs. Conceptually, the model is consistent with including this latter group in the collection of entrepreneurs, but they will be excluded (with one exception) in the empirical analysis. The definition is conceptually distinct from "self-employed." A self-employed person need not have any other employees, and the kinds and combinations of skills that are necessary for real entrepreneurship are less important for, say, a self-employed handyman who works alone. At the empirical level, self-employed individuals are entrepreneurs if they view themselves as having started a business.

The model presented below is one where an individual can decide to become an entrepreneur using a variety of skills or to specialize using only one. The model is tested using data on graduates from the Stanford Graduate School of Business. The data combine information on post-graduate work experience and incomes with courses taken and grades obtained when the individuals were attending the Stanford Graduate School of Business.

The primary theoretical predictions are:

1. Individuals with more balanced skill sets are more likely to become entrepreneurs.
2. The supply of entrepreneurs is smaller for production processes that require a higher number of independent skills.
3. Individuals who become entrepreneurs should have a more balanced human capital investment strategy on average than those who become specialists.

A number of the predictions are tested empirically using data on Stanford alumni and are borne out. Specifically, those who end up being entrepreneurs are more likely to have varied backgrounds, both in school and on the job. The probability of being an entrepreneur in an employment spell is positively related to the number of different roles that an individual has had over his or her career and with the generality of the curriculum followed when at school. Some of this reflects balanced skills that the individual possesses before entering the labor market and some reflects the balance that is acquired after entering the workforce. There is also some evidence that risk tolerance enters into the decision to become an entrepreneur. Specifically, those individuals who have displayed willingness to choose risky occupations in the past are more likely to become entrepreneurs.

A Model of the Choice to Become an Entrepreneur

Initially, let there be only two skills, denoted x_1 and x_2 with $x_1, x_2 \geq 0$. An individual can be a specialist, in which case he receives income associated with his best skill, or he can be an entrepreneur, in which case he is limited by his weakest attribute. Thus, for specialists,

$$\text{Specialist income} = \max[x_1, x_2]. \quad (1)$$

Entrepreneurs, however, must be good at many things. Even if they do not do the job themselves, they must know enough about a field to hire specialists intelligently. The jack-of-all-trades aspect of entrepreneurship is captured in the income function

$$\text{Entrepreneur income} = \lambda \min[x_1, x_2], \quad (2)$$

where λ is a parameter that is determined by market equilibrium that establishes the value of an entrepreneur. The value of λ, which is called the market value of entrepreneurial talent, will be derived below.

The income-generating functions described in (1) and (2) may seem special, but they are implied by a general production function that exhibits constant returns to scale and ability distributions that are symmetric. This is shown in the appendix, entitled "The Underlying Economy."[5] Creativity and willingness to take risks are two factors that are often mentioned as affecting the decision to become an entrepreneur.[6] Creativity is suppressed in this model because it is unobservable. Formally, more creative individuals can be thought of as those with larger values of λ. They have higher market values for entrepreneurial talent because a given amount of raw skill translates into more entrepreneurial output. Risk preference is simply ignored in this model where everything other than endowment of x_1 and x_2 is deterministic.[7] Risk can be thought of as entering through a stochastic λ.

[5] The perfect substitutes/perfect complements income function is extreme, but it is an expository convenience. Any production process that has complementarity of skills for entrepreneurs and substitution of skills for specialists would be consistent with the intuition.

[6] Kihlstrom and Laffont (1979) were the first to argue that entrepreneurs tend to be less risk-averse than others in society. Iyigun and Owen (1998) suggest that entrepreneurship is risky and risk-averse agents are less likely to go into entrepreneurship in a developed economy where a larger selection of safer (insured) jobs exists.

[7] Becker and Murphy (1992) use a similar notion of specialization. Becker and Mulligan (2002) apply a technology somewhat like the one in this article to discuss the difference between market (specialized) and household (generalized) work.

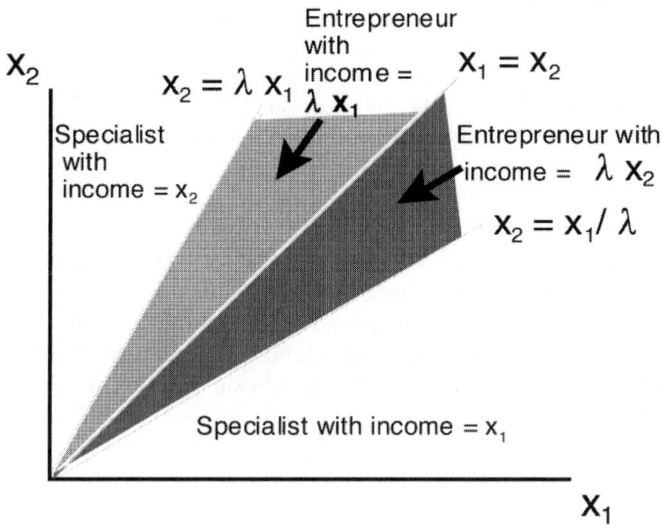

FIG. 1

Who Becomes an Entrepreneur?

Who decides to become an entrepreneur and who decides to become a specialist? The decision is straightforward. Think of individuals as being endowed with a pair of skills (x_1, x_2).[8] The joint density on x_1 and x_2 is given by $g(x_1, x_2)$. The individual chooses to become an entrepreneur if and only if

$$\lambda \min[x_1, x_2] > \max[x_1, x_2]. \tag{3}$$

It is easiest to see this graphically. A given individual is endowed with x_1 and x_2, shown as a point in figure 1. For all points below the 45° line, $x_1 > x_2$, so that a specialist whose endowment lies below the 45° line would always choose to specialize in x_1 and would have income given by x_1; x_2 is irrelevant to this specialist. In order for that individual to prefer being an entrepreneur to being a specialist, it is necessary that

$$\lambda \min[x_1, x_2] > \max[x_1, x_2],$$

[8] Lucas (1978) offers a model in which an individual can choose to work for someone or to be an entrepreneur. The difference between Lucas's model and this one is that, in Lucas, managerial talent is distinct from labor talent. Here, workers and managers have the same two skills, just in different combinations. The complementarity between skills that is the essence of this story is absent from Lucas. Still, Lucas derives implications for the size distribution of firms that are similar to those derived below.

which here requires that

$$\lambda x_2 > x_1$$

because $\min[x_1, x_2] = x_2$ and $\max[x_1, x_2] = x_1$.

Thus, for individuals for points below the 45° line, the condition for entrepreneurship is

$$x_2 > x_1/\lambda. \tag{4}$$

This is shown as the shaded area on the diagram between the lines $x_1 = x_2$ and $x_2 = x_1/\lambda$. The area below the line $x_2 = x_1/\lambda$ corresponds to points where the individual specializes and receives income x_1.

Above the 45° line, the converse is true. Here, $x_2 > x_1$ so the specialist receives income x_2. In these cases, the condition for entrepreneurship, that

$$\lambda \min[x_1, x_2] > \max[x_1, x_2]$$

becomes

$$\lambda x_1 > x_2,$$

so an individual for whom x_2 exceeds x_1 becomes an entrepreneur when

$$x_2 < \lambda x_1. \tag{5}$$

This is shown as the cross-hatch area in the diagram. The region in the northwest corner corresponds to individuals who have sufficiently high values of x_2 relative to x_1 that it pays for them to specialize in x_2 and to receive income x_2.

The probability of becoming an entrepreneur for any λ is given by the probability that the pair of skills lies in one of the two shaded areas in figure 1 or

$$\text{prob of entrepreneur} = \int_0^\infty \int_{x_1/\lambda}^{\lambda x_1} g(x_1, x_2) dx_2 dx_1. \tag{6}$$

It is now possible to derive and explain intuitively how the entrepreneurial decision varies with a number of different parameters. First, consider λ, the market value of entrepreneurial talent. Differentiate (6) with respect to λ to obtain

$$\frac{\partial \text{ prob}}{\partial \lambda} = \int_0^\infty \left[g(x_1, \lambda x_1) x_1 + g\left(x_1, \frac{x_1}{\lambda}\right)\frac{x_1}{\lambda^2}\right] dx_1,$$

which is positive.

The higher is λ, the more likely is the individual to become an entrepreneur. Diagrammatically, as λ increases, the shaded areas become larger because the borders move toward the axes. If λ were infinity, everyone would become an entrepreneur since for any positive values of x_1 and x_2 entrepreneurial income would be infinite. As λ goes to one, the shaded areas get pinched. When $\lambda = 1$, the borders of the shaded area are the line $x_1 = x_2$, and there are no entrepreneurs. Obviously, if $\lambda = 1$, it is impossible for condition (3) to hold since the min of something can never exceed the max of that same thing.

This result is important for equilibrium. The market value of entrepreneurial talent, λ, is a parameter that determines the supply of entrepreneurs in an economy. As λ rises, everyone chooses to become an entrepreneur. As λ falls to one, no one opts for entrepreneurship. This will guarantee an interior solution for λ and will ensure that there is a finite number of individuals wanting to enter entrepreneurship.

It is also possible to think of λ as being person specific. Some individuals have a comparative advantage in entrepreneurship. This might relate to creativity or other skills, but it is reflected in high values of λ. Since such talents are generally unobservable, not much more is said about the idiosyncratic variation in λ.[9]

If λ is person specific, it might also be unknown. This would explain why some people try entrepreneurship and then switch back to working for others. Perhaps individuals must try being an entrepreneur before they know whether they have a natural talent for running their own business. There is some evidence of this. There is a significant fraction of individuals who move back to working for others after having been an entrepreneur, but the probability is about 10 times as high that a year as an entrepreneur is succeeded by another year as an entrepreneur than by a year working for someone else.

Balance

In what follows, it is shown that as the correlation between x_1 and x_2 rises, the supply of entrepreneurs increases. Before deriving this formally, we state the intuition. Since entrepreneurial output and income is determined by the weakest link, it does little good to have a high value of x_1 if x_2 is not also high. Under such circumstances, it is necessary that x_2 be high whenever x_1 is high or there is little chance that an individual

[9] One of the skills can be interpreted as the ability to raise capital. This argument is central to Evans and Jovanovic (1989). Holtz-Eakin, Joulfaian, and Rosen (1994) show that capital is important in starting a business by linking the receipt of an inheritance to the likelihood of starting a business. Recent work by Gentry and Hubbard (2002) explores the relation of saving to entrepreneurial investment. Their motivation is growth and macroeconomic factors, but the results are relevant to this study as well. They find that there is an interdependence between entrepreneurial saving and investment.

will become an entrepreneur. Diagrammatically, for any given λ, a larger proportion of the population prefers to be entrepreneurs, the more points lie in the shaded area of figure 1. The shaded area consists of points where x_1 and x_2 are close in value. For small values of λ, only points very close to the $x_1 = x_2$ line result in choosing to become an entrepreneur. If most of the mass of the distribution lies close to the axes, then individuals will be inclined to specialize in one or the other skill because they have a strong absolute advantage in one skill. Entrepreneurs are jacks-of-all-trades, which means that they must be relatively good (or relatively bad) at everything.[10]

Formally, let x_2 be defined in terms of x_1 as follows:

$$x_2 = \rho x_1 + (1 - \rho)\nu,$$

where x_1 has density $f(x_1)$ and ν has density $h(\nu)$. When $\rho = 1$, x_1 and x_2 are perfectly correlated. When $\rho = 0$, they are uncorrelated. In fact, ρ is the correlation coefficient between x_1 and x_2. The probability of being an entrepreneur in (6) can be rewritten as

$$\text{prob of entrepreneurship} = \int_0^\infty \int_{[(x_1/\lambda)-\rho x_1]/(1-\rho)}^{(\lambda x_1 - \rho x_1)/(1-\rho)} f(x_1)h(\nu)d\ \nu dx_1 \qquad (7)$$

by using a standard change of variables and altering the limits of integration appropriately.

Next, differentiate (7) with respect to ρ to obtain

$$\frac{\partial}{\partial \rho} = \int_0^\infty \left[h(UL)\frac{\partial UL}{\partial \rho} - h(LL)\frac{\partial LL}{\partial \rho}\right] f(x_1)dx_1,$$

where UL and LL stand for upper and lower limits of the inside integral in (7). After substitution, this becomes

$$\frac{\partial}{\partial \rho} = \int_0^\infty \left[h(UL)\frac{x_1(\lambda - 1)}{(1-\rho)^2} + h(LL)\frac{x_1(1 - 1/\lambda)}{(1-\rho)^2}\right] f(x_1)dx_1,$$

which is positive since density functions are always positive and since $\lambda > 1$ for there to be any entrepreneurs in the economy at all. Thus, as

[10] Stopford and Baden-Fuller (1994) list five components (proactiveness, team orientation, dispute resolution skills, being innovative, and ability to learn) that are important in entrepreneurship. Thus, an entrepreneur might be someone who was highly endowed with each of the five factors.

correlation increases between the two variables, the proportion of entrepreneurs rises.

The idea that balance is important suggests that the supply of entrepreneurship may vary by industry. For example, consider insurance agencies. The ability to understand complex insurance policies is a skill that is likely to be correlated with the accounting and management skills necessary to run a business. As a result, there are many who are well suited to running their own agencies and so the number of agencies should be great and their average size small.

An alternative example involves artists. Because art and business are quite distinct skills, there is no reason to expect strong positive correlation between artistic talent and business skills. As long as both artistic and business skills are relevant for production in the art business, then few will have high enough levels to avoid specializing in one or the other aspect of the business. Thus, the supply of entrepreneurial talent in art would be expected to be low, so most artists must be managed by others. The prediction is that there would be very few artists who run their own studios and publicize their own work.

The empirical statements are verifiable by looking at real world data.[11] In situations where entrepreneurs are rare, a few must run the whole industry, driving up concentration ratios. In situations where many opt to be entrepreneurs, the concentration ratios should be low. Of course, other technological considerations are key here and must be held constant. If scale economies are more important in some industries (e.g., automobiles) than in others (e.g., restaurants), the concentration ratios are likely to be higher in the former than the latter, independent of entrepreneurial supply considerations.

Complexity of Production

Some production processes are very complex, requiring many skills in order to produce output. Others are relatively straightforward. As the world has become complex, a larger variety of skills may be required to be an entrepreneur. In an agrarian society, a farmer did not require too many business skills to run his small farm and get his produce to market. The founders of the modern corporation are a different breed. They are more than competent technicians; they must understand how to create a worldwide business.

What happens to the supply of entrepreneurs as the number of factors increases? Without being more specific about the distribution of the factors, it is impossible to make qualitative statements. However, it is possible

[11] To make statements about groups, it is necessary to show that the propositions are true in a statistical sense at the level of the population. This is derived in the appendix.

to show that the introduction of independent factors always reduces the supply of entrepreneurs.

Consider the original joint density $g(x_1, x_2)$. Now introduce a third factor, x_3, and let the density of the three be denoted $k(x_1, x_2, x_3)$. If x_3 is an independent factor with (marginal) density $m(x_3)$, then it is possible to write

$$k(x_1, x_2, x_3) = \int m(x_3)\left\{\int \int g(x_1, x_2)dx_2 dx_1\right\}dx_3.$$

The condition necessary to ensure an entrepreneur for two variables must still hold. For any given x_3 and for a given λ, if the projection onto the (x_1, x_2) plane does not lie in the entrepreneurial area, the individual will not choose to be an entrepreneur. That is, if

$$\lambda \min [x_1, x_2] < \max [x_1, x_2],$$

the individual becomes a specialist, irrespective of x_3. In addition, there are some potential cutoff values, x_3^* and x_3^{**}, that are also required for entrepreneurship. So the probability of being an entrepreneur cannot exceed

$$\int_{x^*_3}^{x_3^{**}} m(x_3)\left\{\int_0^{\infty} \int_{x_1/\lambda}^{\lambda x_1} g(x_1, x_2)dx_2 dx_1\right\}dx_3,$$

which can be written as

$$\{M(x_3^{**}) - M(x_3^*)\} \int_0^{\infty} \int_{x_1/\lambda}^{\lambda x_1} g(x_1, x_2)dx_2 dx_1.$$

Since the first term cannot exceed one, the probability of being an entrepreneur cannot be higher with three factors than with two and, in general, must be lower.

The proof can be repeated, adding one factor at a time. Therefore, the supply of entrepreneurs falls as the production process requires more independent skills. More individuals can run family farms than multinational, multiproduct conglomerates.

There is some relation between correlation of skills and complexity. For example, it is possible to think of being an accountant as requiring only one skill, namely, accounting, which can be used to serve clients or run the business. A skill is necessarily perfectly correlated with itself, so the probability that individuals possess all the skills necessary to run an accounting business, other things equal, might be thought to be higher than that associated with running an art business, where two distinct skills are required.

The Premium to Entrepreneurship

It is now straightforward to show that an equilibrium λ always exists. To make things simple, but without loss of generality, suppose that there are a fixed number of firms in an economy and each firm requires one and only one entrepreneur. Then the demand for entrepreneurs is perfectly inelastic at q^*, where q^* is the number of entrepreneurs demanded. Let the number of individuals in the labor force be given by N. Then, using (6), which defines the probability of being an entrepreneur as a function of λ, the supply of entrepreneurs is simply

$$N\int_0^{\infty}\int_{x_1/\lambda}^{\lambda x_1} g(x_1, x_2)dx_2dx_1.$$

Market equilibrium occurs when λ is set such that

$$N\int_0^{\infty}\int_{x_1/\lambda}^{\lambda x_1} g(x_1, x_2)dx_2dx_1 = q^*. \tag{8}$$

Equation (8) is one equation in one unknown, namely λ, which determines the equilibrium value of entrepreneurship. The market value of entrepreneurial talent adjusts to induce enough individuals to become entrepreneurs so that demand is satisfied. When $\lambda = 1$, no one chooses to be an entrepreneur. As $\lambda \to \infty$, all choose to be entrepreneurs so there must be λ that sustains N^* as the equilibrium number of entrepreneurs. This is true for any demand for entrepreneur function. There is always an intersection of demand with supply, although corner solutions are possible.

Investment

So far, x_1 and x_2 have been taken as given. But much of economic activity as it relates to labor markets involves investment in skills. It is important to take investment in skills into account both for the purposes of completing the theory and in order to allow predictions for empirical analysis.

Augment the previous model by defining x_1^0 as the initial stock of skill x_1, x_2^0 as the initial stock of skill x_2, and x_1 and x_2 as the (final) attained level. Let a particular individual, with endowed skills (x_1^0, x_2^0), obtain levels of x_1, x_2, according to the cost function

$$C(x_1, x_2),$$

with C_1, $C_2 > 0$, $C_{ii} > 0$.

Define x_1 to be the skill with which the individual is endowed the largest amount. This means that a worker who chooses to specialize is likely to specialize in x_1 and will solve

$$\max_{x_1} \; x_1 - C(x_1, x_2)$$

with first-order condition

$$1 - C_1(x_1, x_2) = 0.$$

Someone who is going to specialize will invest in only one of the two skills. There is no value to augmenting a skill that will not be used. It is possible that C_2 is sufficiently low relative to C_1 that the individual will ignore his higher endowment of x_1 and instead specialize in x_2. This is of little importance. Essential here is that the specialist invests in one or the other, but not both.

Now consider an individual who is going to become an entrepreneur. His constraint is the minimum skill, defined to be x_2. Should the aspiring entrepreneur invest in x_1, in x_2, or in both?

Since the constraint is x_2, there is no point in investing in x_1 unless x_2 is brought up at least to the level of x_1. If there is an interior solution for x_2, then it satisfies

$$\lambda - C_2(x_1, x_2) = 0.$$

There are three possibilities, but they can be dealt with quickly. If $C_2(x_1^0, x_2^0) > \lambda$, then it does not pay for the individual to increase his stock of x_2 and so no investment occurs. (It surely does not pay to increase x_1 since there is already an excess of x_1 at x_1^0.) If $C_2(x_1^0, x_2^0) < \lambda$, but $C_2(x_1^0, x_1^0) > \lambda$, the individual will invest only in x_2 because it does not pay even to bring x_2 up to the endowed level of x_1. (There is no advantage to augmenting x_1 until x_2 has reached the level of x_1.) In this case, the individual specializes in investment in x_2 and behaves identically to a specialist, except that he invests in the skill in which he is weak instead of the skill in which he is strong, which is the more common case for the specialist. Finally, if $C_2(x_1^0, x_1^0) < \lambda$, then it pays for the individual to exceed x_1^0 in attained x_2. But now x_1 becomes the constraint. As long as $C_1(x_1^0, x_1^0) < \lambda$, the individual benefits by increasing his investment in x_1 as well and continues to do so, but the optimum must have $x_1 = x_2$ in this case. What is important, however, is that, in this situation, the individual does not look like a specialist; he invests in more than one skill.

Investment can take a number of forms, the most important of which is formal schooling and on-the-job training. Thus, those who eventually become entrepreneurs should not specialize in skill acquisition, and this might be reflected in taking a wide variety of courses.

Additionally, individuals who will eventually become entrepreneurs should take on a variety of jobs to acquire the skills necessary to become an entrepreneur. Thus, an individual might spend some time working in a financial role, some time in human resources, some time as a manager, some time as a skilled staff worker, and so forth. Having a large variety of roles is a standard way to acquire a variety of skills and is the method used for workers where the intention is to create a multiskilled workforce.

To summarize, those who are going to specialize invest in only one skill. Those who become entrepreneurs may invest in one skill, but if they do so, it will be the skill in which they are weak. But entrepreneurs are the only individuals who may invest in more than one skill. To put this in somewhat less stark terms, individuals who become entrepreneurs should have a more balanced investment strategy on average than those who end up specializing as wage and salary workers.

Innovation

No reference has been made to innovation. When thinking of the truly successful entrepreneurs, individuals who had some new idea usually come to mind. Even in a traditional industry such as retail, a founder like Sam Walton of Walmart used a new business process that allowed his firm to undercut the competition. The generalist view of entrepreneurship deemphasizes innovation, although it is not inconsistent with it. One interpretation of the value of having multiple skills is that it is easier to innovate when the entire situation can be seen. A technical engineer may be superb at creating a new device, but that device may not have any business value. The innovator who succeeds is the one who can come up with something that is not only technically sound but business relevant as well.

Empirical Analysis

There are a number of implications that have been suggested in the theory section above. Before going to direct tests of the model, it is useful to provide some information on the composition of entrepreneurs.

Who Are the Entrepreneurs?

To examine the issues discussed in the theory section more directly, a unique data set will be used. In the late 1990s, Stanford surveyed its Graduate School of Business alumni (from all prior years). The primary focus of the survey was compiling a job history for each of the graduates, with special emphasis on information about starting businesses.[12] This resulted in a sample of about 5,000 respondents. In addition to the detailed job histories, these data were matched with the student transcripts so that it is possible to see which courses were taken by those who went on to be entrepreneurs and which by those who became specialists. Additionally, the grade obtained in each of the courses taken is reported in the data.

The Stanford MBA data make clear that at least among this population, most entrepreneurs are not in technical fields. In table 1, the industries in which Stanford MBAs are found are dominated by construction and real estate, retail trade, and business consulting of some form. The technical fields are much less important. This finding is less surprising in these data, given that the more technical people would likely be found in engineering rather than MBA programs at Stanford. The Stanford MBA program recruits a significant fraction of engineers and students with technical backgrounds but fewer than would be found in those technical departments themselves. However, a comparison with the March 2002 Current Population Survey displays the same pattern. Among incorporated self-employed (which eliminates most household and other service workers who work alone), the dominant industries are, in order, construction, retail trade, professional services, business services, and real estate. Again, technical fields are minor players.

The basic hypothesis is that entrepreneurs are jacks-of-all-trades. In the "Investment" section above, it was shown that individuals who want to become entrepreneurs invest in a broader range of skills than do those who want to become specialists. Going into any job, individuals with a broader range of skills, acquired either through investment or through endowments, are more likely to be entrepreneurs.[13]

The data allow this hypothesis to be tested. The data set is a job history panel so that each respondent has one row of data corresponding to each employer (including self) that he or she has held. For example, an individual who had six employment spells would have six rows of data, one for each spell. An individual who had four employment spells and one spell of unemployment would have five rows of data. The beginning and ending dates for each job are recorded, as are the beginning and ending salary and size of firm. Additionally, all roles within the employment spell (up to five) are described through a coding system that corresponds to occupational titles. Industry and demographic data are also provided.

[12] The response rate was 40%. Some individuals were very old, and others were no longer alive, which accounts for some of the nonresponses.

[13] Lentz and Laband (1990) find that there is a higher likelihood of self-employment among the children of the self-employed. They interpret this as human capital that is passed from one generation to the next. There are also papers on the link between education and entrepreneurship. See, e.g., Bates (1985, 1990).

Table 1
Industrial Breakdown of Entrepreneurs in the Stanford MBA Sample

Industry Code	Percent
Management Consulting	14.51
Construction/Real Estate Development	8.07
Investment Management	5.87
Retail/Wholesale	5.08
Venture Capital	5.08
Hardware/Software/Systems Services	4.86
Investment Banking/Brokerage	4.23
Real Estate Finance	4.01
High Tech—Computers/Software	3.73
Entrepreneurial Services	3.67
Consumer Products	2.65
Entrepreneurial Manufacturing	2.32
Entertainment/Leisure/Sports	2.20
Food/Lodging	2.03
Marketing Services	1.69
Diversified Financial Services	1.58
Diversified Service	1.58
Printing/Publishing	1.58
High Tech—Other	1.36
Health Care Services	1.30
Extractive Mineral/Natural Resources	1.30
Telecommunications Services	1.24
Foundation/Nonprofit Organizations	1.13
Agriculture	1.02
Medical Instruments and Devices	1.02
Accounting	.90
Import/Export/International Trade	.90
Commercial Banking	.85
Education	.79
Radio/TV/Cable/Film	.79
High Tech—Telecommunications Products	.79
Energy	.73
High Tech—Computers/Hardware	.73
Industrial Equipment	.73
Advertising	.68
Insurance	.68
Transportation Services/Shipping	.68
Legal Services	.62
Multimedia Services	.62
High Tech—Multimedia Products	.56
Rubber/Plastics	.56
Biotechnology	.51
Diversified Manufacturing	.45
Public Relations	.34
Chemical	.34
High Tech—Consumer/Electronics	.34
High Tech—Semiconductors	.34
Architecture	.28
Arts	.28
Environmental/Waste Management/Recycling	.28
Social Services	.28
Unknown	.28
Aerospace	.28
Automotive/Transportation Equipment	.28
High Tech—Optics	.28
Apparel/Textiles	.23
Government	.17
Pharmaceuticals	.17
Utilities	.06
High Tech—Networking	.06

NOTE.—1,771 observations, entrepreneurs only.

Table 2
Variables and Descriptive Statistics

Variable	Observations	Mean	SD	Min	Max
A. Whole sample:					
mbayear	26,901	74.27	14.21	13	97
male	27,283	.83	.37	0	1
age	26,863	50.24	13.59	25	93
white	27,283	.86	.35	0	1
exp	26,778	9.54	9.29	0	63
NPRIOR	27,277	3.26	3.39	0	37
NROLES	4,877	5.23	3.78	0	37
NROLESOTH	4,216	4.73	3.27	1	32
entre	27,283	.07	.25	0	1
numbus	27,283	.39	.79	0	5
avjobten	26,737	2.03	1.70	0	25
specdif	1,996	2.49	1.13	0	9
nafter	27,262	3.76	3.73	0	37
yrleft	26,588	10.74	9.93	0	45
B. Specialists:					
mbayear	25,120	74.34	14.25	13	97
male	25,482	.83	.38	0	1
age	25,081	50.18	13.65	25	93
white	25,482	.86	.35	0	1
exp	25,010	9.14	9.08	0	63
NPRIOR	25,476	3.10	3.26	0	37
NROLES	3,710	4.67	3.39	0	32
NROLESOTH	671	4.60	3.02	1	21
entre	25,482	0	0	0	0
numbus	25,482	.36	.75	0	5
avjobten	24,969	1.97	1.68	0	25
specdif	1,673	2.51	1.14	1	9
nafter	24,457	3.92	3.74	0	37
yrleft	23,879	11.39	10.01	0	45
C. Entrepreneurs:					
mbayear	1,781	73.22	13.62	13	97
male	1,801	.89	.31	0	1
age	1,782	51.12	12.76	26	88
white	1,801	.86	.34	0	1
exp	1,768	15.25	10.21	0	59
NPRIOR	1,801	5.55	4.27	0	35
NROLES	1,167	6.95	4.39	0	37
NROLESOTH	3,545	4.76	3.32	1	32
entre	1,801	1	0	1	1
numbus	1,801	.85	1.19	0	5
avjobten	1,768	2.88	1.75	0	18
specdif	323	2.36	1.05	0	7
nafter	2,805	2.34	3.28	0	25
yrleft	2,709	5.06	7.04	0	41

Number of Prior Roles

Table 2 provides the means and standard deviations of the relevant variables used in subsequent analyses. An entrepreneurial employment spell is defined to be one for which the respondent stated that he or she was "Founder—among those who initially started the business." The standard way to acquire human capital is through formal schooling and on-

Table 3
Proportion of Entrepreneurs by Number of Prior Roles

	A. Roles				
	≤ 3	3–16	> 16		
Proportion of entrepreneurs	.03	.10	.29		
	B. Experience Level				
	P Low NPRIOR	P High NPRIOR	Difference	SE	*N*
Roles:					
3 or fewer	.015	.029	.014	.003	8,163
4–8	.034	.073	.039	.005	6,971
9–13	.073	.11	.037	.009	4,597
14–19	.099	.115	.016	.011	3,415
20–29	.102	.138	.036	.012	2,965
30 or more	.145	.187	.042	.022	1,148

Note.—P = proportion.

the-job training. Therefore, individuals who plan to become entrepreneurs are expected to invest in "general" on-the-job training, where "general" here refers to a variety of skills.

The unit of observation in tables 2 and 3 is the row, which consists of an employment spell for any given individual. The key independent variable, "NPRIOR," is the number of roles in total that the individual has had before the employment spell in question. So if an individual had three previous employers, and held two roles with the first, four roles with the second, and one with the third, then NPRIOR would equal seven. The dependent variable, ENTRE, is a dummy equal to one if the employment spell is one in which the individual founded his or her own business. About 6.6% of all employment spells (rows) are entrepreneurial ones. Taken over the lifetime, however, about 24% of the sample started at least one business.

Table 3 provides the initial evidence on the relation of varied background to entrepreneurship. Panel A of table 3 reports the proportion of entrepreneurial employment spells, broken down by number of prior roles. Only 3% of those who have had fewer than three roles are entrepreneurs, whereas 29% of those with over 16 prior roles are entrepreneurs.[14] Although having more than 16 roles is well above the mean, it is far from the maximum number of roles held by individuals. (See table 2.) The point is that the simplest statistics show that those who undertake

[14] One possibility is that those who have been entrepreneurs in the past list many roles when they are entrepreneurs and that entrepreneurship is serially correlated. To check this, NPRIOR was redefined such that each entrepreneurial employment was given one and only one role. The results were substantially unaltered.

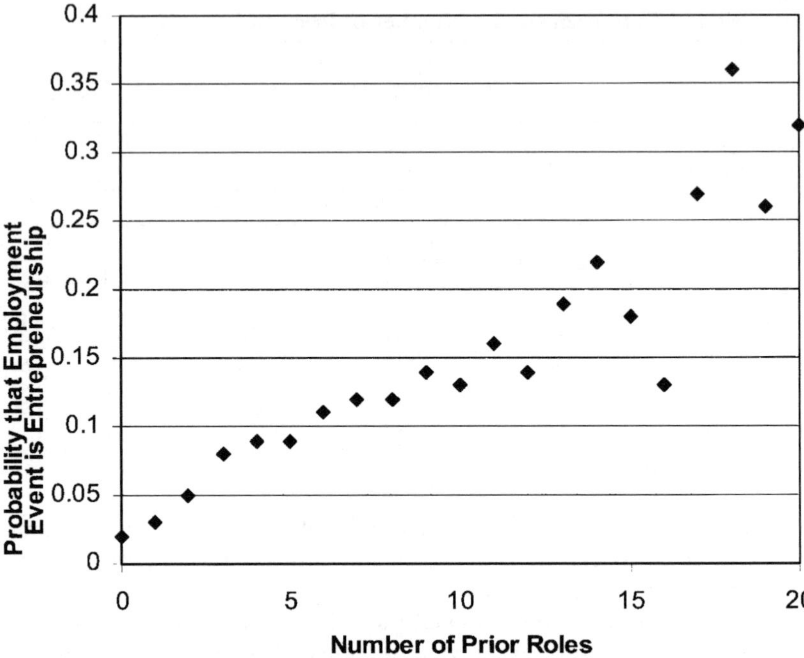

FIG. 2

many different assignments also have much higher probabilities of becoming entrepreneurs. More detail is given by the plot in figure 2. A clear positive relation of entrepreneurship to number of prior roles is apparent.

Because it may be expected that individuals become entrepreneurs after acquiring some on-the-job experience, it is useful to break the data up by experience level. This is done in table 3, panel B. The column "P Low NPRIOR" is the proportion of entrepreneurial spells in the group of individuals who have the median or fewer than the number of prior roles for this experience group. The column "P High NPRIOR" gives the proportion of entrepreneurial spells in the group of individuals who have more than the median number of roles for that experience group. Thus, the first row of panel B shows that those with the median or fewer prior roles were half as likely to be entrepreneurs as those with more than the median number of roles (.015/.029). Since an individual may enter the data set a number of times at different levels of experience, the rows are not independent tests of the hypothesis.[15] But the pattern is clear. Those who have had more roles in the past are more likely to be entrepreneurs.

[15] In fact, because experience brackets contain more than 1 year of experience, it is possible that an individual might enter even the same calculation more than

There are two interpretations of the NPRIOR variable, both of which are consistent with the jack-of-all-trades hypothesis. The first is that those who are endowed with high levels of multiple skills (or have acquired them by the time they reach the labor market) are able to perform many roles and NPRIOR proxies a person-specific effect—high NPRIOR are endowed with a balanced skill set. The second interpretation is that those who want to be entrepreneurs intentionally choose to perform a number of roles in order to acquire a balanced skill set. This is the investment route emphasized in the investment section above. Either interpretation is consistent with the model. The point is that a general set of skills is required for entrepreneurship. The existence of multiple roles is evidence for the prior existence or acquisition of the broad skill set. Below, some attempt will be made to distinguish between the two versions.

A Deeper Analysis

Table 4 contains the results of a deeper examination of the likelihood of starting a business to varied background. The basic result is contained in column 1. Each observation is a row in the work history matrix, referring to one employment spell for one individual. The data are stacked. Two spells for one individual count the same as one spell for each of two individuals. Columns 2 and 3 do not assume that same statistical structure, but more on that later.

The key result is that the likelihood that an employment spell is entrepreneurial is positively and strongly associated with the number of prior roles. The partial effect is .0851 in the logit. This translates into a large effect. A one standard deviation increase in the number of prior roles implies an increase in the probability of being an entrepreneur by about .018, which is about one-fourth the probability that an employment spell is an entrepreneurial one. The finding is again consistent with the view that entrepreneurs have more varied backgrounds, at least in terms of the roles that they have played at work, than nonentrepreneurs.

Although not proof of the theory, this evidence can be viewed as consistent with the premise on which the model is based. That premise is that the entrepreneurial production technology is one that exhibits strong complementarity among skills. The extreme formulation used for expository purposes is that entrepreneurial output is proportional to $\min[x_1, x_2]$. Given that technology, the implication is that individuals who possess more varied skills are those most likely to be entrepreneurs. That prediction is borne out by the data in the sense that those with general backgrounds are more likely to start companies.

Experience is also important. Five years of additional experience in-

---

once, say, once with 5 and once with 7 years of experience. Thus, there is not complete independence even within a row of the table.

Table 4
Logit Analysis Panel Data

Variable	Correlation Structure						
	Independent (1)	AR-1 (2)	Unstructured (3)	Independent (4)	Independent (5)	Year Basis – Unstructured (6)	Independent (7)
EXP	.0452	.0502	.0534	.0070	.0205	.0341	.0430
	(.0036)	(.0039)	(.0038)	(.0044)	(.0048)	(.0058)	(.0037)
NPRIOR	.0851	.0769	.0706	.0808	.1166	.0875	.0764
	(.0079)	(.0088)	(.0092)	(.0082)	(.0088)	(.0117)	(.0078)
MALE	.4757	.4562	.4565	.5266	.4769	.6021	.4605
	(.0843)	(.0950)	(.0998)	(.0865)	(.0846)	(.1376)	(.0867)
MBAYEAR	−.0070	−.0044	−.0054	−.0214	−.0094	−.0136	−.0218
	(.0074)	(.0084)	(.0090)	(.0076)	(.0074)	(.0110)	(.0074)
AGE	−.0265	−.0250	−.0256	−.0117	−.0281	−.0299	.0024
	(.0078)	(.0089)	(.0095)	(.0080)	(.0078)	(.0112)	(.0079)
NAFTER				.0186			
				(.0119)			
YRLEFT				.0786			
				(.0060)			
AVJOBTEN					.1218		
					(.0152)		
SDFIRST							.00090
							(.00020)
Constant	−2.1085	−2.3626	−2.2253	−.7957	−1.9849	1.583	−5.499
	(.9397)	(1.0669)	(1.1369)	(.9587)	(.9330)	(1.37)	(.9904)
Wald χ_2	842	706	786	944	876	281	617
Number of observations	26,819	26,663	26,819	26,163	26,779	62,229	21,956

NOTE.—Dependent variable: 1 if employment spell is entrepreneurial. Standard errors are in parentheses.

Table 5
Multinomial Logit Analysis of C-Levels, Employees, and "True" Entrepreneurs

	1		2		3	
	Other	C-Levels	Other	C-Levels	Other	C-Levels
EXP	−.049	.008	−.025	.028	−.049	.008
	(.003)	(.005)	(.004)	(.005)	(.004)	(.005)
NPRIOR	−.090	.0004			−.092	−.001
	(.008)	(.098)			(.008)	(.010)
MALE	−.517	.575	−.480	.602	−.515	.577
	(.090)	(.143)	(.091)	(.143)	(.090)	(.143)
MBAYEAR	−.010	.027	.010	.046	−.013	.024
	(.008)	(.011)	(.008)	(.011)	(.008)	(.011)
AGE	.011	.030	.029	.048	.007	.026
	(.008)	(.011)	(.009)	(.011)	(.008)	(.011)
NROLES			−.175	−.069		
			(.007)	(.009)		
SDFIRST					−.00075	−.00084
					(.00021)	(.00032)
Log likelihood	−10,540		−10,325		10,533	
Number of observations	20,920		20,920		20,920	

NOTE.—Omitted category: "true" entrepreneurs.

crease the probability of being an entrepreneur by about .014, or about one-fifth the probability that an employment spell is an entrepreneurial one. Men are more likely to start companies than women and younger individuals, for a given amount of experience, start more companies than older ones.

A statistical issue is that the rows are not independent because they belong to the same individual. For example, individuals who have already started one company are more likely than the average individual to start another at the next employment effect. Treating each row (employment spell) as independent may not be appropriate under these circumstances and the standard errors will be inconsistent. To run logit, taking into account correlation between rows due to the panel nature of the data, the approach of generalized estimating equations is used. The results are reported in table 4.

Columns 2 and 3 of table 5 introduce different forms of dependence across observations to allow for the fact that the data consist of a panel in which more than one row of data may belong to one individual. In column 2, an AR-1 structure is assumed so that the correlation between rows s and t for a given individual is

$$\rho^{|t-s|}.$$

In column 3, the correlation matrix is unstructured, and each correlation is allowed to be free and is estimated by the data.

The key result is that the coefficient on NPRIOR is not very sensitive to the correlation structure assumed. Even when nonindependence is taken into account, the relation of number of roles on entrepreneurial activity is large and significant.[16]

There is another potential statistical structure. Employment spells vary in duration. To address this, let us define a person-year as the relevant unit of analysis for the purpose of correlation structure. Every year of a person's career constitutes a separate observation during which he or she is an entrepreneur or working for someone else. Thus, someone who has been in the labor force for 15 years contributes 15 observations to the analysis. The within-job, between-year correlation structure is taken into account in estimating the logit so that standard errors are estimated consistently. Results are reported in column 6 of table 4. It is evident that NPRIOR maintains the same relative importance, has almost identical magnitude, and is statistically significant.

Which Interpretation?

Already mentioned is that there are (at least) two possible versions of the jack-of-all-trades view. One is that people are endowed with skills and that their labor market behavior merely reflects endowed skills or those acquired in school before they enter the labor market. An alternative is that individuals choose to perform a variety of roles as an investment in acquiring the broad skills necessary to become an entrepreneur. To get at this, the panel data are helpful.

If endowed characteristics simply enable a generalist to perform many roles, then the timing of those roles should not matter. That is, roles held after any particular employment spell should have the same "effect" on the likelihood that the spell is an entrepreneurial one as a role held before. However, if performing many labor market roles actually enhances the individual's ability to become an entrepreneur, then roles before any particular employment spell should be more salient than roles held after an employment spell. Therefore, another variable, NAFTER, was created, which counts the number of roles that an individual had in all employment spells following the one in a given row. It is analogous to NPRIOR, but it is defined in reverse. Rather than counting the roles held up to that point, it counts the roles after that point. Also, just as it was necessary to correct for experience, it is also necessary to correct for number of years after or number of roles will pick up life cycle effects. (Those who

[16] The sorting effect of individuals into different occupations picks up comparative advantage. If entrepreneurs have a comparative disadvantage as specialists, and not merely an absolute advantage, then those who become entrepreneurs should, when they are in training in nonentrepreneurial jobs, earn less than those who choose to be specialists. A regression of earnings in nonentrepreneurial jobs on EXP and a dummy equal to one if the individual ever founds a business in the career yields a negative, albeit insignificant, coefficient on the dummy.

have a long career in the data after a particular spell will have higher values of NAFTER, irrespective of their general skills.) YRLEFT measures the number of years remaining until the end of the career record at the point at which the employment spell in question occurs.

Results are reported in column 4 of table 4. Although NAFTER enters positively, it is less than one-fourth as important in magnitude as NPRIOR. This is evidence that multiple roles are actually productive in preparing a person to be an entrepreneur and that they are not merely a proxy for unobserved general skills.[17] One caveat: NAFTER is defined only for those who have a subsequent employment spell so NAFTER might relate to a preference for changing employers. This is unlikely, because, as discussed below, those who change employers frequently are less likely, rather than more likely, to become entrepreneurs.

Alternative Explanations

Might the results reflect something other than the jack-of-all-trades view? One possibility is that entrepreneurs have more roles when they are entrepreneurs than others and that the results are spurious. Indeed, there is some evidence that this is the case. On average, entrepreneurs have on average 2.4 roles during their entrepreneurial job compared to an average of 1.2 roles for all nonentrepreneurial spells.

To deal with this, a variable like NPRIOR was defined, except that each entrepreneurial spell is allowed to add only one role to NPRIOR, irrespective of the number of roles claimed during the entrepreneurial employment spell. This approach intentionally biases the results against finding a positive relation of entrepreneurial activity to the number of roles because one role is below the mean number of roles per employment spell. The results are qualitatively identical to those reported. The coefficient on the redefined variable is about half that on NPRIOR, but otherwise similar.

It is also possible that those with more roles received more promotions with their previous employer. To check this, the same models were run including a variable that measured the final salary on the last job. This was done only for those with at most one entrepreneurial spell so that the prior salary would be unambiguously defined. Those with higher final salaries do have a slightly higher probability of being entrepreneurs, but

[17] The evidence is not dispositive, however. Even if large effects of NAFTER were found, it could be that being an entrepreneur prepares a worker to take on many roles in subsequent employment, precisely because entrepreneurs perform many tasks. Conversely, if entrepreneurship is an absorbing state or close to it, an entrepreneurial spell could be associated with fewer roles after because there are fewer jobs. This is unlikely, however, because entrepreneurs have more roles per year after a given employment spell on average than nonentrepreneurs.

the coefficient is not significant in the logits with ENTRE as the dependent variable.

One theory, that entrepreneurs are those who are impatient, want variety, or fall into entrepreneurship accidentally because they are the individuals who change jobs often, is not borne out by the data. To examine this hypothesis, "average employment tenure" defined as experience/number of employment spells was included in the logit. The result, in column 5 of table 4, is that those who have longer employment spells are more likely to be entrepreneurs. If entrepreneurs were impatient types, the relation of entrepreneurship to average job tenure (AVJOBTEN) should be negative, not positive. There is surely some version of a taste argument that can be made, but the most straightforward ones do not seem to be consistent with the sign on AVJOBTEN.

The most frequently offered explanation of entrepreneurship is that entrepreneurs are those who are willing to take risks. Taste explanations are never particularly appealing unless they can be related to something measurable. There is some hope of doing just that in the data set. Most entrepreneurs work in other jobs before starting their own companies. The jobs that they choose before becoming entrepreneurs can provide some information on their preference for risk. Specifically, the data provide detailed industry/occupation codes that correspond to fields like accounting, education, insurance, chemical manufacturing, and so forth. There are 74 different industry/occupation categories. Some industries/occupations are riskier than others, as can be proxied by examining the wage distribution of individuals within these occupations. To measure this, the standard deviation of income in each of the 74 groups was calculated. Then, each individual was classified according to the industry/occupation of his or her first job choice after graduation from Stanford. The standard deviation of income in that category becomes a variable attached to the individual. The idea is that those who are less risk averse should be willing to enter the higher income variance occupations than the most risk averse, so that the standard deviation of income in the first occupation measures their tolerance for risk.

The standard deviation in income of the individual's first industry/occupation, FIRSTSD, is entered as a variable in the entrepreneurship logit. The results are reported in column 7 of table 4.

Note first that the coefficient on FIRSTSD is positive. Those who enter high income variance industries/occupations are more likely to have later entrepreneurial employment spells. This is consistent with the view that entrepreneurs are less risk averse. Of course, it could also reflect other differences across occupations correlated with wage variance. For example, fields with high wage variance like investment banking might attract people who think they are movers and shakers, and the same people might also be inclined to start their own businesses. This is a taste explanation, and there are surely others that cannot be rejected by these data, but the risk aversion view is consistent with the findings.

Second, even though FIRSTSD enters significantly, the coefficient and standard errors on NPRIOR remain virtually unchanged. Even if risk or other taste factors affect the choice to start a business, the generalist argument still seems to apply and is not weakened by this effect.

Defining *Entrepreneur*

Because the definition of *entrepreneur* is somewhat arbitrary, another group was defined to be entrepreneurs. They are those who reported their position as high-level general manager, specifically, "I am responsible for the organization's overall direction, including responsibility for major business functions and personnel decisions (examples: CEO, President, COO, Executive Director)." Although individuals in this category may not assume the same risk as those who found a business, they are senior general managers, so the jack-of-all-trades argument should pertain to them as well.

To determine whether general managers are also jacks-of-all-trades, a multinomial logit was estimated. The dependent variable can either be "entrepreneur," which is the left-out group, "C-level manager," or "other." The results (see table 5, col. 1) show a significant negative coefficient on NPRIOR for those who are neither C-level managers nor entrepreneurs relative to "true" entrepreneurs. But the coefficient on NPRIOR for those who are C-levels is zero relative to the true entrepreneurs. Apparently, prior roles do not distinguish between entrepreneurs and C-level managers. The prior skills seem to be the same. So the jack-of-all-trades story applies well to senior-level managers.

It is possible to distinguish senior managers from true entrepreneurs. The theory suggests that those who start their own businesses must perform many tasks as entrepreneurs that are not required of C-level managers. For example, the chief technology officer need not raise funds for the firm since the chief financial officer generally performs that task. As a result, NROLES, which differs from NPRIOR only in that it includes the roles performed in the current job, should be more important for true entrepreneurs than for chief-level managers. Column 2 of table 5 contains the results of a multinomial logit, where the comparisons are relative to true entrepreneurs.

The multinominal logit results are as predicted. As before, NROLES has a strong negative effect on being an employee throughout the career, that is, it has a strong positive effect on being a true entrepreneur. More important, NROLES also has a negative effect on being a chief-level manager relative to being a true entrepreneur. Although having many roles increases the probability of having been a high-level manager, having many roles makes one even more likely to be a true entrepreneur.

One difference between C-levels and true entrepreneurs is the amount of risk that they face. It is likely that being a true entrepreneur is riskier than being a C-level employee because one's own physical capital is often at stake when starting a business. Also, high risk of failure might create riskiness associated with true entrepreneurship. Perhaps risk preferences help distinguish between being a C-level employee and a true entrepreneur.

Using the earlier approach, where risk aversion is proxied by avoidance of industry/occupations with high income variance, column 3 of table 5 reports the results with SDFIRST included. There is in fact evidence that a difference between C-level managers and entrepreneurs is that true entrepreneurs are more inclined to take risks. The SDFIRST variable enters negatively in both the employee and C-level manager part of the multinomial logit and with almost the same magnitude. Both groups, C-level managers and other employees, are less inclined to start out in high variance industries/occupations than those who become true entrepreneurs.

One interesting aside is that women are more likely to be employees of others than they are to be entrepreneurs, but they are more likely to be entrepreneurs than they are to be high-level managers. Women may have escaped historical discrimination at very high levels of corporations by starting their own businesses.

General and Specialized Curricula

The data on work histories were matched with data from student transcripts. As a result, we have information on the courses taken while the individual was a student at the Stanford Graduate School of Business. The records begin in the mid-1980s, so the transcript-matched data pertain only to those who graduated during approximately the last 15 years. But almost 2,000 records of alumni work history data have been matched with transcript information, so a significant amount of information is contained in the 15 years of records.

Simple relationships can be seen in the comparison of means in table 2. The variable SPECDIF is the difference between the maximum number of courses taken in one field and the average number of courses taken across fields. This is a measure of lopsidedness in the study curriculum. SPECDIF is lower by about one-fifth of a standard deviation for entrepreneurs than it is for specialists. The difference between the average level of SPECDIF between groups suggests that those who later become entrepreneurs took a more balanced set of courses while at Stanford.

Additionally, the simple correlation between SPECDIF and NROLES for lifetime roles is negative and significant, as is the partial correlation, holding constant MBA year. Those who take more specialized curricula

Table 6
Tobit and Logits with Stanford Course Data, Number of Businesses Started (Tobit and Ordered-Probit), and Ever Started a Business (Logit)

Variable	Logit (1)	Tobit (2)	Ordered Probit (3)
EXP	.0259	.0266	.0612
	(.0185)	(.0196)	(.0042)
SPECDIF	−.1458	−.1452	−.0433
	(.0581)	(.0592)	(.0180)
MALE	.6025	.6305	.3857
	(.1511)	(.1531)	(.0482)
MBAYEAR	−.0318	−.0384	.0241
	(.0215)	(.0224)	(.0069)
AGE	.0250	.0264	.0146
	(.0179)	(.1531)	(.0053)
Constant	.0202	.3243	Cut parameters
	(2.4182)	(2.4897)	omitted
Log likelihood	−841	−1,181	
Number of observations	1,952	1,950	

NOTE.—Standard errors are in parentheses.

also take on fewer roles when they enter the labor market, which suggests that generalized formal education and generalized on-the-job training are complements.

The first analysis reported in table 6 presents logits, tobits, and ordered probits, analogous to those in table 4, but only pre-labor-market characteristics are allowed to affect the career. The approach is to assume that school prepares an individual for the labor market and then to observe how differences in the educational experience are reflected in the subsequent career. The jack-of-all-trades theory suggests that those who have large values of SPECDIF should be less likely to become entrepreneurs.

The results confirm the hypothesis. The more specialized is the curriculum, as measured by SPECDIF, the fewer businesses the individual starts and the less likely is the individual to start a business. Once again, it is the generalists, as reflected in generalized course curricula, who end up founding a business after they leave school. Those who want to found a business prepare themselves by taking a variety of different courses that they hope will later prove useful when they start businesses. An alternative view is that those who happen to take a varied set of courses start a business later because the spell has given them the general skills necessary to found a business. Both views are consistent with the jack-of-all-trades view of entrepreneurship. Only if entrepreneurs need general skills will a varied course background be correlated with later entrepreneurial activity.[18]

[18] For example, suppose that some people like variety. They take many different types of courses and also become entrepreneurs. But those who like variety would only become entrepreneurs if entrepreneurship offered a more varied experience, which says that entrepreneurship is a general rather than specialized occupation.

The results of the Stanford course data support the earlier conclusions. Entrepreneurs are jacks-of-all-trades. They have more varied course work while in the MBA program and have many more positions when they are actually in the labor market.

Conclusion

Entrepreneurs are individuals who are multifaceted. Although not necessarily superb at anything, entrepreneurs have to be sufficiently skilled in a variety of areas to put together the many ingredients required to create a successful business. As a result, entrepreneurs tend to be more balanced individuals.

Two kinds of evidence are provided. First, those who have more varied careers, as evidenced by having performed more roles as part of their work experience, are more likely to be entrepreneurs. There are two interpretations of this result, both consistent with the jack-of-all-trades view. The first is that the correlation between number of roles and entrepreneurship reflects endowed differences in general skills across people. Those with more general skills can perform more roles. The second is that the correlation reflects conscious investment, where individuals who plan to become entrepreneurs take on many roles so that they can acquire the varied background necessary to start a business. Each version finds some support, but the investment view seems to dominate.

Second, the pattern of investment that occurs prior to entering the labor market is also consistent with the generalist view of entrepreneurship. In the Stanford MBA data, it is found that those students who study a more varied curriculum are more likely to be entrepreneurs and to start a larger number of businesses over their careers.

Much more can be done, especially at the empirical level, given the richness of the data. The prevalence of entrepreneurship by occupation and industry is predicted by the model. Educational systems differ by country in terms of amount of specialization, and this has implications for the proportion of entrepreneurs by country. The model gives quite specific predictions about these relations, but investigation is left to the future.

Appendix

The Underlying Economy

In this section, the income-generating functions shown in (1) and (2) are derived from a more fundamental production function.[19]

Let there be two raw skills, y_1 and y_2, for example, verbal and quan-

[19] Lucas (1978) uses this production function to discuss income distribution and the size distribution of firms.

titative ability. In any given firm with an entrepreneur who has the ability pair (y_1, y_2), output is given by

$$\text{Output} = Q(\min(y_1, y_2))f(Y_1, Y_2),$$

where $Q(\ )$ and $f(\ )$ are parts of the production function and Y_1 and Y_2 are the amounts of skills employed by the firm in efficiency units. Normalize the price of a unit of output to 1 and let wages (determined by the equilibrium) be given by w_1 and w_2. Then profit is given by

$$\pi(y_1, y_2, w_1, w_2) = Q(\min(y_1, y_2))f(Y_1, Y_2) - w_1Y_1 - w_2Y_2. \quad \text{(A1)}$$

To avoid discussion of number of firms in an industry, which is not central to the analysis, simply assume that each entrepreneur is a local monopolist. Thus, the $Q(\ )$ function incorporates the monopoly price into the measure of output.

Maximization of the profit function in (A1) yields the firm's demand curves for Y_1 and Y_2, which are written as

$$Y_i^d = Y_i^d(\min(y_1, y_2), w_1, w_2) \quad \text{for } I = 1, 2. \quad \text{(A2)}$$

The underlying density of skills in the overall working population is given by $g(y_1, y_2)$. $A_1(w_1, w_2)$ is the set of individuals who choose to specialize in supplying skill y_1. It is given by

$$A_1(w_1, w_2) = \{(y_1, y_2)|w_1y_1 > \max[w_2y_2, \pi(y_1, y_2, w_1, w_2)]\}. \quad \text{(A3)}$$

Analogously, A_2, the set of individuals who choose to specialize in skill y_2 is given by

$$A_2(w_1, w_2) = \{(y_1, y_2)|w_2y_2 > \max[w_1y_1, \pi(y_1, y_2, w_1, w_2)]\}. \quad \text{(A4)}$$

Finally, entrepreneurs are defined as the set

$$E(w_1, w_2) = \{(y_1, y_2)|\pi(y_1, y_2, w_1, w_2) > \max[w_1y_1, w_2y_2]\}. \quad \text{(A5)}$$

Then let $g_i(y_1, y_2)$ be the density function of abilities of individuals in set A_i derived from $g(\ )$ and (A3) and (A4) and $g_e(y_1, y_2)$ be the corresponding density among entrepreneurs, derived from $g(\ )$ and (A5).

The following two supply-equals-demand equations determine the equilibrium values of w_1 and w_2:

$$\int\int y_1 g_1(y_1, y_2) dy_1 dy_2$$
$$= \int\int Y_1^d(\min(y_1, y_2), w_1, w_2) g_e(y_1, y_2) dy_1 dy_2 \quad \text{(A6)}$$

and

$$\int\int y_2 g_2(y_1, y_2) dy_1 dy_2$$
$$= \int\int Y_2^d(\min(y_1, y_2), w_1, w_2) g_e(y_1, y_2) dy_1 dy_2. \quad \text{(A7)}$$

Next, define $x_1 = w_1 y_1$ and $x_2 = w_2 y_2$ where the wages are obtained from the market equilibrium given in (A5) and (A7). Sufficient conditions to derive income-generating functions (1) and (2) are constant returns to scale and symmetry. Specialist income in (1) comes directly from the definition of x_i and the conditions that define sets A_i. This is simply relabeling. Given that specialist income is x_i, it is now shown that entrepreneurial income moves in proportion to $\min(y_1, y_2)$ under the assumptions of constant returns to scale and symmetry. Let production exhibit constant returns to scale such that

$$Q(k \min(y_1, y_2)) f(kY_1, kY_2)$$
$$= kQ(\min(y_1, y_2)) f(Y_1, Y_2), \quad \text{(A8)}$$

and symmetry such that $f(y_1, y_2) = f(y_2, y_1)$ and $g(y_1, y_2) = g(y_2, y_1)$.

We want to show that

$$\pi(y_1, y_2, w_1, w_2) = \lambda \min(x_1, x_2). \quad \text{(A9)}$$

Symmetry guarantees that $Y_1^0 = Y_2^0$ and that $w_1 = w_2 = w$. Let $z = \min(y_1^0, y_2^0)$. Constant returns to scale implies that if entrepreneur of skill z employs y^0 of each type of labor, then an entrepreneur of skill z' employs $(y^0)(z'/z)$ of each type of labor. That, coupled with (A8) implies that profits are equal to $(z'/z)\pi(z, w, w)$. Let $\lambda \equiv \pi(z, w, w)/zw$. Then

$$\begin{aligned} \pi(z', w, w) &= (z'/z)\pi(z, w, w) \\ &= \lambda z'/w \\ &= (\lambda/w) \min(y_1, y_2) \\ &= \lambda \min(x_1, x_2), \end{aligned}$$

which is (2), the entrepreneurial income function.

References

Bates, Timothy. 1985. Entrepreneur human capital endowments and minority business viability. *Journal of Human Resources* 20, no. 4:540–54.

———. 1990. Entrepreneur human capital inputs and small business longevity. *Review of Economics and Statistics* 72, no. 4:551–59.

Becker, Gary S., and Casey Mulligan. 2002. The division of labor and household and market economies. Unpublished manuscript, Department of Economics, University of Chicago.

Becker, Gary S., and Kevin M. Murphy. 1992. The division of labor, coordination costs, and knowledge. *Quarterly Journal of Economics* 57, no. 4:1137–60.

De Meza, David, and Clive Southey. 1996. The borrower's curse: Optimism, finance and entrepreneurship. *Economic Journal* 106, no. 435: 375–86.

Evans, David S., and Boyan Jovanovic. 1989. An estimated model of entrepreneurial choice under liquidity constraints. *Journal of Political Economy* 97, no. 4:808–27.

Evans, David S., and Linda S. Leighton. 1989. Some empirical aspects of entrepreneurship. *American Economic Review* 79, no. 3:519–35.

Gentry, William M., and Glenn Hubbard. 2002. Entrepreneurship and household saving. Unpublished manuscript, Columbia University.

Hamilton, Barton H. 2000. Does entrepreneurship pay? An empirical analysis of the returns of self-employment. *Journal of Political Economy* 108, no. 3:604–31.

Holmes, Thomas J., and James A. Schmitz Jr. 1990. A theory of entrepreneurship and its application to the study of business transfers. *Journal of Political Economy* 98, no. 2:265–94.

Holtz-Eakin, Douglas, David Joulfaian, and Harvey S. Rosen. 1994. Entrepreneurial decisions and liquidity constraints. *RAND Journal of Economics* 25, no. 2:334–47.

Iyigun, Murat F., and Ann L. Owen. 1998. Risk, entrepreneurship, and human-capital accumulation. In "Banking crises, currency crises, and macroeconomic uncertainty." *American Economic Review* 88, no. 2, Papers and Proceedings of the Hundred and Tenth Annual Meeting of the American Economic Association, 454–57.

Kihlstrom, Richard E., and Jean-Jacques Laffont. 1979. A general equilibrium entrepreneurial theory of firm formation based on risk aversion. *Journal of Political Economy* 87, no. 4:719–48.

Landier, Augustin. 2002. Entrepreneurship and the stigma of failure. Thesis, Graduate School of Business, MIT.

Lazear, Edward P. 1995. Introduction. In *Economic transition in Eastern Europe and Russia: Realities of reform*, ed. Edward P. Lazear. Hoover

Institution Press Publication no. 425. Stanford, CA: Hoover Institution Press.

Lentz, Bernard F., and David N. Laband. 1990. Entrepreneurial success and occupational inheritance among proprietors. *Canadian Journal of Economics* 23, no. 3:563–79.

Lucas, Robert E., Jr. 1978. On the size distribution of business firms. *Bell Journal of Economics* 9, no. 2:508–23.

Otani, Kiyoshi. 1996. A human capital approach to entrepreneurial capacity. *Economica* 63, no. 250:273–89.

Stopford, John M., and Charles W. F. Baden-Fuller. 1994. Creating corporate entrepreneurship. *Strategic Management Journal* 15, no. 7: 521–36.

[33]

Is Mobility of Technical Personnel a Source of R&D Spillovers?

Jarle Møen, *Norwegian School of Economics and Business Administration*

Labor mobility is considered to be an important source of knowledge externalities, making it difficult for firms to appropriate returns to research and development (R&D). Interfirm transfers of knowledge embodied in people should be analyzed within a human capital framework. Testing such a framework, I find that the technical staff in R&D-intensive firms pays for the knowledge they accumulate on the job through lower wages early in their career. They later earn a return on these implicit investments through higher wages. This suggests that the potential externalities associated with labor mobility are, at least partially, internalized in the labor market.

> "Don't let your employees do to you what you did to your former boss." The golden rule of protecting trade secrets, as defined by Intel general counsel Roger Borovoy. (Jackson 1997)

This article has benefited greatly from the continuous advice and encouragement of my thesis adviser, Tor Jakob Klette. I am also indebted to Zvi Griliches for suggesting the topic to me and for inviting me to visit Harvard University and the National Bureau of Economic Research while working on the project. Finally, I have received useful comments from James Adams, Pierre Azoulay, Gilles Duranton, Torbjørn Hægeland, Adam Jaffe, Lars Mathiesen, Oddbjørn Raaum, Kjell Salvanes, Scott Stern, and seminar participants in Oslo, Bergen, Trondheim, Aarhus, Jyväskula, and Cambridge. Remaining errors and all opinions expressed are my sole responsibility. The project is financed by the Research Council of Norway.
Contact the author at jarle.moen@nhh.no.

[*Journal of Labor Economics*, 2005, vol. 23, no. 1]

0734-306X/2005/2301-0004$10.00

Introduction

Labor mobility is likely to be a very important source of knowledge diffusion. Surveying 100 founders of companies on the 1989 *Inc.* 500 list of the fastest growing companies in the United States, Bhide (1994, 151) finds that 71% "replicated or modified an idea encountered through previous employment." With respect to technical employees, Almeida and Kogut (1999) demonstrate by an analysis of patent data from the semiconductor industry that ideas are spread through the mobility of key engineers. Evidence of this kind, however, does not justify the common proposition that labor mobility is an important source of knowledge spillovers. Such spillovers (or externalities) are thought to cause underinvestment in private research and development (R&D) because workers have incentives to exploit their employers' research results by setting up or joining a competitor.

The aim of this article is threefold. First, I want to clarify how labor mobility can affect R&D investments. I will argue that there are market mechanisms that may internalize the potential externalities involved. Second, I present a framework to test the existence of such market mechanisms. Third, I present empirical findings suggesting that these mechanisms actually exist. My results are, however, not entirely conclusive with respect to the exact mechanism.

The link between labor mobility and knowledge spillovers dates back to Arrow's (1962) article on the public good aspect of knowledge. Arrow writes that "no amount of legal protection can make a thoroughly appropriable commodity of something so intangible as information" and adds that "mobility of personnel among firms provides a way of spreading information" (1962, 615). Following Arrow's seminal work, a large literature on R&D spillovers has evolved, and economists working in the field have continued to consider labor mobility an important spillover channel. Geroski (1995, 78) formulates what appears to be a common view, writing that "last but not least, spillovers occur when a researcher paid by one firm to generate new knowledge transfers to another firm (or creates a spin-off firm) without compensating his/her former employer for the full inventory of ideas that travels with him or her."[1]

That workers do not make such compensations may seem obvious since they already possess their employers' knowledge when they decide to leave. The timing of events implicit in Geroski's formulation, however,

[1] Jaffe (1996, 6) writes that "knowledge spillovers also occur when researchers leave a firm and take a job at another firm." Stephan (1996, 1228) writes that "future work should also focus on the role mobility within the industrial sector plays in facilitating spillovers." Gersbach and Schmutzler (1997, 3) write that "spillovers arise because employees who change jobs take with them all their knowledge, some of which is not specific to their original firm."

is misleading. To the extent that research work has a general training element, workers may pay for knowledge as it is accumulated. Whether labor mobility actually reduces appropriability and R&D investments, therefore, is an empirical question. The approach I suggest to answer this question is to test key implications of models that assume perfect markets. If one uses standard methodologies for estimating R&D spillovers without first considering such a "benchmark" case, the results of ordinary market exchange may mistakenly be interpreted as R&D spillovers.[2] In that case, public policy will be misguided.[3]

The basic implications of labor mobility follow from classical human capital theory (Mincer 1958; Becker 1962, 1964). To the extent that workers in R&D-intensive firms get access to valuable knowledge on the job, they will expect higher wages in the future. When holding jobs that give access to such knowledge, they should therefore be willing to pay for what they learn by accepting wages below their alternative wage. This hypothesis can be tested using extended Mincer (1974) wage regressions, which is the standard approach in the training literature.

Utilizing a large matched employer-employee data set from the Norwegian machinery and equipment industry, I find that the technical staff in R&D-intensive firms pay for the knowledge they accumulate on the job through lower wages in the beginning of their career and that they later earn a return on these implicit investments through higher wages. Somewhat surprisingly, I find as strong results for workers with secondary technical education as for scientists and engineers. This indicates that R&D intensity is not only a measure of learning associated with doing research but also a proxy for the value of general work experience from high-tech firms. When estimating the price paid for learning separately from the return to research experience, I find that having work experience from R&D-intensive firms is associated with higher wages, while the employers' current R&D intensity reduces wages for workers with less than 20 years experience.[4] Furthermore, as predicted by human capital theory, the youngest workers appear to invest most heavily in on-the-job learning. These findings suggest that the potential externalities associated with labor mobility, at least to some extent, are internalized in the labor market.[5]

[2] See, e.g., Jaffe (1986), Jaffe, Trajtenberg, and Henderson (1993). See Sec. V for a short discussion of the problem with these methodologies in my context.

[3] Weder and Grubel (1993), Zucker, Darby, and Armstrong (1998), Klette and Møen (1999), and Klette, Møen, and Griliches (2000) elaborate on this point. See also Geroski (1995) and Cooper (2001) for contributions that are skeptical as to whether spillovers imply underinvestment in R&D.

[4] I will use "R&D experience" as a shortened term for experience from R&D-intensive firms.

[5] This does not guarantee optimal R&D investments, however, as credit restrictions or risk-averse preferences may reduce workers' willingness to cofinance R&D. I will return to this in Sec. V.

The rest of the article is organized as follows. Section II outlines some relevant theoretical models. Section III discusses the data and the industries analyzed. Section IV derives empirical results regarding R&D investments and wages, and Section V contains my concluding remarks.

R&D Investments and Human Capital Theory

Research is a learning process, and R&D investments, therefore, may increase not only a firm's stock of innovations but also the human capital of research workers. In the literature, however, R&D capital (Griliches 1973) and human capital (e.g., Becker 1964) are rarely discussed together.

Essentially, R&D capital is knowledge that can earn a monopoly rent, and this rent is what motivates investments in R&D. If the results of a research project can be perfectly protected by patents or other intellectual property right instruments, labor mobility is not a concern to firms when it comes to appropriating returns. However, often the intellectual property rights cannot be effectively protected. The R&D capital of firms is then to a large extent embodied in the employees. Such knowledge is what Zucker, Darby, and Brewer (1994, 1998) have called intellectual human capital.[6] Under these circumstances labor mobility is potentially a threat to the firms. Pakes and Nitzan (1983) analyze the investment incentives of entrepreneurs facing such a situation and conclude that it is possible to design labor contracts that solve the problem, as I argue below.

As knowledge diffuses, intellectual human capital will become "ordinary" human capital that can be acquired through schooling or on-the-job training. On-the-job training also has relevance for an analysis of labor mobility and R&D investments. There may be more to learn in firms conducting research because such firms are likely to use the most

[6] Intellectual human capital is human capital that can earn a monopoly rent because the knowledge is not publicly available nor perfectly protected. This distinguishes it from "ordinary" human capital, which is widely diffused knowledge that can be acquired at a cost and earns a normal rate of return on the implied investment. How quickly intellectual human capital depreciates and becomes ordinary human capital depends both on the complexity and tacitness of the knowledge and on whether those who possess the knowledge try to keep it secret. Innovations that can be communicated at no cost represent a limiting case and will not add to anyone's human capital once the ideas are in the public domain. Zucker et al. (1998, 291) stress that scientific discoveries that create intellectual human capital are "characterized by natural excludability" and that this solves the appropriability problem. From the point of view of an individual scientist, this is correct, but not from the point of view of an investing firm or entrepreneur. Zucker, Darby, and Armstrong (1998, 66) recognize this within their setting, writing that "much of the fruits of the biotechnological revolution was much more appropriable by the star scientists than by the universities that (typically) employed them." They also recognize that universities indirectly appropriate returns through the wage mechanism (see n. 36).

up-to-date technology and frequently change their products and production processes. This training may be valuable to other firms. Furthermore, the distinction between intellectual human capital and on-the-job training does not constitute a clear dichotomy. Many innovations are incremental product and process improvements made at the factory floor, and small innovations of this type may as well be considered excellent craftsmanship as innovations. A case where different firms offer different opportunities for on-the-job training is analyzed by Rosen (1972).

The rest of this section will outline the theoretical models of Rosen (1972) and Pakes and Nitzan (1983). Although highly relevant for work on R&D investments, training, and labor turnover, these models have received modest attention in the literature. The main predictions of the models will be discussed and tested in the empirical part of the article.

Rosen's 1972 Model

Rosen (1972, 327) models on-the-job learning using a compensating differential framework and turns it into "an economic theory of occupational mobility." Rosen thinks of jobs as tied packages of work and learning. Workers sell the services of their skills and simultaneously purchase an opportunity to augment those skills. Some jobs provide more learning opportunities than others. The difference between the maximum market rental of a worker's existing skills and the wage that he or she receives in a given job is the implicit price the worker pays for learning. Basic human capital theory suggests that a worker's incentive to accumulate human capital is largest at a young age. As the worker grows older, he or she will have fewer years to collect returns on a given investment, and obviously workers have no incentives to pay for increasing their human capital in the last year before retirement. This implies that the "optimal human capital investment program is implemented by a sequence of job assignments in which workers systematically move and are promoted across jobs that offer successively smaller learning opportunities" (Rosen 1986, 677).

The point of departure in Rosen's model is a net wage equation,

$$y = \omega H - P(k), \tag{1}$$

where y is income; ω is the unit rental price of human capital, H; and k is an index measuring potential learning-by-experience on the job, $k \in [0, \bar{k}]$; and $P(k)$ is an implicit or shadow price function giving the market-equalizing wage differential between a job with no learning potential and a job with learning potential k. The actual amount of learning by individual i is proportional to k and depends on individual i's ability, $\alpha_i \in [0, 1]$ such that

$$\dot{H}_{it} = \alpha_i k. \tag{2}$$

The worker's problem is, then, to choose a sequence of jobs, k_t, over his or her lifetime, T, to maximize the present value of income, that is,

$$\max_{k_t} V = \int_0^T [\omega H_t - P(k_t)]e^{-rt}dt, \tag{3}$$

subject to an initial stock of human capital, H_0, and $\dot{H}_{it} = \alpha_i k$. Optimization requires that at any time, $t \in [0, T]$,

$$\frac{P'(k_t)}{\alpha_i} = \frac{\omega}{r}[1 - e^{-r(T-t)}]. \tag{4}$$

The expression on the left-hand side is the marginal cost of investing in human capital, and the expression on the right-hand side is the discounted marginal return. It seems reasonable to assume that $P'(k) > 0$ and $P''(k) > 0$, that is, that the marginal cost of learning is positive and increasing. Given this, optimality requires k_t to be largest at the time of entry into the labor market and then to decrease monotonically over time.

Note that the marginal cost of a given real investment in human capital decreases with the ability to learn. Hence, workers with higher ability will, all else equal, find it profitable to choose jobs with greater learning potential. In the words of Rosen (1972, 338): "Economic incentives induce more 'able' workers to learn more and to accumulate knowledge more rapidly than the less 'able.'" This will give rise to a potential selection problem (ability bias) in the empirical application of the model.

The Pakes-Nitzan Model

In Rosen (1972), firms differ with respect to learning opportunities, but they do not have market power. Pakes and Nitzan (1983) formalize learning in a different way, downplaying training but emphasizing the strategic aspects of the innovation process. Their point of departure is Arrow's (1962) reference to labor mobility as a source of R&D spillovers and the observation that scientists get access to valuable information.[7] They argue that even though mobility of scientific personnel will spread knowledge produced in industrial laboratories, it need not be a mechanism that reduces the profitability of research projects and employment in such pro-

[7] Note that Pakes and Nitzan (1983) is, strictly speaking, not a model about human capital. Research scientists learn about research results but do not increase their generic productivity. Hence, the model lacks the training perspective crucial in Rosen (1972) .

jects. Both scientists and firms are aware of the fact that working on a research project gives access to valuable information.[8] Once such information is disclosed or developed, scientists, if they are to stay with the firm, will have to receive a wage increase reflecting their new market value. Thus, scientists expect that accepting a research position implies a future wage increase, and consequently they accept an initial wage below their alternative wage.[9] In this respect, the Pakes-Nitzan model resembles a bonding model, having a wage scheme designed to deter workers from moving and taking information with them. Workers "pay" up front for the right to work at an R&D-intensive firm by accepting low starting wages and redeem this implicit bond by receiving higher wages later on.

Next, Pakes and Nitzan (1983) notice that if the innovation makes the firm a true monopolist, it will never be profitable for the firm and the scientist to split, since the sum of rents in a duopolistic market will be less than the monopoly rent.[10] Mobility, therefore, should only be observed when it increases the joint profit of the firm and the scientist. This may happen if the firm cannot prevent other firms from getting access to valuable information and entering the market.[11] The scientist, by setting

[8] Pakes and Nitzan (1983) explicitly model the uncertainty involved in research. This feature of the model does not alter the simple intuition given here, however, because they assume that utility functions are linear in income. Discussing this assumption, they acknowledge that both risk aversion and a lower bound on wages will affect R&D investments.

[9] See Anand and Galetovic (2000) for a model in which the firm cannot commit ex ante to share profits with the researcher. In this setting, underinvestment in R&D may occur. Kim and Marschke (2003) extend Pakes and Nitzan's model by allowing firms to protect innovations by patenting and by showing empirically that higher risk of a scientists' departure is associated with a higher propensity to patent.

[10] Pakes and Nitzan (1983) model only a situation with one entrepreneur and one scientist. If several scientists have equal access to the same critical information, this will complicate the analysis because of potential strategic interaction among the scientists. Combes and Duranton (2001) develop a game theoretic model in which a continuum of workers share the same (exogenously given) strategic knowledge. They show that the "joint profit" effect driving the result in Pakes and Nitzan's model is not robust to this variation. Pakes and Nitzan dismiss the case with a large number of workers sharing exactly the same strategic knowledge about a firm as being of little relevance. They do not present strong arguments, but it seems reasonable to assume that if several scientists work on the same project, their knowledge is more often complementary than substitutable. See Rajan and Zingales's (2001) "horizontal hierarchy" and Rønde (2001) for models with this flavor. Empirical work on this topic will require data on the size of the R&D teams in different firms.

[11] Note that spillovers at this point enter the story, but mobility will be a consequence, not a source, of spillovers. Information can leak out to third parties by reverse engineering, inspection of patent documents, independent research on the same technological problem, etc. See Levin et al. (1987) for a survey of the

up a rival, will then break into profits that otherwise accrue to third parties, and since this profit will be part of the scientists' alternative wage in "period 2," it is possible for the firm to extract this rent when setting the "period 1" wage. Another situation that may induce the scientist to join or set up a rival is when the research project creates "spin-offs," some of which are better exploited in a separate firm because of coordination costs.[12] Summarizing the insight of their model, Pakes and Nitzan (1983, 359–60) write that "mobility of scientific personnel is not, in itself, a source of concern to entrepreneurs. . . . An optimizing entrepreneur who is free to choose among alternative contracts will always choose one which only induces the scientist to leave and join a rival if the sum of the benefits to the two agents increases as a result of the scientist's leaving. Contracts which specify labor payment in the form of a flat wage and stock option (or other profit sharing agreement) ought to be able to induce close approximation to this behavior."

Balkin and Gomez-Mejia (1985) provide empirical evidence in support of Pakes and Nitzan's prediction. Surveying 105 companies in the Route 128 region around Boston, they find that incentive pay programs are far more common in high-tech firms than in other firms and that such programs are used for broad levels of technical employees. In addition, key scientists and engineers who help form the companies at an early stage are given long-term stock options.

Comparing Rosen (1972) and Pakes and Nitzan (1983)

Narrowly interpreted, as a 2-period model about information, Pakes and Nitzan (1983) predict a wage discount when a scientist enters a research firm and a wage rise thereafter, regardless of when in the career this happens. The wage profile is driven by a rent that exists as long as the research results are not completely diffused in the industry. In Rosen's (1972) model, by contrast, a steeper wage profile is associated with high-tech or research experience rather than research results. Since the experience gained by working on a new technology may be of value after the rent associated with the technology is competed away, the human capital investments in Rosen's model are likely to depreciate more slowly than the "intellectual" human capital investments in Pakes and Nitzan's model. While young workers thus have a stronger incentive to invest in ordinary

importance of various information channels. Labor mobility receives a middle score in their study.

[12] See Franco and Filson (2000) for a model focusing particularly on spin-off firms, but looking at process innovations in a homogeneous product industry. Franco and Filson get results similar to Pakes and Nitzan (1983) in that knowledge spillovers are internalized in the labor market, but they do not endogenize mobility by considering the potential "joint profit" resulting if a spin-off firm is not established.

human capital than do old workers, workers of different ages may have more similar incentives with respect to investing in access to research results. In principle, the two models could be tested against each other based on this difference. However, when going beyond stylized versions of the models, the effects of research experience and research rents are complementary rather than alternative explanations for expecting a steeper wage profile in research firms. Clearly, doing research has a training element in addition to giving workers insight in particular research results, and given that R&D is a cumulative process where detailed knowledge about the current technology is an important input in the development of the new technology, it is highly unlikely that an old worker will get a high-tech job without being on a high-tech career track already. Taking account of this cumulative aspect of R&D and thinking in terms of a "multiperiod" Pakes and Nitzan model, a worker who continuously invests part of his or her wage in relatively short-term R&D projects is likely to have a steadily increasing wage profile over the whole career.[13] This is because R&D investments on average should have a higher return than financial savings because of the higher risk, and because workers may become exposed to more and more valuable research results as they gain general research experience.

Predictions Regarding Mobility

At first sight, Rosen (1972) and Pakes and Nitzan (1983) seem to have specific predictions not only with respect to wage profiles but also with respect to mobility patterns. A main prediction of Rosen's model is that workers consistently move to jobs with less learning opportunities. In my context, that may imply that workers move from more to less R&D-intensive firms, but as pointed out by Rosen himself, there is heterogeneity with respect to the learning content of jobs not only across, but also within, firms. Hence, a clear prediction cannot be deduced.

Pakes and Nitzan (1983) predict that R&D firms are able to avoid turnover, and thereby spillovers, by sharing the monopoly rent at stake with the workers. In the presence of spin-off innovations or sources of spillovers other than labor mobility, however, they show that mobility actually can be a way of appropriating returns. The model, therefore, like Rosen's, fails to give clear predictions with respect to worker mobility between firms. Furthermore, Pakes and Nitzan do not consider firm-specific knowledge. If firms with different levels of R&D intensities differ

[13] Changes in R&D intensity within a research career will, however, reduce the smoothness of such a wage profile. Fosfuri and Rønde (2004) formally extend the model of Pakes and Nitzan (1983) to a 3-period setting with a focus on cumulative innovation.

with respect to firm-specific human capital, this will also influence the relationship between turnover and R&D investments.[14]

Because of the lack of strong predictions, empirical mobility patterns cannot be used directly to test the theories above. In the empirical part of the article, I will therefore focus on the effect of R&D on wages. A descriptive analysis of R&D and turnover can be found in the working paper version of this article (Møen 2000).[15]

Data and Industry Description

Data Description

The data used in this study come from three main sources: governmental administrative records prepared by Statistics Norway, the annual manufacturing census of Statistics Norway, and the biennial R&D survey of Statistics Norway supplemented with other surveys of immaterial investments and innovation done by the same bureau. The Norwegian data are extraordinary in the senses that the entire working population can be followed over a number of years for which extremely rich information is available both about the workers and about their employers. When analyzing wage profiles and labor mobility, the extensive coverage offered by the Norwegian data is a great advantage.

I have chosen to focus on the technical staff in the machinery and equipment industries as these industries have many high-tech firms and have a fairly complete coverage in the R&D surveys.[16] The matched employer-employee data set covers the years 1986–95, and I have only included men employed full-time in the analysis below. Women do not constitute a large share of the labor stock in these industries, and they are known to have different career patterns and preferences than men. Roughly speaking, the main sample has annual observations of about 30,000 workers in 750 plants.

Both the (normalized) length of the highest attained education and the

[14] In the training literature, the effect of training on turnover propensities has been used to assess whether the human capital built up is general or firm specific; see, e.g., Loewenstein and Spletzer (1999) and Parent (1999). For a theoretical model of knowledge diffusion with partly firm specific human capital, see Fosfuri, Motta, and Rønde (2001), who analyze a firm's decision to invest in production facilities abroad.

[15] I find that the turnover rate in the Norwegian machinery and equipment industry is about 20% and that turnover is not significantly related to firms' R&D intensity. Excess labor turnover, by contrast, i.e., hires and quits over and above the level necessary to accomplish changes in the number of employees, is less in R&D-intensive firms. This effect is particularly pronounced for workers with secondary technical education.

[16] I define the technical staff as workers with secondary technical education and workers with higher technical or scientific education. I refer to the latter group as scientists and engineers.

type of education are recorded in the data. Occupation, however, is not available. Hence, it is not possible to look specifically at researchers, and workers' learning will be proxied by the employers' R&D intensity. I measure R&D intensity as R&D man-years per employee at the three-digit line of business level within firms.[17] If all workers within a firm participate equally in the firm's R&D efforts, R&D man-years per employee will measure the share of time that each worker uses to perform R&D. Since R&D work obviously is not shared equally among the employees, R&D intensity is a noisy proxy for what we want to capture. Measurement errors in the R&D variable add to this noise.

Table 1 gives descriptive statistics for the workers. Further information about the data is given in the data appendix.

Industry Description

The Norwegian machinery and equipment industry constitutes about 36% of Norwegian manufacturing in terms of employment, and it performs about 55% of total manufacturing R&D.[18] Both shares are somewhat below the Organization for Economic Cooperation and Development (OECD) average. Close to 40% of the production value is exported. The OECD average is about 25%. Compared with other OECD economies, the Norwegian machinery and equipment industry has a relative strength within shipbuilding and nonelectrical machinery, particularly machinery and tools related to the petroleum sector and to hydro-electric power production. Relative to other OECD economies, it has a modest motor vehicle industry. The share of professional goods and electronics is also somewhat below the OECD average, despite the Norwegian government (like most OECD governments) having actively promoted the electronics industry since the 1960s. The share of office machinery and computers was right on the OECD average in the years analyzed, and it was more R&D-intensive than the OECD average. The main company in this sector was Norsk Data. For many years this company was a successful producer of minicomputers, but it run into considerable difficulties adapting to the PC revolution of the late 1980s. It went out of

[17] This means that R&D intensity is measured at a level "in between" the firm and the plant. I will use the term "firm-level R&D intensity" in what follows. If R&D man-years were not reported, the value has been imputed based on the firms' R&D spending. I have censored the R&D intensity variable at 0.8 in order to reduce the influence of outliers. This affects 0.4% of the observations with positive R&D intensity.

[18] The description given here is based on average numbers for the years 1986–92 computed from the OECD, Directorate for Science, Technology and Industry, Structural Analysis Industrial Database (1995), and the Analytical Businesss Enterprise Research and Development Database (1996).

Table 1
Worker Characteristics by Education

	Mean	SD	Tenth Percentile	Ninetieth Percentile
Higher technical or scientific education:				
Years of education	14.5	1.7	13.0	17.0
Years of experience	17.9	11.6	4	35
Years of tenure[a]	6.4	5.1	1.3	13.2
Wage in 1995 NOK	355,100	127,800	242,400	480,800
Working at R&D performing plant	78%	.17	0	1
R&D intensity if at R&D performing plant	.123	.129	.006	.266
Union membership	28%	.20	0	1
Secondary technical education:				
Years of education	11.2	.9	10.0	12.0
Years of experience	17.2	11.9	3	35
Years of tenure[a]	6.6	5.6	1.2	13.2
Wage in 1995 NOK	246,100	70,900	177,700	336,400
Working at R&D performing plant	55%	.25	0	1
R&D intensity if at R&D performing plant	.054	.078	.001	.147
Union membership	45%	.25	0	1

NOTE.—NOK = Norwegian kroner. The numbers are based on all worker-year observations of men in the machinery and equipment industry in Norway, 1986–95, working full time, whose current R&D intensity is known, and whose R&D experience in at least 1 previous year is known (see table A1). There are 65,422 observations of workers with higher technical education and 227,418 observations of workers with secondary technical education in the sample. An R&D plant is a plant belonging to a firm that conducts some R&D within the plant's three-digit international standard industrial classification industry. The R&D intensity is measured as R&D man-years per employee at the three-digit line of business level within firms. Wage in 1995 NOK is rounded to the nearest 100.

[a] 17% of the observations have the job starting date censored at April 30, 1978.

business during the severe downturn in the Norwegian economy in the early 1990s.

The Effect of R&D Investments on Wages

Pakes and Nitzan (1983) predict lower starting wages and higher wage growth for workers doing research, and Rosen (1972) predicts the same pattern more generally for workers having jobs with a high learning potential. A key assumption behind both models is that workers mainly acquire general knowledge on the job. Testing these models, that is, testing to what extent different firms offer different learning opportunities, and to what extent workers pay for their knowledge accumulation, we would like to estimate equation (1), which is Rosen's point of departure. In principle this is possible. Human capital, H, can be decomposed, and the price or relative weight of its various components can be estimated using a standard log-linear hedonic wage regression. Furthermore, potential learning-by-experience on the job, k, may be proxied by the employer's R&D intensity as it seems reasonable to assume that workers in high-

Table 2
Worker Mobility between Plants with Known R&D Intensity

	With No R&D	With R&D Intensity ∈ ⟨0, .05]	With R&D Intensity ∈ ⟨.05, .2]	With R&D Intensity > .2	Total No. of Separations
Left a non-R&D plant and joined a plant	65.5	27.3	5.7	1.5	3,168
Left a plant with R&D intensity ∈ ⟨0,.05] and joined a plant	27.8	61.1	8.9	2.2	3,330
Left a plant with R&D intensity ∈ ⟨.05,.2] and joined a plant	11.9	42.6	40.4	5.1	2,841
Left a plant with R&D intensity > .2 and joined a plant	12.9	9.3	13.9	64.0	497
% of jobs in plants	34.6	42.3	17.2	5.9	

NOTE.—The numbers are percentage shares of total separations from each category of plants and sum to 100 horizontally. Transitions out of the sample have been excluded.

tech, R&D-intensive firms learn more than workers in low-tech firms. However, some problems are immediately evident. Work experience needs to be decomposed according to the training or research content of the jobs that workers have had at different stages of their career, but complete information about the workers' career histories is not available. Furthermore, it is far from obvious how one can summarize what is known about the workers' experience from different firms into a good measure of human capital. In what follows, I will suggest several solutions to these problems.

A First Look at the Effect of R&D on the Earnings Profile

One way to get around the missing career data is to assume that workers' career trajectories are such that the R&D intensity is constant over their career. Table 2, reporting a transition matrix of job changes for technical employees between plants with known R&D intensities, shows that this assumption is valid as an approximation. Of workers leaving a firm that does not conduct R&D (within the plant's line of business), 65.5% move to another firm that does not conduct R&D, even though jobs in such firms account only for 34.6% of all jobs. Of workers leaving a firm with R&D intensity above 0.2, 64.0% move to another firm with R&D intensity above 0.2, although such firms only account for 5.9% of all jobs. The pattern is the same for workers leaving firms with intermediate levels of R&D intensity.[19] One explanation for the observed sta-

[19] The correlation coefficient between R&D intensity in year t and $t-1$ is 0.84. It falls somewhat when the time interval is increased, but the coefficient is still 0.57 and highly significant between year t and $t-9$. This is the longest observable time span. The correlation coefficients are downward biased because of measurement errors in the R&D intensity.

bility in R&D intensity across jobs may be that there is some specificity associated with a given technology level within the industry. As workers grow older, they will then prefer jobs with less learning within firms at the same level of R&D intensity as those for which they have previously worked. Another explanation may be that workers have preference for work at a given technology level. The work of Almeida and Kogut (1999), Stern (1999), and others suggests that scientists and technical personnel have preferences regarding the technological environment in which they work.

Assuming career trajectories with a constant R&D intensity, we can utilize the structural relationship between k and H, given in equation (2) together with the optimal time path for learning investments implicit in equation (4). Under this assumption, the R&D intensity will at each point of time reveal information both about k and about the component of H representing accumulated R&D experience. More specifically, the estimated joint effect will give the returns to R&D experience minus the cost of learning. Working for a highly R&D-intensive employer should cause a large negative wage premium early in the career, reflecting the implicit price paid for the R&D experience. At the same time, this experience has not had much time to affect the stock of human capital. As time goes by, workers' willingness to pay for human capital accumulation decreases and approaches zero, but differences in previous R&D experience will translate into differences in human capital. Workers who are in R&D-intensive firms and have a long R&D-intensive career behind them will therefore have a large positive wage premium reflecting the human capital accumulated.

Table 3, in columns 1 and 3, reports the results of ordinary least squares wage regressions where cross-terms between experience and current R&D intensity are added to test the hypothesis that employees with a career in R&D-intensive firms have a steeper experience-earnings profile than other workers. Note that my data include multiple observations on the same worker and that the estimated coefficients do not reflect any correction for unobserved heterogeneity.[20] I return to this problem in the next section. Additional control variables included are years of schooling, seven experience dummies, a quadratic in plant number of employees, and year dummies.[21] An alternative illustration is given in figure 1, where earnings-experience profiles for workers in firms with no R&D and in

[20] Huber-White robust standard errors allowing for clustering of errors by individuals, as implemented by Rogers (1993) in STATA, are in parentheses.

[21] I have chosen to use experience dummies rather than a higher order polynomial in the main specification because the tabulation of cross-terms between R&D intensity and a higher order polynomial is difficult to interpret.

Table 3
The Effect of Current R&D and Previous R&D Experience on Earnings

	Secondary Technical Education		Higher Technical or Scientific Education	
	(1)	(2)	(3)	(4)
Current R&D intensity				
× less than 1 year experience	−.204**	−.204**	−.134**	−.130**
	(.050)	(.050)	(.044)	(.044)
× 1–2 years experience	−.314**	−.342**	−.131**	−.266**
	(.062)	(.073)	(.030)	(.038)
× 3–5 years experience	−.190**	−.233**	−.058*	−.123**
	(.031)	(.039)	(.023)	(.027)
× 6–10 years experience	−.177**	−.208**	.010	−.097**
	(.028)	(.032)	(.023)	(.025)
× 11–15 years experience	−.080*	−.084*	.014	−.062*
	(.032)	(−.037)	(.027)	(.069)
× 16–20 years experience	−.069⁺	−.079⁺	.031	−.022
	(.036)	(.041)	(.034)	(.038)
× 21–35 years experience	.089**	.038	.094**	.039
	(.029)	(.030)	(.030)	(.032)
× more than 35 years experience	.234**	.285**	.222**	.265**
	(.048)	(.053)	(.056)	(.062)
Average R&D intensity over previous career				
× 1–2 years experience		.055		.261**
		(.093)		(.051)
× 3–5 years experience		.094		.142**
		(.057)		(.038)
× 6–10 years experience		.070		.186**
		(.055)		(.036)
× 11–15 years experience		.010		.167**
		(.062)		(.038)
× 16–20 years experience		.028		.134**
		(.059)		(.052)
× 21–35 years experience		.171**		.165**
		(.060)		(.053)
× more than 35 years experience		−.158*		−.149
		(.072)		(.103)
Sample size	227,418	227,418	65,422	65,422
R^2	.19	.19	.28	.28

NOTE.—The dependent variable is ln (real annual earnings). Control variables included in the regression, but not reported, are seven experience dummies, years of schooling, a quadratic in plant number of employees, and year dummies. The coefficients are estimated using ordinary least squares. Huber-White robust standard errors allowing for clustering of errors by individuals are in parentheses.

⁺ Significant at the 10% level.
* Significant at the 5% level.
** Significant at the 1% level.

firms with R&D intensity 0.2 is graphed, based on a specification continuous in experience.

The results support the main theoretical prediction of Rosen (1972) and Pakes and Nitzan (1983). Early in the career both workers with secondary technical education and scientists and engineers accept a significant wage discount when working for R&D-intensive firms, but over

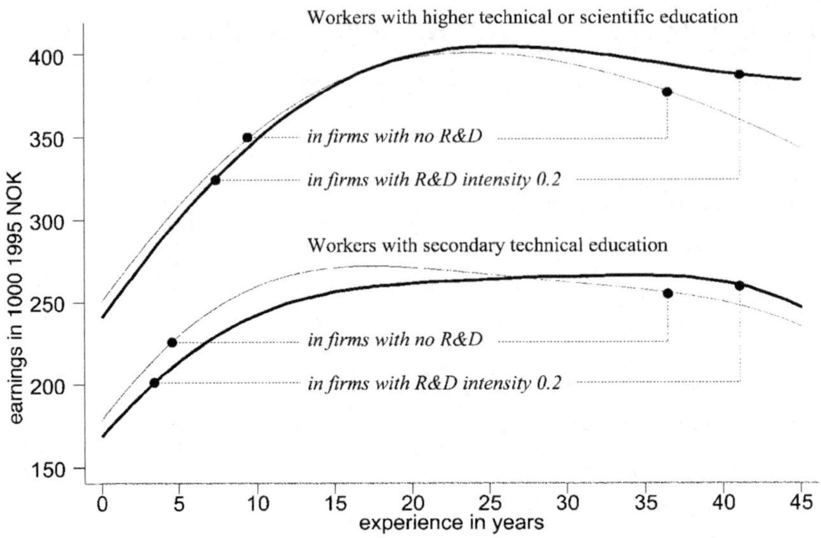

FIG. 1.—The graphs are based on regressions similar to those in table 3, cols. 1 and 3, except that experience dummies interacted with R&D intensity is exchanged with a quartic in experience interacted with a quadratic in R&D intensity. The graphs for workers with higher education are based on 15 years of education and a firm with 100 employees. The graphs for workers with secondary education are based on 12 years of education and a firm with 100 employees. Business cycle conditions are assumed to be like 1995.

time this discount is changed into a significant wage premium. The pattern strongly suggests that R&D investments of firms translate into general human capital and that workers both pay and get paid for the knowledge they accumulate. Note that both the discounts and the premia are biased toward zero because of measurement errors in the R&D variable. (Further biases are discussed below.) Despite this, the coefficients translate into numbers that are of economic significance. Looking, for example, at a firm with 0.2 R&D intensity, scientists and engineers in their first year have on average 2.7% lower wages than colleagues starting in firms with no R&D. Scientists and engineers with more than 35 years of experience have on average 4.4% higher wages than similarly experienced colleagues in firms with no R&D. About 18% of the scientists and engineers in the sample work in firms with R&D intensity above 0.2. Using an alternative specification, interacting experience with a dummy that is one if the R&D intensity is 0.2 or above, I have estimated the average discount in the beginning of the career for this group of workers to be 6.1% and the premium at the end of the career to be 6.8%.[22]

[22] In this regression, workers in firms with R&D intensity between 0.2 and 0.05 were excluded. This was done partly to facilitate a sharper comparison between

The magnitude of the discounts and premia are similar for workers with secondary technical education in R&D-intensive firms. According to the estimates in table 3, column 1, they have a 4.1% wage discount in the beginning of their career and a 4.7% premium in the end of their career if working in a firm with 0.2 R&D intensity.

One way to check the plausibility of the estimates is to calculate the internal rate of the return to choosing an R&D-intensive career. For a worker with secondary technical education, the internal rate of return is 5.7%, and for workers with higher technical or scientific education, it is 3.6%.[23] These numbers should be considered rough estimates, but they are in a reasonable range. Figure 1, which is based on a specification with a quartic in experience interacted with a quadratic in R&D intensity, gives the same qualitative results as table 3. In figure 1, however, the premium late in the career is particularly evident for scientists and engineers. The wage discount early in the career, by contrast, is most evident for workers with secondary technical education.[24]

Given that research is mostly performed by workers with higher education, the significant effect of R&D on wages for workers with secondary technical education, brought out both in table 3 and in figure 1, is somewhat surprising. Presumably, R&D intensity not only measures direct research exposure but works as a general proxy for technological training at all levels within firms. This suggests that the estimated coefficients are driven more by the long-term value of high-tech experience, that is, the accumulation of ordinary human capital as modeled by Rosen (1972), than by knowledge about specific research results (intellectual human capital) as modeled by Pakes and Nitzan (1983).

---

workers in firms with high and low R&D intensity and partly because R&D intensity is a poorer proxy for each individual's R&D exposure in firms with intermediate levels of R&D intensity. The results from this alternative specification are reported in the working paper version of this article (Møen 2000).

[23] I assume that the workers are employed in a firm with 100 employees and that the business cycle is as it was in 1995. Workers with secondary education are assumed to have 12 years of schooling and to work for 45 years. Workers with higher education are assumed to have 15 years of schooling and to work for 42 years. An R&D-intensive career is defined as a career in a firm with R&D intensity 0.2 or above. The benchmark is a firm with R&D intensity 0.05 or below.

[24] A slight anomaly in fig. 1 is the increase in the wage discount for workers with secondary education over the first 10 years of their careers. Such an increase in the very first years is also found in tables 3 and 4 and is not inconceivable, as discussed in n. 28. Whereas the specifications with experience dummies are completely flexible with respect to the length of such an effect, the specification behind fig. 1 causes some interdependence in discounts and premia across time. This may explain why the effect is most visible there.

Potential Biases

One major obstacle to identifying compensating differentials, whether associated with training or other job amenities, has been the potential correlation between job amenities and unobserved individual characteristics. This problem is also present in my analysis. In addition, there are potential biases caused by imprecise and incomplete data. In this section I will discuss these biases and to what extent they can be alleviated.

Two specific biases have already been mentioned: measurement error and a potential ability bias. Standard measurement error in the R&D variables will bias all coefficients of interest toward zero. The ability bias is more complex. A bias arises in Rosen's model because highly talented workers have a lower cost of learning and absorb more knowledge in a job with a large potential for learning than less-talented workers. This implies a tendency for talented workers to self-select into R&D-intensive firms. A steeper wage profile in R&D-intensive firms will then partly be driven by the greater learning opportunities in these firms and partly by "fast-learning" workers being attracted by these greater learning opportunities. Note that this is not the "traditional" ability (to produce) bias associated with unobserved differences in skills and thoroughly studied in the returns to education literature. The traditional ability bias is a fixed effect affecting the wage level. Learning ability as modeled in Rosen (1972) affects the wage slope. Although not explicitly modeled by Rosen (1972), ability to produce is likely to be correlated with learning ability. An ability to produce bias will, if present, bias the estimated wage level upward, causing the wage discount in the beginning of the career to be underestimated and the wage premium at the end of the career to be overestimated.

A fourth bias is present in table 3 to the extent that workers switch between employment in high-tech and low-tech firms. Although table 2 indicates that this kind of behavior is not very common, it clearly does happen. A bias arises because the specification used assumes that we can compare experienced workers in R&D-intensive firms with experienced workers in less R&D-intensive firms and learn how much more human capital is accumulated in R&D-intensive firms. Workers who transfer out of R&D-intensive firms, however, will increase the wage level of the "comparison group" in the less R&D-intensive firms and cause a downward bias on the estimated gain from working in R&D-intensive firms. In the same way, workers who transfer from firms that do not invest much in R&D to firms that do have less human capital than those who have been in R&D-intensive firms for their entire career.

Hence, they will reduce the average wage level in R&D-intensive firms and add to the bias. The result is that the wage premia associated with the last periods of a "high-tech career" are underestimated, that is, we will underestimate the steepness of the experience earnings profile.[25]

Finally, there is a fifth bias or, rather, possibility for misinterpretation. The work of Almeida and Kogut (1999), Stern (1999), and others suggests that scientists and technical personnel have preferences regarding the technological environment in which they work. Stern, in particular, finds that firms that allow their workers to do "open" science can pay lower wages for a given quality of their workforce. This implies that the wage discount for young workers in R&D firms observed in table 3 may partly be caused by a preferences for working in such an environment, rather than itself being payment for greater learning opportunities. The wage discount when interpreted as a price paid for learning will then be overestimated, and the corresponding wage premium at the end of the career will be underestimated.

To sum up the above discussion: first, measurement error in R&D will bias all coefficients of interest toward zero. Given what is known about the data quality, there is no doubt that this bias is present. Second, a traditional ability bias (ability to produce) may bias the estimated wage level in R&D firms upward, and a "preference for R&D" bias may bias the estimated wage level in R&D firms downward. Third, a learning ability bias, as modeled by Rosen (1972), may bias the estimated wage growth in R&D firms upward, and a "switching bias" may bias the estimated wage growth in R&D firms downward.

Estimates Based on Wage Growth

A simple way to avoid the traditional ability bias, the preference for R&D bias, and the switching bias is to estimate a wage equation in first differences, that is, to investigate how firms' R&D intensities affect wage growth. This is done in table 4 on workers staying with the same employer over the years $t - 2$ to t.[26] A drawback of this specification is that it does

[25] By reducing the steepness of the experience earnings profile for workers with a high-tech career, this bias could explain why the estimated net return does not become positive until the workers have somewhere between 10 and 20 years experience. The bias is eliminated when current and previous R&D are included separately in the regression; see table 3, cols. 2 and 4, discussed below.

[26] Møen (2000), table 5, reports the same regression on a sample including separators, trying to utilize the change in R&D intensity between their present and previous employer. The results did not come out particularly well. One reason for this is presumably measurement error in the change in R&D intensity; see the discussion in n. 27 below.

Table 4
The Effect of R&D on 2-Year Earnings Growth for Workers Staying with Same Employer from Year $t-2$ to Year t

	Secondary Technical Education	Higher Technical or Scientific Education
Average R&D intensity over the years $t-2$, $t-1$, and t		
× 2 years experience	−.224	.026
	(.145)	(.044)
× 3–5 years experience	.084*	.060**
	(.040)	(.020)
× 6–10 years experience	.078**	.050**
	(.017)	(.013)
× 11–15 years experience	.102**	.040*
	(.017)	(.016)
× 16–20 years experience	.092**	.048**
	(.015)	(.015)
× 21–35 years experience	.063**	.008
	(.009)	(.011)
× more than 35 years experience	.047**	.007
	(.016)	(.020)
Sample size	135,394	40,809
R^2	.06	.07

NOTE.—The dependent variable is the first difference of ln (real annual earnings) between year t and year $t-2$. Control variables included in the regression, but not reported, are cash flow before wage payments per employee, seven experience dummies, the change in plant size measured by number of employees, and year dummies. The coefficients are estimated using ordinary least squares. Huber-White robust standard errors allowing for clustering of errors by individuals are in parentheses.
$^+$ Significant at the 10% level.
* Significant at the 5% level.
** Significant at the 1% level.

not identify the learning effect of R&D on the wage level.[27] Furthermore, we are still left with the measurement error bias and the learning ability bias. This is not a devastating problem, however, as these biases cannot drive the results. Measurement error represents a bias against the predictions I want to test, and the learning ability bias as modeled by Rosen will not be present unless R&D firms do offer greater learning opportunities. If present, the learning ability bias will, however, go in the opposite direction of the measurement error bias and increase the coefficient

[27] The wage level is in principle identified if using a fixed effects specification, i.e., a within estimator. Such a specification does not produce results in line with those in tables 3 and 4, however, and the estimated coefficients become highly unstable. At first this may seem puzzling, as there is a close relationship between first difference estimators and the fixed-effects estimators. The relationship estimated in table 4 is not, however, a complete first difference transformation of the relationship estimated in table 3. Table 3 relates wage levels to R&D investments, while table 4 relates wage growth to R&D investments, not to the growth in such investments. A within or first difference transformation of R&D magnifies the effect of measurement error greatly, as there are both severe measurement errors in the level of R&D investments and little within-firm variation.

on the interaction between experience and R&D intensity. Without an instrument that is correlated with the propensity to choose an R&D-intensive career, but not otherwise with the workers' wage growth, it is not possible to solve this problem. Such an instrument is not available.

Looking now at the results in table 4, using first differences, the broad picture emerging is that workers with technical or scientific education in R&D-intensive firms staying with the same employer have higher wage growth throughout their career.[28] This is consistent with the previous finding that R&D translates into human capital on which workers earn a return.[29] Wage growth also appears to level off toward the end of the career, consistent with workers having less incentive to accumulate human capital when getting closer to retirement age.

Estimating the Price of Learning and the Return to R&D Experience Separately

Table 3, in columns 1 and 3, utilizes cross sectional information only and estimates in one coefficient the return to previous R&D experience minus the price paid for current learning opportunities. Utilizing the longitudinal dimension of the data set, it is possible to specify these two components separately. The learning opportunity that a worker faces depends only on current R&D intensity, while average R&D intensity in previous years reveals information about the workers' R&D experience. Note, however, that the stability in R&D intensity over the workers' careers, evident in table 2 (and n. 19), makes current and previous R&D intensities somewhat collinear. A high level of precision can therefore not be expected when including both variables.

Table 3, in columns 2 and 4, reports the results of interacting current R&D intensity and the average of previously observed R&D intensities separately with experience dummies. The first thing to notice is that the coefficients on the average of previously observed R&D intensities, that is, the return to R&D experience, are mostly positive, while the coefficients on current R&D intensity, that is, the implicit price paid for learning

[28] Wage growth for workers with secondary technical education is negatively correlated with the employers' R&D intensity in the first 2 years of the career. This is also evident in table 3, and in fig. 1; see n. 24. This evidence may reflect that it takes some time to "absorb" the complexity of R&D-intensive firms or that workers, because of imperfect information about the quality of the training, are unwilling to pay the full cost of the training at once, but that firms are able to extract this premium through lower wage growth during the very first years of the workers' career.

[29] Cash flow before wage payments per worker is included to control for the rent-sharing effect of successful innovations found by van Reenen (1996). Such a rent-sharing effect is present in the data, but it does not dominate the effect of R&D.

opportunities, are mostly negative. Note also that current R&D intensity has a more negative impact when previous R&D experience is included (see cols. 1 and 3).

The price paid for learning decreases over time as predicted by theory, but the data do not bring out the expected wage increase over time that should be associated with R&D experience. Furthermore, the coefficients on current R&D, that is, learning, become positive late in the career. These two features seem connected. The employer's current R&D intensity appears to be a better proxy for old workers' human capital than the average of previously observed R&D intensities. This could result from some selection process where workers whose technological experience has become obsolete move out of, or are displaced from, R&D-intensive firms.

In order to assess the importance of learning for the industry on an aggregate level, I have summarized the estimated wage discount for all R&D firms. This sum amounts to 0.7% of the total wage bill for technical personnel in all R&D performing firms and 2.6% of industry R&D investments. Looking only at firms with R&D intensity above 0.2, the wage discount represent 3.0% of their total wage bill and 2.5% of their R&D investments. These numbers are not very big, nor are they negligible.

The Value of R&D Experience from the Current Employer versus Previous Employers

Loewenstein and Spletzer (1998, 1999) and Lengermann (1999), who study the effect of formal on-the-job training, find that the return to training received from previous employers exceeds the return to training received from the current employer. Loewenstein and Spletzer argue that this may reflect the fact that employers extract some returns to general training and that workers do not realize the full returns until they change jobs. If something similar applies to the value of experience from R&D-intensive firms, it would imply that the potential R&D spillovers involved when workers in R&D-intensive firms change employers is only partially internalized in the labor market. In order to investigate this possibility, I have, for each employee where sufficient career information is present, calculated the average observed R&D intensity in previous years when working for the current employer and the average observed R&D intensity in years working for previous employers.

With a smaller sample size and three R&D measures, an extension of the specification with experience dummies interacted with R&D intensities, used in table 3, is not feasible. It is necessary to put more restrictions on the specification, and I have chosen to approximate the price paid for learning opportunities with current R&D intensity and its interaction with years of overall work experience. The R&D experience built up with the current employer is proxied with the average observed R&D intensity in

Table 5
The Effect of Current R&D, R&D Experience from the Current Employer, and R&D Experience from Previous Employers

	Secondary Technical Education		Higher Technical or Scientific Education	
	(1)	(2)	(3)	(4)
Current R&D intensity	−.316**	−.480**	−.104**	−.157**
	(.042)	(.063)	(.034)	(.052)
× experience	.017**	.038**	.003	.008
	(.002)	(.008)	(.002)	(.007)
× experience^2		−.001**		−.0001
		(.0002)		(.0002)
Mean R&D intensity in previous years with current employer				
× tenure	.028**	.124**	.032**	.089**
	(.008)	(.018)	(.007)	(.014)
× tenure^2		−.014**		−.008**
		(.002)		(.002)
Mean R&D intensity in years with previous employer(s)				
× (experience − tenure)	.008*	.005	.005*	.013*
	(.003)	(.008)	(.002)	(.006)
× (experience − tenure)^2		.0001		−.0004
		(.0003)		(.0003)
Sample size	62,243	62,243	17,675	17,675
R^2	.23	.24	.33	.33

NOTE.—The dependent variable is ln (real annual earnings). Control variables included in the regression, but not reported, are years of schooling, a quadratic in plant number of employees, a quartic in experience, a quadratic in tenure, a dummy for having changed employer at least once, year dummies, and a dummy variable for job relationships whose starting date is censored at April 30, 1978, together with its interactions with all tenure variables. The coefficients are estimated using ordinary least squares. Huber-White robust standard errors allowing for clustering of errors by individuals are in parentheses. Mean R&D intensity is calculated over the years where information about the R&D intensity is available. Workers in firms where R&D information is not available in the sample year and in at least 1 prior year, and workers who have had previous employment without R&D intensity being known in at least 1 year have been excluded.

* Significant at the 5% level.
** Significant at the 1% level.

previous years working for this employer times years of tenure with this employer. The R&D experience built up with previous employers is proxied with the average observed R&D intensity while working for previous employers times years of experience prior to the current employment relationship. These measures, resembling sums of R&D intensities, are consistent with equation (2).

Table 5, in columns 1 and 3, reports the results. Columns 2 and 4 report a slightly less restrictive specification where nonlinear interactions with experience and tenure are allowed. All regressions confirm the previous finding that current R&D intensity has a significantly negative impact on wages early in the career. The positive cross-term with experience also confirms that this negative impact, interpreted to be the price paid for learning opportunities, diminishes over time. With respect to the R&D

experience built up over the career, both R&D experience from the current employer and R&D experience from previous employers have a positive and significant impact on wages. The R&D experience from the current employer, however, seems to be more highly valued. Unfortunately, this result is more suggestive than conclusive. In order to construct the variables needed, all years working with the current employer must be included in the sample, while information about previous employers can be less complete. Hence, the average R&D intensity in years working for previous employers is measured with greater error than average R&D intensity in years working for the current employer, and coefficients on variables involving the former measure will therefore be more biased toward zero.[30] In addition, the coefficient on R&D experience with the current employer could be upward biased. This would happen if recent R&D experience shows that knowledge accumulated earlier in the career has not become obsolete. The results for old workers in table 3, columns 2 and 4, indicate that this may be the case.

Robustness and Econometric Issues

A number of alternative specifications have been tried to assess the robustness of the results.[31] In one specification, more than 30 additional

[30] If the sample is restricted to workers whose complete career is known, the return to R&D experience from previous employers appears to be above the return to R&D experience from the current employer for workers with higher education, while both coefficients become insignificant for workers with secondary education. For these workers, the coefficient on previous R&D experience even has a negative sign.

[31] In addition to trying out different specifications within the sample of workers with technical education, I have also run the basic regressions on workers with nontechnical education. The effect of R&D experience on workers with nontechnical secondary and higher education resembles the effect on workers with technical education in that they seem to have a steeper experience-earnings profile if working in R&D-intensive firms. The results are fairly strong for workers with secondary nontechnical education but less evident for workers with higher nontechnical education. It is not clear why these workers should be affected by the R&D intensity of their employers, but several explanations are possible. First, many workers with a general secondary degree are in fact production workers and, hence, quite comparable with those with secondary technical education included in the main sample. Second, R&D-intensive firms may be advanced along many dimensions and may also offer valuable work experience to the nontechnical staff. Third, R&D-intensive firms also appear to be intensive in formal training. In years in which the data set includes measures of both R&D investments and formal training, these measures are significantly, positively correlated. Fourth, it is possible that there is positive correlation between the value of firms' technical and nontechnical trade secrets or that not only the technical staff but also administrative managers in R&D-intensive firms have access to sensitive technological information. Then the Pakes and Nitzan (1983) model applies to this group as well as to the technical employees, and it is in any case conceivable that R&D-

control variables were included, such as proxies for hours worked, the capital-to-labor ratio, the Herfindal index, the market share of the firm, the union density, and four-digit industry dummies. This did not change the quantitative results. The results are also robust to including firm size, years of education, and union density in interaction with experience. Dividing the sample into different time periods, however, reveals that the effect of R&D on the wage-experience profile is more pronounced in the 1980s than in the 1990s. This may be related to the severe recession in the Norwegian economy starting in the late 1980s, causing a restructuring of, and a decline in, some of the most innovative subindustries. The decline in the profitability of high-tech firms is likely to have affected both the returns to previously accumulated human capital and workers' willingness to pay for access to new knowledge. Another interesting finding is reported in Møen (2002), in which I find that the wage discount and steeper wage growth in R&D firms is more strongly associated with IT-related R&D than with other types of R&D. This finding is consistent with the idea that IT represents a "general purpose technology" and, hence, that IT projects generate more general human capital than other R&D projects.

All regressions reported in tables 3–5 allow for correlated error terms across observations of the same individual in different years (see n. 20). However, one could also argue that error terms for workers belonging to the same firm may be correlated. Allowing for such correlations when computing the standard error of the estimated coefficients reduces their significance, but the qualitative results are even robust to including firm-specific fixed effects in the regressions.

A Comparison with the Training Literature

It may be worthwhile to compare the overall results of my analysis to similar analyses of on-the-job training. Although this article, as far as I know, is the first to look at the effect of R&D on wages, there exists a large literature on the effect of formal training. In this literature, a number of authors have found training to be correlated with wage growth, but finding support for a negative effect on starting wages, such as human capital theory predicts, is unusual.[32] Common interpretations are that

intensive firms to a larger extent than other firms use stock options and similar compensation schemes for their managers, e.g., because of cash constraints. Finally, the Norwegian economy is strongly unionized. Unions often demand similar earnings plans for all employees in a firm.

[32] Barron, Black, and Loewenstein (1989), Lynch (1992), or Barron, Berger, and Black (1999). One exception is Autor (2001, 1411). Studying temporary help firms, he finds that "wages are lower at firms offering training by a modest, but statistically significant magnitude." Lynch (1992) find a negative effect of uncompleted training for workers with less than a high school education, but not for workers with a high school degree or some college education.

workers do not pay for general training or that the implicit price is masked by a positive ability bias. In this perspective, the strong negative effect of R&D on starting wages present in this sample is remarkable. It suggests that firms' technology levels are more important to wages than formal on-the-job training. One explanation for this could be that while most formal training is short-term, working in a technologically challenging environment affects human capital accumulation for the entire duration of a job.[33]

Concluding Remarks

Labor mobility is often considered to be an important source of knowledge externalities, making it difficult for firms to appropriate returns to R&D investments. Pakes and Nitzan (1983), however, analyze the problem formally and find that labor turnover should not be a problem for R&D firms. Both scientists and firms are aware of the fact that working on a research project gives access to valuable information. Once such information is disclosed or developed, scientists, if they are to stay with the firm, will have to receive a wage increase reflecting their new market value. Thus, scientists expect that accepting a research position implies a future wage increase, and consequently they can accept an initial wage below their alternative wage, without experiencing a welfare loss.

Research firms are likely to use the most up-to-date technology and frequently change their products and production processes. Because of this, one would think that even workers who do not have direct access to the results of the R&D projects learn more in these firms. Rosen (1972) provides a model in which different firms offer different opportunities for on-the-job learning, and he derives implications with respect to wages that resemble those of Pakes and Nitzan (1983). Rosen thinks of jobs as tied packages of work and learning. Workers sell the services of their skills and simultaneously purchase an opportunity to augment those skills.

I have argued that interfirm transfers of R&D results embodied in people should be analyzed within a human capital framework similar to the models of Rosen (1972) and Pakes and Nitzan (1983).[34] Testing such a framework using matched employer-employee data from the Norwegian machinery and equipment industries, I find that the technical staff in R&D-intensive firms indeed pay for the knowledge they accumulate on the job through lower wages in the beginning of their career. Later in the

[33] This view is consistent with the results in Grooth and Mekkelholt (1995) and in Parent (2002, 376), who in a recent study of earnings finds "strong evidence of a negative correlation between job-specific wage growth and the job-specific intercept."

[34] There exists a related literature analyzing the transfer of embodied human capital from universities to firms; see Jensen and Thursby (2001) and Zucker, Darby, and Torero (2002) for recent contributions.

career they earn a return on these implicit investments through higher wages.[35] These results appear despite the existence of several biases against finding different experience-earnings profiles in R&D-intensive and non-R&D-intensive firms. This suggests that potential externalities associated with labor mobility out of research firms, at least to some extent, are internalized in the labor market.[36]

An alternative explanation for the steeper experience-earnings profile in R&D-intensive firms has been suggested to me by several discussants. Inspired by Freeman (1977), one might imagine that workers and firms are uncertain about the workers' ability.[37] One might also imagine that high-ability workers have higher productivity in research firms than in firms using well-known technologies demanding less problem solving.[38] All workers are in this case willing to accept a lower wage in R&D firms because the expected wage growth is higher. As information about ability is revealed, ex post high-ability workers in research firms perform well, while others, ex post low-ability workers, exit having lost in the "lottery."[39] Such a model, which does not involve any kind of human capital accumulation, is consistent with the results in tables 3, columns 1 and 3, but not with the results in columns 2 and 4, where previous R&D experience is introduced in addition to current R&D. In a model like the one sketched above, current R&D should be associated with a wage premium as soon as the firm and the worker have learned about the worker's ability and should be stable thereafter, while previous R&D experience should not affect wages. This is not the case.

An important question that this analysis has not addressed is whether

[35] In order to make separate evaluations of the models of Rosen (1972) and Pakes and Nitzan (1983), I believe that detailed survey data on wage contracts for key scientists is necessary. Gathering and analyzing such data are left for future work. The significant finding in this study is that workers seem willing to do intertemporal wage trade-offs that can internalize potential R&D spillovers.

[36] The only related finding I know of in the previous literature is Zucker, Darby, and Armstrong (1998, 80), who, in an academic setting, claim that "competitive university salaries are lower, other things equal, in areas where faculty expect the possibility of receiving substantial outside income or wealth as a result of skills developed doing research at the university."

[37] Freeman's (1977, 421) point of departure is that "not all prospective research workers prove equally adept. Moreover, no individual can be certain in advance just how well he will do. As time passes, his capabilities become clearer to himself as well as to his employer."

[38] An alternative formulation is that there is uncertainty about the quality of the match (see, e.g., Jovanovic 1979) and that match quality is more important in research firms than in other firms with more "routine jobs."

[39] Freeman (1977) focuses on risk-averse workers' demand for a wage contract with insurance, but it follows from his model that if workers are not completely risk averse they will accept a beginning wage below their alternative wage in order to participate in the implicit "lottery."

workers pay for the full value of the knowledge they accumulate in R&D-intensive firms. The conceptual distinction between knowledge diffusion and true externalities motivated my analysis, but identifying possible true externalities is challenging. Building on previous methodologies such as that of Jaffe (1986), one could construct a spillover pool for each firm based on R&D conducted in other firms from which the firm in question has recruited. This measure could be inserted in a knowledge production function. The problem is that it is not possible to know whether the coefficient on such a spillover pool represents an externality or whether the knowledge is "bought" in the labor market. Inserting the spillover pool in a profit function will not solve the problem either. In equilibrium, labor mobility should affect the profit of all firms, and only by comparing different equilibria with exogenous variation in the degree of labor mobility could an effect of mobility-induced externalities be identified. This is obviously difficult. I believe that the best way to proceed is to model explicit mechanisms that might cause externalities and to derive testable implications from such specific models. Case studies of firms that have lost or hired workers with strategic knowledge would also be valuable.

From a theoretical point of view it is conceivable that labor mobility does create some externalities. If firms have limited ability to commit themselves to share future profits with their employees, or if several workers have access to exactly the same research results, this may undermine the "joint profit" effect underlying optimal R&D investments in Pakes and Nitzan's (1983) model.[40] Furthermore, information asymmetries and other barriers to mobility may enhance the firms' ability to appropriate rents while at the same time reduce workers' incentives to pay for knowledge accumulation.[41] Mechanisms that induce employers to pay for general human capital accumulation create a positive externality to the worker's future employer if the worker decides to quit or if the firm goes out of business. A complete welfare analysis must also incorporate the fact that, even if workers pay for all the knowledge they accumulate, this "solution"

[40] See nn. 9 and 10.

[41] See Acemoglu and Pischke (1999), although they do not write with reference to R&D investments. A particularly important kind of imperfection may be distortions in the wage structure, which makes a wedge between wages and marginal productivity increase with the workers' human capital. Firms then have an incentive to invest in R&D-producing general human capital because they get a share of the return. A simple mechanism that could cause such wage compression is that firms receive a fraction of the productivity of the workers as profit because of matching, search costs, or other sorts of labor-market friction. If the employer receives a fraction of the workers' productivity, the employer's level of profit will increase with the workers' productivity and, therefore, with their human capital. Another possible mechanism is complementarity between firm-specific and general human capital. In this case, the alternative wage for the scientist will increase less than his or her productivity as he or she receives training.

to the spillover problem does not guarantee optimal R&D investments. If workers cofinance R&D through lower wages, and if the value of the knowledge they accumulate depends on the outcome of the R&D project, they become exposed to the risk associated with the project. Risk aversion among workers may then become a new source of distortion since human capital investments cannot be diversified.[42] Liquidity constraints making workers unwilling to trade off current wage for future wage on a large scale may also create problems.[43]

Appendix

Data Issues

Information about individual workers comes from a number of governmental administrative records, which are prepared by Statistics Norway for research use. Barth and Dale-Olsen (1999, app. 2), give some details on the various registers included in the database. The available registers cover the years 1986–95.

Plant-level information about employers comes from the annual manufacturing census of Statistics Norway. Microdata are available from 1966.[44] Information about R&D at the line of business level within firms are collected from R&D surveys and other surveys of immaterial investments and innovation. Microdata are available for 1970, biennially 1975–81, annually 1981–85, and biennially 1985–95.[45] In the machinery and equipment industries utilized in this study, the R&D surveys have close to full coverage for firms with more than 20 employees. For years and firms not covered by the R&D surveys, three other data sets have been utilized. Two surveys of immaterial investments cover the years 1986–90, and an innovation survey covers the year 1992.[46]

[42] See n. 8.

[43] See Fosfuri et al. (2001) for a model emphasizing this in a developing country context. Aghion and Tirole (1994) make a similar point.

[44] The census is documented in the series *Manufacturing Statistics*, Official Statistics of Norway, Statistics Norway, Oslo. The microdata are documented in a mimeograph from 1991 by Halvorsen, Jenssen, and Foyn (1991) in Statistics Norway.

[45] The R&D surveys are documented in the series *FoU-statistikk Forsknings- og utviklingsarbeid Utgifter og personale* (1970–1991 by Forskningsrådenes samarbeidsutvalg and Norges Forskningsråd); *FoU-statistikk og indikatorer* (1995 by Utredningsinstituttet for forskning og høyere utdannelse); and *Det norske forskningssystemet—statistikk og indikatorer* (1997 by Norges forskningsråd). More details are given in the series *FoU virksomhet Utgifter og personale i næringene industri, bergverksdrift og anleggsvirksomhet*, by NTNF and in Report 96/14 from Statistics Norway by Skorge, Foyn, and Frengen. All publications have summaries in English.

[46] The former two surveys are documented by Frenger in Interne notater 90/11 and Notater 93/14 from Statistics Norway. The latter survey is documented in the reports 95/7 and 95/26 from Statistics Norway by Frengen, Foyn, and Ragnarsøn. Report 95/26 is in English.

Table A1
Sample Size and Trimming Procedures

	Sample Size
Total no. of observations in the machinery and equipment industries, 1986–95:	810,559
Women	125,111
Part-time workers	11,314
Workers with unknown education	8,968
Workers with primary education only	141,216
Workers with secondary or higher nontechnical-nonscientific education	94,325
Total no. of observations of full-time working male technical staff:	429,625
Workers in firms that cannot be matched to the time series files of the manufacturing statistics	39,527
Workers in firms where R&D information is not available	46,744
Workers not working for the whole year because they are entering the labor force	9,982
Workers not working for the whole year because they are leaving the labor force	14,044
Workers with secondary technical education and earnings below NOK 75.000 (1995 value)	2,723
Workers with higher technical or scientific education and earnings below NOK 150.000 (1995 value)	566
Workers without any information about R&D in previous years	23,189
Sample (trimmed)	292,840

NOTE.—Each entry refers to the number of observations deleted among the observations left after the deletions in the rows above have been conducted. Workers with secondary technical education and earnings below NOK 75,000 (1995 value) and workers with higher technical or scientific education and earnings below NOK 150,000 (1995 value) have been excluded because such low earnings suggest that they have not worked full-time for an entire year.

I have used the following procedure when constructing the R&D database. First, I have linked the R&D surveys to the manufacturing census. Next, for firms and years not included in the R&D surveys, I have used R&D information from the surveys of immaterial investments and from the innovation survey. For firms and years where R&D information is still missing, I have used survey information about planned R&D 1 and 2 years ahead and information about previous R&D. In the final stage, missing R&D variables were imputed by linear interpolation and by extrapolating the first observed R&D intensity backward in time and the last observed R&D intensity forward in time, firm by firm. Firms' R&D investments are known to be stable over time. Imputing missing information when possible, therefore, seems preferable to deleting the observations. Of the worker-year R&D variables, 80% are from surveys, 5% are imputed by interpolation, and 15% are imputed by extrapolation; 65% of the imputed R&D intensities are zero.

Even though the data set is rich, I do not have complete information about the workers' careers. First, the individual records start in 1986. Second, despite my effort to collect R&D information as described above, small firms are not necessarily covered. Third, the match between the different data sources is not perfect. Because of the second and third problem, R&D information is missing for approximately 20% of the worker-year observations. On the positive side, however, I can extract some information about workers' careers prior to 1986, the first year included in the matched data set. I know when the workers started the job they held in 1986, and this can be combined with information about the employers' R&D investments.

Earnings is measured as taxable labor income. I have often referred to this as the workers' wage. Observations with unreliable earnings measures have been excluded. Trimming based on earnings reduces the sample by 8%. Experience is measured as real work experience for the youngest cohorts, years since graduation for older cohorts, and potential work experience for cohorts graduating before November 1970. Both experience and tenure are measured in years (with decimals) completed at the beginning of the calendar year. Dummy variables for experience are based on the integer of experience. Detailed information about the sample size and trimming procedures is given in table A1.

References

Acemoglu, D., and J.-S. Pischke. 1999. Beyond Becker: Training in imperfect labor markets. *Economic Journal* 109, no. 453:F112–F142.

Aghion, P., and J. Tirole. 1994. The management of innovation. *Quarterly Journal of Economics* 109, no. 4:1185–1209.

Almeida, P., and B. Kogut. 1999. Localization of knowledge and the mobility of engineers in regional networks. *Management Science* 45, no. 7:905–17.

Anand, B. N., and A. Galetovic. 2000. Weak property rights and holdup in R&D. *Journal of Economics and Management Strategy* 9, no. 4: 615–42.

Arrow, K. J. 1962. Economic welfare and the allocation of resources for invention. In *The rate and direction of inventive activity: Economic and social factors*, vol. 13, ed. R. R. Nelson, 609–25. NBER Special Conference Series. Princeton, NJ: Princeton University Press.

Autor, D. H. 2001. Why do temporary help firms provide free general skills training? *Quarterly Journal of Economics* 116, no. 4:1409–48.

Balkin, D. B., and L. R. Gomez-Mejia. 1985. Compensation practices in high-technology industries. *Personnel Administrator* 30 (June): 111–23.

Barron, J. M., M. C. Berger, and D. A. Black. 1999. Do workers pay for on-the-job training? *Journal of Human Resources* 34, no. 2:235–52.

Barron, J. M., D. A. Black, and M. A. Loewenstein. 1989. Job matching and on-the-job training. *Journal of Labor Economics* 7, no. 1:1–19.

Barth, E., and H. Dale-Olsen. 1999. The employer's wage policy and worker turnover. In *The creation and analysis of employer-employee data*, ed. J. C. Haltiwanger, J. I. Lane, J. R. Spletzer, J. J. M. Theeuwes, and K. R. Troske, 285–312. Contributions to Economic Analysis no. 241. Amsterdam: North Holland.

Becker, G. S. 1962. Investment in human beings. *Journal of Political Economy* 70, no. 5:9–49.

———. 1964. *Human capital.* New York: Columbia University Press.

Bhide, A. 1994. How entrepreneurs craft strategies that work. *Harvard Business Review* 72, no. 2:150–61.

Combes, P. P., and G. Duranton. 2001. Labor pooling, labor poaching and spatial clustering. CEPR Discussion Paper no. 2975, Centre for Economic Policy Research, London.

Cooper, D. P. 2001. Innovation and reciprocal externalities: Information transmission via job mobility. *Journal of Economic Behavior and Organization* 45, no. 4:403–25.

Fosfuri, A., M. Motta, and T. Rønde. 2001. Foreign direct investment and spillover through workers' mobility. *Journal of International Economics* 53, no. 4:205–22.

Fosfuri, A., and T. Rønde. 2004. High-tech clusters, technology spillovers and trade secret laws. *International Journal of Industrial Organization* 22, no. 1:45–65.

Franco, A. M., and D. Filson. 2000. Knowledge diffusion through employee mobility. Research Department Staff Report no. 272, Federal Reserve Bank of Minneapolis.

Freeman, S. 1977. Wage trends as performance displays productive potential: A model and application to academic early retirement. *Bell Journal of Economics* 8, no. 2:419–43.

Geroski, P. A. 1995. Do spillovers undermine the incentive to innovate? In *Economic approaches to innovation*, ed. S. Dowrick, 76–97. Aldershot: Elgar.

Gersbach, H., and A. Schmutzler. 1997. Endogenous spillovers, the market for human capital, and incentives for innovation. Discussion Paper no. 248, Universität Heidelberg, Wirtschaftswissenschaftliche Fakultät.

Griliches, Z. 1973. Research expenditures and growth accounting. In *Science and technology in economic growth*, ed. B. R. Williams, 59–95. London: Macmillan.

Grooth, W., and E. Mekkelholt. 1995. The rate of return to investments in on-the-job-training. *Applied Economics* 27, no. 2:173–81.

Halvorsen, Reidun, Reider Jenssen, and Frank Foyn. 1991. Dokumentasjon av industristatistikkens tidsseriebase. Photocopy. Statistics Norway, Oslo.

Jackson, T. 1997. *Inside Intel: Andrew Grove and the rise of the world's most powerful chip company*. New York: Dutton.

Jaffe, A. 1986. Technological opportunity and spillovers from R&D. *American Economic Review* 76, no. 5:984–1001.

———. 1996. Economic analysis of research spillovers: Implications for the advanced technology program. National Institute of Standards and Technology, Report GCR 97-708, National Institute of Standards and Technology, Gaithersburg, MD.

Jaffe, A. B., M. Trajtenberg, and R. Henderson. 1993. Geographic localization of knowledge spillovers as evidenced by patent citations. *Quarterly Journal of Economics* 108, no. 3:577–98.

Jensen, R., and M. Thursby. 2001. Proofs and prototypes for sale: The licensing of university inventions. *American Economic Review* 91, no. 1:240–59.

Jovanovic, B. 1979. Job-matching and the theory of turnover. *Journal of Political Economy* 87, no. 5:972–90.
Kim, J., and G. Marschke. 2003. Labor mobility of scientists, technological diffusion and the firm's patenting decision. Discussion Paper no. 01-03, Department of Economics, State University of New York at Albany.
Klette, T. J., and J. Møen. 1999. From growth theory to technology policy—coordination problems in theory and practice. *Nordic Journal of Political Economy* 25, no. 1:53–74.
Klette, T. J., J. Møen, and Z. Griliches. 2000. Do subsidies to commercial R&D reduce market failure? Microeconometric evaluation studies. *Research Policy* 29, nos. 4–5:471–95.
Lengermann, P. A. 1999. How long do the benefits of training last? Evidence from long term effects across current and previous employers. *Research in Labor Economics* 18:439–61.
Levin, R. C., A. K. Klevorick, R. R. Nelson, and S. G. Winter. 1987. Appropriating the returns from industrial research and development. *Brookings Papers on Economic Activity, Microeconomics,* no. 3, 783–831.
Loewenstein, M. A., and J. R. Spletzer. 1998. Dividing the costs and returns to general training. *Journal of Labor Economics* 16, no. 1:142–71.
———. 1999. General and specific training, evidence and implications. *Journal of Human Resources* 34, no. 4:710–33.
Lynch, L. M. 1992. Private sector training and the earnings of young workers. *American Economic Review* 82, no. 1:299–312.
Mincer, J. 1958. Investment in human capital and personal income distribution. *Journal of Political Economy* 66, no. 4:281–302.
———. 1974. *Schooling, experience and earnings.* New York: Columbia University Press.
Møen, J. 2000. Is mobility of technical personnel a source of R&D spillovers? NBER Working Paper no. 7834, National Bureau of Economic Research, Cambridge, MA.
———. 2002. Spin-offs and spillovers: Tracing knowledge by following employees across firms. Discussion Paper no. 5/02, Department of Finance and Management Science, Norwegian School of Economics and Business Administration.
Pakes, A., and S. Nitzan. 1983. Optimum contracts for research personnel, research employment, and the establishment of "rival" enterprises. *Journal of Labor Economics* 1:345–65.
Parent, D. 1999. Wages and mobility: The impact of employer-provided training. *Journal of Labor Economics* 17, no. 2:298–317.
———. 2002. Matching, human capital and the covariance structure of earnings. *Labour Economics* 9, no. 3:375–404.
Rajan, R. G., and L. Zingales. 2001. The firm as a dedicated hierarchy: A theory of the origins and growth of firms. *Quarterly Journal of Economics* 116, no. 3:805–51.
Rogers, W. H. 1993. Regression standard errors in clustered samples. *Stata Technical Bulletin* 13:19–23.
Rønde, T. 2001. Trade secrets and information sharing. *Journal of Economics and Management Strategy* 10, no. 3:391–417.

Rosen, S. 1972. Learning and experience in the labor market. *Journal of Human Resources* 7, no. 3:326–42.

———. 1986. The theory of equalizing differences. In *Handbook of labor economics*, vol. 1, ed. O. Ashenfelter and R. Layard, 641–92. Amsterdam: North-Holland.

Stephan, P. 1996. The economics of science. *Journal of Economic Literature* 34, no. 3:1199–1235.

Stern, S. 1999. Do scientists pay to be scientists? NBER Working Paper no. 7410, National Bureau of Economic Research, Cambridge, MA.

Van Reenen, J. 1996. The creation and capture of rents: Wages and innovation in a panel of UK companies. *Quarterly Journal of Economics* 111, no. 1:195–226.

Weder, R., and H. G. Grubel. 1993. The new growth theory and Coasean economics: Institutions to capture externalities. *Weltwirtschaftliches Archiv* 129, no. 3:488–513.

Zucker, L. G., M. R. Darby, and J. Armstrong. 1998. Geographically localized knowledge: Spillovers or markets? *Economic Inquiry* 36, no. 1:65–86.

Zucker, L. G., M. R. Darby, and M. B. Brewer. 1994. Intellectual capital and the birth of U.S. biotechnology enterprises. NBER Working Paper no. 4653, National Bureau of Economic Research, Cambridge, MA.

———. 1998. Intellectual human capital and the birth of U.S. biotechnology enterprises. *American Economic Review* 88, no. 1:290–306.

Zucker, L. G., M. R. Darby, and M. Torero. 2002. Labor mobility from academe to commerce. *Journal of Labor Economics* 20, no. 3:629–59.

[34]

MANAGEMENT SCIENCE
Vol. 53, No. 6, June 2007, pp. 919–933
ISSN 0025-1909 | EISSN 1526-5501 | 07 | 5306 | 0919

informs®
DOI 10.1287/mnsc.1060.0648

When Do Employees Become Entrepreneurs?

Thomas Hellmann
Sauder School of Business, University of British Columbia, Vancouver, British Columbia V6T 1Z2, Canada
hellmann@ubc.sauder.ca

This paper examines an economic theory of when employees become entrepreneurs. It jointly addresses the two fundamental questions of when employees generate innovations, and whether these innovations are developed as internal ventures or outside the firm. The model shows that if generating innovations distracts employees from their assigned tasks, firms may discourage innovation. Firms may reject profitable opportunities that fall outside of their core activities. If employees own the intellectual property (IP), they may leave to do a start-up. The allocation of IP rights also affects the generation of innovation. The external entrepreneurial environment is a complement to firm-internal innovation. If the external environment is particularly good, firms may embrace employee innovation and take advantage of it through spin-offs.

Key words: entrepreneurship; innovation; intellectual property rights; strategic management; employment policy
History: Accepted by Scott Shane, technological innovation, product development, and entrepreneurship; received April 11, 2006. This paper was with the author 5 months for 2 revisions.

1. Introduction

Where do entrepreneurs come from? Employees of established companies turn out to be one of the most important sources for entrepreneurship. The semiconductor industry, for example, has an impressive genealogy, where, generation after generation, employees left their parent company to launch the next entrant.[1] In a sample of fast-growing private companies, Bhidé (1994, p. 151) finds that "71 percent of all founders had replicated or modified an idea encountered through previous employment." Similarly, Cooper (1985) reports that in a broad cross section of industries 70 percent of all founders pursued opportunities closely related to their previous employment.

A popular complaint is that established firms simply miss out on all these great opportunities,[2] yet this hardly constitutes a systematic explanation. This paper argues that the departure of employees to become entrepreneurs can be understood as a natural equilibrium outcome in the process of innovation. In addition, we stress that it is not the only possible equilibrium outcome. While some employees leave, others stay to develop their innovations internally, turning employees into intrapreneurs.[3] Other firms use their employees' innovations to found spin-offs.[4] To get a better understanding of the supply of entrepreneurs, we need to ask how companies manage employee innovations.

This paper provides a theory of how firms formulate polices that affect the generation and development of employee innovations. We consider a model where a wealth-constrained employee sometimes obtains a new idea that is unrelated to her assigned employment task. She faces a multitask choice of either focusing on her task, or exploring the new idea. The incentive to remain focused depends on how likely she is to succeed with the assigned task, and how well the task is compensated. The attraction of pursuing a new idea is that it may generate far greater returns (including private benefits) than the employee could ever hope to earn on her assigned task. However, pursing an idea is risky, and whether it generates returns depends on how it can be commercialized.

In managing the employee's incentives, the firm faces a fundamental trade-off between exploitation and exploration. The firm wants the employee to

[1] The "traitorous eight" left Shockley Labs to create Fairchild Semiconductor, which itself saw its employees starting, among others, National Semiconductor, Intel, AMD, and LSI Logic, which in turn became parents to Cypress, Zilog, Sierra Semiconductor, and many other semiconductor companies.

[2] This notion has been popularized by Smith and Alexander (1999), who describe how Xerox "fumbled the future" by not capitalizing on some of its inventions for the personal computer, such as the mouse. Interestingly, Smith and Alexander put little emphasis on the fact that Xerox's main line of business was photocopiers, with which it was a highly successful and profitable firm during this period. Chesbrough (2002) provides a more thoughtful analysis of the Xerox experience.

[3] Bartlett and Mohammed (1995) report that 3M allows employees to use 15 percent of their time to develop new ideas. The company also has an objective of generating 25 percent of its sales from products that are less than five years old.

[4] Allen (1998) discusses Thermo-Electron as an example of a company that makes extensive use of spin-offs.

focus on the core task, yet the employee is attracted to idea exploration, which holds promise of greater private benefits. The firm can offer bonuses for core tasks and define policies for developing employee innovations. An important assumption is that the firm is able to commit to policies, even if they are ex post inefficient. The analysis generates four types of equilibria. First, in a *focused* equilibrium, the firm provides sufficient bonuses to keep all employees focused on their assigned tasks. The firm has a policy of refusing to develop any innovations that fall outside of the core activities, although in equilibrium employees never generate innovations. Second, the *stubborn* equilibrium is similar to the focused equilibrium, except that some employees (those with ideas and poor prospects for their assigned task) explore their ideas and sometimes generate profitable innovations. The firm sticks to its policy of refusing to support innovations. In equilibrium the employee leaves to become an entrepreneur (this requires that the employee owns the (intellectual property (IP)). In the first two equilibria, the firm actively discourages idea exploration. In the next two equilibria the firm accepts that it cannot "stem the tide" of its employee's ambitions, and decides to "go with the flow." In the *intrapreneurial* equilibrium, employees excessively explore their ideas, and the firm makes the best of their innovations by developing them as internal ventures. The *entrepreneurial* equilibrium is similar, except that development occurs through start-ups (if the employee owns the IP) or spin-offs (if the firm owns the IP).

The model generates several key insights. In the first two equilibria firms actively discourage employee innovation. In the stubborn equilibrium this even leads to a "dramatic" sequence of events, where employees prefer to develop their innovations inside the firm, the firm refuses (even though this is ex post inefficient), and the employees leave to become entrepreneurs. The model explains the curious empirical phenomenon that some entrepreneurs start their companies only after being rejected by their employers. While this may lead to a perception of "firms missing opportunities," the analysis shows that the firm's intransigence may be part of an optimal ex ante policy to promote greater focus on core tasks.

The model also generates new insights about IP rights and the importance of the external environment. If the employee owns the IP, as the external environment becomes increasingly attractive, the outcomes go from the focused, to the stubborn, to the intrapreneurial, to the entrepreneurial equilibrium. Contrast this with the case in which the firm owns the IP: For most parameter values the firm implements the focused equilibrium; only if the external environment becomes very attractive does the firm switch to the entrepreneurial equilibrium. Allocating IP rights to the employee thus increases employee innovation. The key intuition is that if the firm owns the IP, then the external environment only constitutes an opportunity for the firm. If the employee owns the IP, then the external environment also constitutes a threat, and the firm has to accommodate employee innovation more.[5]

The paper differs from the prior literature in several important respects. Most of the literature on innovation uses a single-task model (e.g., Aghion and Tirole 1994, Anand and Galetovic 2000, Gans and Stern 2000) and focuses on underinvestment in innovation. This paper considers a multitask model where incentives for innovation may interfere with incentives for core tasks. The prior literature explains employee departures as efficient outcomes (e.g., Cassiman and Ueda 2006, Lewis and Yao 2001, Pakes and Nitzan 1983), whereas our analysis can account for employee departures that are ex post inefficient. This is because the firm can commit to an optimal ex ante policy that trades off the benefits of greater employee focus with the costs of inefficient employee departures. Moreover, unlike some of the previous literature, the allocation of IP rights matters in this model. The reason that the Coase theorem does not apply is that firms can commit to ex post inefficient outcomes. The paper is related to several other literatures, not only economic theory, but also empirical research, ranging from ethnographic observation, to legal study, to econometric data analysis. Rather than attempting to summarize all this in the introduction, we defer the full literature discussion until after we have developed the theory.

The remainder of this paper is structured as follows. Section 2 derives the theory model. Section 3 provides an extensive discussion, including a literature review. It is followed by a brief conclusion. All proofs are given in the online appendix (provided in the e-companion).[6]

2. The Theory

2.1. The Base Model

Suppose there is a firm and a wealth-constrained employee. At date 0 an employment contract is drafted. At date 1 the employee may get a new idea that is unrelated to her main employment task. She has a

[5] The analysis takes the legal regime as exogenous, and assumes that the firm and the employee cannot privately reallocate property rights. This case is realistic if the law forbids such private recontracting. In California, for example, the law does not enforce certain private contracts that reallocate IP rights back to the firm. See Gilson (1999) and Hyde (1998).

[6] An electronic companion to this paper is available as part of the online version that can be found at http://mansci.journal.informs.org/.

private choice about whether to explore her idea or to focus on her assigned task. If she sticks to her task, she generates some verifiable value at date 3. If she explores her idea, she discovers at date 2 whether or not the idea is feasible. If it is feasible, the idea can be developed inside or outside the firm, or else it can be shelved. All returns accrue at date 3. All parties are risk-neutral profit maximizers. There is no discounting. For convenience, Table EC.1 in the online appendix provides a time line and Table EC.2 summarizes the notation used in the paper.

Let us begin with the idea exploration stage at date 1. As part of her regular job activities, the employee may serendipitously get a new idea that falls outside her main task. Let λ be the probability that she obtains such an idea. When she gets an idea, she can choose to ignore or explore it. We model this as a multitask problem, where she can focus either on Project *A* or *B*.

Project *A* denotes the assigned employment task, which may include a large variety of activities that improve the firm's current core businesses. If successful, the assigned task generates a value a for the firm. At the time of receiving the idea, the employee also receives a private signal about the likelihood of success: Either the prospects for the assigned task are "strong," so that success is certain; or they are "weak," so that the probability of success is given by $\phi \in (0, 1)$. The probability that the core prospects are strong is denoted by $\gamma \in (0, 1)$.[7]

Project *B* represents exploring a new idea. There are two stages to turning an idea into a successful venture: exploring the idea and developing the innovation. If the employee explores her idea, she finds out, at date 2, whether it is feasible (which occurs with probability p) or not. If it is not feasible, all parties receive zero utility. If it is feasible, we call the idea an innovation. A development effort is then required to generate value from the innovation.[8] The employee is indispensable for development.[9]

If the employee has an idea, she has to choose between focusing on her assigned task (*A*) or exploring her idea (*B*). We distinguish three levels of exploration:

σ^-: No exploration (employee never explores ideas)
σ^*: Efficient exploration (employee explores ideas only if core prospects weak)
σ^+: Excessive exploration (employee explores ideas irrespective of core prospects)

Condition (1) below justifies the use of the word "efficient" and "excessive."

Consider now the events at date 2. If the employee has chosen *A*, then nothing happens until date 3. If the employee has chosen to pursue an idea that turns out to be feasible, there are three possible development choices:

δ^0: No Development (Shelving)
δ^y: Internal Development (Internal ventures)
δ^x: External Development (Start-ups, Spin-offs)

If the innovation is not developed (δ^0), there are no returns. If the innovation is developed internally (δ^y), it generates a total utility y. If the innovation is developed externally (δ^x), it generates a total utility x. We assume that x and y are exogenous and symmetrically known.[10]

The utilities x and y include not only financial returns, but also any private benefits that the employee derives from developing her innovation. Private benefits might represent an increase in the value of the employee's human capital, and/or the satisfaction of developing her innovation, as well as monetary returns that the employee is always able to extract.[11] Private benefits allow us to capture the notion that employees are attracted to exploring ideas, rather than performing core tasks. This introduces a natural source of disagreement between the employee and the firm. Let β_x be the fraction of the total utility x that constitutes a private benefit, i.e., $\beta_x x$ is the expected (nontransferable) private benefit and $(1 - \beta_x)x$ is the expected (transferable) financial return. It is the same for β_y. For expositional convenience, we focus on the

[7] We assume that λ and γ are independent. In an earlier version of the paper we found that relaxing this assumption generated the same results, but made the analysis less transparent.

[8] We assume that a successful innovation does not cannibalize, nor complement, the firm's other profits. Adding such effects to the model is straightforward, but generates few additional insights.

[9] This assumption eliminates the issue of whether the idea leads to the creation of one or two ventures. Anton and Yao (1994, 1995) provide an extensive analysis of this problem.

[10] Section 3 discusses the literature that compares the efficiency of internal and external ventures. This literature focuses on differences in information, incentives, the availability of complementary assets inside the firm, the risk of appropriation outside the firm, and a variety of other aspects. Here we do not derive underlying differences between x and y, but instead focus on how such differences affect the generation and development of innovations. We also do not provide an explicit model of the development process, but note that x and y represent general expected utilities of this process.

[11] The notion of private benefits is broad. It includes the usual nonmonetary benefits, such as satisfaction of doing a venture, etc. (Incidentally, these benefits can be unrelated to the ultimate success of the venture, i.e., the employee may enjoy private benefits from development alone.) Moreover, private benefits can also represent nontransferable financial returns. These are financial returns that the employee can always extract, such as through efficiency wages, information rents, or hold-up power. They are not transferable because the employee is wealth constrained, and cannot commit not to extract them once development has been initiated. An earlier version of the paper explicitly modeled private benefits as a hold-up rent, but the current specification is more general.

case where $\beta_x = \beta_y$. The online appendix shows that relaxing this assumption changes very little in the model.

We want to focus on the parameter constellation where there can be either too much or too little exploration. For this, we assume that it is efficient to explore if the core prospects are weak, but not when they are strong. Formally, this means

$$\phi a < py < a. \tag{1}$$

We are now in a position to describe the contractual choices at date 0. We assume that all value realizations at date 3 are verifiable, so that contracts can be made contingent on them. Contracts are complete in the sense that we allow our contracting parties to use all verifiable information. If the employee generates value in the core, the firm may reward this with w_a. If she generates value from an internal development, the firm may reward this with w_y.[12] The firm can also set a reward w_x if the employee develops an innovation externally. The employee is wealth constrained. This imposes an important constraint on the set of contracts that the firm can offer, namely $w_a, w_y, w_x \geq 0$.[13]

There are two regimes of IP rights: the employee-friendly regime (denoted by EIP) where innovations belong to the employee, and the firm-friendly regime (denoted by FIP) where innovations belong to the firm. The legal regime is exogenous, and we assume that the firm and the employee cannot privately reallocate property rights.[14]

We are interested in how firm policies affect the process of innovation. For this, we consider the two polar cases where the firm is either able or unable to commit itself. We can think of commitment as a deliberate policy of top management, and/or a corporate culture among middle management (who may be responsible for development decisions) that is sanctioned by top management. For simplicity, we treat the ability to commit as exogenous, but one can use standard economic reasoning to derive commitment endogenously from a reputation condition in an infinitely repeated game. The firm is able to commit to a policy as long as the long-term benefit of maintaining its reputation outweighs the short-term benefit of reneging on its commitment. Long-term benefits are higher, the greater the number of employees, and the lower the firm's discount rate.[15] For the commitment case, the firm can specify at date 0 how it will behave at date 2. If the firm owns the IP, it can commit to choose either δ^0, δ^y, or δ^x. If the employee owns the IP, then the firm can only make a commitment about whether or not to do an internal development δ^y.[16]

2.2. The Employee-Friendly Regime

We begin our analysis with the case in which the employee owns the IP. We first focus on the case of $x < y$, where start-ups are inefficient compared to internal ventures. Even though one might expect no start-ups to arise under these circumstances, we will see otherwise.

Consider an employee with an innovation at date 2. If she leaves the firm, she obtains x, which includes the financial value as well as private benefit. Because leaving is inefficient for $x < y$, the firm may try to retain her with an offer w_y, giving her a utility of $\beta_y y + w_y$. For $x < \beta_y y$ the firm sets $w_y = 0$, but for $x > \beta_y y$ the firm needs to offer $w_y = x - \beta_y y$. The employee's utility of staying with the firm is thus given by $z \equiv \text{Max}[\beta_y y, x]$.

Consider now the multitask choice at date 1. The employee's motivation for pursuing an idea includes not only financial returns, but also private benefits. By contrast, the motivation for performing the assigned task is purely financial. If the employee has an idea, she explores it whenever $pz > w_a$ for a strong core, and whenever $pz > \phi w_a$ for a weak core. This assumes that the firm makes an offer for internal venturing. If the firm has a policy of no internal ventures, the employee explores her idea whenever $px > w_a$ for a strong core, and whenever $px > \phi w_a$ for a weak core. These four equations constitute the employee's incentive constraints.

The firm has two levers for influencing the employee's choice. First, to encourage the employee to focus more on her assigned task, the firm can increase w_a. Unfortunately, this increases the compensation of all employees, including all those with no ideas, so it can become expensive for the firm. Intuitively, increasing w_a works well for higher values of λ, but becomes costly for lower values of λ.

[12] Note that in this model it makes no difference whether the payment w_y is made at date 2 or date 3. That is, it makes no difference whether the contract is made contingent on the outcome y or on the development decision δ^y. It is the same for w_x. Note also that if the employee generates no value at date 3, the firm could reward this with w_0. It is easy to see that the firm never benefits from rewarding the employee for doing nothing, so that $w_0 = 0$ is always optimal.

[13] One may wonder if the employee cannot relax the wealth constraint by pledging some of the returns from the innovation. For an internal venture, all of the employee's wealth comes from w_y, so that $w_y \geq 0$ remains valid. If the development is external, then the firm cannot bond its employees, implying $w_x \geq 0$.

[14] Note that even if reallocation were allowed, it might not occur because of the employee's wealth constraint. An earlier working paper examined this in greater detail.

[15] Simon (1951) and Kreps (1990) argue that an employment relationship can be thought of as an infinitely repeated game, where the firm is a long-run player, facing successive generations of employees who are short-term players.

[16] The firm could also "bribe" the employee to shelve the innovation. It is easy to show that this is never optimal in this model.

The firm's second lever is its development policy, specifically, whether or not to engage in internal ventures. Ex post (i.e., at date 2), internal venturing is profitable because the firm receives a profit of $y-z$. Ex ante (i.e., at date 1), however, it may encourage too much idea exploration. For higher values of x, there is no ex ante incentive effect to internal venturing. Specifically, for $x > \beta_y y$, we have $z = x$, so that the employee gets the same utility irrespective of whether he stays or leaves. For lower values of x, namely for $x < \beta_y y$, the employee prefers internal venturing. A policy of refusing internal ventures therefore has the ex ante effect of discouraging idea exploration. Intuitively, the ex ante incentive effect is stronger for lower values of x, because the refusal leaves the employee with a weak outside option. It is stronger for higher β_y, because the employee enjoys internal ventures. Moreover, the firm cares more about focus if core prospects are strong, so that the incentive effect matters more for higher values of γ.

This provides an intuitive overview of the firm's choices. We can now turn to the more formal derivation of the firm's optimal policies. As a benchmark, consider the case where the firm is unable to commit to inefficient ex post actions. In this case, its development policy is always δ^y (for $x < y$). At date 1, the employee chooses σ^- if $pz \leq \phi w_a$; chooses σ^* if $\phi w_a \leq pz \leq w_a$; and chooses σ^+ if $w_a < pz$. To implement (σ^-, δ^y) as an equilibrium, the firm chooses the lowest w_a that satisfies $pz \leq \phi w_a$, given by $w_a = pz\phi^{-1}$. This gives the firm an ex ante utility of

$$U(\sigma^-, \delta^y) = (\gamma + \bar{\gamma}\phi)(a - pz\phi^{-1}). \tag{2}$$

Throughout the paper, a bar above a probability parameter denotes its complement, i.e., $\bar{\gamma} = 1-\gamma$. To implement (σ^*, δ^y), the firm sets $w_a = pz$ so that

$$U(\sigma^*, \delta^y) = \lambda\bar{\gamma}(py - pz) + (\gamma + \bar{\lambda}\bar{\gamma}\phi)(a - pz). \tag{3}$$

To implement σ^+, the firm simply sets $w_a = 0$ so that

$$U(\sigma^+, \delta^y) = \lambda(py - pz) + \bar{\lambda}(\gamma + \bar{\gamma}\phi)a. \tag{4}$$

Straightforward calculations reveal that $U(\sigma^*, \delta^y) > U(\sigma^-, \delta^y)$, i.e., (σ^-, δ^y) is a dominated policy. Moreover, after transformations we obtain $U(\sigma^+, \delta^y) > U(\sigma^*, \delta^y) \Leftrightarrow \lambda < (\gamma + \bar{\gamma}\phi)pz/(\gamma(a-py) + (\gamma + \bar{\gamma}\phi)pz)$. We want to focus on the model where this condition holds for all values of x. We therefore introduce the following parameter restriction:

$$\lambda < \hat{\lambda} \equiv \frac{(\gamma + \bar{\gamma}\phi)p\beta_y y}{\gamma(a-py) + (\gamma + \bar{\gamma}\phi)p\beta_y y}.$$

This condition simply says that it is sufficiently rare that employees get ideas outside of their assigned task. The assumption reflects our emphasis on ideas that are unconventional and unplanned (see also §3.1). In the online appendix we discuss what happens when we relax this condition. We have thus derived our first proposition.

Proposition 1. *Suppose that $x < y$, $\lambda < \hat{\lambda}$, and that the employee owns the IP.*

- *If the firm is unable to commit to a development policy, the intrapreneurial equilibrium (σ^+, δ^y) obtains. The employee excessively explores her ideas, and the firm makes the best of it by developing the innovations internally.*

Proposition 1 provides a benchmark for all subsequent propositions. It says that in the absence of commitment, the firm accepts that the employee likes to explore ideas. The firm makes the best of it through internal ventures. In equilibrium there is excessive exploration, i.e., the employee explores her ideas even if the core prospects are strong. We call this the intrapreneurial equilibrium.

Now consider the case in which the firm can commit to a development policy. As discussed above, the firm may adopt a policy of no internal ventures to improve ex ante incentives. The employee's remaining option is to do a start-up. This corresponds to a development decision δ^x. The firm can still use bonuses to affect the employee's choices. The employee chooses σ^- if $px \leq \phi w_a$, chooses σ^* if $\phi w_a \leq px \leq w_a$, and chooses σ^+ if $w_a \leq px$. To implement (σ^-, δ^x), the firm sets $w_a = px\phi^{-1}$, which gives

$$U(\sigma^-, \delta^x) = (\gamma + \bar{\gamma}\phi)(a - px\phi^{-1}). \tag{5}$$

To implement (σ^*, δ^x), it sets $w_a = px$ so that

$$U(\sigma^*, \delta^x) = (\gamma + \bar{\lambda}\bar{\gamma}\phi)(a - px). \tag{6}$$

To implement σ^+, it simply sets $w_a = 0$ so that

$$U(\sigma^+, \delta^x) = \bar{\lambda}(\gamma + \bar{\gamma}\phi)a. \tag{7}$$

We immediately note that $U(\sigma^+, \delta^y) > U(\sigma^+, \delta^x)$, i.e., (σ^+, δ^x) is a dominated policy. With this, we proceed to Proposition 2. The proof in the online appendix derives the critical values $\hat{x}_1, \hat{x}_2$ (where $0 < \hat{x}_1 < \hat{x}_2 < y$) and $\hat{\gamma}_1$ (where $0 < \hat{\gamma}_1 < 1$).

Proposition 2. *Suppose that $x < y$, $\lambda < \hat{\lambda}$, $\gamma > \hat{\gamma}_1$, that the employee owns the IP, and that the firm is able to commit to a development policy.*

- *With a weak external environment ($x < \hat{x}_1$) the focused equilibrium (σ^-, δ^x) obtains; the firm refuses internal venturing; the employee never explores her ideas.*
- *With an intermediate external environment ($\hat{x}_1 < x < \hat{x}_2$) the stubborn equilibrium (σ^*, δ^x) obtains; the firm refuses internal venturing; the employee explores ideas if the core prospects are weak; if she finds an innovation she becomes an entrepreneur, even though internal venturing would be preferred.*
- *With a strong external environment ($\hat{x}_2 < x < y$) the intrapreneurial equilibrium (σ^+, δ^y) obtains; the firm welcomes internal ventures; the employee explores her*

ideas, even if the core prospects are strong; if she finds an innovation she becomes an intrapreneur.

• *The critical boundary $\hat{x}_1$ is increasing in a and ϕ, and decreasing in p. The critical boundary $\hat{x}_2$ is increasing in a and decreasing in p.*

Proposition 2 shows that the equilibrium outcome is affected by the external environment. We distinguish between three outcome regions. Consider first an environment where the employee faces weak outside options in terms of doing a start-up. In this case, the firm's refusal to implement internal ventures puts a real damper on the employee's desire to explore new ideas. We call (σ^-, δ^x) the focused equilibrium, because the firm's refusal has the intended effect that the employee always focuses on her assigned task. In addition to refusing internal ventures, the firm also provides a bonus of $w_a = px\phi^{-1}$ for the core task. The focused equilibrium works well for the firm as long as the external environment constitutes a low threat.

As x rises, the cost of providing a bonus for the core task becomes too onerous. At this point, the firm switches to a slightly different policy: It continues to refuse internal ventures, but resigns itself to the fact that some employees will explore their ideas, namely, when they face poor prospects on their assigned core task. In order not to further encourage idea exploration, the firm sticks to its commitment not to engage in internal ventures. As a result, the employee does a start-up, even though she would prefer to do an internal venture. This is an important and perhaps surprising result. We call (σ^*, δ^x) the stubborn equilibrium because the firm stubbornly refuses to do internal ventures, even though in equilibrium it loses out on internally generated opportunities. While the firm may appear to be too inflexible, its policy of refusing some of the internally generated opportunities is actually part of an optimal policy of encouraging employees to focus on their core activities.

As x rises further, even the stubborn policy becomes too costly for the firm. For $x > \hat{x}_2$ the firm no longer wants to stem the flow of entrepreneurial ideas, but instead decides to go with the flow. It lets the employee pursue all her ideas, and develops her innovations through internal ventures. Note that this is the same as in Proposition 1.

The comparative statics of the equilibrium regions are very intuitive. A higher return on core activities (higher a) increases the ranges of the focused and stubborn equilibria. This is because a higher return encourages the firm to focus its employees. Better odds of idea exploration (higher p) have the exact opposite effect. If the odds in the case of a weak core become more favorable (higher ϕ), this increases the range of the focused equilibrium. Note that this is the only equilibrium where the employee refrains from exploring ideas with a weak core.

Proposition 2 also uses the condition $\gamma \geq \hat{\gamma}_1$, which requires that strong core prospects occur sufficiently often. We already noted above that for higher values of γ the firm cares more about focusing its employees. The model with higher values of γ characterizes a firm that needs to deal with the fact that employees occasionally face poor prospects on their assigned task. The model with small γ characterizes a less-fortunate situation where employees have badly designed jobs in the first place, in the sense that their assigned task has a poor chance of success most of the time. This is probably a less realistic case, and we discuss it in the online appendix.

2.3. The Firm-Friendly Regime

So far, we have assumed that the employee owns the IP. If the firm owns the IP, it has an additional policy at its disposal, namely, to shelve an innovation (i.e., not to develop it at all). The traditional motive for shelving an innovation, cannibalization, is purposely assumed away in this model. Instead, here the motive is that shelving may improve ex ante incentives. The threat of shelving is much more powerful than the refusal to do internal ventures, because shelving implies that an employee who explores an idea always obtains zero utility. This makes it easy for the firm to prevent idea exploration. In addition, the firm hardly needs to pay a bonus for the core activity. Any arbitrarily small bonus ($w_a \to 0$) keeps the employee focused on her assigned task, irrespective of whether core prospects are strong or weak. We denote this policy as (σ^-, δ^0), and note that it simply yields

$$U(\sigma^-, \delta^0) = (\gamma + \bar{\gamma}\phi)a.$$

For the next proposition, the online appendix derives the critical value $\hat{\gamma}_2$, and shows that $\hat{\gamma}_2 < \hat{\gamma}_1$.[17]

Proposition 3. *Suppose $x < y$, $\lambda < \hat{\lambda}$, $\gamma > \hat{\gamma}_2$, and that the firm owns the IP.*

• *If the firm is unable to commit to a development policy, then the intrapreneurial equilibrium (σ^+, δ^y) obtains.*

• *If the firm is able to commit to a development policy, then the focused equilibrium (σ^-, δ^0) obtains. The firm threatens to shelve all innovations, so the employee never explores her ideas.*

Proposition 3 constitutes a simple but powerful result. First, if the firm cannot commit, then the equilibrium is always intrapreneurial, just as in Proposition 1.[18] The power to shelve innovations is useless for the firm if it cannot commit to use this power.

[17] In fact, the online appendix shows that for some parameter values we have $\hat{\gamma}_2 < 0$, in which case the condition $\gamma > \hat{\gamma}_2$ is never binding anyway.

[18] There is one minor difference, namely, that the firm now always sets $w_y = 0$, whereas in Proposition 1, it had to offer $w_y = x - \beta_y y$ for $\beta_y y < x < y$.

With commitment, however, giving the firm the IP dramatically changes the equilibrium. The key intuition is that the external environment no longer constitutes a threat to the firm. The firm finds it easy to implement a focused equilibrium. It commits to shelve all employee innovations. This discourages idea exploration, so that employees always focus on the core business.

Why does the firm choose such an extreme policy? In principle it would like its employees to pursue new ideas when the core is weak, but not when the core is strong. The problem is that inducing such a separating outcome (σ^*) is costly to the firm, because it requires paying a bonus for the core task, even when the employee doesn't have an idea. If ideas are sufficiently rare ($\lambda < \hat{\lambda}$), and ex ante incentives matter sufficiently ($\gamma > \hat{\gamma}_2$), then the firm prefers to forgo the benefits of innovation by refusing any type of development (δ^0).

Proposition 3 (for the FIP model) calls the equilibrium focused, just as in Proposition 2 (for the EIP model). This is because in both cases the employee never explores her ideas. The only difference is that in the EIP model the firm offers a bonus $w_a = px\phi^{-1}$, whereas in the FIP model the firm doesn't even need to do that, i.e., $w_a = 0$. Note that for $x \to 0$, the two policies become identical.

Comparing Propositions 2 and 3, we recognize that the allocation of IP rights can have a dramatic effect both on the generation and development of innovations. In terms of a social welfare comparison, one may be inclined to argue that the firm-friendly regime is less desirable because it undermines innovation. However, a comparison of the two regimes requires that we also take into account the value of core tasks. It is easy to verify that neither regime Pareto-dominates the other: The employee is always better off under EIP, and the firm is always better off under FIP. An earlier working paper (Hellmann 2005) provides a full comparison of the two regimes using the standard welfare criterion of maximizing the sum of the employee's and the firm's utilities. The main finding is that neither regime clearly dominates the other, precisely because there is a nonobvious tradeoff between favoring innovation in the EIP regime versus favoring value creation from core activities in the FIP regime.

Our model focuses on the two polar cases where either the employee or the firm owns the IP. The earlier working paper version also examined a model with imperfect IP rights. Even if the firm owns the IP in principle, the employee can "go alone," i.e., leave and develop the innovation without formally owning the IP. The imperfect IP rights model convexifies the two models, so that outcomes converge to the EIP (FIP) model for sufficiently high (low) returns of pursuing a go-alone policy. In this model, the firm may also want to decrease the returns to a go-alone, such as by committing to sue departing employees.[19]

2.4. When External Development Is Better Than Internal

We now turn to the analysis where the external environment dominates internal venturing ($x > y$). In this case, external venturing constitutes not only a potential threat, but also an opportunity. The online appendix derives a critical value $\hat{x}_3$ that satisfies $\hat{x}_3 > y$. With this we obtain the following results.

PROPOSITION 4. *Suppose $x > y$, $\lambda < \hat{\lambda}$, and $\gamma > \hat{\gamma}_2$.*

- *If the employee owns the IP, the entrepreneurial equilibrium (σ^+, δ^x) obtains, irrespective of whether the firm is able or unable to commit.*
- *If the firm owns the IP and is unable to commit, the entrepreneurial equilibrium (σ^+, δ^x) obtains.*
- *If the firm owns the IP and is able to commit, then the focused equilibrium (σ^-, δ^0) obtains for $x < \hat{x}_3$, and the entrepreneurial equilibrium (σ^+, δ^x) obtains for $x > \hat{x}_3$. The critical boundary $\hat{x}_3$ is increasing in a and ϕ, and decreasing in p.*

If the employee owns the IP, she simply leaves the firm every time she has an innovation. We call this the entrepreneurial equilibrium. The firm finds it too costly to provide incentives against idea exploration, and resigns itself to the fact that the employee explores excessively. In this case, the external environment is all threat and no opportunity for the firm. Because the firm's optimal policy does not involve any inefficient ex post decisions, the optimal policy is the same with or without commitment.

This result depends on the allocation of IP rights. If the firm owns the IP, preventing idea exploration is easier. Indeed, for $x < \hat{x}_3$, the firm's optimal policy remains the same as in Proposition 3: The firm threatens not to develop any innovations (δ^0), which convinces the employee not to explore her ideas (σ^-). For larger values of x, however, the external environment becomes so attractive that the opportunity cost of the focused equilibrium becomes too high. For $x > \hat{x}_3$ the firm changes its optimal policy. It welcomes all idea exploration, even if the core prospects are strong. It makes the best use of employee innovations by ensuring their external development. We call this an entrepreneurial equilibrium, but note that there is an important difference between the EIP and the FIP model. In the EIP model, the employee leaves to do a start-up, without sharing any of the financial rewards with her employer. In the FIP model, the

[19] Jackson (1998) describes how Intel takes an aggressive approach of protecting its IP, harassing departing employees with frivolous lawsuits.

Hellmann: *When Do Employees Become Entrepreneurs?*
Management Science 53(6), pp. 919–933, ©2007 INFORMS

Figure 1 Overview of Equilibria

Panel A: Equilibrium outcomes

<table>
<tr><td>External environment</td><td>Weak</td><td>Intermediate</td><td>Strong</td><td>Slightly superior</td><td>Strongly superior</td></tr>
<tr><td>Employee-friendly IP</td><td>Focused</td><td>Stubborn</td><td>Intrapreneurial</td><td colspan="2">Entrepreneurial</td></tr>
<tr><td>Firm-friendly IP</td><td colspan="4">Focused</td><td>Entrepreneurial</td></tr>
<tr><td>No commitment case</td><td colspan="3">Intrapreneurial</td><td colspan="2">Entrepreneurial</td></tr>
</table>

Panel B: Equilibrium properties

Equilibrium	Focused	Stubborn	Intrapreneurial	Entrepreneurial
Bonus for success with core task	High under EIP None under FIP	Moderate	None	None
Firm stance on internal ventures	Refuse	Refuse	Encourage	External ventures dominate internal ventures
Employee exploration level	None	Efficient	Excessive	Excessive
Equilibrium venturing activities	None	"Inefficient" start-ups	Internal ventures	Start-ups under EIP Spin-offs under FIP

firm hires the employee to manage a spin-off. The employee enjoys the private benefits $\beta_x x$, but does not reap any of the transferable financial returns $(1-\beta_x)x$.

The comparative statics of $\hat{x}_3$ are intuitive. A higher return on core activities (higher a, higher ϕ) increases the range of the focused equilibrium, while better odds for idea exploration (higher p) has the opposite effect.

The policy of deterring idea exploration requires commitment not to develop profitable innovation. Without commitment, the firm's threat to shelve innovations is not credible. In this case the entrepreneurial equilibrium obtains for all x with $x > y$.

2.5. Key Insights and Robustness

Propositions 1 through 4 constitute the core of our theoretical analysis. To provide a better overview, Figure 1 summarizes these key results. Panel A shows how the equilibrium changes as a function of the external environment, the allocation of IP rights, and commitment. Panel B provides a description of the properties of the four types of equilibria.

In §3 we will discuss how our results relate to a variety of literatures. As a preamble, it is useful to summarize the key insights of the model and discuss how robust they are.

This paper proposes the multitask incentive problem as a framework for analyzing firm policies towards employee innovation. Within this framework the model generates the following key insights.

(1) A firm may discourage employee innovation in order to focus its employees on their core task.

(2) To focus its employees, a firm may refuse to implement profitable employee innovations. In equilibrium, employees with innovations may leave to do start-ups, even if they would prefer to do internal ventures.

(3) Allocating IP rights to the employee instead of the firm leads to more innovation, more internal ventures, and more start-ups.

(4) The effect of the external environment is as follows:

- If the employee owns the IP, then a better external environment encourages employees to generate more innovations and develop them as internal ventures or start-ups.
- If the firm owns the IP, then it prefers to focus its employees on their core tasks unless the external

environment is very strong, in which case it supports employee innovation through spin-offs.

(5) If a firm is unable to commit to an innovation policy, then the allocation of IP rights is irrelevant for the outcome. The firm supports all employee innovations, developing them internally or externally, whichever way is more profitable.

How general are these key insights? We would argue that the first insight is very general, and survives many model permutations. It captures a fundamental trade-off that emerges in a multitask environment. In §3.1 we explain how the multitask model captures a bottom-up perspective for understanding the process of innovation.

The second insight may seem model specific at first, but we would argue that the result is fairly general. In terms of the underlying intuition, all that is required is that the firm cares more about ongoing (ex ante) incentives of its employees, rather than the one-time (ex post) opportunity of pursuing an employee innovation. In terms of a formal model, the simplification that there are only two core states—strong ($\tilde{\phi} = 1$) and weak ($\tilde{\phi} = \phi$)—understates the generality of the stubborn equilibrium. In a more general model where $\tilde{\phi}$ is a random variable distributed over $[0, 1]$, it is easy to see that for any finite bonus w_a there always exist some employees who explore ideas, namely, those with $\tilde{\phi} w_a < px$. As long as the firm wants to focus its employees, it will refuse to implement those employee's innovations, and they will leave to do their own start-ups.

The third and fourth insights are interrelated. Their basic insight is also quite general. The key is that if the firm owns the IP, the external environment constitutes an opportunity, but never a threat. The firm has an easy way of focusing its employees, namely, by shelving all innovations. Without the IP, however, the external environment also constitutes a threat for the firm, which forces it to be more accommodating towards employee innovation.

The fifth key insight about the role of commitment is also fairly general. The main issue is that without commitment, the firm cannot have a deliberate policy on employee innovation. Our analysis suggests that IP rights matter precisely because they influence the firm's commitment to inefficient development decisions. Without commitment, the full force of the Coase theorem comes back, predicting efficient development decisions.[20]

[20] In the online appendix we note one exception to this. For $\beta_x > \beta_y$ there is a small region around y, where the employee prefers to do a start-ups if she owns the IP, but the firm prefers to do an internal venture if it owns the IP. IP rights matter here because the employee's wealth constraint prevents renegotiation to the jointly optimal outcome. A similar argument applies for $\beta_x < \beta_y$.

This verbal analysis suggests that the main intuitions of the model are quite general. The online appendix also contains a discussion of how our formal analysis depends on several parameter restrictions.

3. Discussion of Model

The previous section explains the theoretical model. In this section we discuss how the model contributes to a variety of theoretical and empirical literatures. We interpret the main results of the model, and derive a variety of empirical implications. Some of the empirical predictions relate to existing work in the literature, others are essentially new and will hopefully provide stimulus for future research. Rather than breaking out the interpretation of the model, the literature review, and the empirical predictions into separate sections, we discuss them jointly. We structure our discussion around a number of themes.

3.1. How Do Firms Structure Incentives for Employee Innovation?

At the core of our model is the firm's choice about what kind of incentives it wants to provide to its employees. Aghion and Tirole (1994) use an incomplete contracts model, where both the employee and employer make relationship-specific investments. Similar to Grossman and Hart (1986), their model focuses on the ex post hold-up problem, and examines how the allocation of property rights might affect efficiency. The use of an incomplete contracts model excludes firm policies such as incentive compensation, because, by assumption, only property rights are ex ante contractible.

Following Aghion and Tirole, a small literature examines under what circumstances firms are better than independent financiers (such as venture capitalists) at financing innovations. Anand and Galetovic (2000), Anand et al. (2004), and Gans and Stern (2000, 2003) emphasize hold-up problems in the presence of weak property rights, while Hellmann (2002) emphasizes incentive conflicts with the firm's core business. All of this literature assumes that the employee already has a well-defined project, and asks how to finance it optimally. De Bettignies and Chemla (2006) consider a task allocation problem, where core tasks have a low elasticity of effort, but innovative tasks have a high elasticity of effort. They show that for better employee outside options, the firm is more interested in internal ventures, which have a higher effort elasticity. In their model, it is the firm, not the employee, who decides on the employee's choice of project.

We focus on a multitask incentive problem, where the employee chooses what activities to focus on (Holmström and Milgrom 1994). At the core of our multitask problem is a trade-off between exploration

Hellmann: When Do Employees Become Entrepreneurs?
Management Science 53(6), pp. 919–933, ©2007 INFORMS

and exploitation (March and Olsen 1991). Exploitation refers to the pursuit of profits in the core business, while exploration concerns the generation of new business opportunities. The most common way of thinking about this trade-off is to take a "top-down" perspective, where senior management decides on the optimal mix of exploration and exploitation. Instead, our model proposes a bottom-up perspective, where the employee's choices ultimately determine what the firm does. That is, the multitask model allows us to focus on how employees determine firm activities. Top-down firm policies continue to matter, because employees respond to the firm's incentive environment.

Our multitask model appears to differ from the abovementioned single-task models in one important way. The single-task literature typically finds that the employee innovates too little, whereas in our model the firm's problem is that the employee wants to explore too much. Casual observation might also suggest that many firms would like their employees to be more innovative. Our multitask approach prompts us to go deeper and ask what type of innovations firms want: innovations that fit with the firm's core activities and generate rents for the firm, or innovations that fall outside of the firm's core activities and mostly generate rents for the employee? The single-task literature implicitly allows only for the first type of innovation. Our multitask model classifies the first type as part of the assigned core task (i.e., part of the employee's job is to come up with improvements for doing the assigned task), and distinguishes it from the second type of innovation, which the firm considers a distraction.

This distinction between different types of innovations builds on the ethnographic studies of Burgelman (1983, 2002) and Burgelman and Sayles (1986). This line of research emphasizes the coexistence of a planned and an unplanned innovation process. It shows how many important innovations emerge from bottom-up "skunks work," where employees explore unconventional ideas before having their research sanctioned by senior management. Burgelman emphasizes that firms face important choices as to how much they tolerate this unplanned process. His most prominent example is Intel, which greatly benefited from tolerating unplanned innovation: In fact, this is how Intel found its way into the semiconductor business. However, when Intel's semiconductor business was doing particularly well (under the leadership of the enigmatic Andy Grove), the company became extremely focused, and lost its tolerance for unplanned innovation. The work of Burgelman is consistent with our approach. In particular, it validates the importance of a bottom-up model, where employees choose whether to explore their ideas versus remaining focused on their assigned tasks. Moreover, Intel's experience is consistent with the model prediction that firms become less tolerant towards exploration when they face a boom in their core activities.[21]

3.2. Why Do Innovative Employees Leave?

An interesting question is why employees sometimes leave their employer to do a start-up, rather than developing their innovation as an internal venture. Some prior work points out why firms should be investing in their employees' ideas. The work of Teece (1986) explains why firms may have a comparative advantage at developing and commercializing their employees' innovations, focusing on the importance of complementary assets. Bankman and Gilson (1999) explain that firms may have tax advantages relative to outside financiers. This makes it all the more surprising when firms reject their employees' ideas.

A seminal theory paper on whether employees with innovations leave or stay is Pakes and Nitzan (1983). Their model is essentially a neoclassical cost-benefit analysis of cannibalization. In their model, the employee leaves if and only if leaving is the efficient decision. Lewis and Yao (2001) also provide an efficiency explanation for employee departures. They argue that a firm may deliberately embrace mobility of its employees in order to be considered an attractive employer. Cassiman and Ueda (2006) provide a dynamic model that also has the feature that the employee leaves if and only if leaving is efficient. Their model assumes that a firm has limited capacity for internal ventures. The firm trades off the returns of an employee's innovation not only against cannibalization, but also against the option value of waiting for better projects in the future. In their environment, it may look like the firm rejects a profitable project, yet there is no inefficiency. The firm does not look at the employee's project in isolation, but instead looks at it from the perspective of optimizing a dynamic project portfolio.

Our model abstracts from cannibalization, not because it is not important, but because this allows us to focus on a different rationale for why firms reject employee innovations. In the stubborn equilibrium, employees depart involuntarily, in the sense that they would have preferred to develop their innovation as an internal venture. We believe that this is a novel result in the literature. To understand its significance, let us examine some of the puzzling evidence about the frustrations experienced by departing

[21] The legal literature also makes a clear distinction between innovations that come out of planned R&D activities versus unplanned innovations. If an employee is hired to invent, then an innovation is considered part of regular output and belongs to the firm. This is not necessarily true for ideas that are largely unrelated to the task for which the employee was hired.

employees. While there is no systematic measure for whether employees depart voluntarily or not, there is anecdotal evidence that suggests that entrepreneurs often offer their innovations to their employer, and decide to start their own companies only after being rejected by their employer. A famous example is Gene Amdahl, who proposed his venture to his employer (IBM). After being rejected by IBM several times, he reluctantly started his company (Amdahl Computer). The founders of Data General also offered their project to their employer (Digital Equipment Corporation (DEC)) in vain before venturing out on their own.

There are several reasons why employees may become frustrated with the firm. Amador and Landier (2003) note that while the employee may be optimistic about the prospects of the innovation, the firm may not share that optimism. They consider an employee with optimistic beliefs about the success of an innovation, and show that a firm with less optimistic beliefs might be unwilling to finance the venture at terms that would be acceptable to the employee. In a related vein, employees may worry that the employer will steal their ideas. Anton and Yao (1995) examine a model where employees leave because their employer cannot commit not to appropriate ideas upon disclosure.

In our model, the frustration of departing employees stems from the fact that the firm forgoes a profitable opportunity for reasons of strategic commitment. This can be interpreted more broadly as rigidity, which prevents the firm from pursuing opportunities outside its defined core business. This is related to the notion of a narrow business strategy, as in Rotemberg and Saloner (1994). One may also interpret the stubbornness of the firm as organizational inertia, where middle managers are not encouraged to venture beyond accepted boundaries.[22]

Our model shows that employees are more likely to leave when bonuses for the assigned task are low. Subramanian (2005) reaches a similar conclusion in a model that bears some similarities to ours. He considers a dynamic model of innovation, where the employee always owns the IP and the firm is never able to commit. Of particular interest here is his empirical finding. He finds that firms that have high levels of employee departures also have relatively lower levels of employee stock options.

3.3. What Explains the Level of External Venturing Activity?

Our theory makes predictions about when employee ideas are developed outside the firm's boundary. Unlike the previous two themes, this is a theme that has benefited from considerable empirical work. In our theory we make a clear distinction between start-ups, where the employee owns the IP, and spin-offs, where the employer owns the IP and maintains some financial interest. Unfortunately, the empirical literature typically does not make that distinction, neither in the data, nor in the use of terms. To avoid further confusion, we maintain our distinction between start-ups and spin-offs, and use the term "external ventures" for both.

Our model predicts that external ventures occur for opportunities that do not fit well with the firm's current core activities. Klepper and Sleeper (2005) provide detailed evidence on external ventures in the laser industry. They show that the external ventures were often based on technologies similar to those of their parent companies. At first sight, this appears inconsistent with our model. However, these external ventures typically targeted different customer segments than those of their parent companies. This finding echoes earlier work by Christensen (1997), who argues that firms develop blind spots to new opportunities when the new customers are different from the firm's current customer base. Upon closer examination, Klepper and Sleeper's evidence thus supports the notion that external ventures typically do not fit with their parents' strategic vision.

Gompers et al. (2005) provide further evidence on the question of how external venturing relates to parent characteristics. They examine venture capital-backed companies in a broad cross section of industries, and link them to their founder's prior employers. They find that parent companies spawn more external ventures when they pursue a more focused business strategy. This is consistent with our stubborn equilibrium, where employees leave because their parent firms pursue a narrow business strategy.

Our model predicts more external ventures when a company is facing weaker prospects in its core business. Gompers et al. (2005) provide two pieces of evidence on this: First, firms with higher earnings spawn less, and second, spawning increases when a company's performance declines. Both of these findings are consistent with our theory.[23]

One of the questions that has preoccupied empirical researchers is whether firms with better technology also spawn more external ventures. Franco and Filson (2006) find that better technical and market know-how increase the external venturing rate.

[22] See also Milgrom and Roberts' (1990) work on influence activities. In addition, note that firm inertia is ex ante optimal in this model. This contrasts with some of the sociological theories, where inertia is an aspect of evolutionary behavior (Hannan and Freeman 1984).

[23] In an unpublished manuscript, Brittain and Freeman (1986) examine external venturing in the semiconductor industry. They also find that slower growth for a parent company increases its spawning rate.

Klepper and Sleeper (2005) find that firms with more patents have more external ventures. At first glance, these results seem to contradict the previous results that better-performing firms generate fewer external ventures. However, better technologies may give rise not only to a stronger core performance, but also to greater potential for innovation. Better technology may thus increase the value of both exploitation and exploration. One paper that is particularly helpful here is Agarwal et al. (2004), who study the disk-drive industry. They distinguish between technological know-how and what they call market-pioneering know-how, which essentially measures a first-mover advantage in the product market. They find that firms that are strong in both types of know-how generate fewer external ventures than firms with strength in only one of the two types of know-how. This is consistent with the notion that technology know-how raises the opportunities for new ideas, but that firms with strong market opportunities rein in their employees' idea exploration in order to focus them on the opportunities in the core business.[24]

Our analysis suggests some venues for further empirical work. With the exception of Agarwal et al. (2004), the empirical literature is silent on how often (and how!) parents participate in external ventures. Our model suggests that spin-offs occur mainly when the firm has strong property rights and the external environment is strong ($x > x_3$). By contrast, start-ups occur when the firm has weaker property rights, and they may occur in considerably weaker external environments (such as $x_1 < x < x_2$).

3.4. What Role Do IP Rights Play?

Our model challenges two widely held beliefs about IP rights. The first is that IP rights predetermine the type of development, so that one should always observe internal ventures if the firm has the IP rights, and external ventures when the employee has the rights. Our model contradicts this notion, because we can get internal ventures when the employee has the rights, and spin-offs when the firm has the rights. The second widely held belief is based on the Coase theorem and holds that the allocation of IP rights is irrelevant. In our model, the Coase theorem applies when there is no commitment. In this case the equilibrium always has excessive idea exploration (σ^+), and the firm always chooses the ex post efficient type of development (internal ventures for $x < y$ and external ventures for $x > y$). However, the Coase theorem no longer applies when the firm is able to commit. Commitment may lead to inefficient ex post outcomes, such as in the stubborn equilibrium. Moreover, the threat of an inefficient ex post decision affects the ex ante incentives. Importantly, the optimal use of commitment power depends on the allocation of IP rights. Indeed, one of the most striking results of the model is that giving the IP rights to the employee dramatically increases the range of external environments (x) for which the employee engages in idea exploration.

How do these predictions stack up against prevailing views in the legal literature? On the normative side, Sterk (1993) argues that the law should not impose restrictions on the types of contracts that an employer can offer, including restrictions that preclude firms from owning the IP generated by their employees. Sterk's argument is based on an efficient market perspective that believes in the supremacy of private contracting over legal restrictions. Talley (1998) counters that informational asymmetries can lead to market inefficiencies, showing how legal rules can help with the generation and disclosure of employee innovations. Merges (1999) further elaborates on the trade-off between employee-friendly and employer-friendly legal regimes, emphasizing bargaining imperfections and hold-up problems.

Turning to the positive legal literature, Gilson (1999) explains how California law favors employees by refusing to enforce noncompete agreements (except in a limited set of circumstances).[25] Gilson proceeds to argue that Silicon Valley's high level of entrepreneurial activity and labor mobility is partly explained by this legal setting. Hyde (1998) provides a complementary analysis of how California courts take an employee-friendly stance in trade-secret cases.

In their empirical studies, Gompers et al. (2004) and Klepper and Sleeper (2005) include a control for whether the parent company is based in California or not. Both papers find significantly higher spawning rates in California. These studies, however, cannot disentangle the effect of IP law from a variety of other California-specific effects. Stuart and Sorenson (2003) propose a novel identification strategy. Their empirical design is slightly different in that they examine how entrepreneurial activity responds to disruptive shocks, notably, acquisitions and IPOs of biotechnology companies. They find evidence of more entrepreneurial activity as a response to shocks, not only in California, but in all U.S. states that have a legal treatment of noncompete clauses similar to that of California.

[24] The empirical literature also examines performance differences among the various types of ventures (internal, start-up, spin-off, etc.), both at the level of the venture and the parent company. Because the theoretical model avoids modeling heterogeneity among parent firms and external ventures, it is not well suited to make predictions about performance differences.

[25] Strictly speaking, noncompete agreements pertain to employment law rather than IP law, but they have a similar effect, because they limit an employee's ability to leave and start a new firm.

One interesting area for future research is how legal rules affect firm policies. Is there any evidence that employee-friendly laws induce firms to adopt less-restrictive policies? Do multilocation firms choose different policies in different legal constituencies? And do IP laws affect where firms locate different employment activities?

3.5. How Does the External Environment Affect Employee Innovation?

Our model shows how the firm's optimal policies towards employee innovation change with external circumstances. The external environment poses both a threat and an opportunity for the firm. Conceptually, the parameter for the external environment (x) measures the returns that can be made from external venturing. Empirically, this may depend on a large number of factors. One factor is the ease of setting up a new company. Djankov et al. (2002) document cross-country differences in how cumbersome it is to start a new business, just from a bureaucratic standpoint. The work of Anton and Yao (1994, 1995), Gans and Stern (2000, 2003), and Gans et al. (2002) emphasizes the appropriation problem in external ventures. Other important factors are the availability of money (e.g., venture capital), management, qualified employees, assets, etc.

There obviously exists a broad literature on how and why different locations witness different levels of entrepreneurial activities. Our model is partly inspired by Saxenian (1994), who notes that one of the important differences between Silicon Valley and Boston's Route 128 was that established Silicon Valley companies, such as HP, had more open cultures, whereas Massachusetts' established companies, such as DEC, tended to have more closed cultures. Saxenian attributes the success of Silicon Valley to these differences in firm behavior and firm culture. Obviously, all of these factors interact with each other, so that it is difficult to separate endogenous and exogenous drivers.

Endogeneity may be a thorn in the empiricist's eye, but can cause a gleam in the theorist's eye. For the purpose of the main model we take the external environment as given, but in the working-paper version we also developed an extension that endogenizes it. For this we considered a model where venture capitalists make endogenous decisions about whether to locate in certain geographical areas, or to develop expertise in certain technological areas. The main insight from endogenizing the presence of venture capital is that there may be multiple equilibria. If employees rarely leave firms, few venture capitalists take interest in that location or technology sector. This makes it harder for employees to leave the firm, and easier for firms to focus their employees; as a consequence, employees rarely leave. This is a self-enforcing low-entrepreneurship equilibrium. There also exists a self-enforcing high-entrepreneurship equilibrium, where there are many employees who leave their firms, and many venture capitalists to finance their innovations.

This model extension is related to the work of Landier (2005), who develops a multiple-equilibria model based on a stigma of failure. If few competent individuals are willing to take entrepreneurial risks, then entrepreneurial failure signals incompetence, which justifies why competent individuals avoid entrepreneurship. This is the low-entrepreneurship (high stigma of failure) equilibrium. However, if failure is not perceived as a negative signal, then competent individuals take entrepreneurial risks, and in equilibrium failure no longer signals incompetence. Gromb and Scharfstein (2001) examine a related model that introduces the firm as an alternative governance mechanisms. In their model, intrapreneurial activity emerges when the entrepreneurial labor market fails because of a stigma of failure. This contrasts with our model, where intrapreneurial activity may increase with greater external entrepreneurial activity. In a related vein, Hellmann and Perotti's (2006) model also has the feature that innovations inside firms complement external entrepreneurial activity.

4. Conclusion

This paper starts with the observation that many entrepreneurs receive their ideas while working for firms in related industries. It builds a model that explains when employees become entrepreneurs. A critical step of the analysis is to recognize that firms may respond to their employees' innovations in a variety of ways. Some firms want to co-opt their employees' innovations through internal ventures or spin-offs, whereas other firms want to discourage development. Some firms may seem too inflexible when they lose out on the opportunities of their departing employees, but the model suggests that this may be the outcome of an optimal firm policy. This is because firms face a trade-off between focusing employees on their core tasks versus providing them with incentives for innovative ideas. The analysis also highlights the importance of IP rights and the entrepreneurial environment. The paper thus provides a theory for the supply of entrepreneurial activity that incorporates a role for established companies. Further exploring the interactions between the policies of established firms and the process of entrepreneurial firm creation remains a fruitful topic for future research.

5. Electronic Companion

An electronic companion to this paper is available as part of the online version that can be found at http://mansci.journal.informs.org/.

Acknowledgments
The author gratefully acknowledges financial support from the Stanford Center for Entrepreneurial Studies, the W. Maurice Young Entrepreneurship and Venture Capital Research Centre at the University of British Columbia, and Social Sciences and Humanities Research Council of Canada. He thanks the anonymous associate editor, three anonymous referees, Wouter Dessein, Denis Gromb, Hans Hvide, Kevin Murdock, John Roberts, Garth Saloner, Kathy Spier, Scott Stern, Dennis Yao, Michael Waldman, Julie Wulf, and seminar participants at Cornell University (Johnson School), Dartmouth University (Tuck), London School of Economics (FMG), NBER (Organizational Economics), Northwestern University (Kellogg), Stanford University (GSB), Stockholm (SIFR), University of British Columbia (Vancouver), University of New South Wales (Sydney), and University of Pennsylvania (Wharton) for their helpful comments. All errors are the author's.

References
Agarwal, R., R. Echambadi, A. M. Franco, M. B. Sarkar. 2004. Knowledge transfer through inheritance: Spin-out generation, growth, and survival. *Acad. Management J.* **47**(4) 501–522.

Aghion, P., J. Tirole. 1994. The management of innovation. *Quart. J. Econom.* **109** 1185–1209.

Allen, J. W. 1998. Capital markets and corporate structure: The equity carve-outs of Thermo Electron. *J. Financial Econom.* **48** 99–124.

Amador, M., A. Landier. 2003. Entrepreneurial pressure and innovation. Mimeo, Graduate School of Business, Stanford University, Stanford, CA.

Anand, B., A. Galetovic. 2000. Weak property rights and hold-up in R&D. *J. Econom., Management Strategy* **9**(4) 615–642.

Anand, B., A. Galetovic, A. Stein. 2004. Incentives versus synergies in the market for talent. Mimeo, Harvard Business School, Boston, MA.

Anton, J., D. Yao. 1994. Expropriations and inventions: Appropriable rents in the absence of property rights. *Amer. Econom. Rev.* **84** 190–209.

Anton, J., D. Yao. 1995. Start-ups, spin-offs, and internal projects. *J. Law, Econom. Organ.* **11**(2) 362–378.

Bankman, J., R. Gilson. 1999. Why start-ups. *Stanford Law Rev.* **51** 289–308.

Bartlett, C. A., A. Mohammed. 1995. 3M: Profile of an innovating company. HBS Case No. 9-395-016, Harvard Business School, Boston, MA.

Bhidé, A. 1994. How entrepreneurs craft strategies that work. *Harvard Bus. Rev.* **72**(2) 150–161.

Brittain, J., J. Freeman. 1986. Entrepreneurship in the semiconductor industry. Mimeo, University of California–Berkeley, Berkeley, CA.

Burgelman, R. 1983. A process model of internal corporate venturing in the diversified major firm. *Admin. Sci. Quart.* **28**(2) 223–244.

Burgelman, R. 2002. Strategy as vector and the inertia of coevolutionary lock-in. *Admin. Sci. Quart.* **47**(2) 325–357.

Burgelman, R., L. Sayles. 1986. *Inside Corporate Innovation*. The Free Press, New York.

Cassiman, B., M. Ueda. 2006. Optimal project rejection and new firm start-ups. *Management Sci.* **52**(2) 262–275.

Chesbrough, H. 2002. Graceful exits and forgone opportunities: Xerox's management of its technology spin-off organizations. *Bus. Hist. Rev.* **76**(4) 803–838.

Christensen, C. 1997. *The Innovator's Dilemma: When New Technologies Cause Great Firms to Fail*. Harvard Business School Press, Boston, MA.

Cooper, A. 1985. The role of incubator organizations in the founding of growth-oriented firms. *J. Bus. Venturing* **1** 75–86.

de Bettignies, J., G. Chemla. 2006. Corporate venture capital: The upside of failure and competition for talent. Mimeo, University of British Columbia, Vancouver, British Columbia.

Djankov, S., R. La Porta, F. Lopez-De-Silanes, A. Shleifer. 2002. The regulation of entry. *Quart. J. Econom.* **117**(1) 1–37.

Franco, A., D. Filson. 2006. Spin-outs: Knowledge diffusion through employee mobility. *RAND J. Econom.* Forthcoming.

Gans, J., S. Stern. 2000. Incumbency and R&D incentives: Licensing in the gale of creative destruction. *J. Econom., Management Strategy* **9**(4) 485–511.

Gans, J., S. Stern. 2003. The product market and the "market for ideas:" Commercialization strategies for technology entrepreneurs. *Res. Policy* **32**(2) 333–350.

Gans, J., D. Hsu, S. Stern. 2002. When does start-up innovation spur the gale of creative destruction? *RAND J. Econom.* **33** 571–586.

Gilson, R. 1999. The legal infrastructure of high technology industrial districts: Silicon Valley, Route 128, and covenants not to compete. *New York Univ. Law Rev.* **74**(3) 575–629.

Gompers, P., J. Lerner, D. Scharfstein. 2005. Entrepreneurial spawning: Public corporations and the genesis of new ventures: 1986–1999. *J. Finance* **60**(2) 577–614.

Gromb, D., D. Scharfstein. 2001. Entrepreneurial activity in equilibrium. Mimeo, Massachusetts Institute of Technology, Boston, MA.

Grossman, S., O. Hart. 1986. The costs and benefits of ownership: A theory of lateral and vertical integration. *J. Political Econom.* **94**(4) 691–719.

Hannan, M., J. Freeman. 1984. Structural inertia and organizational change. *Amer. Sociol. Rev.* **49**(2) 149–164.

Hellmann, T. 2002. A theory of strategic venture investing. *J. Financial Econom.* **64**(2) 285–314.

Hellmann, T. 2005. Why do employees become entrepreneurs? Working paper version, mimeo, University of British Columbia, Vancouver, British Columbia, http://strategy.sauder.ubc.ca/hellmann/.

Hellmann, T., E. Perotti. 2006. The circulation of ideas: Firms versus markets. Mimeo, University of British Columbia, Vancouver, British Columbia.

Holmström, B., P. Milgrom. 1994. The firm as an incentive system. *Amer. Econom. Rev.* **84**(4) 972–991.

Hyde, A. 1998. Silicon Valley's high-velocity labor market. *J. Appl. Corporate Finance* **11**(2) 28–37.

Jackson, T. 1998. *Inside Intel*. Penguin Putnam Inc., New York.

Klepper, S., S. Sleeper. 2005. Entry by spin-offs. *Management Sci.* **51**(8) 1291–1306.

Kreps, D. M. 1990. Corporate culture and economic theory. J. Alt, K. Shepsle, eds. *Rational Perspectives on Political Economy*, Cambridge University Press, Cambridge, UK, 90–143.

Landier, A. 2005. Entrepreneurship and the stigma of failure. Mimeo, New York University, New York.

Lewis, T., D. Yao. 2001. Innovation, knowledge flow, and worker mobility. Mimeo, Wharton School, University of Pennsylvania, Philadelphia, PA.

March, J., J. Olsen. 1991. Exploration and exploitation in organizational learning. *Organ. Sci.* **2** 71–78.

Merges, R. 1999. The law and economics of employee inventions. *Harvard J. Law Tech.* **13**(1) 1–63.

Milgrom, P., J. Roberts. 1990. The efficiency of equity in organizational decision processes. *Amer. Econom. Rev.* **80**(2) 154–159.

Pakes, A., S. Nitzan. 1983. Optimum contract for research personnel, research employment, and the establishment of rival enterprises. *J. Labor Econom.* **1**(4) 345–365.

Rotemberg, J., G. Saloner. 1994. Benefits of narrow business strategy. *Amer. Econom. Rev.* **84**(5) 1330–1349.

Saxenian, A. 1994. *Regional Advantage: Culture and Competition in Silicon Valley and Route 128*. Harvard University Press, Cambridge, MA.

Simon, H. 1951. A formal theory of the employment relationship. *Econometrica* **19** 293–305.

Smith, D., R. Alexander. 1999. *Fumbling the Future: How Xerox Invented, Then Ignored the First Personal Computer*. William Morrow & Company, New York.

Sterk, S. 1993. Restraints on alienation of human capital. *Virginia Law Rev.* **79** 383–460.

Stuart, T., O. Sorenson. 2003. Liquidity events and the geographic distribution of entrepreneurial activity. *Admin. Sci. Quart.* **48** 175–201.

Subramanian, N. 2006. The economics of intrapreneurial innovation. *J. Econom. Behav. Organ.* **58**(4) 487–510.

Talley, E. 1998. Turning servile opportunities to gold: A strategic analysis of the corporate opportunities doctrine. *Yale Law J.* **108** 277–375.

Teece, D. 1986. Profiting from technological innovation: Implications for integration, collaboration, licensing and public policy. *Res. Policy* **15** 285–305.

[35]

The Economic Journal, **119** (*July*), 1010–1035. © The Author(s). Journal compilation © Royal Economic Society 2009. Published by Blackwell Publishing, 9600 Garsington Road, Oxford OX4 2DQ, UK and 350 Main Street, Malden, MA 02148, USA.

THE QUALITY OF ENTREPRENEURS*

Hans K. Hvide

What determines the quality of entrepreneurs? The article proposes a model of the interaction between individual workers' decision to become entrepreneurs and employers' efforts to keep their best workers and ideas. The main prediction from the model is that larger firms produce entrepreneurs of higher quality than smaller firms. Using novel and unique Norwegian data, I find that previous employer size exerts a significant influence on entrepreneurial performance. For example, increasing previous employer size from the 25 percentile to the 75 percentile increases yearly profitability on assets by 6 percentage points.

Due to their role in fostering employment opportunities and growth, entrepreneurs are viewed as essential to long-run economic performance. For example, the US emphasis on the role of the entrepreneur has been viewed as a main reason for why the US outperformed Europe in the twentieth century. In spite of the economic importance attached to entrepreneurs, we have remarkably little systematic knowledge about them. In particular, we know almost nothing about who become successful entrepreneurs. This question is the focus of the article.

My starting point is broad empirical evidence suggesting that founders of new firms tend to be experienced workers who pursue opportunities closely related to their previous employment. Cooper (1985) finds that 70% of the founders of new firms in a broad cross-section of US industries were previously employed in the same industry, and Bhide (2000) reports that 71% of the entrepreneurs in his sample of fast-growing companies '[...] replicated or modified an idea encountered through previous employment'. Klepper and Thompson (2005) summarise studies from the automobile, laser, disc drive and semiconductor industries and find that about 20% of entrants in these industries have their basis in established firms from the same industry. This evidence suggests that entrepreneurs do not come from out of the blue but build their human intellectual capital through work experience in established firms.

Established firms can be expected to respond actively to the possibility of key workers leaving, by offering contingency wages or job security, by including no-compete clauses in the employment contract, or even by suing leaving workers. Still, there are limitations to these measures; the employer might lack good information about the productivity of individual employees, no-compete clauses must be of limited scope to be legally binding (Posner and Triantis, 2001; Stone, 2002), and even for established innovations, the effective protection of patents is variable (Cohen *et al.*, 2000).

These observations suggest that a theory of the quality of entrepreneurs should address the interaction between workers' entrepreneurship decisions and firms' effort to protect their intellectual property. I suggest such a theory. In the model, a principal

* This research has been supported by the ESRC, grant no RES-000-22-2080. Thanks to Oliver Hart, Eirik Kristiansen, Ed Lazear, Jim March, Lars Mathiesen, Jarle Møen, Paul Romer, three referees, and seminar participants at Aberdeen, Carnegie-Mellon, Columbia-NYU, Harvard-MIT, IBM, IZA, NHH, Stanford, USC, Zurich, Econometric Society world meeting 2005, European Science days at Steyr and J.J. Laffont Memorial conference for comments and suggestions. Special thanks to Stanford GSB for its hospitality.

(owner) hires a worker (specialist) to engage in a project. The output of this project is an idea of uncertain quality *ex ante*, with imperfect property rights attached to it, meaning that the firm to a limited extent can use sticks to keep the worker from leaving to capture (a fraction of) the value of the idea for himself. Instead, the firm must use an attractive continuation wage as a carrot. I consider two different possible firm organisations, 'small' and 'large'. The main assumption made is that a large firm is from a technical perspective more efficient, in that it can exploit returns to scale due to, for example, worker specialisation, but also less informed about the progress and content of the projects the workers are engaged in than in a small firm.

It is a commonplace to state that the reason that a firm with superior technology does not grow infinitely large (or supply the entire market) is because of limited 'span of control' by the manager but it is rarely stated exactly how this limited span of control will decrease firm productivity as size grows large. Here the idea is that reduced monitoring ability will lead to the departure of the employees with the best projects, because these projects cannot be recognised.

The main insights from the model can be summarised as follows. Small firms can implement a wage policy that is fine-tuned to workers' outside option and workers leave a small firm to become entrepreneurs only to the extent that their private benefits, such as more flexible work hours or a sense of freedom and pride, from doing so are sufficiently high.[1] On the other hand, large firms have a more rigid wage policy and as a consequence lose the best workers and ideas. Entrepreneurs emerging from large firms will therefore have higher quality than entrepreneurs emerging from small firms. Second, recall that the downside from being large is that the principal obtains inferior information about the value of individual employees (or projects). When property rights protection is weak, the value of information is high, since the worker leaving with the idea is more of a threat, and being small becomes more beneficial. On the other hand, when property rights protection is strong, the value of information is low, since the worker leaving is not a serious threat, and the principal can focus on technical efficiency and commercialisation potential. Thus improved property rights make a larger firm more likely to be optimal and hence increases the quality of entrepreneurs.

In Section 6, I use novel and unique data from Norway to assess the empirical validity of the main hypothesis obtained from the theoretical part, that the quality of entrepreneurs increases in the size of the previous employer.[2] I obtain results that are consistent with previous employer size exerting a noticeable influence on entrepreneurial performance. For example, increasing previous employer size from the 25 percentile to the 75 percentile increases yearly profitability on assets by 6 percentage points.

[1] To let private benefits play a serious role in determining the entrepreneurship decision could sound overly *ad hoc*. Emphasising private benefits for entrepreneurs goes back at least to John Stuart Mill and Alfred Marshall: 'Concerning the entrepreneur [J. S. Mill] was explicit when he stated that small producers often value so highly the feeling of being their own masters that they consume their small capital in an unsuccessful struggle for independence; "they either sink into the condition of hired labourers, or become dependent upon others for support".' (Evans, 1949, p. 339). And Marshall wrote that '"Organizers of improved methods and appliances are stimulated by a noble emulation more than by any love of wealth for its own sake," and the entrepreneur "often holds his own great tenacity even under considerable disadvantages; for the freedom and dignity of his position are very attractive to him."' (Evans, 1949, p. 339).

[2] As discussed in the working paper version of the article, Hvide (2005), the available evidence supports the notion that firm size increases in property rights protection.

1. Related Literature

The main novelty of the article is to ask what determines the quality of entrepreneurs, a question I am not aware has been addressed head-on by established work.

On the background of entrepreneurs, in a typical economic theory of business start-ups, the defining feature of the entrepreneur – a business idea of some novelty – is simply assumed to exist (Kihlstrom and Laffont, 1979; Evans and Jovanovic, 1989; Rajan and Zingales, 1998, 2001; Landier, 2001). The process in which the entrepreneur obtains these ideas and his human capital is usually ignored.[3] Lazear (2005) models and finds empirical support for the idea that entrepreneurs are jacks-of-all-trades rather than specialists. Lazear (2005) does not focus on the transition to entrepreneurship (nor is explaining entrepreneurial quality his focus). Hellmann (2007) and Subramanian (2005) consider the multi-tasking problem that ensues if a worker can engage in 'private activities' on the job with the intention of creating a start-up later. Hellmann (2007) shows that entrepreneurship may come about through the firm committing not to reward such private activities (it could be *ex post* efficient to support them). Subramanian (2005) finds that the level of private activities inside the firm should be positively correlated with pay-for-performance contracts. In contrast to these papers, I assume that the principal can stop the worker from engaging in private activities but cannot stop the worker from leaving with ideas that are generated through his legitimate work.[4] Finally, Klepper and Thompson (2005) model a situation where an employee gets frustrated and leaves a firm if his opinion about the firm's best course of action disagrees too much with the prevailing opinion. Their model predicts a non-monotonic relation between the size of the employer and the quality of the start-up.

Since the article focuses on the interaction between entrepreneurship and the organisation of established firms, there is a related literature within Organisational Economics. Aghion and Tirole (1994) consider the management of innovations in a setting based on Grossman and Hart (1986). In their setting, there is no entrepreneurship in equilibrium due to ownership rights being fully contractible.[5] In the same tradition, but with imperfect property rights, Rajan and Zingales (2001) considers the optimal choice of organisation for a firm with an exogenously given 'critical resource' controlled by an owner. In my setting, the employee has informal control in that only he has the knowledge and human capital necessary for the innovation. Moreover, there

[3] Shane (2000) argues that this also holds for the Psychology and Management literatures on entrepreneurship.

[4] Hvide and Kristiansen (2006) builds on the current model's large firm case to analyse a moral hazard setting where the employer can litigate the worker if the latter leaves with an idea. That paper focuses on the interaction between the firm's suing decision and the potential entrepreneur's exertion of effort. Some other literature, Dunn and Holtz-Eakin (2000), among others, suggests that self-employed build their human capital through learning spillovers from their parents. Several papers, Evans and Jovanovic (1989), Evans and Leighton (1989) and Holtz-Eakin *et al.* (1994*a,b*) find evidence that financial constraints affect whether an individual becomes an entrepreneur. Gentry and Hubbard (2002) investigate the macro-level implications of the relation between savings and investments made by entrepreneurs. De Meza and Southey (1996), Amador and Landier (2003) and Landier and Thesmar (2009) study the role of entrepreneurial overconfidence, while Bates (1990) studies the link between education and entrepreneurship.

[5] Anton and Yao (1994) consider the effects of weak property rights in a market for ideas. Anton and Yao (1995) build on Anton and Yao (1994) to study whether an idea generated inside a firm leads to creation of one or two ventures. This question is eliminated in the current setting by assuming that the participation of the employee is essential to develop the idea.

is no start-up activity by workers in equilibrium in Rajan and Zingales (2001), due to information being complete. It is interesting to note, however, that both frameworks obtain the prediction that optimal firm size increases in the strength of property rights. Pakes and Nitzan (1983) were the first to consider the problems that arise where agents can appropriate part of their output. In their setting, there are no property rights and the principal is fully informed about the value of individual workers and therefore only efficient separations will occur, if any.

The idea that supervisors in larger firms have inferior information about projects than supervisors in small firms goes back to work by Simon (1973), who emphasised that as firm size grows, managerial attention becomes a scarce resource. A version of this idea was later used by Williamson (1981) and by Rosen (1982). These papers do not consider the implications of limited managerial resources for the design of compensation contracts, nor for turnover.

On the empirical side, previous work on entrepreneurial performance has focused on the effect of liquidity constraints and, in particular, whether more wealthy individuals do better as entrepreneurs. Evans and Jovanovic (1989) find that a doubling of founder wealth is associated with a 14% increase in self-employed earnings (Table 3, p. 820) but they do not control for investment level nor for the fact that wealthier founders tend to have higher effective human capital.[6] As shown by Hvide and Møen (2007), who use the same dataset as in the present article, lack of such controls can lead to seriously upward biased estimates of the effect of wealth on performance.

2. The Framework

A principal (owner) identifies a research area and hires a worker (specialist) at time 0 to generate ideas in that area. I shall assume that the firm has only this project and that the worker is hired from a homogeneous pool of risk neutral agents with reservation utility $\bar{U} = 0$. As explained later, the latter assumption means that I can safely assume that the worker waives any formal ownership rights for future output.

Upon signing the contract, the agent is paid a fixed wage F and supplies one unit of effort. Inelastic effort supply is meant to capture a situation where it is easier to monitor the input of the agent in terms of time spent and effort exerted than assessing his output. At time 1, he produces an idea with value x, where x is stochastic and non-negative (a bad idea can be viewed as no idea, i.e., $x = 0$). By the complex and often fuzzy nature of early stage innovations, x is assumed to be non-contractible (contracts are not enforceable by courts due to verifiability problems). The distribution of x follows $G(x)$ with density $g(x)$ and support on [0,1]. For convenience, it is assumed that $G(\cdot)$ is twice differentiable and with $g(1) = 0$. Although x is referred to as the quality of an idea produced by the agent, one can also think of x as the value of the worker's human capital if employed by the firm after time 1.

[6] Holtz-Eakin *et al.* (1994*a*) use data on the tax receipts of a sample of entrepreneurs that had received inheritances from large estates and find that a $150,000 inheritance increase will increase the survival probability by 1.3 percentage points and, if the enterprise survives, its receipts by almost 20%. That paper suffers from a similar problem to Evans and Jovanovic (1989) in that the positive relation between liquidity and receipts could just reflect that founders that receive a large inheritance are more likely to invest in the enterprise and thus be more highly capitalised.

I assume that x is known to the worker upon its realisation at time 1, while the principal receives a signal $s \in S$ about x. The precision of the signal is inversely related to the firm size in a particularly simple manner; in a small firm $s = x$ and in a large firm s is uninformative. This main assumption is discussed in more detail in Section 3.2. Conditional upon s, the firm makes a continuation wage offer B to the agent. Whether the firm can commit to a continuation wage schedule $B(s)$ at time 0 or not is immaterial to the analysis, as long as $\bar{U} = 0$, but I shall speak as if commitments were not possible.[7] Agents are protected by limited liability so that both F and B must be non-negative.

The agent decides whether to accept or reject the offer B (communication between the agent and the principal is discussed in Section 5). Accepting B means signing an extension of the employment contract and the final (time 2) payoffs become $x - B$ to the firm and B to the agent, respectively.[8] If the worker rejects the offer B, he quits the firm and develops a start-up based on x at time 1. An 'entrepreneur' is such an individual. The time 2 payoff from becoming an entrepreneur, $U(x,.)$, depends on the fixed costs (or benefits) from starting up a firm and the degree to which the idea generated is possible to carry away. I formalise this as the entrepreneurial payoff being equal to,

$$U(x) = \max(0, \eta x - c). \quad (1)$$

Since this is a key equation of the model setup, I discuss it in some detail. The parameter c is meant to capture the sum of financing costs and the private benefits the agent enjoys from being entrepreneur. For example, if a start-up requires a high investment cost (think of a pharmaceutical laboratory) or is located in region with a dearth of venture capitalists, then the financing costs are high.[9] For simplicity, it is assumed that the worker can finance this cost from own funds (the case with external financing is an interesting extension discussed informally in Section 5). Private benefits, on the other hand, are the monetary equivalent the entrepreneur is willing to sacrifice in order to start a venture (be his own boss, a sense of pride etc.). Since empirical evidence indicates that private benefits can be substantial (Hamilton, 2000), both the cases $c > 0$ and $c < 0$ are analysed. For simplicity it is assumed that c is known to both the worker and the employer at time zero, and is constant across all workers in the labour pool.[10]

The parameter η reflects the degree to which the idea generated by the agent has value to him if he leaves; the higher η the higher fraction of value the agent can carry

[7] If $\bar{U} > 0$ then under pre-commitment there will be an indeterminancy in the optimal contract, as in Pakes and Nitzan (1983), since F and B will be substitutable, but that will have no impact on the results. By the assumption $\bar{U} = 0$, then $F = 0$ in optimum and the optimal $B(s)$ schedule is uniquely determined.

[8] One might ask why the firm would not write a longer contract in the first place to avoid many problems. However, contracts of unbounded duration would not be acceptable according to labour laws in most countries and, for any contract with finite length, the problems the article discusses can occur.

[9] Evidence from Hellmann and Puri (2002) suggests that venture capitalists may boost the gross entrepreneurial payoff $U(x)$, in addition to providing funding. As long as the firm has the option of hiring the venture capitalist as a consultant, or a person with similar expertise, it seems unlikely that this possibility would change our results qualitatively.

[10] Since c does not interact with x, letting c be unknown to both parties prior to contracting (but learned at time 1) would give the same type of results. Letting c be privately known to the worker prior to signing the employment contract, would open for the firm screening for workers with a high c. When precommitment to $B(s)$ is not possible, such screening is not possible, since all worker types will prefer a high F. When precommitment to $B(s)$ is possible such screening contracts would give an advantage to small firms relative to large firms but apart from that make no qualitative difference to our results.

with him. My focus is on how the strength of property rights protection affects η (how technological complementarities can affect η is discussed in Section 5).

There are at least two ways in which η depends on the legal environment; through non-compete clauses and through patenting. Let us discuss these two mechanisms in turn. Non-compete clauses impose a direct restriction on the worker's legal ability to start up a business whose activities are close or similar to those engaged in at his previous employer. The ease to which non-compete clauses can be enforced varies between industries, one reason being that what is meant by a related business can be much harder to define in say the management consulting industry than in the electronics industry (Posner and Triantis, 2001; Stone, 2002). Varying η can also reflect differences in the strength of patent protection across industries, where the sources of variation are technological that can be taken as exogenous to the model.[11] This variable therefore is well-suited for comparative statics exercises.

I assume that the worker after accepting B can reveal x to the principal. This is a fairly innocuous assumption, as the agent would have no incentives to misrepresent x at this point and (non-verifiable) cheap talk will be credible. As a consequence, the quantity x can be interpreted as the value of the idea given its best use by the firm, either through development inside or through a spin off (creation of a new division organised as a separate firm). This interpretation will be explained further in Subsection 3.3. Since a spin-off can plausibly imitate a start-up along all technical dimensions (recall the assumption that the worker has committed to work for the firm) and there will be no efficiency loss associated with legal costs or other hassles, I shall assume throughout that $\eta \in (0,1)$.

The last thing to specify in the model is the payoff to the firm if the worker decides to leave. There are at least two, counteracting, effects. First, the employee leaving could involve some loss in profits for the firm due to increased product market competition. Second, the employee starting up a firm based on x could open up for the possibility of suing the agent, with some positive payoff if successful. To simplify the analysis, I assume that these effects are small, or cancel each other out, and that the employee leaving is, from the firm's perspective, equivalent to the employee leaving the workforce.

3. Quality of Entrepreneurs

This Section analyses the interplay between the entrepreneurship decision made by workers and the optimal wage offers made by firms, taking firm size as given. Section 4 considers the optimal firm size.

3.1. *Small Firm*

Recall that in a small firm, the owner knows x, i.e., $s = x$. If $c > 0$ then $U(x) < x$, it will be efficient that the worker stays and the firm will offer a bonus that is sufficient to retain the worker. If $c < 0$, however, then $U(x) > x$ for sufficiently low x. Without loss of

[11] For example, evidence from Cohen *et al.* (2000) indicates that innovations in the drugs (pharmaceutical), computers and chemicals industries are easier to patent than innovations in the food or steel industry.

generality, assume that the worker stays in the firm if $B_S = U(x)$. Since there is no reason to leave rents on the table, the optimal bonus offer by the firm will be,

$$B_S^* = \begin{cases} 0 & \text{if } U(x) < 0 \text{ or } U(x) > x \\ \eta x - c & \text{if } 0 < U(x) < x. \end{cases} \quad (2)$$

The leaving option implies that the bonus payment to the agent will be equivalent to a call option on (a fraction of) the value of the project. Incidentally, this payment scheme resembles pay structure in small technologically oriented firms, which commonly pay their workers with stocks or options.[12]

From the optimal bonus offer by the firm I get the following on the quality of entrepreneurs from a small firm.

PROPOSITION 1. *There will be entrepreneurs emerging from small firms only when private benefits from a start-up outweigh the financing costs* ($c < 0$). *The entrepreneurs will be of low quality.*

Proof. For $c > 0$ then $U(x) < x$ and the firm will always pay a sufficient amount to keep the worker. When $c < 0$, the worker leaves in equilibrium if $\eta x - c > x$, or in other words if $x < P$, where $P \equiv -c/(1 - \eta) > 0$. Hence, from a small firm, workers with ideas on the interval $[0,P]$ will become entrepreneurs.

In a small firm there cannot be entrepreneurship unless there are private benefits associated with entrepreneurship. Moreover, entrepreneurs from small firms will tend to have low quality ideas, i.e., the value of their ideas are not a random sample from the distribution of x but a random sample below some threshold. Note that there will be a socially optimal level of separations in equilibrium, since the firm extracts all the surplus created when it sets the separation threshold through B_S^*.

Assuming that $c < 0$, the comparative statics properties on the quality and quantity of entrepreneurs emerging from a small firm are straightforward.

REMARK 1. *The quality and quantity of entrepreneurs from a small firm decreases in* c *and increases in* η.

Proof. The remark follows from differentiating P with respect to c and η.

An increased financing cost or decreased private benefit, i.e., an increased c, makes entrepreneurship less efficient and only workers with ideas of very low quality will leave to start-up their own firm. Weaker property protection makes the quality of entrepreneurs go up.

3.2. *Large Firm*

I assume that in a large firm, the owner gets less precise information about x than the owner of a small firm. Since this is a key assumption, let me spend some time justifying it before moving on to the analysis.

[12] Oyer and Schaefer (2005) provide broader evidence that firms compensate employees in order to retain them, rather than to elicit effort.

The idea that smaller firms are better monitors has a long theoretical tradition, going back to at least Simon's (1973) work on managerial overload. Later, it was emphasised by, e.g., Rosen (1982) in his famous paper on authority and control: 'It is the supervisory activity that congests management scale economies and produces determinate firm sizes.' And 'As suggested by Williamson (1981) and others, fulfilment of plans requires monitoring of workers in adjacent ranks, because transaction costs raise possibilities for opportunistic behavior of subordinates and impose limitations on the information transfer within the firm.' (Rosen, 1982, p. 313).

On the empirical side, the assumption that smaller firms are better monitors is supported by evidence from Agell (2004), who interviewed the pay managers of 1,200 Swedish companies about the ease to which pay can be tied to performance for individual workers. Agell (2004) reports that pay managers at larger firms find it significantly more difficult to evaluate the performance of individual employees than pay managers at smaller firms. At a more anecdotal level (Shockley Semiconductors, Xerox Parc) large corporations are often criticised for being slow to adapt and to pose bureaucratic hurdles to employees with ideas and initiative. I suggest a reason why it might be so: large corporations are sluggish because they lack information about the right course of action.

Given the informational benefit from being small, there must be some benefit from being large for there to be an interesting choice of firm size. I formalise this by assuming that if the worker stays, being large has scale advantages that boost the value of the idea by a factor $\alpha > 1$: an idea of quality x is worth x to a small firm and αx to a large firm.[13]

The model's distinction between a small and a large firm can be seen as a reduced-form specification of several plausible mechanisms:

- Similar to Stein (2002), large can mean hiring a bureaucrat, or middle manager, in addition to a worker. The upside of hiring a bureaucrat is that he can boost the value of the project through, e.g., facilitating commercialisation. The downside of hiring a bureaucrat is that he acts as a veil between the agent and the owner, either because of lack of competence in assessing x or because he has a private agenda.[14]
- One can interpret a large firm as one with more workers than a small firm. The upside of employing more workers is to exploit scale advantages or complementarities between the worker's skills, and the downside with team production that it becomes harder to evaluate a particular worker's marginal contribution or value (Alchian and Demsetz, 1972).
- One can think of a large firm as having more projects than a small firm. On one hand, this can create synergies through, e.g., knowledge spillovers[15] but on the other hand more projects will tend to increase managerial overload by reducing the time available for the principal on each project (Simon, 1973).

[13] To have the 'boosting factor' α as a constant makes the analysis more convenient but is not necessary. It is sufficient for the same type of results to go through that $\alpha(x)x > x$ for all x.

[14] For example, as in Stein (2002), the bureaucrat cannot be trusted to make true announcements about x due to private benefits from keeping the project alive.

[15] There is empirical evidence supporting the idea of such synergies in Cockburn and Henderson (1996), who find that larger research efforts in the biotech industry are more productive, ascribed to knowledge spillovers.

Let us now move to the analysis. Recall that in a large firm, the principal knows $g(\cdot)$. The optimal offer by the firm is denoted by B_L^*, and the analysis is confined to the case $B_L^* > 0$, which occurs when $|c|$ is not too large (if $c << 0$ then all workers become entrepreneurs and when $c >> 0$, no workers become entrepreneurs even for $B_L = 0$, neither case very interesting). Given an offer $B_L > 0$, then a worker leaves iff $U(x) = \eta x - c > B_L$, or in other words if x exceeds a cutoff z, where

$$z = (B_L + c)/\eta. \tag{3}$$

Taking into account workers' best response, and viewing z as the choice variable for the firm, the profit of a large firm equals,

$$\Pi_L = \int_0^z (\alpha x - \eta z + c)g(x)\mathrm{d}x. \tag{4}$$

The firm's profit maximisation problem can be written as finding an optimal cutoff z^*, where

$$z^* = \arg\max_{z\in[0,1]} \left\{ \int_0^z (\alpha x - \eta z + c)g(x)\mathrm{d}x \right\}. \tag{5}$$

The first order derivative of Π_L equals,

$$\Pi_L' = (\alpha z - \eta z + c)g(z) - \eta G(z). \tag{6}$$

This equation reflects the trade-off faced when deciding upon a cutoff z. Setting a higher z will decrease leaving at the margin but also increase the wage payment to all worker types that stay.[16] To see that z^* must be less than unity, observe that $\Pi_L'(1) = (\alpha - \eta + c)g(1) - \eta G(1) = -\eta < 0$. I then have the following result.

PROPOSITION 2. *There will always be entrepreneurs emerging from a large firm and the entrepreneurs will be of high quality.*

Entrepreneurs from large firms are from the top of the distribution of x simply because $U(x)$ slopes upwards in x and therefore workers with the best ideas reject the fixed bonus offer.

Assume for convenience that $\Pi_L' = 0$ has a unique solution; some conditions for uniqueness were considered in the working paper version, Hvide (2005). In that case, the first order condition (5) implicitly defines a differentiable function $z^*(\alpha,\eta,c)$ with the following comparative statics properties.

REMARK 2. *The quality (quantity) of entrepreneurs from a large firm,*

Increases (decreases) in α
Increases (decreases) in c
Decreases (increases) in η

[16] The second order condition for profit maximum is,

$$\Pi_L'' = (\alpha - 2\eta)g(z^*) + (\alpha z^* - \eta z^* + c_E)g'(z^*) < 0.$$

Proof. Implicitly differentiate the first order condition for profit maximum to obtain,

$$\begin{aligned}\frac{dz^*}{d\alpha} &= -z^* g(z^*)/\Pi_L'' > 0\\ \frac{dz^*}{dc} &= -g(z^*)/\Pi_L'' > 0 \\ \frac{dz^*}{d\eta} &= z^* g(z^*) + G(z^*)/\Pi_L'' < 0.\end{aligned} \tag{7}$$

These results are intuitive. Increased technical efficiency α makes the marginal employee more valuable to the firm, without affecting $U(\cdot)$, and the firm increases z (by increasing B_L). This reduces the probability of entrepreneurship but increases the average quality of entrepreneurs. An increased financing cost or decreased private benefit, i.e., an increased c, makes it cheaper for the firm to keep the marginal employee and z increases. Finally, weaker property rights protection makes it more expensive for the firm to keep the marginal employee and z decreases.

To sum up, a large firm is constrained by low quality on the information about the ideas generated and therefore has a rigid wage policy. The workers with the best ideas will tend to leave the firm to start their own venture. Therefore, entrepreneurs from large firms do not carry with them a random draw from the distribution of x but the distribution of x above some threshold.

3.3. *Some Interpretational Issues*

I have assumed that the firm has the formal ownership rights to the project. This is an innocuous assumption given that the reservation utility of a worker is zero ($\bar{U} = 0$), since the worker makes rents and therefore will be willing to waive any formal ownership rights at the time of employment. If $\bar{U} > 0$, however, then conceivably it would be optimal to sign a contract that gives some ownership rights of the project to the agent. To see why this would not be the case, suppose that contracts on formal ownership can be written to induce any η on the interval $[\eta_L,1]$. Then the agreed-upon η would equal the minimum point η_L. The reason is simple; if the agent's expected utility needs to be increased to ensure participation, it will be more efficient to increase it via increasing the fixed wage F rather than increasing η, since a higher η means a greater efficiency loss (due to an increased probability of entrepreneurship). Therefore, the only consequence of $\bar{U} > 0$ in the model would be to increase the fixed wage F paid by the firm. This is convenient because it means that the model can be encompassed in a general equilibrium type of framework (with endogenous $\bar{U}$) without any of the basic results being reversed.

Second, the spin-off option has been implicit in the firm's production function. Look more into this possible venue of developing ideas by considering the case of a small firm (the case of a large firm is practically speaking identical). A reasonable feature of spin-offs is that they have higher fixed costs than internal development. For spin-offs to occur at all, there must consequently be some technological advantage to it, such as negative synergies between the project and other existing projects of the firm (a spin-off occurs if the innovation does not fit existing business areas, or it might be easier to incentivise the agent in a separate unit). We can formalise this by assuming that developing an idea with 'intrinsic' quality y in-house gives it economic value equal to py, while developing it through a spin-off gives it value $\theta y - f$, where $f > 0$, and $\theta > p$. Conditional on the firm developing ideas optimally, it follows that $x = \max(py, \theta y - f)$. We then have that,

REMARK 3. *Only innovations of sufficiently high quality will be developed through a spin-off.*

This result is consistent with evidence of a positive stock market reaction to spin-off decisions, e.g., Desai and Jain (1999) for US evidence and Veld and Veld-Merkoulova (2004) for European evidence.[17]

Third, I have stressed the interpretation of η as reflecting the extent to which property rights are enforceable. The flexibility of the model allows for other interpretations. If we choose to interpret x as the human capital the agent acquires from working on the project, one interpretation of η is the extent to which this human capital is firm-specific; where a high degree of firm specificity implies that η is low. Although a firm may be able to affect the firm specificity of human capital acquisition when designing training projects, it seems reasonable to assume that firm specificity in the human capital acquired from working on a research project to a large extent is driven by technological factors outside the decision sphere of the firm. This justifies treating η as exogenous also under this interpretation of x.

Finally, while η is a constant in (1), the findings can easily be generalised to the case where η is a function of x,

$$U(x) = \max[0, \eta(x)x - c], \tag{8}$$

where the only restriction put on $\eta(x)$ is that $\mathrm{d}U/\mathrm{d}x = \eta'(x)x + \eta > 0$, i.e., the agent prefers a higher x to a lower x. This more flexible formalisation allows for complementarities that can vary upon the quality of the idea. Also it covers, for example, cases where it is easier for firms to enforce property rights when x becomes bigger.

4. Endogenous Firm Size

This Section studies how the property rights regime affects the trade-off between technical efficiency and quality of information, and determines the optimal firm size.

To solve for the optimal firm size, I need to calculate the difference in 'indirect' expected profits between organising as a large firm and organising as a small firm,

$$\begin{aligned}\Delta^*[\alpha, c, \eta, z^*(\alpha, c, \eta)] &= \Pi_L^* - \Pi_S^* = \alpha \int_0^{z^*} xg(x) - G(z^*)B_L^* - \Pi_S^* \\ &= (\alpha - 1)\mathrm{E}(x) - \alpha \int_{z^*}^1 xg(x) + \int_0^{I(c)P} xg(x) - [G(z^*)B_L^* - \mathrm{E}(B_S^*)]. \end{aligned} \tag{9}$$

[17] An interesting idea that will not be pursued further here is that a spin-off may make it easier to write a good compensation contract to the worker. At the extreme, we could think of the project organised in a spin-off to make x verifiable to courts. Even in that case the current analysis holds through as long as the fixed cost of setting up a spin-off, f, is not too low.

The first term gives the (hypothetical) productivity difference between a large and a small firm if there were no entrepreneurship, the second term gives the reduction in productivity for large firms due to workers leaving, and the third term gives the reduction in productivity in small firms due to workers leaving, where $I(c)$ is an indicator function which equals 1 if $c < 0$ and 0 otherwise. The final term gives the difference in expected wage bill for a large and a small firm. The Δ^*-function is continuous since both Π_L^* and Π_S^* are continuous.

We can note the following results,

REMARK 4. *When*

(*i*) *α is sufficiently high or*
(*ii*) *η is sufficiently small, then a large firm is optimal.*

Proof. To prove (*i*) I show that for $\alpha \geq [\mathrm{E}(x) + \eta - c)]/\mathrm{E}(x)$ then a large firm dominates for all (c,η). The argument is simple. Suppose that the large firm pays a bonus equal to $\eta - c$ to all workers. It would then retain all and have profits equal to $\alpha\mathrm{E}(x) - \eta + c$. A small firm can never make a profit higher than $\mathrm{E}(x)$. Hence the difference in profits between a large firm and a small firm must be at least $(\alpha - 1)\mathrm{E}(x) - \eta + c$. This expression is greater than zero for $\alpha > [\mathrm{E}(x) + \eta - c)]/\mathrm{E}(x)$. To prove (*ii*), there are two cases to consider, $c < 0$ and $c > 0$. For $c < 0$, consider the limit case $\eta = 0$. In that case it is either optimal to pay all agents c (in which case all stay) or to pay them all zero (in which all leave). Assuming that $\mathrm{E}(x) - c > 0$, paying all agents c is optimal. The profits of a large firm then equals $\alpha\mathrm{E}(x) - c$ and the profits of a small firm equals $\mathrm{E}(x) - c$. Therefore $\Delta^* = (\alpha - 1)\mathrm{E}(x) > 0$ for $c < 0$ and $\eta = 0$. By continuity, an interval $[0,\bar{\eta}]$ must exist such that a large firm is optimal for any $\eta \in [0, \bar{\eta}]$. For $c > 0$, the worker stays even when $B_L = 0$ if $\eta \in [0,c]$. It follows directly that $\Delta^* > 0$ for $\eta \in [0,c]$.

I now have the following result.

PROPOSITION 3. *For any c, then for α not too large there exists a cutoff η^* such that a large firm dominates for $\eta < \eta^*$ and a small firm dominates for $\eta > \eta^*$.*

Proof. See Appendix A.

Recall that the optimal firm size is a trade-off between improved technical efficiency (parameterised by $\dot{\alpha} > 1$) and inferior quality of information. When property rights are strong (η low) the leaving option is less of a threat, information about the quality of individual projects therefore of relatively small value, and a large firm is optimal. On the other hand, when property rights are weak (η high), the leaving option is more of a threat, the value of information high, and a small firm becomes optimal. Hence worsened property rights makes a small firm more likely to be optimal.[18]

[18] There is a qualification to this result. When the property rights are very weak (η close to 1), the surplus that can be extracted from the worker is low, and information can become of little value. Hence the relationship between weakened property rights protection and the value of information can be non-monotonic. Proposition 3 excludes this possibility to have an impact on the optimal firm size for α sufficiently close to 1. Restricting η to be bounded away from 1 will have the same effect.

Finally I derive some additional comparative statics results.

REMARK 5. *A large firm is more likely to be optimal the higher α. For the absolute value of c sufficiently small, a large firm is less likely to be optimal the higher c.*

Proof. See Appendix A.

While the first result is not surprising, the second result essentially says that increasing c will be more beneficial to the profits of small firms than to the profits of large firms. The intuition behind this result is that a higher c benefits a small firm for all (or approximately all) workers, while it benefits large firms only for workers with $x \in [0,z]$.

5. Discussion

The purpose of the analysis has been to clarify our understanding of the relationship between an entrepreneur's background and his likelihood of obtaining success as an entrepreneur. Here I discuss three aspects of the model; financing of entrepreneurs, the absence of contingent offers from a large firm and a possible interaction between property rights protection and observability.

I have assumed that workers can self-finance if they decide to become entrepreneurs. Suppose instead that a start-up requires external funding, and eventual repayment according to a debt contract. If financiers are risk averse or there is some cost of bankruptcy, entrepreneurs from small firms will then get worse financing conditions than entrepreneurs from large firms (i.e., have a higher c). The reason is simple: entrepreneurs from small firms carry (more) credit risk and financiers will only be willing to tolerate this if they are compensated through a higher effective interest rate.[19] Interestingly, this argument suggests a reason why venture capitalists may be able to better capture value from ideas than established firms: the fact that a worker has left (and whether he has left a small or a large firm) will provide the venture capitalist with a better-informed position than (a large) employer had.

I have assumed that profit sharing schemes or contingent pay to the agent in a large firm are not feasible, since x is non-contractible. Given that x is not directly contractible, there are two versions of profit sharing that can be taken into account. First, the large firm might offer the worker shares in the company if he stays on. This could possibly solve the efficiency loss problems due to turnover since the agent's payoff would depend on the quality of the idea also if he stays on in the firm (a higher stock price of the firm if the idea is valuable). However, even if profit sharing through stocks, stock options or group rewards might give the agent stronger incentives to stay on in the firm, they would typically only marginally do so, since the increased value due to a good project would have to be shared with the other shareholders or project members (Merges, 1999; Kim and Marschke, 2005).[20] A better solution than awarding the worker

[19] This will have the qualitative effect of shifting the Δ^* function to the left.

[20] If the firm is small, the worker might get a substantial share of the firm if he stays. This is consistent with anecdotal evidence of key workers in Silicon Valley firms having significant ownership stakes in their companies.

stocks might therefore be for the firm to allow the worker to set up his own company and get a share of the profits of the new company. Although such worker-initiated spin-offs are quite common (for example spin-offs from the MIT Engineering School) there are limitations to this solution. It is for example not clear what would stop the worker from neglecting the spin-off and to develop the idea inside a different company (Anton and Yao, 1995). In line with this argument, Anand and Khanna (2000) find in their sample of UK firms that there are few licensing contracts in markets were it is 'difficult to clearly specify the content and boundaries of knowledge and other intangible assets' (p. 126). Arora and Ceccagnoli (2006) reports a similar finding with the use of US data. If it is difficult to use profit sharing or other incentive contracts, we are back in a similar situation to the one focused on in the model.

Instead of profit sharing, an alternative way to let the agent's payoff depend upon the quality of the idea would be to let the informed worker demand a continuation wage rather than having the (uninformed) firm offer a wage. Obviously, such communication would have no effect on the small firm solution. Potentially, however, communication could enable the agent in a large firm to reveal the value of the idea to the principal, and only efficient separations would occur. Some versions of this argument seem sound, others not. First, suppose that the worker has a cheap means of credibly communicating the content (and value of) the idea. Such revelation would be unlikely, however, since it may drastically reduce the value of the worker's position (Anton and Yao, 1995), by making it easier for the firm to claim its property right. Second, if the agent can only make non-verifiable statements about the value of the idea, talk would be utterly cheap since the worker would have incentives to inflate x and thereby get a higher B. Third, and more interestingly, we could turn the bargaining table around and let the worker make a wage demand instead of the firm making a wage offer. Such a version of the incomplete information bargaining, which is analysed in Appendix B, opens up the interesting possibility that the worker can credibly reveal information about x through his demand (in fact, I construct a separating equilibrium where the demand fully reveals x). Apart from this, there must be workers leaving in equilibrium, and the probability of leaving increases in x, so that the main results on entrepreneurship are unchanged.[21]

I assumed that the information advantage of a small firm relative to a large firm is independent of the strength of property rights protection η. It is however, worth thinking more about what causes property rights to be weak. Presumably, this occurs due to problems with unverifiability, so that one could expect a positive correlation between the strength of property rights and the extent to which small firms would have an information advantage. This argument is complementary to the model, in that there would be even greater reasons to choose a large firm if both ownership rights are strong and information is good, and, vice versa, there would be even greater reason for organising as a small firm if both ownership protection is weak and the informational

[21] Of course, there is a wide range of other formalisations of the incomplete bargaining game between the worker and the firm. I am not aware of any such formulation, however, without the property that bargaining is more likely to break down (i.e., entrepreneurship to occur) the higher x, which is the essential feature needed.

advantage versus large firms becomes greater.[22] Finally, I have assumed that the principal can monitor the input (effort, time) of agents in a perfect manner. What if (a portion of) effort had to be induced via incentives? Since a small firm can condition its pay on a less noisy measure of x than a large firm can, it can link pay and performance in a more direct fashion, and therefore elicit effort more easily than a large firm. Modelling such a consideration would tend to shift the Δ^*-function to the left in the model.

6. Empirical Analysis

The theoretical analysis predicts that entrepreneurs originating from larger firms start up companies of higher quality than entrepreneurs originating from smaller firms. This section confronts this hypothesis with data from Norway.

6.1. *Data*

The dataset consists of incorporated companies founded between 1994 and 2002 in Norway.[23] The dataset is compiled from three different sources. First, accounting information from Dun and Bradstreet's database of accounting figures based on the annual tax reports. These data include variables such as organisation identification number, industry codes, sales, assets, profits and so on for the years 1994–2005. Second, data on individuals are prepared by Statistics Norway. These records include individual identification number and yearly sociodemographic variables such as age, education, wealth, and earnings split into labour income and capital income, for 1986–2004. The Statistics Norway data is used to construct measures of founder background. Third, founding documents submitted by new firms to the government agency *Brønnysundregisteret.* These data include the organisation identification number for the start-up, personal identification number of the founders, total capitalisation of the company and each founder's respective ownership share. Using these data, I define an 'entrepreneur' as an individual with a majority stake, i.e., more than 50% of the total shares, in a newly established incorporated company.[24]

Many start-ups in Norway, particularly in real estate, are tax-shelters or have minimal activity. The sampling procedure dealt with this in two ways.[25] First, by oversampling

[22] Related, we have assumed that the principal knows x in a small firm and knows nothing about x in a large firm (except its distribution). Allowing for an informative signal received by the principal in a large firm will open up for the possibility that the set of entrepreneurs in a large firm will be a non-connected set. This can mean a lower quality of entrepreneurs from a large firm than presently but it will still be true that the average quality of entrepreneurs will be higher from a large firm than from a small firm.

[23] The main advantage of focusing on incorporated companies rather than on the self-employed is that incorporated companies trace out much more detailed data on the business unit, such as investment levels and industry code classification. Moreover, since setting up an incorporated company carries tax benefits relative to being self-employed (e.g., more beneficial write-offs for expenses such as home office, company car and computer equipment) incorporation status will be more tax efficient than self-employed status except for the smallest projects. Also, incorporated companies are required to have an external auditor certifying the accounting statements in the annual reports, which makes the reliability of the data higher than for self-employed.

[24] Restricting the sample to majority owners avoids the problem of defining founder background characteristics when dealing with multiple founders. Other advantages include avoiding double counting of companies and nominal founders such as 'sleeping spouses'.

[25] Avoiding sampling empty companies was an important concern since the incorporation documents had to be hand-collected by research assistants at a considerable cost.

start-ups in manufacturing and IT, since tax shelters are less likely to occur in these industries. As an added benefit, these industries exhibit high variation in perceived capital-intensity. Second, to avoid sampling empty shells, start-ups sampled were required to have at least NOK 500,000 in sales and at least two persons employed during one of the first two years of operation.[26]

For each start-up identified in Dun and Bradstreet's database, a list of founders was compiled based on the founding documents. Next, the founders' associated socio-demographic information from the public registers supplied by Statistics Norway was matched in. This procedure gave a sample of about 1000 entrepreneurs and 5000 entrepreneur-years.

We can note several unique features of the dataset. Most importantly, it contains detailed information about the employment and wage history of the entrepreneurs. This enables me to control for human capital and the entrepreneur's opportunity cost, and to construct measures of the size of the previous employer. Second, the dataset includes the industry codes of the start-ups (in contrast to datsets on self-employed I am aware of), enabling industry-specific effects related to employer size or profitability to be controlled for. Third, the detailed data allow me to construct several different performance measures for the start-ups which captures different aspects of start-up success.

6.2. *Empirical Strategy*

I use operating returns on assets (OROA) as the benchmark performance measure. OROA is the standard performance measure in a large accounting and financial economics literature; see e.g. Bennedsen *et al.* (2007) and the references therein. OROA is defined as the ratio of earnings before interest and taxes (EBIT) to the total asset base used to generate them. Unlike returns on equity or returns to capital employed, OROA compares firm profitability relative to total assets.[27] In contrast to net income-based measures such as return on assets, OROA is not affected by capital structure or dividend policy differences across firms. Following Frank and Goyal (2003), I use three-year average OROA, weighted by the level of assets. This measure weighs profitability in years where the start-up is larger more heavily than profitability in years where the start-up is smaller.

OROA does not necessarily capture the success of companies whose asset base grows fast relative to profitability. To capture such growth companies, I therefore use asset growth as an alternative performance measure to OROA. As a second alternative performance measure I use start-up survival.[28] Focusing on survival is useful, for example, to accommodate the possibility that growth-oriented companies might also have a higher probability of failure. To have a reasonable sample size, I focus on four-year survival.

[26] The final sample consisted of about 35% start-ups within manufacturing (NACE 23–35), 50% within IT (NACE 72) and 15% drawn randomly from other non-financial sectors.

[27] The asset base I use to compute yearly OROA is the average of assets at the beginning and the end of the calendar year. This is in line with the accounting literature.

[28] The data do not contain information on whether a start-up ceases to exist due to a liquidiation or due to a buy-out. This reduces the reliability of the survival measure I construct. Buy-outs occur rarely for very young firms so this lack of detail is unlikely to cause much bias.

As outlined in Section 3.2, the measure of firm size best aligned with the theoretical model is the number of employees. Employee supervision typically occurs at the plant level. Therefore, the number of employees at plant level captures the difficulty to which principals can assess individual productivity, and I therefore measure employer size by number of employees at plant level.[29] Since the data gives access to yearly plant identification number for all workers is Norway, I was able to construct exact yearly measures of plant size for the majority of the future entrepreneurs. I use the average plant size over three years before the venture's first year of operations.[30] Reasonably, the informational effects of larger employer size should be declining at the margin. To accommodate this concavity, I use log number of employees as the benchmark employer size measure.

A positive correlation between previous employer size and performance need not be due to the informational reasons outlined in the model. It is well known from the labour economics literature that larger firms tend to recruit workers with stronger formal credentials such as education level, so that omitting controls for such credentials could create a spurious relation between employer size and entrepreneur quality. I therefore include controls for age and years of education, and these variables squared. It is also well known from labour economics (Fox, 2004) that large firms pay higher wages than small firms even after controlling for education level and experience. Although there is no consensus on what creates these pay differences, they may well be due to differences in ability at the hiring stage (or through subsequent training). To accommodate ability differences not captured by education or experience, I include a control for the wage earned by the entrepreneur while still being an employee. As a benchmark measure of previous wage, I use the average wage over three years before the venture's first year of operations.

More wealthy founders tend to start up companies that are larger (see e.g., Holtz-Eakin *et al.* (1994*b*), or, using the same data as here, Hvide and Møen, 2007), so that if entrepreneurs from larger firms are wealthier then they may perform better for liquidity or size reasons not related to wage. To accommodate possible liquidity and size effects, I include founder taxable gross wealth and company size at start-up date as measured by the total equity infusion as controls. Finally, I include dummy variables for start-up year and 2-digit industry codes to capture business cycle and industry effects.

6.3. *Descriptive Statistics*

Table 1 presents summary statistics of the founders and firms in the sample. As expected, founders tend to be experienced workers and the start-ups are generally small. There is substantial variation in the size of previous employer and a majority of firms survive the first four years.

[29] Plants are defined through having an independent business address.

[30] Because of the tax benefits of incorporation status, the start-ups could be continuations of existing sole proprietorships. We avoided this problem by defining plant size as an average only over years where the founder was not self-employed (the same applies to the previous wage measure defined below).

Table 1

Summary Statistics ($n = 988$)

Variable	Description	Median	Mean (St.Dev)
Age	Founder age at start-up date	39	40.1 (8.6)
Education	Founder education in years, at start-up date	12	13.1 (2.6)
Wealth	Founder taxable wealth, 3-year average before start-up date	454.2	1200.0 (5012.1)
Wage before	Founder wage income, 3-year average before start-up date	428.5	477.6 (297.0)
Prevsize	Employer size in number of employees, 3-year average before start-up date	28.8	232.0 (1097.0)
Start-up year	First reporting year	1999	1998.5 (2.8)
Equity	Start-up equity at start-up date	100	185.2 (1124.5)
Sales	Start-up sales, first year	957.1	2636.7 (7740.1)
Assets	Assets, end of first year	707.0	2184.0 (7480.0)
OROA	Start-up return on assets, asset–weighted 3-year average	0.123	0.155 (0.288)
Growth	Yearly asset growth	0.057	0.228 (0.696)
Survival	4-year survival	1	0.74 (0.441)

Note. Krone values are expressed in (1000) 2002 kroner. 1000 kroner is equivalent to about 140 Euro. The formal capital requirement for registering an incorporated company was NOK 50,000 in equity until 1998 and NOK 100,000 thereafter. The asset base used to compute yearly OROA is the average of assets at the beginning and at the end of the calendar year. OROA is based on yearly 5% winsorised values.

6.4. *Previous Employer Size and Performance*

I now turn to the main question: do start-ups where the entrepreneur was previously employed at a larger firm perform better? As discussed above, I use three complementary performance measures; OROA, asset growth and 4-year survival. I start out looking at the effect of employer previous size on OROA (Table 2).

Regressing OROA on employer size without controlling for previous wage, start-up size or industry codes, I find a statistically significant relationship, as depicted in column (I). A one log increase in previous size (which approximately equals a doubling of previous size) gives a 2.4 percentage points increase in OROA. Introducing a set of controls for human capital in column (II) we see that the size of the estimated effect is weakened but only slightly. The economic magnitude of a change in previous size seems quite considerable. For example, by moving from the 25 percentile to the 75 percentile in previous employer size (from 8 to 119 employees) gives a 6 percentage points increase in OROA. In comparison, an increase from the 25 percentile to the 75 percentile in previous wages (315K to 570K) gives 3 percentage points increase in OROA. Hence the relation between previous employer size and OROA seems quite strong both in absolute terms and relative to the strength of the relationship between founder's previous wage and OROA.

Table 2

Entrepreneurial Performance

	I	II	III	IV	V	VI	VII	VIII
	OROA	OROA	Growth	Growth	Survival	Survival	Wages	Wages
Ln(previous size)	0.024***	0.021***	0.019***	0.018***	−0.031	−0.014	0.027***	0.012
	(0.005)	(0.005)	(0.005)	(0.006)	(0.028)	(−0.030)	(0.008)	(0.009)
Ln(previous wage)		0.054**		0.035		0.075		0.355***
		(0.023)		(0.023)		(0.121)		(0.068)
Age, education	Yes	Yes	Yes	Yes	Yes	Yes	Yes	Yes
Equity, wealth	No	Yes	No	Yes	No	Yes	No	Yes
Industry dummies	No	Yes	No	Yes	No	Yes	No	Yes
Year dummies	Yes	Yes	Yes	Yes	Yes	Yes	Yes	Yes
N	988	924	4,786	4,166	730	638	987	919
R^2	0.084	0.141	0.072	0.090	0.019	0.052	0.126	0.311

***, **, * significant coefficients at 1%, 5% and 10% levels. Standard errors in parenthesis.
The dependent variable in columns (I) and (II) is the OROA of the start-up, calculated as a weighted average over the three first years excluding the first year. The dependant variable in columns (III) and (IV) is yearly asset growth. The dependent variable in the probit regression of columns (V) and (VI) is a dummy which equals 1 if the firm is in operation after four years and zero otherwise. This regression has a lower number of observations because some firms in the sample were founded after 2001. The dependent variable in columns (VII) and (VIII) is the log of 4-year average wage after the start-up year. In (VII) and (VIII), I add a control on whether the founder works for the start-up for at least one year or not, and in regressions (III) and (IV), I include dummies for start-up age in years. In the OROA, wage and survival regressions I report robust standard errors. In the growth regression I report Huber-White robust standard errors allowing for clustering of errors by individuals. (***) denotes significant at the 1% level, (**) significant at the 5% level, and (*) significant at the 10% level. Standard errors in parenthesis.

Moving on to relationship between previous employer size and asset growth in column (III) and (IV), the results are somewhat similar. A one log increase in previous size gives about 2 percentage points higher yearly asset growth. An increase from the 25 percentile to the 75 percentile in previous size gives 5 percentage points higher asset growth, while an increasing previous wage from the 25 percentile to the 75 percentile gives 2 percentage points higher asset growth. In columns (V) and (VI), I investigate 4-year survival. Here, somewhat surprisingly, neither human capital as measured by previous wage nor previous employer size are statistically significant.[31]

As a robustness check,[32] I undertook the same regression as in column (II) and (IV) but defining employer size with the log of number of employees at the company level rather than at the plant level. The results were almost identical. The positive relation between OROA and previous employer size could be offset by a negative relation between entrepreneurial wage and previous employer size. Analysis of the relation between previous employer size and performance as measured by entrepreneurial wages in columns (VII) and (VIII), however, suggested a weak positive relation between previous employer size and entrepreneur wage.

[31] Start-up size as measured by log total equity and founder wealth turn out statistically signficant. As shown by Hvide and Møen (2007), although total equity and founder wealth is statistically significant in explaining survival, the economic magnitude is small. One reason for these somewhat puzzling results is that I am unable to separate liquiditations from buy-outs when defining survival.

[32] In addition to the checks reported below, I also ran median and robust regressions. These gave results close to those reported in Table 2.

To accommodate non-linear effects from previous size other than those dictated by the log formulation, I used two alternative specifications: a dummy specification where the previous size variable was replaced with 10 dummy variables defined by previous size deciles, and a fourth-order polynomial specification. Both these formulations suggested that the positive effect of previous size is not confined to a particular interval of the previous size variable (consistent with the log formulation in Table 2, the effect of previous size in these alternative formulations was monotonically positive and declining). The levels of statistical and economic significance of previous size in these regressions were very similar to those in (II).

The empirical results reported in Table 2 are consistent with the theoretical prediction of the model. Since I do not have access to natural experiments or instruments that could establish causality, it is impossible to exclude the possibility that other mechanisms than outlined in the theoretical model drives the empirical results. Let me discuss some of them.

First, in a small firm the entrepreneur may be more likely to be one of the founders, and this may increase his incentives to stay irrespective of the firm's ability to monitor. This mechanism, which would tend to stop promising ideas from leaving small firms, would seem more plausible for the very smallest previous employers. Since I find a monotonic relationship between previous employer size and start-up performance, however, this mechanism seem unlikely to explain the empirical results.

Second, a common strategy for entrepreneurs is to consult or subcontract for former employers. Such a strategy could be more lucrative for larger or more productive former employers, even controlling for the wage earned by the entrepreneur while being an employee. Since I do not know the identity of the main customers of the start-up, it is hard to test this hypothesis directly. As an indirect test, one can investigate the IT industry, where such subcontracting through consulting services seems perhaps more likely than in manufacturing. To this end, I ran the same regressions as in Table 2 but splitting the sample into IT and non-IT companies. The results conveyed minimal differences in the effect of previous size for IT and non-IT companies. Moreover, I included the profitability and sales volume of the previous employer as additional controls. Including these variables gave a stronger effect of previous firm size on OROA, so the results do not seem to be explained by larger former employers being more profitable.

Third, if large firms tend to defer wages more than small firms (as suggested by Lazear and Moore, 1984) then for a given wage young workers in large firms will tend to be more productive than young workers in small firms. Given that the entrepreneurs in the sample are quite mature, with an average age of above 40 years, this alternative explanation seems unlikely.[33]

Finally, if workers anticipate the likelihood that they will have an idea, this could affects their initial choice of firm size. More creative employees should self-select into small firms if their creativity is more observable in such settings (which might explain why many top engineering graduates prefer to work for small firms rather than for large established firms). Self-selection based on creativity then arguably suggests that

[33] I experimented with other measures of previous wage, such as including yearly wages three years before the start-up date, to capture differences in wage evolution. These specifications gave unchanged results with respect to the effect of firm size.

entrepreneurs from small firms are of higher quality than entrepreneurs from large firms, which stands in contrast to the empirical findings.

7. Conclusion

Understanding the determinants of entrepreneurship is a key question to policy makers, economists and students of business. Still progress in the area has been limited; for example, microeconomics textbooks' notion of an entrepreneur is synonymous to an exogenously given production possibility frontier; courses on entrepreneurship are largely case-based and normatively (or perhaps rather romantically) tilted, and commonly taught by practitioners rather than academics. This state of affairs justifies an aggressive research effort by economists towards a better understanding of entrepreneurship.

The starting point of the article is the stylised fact that entrepreneurs are experienced workers who follow start-up opportunities closely related to their previous work experience. From this, I suggested an imperfect property rights framework that attempts to capture the interaction between the worker's leaving decision and an established firm's innovation strategy. The reduced-form nature of the model makes it consistent with several underlying mechanisms and forces a focus on results that are robust across these mechanisms.

The main insights were twofold. First, due to large firms having less precise information about the value of individual employees, large firms will pay less flexible wages and produce entrepreneurs of higher quality than small firms. Second, by affecting the trade-off between information quality and returns to scale, a large firm is more likely to be optimal the stronger property rights protection. Evidence on the quality of entrepreneurs from Norway gave encouraging support to the first hypothesis.

The main advantage of the current framework is that it is simple, and can be extended in various directions. One example would be to capture aspects the established firm's choice of R&D strategy (such as project choice and the extent to which workers are delegated decisions at project level) and labour market competition for workers.

Appendix

A. Proofs on $\Delta(\cdot)$.

I start out by proving Proposition 4 in several steps. For brevity I skip the * notation, and throughout the Appendix I assume interior solutions to the large firm's profit maximisation, i.e., $z \in (0.1)$. To be able to prove the proposition for both $c < 0$ and $c > 0$ it is convenient to define the function,

$$\Omega(c,\eta) = \begin{cases} P & \text{if } c < 0 \\ A & \text{if } c \geq 0, \end{cases} \tag{10}$$

where $P \equiv -c/(1-\eta)$ and $A \equiv c/\eta$, as before. Clearly the function $\Omega(c,\eta)$ slopes downward in c for $c < 0$ and upwards in c for $c > 0$. $\Omega(\cdot)$ decreases in η since $\mathrm{d}P/\mathrm{d}\eta = -c/(1-\eta)^2 < 0$ and $\mathrm{d}A/\mathrm{d}\eta = -c/\eta^2 < 0$. $\Omega(\cdot)$ is continuous in c since $\lim_{c\to 0^+} A = 0 = \lim_{c\to 0^-} P$.

Recall that,

$$\Delta[\alpha, c, \eta, z(\alpha, c, \eta)] = \Pi_L - \Pi_S = \alpha \int_0^z xg(x) - G(z)B_L - \Pi_S. \tag{11}$$

We start out with the following useful application of the envelope theorem.

LEMMA 1.

$$\frac{d\Delta}{d\eta} = \frac{\partial\Delta}{\partial\eta} = \int_\Omega^1 xg(x) - G(z)z. \tag{12}$$

Proof. Recall that $\Delta = \alpha \int_0^z xg(x) - G(z)B_L - \Pi_S$, where $B_L = \eta z - c$. Therefore,

$$\frac{d\Delta}{d\eta} = \frac{\partial\Delta}{\partial\eta} + \frac{\partial\Delta}{\partial z}\frac{\partial z}{\partial\eta}. \tag{13}$$

But by the construction of Δ, we have that $\partial\Delta/\partial z = 0$. Recall that $B_L = \eta z - c$ and $\partial B_L/\partial\eta = z$. Therefore,

$$\frac{d\Delta}{d\eta} = \frac{\partial\Delta}{\partial\eta} = -G(z)\frac{\partial B_L}{\partial\eta} + \frac{\partial\Pi_S}{\partial\eta} = \int_\Omega^1 xg(x) - G(z)z.$$

The next Lemma is also on a property of Δ.

LEMMA 2. *Δ is convex in η.*

Proof. Write for brevity $d\Omega/d\eta = \Omega'$ and recall that $\Omega' < 0$. Using the envelope theorem again,

$$\begin{aligned}\frac{d^2\Delta}{d\eta^2} &= \frac{\partial^2\Delta}{\partial\eta^2} + \frac{\partial^2\Delta}{\partial z\partial\eta}\frac{\partial z}{\partial\eta}\\ &= -\Omega'\Omega g(\Omega) - [g(z)z + G(z)]\frac{\partial z}{\partial\eta} > 0.\end{aligned} \tag{14}$$

Since $\partial z/\partial\eta < 0$.

Now define,

$$\Delta^1(\cdot) = \Delta(\eta = 1, .). \tag{15}$$

Since Δ is convex in η from the previous Lemma, the following obvious but very useful observation follows directly.

LEMMA 3. $\Delta^1(\cdot) < 0$ implies both existence and uniqueness of η^*.

Recall that,

$$\Delta^1(\cdot) = \alpha \int_0^z xg(x) - G(z)B_L - \Pi_S. \tag{16}$$

Now define $\bar{\alpha}$ as a value of α which makes $\Delta^1(\cdot) = 0$. Since $d\Delta^1(\cdot)/d\alpha > 0$ there is at most one $\bar{\alpha}$ for each c, and hence there exists a function f such that $\bar{\alpha} = f(c)$, whose graph demarcates the values of (c,α) where $\Delta^1(\cdot) < 0$ from values of (c,α) where $\Delta^1(\cdot) > 0$. The area north of the graph would have $\Delta^1(\cdot) > 0$ and the area south of the graph would have $\Delta^1(\cdot) < 0$ since $d\Delta^1(\cdot)/d\alpha > 0 > 0$. So one way of completing the proof is to show that $f(c) > 1$ for all c. It turns out, however, that there is a simpler way to complete the proof. Specifically, set $\Delta^1(\cdot) = 0$ and rearrange to form a fixed-point definition of $\bar{\alpha}$,

$$\bar{\alpha} = h[\bar{\alpha}, c, z(\bar{\alpha}, c)] = \frac{G(z)B_L + \Pi_S}{\int_0^z xg(x)}.$$

It is sufficient to prove the Proposition to show that any solution to this equation must have $\bar{\alpha} > 1$ for any c. Note first that $h(\cdot)$ converges to $(1 - c) + \Pi_S$ when $\bar{\alpha}$ goes to infinity, since z converges to 1 and B_L converges to $1 - c$. The reason is simple; when α becomes large, the large firm maximises profits by keeping all workers and must pay them $1 - c$. It is therefore sufficient for existence and uniqueness of a fixed point above 1 that $h[1,c,z(1,c)] > 1$. But this is trivial to see. Recall that $\Pi_L = \int_0^z xg(x) - G(z)B_L$.

I therefore have that,

$$h(\bar{\alpha} = 1) = \frac{\int_0^z xg(x) + \Pi_S - \Pi_L}{\int_0^z xg(x)}. \tag{17}$$

But the right hand side of this expression exceeds 1, since a small firm must make weakly higher profits than a large firm for any c when $\alpha = 1$.[34] Therefore $h[1,c,z(1,c)] > 1$, and the proposition is proved. One may note that this result does not depend on the firm getting nothing if the worker leaves, since if $\eta = 1$ then the firm gets nothing anyway, given that Δ is convex.

Proof of Remark 5. Let me first prove the second statement. We have that,

$$\frac{\mathrm{d}\Delta}{\mathrm{d}c} = \frac{\partial \Delta}{\partial c} + \frac{\partial \Delta}{\partial z}\frac{\partial z}{\partial c}. \tag{18}$$

But by the construction of Δ, the envelope theorem gives $\partial\Delta/\partial z = 0$. Therefore

$$\frac{\mathrm{d}\Delta}{\mathrm{d}c} = \frac{\partial \Delta}{\partial c} = -G(z)\frac{\partial B_L}{\partial c} - \frac{\partial \Pi_S}{\partial c} = G(z) - \int_\Omega^1 xg(x)\mathrm{d}x = G(z) + \int_0^\Omega xg(x)\mathrm{d}x - 1. \tag{19}$$

Using the envelope theorem and the implicit function rule,

$$\frac{\mathrm{d}\eta^*}{\mathrm{d}c} = -\frac{\Delta_c^*}{\Delta_\eta^*} = \frac{1 - G(\Omega) - G(z)}{\Delta_\eta^*}. \tag{20}$$

Note that by uniqueness of η^* then $\Delta_\eta^* < 0$. For c close to zero, then $G(\Omega)$ is close to zero, and therefore $\mathrm{d}\eta^*/\mathrm{d}c < 0$. To find $\mathrm{d}\eta^*/\mathrm{d}\alpha$, I use the same method to obtain,

$$\frac{\mathrm{d}\eta^*}{\mathrm{d}\alpha} = \frac{\partial \eta}{\partial \alpha} = -\frac{\Delta_\alpha^*}{\Delta_\eta^*} = -\frac{\int_0^z G(x)}{\Delta_\eta^*}. \tag{21}$$

B. Alternative Bargaining Formulation

In the main analysis, the firm makes a take-it or leave-it offer to the agent at time 1. Now consider the opposite formulation, i.e., when the worker can give a take-it or leave-it wage demand to the principal and leaves the firm if these conditions are not met. Since the principal in practice may not know at which point in time the agent learns about x, this formulation may be more realistic and opens up for the interesting possibility that the agent through the wage demand can reveal information about x (if employed in a large firm).

Let us begin by considering a small firm (symmetric information). First suppose that $c > 0$. In that case, it is always efficient that the agent stays in the firm, and the optimal take-it or leave-it

[34] Recall that the large firm pays the same wage to all agents that stay in the firm. The small firm can replicate whatever wage policy the large follows but pay less to the agents with a low x (it can do even better but that is beside the point). Therefore $\Pi_S - \Pi_B > 0$ when $\alpha = 1$, $\eta = 1$, and $c < 1$. When $c = 1$, the profits of both types of firm equals $E(x)$, in which case the only solution to the fixed point equation is for $\bar{\alpha} = 1$. Therefore the curve defined by the function $f(c)$ can hit the c axis only in the point $c = 1$.

demand, denoted by w, clearly equals x. That leaves the firm's profit at zero. In the case with $c < 0$, the worker will take x if $x > c$ but leave the firm if $x < c$. Hence I can express the optimal take-it or leave-it demand as,

$$w = \begin{cases} 0 & \text{if } x < c \\ x & \text{if } x \geq c. \end{cases} \tag{22}$$

Note that separations are efficient also in this case, since the worker takes the social benefits and costs of a separation fully into account. *A large firm.* To illustrate how the agent can convey information through his demand, and how this may affect who becomes entrepreneurs, I focus on the equilibrium which conveys the maximal amount of information about the true x. This turns out to be a separating equilibrium where the agent fully reveals x through his demand.[35]

Let the agent's strategy be described by a (possibly random) function $w(x)$, which describes the wage he demands as a function of x, and let the firm's strategy be represented by an accept function $Q(w)$. The interpretation of the second function is that for any demand w, the firm accepts the demand with probability $Q(w)$ and rejects the demand (in which case the worker becomes an entrepreneur) with probability $1 - Q(w)$.

Suppose that a separating equilibrium exists where the agent fully reveals x through his demand w. By standard reasons, there cannot exist such an equilibrium where the firm's strategy is deterministic (i.e., only takes the values 0 or 1).[36] I therefore confine attention to the firm playing a mixed strategy.[37] For a firm playing a mixed strategy, it must be indifferent between accepting the worker's demand or not, and consequently $w(x) = x$, i.e., the worker takes all the rents. For a given x and $Q(\cdot)$, the agent's utility from demanding w equals,

$$U(w,x) = Q(w)w + [1 - Q(w)](\eta x - c). \tag{23}$$

The first term reflects that the agent gets the continuation wage w if the firm accepts the demand. The second term reflects that the worker gets the entrepreneurial payoff $U(x)$ if the demand is rejected. Without loss of generality I can assume that $Q(\cdot)$ is continuous and differentiable. Writing $Q(w)$ as Q, and dQ/dw as Q', I then have that,

$$\frac{dU}{dw} = Q'(w - \eta x + c) + Q. \tag{24}$$

For a separating equilibrium to exist, $w = x$ must be the optimal wage demand for all x,

$$Q'(x - \eta x + c) + Q = 0. \tag{25}$$

This is a linear partial differential equation in x, with the solution,

$$Q(x) = K[(1-\eta)x + c]^{-\frac{1}{1-\eta}}. \tag{26}$$

where K is an integration constant. Since it is strictly optimal for the firm to accept a zero demand, I can impose the corner condition $Q(0) = 1$ to arrive at,

$$Q(x) = c^{\frac{1}{1-\eta}}[(1-\eta)x + c]^{-\frac{1}{1-\eta}}. \tag{27}$$

As can easily be verified, this function is decreasing and convex in x.[38] I therefore have that,

[35] This Section has formal resemblence to the financier-entrepreneur costly state verification repayment game modelled in Hvide and Leite (2004). As discussed there, the separating equilibrium has some attractive equilibrium selection properties. Issues of renegotiation are also discussed in that paper.

[36] Suppose that the firm plays a pure strategy, and let $\tilde{w}$ be the highest offer that the firm would accept with probability 1. Then all worker types would strictly prefer to offer $\tilde{w}$, which would contradict the assumption that there exists a fully separating equilibrium.

[37] One interpretation of such a mixed strategy is that there are many possible firm types, and that each of these play a pure strategy. In aggregate, however, they must play a mixed strategy described by $Q(.)$.

[38] That $Q(\cdot)$ is convex intuitively means that the principal is particularly sensitive to the agent lying when making small offers w, which is economically plausible.

PROPOSITION 4. *When the agent makes the take-it or leave-it demand, the firm makes zero profits, and the probability of the worker becoming an entrepreneur increases in x.*

Proof. This follows directly from $Q(x)$ decreasing in x.

As can be shown, this equilibrium has the same qualitative comparative statics properties to that in the main case (Remark 4).

University of Aberdeen CEPR and the Norwegian School of Economics and Business

Submitted: 21 March 2006
Accepted: 29 May 2008

References

Agell, J. (2004). 'Why are small firms different? Managers' views', *Scandinavian Journal of Economics*, vol. 106, pp. 437–52.

Aghion, P. and Tirole, J. (1994). 'The management of innovation', *Quarterly Journal of Economics*, vol. 113, pp. 1185–209.

Alchian, A.A. and Demsetz, H. (1972). 'Production, information costs, and economic organization', *American Economic Review*, vol. 62, pp. 777–95.

Amador, M. and Landier, A. (2003). 'Entrepreneurial pressure and innovation', available at http://www.ssrn.com

Anand, N.B. and Khanna, T. (2000). 'The structure of licensing contracts', *Journal of Industrial Economics*, 48, pp. 103–35.

Anton, J. and Yao, D. (1994). 'Expropriations and inventions: appropriable rents in the absence of property rights', *American Economic Review*, vol. 84, pp. 190–209.

Anton, J. and Yao, D. (1995). 'Start-ups, spin-offs and internal projects', *Journal of Law, Economics and Organization*, vol. 11(2), pp. 362–78.

Arora, A. and Ceccagnoli, M. (2006). 'Patent protection, complementary assets, and firms' incentives for technology licensing', *Management Science*, vol. 52, pp. 293–308.

Bates, T. (1990). 'Entrepreneur human capital inputs and small business longevity', *Review of Economics and Statistics*, vol. 72, pp. 551–59.

Bennedson, M., Nielsen, K., Pérez-González, F. and Wolfenzon, D. (2007). 'Inside the family firm: the role of families in succession decisions and performance', *Quaterly Journal of Economics*, vol. 122, pp. 647–91.

Bhide, A.V. (2000). *The Origin and Evolution of New Businesses*, Oxford: Oxford University Press.

Cockburn, I. and Henderson, R. (1996). 'Scale, scope, and spillovers: the determinants of research productivity in drug discovery', *The RAND Journal of Economics*, vol. 27, pp. 32–59.

Cohen, W.M., Nelson, R.P. and Walsh, J.P. (2000). 'Protecting their intellectual assets: appropriability conditions and why U.S. manufacturing firms patent (or not)', NBER Working Paper No. 7552.

Cooper, A.C. (1985). 'The role of incubator organizations in the founding of growth-oriented firms', *Journal of Business Venturing*, vol. 1, pp. 75–86.

De Meza, D. and Southey, C. (1996). 'The borrower's curse: optimism, finance and entrepreneurship', ECONOMIC JOURNAL, vol. 106, pp. 375–86.

Desai, H. and Jain, P.C. (1999). 'Firm performance and focus: long-run stock market performance following spinoffs', *Journal of Financial Economics*, vol. 54, pp. 75–101.

Dunn, T. and Holtz-Eakin, D. (2000). 'Financial capital, human capital, and the transition to self-employment: evidence from intergenerational links', *Journal of Labor Economics*, vol. 28, pp. 282–305.

Evans, D.S. and Jovanovic, B. (1989). 'An estimated model of entrepreneurial choice under liquidity constraints', *Journal of Political Economy*, vol. 97, pp. 808–27.

Evans, D.S. and Leighton, L.S. (1989). 'Some empirical aspects of entrepreneurship', *American Economic Review*, vol. 79, pp. 519–35.

Evans, G.H. Jr (1949). 'The entrepreneur and economic theory: a historical and analytical approach', *American Economic Review, Papers and Proceedings*, vol. 39, pp. 336–48.

Fox, J. (2004). 'Employer size wage gap revisited', mimeo, University of Chicago Economics Department.

Frank, M.Z. and Goyal, V.K. (2003). 'Capital structure decision: which factors are reliably important?', Working Paper, UBC and HKUST.

Gentry, W. and Hubbard, R. (2002). 'Entrepreneurship and household saving', *Advances in Economic Analysis & Policy*, vol. 4, p. 1053.

Grossman, S.J. and Hart, O.D. (1986). 'The costs and benefits of ownership: a theory of vertical and lateral integration', *Journal of Political Economy*, vol. 94, pp. 691–719.

Hamilton, B. (2000). 'Does entrepreneurship pay? An empirical analysis of the returns to self-employment', *Journal of Political Economy*, vol. 108, pp. 604–31.

Hellmann, T. (2007). 'When do employees become entrepreneurs?', *Management Science*, vol. 53, pp. 919–33.

Hellmann, T. and Puri, M. (2002). 'Venture capital and the professionalization of start-up firms: empirical evidence', *Journal of Finance*, vol. 57, pp. 169–97.

Holtz-Eakin, D., Joulfaia, D. and Rosen, H.S. (1994*a*). 'Sticking it out: entrepreneurial survival and liquidity constraints', *Journal of Political Economy*, vol. 102, pp. 53–75.

Holtz-Eakin, D., Joulfaia, D. and Rosen, H.S. (1994*b*). 'Entrepreneurial decisions and liquidity constraints', *Rand Journal of Economics*, vol. 25, pp. 334–47.

Hvide, H.K. (2005). 'Firm size and the quality of entrepreneurs', CEPR Working Paper No. 4979, available at http://www.ssrn.com.

Hvide, H.K. and Kristiansen, E.G. (2006). 'Management of knowledge workers', available at http://www.ssrn.com.

Hvide, H.K. and Leite, T. (2004). 'Capital structure under imperfect enforcement', CEPR Working Paper No. 4757.

Hvide, H.K. and Møen, J. (2007). 'Liquidity constraints and entrepreneurial performance', available at http://www.ssrn.com.

Kihlstrom, R. and Laffont, J.J. (1979). 'A general equilibrium entrepreneurial theory of the firm based on risk aversion', *Journal of Political Economy*, vol. 87, pp. 719–48.

Klepper, S. and Thompson, P. (2005). 'Spinoff entry in high-tech industries: motives and consequences', mimeo, Dept. of Decision and Social Sciences, Carnegie Mellon University.

Kim, J. and Marschke, G. (2005). 'Labor mobility of scientists, technological diffusion, and the firm's patenting decision', *RAND Journal of Economics*, vol. 36, pp. 298–317.

Landier, A. (2001). 'Endogenous entrepreneurial risk: the stigma of failure', mimeo, MIT.

Landier, A. and Thesmar, D. (2009). 'Financial contracting with optimistic entrepreneurs: theory and evidence', *Review of Financial Studies*, vol. 22, pp. 117–50.

Lazear, E.P. (2005). 'Entrepreneurship', *Journal of Labor Economics*, vol. 23, pp. 649–80.

Lazear, E.P. and Moore, R.L. (1984). 'Incentives, productivity, and labor contracts', *Quarterly Journal of Economics*, vol. 99, pp. 275–96.

Merges, R.P. (1999). 'The law and economics of employee innovations', *Harvard Journal of Law & Technology*, vol. 13, pp. 1–55.

Oyer, P. and Schaefer, S. (2005). 'Why do some firms give stock options to all employees? An empirical examination of alternative theories', *Journal of Financial Economics*, vol. 76, pp. 99–133.

Pakes, A. and Nitzan, S. (1983). 'Optimum contracts for research personnel, research employment, and the establishment of "rival" enterprises', *Journal of Labor Economics*, vol. 1, pp. 345–65.

Posner, E. and Triantis, G.G. (2001). 'Covenants not to compete from an incomplete contracts perspective', Working Paper, Chicago Law School, available at http://www.ssrn.com.

Rajan, R.G. and Zingales, L. (1998). 'Power in a theory of the firm', *Quarterly Journal of Economics*, vol. 113, pp. 387–432.

Rajan, R.G. and Zingales, L. (2001). 'The firm as a dedicated hierarchy: a theory of the origin and growth of firms', *Quarterly Journal of Economics*, vol. 116, pp. 805–51.

Rosen, S. (1982). 'Authority, control, and the distribution of earnings', *Bell Journal of Economics*, vol. 13, pp. 311–23.

Shane, S. (2000). 'Prior knowledge and the discovery of entrepreneurial opportunities', *Organization Science*, vol. 11, pp. 448–69.

Simon, H.A. (1973). 'Theories of bounded rationality', in (C.B. McGuire and R. Radner eds.), *Decision and Organization*, pp. 161–76, Amsterdam: North-Holland.

Stein, J.C. (2002). 'Information production and capital allocation: decentralized versus hierarchical firms', *Journal of Finance*, vol. 57, pp. 1891–921.

Stone, K.V.W. (2002). 'Knowledge at work: disputes over the ownership of human capital in the changing workplace', *Connecticut Law Rewiew*, vol. 34, pp. 721–63.

Subramanian, N. (2005). 'The economics of intrapreneurial innovation', *Journal of Economic Behavior and Organization*, vol. 58, pp. 487–510.

Veld, C. and Veld-Merkoulova, Y.V. (2004). 'Do spin-offs really create value? The European case', *Journal of Banking and Finance*, vol. 28, pp. 1111–35.

Williamson, O.E. (1981). 'The modern corporation: origins, evolution, attributes', *Journal of Economic Literature*, vol. 19, pp. 1537–68.